# Lecture Notes in Physics

## The Lecture Notes in Physics

The series Lecture Notes in Physics (LNP), founded in 1969, reports new developments in physics research and teaching – quickly and informally, but with a high quality and the explicit aim to summarize and communicate current knowledge in an accessible way. Books published in this series are conceived as bridging material between advanced graduate textbooks and the forefront of research and to serve three purposes:

- to be a compact and modern up-to-date source of reference on a well-defined topic
- to serve as an accessible introduction to the field to postgraduate students and nonspecialist researchers from related areas
- to be a source of advanced teaching material for specialized seminars, courses and schools

Both monographs and multi-author volumes will be considered for publication. Edited volumes should, however, consist of a very limited number of contributions only. Proceedings will not be considered for LNP.

Volumes published in LNP are disseminated both in print and in electronic formats, the electronic archive being available at springerlink.com. The series content is indexed, abstracted and referenced by many abstracting and information services, bibliographic networks, subscription agencies, library networks, and consortia.

Proposals should be sent to a member of the Editorial Board, or directly to the managing editor at Springer:

Christian Caron
Springer Heidelberg
Physics Editorial Department I
Tiergartenstrasse 17
69121 Heidelberg / Germany
christian.caron@springer.com

Roland Diehl · Dieter H. Hartmann ·
Nikos Prantzos (Eds.)

# Astronomy with Radioactivities

Springer

Roland Diehl
MPI für extraterrestrische Physik
Giessenbachstr. 1
85740 Garching
Germany

Dieter H. Hartmann
Clemson University
Dept. Physics & Astronomy
29634 Clemson South Carolina
USA

Nikos Prantzos
Institut d'Astrophysique de Paris
Boulevard Arago 98bis
75014 Paris
France

Diehl, R., Hartmann, D.H., Prantzos, N. (Eds.): *Astronomy with Radioactivities*, Lect. Notes Phys. 812 (Springer, Berlin Heidelberg 2011), DOI 10.1007/978-3-642-12698-7

Lecture Notes in Physics ISSN 0075-8450 e-ISSN 1616-6361
ISBN 978-3-642-12697-0 e-ISBN 978-3-642-12698-7
DOI 10.1007/978-3-642-12698-7
Springer Heidelberg Dordrecht London New York

Library of Congress Control Number: 2010930266



*Cover design:* Integra Software Services Pvt. Ltd., Pondicherry

Printed on acid-free paper

Springer is part of Springer Science+Business Media (www.springer.com)

# Preface

Nuclear energy has been recognized almost a 100 years ago as the energy source which makes the stars shine. It became clear that stars, supernovae, and other cosmic sites of nuclear fusion reactions are responsible to process the primordial elemental abundances, which consist of mainly hydrogen and helium, into the rich variety of 92 natural chemical elements through most of the 3,100 known isotopes.

As part of this *cosmic nucleosynthesis*, unstable isotopes are created alongside. These lead to unique astronomical signatures about the cosmic nucleosynthesis process and its environments, which need to be captured and unraveled to teach us about the astrophysics of cosmic nucleosynthesis and related processes.

It is the goal of this book to introduce this specific kind of *astronomy*, i.e. the acquisition and reading of measurements on unstable isotopes in different parts of the universe, and to relate other astronomical and nuclear-physics measurements of very different kinds and corresponding theories to the *astrophysical studies on nuclear-physics aspects of cosmic objects and evolutionary processes*. This field is an important part of *nuclear astrophysics*.

Pursuing this goal, we describe stars of different types, stellar explosions which terminate stellar evolutions, and other explosions triggered by mass transfers and instabilities in binary stars. We also address nuclear reactions and transport processes in interstellar space. The contexts of cosmic rays and in particular of chemical evolution are exposed. A special chapter is dedicated to the solar system, where we have our best material samples of one star formed 4.6 Gigayears (Gy) ago with its planets and our Earth. We add to this book a description of key tools which astrophysicists employ in those particular studies.

This book grew out of a series of small-scale, bi-annual workshops, which started in 1996 in Clemson, South Carolina (USA) as *The Radioactive Galaxy*, and was later continued mostly at the Max Planck Society's "Ringberg" castle in the Bavarian alps (Germany) under the title of *Astronomy with Radioactivities*. During these workshops, the variety of themes of nuclear astrophysics were discussed among experimenters in meteoritics, in astronomy with spectral lines from gamma-rays to radio wavelengths, and in cosmic rays, and also with and among theorists of stellar evolution and explosions, of cosmic chemical evolution, and with nuclear reaction experts. These multi-facetted discussions are reflected in this book, for the benefit and pleasure of all people interested in the science of how nature made the elements we are composed of.

Garching, Germany
Clemson, South Carolina
Paris, France
25 March 2010

Roland Diehl
Dieter H. Hartmann
Nikos Prantzos

# Acknowledgements

Many colleagues have contributed to this field, yet are not explicitly co-authors of chapters in this book. The choice of authors grew out of opportunities around 2008, being aware that there were many more who could (and probably would) have joined this adventure. The authors of this book are grateful for stimulating discussions and for collaborative work and publications shared with many dear friends and colleagues. We explicitly would like to mention here Sachiko Amari, Peter von Ballmoos, John Beacom, Peter Biermann, Bob Binns, Andreas Burkert, Roger Chevalier, John Cowan, John Danziger, Iris Dillmann, Mounib El Eid, Bruce Elmegreen, Brian Fields, Claes Fransson, Roberto Gallino, Neil Gehrels, Matthieu Gounelle, Alexander Heger, Wolfgang Hillebrandt, Rob Hoffman, Peter Hoppe, Christian Illiadis, Anatoli Iyudin, Thomas Janka, Franz Käppeler, Jürgen Knödlseder, Gunther Korschinek, Karl-Ludwig Kratz, Pavel Kroupa, Karlheinz Langanke, John Lattanzio, Bruno Leibundgut, Mark Leising, Marco Limongi, Günther W. Lugmair, Gabriel Martinez-Pinedo, Bradley Meyer, Georges Meynet, Peter Milne, Yuko Mochizuki, Thierry Montmerle, Nami Mowlavi, Ewald Müller, Ken'ichi Nomoto, Uli Ott, Etienne Parizot, Volker Schönfelder, David Smith, Andrew Strong, Vincent Tatischeff, Lih-Sin The, James Truran, Jacco Vink, Stan Woosley, Hans Zinnecker, and Ernst Zinner.
Thank you all.

# Contents

# Part I
# The Role of Radioactivities in Astrophysics

In the following two Chapters we introduce basic astrophysical considerations related to radioactive isotopes and their decay. A general introduction to the terminology of radioactivity and its physical meaning is followed by a personal account from one of the fathers of this field. The latter presents the development of the science of cosmic radioactivity and discusses the fundamental questions that were raised in the early years and those challenging its present day practitioners.

# Chapter 1
# Introduction to Astronomy with Radioactivity

**R. Diehl**

## 1.1 The Origin of Radioactivity

The nineteenth century spawned various efforts to bring order into the elements encountered in nature. Among the most important was an inventory of the *elements* assembled by the Russian chemist Dimitri Mendeleyev in 1869, which grouped elements according to their chemical properties, their *valences*, as derived from the compounds they were able to form, at the same time sorting the elements by atomic weight. The genius of Mendeleyev lay in his confidence in these sorting principles, which enforced gaps in his table for expected but then unknown elements, and Mendeleyev was able to predict the physical and chemical properties of such elements-to-be-found. The tabular arrangement invented by Mendeleyev (Fig. 1.1) still is in use today, and is being populated at the high-mass end by the great experiments in heavy-ion collider laboratories to create the short-lived elements predicted to exist. The second half of the nineteenth century thus saw scientists being all-excited about chemistry and the fascinating discoveries one could make using Mendeleyev's sorting principles. Note that this was some 30 years before sub-atomic particles and the atom were discovered. Today the existence of 112 elements is firmly established (the currently heaviest) element no. 112 was officially named *Copernicium (Cn)* in February 2010 by IUPAC, the international union of chemistry.

In the late nineteenth century, scientists also were excited about new types of penetrating radiation. Conrad Röntgen's discovery in 1895 of *X-rays* as a type of electromagnetic radiation is important for understanding the conditions under which Antoine Henri Becquerel discovered radioactivity in 1896. Becquerel also was engaged in chemical experiments, in his research on phosphorescence exploiting the chemistry of photographic-plates materials. At the time, Becquerel had prepared some plates treated with uranium-carrying minerals, but did not get around to make the planned experiment. When he found the plates in their dark storage some time later, he accidentally processed them, and was surprised to find an image of a

R. Diehl (✉)
Max Planck Institut für extraterrestrische Physik, 85748 Garching, Germany, rod@mpe.mpg.de

Diehl, R.: *Introduction to Astronomy with Radioactivity*. Lect. Notes Phys. **812**, 3–23 (2011)
DOI 10.1007/978-3-642-12698-7_1 

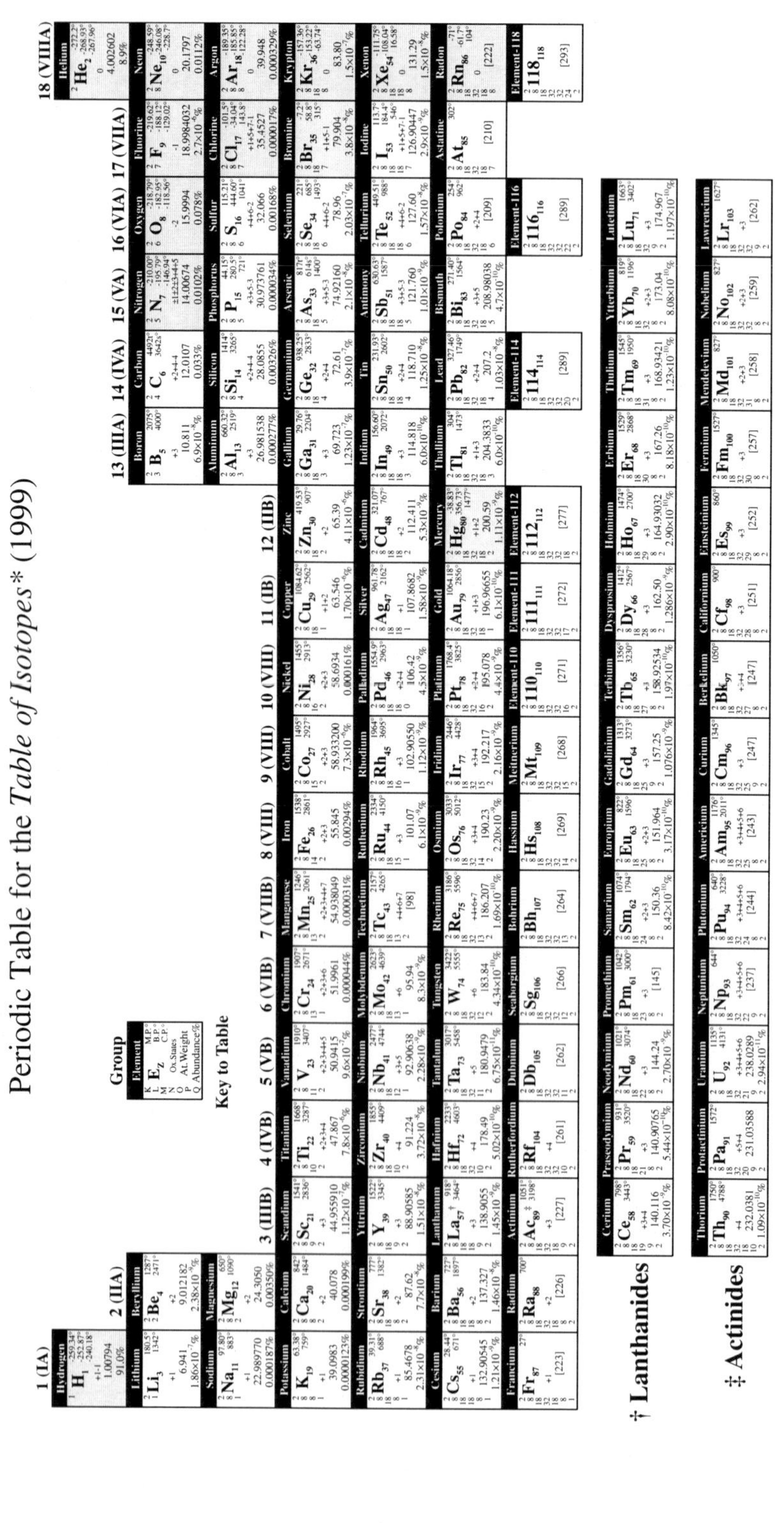


**Fig. 1.1** The periodic table of elements, grouping chemical elements according to their chemical-reaction properties and their atomic weight, after Mendeleyev (1869)

coin which happened to have been stored with the plates. Excited about X-rays, he believed he had found yet another type of radiation. Within a few years, Becquerel with Marie and Pierre Curie and others recognized that the origin of the observed radiation were elemental transformations of the uranium minerals, and the physical process of *radioactivity* had been found. The revolutionary aspect of elements being able to spontaneously change their nature became masked at the beginning of the twentieth century, when sub-atomic particles and the atom were discovered. But it is worth emphasizing that well before atomic and quantum physics was known, the physics of weak interactions had been discovered in its form of *radioactivity*.

In an ensemble consisting of a large number of identical radioactive isotopes, it is well-known that the number remaining declines exponentially with time: Radioactive decay is described by

$$\frac{dN}{dt} = -\lambda \cdot N \tag{1.1}$$

Here $N$ is the number of isotopes, and the *radioactive-decay constant* $\lambda$ is the inverse of the decay time $\tau$. The decay time $\tau$ is the time after which the number of isotopes is reduced by decay to $1/e$ of the original number:

$$N = N_0 \cdot \exp\frac{-t}{\tau} \tag{1.2}$$

The radioactive half-life $T_{1/2}$, correspondingly, is defined as the time after which the number of isotopes is reduced by decay to $1/2$ of the original amount, with

$$T_{1/2} = \frac{\tau}{\ln(2)} \tag{1.3}$$

Depending on the astrophysical objective, radioactive isotopes may be called *short-lived* or *long-lived* for identical lifetimes: The relation of the radioactive lifetime to astrophysical time scales of interest ist what matters. Examples are the utilization of $^{26}$Al and $^{60}$Fe ($\tau \sim$ My) diagnostics of the early solar system (Chap. 6) or of nucleosynthesis source types (Chaps. 3, 4, and 5).

An *isotope* is defined by the number of its two types of nucleons, *protons* (the number of protons defines the charge number Z) and *neutrons* (the sum of the numbers of protons and neutrons defines the mass number A), written as $^{A}$X for an element “X”. Note that some isotopes may exist in different nuclear quantum states which have significant stability by themselves, so that transitions between these configurations may liberate the binding energy differences; such states of the same isotope are called *isomers*.

The above exponential decay law is a consequence of a surprisingly simple physical property: The probability per unit time for a single radioactive nucleus to decay is independent of the age of that nucleus. Unlike our commonsense experience with living things, decay does not become more likely as the nucleus ages. Radioactive decay is a nuclear transition from one set of nucleons constituting a nucleus to a different and energetically-favored set with the same number of nucleons. Different types of interactions can mediate such a transition (see below). In *$\beta$-decays* it is mediated by the *weak transition* of a neutron into a proton, or more generally, nucleons of one type into the other type[1]:

$$n \longrightarrow p + e^- + \overline{\nu_e} \tag{1.4}$$

If such a process occurs inside an atomic nucleus, the quantum state of the nucleus is altered. Depending on the variety of configurations in which this new state may be realized (i.e. the *phase space* available to the decaying nucleus), this change may be more or less likely, in nature's attempt to minimize the total energy of a composite system of nucleons. The decay probability $\lambda$ per unit time for a single radioactive nucleus is therefore a property which is specific for each particular type of isotope. It can be estimated by Fermi's *Golden Rule* formula though time-dependent perturbation theory (e.g. Messiah, 1962). When schematically simplified to convey the main ingredients, the decay probability is:

$$\lambda = \frac{4\pi^2}{h} V_{fi}^2\, \rho(W) = \frac{1}{\tau} \tag{1.5}$$

where $\rho(W)$ is the number of final states having suitable energy $W$. The full theory involves an integral over the final kinematic states, suppressed here for simplicity. The matrix element $V_{fi}$ is the result of the transition-causing potential between initial and final states. The parameter $\tau$, defined as $1/\lambda$, can easily be shown to be the mean lifetime of the nucleus against radioactive decay.

In general, transitions occur from the ground state of the parent nucleus to an excited state of the daughter nucleus. But quantum mechanical transition rules may allow and even prefer other initial and final states. Excess binding energy will be transferred to the end products, which are the daughter nucleus, but also emitted (or absorbed, in the case of electron capture transitions) leptons (electrons, positrons, neutrinos) and $\gamma$-ray photons.

The occupancy of states is mediated by the *thermal* excitation spectrum of the *Boltzmann distribution* of particles at different energies according to:

$$\frac{dN}{dE} = G_j \cdot e^{-\frac{E}{k_B T}} \tag{1.6}$$

Here $k_B$ is Boltzmann's constant, $T$ the temperature of the particle population, $E$ the energy, and $G_j$ the statistical weight factor of all different possible states $j$ which

[1] In a broader sense, nuclear physics may be considered being similar to chemistry: elementary building blocks are rearranged to form different species, with macroscopically-emerging properties such as characteristic, well-defined energy releases in such transitions.

correspond to a specific energy $E$.[2] In natural environments, particles will populate different states as temperature dictates. Transition rates among states thus will depend on temperature. Inside stars, and more so in explosive environments, temperatures can reach ranges which are typical for nuclear energy-level differences. Therefore, in cosmic sites, radioactive decay constants may be significantly different from what we measure in terrestrial laboratories on *cold* samples (see Sect. 1.2 for more detail).

The atomic-shell environment of a nucleus may modify radioactive decay, in particular if a decay involves *capture of an electron* to transform a proton into a neutron. Such decays are inhibited in fully-ionized plasma conditions, due to the non-availability of electrons. Which radioactive decays are to be expected, in a mix of nuclei such as could be imagined to exist in a thermal bath of nuclei? What are stable configurations of nucleons inside a nucleus? This involves an understanding of the nuclear forces, an area of current research and far from being fully understood. Nevertheless, a few general ideas appear well established. These can be summarized in the expression for nuclear masses (Weizsäcker, 1935):

$$m(Z,A) = ZM_p + (A - Z)m_n - BE \tag{1.7}$$

with

$$BE = a_{\text{volume}} A - a_{\text{surface}} A^{2/3} - a_{\text{coulomb}} \frac{Z^2}{A^{1/3}} - a_{\text{asymmetry}} \frac{(a - 2Z)^2}{4A} - \frac{\delta}{A^{1/2}} \tag{1.8}$$

This description emphasizes that the total *binding energy* (BE) is another free parameter for a system of nucleons, allowing them to adopt bound states of lower energy than the sum of the free nucleons. Thus, in a thermal mixture of nucleons, bound nuclei will be formed, and their abundance depends on their composition and shape, and on the overall system temperature, defining how the total available phase space of internal and kinetic energy states is populated.

The key gross features characterizing the bound systems with local maxima of binding energy are (1) the *odd-even* effect described by the last term, which results in odd-nucleon nuclei being less favored that even-nucleon nuclei, and (2) the neutron excess favored by the asymmetry term, which results in heavier nuclei being relatively more neutron rich.

As an illustration of the available nuclear configurations, Fig. 1.2 shows the stable (black) and unstable isotopes, the latter decaying by $\beta^-$-decay (blue) and $\beta^+$-decay (orange). The *ragged* structure signifies that there are systematic variations of nuclear stability with nucleon number, some nucleonic numbers allowing for a greater variety of stable configurations of higher binding energy. These are, in particular, *magic numbers* of protons and neutrons of 2, 8, 20, 28, 50, and 82.We now

[2] States may differ in their quantum numbers, such as spin, or orbital-momenta projections; if they obtain the same energy $E$, they are called *degenerate*.

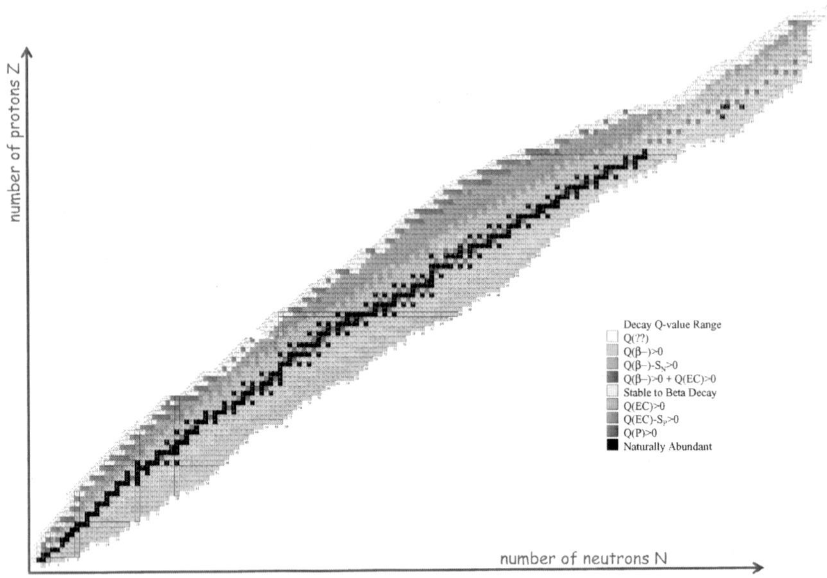


**Fig. 1.2** The table of isotopes, showing nuclei in a chart of neutron number (abscissa) versus proton number (ordinate). The stable elements are marked in *black*. All other isotopes are unstable, or radioactive, and will decay until a stable nucleus is obtained

know approximately 3,100 such *isotopes* making up the 114 now-known chemical elements, but only 286 of these isotopes are considered stable. (The latest (7th) edition of the Karlsruher Nuklidkarte (Pfennig et al., 2007) lists 2,962 experimentally-observed isotopes and 652 isomers, its first edition (1958) included 1,297 known isotopes of 102 then-known elements. At the time of writing, elements 115 and 113 are the most massive superheavy elements which have been synthesized and found to exist at least for short time intervals, although more massive elements may exist in an island of stability beyond).

Unstable isotopes, once produced, will be *radioactive*, i.e. they will transmute to other isotopes through nuclear interactions, until at the end of such a decay chain a stable isotope is produced. Weak interactions will mediate transitions between protons and neutrons and lead to neutrino emission, involvements of atomic-shell electrons will result in X-rays from atomic-shell transitions after electron capture and internal-conversion transitions, and $\gamma$-rays will be emitted in electromagnetic transitions between excitation levels of a nucleus.

It is the subject of this book to explain in detail the astrophysical implications of this characteristic process of nuclear rearrangements, and what can be learned from measurements of the messengers of radioactive decays. But first we describe the phenomenon of radioactivity in more detail.

## 1.2 The Processes of Radioactivity

After Becquerel's discovery of radioactivity in 1896, Rutherford and others found out in the early twentieth century that there were different types of radioactive decay (Rutherford, 1903). They called them *$\alpha$ decay, $\beta$ decay* and *$\gamma$ decay*, terms which are still used today. It was soon understood that they are different types of interactions, all causing the same, spontaneous, and time-independent decay of an unstable nucleus into another and more stable nucleus.

*Alpha decay*: This describes the ejection of a $^4$He nucleus from the parent radioactive nucleus upon decay. $^4$He nuclei have since been known also as *alpha particles* for that reason. This decay is intrinsically fast, as it is caused by the *strong* nuclear interaction quickly clustering the nucleus into an alpha particle and the daughter nucleus. Since $\alpha$-nuclei are tighly-bound, they have been found as substructures even within nuclei. In the cases of nuclei much heavier than Fe, a nucleus thus consisting of many nucleons and embedded $\alpha$ clusters can find a preferred state for its number of nucleons by separation of such an $\alpha$ cluster liberating the binding-energy difference.[3] In such heavy nuclei, Coulomb repulsion helps to overcome the potential barrier which is set up by the strong nuclear force, and decay can occur through emission of an $\alpha$ particle. The $\alpha$ particle *tunnels*, with some calculable probability, through the potential barrier, towards an overall more stable and less-energetic assembly of the nucleons.

An example of $\alpha$ decay is $_{88}\mathrm{Ra}^{226} \Rightarrow {}_{86}\mathrm{Rn}^{222} + {}_2\mathrm{He}^4$, which is one step in the decay series starting from $^{238}$U. The daughter nucleus, $_{86}\mathrm{Rn}^{222}$, has charge $Z - 2$, where $Z$ is the original charge of the radioactive nucleus ($Z = 88$ in this example), because the $\alpha$ particle carried away two charge units from the original radioactive nucleus. Such decay frequently leads to an excited state of the daughter nucleus. Kinetic energy $E_\alpha$ for the $\alpha$ particle is made available from the nuclear binding energy liberation expressed by the *Q-value* of the reaction if the mass of the radioactive nucleus exceeds the sum of the masses of the daughter nucleus and of the helium nucleus[4]:

$$Q_\alpha = [M(_{88}\mathrm{Ra}^{226}) - \mathrm{M}(_{86}\mathrm{Rn}^{222}) - \mathrm{M}(_2\mathrm{He}^4)]\mathrm{c}^2 \tag{1.9}$$

The range of the $\alpha$ particle (its stopping length) is about 2.7 cm in standard air (for an $\alpha$ particle with $E_\alpha$ of 4 MeV), and it will produce about $2\times10^5$ ionizations before being stopped. Even in a molecular cloud, though its range would be perhaps $10^{14}$ times larger, the $\alpha$ particle would not escape from the cloud. Within small solids (dust grains), the trapping of radioactive energy from $\alpha$ decay provides a source of heat which may result in characteristic melting signatures.[5]

---

[3] The binding energy *per nucleon* is maximized for nucleons bound as a Fe nucleus.

[4] These masses may be either nuclear masses or atomic masses, the electron number is conserved, and their binding energies are negligible, in comparison.

[5] Within an FeNi meteorite, e.g., an $\alpha$ particle from radioactivity has a range of only $\sim$10 μm.

*Beta decay*: This is the most-peculiar radioactive decay type, as it is caused by the nuclear *weak interaction* which converts neutrons into protons and vice versa. The neutrino $\nu$ carries energy and momentum to balance the dynamic quantities, as Pauli famously proposed in 1930 (Pauli did not publish this conjecture until 1961 in a letter he wrote to colleagues). The $\nu$ was given its name by Fermi, and was discovered experimentally in 1932 by James Chadwick, i.e. *after* Wolfgang Pauli had predicted its existence. Neutrinos from the Sun have been discovered to *oscillate* between flavors. Beta decays are being studied in great detail by modern physics experiments, to understand the nature and mass of the $\nu$. Understanding $\beta$ decay challenges our mind, as it involves several such unfamiliar concepts and particles.

There are three types[6] of $\beta$-decay:

$$^A_Z X_N \longrightarrow \ ^A_{Z-1}X_{N+1} + e^+ + \nu_e \tag{1.10}$$

$$^A_Z X_N \longrightarrow \ ^A_{Z+1}X_{N-1} + e^- + \overline{\nu_e} \tag{1.11}$$

$$^A_Z X_N + e^- \longrightarrow \ ^A_{Z-1}X_{N+1} + \nu_e \tag{1.12}$$

In addition to Eq. (1.4) ($\beta^-$ *decay*), these are the conversion of a proton into a neutron ($\beta^+$ *decay*), and *electron capture*. The weak interaction itself involves two different aspects with intrinsic and different strength, the vector and axial-vector couplings. The $V^2_{fi}$ term in Eq. (1.5) thus is composed of two terms. These result in *Fermi* and *Gamow-Teller transitions*, respectively (see Langanke and Martínez-Pinedo, 2003, for a review of weak-interaction physics in nuclear astrophysics).

An example of $\beta$ decay is $^{13}_{7}\mathrm{N} \longrightarrow \ ^{13}_{6}\mathrm{C} + \mathrm{e}^+ + \nu$, having mean lifetime $\tau$ near 10 min. The kinetic energy $Q$ of the two leptons, as well as the created electron's mass, must be provided by the radioactive nucleus having greater mass than the sum of the masses of the daughter nucleus and of an electron (neglecting the comparatively-small neutrino mass).

$$Q_\beta = \left[M\left(^{13}_{7}\mathrm{N}\right) - \mathrm{M}\left(^{13}_{6}\mathrm{C}\right) - \mathrm{m_e}\right]\mathrm{c}^2 \tag{1.13}$$

where these masses are nuclear masses, not atomic masses. A small fraction of the energy release $Q_\beta$ appears as the recoil kinetic energy of the daughter nucleus, but the remainder appears as the kinetic energy of electron and of neutrino.

Capture of an electron is a *two-particle* reaction, the bound atomic electron $e^-$ or a free electron in hot plasma being required for this type of $\beta$ decay. Therefore, depending on availability of the electron, electron-capture $\beta$ decay lifetimes can be very different for different environments. In the laboratory case, electron capture usually involves the 1s electrons of the atomic structure surrounding the radioactive nucleus, because those present their largest density at the nucleus.

---

[6] We ignore here two additional $\beta$ decays which are possible from $\nu$ and $\overline{\nu}$ captures, due to their small probabilities.

In many cases the electron capture competes with $e^+ + \nu$ emission. In above example, $^{13}$N can decay not only by emitting $e^+ + \nu$, but also by capturing an electron: $^{13}_{7}\mathrm{N} + \mathrm{e}^- \longrightarrow ^{13}_{6}\mathrm{C} + \nu$. In this case the capture of a 1s electron happens to be much slower than the rate of $e^+$ emission. But cases exist for which the mass excess is not large enough to provide for the creation of the $e^+$ mass for emission, so that only electron capture remains to the unstable nucleus to decay. Another relevant example is the decay of $^7$Be. Its mass excess over the daugther nucleus $^7$Li is only 0.351 MeV. This excess is insufficient to provide for creation of the rest mass of an emitted $e^+$, which is 0.511 MeV. Therefore, the $^7$Be nucleus is stable against $e^+ + \nu$ emission. However, electron capture adds 0.511 MeV of rest-mass energy to the mass of the $^7$Be nucleus, giving a total 0.862 MeV of energy above the mass of the $^7$Li nucleus. Therefore, the $e^-$ capture process (above) emits a monoenergetic neutrino having $E_\nu = 0.862$ MeV.[7]

The situation for electron capture processes differs significantly in the interiors of stars and supernovae: Nuclei are ionized in plasma at such high temperature. The capture lifetime of $^7$Be, for example, which is 53 days against 1s electron capture in the laboratory, is lengthened to about 4 months at the solar center (see theory by Bahcall, 1964; Takahashi and Yokoi, 1983), where the free electron density is less at the nucleus.

The range of the $\beta$ particle (its stopping length) is small, being a charged particle which undergoes Coulomb scattering. An MeV electron has a range of several meters in standard air, during which it loses energy by ionizations and inelastic scattering. Energy deposit or escape is a major issue in expanding envelopes of stellar explosions, supernovae (positrons from $^{56}$Co and $^{44}$Ti) and novae (many $\beta^+$ decays such as $^{13}$N) (see Chaps. 4, 5, and 7 for a discussion of the various astrophysical implications). Even in small solids and dust grains, energy deposition from $^{26}$Al $\beta$-decay, for example, injects 0.355 W/kg of heat. This is sufficient to result in melting signatures, which have been used to study condensation sequences of solids in the early solar system (see Chap. 6).

*Gamma decay*: In $\gamma$ decay the radioactive transition to a different and more stable nucleus is mediated by the *electromagnetic interaction*. A nucleus relaxes from its excited configuration of the nucleons to a lower-lying state of the same nucleons. This is intrinsically a fast process; typical lifetimes for excited states of an atomic nucleus are $10^{-9}$ s. We denote such electromagnetic transitions of an excited nucleus *radioactive $\gamma$-decay* when the decay time of the excited nucleus is considerably longer and that nucleus thus may be considered a temporarily-stable configuration of its own, a *metastable* nucleus.

How is stability, or instability, of a nuclear-excited state effected? Electromagnetic transitions

$$A^\star \longrightarrow A^{g.s.} + \gamma \tag{1.14}$$

[7] This neutrino line has just recently been detected by the Borexino collaboration arriving from the center of the Sun (Arpesella et al., 2008).

must satisfy spin (angular momentum) conservation in the transition. The spin state of a nuclear state is a property of the nucleus, and reflects how protons and neutrons are spread over the quantum-mechanically allowed *shells* or nucleon wave functions (as expressed in the *shell model* view of an atomic nucleus). The photon (*$\gamma$ quantum*) emitted (Eq. 1.14) will thus have a *multipolarity* dictated by the initial and final states of the nucleus. Dipole radiation is most common and has multipolarity 1, emitted when initial and final state have angular momentum difference $\Delta l = 1$. Quadrupole radiation (multipolarity 2, from $\Delta l = 2$) is ~6 orders of magnitude more difficult to obtain, and likewise, higher multipolarity transitions are becoming less likely by similar probability decreases (the *Weisskopf estimates* (see Weisskopf, 1951)). This explains why some excited states in atomic nuclei are much more long-lived (*meta-stable*) than others; their transitions to the ground state are considered as *radioactivity* and called *$\gamma$ decay*.

The range of a $\gamma$-ray (its stopping length) is typically about 5–10 g/cm$^2$ in passing through matter of all types. Hence, except for dense stars and their explosions, radioactive energy from $\gamma$ decay is of astronomical implication only.[8]

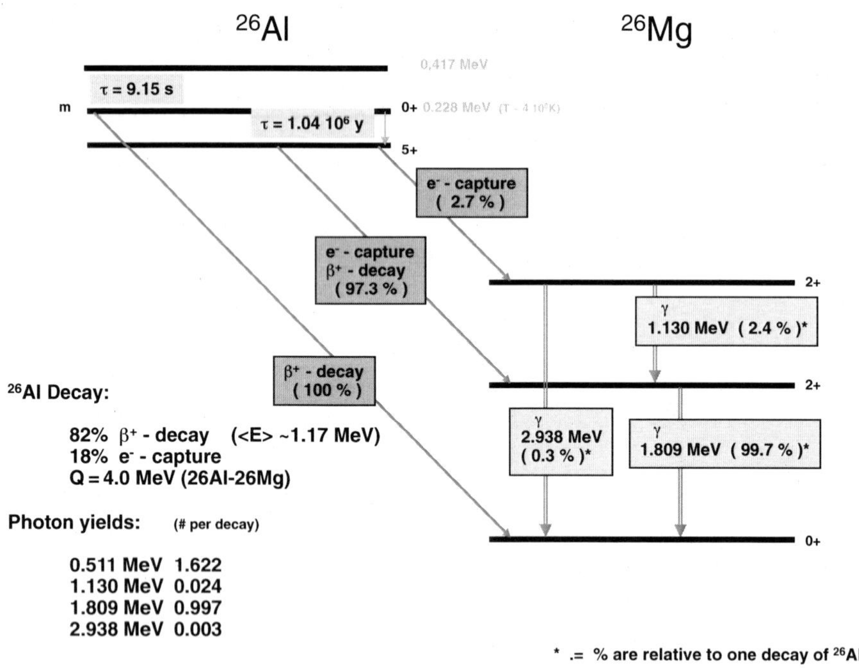


**Fig. 1.3** $^{26}$Al decay. The $^{26}$Al nucleus ground state has a long radioactive lifetime, due to the large spin difference of its state to lower-lying states of the daughter nucleus $^{26}$Mg. An isomeric excited state of $^{26}$Al exists at 228 keV excitation energy. If thermally excited, $^{26}$Al may decay through this state. Secondary products, lifetime, and radioactive energy available for deposits and observation depend on the environment

[8] Gamma-rays from nuclear transitions following $^{56}$Ni decay (though this is a $\beta$ decay by itself) inject radioactive energy through $\gamma$-rays from such nuclear transitions into the supernova envelope,

An illustrative example of radioactive decay is the $^{26}$Al nucleus. Its decay scheme is illustrated in Fig. 1.3. The ground state of $^{26}$Al is a 5+ state. Lower-lying states of the neighboring isotope $^{26}$Mg have states 2+ and 0+, so that a rather large change of angular momentum $\Delta l$ must be carried by radioactive-decay secondaries. This explains the large $\beta$-decay lifetime of $^{26}$Al of $\tau \sim 1.04\ 10^6$ year. In the level scheme of $^{26}$Al, excited states exist at energies 228, 417, and 1,058 keV. The 0+ and 3+ states of these next excited states are more favorable for decay due to their smaller angular momentum differences to the $^{26}$Mg states, although $\Delta l = 0$ would not be *allowed* for the 228 keV state to decay to $^{26}$Mg's ground state. This explains its relatively long lifetime of 9.15 s, and it is a *metastable* state of $^{26}$Al. If thermally excited, which would occur in nucleosynthesis sites exceeding a few $10^8$K, $^{26}$Al may decay through this state without $\gamma$-ray emission as $^{26}\mathrm{Al}^{\mathrm{g.s.}} + \gamma \longrightarrow ^{26}\mathrm{Al}^{\mathrm{m}} \longrightarrow ^{26}\mathrm{Mg} + e^+$, while the ground state decay is predominantly a $\beta^+$ *decay* through excited $^{26}$Mg states and thus including $\gamma$-ray emission. Secondary products, lifetime, and radioactive energy available for deposits and observation depend on the environment (See also Sect. 7.4).

## 1.3 Radioactivity and Cosmic Nucleosynthesis

Nuclear reactions in cosmic sites re-arrange the basic constituents of atomic nuclei (neutrons and protons) among the different configurations which are allowed by Nature. Throughout cosmic history, such reactions occur in different sites, and lead to rearrangements of the relative abundances of cosmic nuclei, a process called *cosmic chemical evolution*.[9]

The *cosmic abundance* of a specific isotope is expressed in different ways, depending on the purpose. Counting the atoms of isotope $i$ per unit volume, one obtains $n_i$, the number density of atoms of species $i$ (atoms/cm$^3$). The interest of cosmic evolution and nucleosynthesis lies in the fractional abundances of species $i$ related to the total, and how it is altered by cosmic nuclear reactions. Observers count a species $i$ and relate it to the abundance of a reference species. For astronomers this is hydrogen. Hydrogen is the most abundant element throughout the universe, and easily observed through its characteristic atomic transitions in spectroscopic astronomical measurements. The astronomical abundance scale is set from most-abundant cosmic element Hydrogen to $\log(X_H) = 12$ (Fig. 1.4), but mineralogists and meteoriticians use *Si* as their reference element and set $\log(X_{Si}) = 6$. Using the definition of Avogadro's constant $A_{Av}$ as the number of atoms which make up $A$ grams of species $i$ (i.e., one mole), we can obtain abundances *by mass*; $A_{Av} = 6.02214\ 10^{23}$ atoms/mole. The mass contained in a species $S$ results from

where it is absorbed in scattering collisions and thermalized. This heats the envelope such that thermal and optically bright supernova light is created. Deposition of $\gamma$-rays from nuclear transitions are the engines which make supernovae to be bright light sources out to the distant universe, used in cosmological studies (Leibundgut, 2000) to, e.g., support evidence for *dark energy*.

[9] We point out that there is no chemistry involved; the term refers to changes in abundances of chemical elements, which are a result of the changes in abundances of isotopes.

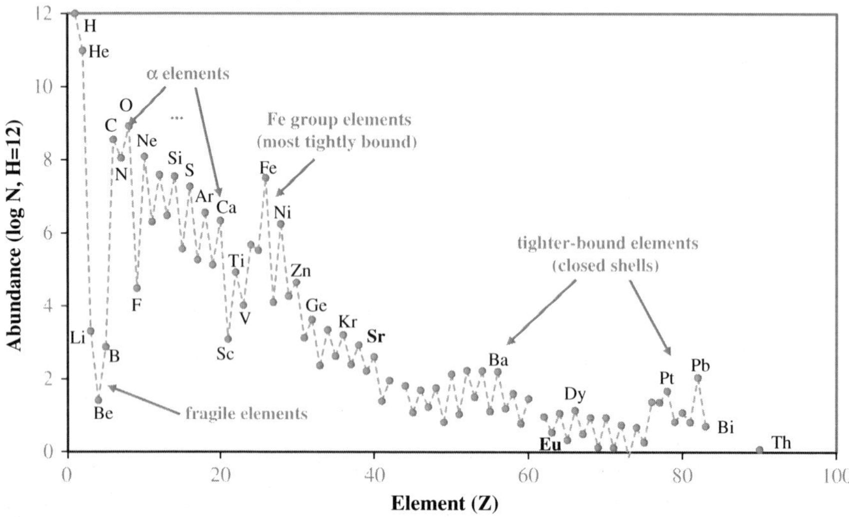


**Fig. 1.4** The abundance of elements in the present-day nearby universe. Abundances (by number) are shown in a logarithmic scale, and span 12 orders of magnitude. The interplay of nuclear properties (several are indicated in the graph) with environmental conditions in cosmic nucleosynthesis sites has created this complex abundance pattern during the course of cosmic history

scaling its abundance by number $N_s$ with its atomic weight $A$. We can get a global measure for cosmic evolution of the composition of matter by tracing how much of the total mass is in hydrogen, helium, and the remainder of elements called *metals*,[10] calling these quantities $X$ for hydrogen abundance, $Y$ for helium abundance, and $Z$ for the cumulative abundance of all nuclei heavier than helium. We call these *mass fractions* of hydrogen $X$, helium $Y$, and metals $Z$, with $X+Y+Z=1$. The metallicity Z is a key parameter used to characterize the evolution of elemental and isotopic composition of cosmic matter. The astronomical abundance scale is set from most-abundant cosmic element Hydrogen to $\log(X_H) = 12$ (Fig. 1.4), but mineralogists and meteoriticians use Si as their reference element and set $\log(X_{Si}) = 6$.

We often relate abundances also to our best-known reference, the solar system, denoting *solar-system* values by the $\odot$ symbol. Abundances of a species $S$ are then expressed in *bracket notation* as

$$\left[\frac{S}{H}\right] \equiv \log\left(\frac{X_S}{X_H}\right)_{\star} - \log\left(\frac{X_S}{X_H}\right)_{\odot} \tag{1.15}$$[11]

---

[10] This nomenclature may be misleading, it is used by convenience among astrophysicists. Only a part of these elements are actually metals under normal terrestial conditions.

[11] Deviations from the standard may be small, so that $[S_1/S_2]$ may be expressed in $\delta$ units (parts per mil), or $\varepsilon$ units (parts in $10^4$), or ppm and ppb; $\delta(^{2}9Si/^{2}8Si)$ thus denotes excess of the $^{2}9Si/^{2}8Si$ isotopic ratio above solar values in units of 0.1%.

Depending on observational method and precision, our astronomical data are *metallicity*, elemental *enrichments* with respect to solar abundances, or isotopic abundances. Relations to nuclear reactions are therefore often indirect. Understanding the nuclear processing of matter in the universe is a formidable challenge, often listed as one of the *big questions* of science.

After nucleosynthesis during the Big Bang, most-abundant were hydrogen (protons) and helium; the total amount of nuclei heavier than He (the *metals*) was less than $10^{-9}$ (by number, relative to hydrogen (Steigman, 2007)). Today, the total mass fraction of metals in *solar abundances*, our local reference which seems remarkably universal, is $Z = 0.0134$ (Asplund et al., 2009), compared to a hydrogen mass fraction of $X = 0.7381$.[12] This growth by about seven orders of magnitude is the effect of cosmic nucleosynthesis. Nuclear reactions in stars, supernovae, novae, and other places where nuclear reactions may occur, all contribute. But it also is essential that the nuclear-reaction products will be made available to observable cosmic gas and solids eventually. This book will also discuss our observational potential for cosmic isotopes, and we address the constraints and biases which limit our ability to draw far reaching conclusions.

The growth of isotopic and elemental abundances from cosmic nucleosynthesis does not occur homogeneously. Rather, the cosmic abundances observed today span a dynamic range of twelve orders of magnitude between abundant hydrogen and rare heavy elements (Fig. 1.4). Moreover, the elemental abundance pattern already illustrates clearly the prominent effects of nuclear structure (see Fig. 1.4): Iron elements are among the most-tightly bound nuclei, and locally elements with even numbers of nucleons are more tightly bound than elements with odd numbers of nuclei. The Helium nucleus (*$\alpha$-particle*) also is more tightly bound than its neighbors in the chart of nuclei, hence all elements which are multiples of $\alpha$'s are more abundant than their neighbors.

Towards the heavier elements beyond the Fe group, abundances drop by about five orders of magnitude again, signifying a substantially-different production process than the mix of charged-particle nuclear reactions that produced the lighter elements: *neutron capture* on Fe *seed nuclei*. The two abundance peaks seen for heavier elements are the results of different environments for cosmic neutron capture reactions (the *r-process* and *s-process*), both determined by neutron capture probabilities having local extrema near *magic numbers*. The different peaks arise from the particular locations at which the processes' reaction path encounters these *magic nuclei*, as neutron captures proceed much faster (slower) than beta decays in the *r*-process (*s*-process).

The issues in cosmic nucleosynthesis research are complex, and cover the astrophysics of stars, stellar explosions, nuclear reactions on surfaces of compact stars and in interstellar space. For each of the potential nuclear-reaction sites, we need to understand first how nuclear reactions proceed under the local conditions, and

---

[12] This implies a *metallicity* of solar matter of 1.4%. Earlier than ~2005, the commonly-used value for solar metallicity had been 2%.

then how material may be ejected into interstellar space from such a source. None of the nucleosynthesis sites is currently understood to a sufficient level of detail which would allow us to sit back and consider cosmic nucleosynthesis *understood*. For example, for the Sun, where one would assume we know most detail, solar neutrino measurements have been a puzzle only alleviated in recent years with the revolutionary adoption of non-zero masses for neutrinos. Even if we consider this a sufficient explanation, solar elemental abundances have recently been revised by almost a factor two based on consistency arguments between three-dimensional models of the solar photosphere and measured elemental line profiles from this photosphere; but helio-seismological measurements for the Sun are in significant conflict with the underlying model which describes the interior structure of the Sun in these three-dimensional models (Asplund et al. 2009). As another example, there are two types of supernova explosions, core-collapse supernovae marking the final gravitational collapse of a massive star once its nuclear fuel is exhausted, and thermonuclear supernovae, thought to originate from detonation of degenerate stars once they exceed a critical threshold for nuclear burning of Carbon. For neither of these supernova types, a *physical* model is available, which would allow us to calculate and predict the outcome (energy and nuclear ashes) from such an explosion under given, realistic, initial conditions (see Chaps. 4 and 5). Much research remains to be done in cosmic nucleosynthesis.

One may consider measurements of cosmic material in all forms to provide a wealth of data, which by now has been exploited to understand cosmic nucleosynthesis. Note, however, that cosmic material as observed has gone through a long and ill-determined journey. We need to understand the trajectory in time and space of the progenitors of our observed cosmic-material sample if we want to interpret it in terms of cosmic nucleosynthesis. This is a formidable task, necessary for distant cosmic objects, but here averaging assumptions help to simplify studies. For more nearby cosmic objects where detailed data are obtained, astrophysical models quickly become very complex, and also need simplifying assumptions to operate for what they are needed. It is one of the objectives of cosmic nucleosynthesis studies to contribute to proper models for processes in such evolution, which are sufficiently isolated to allow their separate treatment. Nevertheless, carrying out *well-defined experiments* for a source of cosmic nucleosynthesis remains a challenge, due to this often ill-constrained history (see Chaps. 6 and 7).

The special role of radioactivity in such studies is contributed by the intrinsic decay of such material after it has been produced in cosmic sites. This brings in a new aspect, the clock of the radioactive decay. With such additional information, changes in isotopic abundances with time will occur naturally, and will leave their imprints in observable isotopic abundance records. For example, the observation of unstable technetium in stellar atmospheres of AGB stars was undisputable proof of synthesis of this element inside the same star, because the evolutionary time of the star exceeds the radioactive lifetime of technetium. Another example, observing radioactive decay $\gamma$-ray lines from short-lived Ni isotopes from a supernova is clear proof of its synthesis in such explosions; measuring its abundance through $\gamma$-ray brightness is a prominent goal for direct *calibration* of processes in the supernova interior. A last example, solar-system meteorites show enrichments in daughter products of characteristic radioactive decays, such as $^{26}$Al and $^{53}$Mn; the fact that these radioactive elements were still not decayed when those solids formed sets important constraints to the time interval between the latest nucleosynthesis event near the forming Sun and the actual condensation of solid bodies in the young solar system. This book will discuss these examples in detail, and illustrate the contributions of radioactivity studies to the subject of cosmic nucleosynthesis.

## 1.4 Observations of Cosmic Radioactive Isotopes

Astronomy has expanded beyond the narrow optical band into *new astronomies* in the past decades. By now, we are familiar with telescopes measuring radio emission through infrared emission towards the long wavelength end, and ultraviolet, X-ray, and $\gamma$-ray emission towards the short wavelength end (see Fig. 1.5). The physical origins of radiation are different in different bands. Thermal radiation dominates emission from cosmic objects in the middle region of the electromagnetic spectrum, from a few 10 K cold molecular clouds at radio wavelengths through dust and stars up to hot interstellar gas radiating X-rays. Non-thermal emission is characteristic for the wavelength extremes, both at radio and $\gamma$-ray energies. Characteristic spectral lines originate from atomic shell electrons over most of the spectrum; nuclear lines are visible only in roughly two decades of the spectrum at 0.1–10 MeV. Few exceptional lines arise at high energy from annihilations of positrons and pions. Thus, cosmic *elements* can be observed in a wide astronomical range, while *isotopes* are observed almost exclusively through $\sim$MeV $\gamma$-rays (see Fig. 1.6 for exceptions). Note that nucleosynthesis reactions occur among isotopes, so that this is the

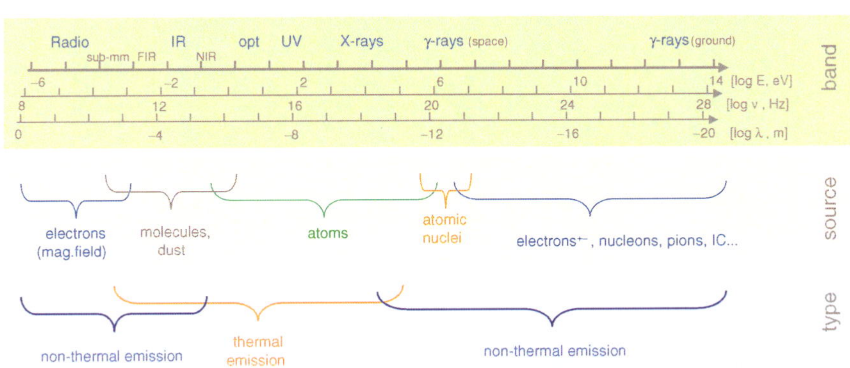


**Fig. 1.5** The electromagnetic spectrum of candidate astronomical measurements ranges across more than twenty orders of magnitude. Not all are easily accessible. Information categories of thermal and non-thermal, and of molecular, atomic, nuclear, and elementary-particle physics origins of cosmic radiation extends over different parts of this broad spectrum. Nuclear physics is accessible in a small band (0.1–10 MeV) only, and is made difficult by penetrating radiation and by large instrumental backgrounds from cosmic-ray interactions

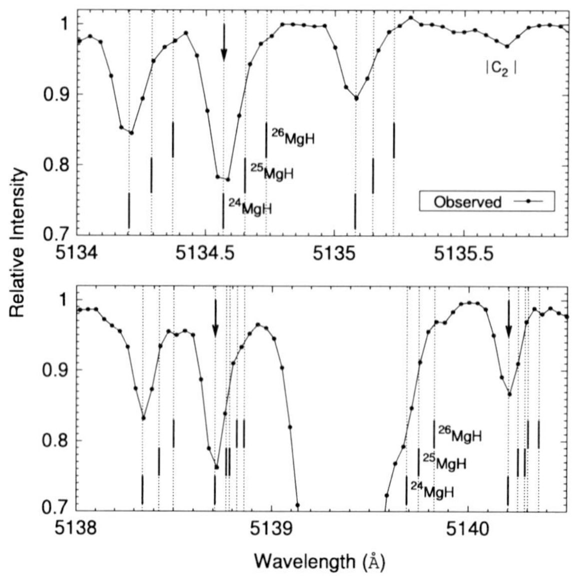


**Fig. 1.6** Example of an absorption-line spectrum of a cool star with a present-generation optical telescope, such as shown in Fig. 1.7. Molecular lines have isotopic shifts, which can be recognized as changes in line shapes, as resulting from the isotopic abundance ratio. (from Yong et al. 2004)

**Fig. 1.7** The Very Large Telescope (VLT) on Mount Paranal in Chile, with four telescopes (*lower right*), is one of the modern optical instruments. Equipped with high-resolution spectrographs such as FLAMES (insert *lower right*), absorption-line spectroscopy of stars in nearby galaxies can be made. (Figures ESO)

information of interest when we wish to investigate cosmic nucleosynthesis environment properties.

Only few elements such as technetium (Tc) do not have any stable isotope; therefore, elemental photospheric absorption and emission line spectroscopy, the backbone of astronomical studies of cosmic nucleosynthesis, have very limited application in astronomy with radioactivities. This is about to change currently, as spectroscopic devices in the optical and lower-energy regime approach resolutions sufficient to resolve fine structure lines, thus enabling isotopic abundance studies. Observational studies of cosmic radioactivities are best performed by techniques which intrinsically obtain isotopic information. These are:

- precision mass spectroscopy in terrestrial laboratories, which has been combined with sophisticated radiochemistry to extract meteoritic components originating from outside the solar system
- spectroscopy of characteristic $\gamma$-ray lines emitted upon radioactive decay in cosmic environments

Both these *astronomical disciplines* have a relatively young history. They encounter some limitations due to their basic methods of how astronomical information is obtained:

- Precision mass spectrometry of meteorites for astronomy with radioactivity began about 1960 with a new discovery of now extinct radioactivity within the young solar system. By heating a bulk meteorite sample the presence of excess $^{129}$Xe was clearly demonstrated, and attributed to trapped gas enriched in $^{129}$I at the time of formation of this meteorite, which from mineralogical arguments is determined to be the early solar system epoch about 4.6 Gigayears (Gy) ago (Reynolds, 1960). This was the first evidence that the matter from which the solar system formed contained radioactive nuclei whose half-lives are too short to be able to survive from that time until today ($^{129}$I decays to $^{129}$Xe within $1.7 \times 10^7$ year). Isotopic anomalies found in such *extra-solar* inclusions, e.g. for C and O isotopes, range over four orders of magnitude for such *star dust* grains as shown in Fig. 1.8 (Zinner, 1998), while isotopic-composition variations among bulk meteoritic-material samples are a few percent at most. The measurements are characterized by an amazing sensitivity and precision, clearly resolving isotopes and counting single atoms at ppb levels to determine isotopic ratios of such rare species with high accuracy. This astronomy in terrestrial laboratories is now an established part of astrophysics (see Clayton and Nittler, 2004, for a recent review). Studies are limited only by sample preparation and by the extraction techniques evaporizing dust grain surfaces for subsequent mass spectrometry. Substantial bias of the technique arises from the preparation of suitable samples, extracting meteoritic material with increased fractions of non-solar material and thus increasing the signal against background from solar-system isotopic composition. In general, this favors the hardest and least-resolvable meteoritic inclusions of most-refractory minerals, and noble gases included in microscopic cavities. Furthermore, the origin of non-solar dust, i.e. the journey from its formation

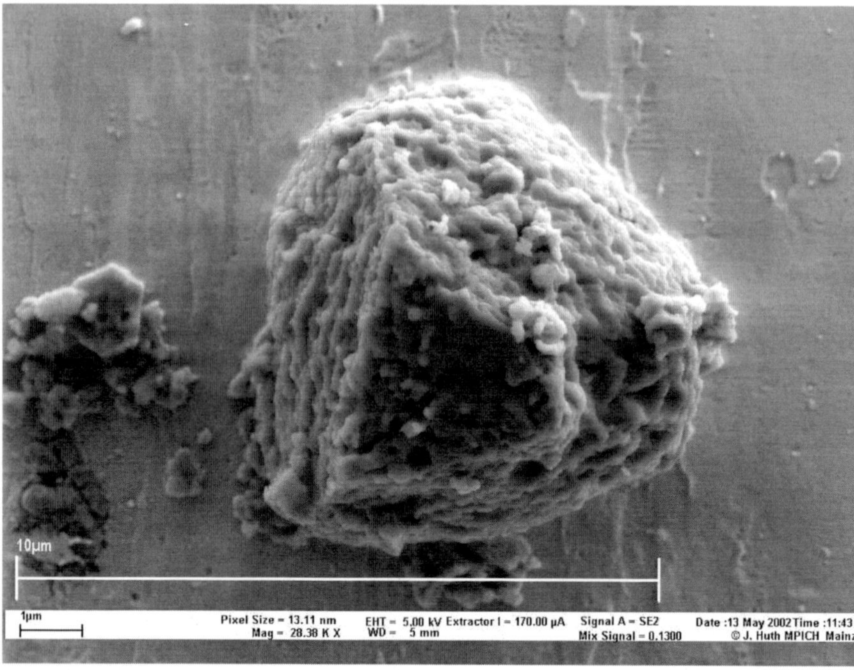


**Fig. 1.8** Meteoritic inclusions such as this SiC grain are recognized as dust formed near a cosmic nucleosynthesis source outside the solar system, from their large isotopic anomalies, which cannot be explained by interstellar nor solar-system processing but are reminiscent of cosmic nucleosynthesis sites. Having condensed in the envelope of a source of new isotopes, laboratory mass spectroscopy can reveal isotopic composition for many elements, thus providing a remote probe of one cosmic nucleosynthesis source

in stellar envelopes or interstellar gas up to inclusion in meteorites which found their way to Earth, remains subject to theoretical modeling based on the observed grain composition and morphology plus (uncertain) theories of cosmic dust formation (Zinner, 1998).

- Characteristic $\gamma$-ray lines from cosmic sources were not known until the 1960s, when spaceflight and its investigations of the near-earth space radiation environment had stimulated measurements of $\gamma$-rays. The discovery of a cosmic $\gamma$-ray line feature near 0.5 MeV from the direction towards the center of our Galaxy in 1972 (Johnson et al., 1972) stimulated balloon and satellite experiments for cosmic $\gamma$-ray line spectroscopy. By now and with current experiments such as the INTEGRAL mission of ESA shown in Fig. 1.9, this technique has established an astronomical discipline of its own, the window of electromagnetic radiation clearly attributed to specific isotopes. Decay of the isotopes $^{26}$Al, $^{60}$Fe, $^{44}$Ti, $^{57}$Ni, and $^{56}$Ni in distant cosmic sites is an established fact, and astrophysical studies make use of such measurements. The downsides of those experiments is the rather poor resolution by astronomy standards (on the order of degrees), and the sensitivity limitations due to large instrumental backgrounds, which effectively

**Fig. 1.9** Example of a present-generation space-borne $\gamma$-ray telescope. The INTEGRAL satellite (artist view picture, ESA) has as one of its two main telescopes a spectrometer SPI, shown at the lower-right schematically with its 19-detector Ge camera and the tungsten mask for imaging by casting a shadow onto the camera. Space-based instruments of this kind have been used to directly record characteristic $\gamma$-ray lines from the decay of unstable isotopes near sites of current-epoch cosmic element formation

only shows the few brightest sources of cosmic $\gamma$-rays until now (see Diehl et al., 2006, for a discussion of achievements and limitations).

Despite their youth and limitations, both methods (described in Chap. 10) to address cosmic radioactivities share a rather direct access to isotopic information, unlike other fields of astronomy. From a combination of all available astronomical methods, the study of cosmic nucleosynthesis will continue to advance towards a truly astrophysical decomposition of the processes and their interplays. This book describes where and how specific astronomical messages from cosmic radioactivity help to complement these studies.

## 1.5 The Structure of this Book

This book begins with a basic introduction in Chap. 2, written by D.D. Clayton, one of the founders of the field, who also contributed a fundamental textbook on stellar astrophysics and many stimulating ideas that determined the course of this field, in particular the need for a concerted observational effort to understand nucleosynthesis from both measurements of electromagnetic radiation and meteoritic material. This Introduction and Overview is followed by discussions of the specific sources of cosmic radioactivity in Chaps. 3, 4, and 5, i.e. stars in their hydrostatic interiors, massive stars and their core-collapse supernova explosions, and binary-star interactions including thermonuclear supernova explosions. These chapters describe our current theoretical models and understanding of each of the cosmic sources of radioactivities, and guide these through references to key observations. Then we turn our perspective directly towards the observational side, and present prominent locations of radioactivities as observed (Chapter 6 addresses the Solar System, Chap. 7 more broadly then the different diffuse radioactivities seen in interstellar gas), and discuss how the data on cosmic radioactivities shape our astrophysical modeling of cosmic nucleosynthesis. The book is completed by a survey of a few tools which are characteristic of the field (in Chaps. 8, 9, and 10), and Appendices holding useful tables, a timeline of milestones of the field, and a glossary of key terms of Astronomy with Radioactivities to guide readers through the basic terminology. Beyond the general guidance of Chap. 2, subsequent chapters can be read in any sequence suitable to the interests and curiosities of the reader, as we provide cross references throughout the book. Enjoy the ride!

## References

Arpesella, C., Back, H.O., Balata, M., et al.: Direct measurement of the Be7 solar neutrino flux with 192 days of Borexino data. Phys. Rev. Lett. **101**(9), 091, 302 (2008). doi:10.1103/PhysRevLett.101.091302, 0805.3843

Asplund, M., Grevesse, N., Sauval, A.J., Scott, P.: The chemical composition of the sun. ARA&A **47**, 481–522 (2009). doi:10.1146/annurev.astro.46.060407.145222, 0909.0948

Bahcall, J.N.: Electron capture in stellar interiors. ApJ **139**, 318 (1964). doi:10.1086/147755

Clayton, D.D., Nittler, L.R.: Astrophysics with presolar stardust. ARA&A **42**, 39–78 (2004). doi:10.1146/annurev.astro.42.053102.134022

Diehl, R., Prantzos, N., von Ballmoos, P.: Astrophysical constraints from gamma-ray spectroscopy. Nucl. Phys. A **777**, 70–97 (2006). doi:10.1016/j.nuclphysa.2005.02.155, arXiv:astroph/0502324

Johnson, W.N. III, Harnden, F.R. Jr, Haymes, R.C.: The spectrum of low-energy gamma radiation from the galactic-center region. ApJ **172**, L1 (1972). doi:10.1086/180878

Langanke, K., Martínez-Pinedo, G.: Nuclear weak-interaction processes in stars. Rev. Modern Phys. **75**, 819–862 (2003). doi:10.1103/RevModPhys.75.819, arXiv:nucl-th/0203071

Leibundgut, B.: Type Ia Supernovae. A&A Rev. **10**, 179–209 (2000). doi:10.1007/s001590000009, arXiv:astro-ph/0003326

Messiah, A.: Quantum Mechanics. North-Holland Publication 1961–1962, Amsterdam (1962)

Pfennig, G., Klewe-Nebenius, H., Seelmann-Eggebert, W.: Karlsruher nuklidkarte – chart of nuclides, 7th edition. (2007) ISBN 3-92-1879-18-3

Reynolds, J.H.: Determination of the age of the elements. Phys. Rev. Lett. **4**(1), 8 (1960). doi:10.1103/PhysRevLett.4.8

Rutherford, E.: Radioactive processes. Proc. Phys. Soc. London **18**, 595–600 (1903). doi:10.1088/1478-7814/18/1/360

Steigman, G.: Primordial nucleosynthesis in the precision cosmology era. Ann. Rev. Nucl. Particle Sci. **57**, 463–491 (2007). doi:10.1146/annurev.nucl.56.080805.140437, 0712.1100

Takahashi, K., Yokoi, K.: Nuclear b-decays of highly ionized heavy atoms in stellar interiors. Nucl. Phys. A **404**, 578–598 (1983). doi:10.1016/0375-9474(83)90277-4

Weisskopf, V.F.: Radiative transition probabilities in nuclei. Phys. Rev. **83**, 1073–1073 (1951). doi:10.1103/PhysRev.83.1073

Weizsäcker, C.F.V.: Zur Theorie der Kernmassen. Zeitschrift fur Physik **96**, 431–458 (1935). doi:10.1007/BF01337700

Yong, D., Lambert, D.L., Allende Prieto, C., Paulson, D.B.: Magnesium isotope ratios in hyades stars. ApJ **603**, 697–707 (2004). doi:10.1086/381701, arXiv:astro-ph/0312054

Zinner, E.: Stellar nucleosynthesis and the isotopic composition of presolar grains from primitive meteorites. Ann. Rev. Earth and Planetary Sci. **26**, 147–188 (1998). doi:10.1146/annurev.earth.26.1.147

# Chapter 2
# The Role of Radioactive Isotopes in Astrophysics

D.D. Clayton

## 2.1 History of Nucleosynthesis and Radioactivity

Radioactive nuclei can not be not infinitely old or they would now be gone. This point had first been made by Lord Rutherford, who concluded in 1929 (Rutherford, 1929) that the elements had been created in the sun 100 million years ago and had somehow got from sun to earth! Terrible astrophysics, but his argument showed profound appreciation of the conundrum of live radioactivity on earth; namely, the nuclei must be created. By the middle of the twentieth century, the consequences had progressed no further. The question, "Why aren't they all gone?" went largely unspoken.

Radioactivity had been discovered by Henri Becquerel in 1896 when he showed that the element uranium emitted radiation that would develop images of the uranium salts on an adjacent photographic plate. Marie Curie coined the term *radioactivity* to describe this phenomenon after she demonstrated that other elements possessed the same spontaneous property. Rutherford separated the rays into three types – $\alpha$, $\beta$, and $\gamma$ – by their physical properties. These are described in Chap. 1.

In mid-twentieth century, astronomers discovered metal-poor stars in increasing numbers, especially after the introduction of the CCD chip, which greatly increased the sensitivity of spectroscopic observations of stars. The abundances relative to hydrogen in old low-mass stars reveals the initial composition of the gas from which those stars formed. Observed metallicities ranged over a continuum of values from $(Fe/H)/(Fe/H)_\odot = 10^{-4}$ to 1 and even greater, so the synthesis of metals must have occurred after the Big Bang. Today we know that creation of the chemical elements occurs in stars by synthesis of them from initial H and He. By the same token, natural radioactivity must be the product of nucleosynthesis in stars.

D.D. Clayton (✉)
Clemson University, Clemson, SC 29634-0978, USA, claydonald@gmail.com

Clayton, D.D.: *The Role of Radioactive Isotopes in Astrophysics*. Lect. Notes Phys. **812**, 25–79 (2011)
DOI 10.1007/978-3-642-12698-7_2 

### 2.1.1 Two Very Different Pioneers

There could hardly be two more different scientists than the two that made the first significant steps toward nucleosynthesis in stars of the elements heavier than carbon. Both were Europeans disrupted in careers by world war II. The originator of nucleosynthesis in stars arose from non-privileged Yorkshire mill villages. The pioneer in glimpsing how nuclear magic numbers can be observed in isotopic abundances arose from a distinguished Viennese family of geologists.

#### 2.1.1.1 Sir Fred Hoyle (1946, 1954)

In a flash of astrophysical brilliance, Fred Hoyle created the theory of nucleosynthesis in stars with two monumental papers (Hoyle et al., 1946, 1954). The first (Hoyle et al., 1946) demonstrated that stars will naturally evolve to central temperatures $4\times10^9$ K and greater, and that nuclear statistical equilibrium (NSE) at high temperature $T$ creates an abundance peak at $^{56}$Fe that Hoyle interpreted as the explanation of that observed abundance peak. This explains why a graph of the abundances of chemical elements in the sun as a function of atomic weight declines almost monotonically from H to Sc (A = 1–45), whereafter it increases sharply to a broad abundance peak centered on A = 56 (Fe). Readers unfamiliar with the meaning and issues of abundances in the universe can find an eloquent essay in the introductory chapter of Clayton (2003).

Hoyle's second paper (Hoyle et al., 1954) described how the elements from carbon to nickel are synthesized during the advanced evolution of massive stars. This is called *primary nucleosynthesis* because the elements are fused from the initial H and He of the massive stars. Hoyle described how the *ashes* of each thermonuclear burning epoch became, upon contraction and heating, the fuel for the next group of elements synthesized. When its core collapses, much of the overlying new elements is returned to the interstellar medium. Such nucleosynthesis creates new heavy nuclei and is responsible for increasing the metallicity of the interstellar medium as it ages.

Because the origin of our chemical elements is one of the grand theories of science (Clayton, 2007) went to some lengths to create what he called *Hoyle's equation* as determined from careful reading of his 1954 paper (Hoyle et al., 1954). Hoyle's words and calculations pointed clearly to ideas of nucleosynthesis in stars that he was advancing for the first time and that are more sweeping than detail-oriented sequels. Hoyle's discussion was phrased in terms of the mass $\Delta m_{\text{new}}$ of new primary isotopes that are being ejected from massive stars. His basic approach to stellar nucleosynthesis is

$$\frac{dm_{\text{new}}}{dt} = H_{\text{nucl}} \tag{2.1}$$

where

$$H_{\text{nucl}} = B(t')\mathbf{Ev}(t'-t)\Sigma_k \Delta m_k \tag{2.2}$$

is the *Hoyle nucleosynthesis rate*. $B(t')$ is the stellar birthrate of stars having total mass such that they evolve to end their lives at time $t$, $\mathbf{Ev}(t', t)$ is an operator (rather than a number) that expresses the nuclear and stellar evolution during its lifetime from $t'$ to $t$, and $\Sigma m_{\rm k}$ is the mass of isotope $k$ ejected at time $t$. Then a sum over all presolar birthdates $t'$ selects the appropriate stellar masses. Stellar evolution was only dimly perceived in 1953 when Hoyle wrote this paper. The structure of red giants was its current literature frontier, and Hoyle was perhaps the world's leading expert on its ideas, enabling him to discern the more fruitful subsequent evolution that he advanced in this 1954 paper.

Hoyle subtitled this paper *The synthesis of elements from Carbon to Nickel*. These primary isotopes are ejected from massive stars that progress through a series of core evolutions that Hoyle laid out for the first time. Hoyle explained that gravitational contraction causes temperature increases after each nuclear fuel is consumed, and he described the nuclear burning during each advanced core evolution. Because those massive stars all evolve almost instantaneously in comparison with galactic timescale, Hoyle takes $B_{\rm M>}(t)$ to be the birthrate of all such massive stars at time $t$, and it clearly equals their death rate at the same time if the numbers of stars are to change only slowly. The subscript $M>$ characterizes stars too massive to become white dwarfs. For those large-mass stars Hoyle et al. (1954) predicted that collapse of the final central evolved core is inevitable. For those massive stars Hoyle's equation expresses the rate of ejection of new primary isotopes from C to Ni as

$$\frac{dm(C-Ni)}{dt} = B_{M>}(t)\mathbf{Ev}_{\rm nucl}\Sigma_k \Delta m_k \tag{2.3}$$

For these short-lived stars, $\mathbf{Ev}$ is an operator (rather than a number) that expresses the nuclear and stellar evolution during the stellar lifetime. It replaces the initial composition of the star by the composition it has attained at the time when its core collapses. Hoyle attributed the mass and identity $k$ of new primary isotopes ejected per massive star to the following successive core burning phases: $^{12}$C and $^{16}$O from He burning; $^{20}$Ne, $^{23}$Na, and $^{24}$Mg from C burning; additional $^{16}$O and $^{24}$Mg from Ne burning; $^{28}$Si and $^{32}$S from O burning; $^{32}$S, $^{36}$Ar and $^{40}$Ca from photoalpha reactions on $^{32}$S and heavier alpha nuclei during later heating of O-burned matter by the inevitable contraction; and finally $^{52}$Cr, $^{56}$Fe, $^{60}$Ni from subsequent nuclear statistical equilibrium. Hoyle also correctly stated that neutrino emission governs the collapse timescale when core temperature exceeds $T = 3\times10^9$ K.

Hoyle's equation expresses a breathtakingly modern view of the metallicity-increasing nucleosynthesis during galactic history. Hoyle missed only the complete photonuclear rearrangement during Si burning and the $n/p$ ratio within the NSE. But his equation, given above, remains correct today. Like Schroedinger's equation, for which much work is required to determine the appropriate Hamiltonian operator used within it, so Hoyle's equation involves much work to determine the Hoyle nuclear evolution $H_{\rm nucl}$.

Hoyle et al. (1954) also distinguished the idea of secondary nuclei, those whose created abundance derives from initial seed concentrations of primary nuclei that had been created in previous massive stars, thereby seeding the interstellar matter with primary elements. These seed nuclei are required in order that secondary nuclei be produced from them in stars. He emphasized especially $^{14}$N, $^{18}$O, $^{19}$F and $^{22}$Ne in that regard, each of which depends on initial primary C and O nuclei within the initial composition of later-generation massive stars. Their yields of secondary nuclei do not obey Hoyle's equation but instead are proportional to the initial metallicity of each star. Hoyle's paper also first noted that $^{22}$Ne would be a source of free neutrons; indeed, it is today their major source in burning shells of massive stars, although that insight is usually attributed to later emphasis by others.

It is unfortunate that Hoyle did not put to paper the equation he envisioned and described. Had he done so, clearer scientific visibility of his unparalleled achievement would have followed more easily. Research in nucleosynthesis has followed his approach during five decades of countless numerical models evaluating Hoyle's equation. Nonetheless, the astrophysical world began to inappropriately cite a paper to appear, with Hoyle as coauthor, in 1957 for the general origin of nucleosynthesis in stars rather than Hoyle's works. Hoyle's great achievement lay somewhat obscurred until modern times, at least in terms of citations of research works.

#### 2.1.1.2 Hans Suess, Nuclear Shell Model, and Abundances

Hans Suess was born in Vienna in 1909 to a distinguished family of geologists. His chemical training and focus on abundances of the elements in meteorites enabled him to establish *Suess's rules* for the abundances and their isotopes. The elemental abundances were not well known after WWII, but testing of a theory of nucleosynthesis needed that data base. In the late 1940s Suess began to argue that certain regularities of the abundances had to reflect nuclear properties of their isotopes. He used those systematics to coauthor with Hans Jensen a discovery paper (Haxel et al., 1949) for the magic numbers of nuclear shell structure. These are hugely important for nucleosynthesis. The magic numbers are favored numbers of identical nucleons, either protons or neutrons, in the sense that clusters of those numbers of identical nucleons have larger than normal binding energy. They are 2, 8, 14, 20, 28, 50, 82 and 126, and they result from the combined effect of a deep spherical potential binding the nucleons together, coupled with a very strong spin-orbit energy that moderates energy gaps between differing nuclear shells. For that paper Jensen shared the 1963 Nobel Prize in Physics with Maria Mayer for their independent theoretical work on how the spin-orbit force establishes the magic numbers. Suess, working in Hamburg and Heidelberg, had seen the evidence of magic numbers in the abundance regularities shown by nuclei having a magic structure of neutrons or protons. For example, the abundances of the isotopes of barium implicated neutron irradiation in their nucleosynthesis. Suess amazingly divined that the very high abundance of $^{56}$Fe had to reflect the doubly-magic properties of its $^{56}$Ni isobar having $N$=28 for both protons and neutrons. That correct assertion was not accepted for a decade because Hoyle's papers (and also B$^2$FH) had maintained that that abundance peak was established within a nuclear equilibrium having excess of neutrons relative to protons, so that the abundance of Fe isotopes could reflect their own nuclear properties rather than those of Ni. This error persisted for a decade, and had many dead-end astrophysical consequences. These have been intensively reviewed (Clayton, 1999), who also reviews Suess's pioneering papers from the 1940s.

The magic numbers were also to play a pivotal role in the theories for nucleosynthesis of the elements heavier than the Fe peak. Some of the earliest testaments to the correctness of the theoretical ideas lay in their interplay with the magic numbers. Suess himself called for two neutron-addition processes needed to account for magic-number abundances within the heavy elements. After moving to the United States in 1950, Suess coauthored with Harold Urey an immensely influential tabulation (Suess and Urey, 1956) of the abundances of the elements, largely from chemical analyses of meteorites rather than from astronomical observations. That review paper by Suess and Urey became a cornerstone of empirical evidence for nucleosynthesis in stars.

## *2.1.2 The Second Decade*

Hoyle's two papers were not enough to establish the theory of nucleosynthesis in stars. Contributions from other pioneers came together to launch the full theory and to win the acceptance that it enjoys today.

### 2.1.2.1 William A. Fowler and Burbidge, Burbidge, Fowler and Hoyle ($B^2FH$)

William A. Fowler had met Fred Hoyle in 1953 when Hoyle famously predicted the existence and the energy of the $0^+$ second excited state of $^{12}C$. In that first application to nuclear physics of what has come to be called *the anthropic principle*, Hoyle had argued that if such a nuclear state of $^{12}C$ did not exist, neither would we! Fowler said that that prediction was what "really hooked me on nucleosynthesis". Fowler met Hoyle again during his 1955 sabbatical leave in Cambridge UK. Fowler was at this time already the leader of the world effort (which Caltech championed) to determine the rates at which nuclear reactions would occur in stars. This involved the now familiar technique of measuring nuclear interactions at MeV energies with Van de Graff accelerator beams of charged ions, and extrapolating measured data downward in energy to the Gamow-peak energy within a thermal distribution (Clayton, 1968, Chap. 4). Fowler's zest for nuclear astrophysics was boundless, and he soon had the entire Kellogg Radiation Laboratory at work on thermonuclear reactions rates between positive ions. Fowler was awarded a share of the 1983 Nobel Prize in Physics for that pioneering program of research. His efforts had been directed primarily to the question of thermonuclear power in stars; but Fowler became intensely attracted also to the larger question of the origin of the elements. In Cambridge he met Geoffrey and Margaret Burbidge, and these three began a project with Fred Hoyle to write a survey paper on the issues of nucleosynthesis in stars. That paper was written at Caltech in 1956 and published the next year (Burbidge et al., 1957). Soon called simply $B^2FH$, it became one of the most celebrated papers in astrophysics.

The authors of $B^2FH$ contributed creatively and energetically to formulating the *neutron-capture processes* for synthesizing the elements heavier than nickel. Ascribing crucial roles to the magic neutron numbers $N = 50$, 82 and 126 they described environmental situations in stars within which selected isotopes would be abundant, fleshing out ideas that Suess had envisioned earlier for two neutron-capture processes. Suess had, however, been unable to formulate these clearly enough to win contemporary acclaim. The slow neutron-capture process, named *s process* by $B^2FH$, envisioned 100-to-1000 years between neutron captures, so that radioactive isotopes would generally beta decay before capturing another neutron, keeping the capture path trailing along the valley of beta stability. The rapid-neutron-capture process, named *r process* by $B^2FH$, envisioned neutron densities so large that neutron captures occur in tens of milliseconds, faster than beta decays, with the result that the capture path moved into the realm of radioactive neutron-rich nuclei, being halted only when no additional neutrons could be stably added owing to their diminishing separation energies. These created *waiting points* at which the capture flow would halt and wait until beta decay occurs (see also Seeger et al., 1965; Clayton, 1968). Defining the *s* process and the *r* process was the high point of the $B^2FH$ contributions to nucleosynthesis theory. Their Appendix included an inspiring table of all heavy isotopes in which each was characterized as being either *s* process only, mostly *s* process, comparable *s* process and *r* process, mostly *r* process and *r* process only. This can be regarded as the next important step in nucleosynthesis theory.

An important aspect of astronomy with radioactivity lies in the competition between beta decay and neutron capture that ensues when neutron capture by stable isotopes create isotopes that can undergo beta decay. Most do so quickly in comparison with the time required to capture another neutron, but some key branch points are slow to decay. $B^2FH$ had inventively shown that such competition at branch points could, when compared with the actual solar abundances, reveal the time scale and neutron density for the operation of the *s* process. Thermally populated excited states of radioactive nuclei often increase the effective decay rate, a delicate point in that aspect of astronomy with radioactivity.

Two cautions about $B^2FH$ must be made in order to not overly eulogize what they achieved. Firstly, both the *s* process and the *r* process are, as described by them, secondary processes of nucleosynthesis. *Secondary nucleosynthesis* refers to the synthesis of new heavy nuclei from other existing heavy nuclei. It does not increase the galactic metallicity, a goal that Hoyle's founding papers had achieved spectacularly as the primary goal of the astronomy of nucleosynthesis. The $B^2FH$ neutron-capture processes instead change one existing heavy nucleus into another. So those processes did not contribute to increasing metal abundances in the galaxy. Today it is known that, contrary to their description, the *r* process is actually primary, because the collapsed supernova core synthesizes the seed nuclei that rapidly capture its free neutrons. The details of this are still not understood, however. Astronomical observations of old metal-poor stars confirm that the *r* process indeed began earlier than did the *s* process. Secondly, the $B^2FH$ descriptions of the *s* process and the *r* process could not immediately be used for astrophysics calculations because they were time-independent formulations. Both the *s* process and the *r* process were described by the static condition $dN/dt = 0$. The $B^2FH$ descriptions were thus of stationary abundances that could exist within appropriate environments for these processes. This enabled a rough but clear correlation between the nuclear systematics of their process abundances and the solar isotopic abundances. But it did not enable calculation of the temporal growth of these abundances.

Over the next two decades $B^2FH$ nonetheless became the default citation for workers wanting a reference to the general theory of nucleosynthesis in stars, vastly eclipsing the rate of citations to Hoyle's previous papers. One key to its success was its citation of over 100 stars showing nucleosynthesis effects in their spectra and of more than 100 astronomical research papers concerning those stars. This drew the astronomical community into the scientific culture of nucleosynthesis (which was then new to astronomers) as Hoyle's papers could not. Astronomers generally cited $B^2FH$ rather than Hoyle, with the result that Hoyle's papers slipped into relative obscurity.

Fowler himself coauthored several other important works for nucleosynthesis, primarily with Fred Hoyle on supernovae and with Donald Clayton on nucleosynthesis by neutron capture chains; however, his most important and essential role in history was the empirical thermonuclear reaction rates determined in his laboratory. For this he shared the 1983 Nobel Prize in physics. Probably Fowler's most significant subsequent work with Hoyle concerned the puzzling nature of the supernova phenomenon and on nucleosynthesis within them (Hoyle and Fowler, 1960). They divided supernovae into Types I and II based on theory, rather than on the presence of H absorption lines in the spectra. Type I was assigned to low-mass progenitor stars that evolved to white-dwarf stars of degenerate carbon, but then ultimately explode in an exothermic thermonuclear display that results in most of the iron in the universe. Electron-degenerate matter, supported by degeneracy pressure, is violently unstable to thermonuclear runaway. The Type II supernovae, on the other hand, occurs in massive stars whose cores are too massive to form white dwarf structure. In these continuing nuclear burning eventually exhausts the nuclear energy supply, and so the core must collapse. Fred Hoyle had predicted this inevitable collapse in 1954; but their 1960 paper carried the physical picture further. Today the world of astronomy uses this Hoyle-Fowler classification based on the physics of the device. It was a landmark in astrophysics.

#### 2.1.2.2 A.G.W. Cameron, a Parallel Force

The year 1957 also saw the emergence of A.G.W. Cameron as one of the pioneers of nucleosynthesis. The Atomic Energy of Canada at Chalk River published in bound-mimeograph form a series of lectures on nucleosynthesis that Cameron delivered at Purdue University in March/April 1957 (Cameron, 1957). These lectures covered in an independent way essentially the same material as in $B^2FH$. Cameron had constructed his treatments by working alone, first as a new faculty member at Iowa State University and then at Chalk River, with only Hoyle's two papers to guide him as well as his training in nuclear physics from University of Saskatchewan. Owing to his independent treatment, Cameron's Lecture Notes became a valuable source of new ideas in nucleosynthesis. In his emphasis on nucleosynthesis within the separate shells in massive stars, Cameron's approach followed Hoyle's equation and thereby enlarged the ideas in Hoyle (1954) more effectively than did $B^2FH$. For the next two decades Cameron stressed in many publications with research students – ones that he had recruited during a guest lecture course at Yale University – the vast nucleosynthesis changes that occur in a massive star at its time of explosion. His papers establish him as the first great disciple of Hoyle's equation. Cameron's later Yale Lecture Notes (1963), written by his Yale students W.D. Arnett, C.J. Hansen and J.W. Truran, were much improved over Cameron (1957). They probably should have been published, but were not and are therefore not generally available; however, they were a strong influence on history, especially on Cameron's students and on Clayton and his students. Partly owing to their unavailability, Clayton published his own textbook (Clayton, 1968).

Cameron began his nucleosynthesis research, however, with the sources of free neutrons in stars that could be responsible for the observed radioactive Technetium observed by Paul Merrill in stellar atmospheres. Because all isotopes of the element Tc are radioactive, the presence of its absorption lines in stellar spectra argued that it had been created within the star during roughly its last lifetime. In 1955 Cameron proposed that the $^{13}C(\alpha, n)^{16}O$ reaction would liberate the extra bound neutron in the $^{13}C$ nucleus and that the liberated neutron could be captured to create heavier isotopes (Cameron, 1955). Later Cameron (1959) calculated with the same motivation the number of neutrons liberated during carbon thermonuclear reactions following the exhaustion of He. The carbon fusion reactions were part of the sequence of thermonuclear stages within evolving massive stars (as Hoyle et al. (1954) had first described).

In his large subsequent body of work, Cameron established himself as a true polymath. He introduced the speeding up of beta decay rate by thermal population of the excited states of a nucleus, many of which decay more rapidly than the ground state. During 1955–1956 Cameron introduced numerical computation on the first vacuum-tube computers into nucleosynthesis problems, and remained thereafter on the cutting edge of nucleosynthesis computation. Cameron and his students repeatedly blazed new paths by programming nuclear reaction networks onto the latest and newest computers. He became an expert on planetary sciences and an important advisor to NASA. In particular, Cameron is forever famous for his work on the origin of the moon as the result of a Mars-like planetary collision with the young earth, a theory of origin that now seems beyond doubt.

#### 2.1.2.3 Donald D. Clayton and Time-Dependent Heavy Element Nucleosynthesis

Beginning as research student at Caltech with Fowler, Donald Clayton began constructing a time-dependent formulation of the *s* process in 1957, the year of $B^2FH$ publication and of Cameron's Chalk River lecture notes. Clayton's discard of the assumption $dN/dt = 0$, an assumption requiring a constant $\sigma N$ curve for *s*-process abundances, altered profoundly the direction of $s$-process research by focusing on how efficiently seed nuclei could be converted to heavy $s$-process nuclei. $B^2FH$ had not addressed that question. Clayton showed, as $B^2FH$ had surmised, that the iron abundance peak must provide the seed nuclei being transmuted into the large overabundances of barium in stars whose spectra showed Ba/Fe some 20–50 times the solar ratio. More surprisingly his results also showed that, as cumulative neutron fluence increases, none of the sequential abundance distributions that are generated resemble the solar abundances (Clayton et al., 1961). The solar $s$-process abundances were required to be a superposition of differing numbers of Fe-seed nuclei (per Fe nucleus) exposed to differing integrated fluxes of free neutrons. The number exposed must be increasingly smaller for increasingly larger neutron irradiations. Galactic history or history within $s$-process stars is required to bring that superposition about. Therein lay new astrophysics. The solar $s$-process abundances were shown to not be simply a smoothly declining $\sigma N$ curve, as $B^2FH$ had speculated, but a superposition of exposures generating narrow regions of atomic weight near the neutron-magic numbers where the assumption $dN/dt = 0$ is severely violated. Two decades of improved measurements and consequent fitting to solar abundances (Käppeler et al., 1982) would be required before advances in stellar evolution would be able to describe the exponential-like fluence distribution that was required. It was a sophisticated interplay between He-shell pulses and cyclically ignited H burning at the base of the envelope of AGB stars (see Chap. 3).

From the time of these first solutions of the neutron-irradiation superpositions resulting in the $s$-process abundances, new phenomenological aspects of heavy element nucleosynthesis were possible. The theory-based fit yielded all $s$-process abundances with meaningful accuracy. These allowed Clayton and Fowler (1961) to publish the first decomposition of heavy-element abundances into their $s$-process and $r$-process parts. $B^2FH$ had suggested the dominant processes for each isotope in their spectacular appendix; but a quantitative decomposition became possible for the first time. Their initial effort has been redone at least a dozen times as new neutron-capture-cross-section data appeared, most notably first by Seeger et al. (1965) and later by a new measurement program in Karlsruhe (Käppeler et al., 1982). This $s$-$r$ decomposition applied to astronomical spectroscopy of old stars has routinely produced meaningful new knowledge. Observations of old metal-poor dwarf stars indicated that the $r$-process abundances began to grow earlier as star formation first began than did the $s$-process abundances (Truran, 1981; Burris et al., 2000). That result demonstrated that the $r$ process is a primary nucleosynthesis process, rather than secondary as $B^2FH$ had stated. This requires the $r$ process to occur within core-collapse supernovae. Quantitative $s$-$r$ decomposition also inspired unforeseen new techniques for radioactivity-based cosmochronology.

Impressed by the new astrophysics lurking in time-dependence, Clayton advocated time-dependent formulation of the $r$ process as well, again jointly with Fowler and with P.A. Seeger, Fowler's research student in Kellogg Lab. For the $r$ process as for the $s$ process, $B^2FH$ had described only a time-independent steady flow that showed the neutron-rich heavy isotopes did indeed have abundant progenitors in suitable time-independent settings; but they had not been able to address whether the entire $r$-process abundances can be synthesized at one set of conditions. Computers at that time (1963) were not capable of handling a full $r$-process network. Seeger et al. (1965) showed that the full mass range cannot be produced together (unless new seed nuclei are injected during the process). $B^2FH$ had creatively defined the key nuclear physics relationships of the $r$ process but were mute on its dynamics. The time-dependent formulation by Seeger et al. (1965) became a prototype for $r$-process astrophysics. It showed the $r$ process to also be a superposition of differing irradiation histories of seed nuclei. Even after four decades of subsequent computations, the nature of the $r$-process superposition remains a frontier puzzle. It is noteworthy for astronomy with radioactivity that the entire $r$-process reactions occur within the realm of radioactive nuclei. Only after rapid expansive cooling can that neutron-rich radioactive abundance distribution undergo a series of beta decays changing each isobar identity until resting at the most-neutron-rich stable isobar (isobar is an isotope having the same atomic weight). A major research goal of nuclear astrophysics today is better laboratory definition of the parameters defining accurately the properties of the neutron-rich radioactive nuclei.

Understanding the time dependence during silicon burning (silicon photoerosion) was the big scientific challenge of the mid-1960s. Clayton introduced nuclear quasiequilibrium as a physical concept (Bodansky et al., 1968a,b) to clarify how silicon transmutes to an iron abundance peak. Quasiequilibrium explained the only big gap in Hoyle's 1954 theory of primary nucleosynthesis, replacing the ill-formulated $\alpha$ process of $B^2FH$. A temporal sequence of quasiequilibrium states facilitated the calculation of the set $\Delta m_k$ for k = 28–62 to be inserted in Hoyle's equation. The sequence of quasiequilibria again involved relaxing the assumption $dN/dt = 0$. The quasiequilibrium concept was powerful and new, and enriched many subsequent aspects of nucleosynthesis reaction networks (the $r$ process, explosive oxygen burning, the $\alpha$-rich freeze out, the origin of $^{48}Ca$, and others). Just as all nuclear reactions proceed at the same rate as their inverses in full nuclear equilibrium (NSE), during quasiequilibrium one refractory nucleus violates that equilibrium assumption by changing abundance only slowly, while all others maintain equilibrium with it. During silicon burning the $^{28}Si$ nucleus is the slowly changing, refractory post to which the quasiequilibrium distribution is attached (Bodansky et al., 1968a,b). The most abundant isotopes between A = 44 and 62 are, in this quasiequilibrium sequence, created as radioactive progenitors rather than as stable isobars, with important consequences for emerging astronomies of radioactivity.

#### 2.1.2.4 The Sequel

The decade 1956–1966 following Hoyle's pioneering two papers had witnessed profound enlargements of the theory of nucleosynthesis in stars. The years following the publication of $B^2FH$ had been marked by vast improvement and reformulation of its influential processes. Calculable time-dependent descriptions of heavt-element nucleosynthesis processes refocused attention from simple correlations between nuclear properties and abundances to the astrophysical histories and stellar evolution that bring them into existence. The sometimes heard statement by astronomers that not much happened after $B^2FH$ reflects lack of awareness of these historic changes. It may be of interest to note that the beginnings of nucleosynthesis theory was an international innovation. Of the pioneers named, only Fowler and Clayton were born Americans; Hoyle, E.M. Burbidge and G.R. Burbidge were English; Suess was Austrian; and Cameron was Canadian.

Innovations continued and accelerated during the next decade 1967–1977. These will not be reviewed here except to say that the evaluation of Hoyle's equation through numerical computation of the evolution of massive stars yielded repeated insights into the interplay between stellar evolution and nucleosynthesis. The $B^2FH$ neutron-capture processes revealed layers of complexity associated with the stars and with time dependences. The innovative center of this research moved away from Caltech, initially to Yale University and to Rice University, where Cameron and Clayton respectively founded schools evaluating Hoyle's equation. New leaders developed within those schools included especially W. David Arnett from Yale (subsequently also Rice) and Stanford E. Woosley from Rice (subsequently U.C. Santa Cruz). Each has many important publications on the evolution of massive stars and the nucleosynthesis in its shells (Arnett, 1996; Woosley and Weaver, 1995) (see Chap. 4). A European role in this research also experienced rebirth during that decade, especially in Munich. It is also the decade 1967–1977 that sees the emergence of several new observational aspects of astronomy with radioactivity. A description of those developments follows.

### *2.1.3 New Astronomy with Radioactivity*

The existence of natural radioactivity clearly holds implications for the origins of atomic nuclei. The chemical elements could not have always existed if the radioactive nuclei were created along with the stable nuclei. In this way naturally occurring radioactivity is intimately related to nucleosynthesis of the chemical elements. It was evident from the structure of the suggested processes of nucleosynthesis that radioactive nuclei played a large role in each of them. The key role of radioactive isotopes during nucleosynthesis and during thermonuclear power in stars was the first astronomy with radioactivity. Presumably the radioactive nuclei would be ejected from stars along with the new stable nuclei unless they decayed within the stars prior to ejection. Radioactivity plays a large role in the H-burning reactions, the PP chains and the CNO cycle (see Clayton, 1968) responsible for the stellar power capable of keeping the stars from cooling. Hans Bethe was the 1968 Nobel Prize awardee for discovering these H-burning cycles in stars just prior to WWII. So it was evident in mid-twentieth centrury that radioactive nuclei carry significant issues for astronomy. In the 1950s the idea arose of viewing the radioactivity at the solar center by detecting neutrinos arriving from the sun. Raymond Davis Jr. would win a Nobel Prize for spearheading that effort.

What made astronomy with radioactivity so exciting scientifically was the discovery in the 1960s and 1970s of altogether new ways of observing radioactivity in astronomy. These aspects of astronomy with radioactivity generated new interdisciplinary connections to nucleosynthesis. All science needs observations to provide an empirical base, astronomy just as surely as laboratory science. To be sure, the scientific method happens differently in astronomy than in laboratory science. In astronomy there are no experiments that can change the initial conditions as a test of theories. No experiments can be planned to refute a hypothesis. One has instead only observations of natural events. In many cases simple observations can be sufficient to refute a hypothesis. It is fortunate that nature provides so many natural events, so that in many cases contrasting separate events constitutes a type of experimentation, giving a spectrum of observations in which conditions differ in ways that must be inferred but are nonetheless real. The experiments are natural and performed by nature herself, however, rather than by scientists.

New types of observations of the occurrence of natural radioactivity galvanized astronomy by providing new kinds of astronomical data. These first historical advances are reviewed in what follows, as previews to the more detailed chapters that discuss them. They are: cosmochronology, the age of the elements; gamma-ray-line astronomy of radioactivity; radioactive power for light curves of supernovae and of novae; extinct radioactivity measured by excess abundances of daughter nuclei in solids formed in the early solar system; extinct radioactivity measured by excess abundances of daughter nuclei in solids condensed while dust particles condense as gas leaves a site of stellar nucleosynthesis. What follows is a brief description of how these applications were discovered or anticipated.

### 2.1.4 Interpreting Exponential Decay

The familiar exponential law of radioactive decay follows from Eq. (1.1) for an ensemble of $N$ radioactive nuclei from the constancy of decay probablility for a single nucleus. Given $N$ such nuclei the expected number of decays per unit time in the ensemble is given by the product of the number $N$ of nuclei and the decay probablility per unit time for a single nucleus. Accordingly

$$\frac{dN}{dt} = N\lambda = N/\tau \tag{2.4}$$

Because $\lambda$, and therefore $\tau$, is a constant for that nuclear species when it exists outside of stars, this equation has a well-known integral form,

$$N(t) = N_0 \, e^{-t/\tau} \tag{2.5}$$

where $N_0$ is the initial number in the ensemble (at $t = 0$). Similarly, if a number $N_0$ is observed now at $t = t_0$, the number that existed at an earlier time $t_1$ would have been

$$N(t_1) = N_0 \, e^{(t_0 - t_1)/\tau} \tag{2.6}$$

provided that new nuclei have not been added to the ensemble during that time interval. In the same spirit, if $N_1$ were created at earlier time $t_1$ and $N_2$ were added at a subsequent time $t_2$, the number $N_0$ that would exist today at $t_0$ is

$$N_0 = N_1\, e^{-(t_0-t_1)/\tau} + N_2\, e^{-(t_0-t_2)/\tau}. \tag{2.7}$$

Generalizing slightly, let $B_\odot(t)$ represent the fractional age distribution of the primary stable solar nuclei at the time of solar formation at $t = t_\odot$. Then $dB_\odot(t)/dt$ is the number of stable solar nuclei that had been born per unit time at time $t$. It is the age distribution that a radioactive species within an ensemble would have *if it were not decaying*. The age distribution for radioactive parents may be thought of as the age distribution of stable nuclei that were synthesized along with the radioactive nuclei. A mnemonic for the symbol $B_\odot$ is the *birthrate* of those solar system nuclei. $dB_\odot/dt$ is the rate at which the stable nuclei and the primary parent radioactive nuclei were added to the total sample destined for the solar system rather than the rate at which they were produced by nucleosynthesis. In drawing this subtle distinction, Clayton (1988) demonstrated the error of the common practice of equating the age distribution of solar nuclei with the rate of galactic nucleosynthesis. The age distribution of solar-system nuclei is vastly different from the galactic nucleosynthesis rate because the evolution of the ISM is involved. Even if the ISM is instantaneously mixed, many of the old nuclei became trapped within stars, so that the age distribution in the ISM becomes biased toward more recent nucleosynthesis.

Given this definition of $dB_\odot/dt$ as the age distribution of solar-system nuclei, and given that their addition to the solar sample began at time $t_1$ and continued until $t_\odot$, the total number of any stable primary nucleosynthesis product would have been,

$$N_{\text{total}} = \int_{t_1}^{t_\odot} \frac{dB_\odot(t')}{dt'}\, dt' \tag{2.8}$$

Then in analogy with Eq. (2.7), if those nuclei are instead radioactive, the number surviving until solar formation at time $t_\odot$ would be

$$N_\odot = \int_{t_1}^{t_\odot} \frac{dB_\odot}{dt'}\, e^{-(t_\odot-t')/\tau} dt' \tag{2.9}$$

If the total duration $t_0 - t_1$ of production is much greater than the mean lifetime $\tau$, and if $dB_\odot$/dt varies only slowly, Eq. (2.9) reduces approximately to

$$N_\odot = \left(\frac{dB_\odot}{dt}\right)_{t_\odot} \tau \tag{2.10}$$

This useful formula estimates the number of remaining radioactive nuclei to be equal to the number created during the last mean lifetime of that radioactive species. If $dB_\odot/dt$ has jagged temporal structure near $t_\odot$, however, a more complicated evaluation would be required. The Eq. (2.10) is very useful as a first approximation to the numbers of radioactive nuclei within the ISM during continuous nucleosynthesis in stars if the birthrate $B(t)$ is redefined as the birthrate spectrum of those stable ISM nuclei residing in the ISM rather than in stars.

These relationships expressing properties of exponential decay are central to understanding both extinct radioactivity in the early solar system and the numbers of radioactive nuclei within astronomical objects.

## 2.2 Disciplines of Astronomy with Radioactivity

### *2.2.1 Nuclear Cosmochronology*

Today it is self evident that the existence of radioactive nuclei implies that they were created at some estimable moment in the past. Were that not the case, they would have long ago have disappeared. But the full implications were not evident to those engaged in the exciting day-to-day goals of understanding natural radioactivity. Nonetheless, in 1929 Ernest Rutherford wrote what may be the first paper on astronomy with radioactivity (Rutherford, 1929). Today we use the term nuclear cosmochronology to mean the attempt to use natural abundances of radioactive nuclei to compute the age of an astronomical object or of the elements themselves. Determining the age of the chemical elements by assuming them to have the same age as the radioactive nuclei became a goal that has attracted many, most notably William A. Fowler and this writer.

#### 2.2.1.1 Uranium and Thorium on Earth

When he began to think on these things, Ernest Rutherford concluded from the ratios of $^{235}$U/$^{238}$U as measured in his laboratory that uranium was created somehow within the sun and transported to earth. Accepting Jeans's estimate that the age of the sun was an exuberant $7\times10^{12}$ years, he wrote in a prophetic paper addressing the puzzle (Rutherford, 1929):

> ..it is clear that the uranium isotopes which we observe on earth must have been forming in the sun at a late period of its history, namely, about $4\times10^9$ years ago. If the uranium could only be formed under special conditions in the early history of the sun, the actino-uranium on account of its average shorter life would have practically disappeared long ago. We may thus conclude, I think with some confidence, that the processes of production of elements like uranium were certainly taking place in the sun $4\times10^9$ years ago and probably still continue today.

Corrected modern numbers for those that Rutherford used are the observed abundance ratio $^{235}$U/$^{238}$U = $7.25\times10^{-3}$ and the mean lifetimes against alpha decay, $\tau(^{235}\mathrm{U}) = 1.029\times10^9$ year and $\tau(^{238}\mathrm{U}) = 6.506\times10^9$ year. Since each isotopic abundance has been exponentially decaying during the age $A_E$ of the earth, their abundance ratio when earth formed would have been $(^{235}\mathrm{U}/^{238}\mathrm{U})_0 = 7.25\times10^{-3}$ $(e^{A_E/\tau(^{235}\mathrm{U})}/e^{A_E/\tau(^{235}\mathrm{U})})$.

The age of the earth has been reliably measured using the fact that these two isotopes of U come to rest, after a series of alpha decays, as different isotopes of Pb, namely $^{207}$Pb and $^{206}$Pb respectively. Using the measured earth age, $A_E = 4.57 \times 10^9$ year, yields the initial U ratio on earth to have been $(^{235}\text{U}/^{238}\text{U})_O = 0.31$. These facts are beyond doubt.

The implication of great consequence stems from the expectation that the $r$ process that is responsible for the nucleosynthesis of both isotopes, should make more $^{235}$U than $^{238}$U. $^{235}$U has six non-fissioning progenitors whereas $^{238}$U has but three. An $r$-process progenitor is a non-fissioning transuranic nucleus that after a series of relatively fast radioactive decays comes to rest at one of these long-lived isotopes of U. For $^{235}$U those nuclei are $A$ = 235, 239, 243, 247, 251 and 255-totaling six $r$-process progenitors-whereas for $^{238}$U they are $A$ = 238, 242, 246 and 35% of 250-totaling 3.35 progenitors. Taking into account the empirical evidence that production by the $r$-process favors even-$A$ nuclei by a slight 20% over adjacent odd-$A$ nuclei, one expects $^{235}$U production to exceed that of $^{238}$U by a factor near $P(^{235}\text{U})/P(^{238}\text{U}) = 1.79$. This argument reveals that the abundance ratio $^{235}$U/$^{238}$U has declined from near 1.79 at production to 0.31 when the earth formed. This decline takes considerable time, showing that U isotopes were synthesized during pre-earth astrophysical history.

This first calculation of nuclear cosmochronology reveals the nature of the problem, but also its uncertainties. Is the production ratio $P(^{235}\text{U})/P(^{238}\text{U}) = 1.79$ correctly estimated? Were the U isotopes synthesized in one single presolar event, in which case it occurred about 6.6 billion years ago, 2 billion years prior to formation of the earth. Or was their production rate distributed in presolar time? If the age distribution of the solar system's $r$-process nuclei is flat between the time of first production and the earth's formation, the beginning of nucleosynthesis would fall near 13 billion years ago. The true age distribution can hardly be known with any assurance, revealing the severe limitation of this single pair for constraining the time of the beginning of nucleosynthesis in our galaxy.

The technique was extended to the ratio $^{232}$Th/$^{238}$U by Fowler et al. (1960); and Fowler returned to it many times in later years (e.g. Fowler 1972); see also the textbook by Clayton (1968). Relevant numbers used are the observed abundance ratio $^{232}\text{Th}/^{238}\text{U} = 4.0$ and the mean lifetimes against alpha decay, $\tau(^{232}\text{Th}) = 20.04 \times 10^9$ year and $\tau(^{238}\text{U}) = 6.506 \times 10^9$ year. The abundance ratio is much more uncertain than it was for the pair $^{235}$U/$^{238}$U because Th and U are different chemical elements. Any two elements having different fractionation chemistry pose a tough problem when seeking their ratio in the initial solar system (or in the sun). Since each isotopic abundance has been exponentially decaying during the age $A_E$ of the earth, their abundance ratio when earth formed would have been $(^{232}\text{Th}/^{238}\text{U})0 = 2.5$ instead of 4.0. And the production ratio in $r$-process events was inferred from the same counting arguments of odd and even progenitors of $^{232}$Th and $^{238}$U to be $P(^{232}\text{Th})/P(^{238}\text{U}) = 1.73$.

A problem is that this Th/U pair does not give transparently concordant numbers with the U pair. Uncertainties in $r$-process production ratios, in the relative abundances of Th and U, and in the arbitrary parameterizations of galactic chemical evolution that have been used each conspire to yield possible solutions in which nucleosynthesis began anywhere from 2 to 10 Gyr prior to solar birth. Fowler strove repeatedly to circumvent these uncertainties, trying to extract the true answer (for there obviously is a true answer!); but in fact, the data are not adequate and the astrophysical model of the galaxy used is not adequate for the task. Data from other chronological species would be needed, along with a more sophisticated appreciation of galactic chemical abundance evolution.

On a more positive note, Fowler's papers (Fowler et al., 1960; Fowler 1972, and several others); inspired many others to tackle this fascinating aspect of astronomy with radioactivity.

#### 2.2.1.2 Cosmochronology with Extinct Radioactive Abundances

In 1960s a new discovery, excess trapped $^{129}$Xe gas in meteorites (Reynolds, 1960), provided the first evidence that the matter from which the solar system formed contained radioactive nuclei whose halflives are too short to be able to survive from that time until today. These are called extinct radioactivity. Discovery of extinct radioactivity utilized the buildup of the daughter abundance from a radioactive decay as a measure of how abundant that radioactive parent was in the initial solar system. What Reynolds (1960) showed was that the excess $^{129}$Xe gas in meteorites had resulted from $^{129}$I decay and that the initial $^{129}$I abundance was about $10^{-4}$ of the initial $^{127}$I abundance. He realized that this datum allowed an estimate of when the iodine isotopes had been created. By assuming that at the time of their nucleosynthesis the initial $^{129}$I abundance had been equal to that of stable initial $^{127}$I, Reynolds argued from the mean lifetime $\tau(^{129}\mathrm{I}) = 23.5\times10^6$ year (23.5 My) that iodine had been created only about 300 My prior to solar system birth. This was the first new cosmochronology technique since Rutherford and brought Reynolds great fame. Unfortunately, Reynolds' conclusion was no more believable than Rutherford's had been, because his astrophysical model was also quite unrealistic. He mistakenly assumed that all of the iodine abundance in the solar system had been created at the same time, and therefore used Eq. (2.5) to estimate that time. Because $10^{-4}$ requires a decay period equal to about 13 mean lifetimes, he had concluded that the age of the elements was $13\tau(^{129}\mathrm{I}) = 300$ My. Fowler et al. (1960) countered that it was more realistic to assume that the synthesis of iodine isotopes was spread out by multiple *r*-process events, occurring between the time $t_1$ of a first nucleosynthesis event until the time $t_0$ of the last such event prior to solar birth. That concept, called *continuous nucleosynthesis*, required use of the Eq. (2.10) limit of Eq. (2.9) instead of Eq. (2.5). That calculation gave the initial nucleosynthesis epoch to be $104\tau(^{129}\mathrm{I}) = 230$ Gigayears (Gy)–very old indeed, unrealistically old. This picture was then modified to allow for an interval of sequestered interstellar gas, probably within a molecular cloud, of about 100 My during which no new nuclei were added to the solar mix – a so-called *free decay interval* in which radioactive abundances would follow Eq. (2.5). Such an interval would require that 100 My before the solar birth, when that interstellar gas withdrew from new nucleosynthesis, the interstellar abundance ratio would have been $N(^{129}\mathrm{I})/N(^{127}\mathrm{I}) = 2.8\times10^{-3}$ rather than $10^{-4}$.

Equation (2.10) then suggests that nucleosynthesis began 8 Gy prior to solar birth, a reasonable number, obtained, unfortunately, by construction. But that construction had nonetheless established a new paradigm for radioactive abundances in the early solar system. The upshot for the problem of radioactive chronology is that the abundance of now extinct radioactivity in the early solar system can say little about when nucleosynthesis first began. But it provides other equally interesting issues for the astronomy of radioactivity.

Radioactive $^{129}$I is of special historical significance as the first extinct radioactive nucleus to be discovered that can be regarded as part of the average radioactivity in the galactic ISM at the time of solar birth. If so it is in that sense typical of what is expected in the ISM. Other extinct radioactivities may not be capable of such an interpretation but must instead be interpreted as produced by a special event associated with the birth of the solar system. Other subsequently discovered extinct nuclei that can be regarded in the first category of galactic survivors are $^{53}$Mn, $^{107}$Pd, $^{182}$Hf, $^{146}$Sm and $^{244}$Pu.

#### 2.2.1.3 Birthrate Function for Primary Solar Abundances

The history of nuclear cosmochronology reveals widespread confusion between the rate of galactic nucleosynthesis and the age spectrum of solar system nuclei. It is the latter that enters into cosmochronology through Eq. (2.9). So endemic is this confusion between these two fundamental concepts that it is important to define in this section the birthrate spectrum for primary abundances in the solar system.

Let $N_\odot$ be the solar abundance of a primary nucleosynthesis product. One may think of it either as the total number of those atoms in the solar system, which consists of awkwardly large numbers, or as the total number of those atoms in the solar system normalized to a defined abundance for a specified nucleus that sets the abundance scale. Two such normalized scales are very common: one used more in astronomy defines the solar abundance of hydrogen to be $N_H = 10^{12}$; another, used more in solar abundances derived from meteorites, defines the solar abundance of silicon as $N_{Si} = 10^6$. This book will primarily use the latter definition because most studies of nucleosynthesis in the literature use the scale Si = $10^6$ (where the chemical symbol is often used to represent the abundance of that element). Let $B_\odot(t)$ be the cumulative fraction, as a function of time, of stable solar primary nuclei that already existed at time $t$. Letting for convenience the starting time $t = 0$ be the time when nucleosynthesis first began, the function $B_\odot(t)$ starts at 0 at $t = 0$ and rises to $B_\odot = 1$ at $t = t_0$, the time when the solar system formed. Note carefully, $B_\odot(t)$ is not the cumulative birth fraction of that nucleus in galactic nucleosynthesis, but rather the cumulative birth fraction of those primary nuclei that actually entered the solar system inventory. Normalizing this definition of $B_\odot(t)$ as a function that rises to unity at solar birth, the product

$$N_\odot \cdot B_\odot(t) = N(t) \tag{2.11}$$

where $N_\odot$ is the solar abundance of that stable species, $N(t)$ is the number of of those solar-system nuclei that already existed at time $t$. It is equally clear that

$dB_{\odot}(t)/dt$ is the fractional birthrate per unit time of primary solar nuclei. It is the age distribution of solar nuclei, and redefined as here to rise to unity it is approximately the same function for all primary stable nuclei.

A mental experiment is needed to set this birthrate function clearly in mind before considering why it differs so from the nucleosynthesis rate. The reader is warned that the research literature suffers endemic confusion over that distinction, requiring its careful definition. For any primary nucleus within solar abundances, *paint each solar-system atom of it red!* This cannot of course be done in reality, but the thought experiment helps understanding of the birthrate spectrum. On the scale Si = $10^6$ one has by construction $10^6$ *red* Si atoms in the solar system. Then imagine watching a film run backwards in time of these $10^6$ *red* Si atoms. Back in the interstellar medium they are mixed with a much larger number of *unpainted* Si atoms. As the film runs to earlier times, these $10^6$ *red* Si atoms have greatly differing histories. Some, after being created in a supernova, have later been inside a star, survived and come back out. But for each atom, its world line reaches a time when that atom first came into existence. Each of the $10^6$ red solar Si atoms is labeled by that birthdate, born from distinct supernovae at differing times. Then a graph is constructed giving the total number that existed at time $t$. This graph starts at 1 atom when the first solar Si atom is created and reaches $10^6$ by the time the solar system forms, when all $10^6$ have been created. If the curve is then normalized by dividing by $10^6$, one has the birthrate spectrum $B_{\odot}(t)$ of solar primary nuclei, which rises to unity at the time the solar system forms. Equation (2.11) depicts this situation, where $N(t)$ rises to $10^6$ at solar birth, but $B_{\odot}(t)$ rises to unity at that time. Although different primary species have vastly differing abundances, for each the function $B_{\odot}(t)$ is approximately the same function because the abundances of all primary nucleosynthesis products rise approximately (but not exactly) together.

Equation (2.11) is the solution of the differential equation

$$dN/dt = dB_{\odot}(t)/dt \text{ for stable nuclei} \tag{2.12}$$

and

$$dN/dt = dB_{\odot}(t)/dt - \lambda N \text{ for radioactive nuclei} \tag{2.13}$$

One frequently sees these equations written for galactic nucleosynthesis, with $dB/dt$ identified as the galactic nucleosynthesis rate and $N(t)$ as the ISM abundance. This is a fundamental error. Consider why?

The galactic nucleosynthesis rate for Si atoms was very large in the young galaxy, when the rate of formation of massive stars and supernovae was large. But those new Si atoms in the interstellar medium become mostly locked up in the interiors of subsequent low-mass stars, and are not available for incorporation into the solar system. This occurs because it is the fate of most interstellar gas to be locked up in low-mass stars. The initially gaseous galaxy is today about 10% gas, with 90% trapped in low-mass stars. The origin of the Si atoms that entered solar matter is therefore biased towards supernovae that occur relatively shortly before solar birth, and biased against those that were created long earlier. This important conceptual distinction was demonstrated quantitatively by Clayton (1988), who stressed the concept of the age spectrum of solar nuclei and who constructed analytic models of galactic chemical evolution that enable a full disclosure of such issues. As an example, the simplest of all realistic models supports a star formation rate and supernova rate that declines exponentially as the interstellar gas is exponentially consumed; but the age spectrum of solar nuclei in that model is constant – equal numbers from equal times, despite the strong bias of nucleosynthesis of Si toward early galactic times. The student can study Clayton (1988) for many related issues for cosmochronology. Some of these issues involving galactic chemical evolution will be addressed later.

#### 2.2.1.4 Uranium and Thorium in Old Dwarf Stars

An altogether new technique in radioactive chronology became possible as CCD detectors enabled astronomers to measure line strengths for much weaker lines than had been previously possible. Butcher (1987) advanced the first such argument when he was able to accurately measure the abundance of thorium in old dwarf stars. The measured abundances enabled him to argue that the oldest stars are no older than 10 Gy. If they were, Butcher argued, the Th, which has resided in the old stars since their births, would by now have decayed to a smaller abundance than it is observed to have. Because the dwarf stars observed appear to be among the early stars formed in our galaxy, the argument concludes that the galaxy age is negligibly greater than 10 Gy. It will be clear that since the $^{232}$Th mean life is a very long 20 Gy, it can have decayed from its initial abundance by only $e^{-0.5} = 0.6$ during a 10 Gy life of a dwarf star. Therefore the method requires not only an accurate measurement of the Th abundance but also an accurate estimate of its initial abundance when the star formed. It is the reasoning required to obtain that initial abundance that is a controversial aspect of this method. Realize also that the Th in an old star is decaying freely, so that its abundance declines faster than in the ISM where new nucleosynthesis of Th continuously replenishes it.

The method advocated measures the ratio of Th/Nd in the star and observes that the measured ratio is smaller than that seen in younger stars. It assumes that the initial Th/Nd ratio in the star would be the same as the ratio seen in young stars, since both Th and Nd are products of the *r* process and may be hoped to have a constant production ratio there. It was then counter argued (Clayton, 1987, 1988) that a constant production ratio for Th/Nd is not to be expected because almost half of the solar Nd abundance has been created in the *s* process, and, furthermore, the *r* process is primary whereas the *s* process is secondary. There exists ample and exciting evidence from the observed *r*-process pattern of abundances in extremely metal-deficient old stars (Truran, 1981; Gilroy et al., 1988; Beers and Christlieb, 2005) that the *r*-process nucleosynthesis began prior to the beginning of the *s* process. That seemed to support the initial skepticism about Butcher's technique. But astronomers hurried to point out that if one omits the very metal-deficient old stars, the ratio of *r*-to-*s* abundances in stars having more than 10% of solar abundances is observed to be a near constant. Clayton (1988) revisited that larger puzzle of parallel growth of *s* and *r* nuclei, noting that the abundance evidence suggests that the *s* process is primary despite its building upon iron seed nuclei; moreover, he presented a nucleosynthesis argument that showed how the *s* process could in fact resemble primary nucleosynthesis even though it is secondary! See Section 3.1 of Clayton (1988) for that argument, which has proven to be of high significance for nucleosynthesis theory.

The lesson to be taken from this exciting new technique and its controversy is this: radioactive nuclear cosmochronology is vitally dependent upon a correct picture of the chemical evolution of the galaxy. As such it has become less the province of nuclear physics and more the province of astronomy. Tinsley and Clayton both made that point in timely and influential ways. Only when many essential details of the history of our galaxy and of the history of nucleosynthesis within it have been settled can these techniques of radioactive chronology yield a reliable answer for the age of the chemical elements. Nuclear cosmochronology is truly an astronomy with radioactivity.

#### 2.2.1.5 Cosmoradiogenic Chronologies

In the early 1960s a different approach to radioactive chronology became possible. It utilizes the buildup of the daughter abundances of radioactive decay during the history of interstellar matter as a measure of how long that decay had been occurring and, therefore, how long ago the production of radioactivity began. One can imagine the stable daughter of a radioactive decay as a bucket into which all interstellar decay of the radioactive parent has been collected. It integrates past decay rather than focusing on how much radioactivity remains. Such daughter-isotope buildup during the history of the earth was already known as *radiogenic abundance*. It had been applied to the ages of earth and of meteorites, samples in which it could be expected that the other initial isotopic compositions were well known-namely, the solar abundances. But radiogenic abundance collected during the history of interstellar matter, what Clayton (1964) called *cosmoradiogenic abundance*, was not seen as possible data for determining the age of the elements themselves because the interstellar abundances are increased by the processes of nucleosynthesis as well as by any radioactive decay for specific isotopes. Thus the cosmoradiogenic abundance could not be easily disentangled from the direct nucleosynthesis abundance.

What cut through that impasse was a credible *quantitative theory* of heavy-element nucleosynthesis. Suess, $B^2FH$ and Cameron had each contributed importantly to the idea that two distinct neutron-capture processes, the *s* process and the *r* process, had been responsible for the creation of all but the very lightest isotopes of the elements heavier than about $Z = 32$. The disentanglement required two things: firstly, an accurate *s*-process theory and secondly, a parent-radioactivity abundance that could be produced only by the *r* process and a daughter that could be produced only by the *s* process. The radiogenic daughter of such a radioactive *r*-process isotope is called a shielded isotope, because it is shielded from *r*-process production by the radioactive parent.

Clayton (1964) made the key first step by noticing that the solar abundance of $^{187}$Os is about twice as great as it is expected to be from $s$-process nucleosynthesis, and that it cannot be synthesized by the $r$ process because neutron-rich matter at $A = 187$ will, upon decaying toward stability, arrest at $^{187}$Re. The decay cannot reach $^{187}$Os, which is shielded by $^{187}$Re. The $r$-process production flows into $^{187}$Re. The abundance of $^{187}$Os can then be thought of as having two parts; a part produced by the $s$ process and a part produced by the decay of very-long-lived $^{187}$Re. The quantitative $s$-process theory (Clayton et al., 1961) reliably accounted for about half of the $^{187}$Os abundance, so the other half had to be the result of the beta decay of $^{187}$Re during ISM residence:

$$^{187}\mathrm{Re} \longrightarrow ^{187}\mathrm{Os} + \mathrm{e}^- + \nu t_{1/2} = 43\ \mathrm{Gyr}$$

That the halflife exceeds the galaxy's age is useful, because only a modest fraction of the $^{187}$Re can therefore have decayed during presolar history and most of it therefore still exists; but because $^{187}$Re is six times more abundant than the $s$-process amount of $^{187}$Os, a decay of only 1/6th of the $^{187}$Re can have doubled the $s$-process $^{187}$Os abundance. The persistence of live $^{187}$Re is even necessary for the method, because it allows one to equate the total amount of $^{187}$Re nucleosynthesis that was destined for solar incorporation to the sum of the quantity of $^{187}$Re remaining at solar birth and the quantity of $^{187}$Re that had already decayed prior to solar birth. The latter is the difference between the total solar abundance of daughter $^{187}$Os and the quantity produced by the $s$ process. That is,

$$^{187}\mathrm{Os} = ^{187}\mathrm{Os_s} + ^{187}\mathrm{Os_c}$$

where $^{187}\mathrm{Os}_c$ is the cosmoradiogenic part of the $^{187}$Os abundance owing to cosmic $^{187}$Re decay. And the $s$-process part can be obtained from $^{187}\mathrm{Os}_s = {}^{186}\mathrm{Os}$ $(\sigma(186)/\sigma(187))$, where $\sigma$ is the appropriate neutron-capture cross section during the operation of the $s$ process. This last relationship is valid because the entirety of $^{186}$Os is from the $s$ process because it is shielded from $r$-process nucleosynthesis.

The discovery of the Re-Os clock was the key that Clayton (1964) used in presenting solutions to three cosmoradiogenic chronologies. These are the beta decay of $^{187}$Re to $^{187}$Os described above, the similar beta decay of $^{87}$Rb to $^{87}$Sr, and the $\alpha$-decay chains by which $^{235,238}$U decay to $^{207,206}$Pb. Each had been known in the study of meteorites, but by this work they joined studies of the age of the elements themselves.

Cosmoradiogenic lead is the more interesting of the other two chronologies, in part because it couples cosmoradiogenic Pb to the older chronology based on $^{235}\mathrm{U}/^{238}\mathrm{U}$, and in part because the Pb isotopes have three contributions to their abundances:

$$^{206}\mathrm{Pb} = ^{206}\mathrm{Pb}_s + ^{206}\mathrm{Pb}_r + ^{206}\mathrm{Pb}_c$$

$$^{207}\mathrm{Pb} = ^{207}\mathrm{Pb}_s + ^{207}\mathrm{Pb}_r + ^{207}\mathrm{Pb}_c$$

where $^{206}Pb_c$ is the cosmoradiogenic part from $^{238}U$ decay, and where $^{207}Pb_c$ is the cosmoradiogenic part from $^{235}U$ decay. Suffice it to be said here that the *s*-process part and the *r*-process part can both be estimated, and from them the chronological solutions can be displayed. Readers can turn to Clayton (1988) for these solutions.

The cosmoradiogenic chronologies seem to indicate an older galaxy than do the direct-remainder chronologies (Clayton, 1988). But they also are compromised by unique and interesting uncertainties. For the $^{187}Re$ cosmoradiogenic chronology, the main uncertainty is how greatly its decay rate is speeded by its incorporation into stars, where ionization increases its beta decay rate. Yokoi et al. (1983) evaluated this effect within an ambitious galactic chemical evolution model that enabled them to take into account the fraction of $^{187}Re$ that is incorporated into stars and ejected again without nuclear processing and how much time such interstellar $^{187}Re$ spends inside of stars. Their results can be reproduced by a 40% increase of the neutral $^{187}Re$ decay rate. For the Pb cosmoradiogenic chronology, the main uncertainty is an especially interesting set of nucleosynthesis problems associated with both the *s*-process part and the *r*-process part of their abundances. Uncertainty about the Pb/U elemental abundance ratio also suggests some caution.

Certainly it can be said that these chronologies present intricate problems in the astronomy with radioactivity.

### 2.2.1.6 Radioactivity and Galactic Chemical Evolution

In astronomy, the abundances of radioactive species are interpreted through the lens of galactic abundance evolution, frequently called galactic chemical evolution (GCE). Interpretations of nuclear cosmochronology or of initial solar abundances of extinct radioactivities depend on the mean expected abundances in the ISM when solar birth occurred. Galactic Chemical Evolution traditionally concerns the chemical composition of the mean ISM. It features continuous nucleosynthesis between a starting time and the time when solar formation occurred. During that lengthy galactic period radioactive abundances undergo decay and are incorporated into new stars, but they are also replenished by the injection of freshly synthesized radioactivity and are diluted by low-metallicity matter falling onto the disk. Competition among these terms renders simple galactic chemical evolution a showplace for the behavior of radioactivity. One differential equation describing that abundance of those specific galactic atoms (the *red atoms*) that will later be incorporated into the solar system can be thought of as the solution of

$$\frac{dN}{dt} = \frac{dB_{\odot}(t)}{dt} - \frac{N(t)}{\tau} \tag{2.14}$$

where $\tau = 1/\lambda$ is the mean radioactive lifetime of the nucleus in question, N(t) is its time-dependent abundance within those (red) atoms that are destined for inclusion in the solar system, and $B_{\odot}(t)$ is the birthrate spectrum of those solar nuclei, defined such that $\frac{dB_{\odot}(t)}{dt}dt$ is the number of stable solar nuclei that were born between times t and t+dt. The birthrate spectrum of solar nuclei is unknown, however, so

Eq. (2.14) can not be solved. Despite this limitation, workers persisted in attempting to fix the time of the beginning of nucleosynthesis from Eq. (2.9), which students can show does solve the differential equation (2.14) above. Many tried this approach by assuming an easily integrable form for $\frac{dB_\odot(t)}{dt}$, most often $\exp(-Gt)$, where $G$ is taken (erroneously) to be an unknown positive number. That assumption proved seductive to many, because the rate of galactic nucleosynthesis is believed to decrease smoothly with time, so maybe $\frac{dB_\odot(t)}{dt}$ can be assumed to decline as well. Importantly, however, $\frac{dB_\odot(t)}{dt}$ differs greatly from the galactic production rate and actually grows with time. The reason for this surprise requires understanding.

The birthrate spectrum imagines that the nuclei destined for the solar system could be tagged at birth (imagine red paint) and ignores all other ISM nuclei. Because $B_\odot(t)$ is unknown, however, astrophysicists instead calculate the evolution of the interstellar abundances. Their mean values in a well mixed ISM at the time of solar birth provide the expected solar abundances. The conceptual difference between $B_\odot(t)$ and the galactic nucleosynthesis rate has confused many unwary researchers. Many published papers have used their chosen form for $\frac{dB_\odot(t)}{dt}$ and integrated Eq. (2.9) in the hope of fixing the beginning time of galactic nucleosynthesis (the lower limit) by comparing those calculated results with known initial radioactive abundances in the solar system. A common form because Eq. (2.9) is then easily integrated is $\frac{dB_\odot(t)}{dt} = \exp(-Gt)$. In retrospect, however, such poor choices for the form of $\frac{dB_\odot(t)}{dt}$ render their conclusions invalid.

How does GCE clarify the problem of galactic radioactivity? The mass $M_G$ of ISM gas is reduced owing to the rate $\psi(t)$ of its incorporation into new stars, but it is increased by the rate $E(t)$ at which mass is ejected from old stars and by the rate $f(t)$ at which new mass falls onto the galactic disk:

$$\frac{dM_G}{dt} = -\psi(t) + E(t) + f(t) \tag{2.15}$$

If one takes the ejecta rate $E(t)$ from the spectrum of newly born stars to be a fixed return fraction $R$ of the rate $\psi(t)$ at which mass joins new stars, and if one assumes linear models in which the rate of star formation is proportional to the mass of gas $M_G$ , the equations governing the interstellar composition of both stable and radioactive nuclei can be solved analytically within families of choices for the infall rate $f(t)$, as Clayton (1985, 1988) has shown. These analytic solutions for ISM abundances do indeed clarify radioactive abundances. For that purpose Clayton writes

$$\frac{dM_G}{dt} = -\psi(t)\,(1-R)\, + f(t) = -\omega M_G + f(t) \tag{2.16}$$

where $\omega M_G(t) = (1-R)\psi(t)$ is valid for linear models. The constant $\omega$ is the rate of consumption of gas by star formation when compensated by gaseous return from stars. These are called *linear models* because the star formation rate $\psi(t)$ is taken to be proportional to the mass $M_G(t)$ of interstellar gas. That linear assumption is not strictly true; but it is plausible in taking the star formation rate to increase as the mass of gas increases, and to decline as it declines. Furthermore, it is supported by observations of star formation rates in spiral galaxies. From their observations, Gao and Solomon (2004) state, "The global star formation rate is linearly proportional to the mass of dense molecular gas in spiral galaxies." The purpose of models of galactic chemical evolution insofar as galactic radioactivity is concerned is to understand the mean expectation for radioactivity in the ISM. The families of linear analytic models are constructed for that purpose.

Coupled with Eq. (2.16) is an equation for the rate of increase of the concentration $Z$ of each interstellar nucleosynthesis product. The concentration is defined as the mass of species $Z$ in the ISM divided by the total mass $M_G$ of ISM gas and dust. The concentration $Z$, rather than total numbers of atoms, is the quantity traditionally used in chemical evolution studies because it is concentration that astronomers measure. The mass $m_Z = ZM_G$ of interstellar species $Z$ is governed by

$$\frac{dm_Z}{dt} = -Z\psi(t) + Z_E\, E(t) + Z_f\, f(t) \tag{2.17}$$

where $Z_E$ and $Z_f$ are, respectively, the concentration of Z in the spectrum of stellar ejecta, where it is large, and in the infalling gas where it is small. The metallicity in infalling gas $Z_f$ may probably be neglected with good accuracy.

Many workers have shown that it is a good approximation for analytic understanding to assume that the ejecta $E(t)$ is returned at once from the entire spectrum of newly born stars despite the dependence of stellar lifetime on stellar mass. That assumption is called the *instantaneous recycling approximation*, and it is reasonable except late in the life of a galaxy when the gas mass and the star-formation rate have both become very small. But for galaxies in early and middle lifetime, Eq. (2.17) can then be written after some straightforward algebra as

$$\frac{dZ}{dt} = y\,\omega - \frac{Zf(t)}{M_G(t)} - \lambda\, Z \tag{2.18}$$

where $m_Z = ZM_G$, $Z$ being the concentration in the ISM gas (taken to be well mixed), and $y$ is the *yield* of element $Z$, defined as the mass $m_Z$ of new $Z$ ejected from the entire spectrum of newly born stars divided by the mass of stellar remnants left behind by that entire spectrum of stars. The yield $y$ for primary nucleosynthesis products may be taken as a constant despite having small variations in full numerical models. The product $y\omega$ in Eq. (2.18) can be called the *galactic nucleosynthesis rate*, which, through $\omega$, depends explicitly on the star formation rate ($\omega M_G = (1-R)\psi(t)$). Equation (2.18) for the ISM concentration differs from Eq. (2.14) for the history of solar nuclei by the existence of the second term of Eq. (2.18). That term reflects the loss of interstellar metal concentration $Z$ when ISM gas containing $Z$ collapses into new stars and is simultaneously diluted by metal-poor infall.

Comparison of Eq. (2.18) with Eq. (2.14) shows clearly why it is an error to integrate Eq. (2.14) thinking that the solar-nuclei birthrate spectrum can be mentally equated with the spectral rate of galactic nucleosynthesis. Equation (2.18) similarly has the galactic nucleosynthesis rate as the first term and the radioactive decay rate as the last term; but the astrophysical Eq. (2.18) contains the middle term involving both the galactic infall rate and the mass of ISM gas, both of which are time dependent. Integrations of Eq. (2.14) by assuming $B_\odot(t)$ ignore the effects of the ISM and can be correct only if the assumed form $dB_\odot(t)/dt$ actually resembles the number of solar nuclei born per unit time rather than the galactic nucleosynthesis rate. As the birthdate of solar nuclei is usually what is being sought in nuclear cosmochronology, one cannot get the answer from Eq. (2.14) without first knowing and inserting the answer. For that reason nuclear cosmochronology must instead be investigated within the context of galactic chemical evolution. These points are central to the subject of astronomy with radioactivity within solar-system nuclei.

Clayton (1988) showed one way forward. One can integrate Eq. (2.18) analytically with the aid of flexible families of functions that enable analytic integration. This can be accomplished by form-fitting the ratio $f(t)/M_G(t)$ to an integrable family of functions $d\theta/dt$. Such form fitting is much more flexible than it would at first seem to be, because most physically plausible time-dependent behavior can be approximated by a specific form fitting choice. One such useful choice has been called the *Clayton Standard Model*. It takes

$$\frac{d\theta}{dt} = f(t)/M_G(t) = \frac{k}{(t+\Delta)} \tag{2.19}$$

and its explicit functions are given in Appendix A1 of this book.

For the astronomy of radioactivity, the great merit of this approach is exact analytic solutions for $Z(t)$ for both stable and radioactive isotopes while simultaneously yielding exact functional representations of the mass $M_G(t)$ of interstellar gas and of the infall rate $f(t)$. Understanding these analytic models greatly aids understanding the behavior of radioactivity within more general numerical models of galactic chemical evolution. The numerical approach is to place the evolution of the galaxy on a computer, taking into account the evolutionary lifetime of each star formed and the specific nucleosynthesis products to be ejected from each star (e.g. Timmes et al., 1995). Although this approach is undoubtedly correct, it obscures theoretical understanding that can be seen more easily within analytic models. Furthermore, surveys of nuclear cosmochronology can more easily be carried out within analytic models.

All well mixed models of galactic chemical evolution can at best yield only an average expectation for the ISM. The true ISM is inhomogeneous in space and the nucleosynthesis rate is sporadic in time rather than maintaining its steady average. Nonetheless, well mixed models, both analytic and numerical, are important in laying out the results that would be true for a rapidly mixed ISM and a smooth rate of galactic nucleosynthesis. One interprets the observations of radioactive abundances against the backdrop of that expectation.

Another result of great importance for short-lived galactic radioactivity is best illustrated within Clayton's standard model. For short-lived radioactivity, i.e. whenever $\lambda$ is small, the mean concentration in the galactic ISM is

$$Z_\lambda = y\,\omega \left(\lambda + \frac{k}{(t+\Delta)}\right)^{-1} \tag{2.20}$$

From this equation and one for a stable isotope the abundance ratio of a short-lived radioactive nucleus to that of a stable primary isotope in the ISM is larger by the factor $(k + 1)$ than one might estimate without taking galactic chemical evolution into account. As an example, consider the interstellar ratio of radioactive $^{26}$Al to stable $^{27}$Al. The formula

$$\frac{Z(^{26}\mathrm{Al})}{Z(^{27}\mathrm{Al})} = \frac{y(26)}{y(27)}\,(k+1)\,\frac{\tau_{26}}{t_\odot} \tag{2.21}$$

can be derived from such a ratio (Clayton et al., 1993). This result is larger by the factor $(k + 1)$ than an estimate using only Eq. (2.10), which had traditionally been used in oversimplified discussions of the steady-state amount of short-lived interstellar radioactivity. That oversimplification can be found in almost all published papers on short-lived interstellar radioactivity. The extra factor $(k + 1)$, which is an effect of infall of low-metallicity gas, became important when a large mass of interstellar $^{26}$Al was detected by gamma-ray astronomy (see Chap. 7).

### 2.2.2 *Gamma-Ray Lines from Galactic Radioactivity*

One thinks of at least three reasons that the idea of astronomically detecting galactic radioactivity did not occur until the 1960s. In the first place, MeV range gamma rays do not penetrate to the ground, but must be detected above the atmosphere. Secondly, detecting MeV gamma rays and measuring their energies is a quite difficult technology, especially so when the background of cosmic-ray induced events above the atmosphere is so large. Thirdly, even a back-of-the-envelope estimate of expected rate of arrival from a stellar source is discouragingly small.

#### 2.2.2.1 The Rice University Program

In 1964 a new aspect of astronomy with radioactivity arose. Robert C. Haymes was hired by Rice University, and he spoke there with Donald Clayton about the possibility of sources of galactic radioactivity that he might seek with an active anti-coincidence collimation for a NaI detector flown beneath a high-altitude balloon. Burbidge et al. (1957) had speculated (incorrectly) that the exponential 55-night decline of the luminosity of many Type Ia supernovae was the optical manifestation of the decline of the spontaneous-fission radioactive decay of $^{254}$Cf in the ejecta of the supernova, which would quickly have become cold without some heating mechanism that had to decline with a roughly two-month halflife. The $^{254}$Cf nucleus would be synthesized by the $r$ process (assumed to occur in Type Ia supernovae) along with uranium and thorium. The large kinetic energy of its spontaneous-fission fragments would be converted to optical emission by being degraded by atomic collisions within the ejected gas. This was called *the californium hypothesis*. But $B^2FH$ had said nothing about gamma-ray lines. The novel excitement at Rice University was the new idea that concepts of nucleosynthesis could be tested directly if the associated gamma-ray lines from the $r$-process radioactivity could be detected on earth coming from supernova remnants. Haymes estimated rather optimistically that his detector could resolve lines having flux at earth greater than about $10^{-4}$ $cm^{-2}$ $s^{-1}$. The first scientific paper written with that goal was soon published (Clayton and Craddock, 1965). It evaluated the full spectrum of radioactivity by an $r$ process normalized to the yield proposed by the Cf hypothesis. They found several promising lines. For example, the strongest from a 900-year old Crab Nebula would be $^{249}$Cf, presenting $10^{-4}cm^{-2}s^{-1}$ gamma-ray lines having energy 0.39 MeV. It is no accident that the $^{249}$Cf halflife (351 year) is of the same order as the age of the Crab Nebula. Given a broad range of halflives in an ensemble of nuclei, the one giving the largest rate of decay has mean life comparable to the age of the ensemble. It can be confirmed that in a remnant of age $T$ the rate of decay per initial nucleus is $\lambda e^{-\lambda T}$, which is easily shown to be maximal for that nucleus whose decay rate $\lambda = 1/\tau = 1/T$. A photograph (Fig. 2.1) from those Rice years is included.

Realistic expectations were not so sanguine, however. The $r$-process yield required by the Cf hypothesis appeared excessively large (Clayton and Craddock, 1965). The $r$-process nuclei in solar abundances would be overproduced by a factor 100 if all Type Ia supernovae produced that yield. Despite that reservation, an exciting chord had been struck. A new astronomy of radioactivity appeared possible, one having significant scientific payoff, and Haymes's gamma-ray telescope began a program of balloon flights. The program did not discover $r$-process radioactivity, but it did discover apparent positron-annihilation radiation from the center of our galaxy (Johnson et al., 1972) and hard X-rays from several sources. The

**Fig. 2.1** Robert C. Haymes and Donald D. Clayton with Haymes' gondola for his balloon-borne gamma ray telescope at Rice University in 1973. Haymes pioneered gamma-ray astronomy in the MeV region and Clayton developed gamma-ray-line targets for this goal and for nucleosynthesis. This program at Rice University trained two Principal Investigators and two Project Scientists for later experiments on NASA's Compton Gamma Ray Observatory

observational program at Rice University also trained two of the principal investigators (G.J. Fishman and J.D. Kurfess) and two of the project scientists (W.N. Johnson and C. Meegan) on NASA's later *Compton Gamma Ray Observatory*, which went into orbit in 1991. And Clayton began a NASA-sponsored research program for laying out theoretical expectations for nucleosynthesis produced radioactivity and its associated nuclear gamma-ray lines in astronomy. Gamma-ray-line astronomy was in that sense launched by the Rice University program.

At least two other groups followed the Rice University lead and became and remained leaders for the subsequent decades. They constructed differing telescope techniques from that of Haymes NaI scintillator. One was the JPL research group – A.S. Jacobson, J.C. Ling, W.A. Wheaton, and W.A. Mahoney. Using a cluster of four cooled high-purity germanium detectors on NASA's *HEAO* 3, which was launched in 1979, they discovered the first galactic gamma-ray line from radioactive decay. A group at MPI für Extraterrestrische Physik in Munich, V. Schönfelder, P. von Ballmoos and R. Diehl, developed a Compton-scattering telescope capable of good energy resolution and a greater angular view of the sky. Their active results from balloon-borne launches in the 1980s presaged the splendid results of their instrument on NASA's Compton Gamma Ray Observatory. Jacobson's JPL group was not so lucky, as their experiment was removed from *CGRO* as a cost saving. Other groups joined the effort to create this new astronomy with radioactivity.

#### 2.2.2.2 Earliest Predictions of Detectable Gamma-Ray Lines

Inspired by the Rice University balloon program, a series of studies of the nucleosynthesis of radioactivity was undertaken. It may seem surprising in retrospect that this could not have been done by a routine search of the chart of nuclides in conjunction with nucleosynthesis theory. But the data base necessary did not exist. Four decades ago, nucleosynthesis theory was undergoing rapid expansion and clarification, so that it was a series of new insights that laid out each new prospect as it was discovered. Unquestionably the most important of these developments occurred in 1967 when a Caltech group (Bodansky et al., 1968a,b) and a Yale group (Truran et al., 1967) showed with differing research techniques that the *e* process of Hoyle (1946) and Burbidge et al. (1957) for the nucleosynthesis of the iron abundance peak was incorrectly applied. Instead of producing iron isotopes as themselves within a neutron-rich nuclear-statistical equilibrium (NSE), as Hoyle had suggested, they were produced as isobars of radioactive nickel that decayed to iron isobars only after nucleosynthesis and ejection from supernovae. Owing to the increasing strength of the Coulomb energy in nuclei, the most stable nucleus for $4n$ nuclei shifts from $Z = N$ below $A = 40$ to $N = Z + 4$ for $A = 44$–$60$. But insufficient neutron excess required those major nuclei to form as unstable $Z = N$ nuclei. This brought into view the astronomical significance of several important radioactive nuclei. A historical study, *Radiogenic Iron*, was written later to lay out the many significant astrophysical issues that hinged on this new understanding (Clayton et al., 1993).

It is of interest in a historical sense to consider the sequence of subsequent predictions of target radioactivity for astronomy. Clayton et al. (1969) jumped on the newly discovered radioactivity in the revised equilibrium processes (preceeding paragraph). Easily the most important nucleus was $^{56}$Ni, having $Z = N$, whose beta decays to $^{56}$Co and thereafter to $^{56}$Fe, having $N = Z + 4$, are accompanied by numerous gamma ray lines that might be resolved by spectroscopic gamma-ray detectors. G.J. Fishman was a research student in R.C. Haymes' Rice University balloon program when he and Clayton began calculating this spectrum of lines and their time dependences. Their motivation was to test nucleosynthesis theory, especially the fresh new view of explosive nucleosynthesis that had led to $^{56}$Ni nucleosynthesis. S.A. Colgate joined this effort because he had independently begun to investigate whether the energy deposition from the radioactivity could explain the exponentially declining Type Ia supernova light curves (Colgate et al., 1969). Burbidge et al. (1957) had attributed those long lasting light curves to heating by spontaneous fission fragments from freshly produced $^{254}$Cf. As the simplest model (Clayton et al., 1969) evaluated fluxes at earth from the rapid expansion of a solar mass of concrete, much of which had been explosively processed to $^{56}$Ni in the way anticipated by Bodansky et al. (1968a,b). The lifetime of $^{56}$Co was seen as favorable for sufficient expansion for the gamma rays to escape. They neglected a massive envelope, as might overlie Type II cores, realizing that such an envelope would seriously limit the escape of the $^{56}$Ni gamma rays. Clayton et al. (1969) was a very important paper, not only because it was correct in its prediction of the strongest gamma-ray nuclear source, but also because of its galvanizing effect upon experimental teams and on NASA planning. It later was included in the AAS Centennial Volume of seminal papers of the twentieth century. NASA Headquarters suggested funding of a research program at Rice to layout additional prospects for this new astronomy of radioactivity. The NASA grant at Rice was entitled *Prospects for Nuclear Gamma-Ray Astronomy*. But the detection of those $^{56}$Co gamma rays did not become possible until SN1987A occurred in 1987.

Clayton et al. (1969) also first suggested that unknown galactic supernovae (unseen owing to optical obscuration) may be discovered in gamma-ray lines. The lines from radioactive $^{44}$Ti, with 60-year halflife, ought to be detectable from several supernovae that have exploded during the past two centuries, considering that about three per century seem to occur on average within the Milky Way. This exciting idea has great implications for galactic nucleosynthesis and astrophysics; but its hope has been frustrated for astrophysical reasons that are not yet understood.

It was quickly realized (Clayton and Silk, 1969) that the entire universe might be filled with detectable $^{56}$Co gamma rays from the past history of supernova explosions throughout the universe. This work suggested that observable gamma rays owing to the collective effects of many supernovae rather than specific young supernovae might be targeted. Detecting the universe in this way remains a goal of the astronomy of radioactivity.

Two years later it was proposed (Clayton, 1971) that the collective effects of galactic supernovae might allow one to observe gamma-ray lines from long-lived radioactive nuclei whose emission would be too weak from individual supernovae.

That first work in that regard focused on $^{60}$Fe, whose 2.6 million years halflife[1] makes emission from an individual supernova too slow to be observable for the modest number of $^{60}$Fe nuclei produced within a single supernova. The $^{60}$Fe nucleus emits a 59 keV gamma ray upon decay, and its daughter $^{60}$Co emits gamma-ray lines of 1.17 and 1.33 MeV. Reasoning that during its long mean lifetime some 50,000 supernovae occur in the Milky Way, their collective effect should be observable. This reasoning applied equally well thirteen years later to the first interstellar radioactivity to be detected, that of $^{26}$Al.

Surprisingly in retrospect, several years passed before Clayton (1974) realized that when $^{57}$Co, the daughter of 36-h $^{57}$Ni, decays, it also emits favorable gamma-ray lines and that its longer 272-day halflife would cause it to radiate these after the $^{56}$Co was almost extinct. This made $^{57}$Co much more significant than had been appreciated. Clayton (1974) drew attention to the significance of $^{57}$Co and to several new ideas for astronomy with radioactivity. Firstly, given an appropriate galactic supernova one might measure the material thickness overlying the radioactive cobalt by the distinct times at which differing gamma-ray lines reach maximum flux. This happens because the structural opacity for the 2.60 MeV gamma ray is only about half that for the 0.84 MeV gamma ray (both from $^{56}$Co decay), so that the 2.60 MeV gamma ray should peak earlier despite the equal rates of emission. Additionally, the 0.12 MeV gamma ray from the slower $^{57}$Co decay suffers even more absorption and so peaks later yet in time. These time delays have not yet been measurable because a time-dependent gamma line flux requires a bright supernova which would have to occur within our own galaxy. Clayton (1974) also warned that hydrodynamic *instabilities could allow the central material to burst through in streams* and thereby appear earlier than anticipated. Early escape of lines from $^{56}$Co was later detected in SN1987A, which exploded in the Magellanic Clouds, and from which gamma-ray lines from both $^{56}$Co and $^{57}$Co were first detected (see Chap. 3). Earliest of the recorded $^{56}$Co photons were by the gamma ray spectrometer aboard NASA's *Solar Maximum Mission*, which happened to be in orbit when SN1987A exploded and whose sun-pointing spectrometer was reinterpreted as a gamma ray telescope (Leising and Share, 1990). Several balloon-borne gamma-ray spectrometers were flown and also detected $^{56}$Co lines. When $^{57}$Co was detected in SN1987A by Kurfess et al. (1992) with the *Oriented Scintillation Spectrometer Experiment* on NASA's *Compton Gamma Ray Observatory* during summer 1991, 4 years (about six halflives of $^{57}$Co) had elapsed since the explosion, so that the $^{57}$Co abundance had decayed to only about 1.5% of its initial amount. Tension surrounded the hope that it would still be detectable. It was, with 4$\sigma$ significance, implying that the initial abundance ratio $^{57}$Ni/$^{56}$Ni was about twice the ratio $^{57}$Fe/$^{56}$Fe measured in terrestrial iron.

At almost the same time as the $^{57}$Co prediction, Clayton and Hoyle (1974) proposed that gamma rays may be detectable from the more common nova explosions owing to the radioactivity created by the nova outburst. These involve the positron-annihilation line from the hot-CNO burning that powers the nova outburst

[1] At that time, the halflife of $^{60}$Fe was best-known as 1.5 My.

(Starrfield et al., 1972). These 511 keV photons would have to be detected very quickly, within roughly $10^3$ s after the outburst because of the short halflives of CNO radioactive nuclei. They also pointed out the prospect of detecting a 1.274 MeV gamma ray following $^{22}$Na decay. Because of its longer halflife (2.6 year), $^{22}$Na remains alive for the roughly $10^6$ s required for the nova ejecta to become transparent to the gamma ray. These goals have not succeeded yet, but they remain a realistic hope of measuring the thermonuclear power of the nova. The model of the nova makes it a remarkable laboratory for thermonuclear explosions. Detection could confirm the model or rule it out. Some years later the possibility arose that the 478 keV gamma-ray line following the decay of radioactive $^7$Be to $^7$Li might also be detectable from novae (Clayton, 1981). Detectabilty requires the nova envelope accreted from the companion star to be enriched by an order of magnitude in $^3$He, but that is perhaps to be expected. If so, the nuclear reaction $^3\mathrm{He} + {^4\mathrm{He}} \Rightarrow {^7\mathrm{Be}} + \gamma$ creates the radioactive nucleus in the thermonuclear flash. One positive aspect is that the $^7$Be halflife is large enough for the envelope to become transparent.

The radioactive $^{22}$Na nucleus is also produced in supernovae, although the complicated details of its nucleosynthesis therein were not well understood when Clayton (1975c) proposed its detectability within supernovae. It might be observable for a decade, allowing ample time to become transparent to the $^{22}$Na gamma-ray line.

Almost the last good prospect to be predicted turned out to be the first actually observed! That was the radioactive $^{26}$Al nucleus (Ramaty and Lingenfelter, 1977). Like radioactive $^{60}$Fe, the $^{26}$Al nucleus is long-lived and thus decays too slowly to be detectable from individual explosions. It is the cumulative yield of many events over the past Myr or so that was discovered. This is the topic of the next section.

The predictions whose history has been recounted here placed tantalizing targets before the community of experimental physicists. Many of these assembled teams to design, build and fly gamma-ray spectrometers capable of detecting these lines. NASA entertained a grants program pursuing these goals, which create a new wavelength range for astronomy. Moreover, the will to fund the *Compton Gamma Ray Observatory* was strengthened by these specific hopes. That hope was fueled also by the surprising discovery of a gamma-ray line from $^{26}$Al nuclei in the interstellar medium.

#### 2.2.2.3 The Surprise $^{26}$Al

The first detection of an interstellar radioactive nucleus came as a surprise. The history of this radioactive isotope in astrophysics reveals that correct predictions can be made on the basis of inadequate reasoning, that scientists may be blindsided by their own excessive faith in their own pictures and beliefs, and that experimental discovery is the arbiter. The possibility of detecting $^{26}$Al nuclei by observing the 1.809 MeV gamma-ray line emitted following its decay to $^{26}$Mg was suggested by Ramaty and Lingenfelter (1977) and independently by Arnett (1977). Their interesting papers were not quantitative predictions because the grounds for their suggestions were not convincing. They suggested wrongly that interstellar $^{26}$Al nuclei would be detectable if the production ratio in massive stars is

$P(^{26}\text{Al})/P(^{26}\text{Mg}) = 10^{-3}$. However, that production ratio, which was expected by carbon burning in massive stars, was inadequate for detectability. The argument showing that that production ratio was unobservable (Clayton, 1984) was submitted for publication only after interstellar $^{26}$Al was discovered (Mahoney et al., 1982, 1984). The predicted 1.809 MeV flux was too small by a wide margin to be detected by contemporary gamma-ray spectrometers. So something was wrong.

Using the gamma-ray spectrometer on NASA's third *High Energy Astrophysical Observatory* (HEAO 3) Mahoney et al. (1982, 1984) reported a measured flux of 1.809 gamma rays that required about 3 $M_\odot$ of interstellar $^{26}$Al nuclei. That quantity was far larger than the predicted interstellar mass. Nonetheless, both Ramaty and Lingenfelter (1977) and Arnett (1977) had urged, on general grounds and on intuitive arguments, that the 1.809 MeV interstellar gamma-ray line be sought. It was indeed observable, the first to be detected.

What were the conflicts that this discovery illuminated? Using Eq. (2.10) for the number of radioactive nuclei born during the last mean lifetime ($\tau = 1.05\times10^6$year) of $^{26}$Al nuclei, the time-average interstellar abundance ratio for aluminum isotopes was traditionally estimated to be $^{26}\text{Al}/^{27}\text{Al} = P(^{26}\text{Al})/P(^{27}\text{Al})\ \tau/10^{10}\text{year} = 10^{-7}$ if one uses $P(^{26}\text{Al})/P(^{27}\text{Al}) = 10^{-3}$. Since total ISM mass is about $10^{10}\ M_\odot$, it contains about $5.8\times10^5\ M_\odot$ of stable $^{27}$Al. Multiplying by the above isotope ratio, the ISM would carry about 0.06 $M_\odot$ of $^{26}$Al nuclei, woefully inadequate for the observed 3 $M_\odot$ of interstellar $^{26}$Al (Mahoney et al., 1982, 1984). So although Ramaty and Lingenfelter (1977) and Arnett (1977) correctly suggested that $^{26}$Al nuclei might provide a suitable ISM radioactivity to seek, their reasoning did not convincingly justify that hope. The discovery of so much interstellar $^{26}$Al was surprising and meant that some assumptions were quite in error. In fact, several were in error.

Confirming detections of the $^{26}$Al radioactivity were quickly made from balloon-borne gamma-ray spectrometers (von Ballmoos et al., 1987; MacCallum et al., 1987), and the total mass was later measured accurately by the gamma ray spectrometer aboard NASA's *Solar Maximum Mission*, whose sun-pointing spectrometer had been reinterpreted as a gamma ray telescope (Share et al., 1985). The existence of 2–3 $M_\odot$, of $^{26}$Al was then beyond doubt. The 1991 launch of NASA's *Compton Gamma Ray Observatory* with its imaging COMPTEL Compton telescope (see Chap. 10) produced the most detailed data about the spatial distribution of interstellar $^{26}$Al (Diehl et al., 1995).

Stepping back to the history of the mid 1970s, a related issue for the astronomy of $^{26}$Al radioactivity had arisen about one year prior to Ramaty and Lingenfelter's 1977 suggestion; namely, it was discovered (Gray and Compston, 1974; Lee et al., 1977) that the molecular cloud from which the sun had formed apparently carried within it the large isotopic ratio $^{26}\text{Al}/^{27}\text{Al} = 5\times10^{-5}$ at the time the planetary system began to form. Had that ratio been applicable to the ISM as a whole, it would have corresponded to 29 $M_\odot$ of radioactive $^{26}$Al nuclei-very much more than was discovered in 1982 by Mahoney et al. The seeming impossibility of producing such a large quantity by nucleosynthesis prompted the idea (Cameron and Truran, 1977;

Wasserburg and Papanastassiou, 1982) that a supernova within the molecular cloud from which the sun was born had injected the solar $^{26}$Al radioactive nuclei into the collapsing solar portion of the cloud just prior to the solar birth. In that case the solar cloud was quite atypical of the ISM at large. The initial solar $^{26}$Al nuclei in meteorites became the standard bearer for a class of extinct radioactive nuclei that required special local production realted to solar birth. But owing to the huge consequent ratio in the forming sun, it seemed to bear no transparent relationship to the existence of several solar masses of interstellar $^{26}$Al nuclei, which necessarily represents nucleosynthesis of $^{26}$Al nuclei by many thousands of supernovae spread out in time over 1–2 million years prior to today. This discovery from meteorites brought the idea of interstellar $^{26}$Al nuclei to the attention of astrophysicists.

What new ideas brought the expected 0.06 $M_\odot$ of $^{26}$Al nuclei in the ISM into line with the observed 3 $M_\odot$ of $^{26}$Al nuclei? Some suggested that novae (Clayton, 1984; Woosley and Weaver, 1980) or AGB stars, rather than supernovae, were the source of the $^{26}$Al nuclei; but that possibility was ruled out by observations made later with the Compton Telescope (COMPTEL) following the launch of *Compton Gamma Ray Observatory*. Those observations (Diehl et al., 1995) showed spatial concentrations of $^{26}$Al nuclei in ISM regions where star formation is currently active. Spatial correlation of $^{26}$Al nuclei with massive stars was clear. Attention therefore returned to supernovae sources with the realization (Arnould et al., 1980) that hot hydrogen burning shells of massive stars constitutes a significant extra source of $^{26}$Al. The small production ratio in carbon burning ($P(^{26}\mathrm{Al})/P(^{27}\mathrm{Al}) = 10^{-3}$) must be augmented by $^{26}$Al-rich convective shells in massive stars. Those calculations became a major industry with numerical models of presupernova evolution. Because $^{26}$Al/$^{27}$Al ratios of order 0.1 exist in these shells, their contribution calculated by using time-dependent numerical models of presupernova evolution would be needed. Weaver and Woosley (1993) calculated that the average production ratio from a standard spectrum of massive stars was near $\mathrm{P}(^{26}\mathrm{Al})/\mathrm{P}(^{27}\mathrm{Al}) = 0.006$, six times larger than the earlier estimates. Secondly, the estimated ratio of mean interstellar $^{26}$Al/$^{27}$Al must be increased because most of the stable $^{27}$Al is locked up inside old stars whereas the live $^{26}$Al nuclei will still be overwhelmingly in the gaseous ISM. This reasoning augments the expected ratio by the factor (k+1) derived from the standard analytic models of chemical evolution of the mean galaxy. Clayton et al. (1993) presented that argument for the expected interstellar ($^{26}$Al/$^{27}$Al) abundance ratio. Because k = 2–4 seems likely from other astrophysical arguments, this (k+1) correction also increases the interstellar ratio by another factor 3–5. These two effects combined amounted to an increase of a factor 20–30 in the initial expectations, bringing the expected value from 0.06 $M_\odot$ to 1–2 $M_\odot$ of $^{26}$Al nuclei in the ISM. The original conflict with global theory was largely resolved.

Moreover, the largest observed flux from $^{26}$Al nuclei concentrated in ISM regions where star formation is currently active. This directly challenged the theoretical simplification that adopts a star-formation rate that is constant in time and spread smoothly through the spatial ISM. Models of chemical evolution of the galaxy use that simplification in order to be calculable. But the observations of 1.809 MeV gamma rays painted a picture of star formation occurring sporadically whenever and wherever large regions of ISM become vulnerable to prolific star formation. The massive $^{26}$Al emission regions stunned and surprised many experts (including this writer). The fluxes from these are moderated by the inverse square of their respective distances, further complicating discussions of the mass of interstellar $^{26}$Al. The history of $^{26}$Al nuclei in the ISM provides a textbook example of scientific progress into a new discipline. Experts disagree in their predictions; correct predictions often rely on intuition as much as on scientific justification; surprises often attend exploratory experimental surveys; and scientists from many disciplines amplify the relevant ideas and measurements. Astronomy with radioactivity is such a science, and its first discovery of an interstellar gamma-ray line emitter, the surprise of $^{26}$Al, was bathed in confusion.

## 2.2.3 Radiogenic Luminosity

A big problem lay in the path of trying to understand how explosive objects in astronomy could remain bright. Their luminosity was expected to fall rapidly as the objects expand. Suppose a rapidly expanding object consists of an ideal non-degenerate gas and that the expansion is too rapid for gain or loss of heat; that is, the expansion is adiabatic. The internal heat of such an object is rapidly lost to the mechanical work of the expansion. For adiabatic expansion $TV^{\gamma-1}$ = constant during the expansion, where $\gamma$ is the ratio of specific heats at constant pressure $c_P$ to the specific heat at constant volume $c_V$. For a perfect monatomic gas $\gamma = c_P/c_V = 5/3$. Thus $TV^{2/3}$ = constant for such an expanding hot object. If the expanding object can escape gravitational binding, the radius of an exploding object may be approximately $R = vt$, so that volume $V$ is proportional to $t^3$. Thus one would expect that $Tt^2$ is constant during the expansion. The problem is then that if luminosity $L$ is proportional to $R^2T^4$, as in a black body, one expects $L$ to be proportional to $t^{-6}$. However, many supernovae that are bright after one week are still bright after three weeks instead of dimmer by the large expected factor near $3^{-6}$. Similar problems exist for novae expansions. The problem, then, lay in discerning what source of energy could keep the expanding objects hot. The answer is radioactivity. The radiogenic luminosity of exploding objects became one of the major aspects of astronomy with radioactivity.

### 2.2.3.1 Radiogenic Luminosity in Supernovae

It was excellent luck that SN 1987A exploded nearby in the Magellanic Cloud in 1987. It became the most observed explosive astronomical event of all time. Astronomers recorded its emissions in every possible wavelength band, which proved necessary in exposing how radiogenic luminosity worked in that kind of supernova. The observations showed that after July 1987 the total power output of SN 1987A declined exponentially for about two years,accurately tracking the 77.2-day halflife of $^{56}$Co. This confirmed that the energy of the positrons emitted and of the subsequent gamma-ray emission following each $^{56}$Co decay were efficiently converted to optical and infrared luminosity. This old idea for declining supernova light curves was first treated for the identification of $^{56}$Co as the relevant radioactivity by Colgate et al. (1969). The observed value of the total luminosity of SN 1987A required 0.075 $M_{\odot}$ of $^{56}$Ni to have been synthesized within the core matter that escaped the central neutron star. With each scattering of a gamma ray it loses roughly half of its energy to the recoil electron, which is quickly degraded into heat. Once its energy has been scattered below 5 keV the large photoelectric opacity converts its entire remaining energy to electron energy, which is also degraded into heat.

Supernovae come in two main types, with additional structural subtypes. Type Ia and Type II differ in how compact they are, how much radioactivity is produced, how that radioactivity is distributed, and how much envelope overlies the radioactivity. The nucleosynthesis of $^{56}$Ni is easily the largest and most important source of heating power for the radiating gas. Such issues impact the way in which radioactivity provides luminosity for the remnant. To truly judge the degree of scientific understanding of radiogenic luminosity requires astronomers to infer, to the best of their ability, the structure of the exploding object. The luminosity being observed has complicated relationship to the location of the radioactivity. In SN1987A itself, for example, the presupernova star was blue, implying smaller size than the common red presupernova stars. The amount of overlying matter was modest owing to a large extent of mass loss during the prior evolution of the star. These characteristics influenced the relative importance of the radiogenic luminosity to the luminosity caused by the shock heated envelope.

Although $^{56}$Ni provides the largest and most important source of heating power, other radioactive nuclei also play very important roles. The $^{56}$Co daughter of $^{56}$Ni is actually more observable in its effect on radioactive luminosity. This is primarily because its 77.2-day halflife allows it to remain alive after the 6.08-day parent $^{56}$Ni has decayed. Therefore, $^{56}$Co delivers heating power at a later time when the expanding remnant is much larger and has suffered more adiabatic cooling. In these circumstances, quantitative astronomy with radioactivity requires careful evaluation of how, when and where the energy released by radioactivity is deposited in the expanding gas (The et al., 1990).

The nucleosynthesis of $^{57}$Ni also plays a role in radiogenic luminosity because it decays rather quickly to 272-day $^{57}$Co. Owing to its longer halflife, $^{57}$Co is still providing radioactive power when the $^{56}$Co has decayed to negligible abundance. This transition of dominant radioactive power occurs after about 2 years. When *Compton Gamma Ray Observatory* (Kurfess et al., 1992) measured with OSSE the actual mass of $^{57}$Co in SN1987A, it corresponded to twice the solar ratio $^{57}$Fe/$^{56}$Fe. This surprised many who had inferred its abundance to be five times the solar ratio, a large isotopic ratio that had been deduced on the basis of the bolometric luminosity exceeding the instantaneous power from $^{56}$Co decay (Suntzeff et al., 1992; Dwek et al., 1992). The radiogenic luminosity truly deriving from $^{57}$Co decay was demonstrated by the OSSE measurement to have been overestimated by a factor near 2.5. That discovery team (Clayton et al., 1992) then advanced a new theoretical aspect of radiogenic luminosity, what they called *delayed power*. They presented a model showing that when the remnant gets sufficiently dilute owing to its expansion, the rate of electronic recombination cannot keep up with the rate of ionization that had been produced somewhat earlier by residual amounts of $^{56}$Co radioactivity. Because of the high degree of ionization, the radiogenic luminosity from $^{56}$Co radioactivity begins to exceed the instantaneous rate of energy deposition from $^{56}$Co decay. In effect, the radiogenic luminosity reflects the rate of $^{56}$Co decay from a somewhat earlier time than the time of observation. This observed upturn in luminosity had been attributed to $^{57}$Co instead of delayed power from $^{56}$Co decay. Such time-dependence became a new aspect of astronomy with radioactivity.

A more general aspect of radiogenic emission lies in any astronomical investigation in which the gas possesses a higher degree of energy excitation than could be expected in the absence of radioactivity. For example, the hydrogenic lines of $He^+$ in SN 1987A were interpreted as a consequence of the high-energy photons escaping from the supernova interior. Hard X-rays emanting from the expanding supernova 1987A were another clear example of Compton-scattered radioactivity gamma rays escaping from the interior; and only radioactivity can produce so many hard X-rays. Another example might be the detectable presence of doubly ionized species, an ionization state that would make no sense in a purely thermal setting. All such radiogenic possibilities are aspects of radiogenic luminosity, and as such a part of astronomy with radioactivity.

A similar transition between identities of powering radioactive nuclei occurs when the decay power of 60-year $^{44}$Ti exceeds that of $^{57}$Co. And no doubt there exists a similar problem, namely, does the rate of radiant emission of $^{57}$Co power keep up with the rate of $^{57}$Co decay power. The radioactive $^{44}$Ti is created in almost the same location as are the Ni isotopes. This occurs in the explosive burning of oxygen and silicon (Bodansky et al.,1968a,b; Woosley et al.,1973) and reflects the extent to which $^{28}$Si has been decomposed into the silicon-burning quasiequilibrium. But radiogenic luminosity from this $^{44}$Ti nucleus depends even more strongly on how its emissions are converted to luminosity and the extent of the time lag between radioactive decay and the creation of observed photons. Its contributions to delayed radiogenic luminosity require careful assessment of those issues.

Such considerations are significant in the scientific understanding of astronomy with radioactivity. The luminosity of a young supernova remnant is one of the easiest observations of radioactivity in astronomy; but its interpretation requires fairly accurate description of the structure of the exploding object.

#### 2.2.3.2 Radiogenic Luminosity in Novae

The brightening of a nova explosion and the associated ejection of matter from its expanded envelope are also issues that are dependent upon the deposition of radioactive heat to the envelope (Starrfield et al., 1972). Those authors showed that a successful model needs to mix carbon from the surface of the underlying white dwarf into the hydrogen-rich accreted envelope in order to have sufficient radioactivity produced by the thermonuclear flash that triggers the nova event. That burning is the hot CN cycle. They showed that radioactive decay keeps the envelope hot while it expands to larger radiating surface area and therefore increased luminosity. They also showed that without the *radioactive afterburner*, mass ejection would not occur. Truran (1982) gives more nuclear-physics details of the nova explosions.

### 2.2.4 Extinct Radioactivity and Immediate Presolar Nucleosynthesis

#### 2.2.4.1 Xenology Revisited

John Reynolds had discovered in 1959 that extinct radioactive $^{129}$I appeared to have existed initially in meteorites. The observed ratio to stable I is near $^{129}\text{I}/^{127}\text{I} = 10^{-4}$. This stood for decades as a nearly unique example. Then in the mid-1970s new discoveries and ideas greatly enlarged the context of astronomy with radioactivity. Using Eq. (A.12) for the number of radioactive nuclei born during the last mean lifetime ($\tau = 23.5 \times 10^6$ year) of $^{129}$I nuclei, the mean interstellar abundance ratio for iodine isotopes is expected to be

$$^{129}\text{I}/^{127}\text{I} = P(^{129}\text{I})/P(^{127}\text{I})(k+1)\tau_{129}/\text{t}_{\odot} = 0.013 - 0.022$$

using $P(^{129}\text{I})/P(^{127}\text{I}) = 1.5$, $\text{t}_{\odot} = 8$ Gy for the presolar duration of nucleosynthesis, and $k = 2 - 4$ within the standard model of radioactivity in galactic chemical evolution. The measured ratio is very much smaller than this expected ratio. The interpretation was that the solar molecular cloud sat dormant, with no new nucleosynthesis input, for 4–6 mean lifetimes of $^{129}$I to allow decay to reduce its activity to the measured level. As plausible as this *waiting period* seemed to be for the next two decades, it was not correct.

Reynolds' laboratory also discovered that that extinct radioactive $^{244}$Pu appeared to have existed initially in meteorites. It was measured by an anomalous pattern of several Xe isotopes that was consistent with the spontaneous-fission spectrum from $^{244}$Pu. (Hudson et al., 1989) presented more modern data setting its abundance relative to that of $^{232}$Th as $^{244}\text{Pu}/^{232}\text{Th} = 3 \times 10^{-3}$. The many studies of these first two extinct radioactivities came to be called *xenology* because of their reliance on excess isotopic abundances of Xe isotopes in meteorites. With a production ratio $P(^{244}\text{Pu}/^{232}\text{Th}) = 0.5$ and mean lifetime $\tau_{244} = 115 \times 10^6$ year, the expected interstellar ratio would be 0.02–0.03 – larger by tenfold than the solar amount. But in this case a sequestering of the solar cloud to allow $^{129}$I nuclei to decay would not sufficiently reduce the $^{244}$Pu abundance. Trying to resolve the xenology puzzles occupied many meteoriticists for two decades. Then three new developments altered xenology studies.

Firstly, existence of such $^{129}$Xe-rich and fission-Xe-rich interstellar dust was predicted (Clayton, 1975a) by the first argument that dust containing evidence of extinct radioactivity should condense within supernovae and influence the early solar system. This was an exciting new idea in the astronomy of radioactivity. However, it appears that chemical ways of forming meteoritic minerals from dust without loss of the carried xenon isotopes is too implausible; so this new idea was not supported for isotopes of Xe. But the idea of such supernova dust had been firmly implanted among the concepts of astronomy with radioactivity; moreover, the future may yet find evidence of fossil xenon. Several other extinct radioactive nuclei were predicted to exist within supernova dust (Clayton, 1975a,b), and many of these are now demonstrated to exist.

Secondly, the $^{26}$Al extinct radioactivity was discovered in meteorites. But its mean lifetime is too short ($\tau_{26} = 1.05 \times 10^6$year) for any to survive the sequestering that xenology had suggested. This led to the second big change. The seeming impossibility of maintaining the large measured quantity of interstellar $^{26}$Al nuclei by nucleosynthesis prompted the idea that a supernova within the molecular cloud from which the sun was born had injected the solar $^{26}$Al radioactive nuclei into the solar collapse just prior to the solar birth. Such a unique association would render the solar cloud atypical of the ISM at large (Cameron and Truran, 1977; Wasserburg and Papanastassiou, 1982). The substantial solar $^{26}$Al abundance initially in meteorites became the standard bearer for a class of extinct radioactive nuclei that required special local production. Cameron and Truran (1977) suggested that far from being a coincidence, the supernova producing the solar $^{26}$Al nuclei also triggered the collapse of the solar molecular cloud by the overpressure its shock wave brought to bear on that cloud. That model for live short-lived radioactivity in the solar cloud came to be referred to as *the supernova trigger*. But Cameron and Truran went too far. They supposed that the xenology-producing radioactivities ($^{129}$I and $^{244}$Pu) were also injected by that supernova trigger. They ignored that the ISM already was expected to contain a hundred times too much $^{129}$Xe for Reynolds' measurement, not too little, and ten times too much $^{244}$Pu. So injecting those radioactivities seems to not be correct either. Nor does the idea of fossil $^{26}$Mg nuclei from Al-rich interstellar dust rather than live $^{26}$Al in meteorites seem correct. But the puzzles of extinct radioactivity had come more sharply in focus, and these new ideas were instrumental for astronomy with radioactivity.

Thirdly, additional extinct radioactive nuclei were discovered. So diverse were their lifetimes that the idea of exponential decay from a starting abundance became increasingly untenable. Then another new idea of great importance for astronomy with radioactivity appeared. Clayton (1983) pointed out that the concept of a homogeneous ISM was at fault. He argued that supernova ejecta enter a hot phase of an ISM in which matter cycles on average among three phases, but that the sun was born from the cold molecular cloud phase. This distinction greatly modifies the expectation of exponential decay in the ISM and of extinct radioactivities in molecular clouds. Huss et al. (2009) provide a modern list of the many extinct radioactivity for which evidence exists.

The revisiting of xenology and these three new developments left extinct radioactivity research changed conceptually.

#### 2.2.4.2 Neon-E

A different noble gas, neon, provided an early but poorly understood evidence of stardust in the meteorites. As with xenology, the $^{20}$Ne/$^{22}$Ne isotopic ratio can be studied by incremental heating of a meteoritic sample in a mass spectrometer. David Black discovered that in the Ivuna meteorite the measured isotopic ratio $^{20}$Ne/$^{22}$Ne decreased precipitously when the sample heating reached 1,000°C, dropping to a ratio near three, much smaller than the common ratio near ten. Variation by a factor three in a trapped noble-gas isotopic component was unprecedented. This suggested that some unknown mineral releases its neon gas near 1,000°C, and that that mineral contains $^{22}$Ne-rich neon. Black (1972) suggested that this small $^{20}$Ne/$^{22}$Ne isotopic ratio was so bizarre that it can not be accounted for by conventional means within an initially homogeneous, gaseous solar system. Black suggested that interstellar grains that had formed somewhere where the $^{20}$Ne/$^{22}$Ne isotopic ratio was smaller than three had survived the origin of the solar system and imprinted the 1000°C temperature outgassing with its isotopic signature. He called this exotic neon component Ne-E. This was a radical far-reaching conclusion, perhaps the first of its kind based on sound analyzed data rather than on pure speculation.

But where might this dust have formed? Arnould and Beelen (1974) remarked that neon gas ejected from explosive He-burning shells of massive stars could resemble Ne—E, but they offered no suggestion for forming carriers of Ne—E there or of otherwise getting such gas into the early solar system. Clayton (1975b) offered a more concrete proposal, arguing that supernova dust would condense before the ejecta could mix with circumstellar matter, roughly within the first year, and that $^{22}$Na (2.6 year) produced by the explosion would condense as the element sodium and decay to daughter $^{22}$Ne only after the grains had been grown. In that manner nearly isotopically pure $^{22}$Ne could be carried within supernova dust into the forming solar system. This was an early prediction of the supernova-dust phenomenon in meteorites. Clayton and Hoyle (1976) argued that nova dust provided another possible source for Ne—E. Either radioactive venue would amount to a new technique for astronomy with radioactivity.

The leading alternative to radiogenic $^{22}$Ne is condensation of dust in a $^{22}$Ne-rich red-giant stellar atmosphere. This can indeed occur, because during He burning the $^{14}$N residue of the prior CN cycle can be converted to $^{22}$Ne by two successive radiative alpha-particle captures (Arnould and Norgaard, 1978). The question arises whether this Ne is sufficiently $^{22}$Ne-rich. And that required much more study of meteoritic specimens. But the possibility was experimentally strengthened by the discovery (Srinivasan and Anders, 1978) that at least one Ne—E component was associated with $s$-process xenon from red-giant atmospheres.

These were exciting developments in isotopic astronomy and for astronomy with radioactivity. This fresh new set of ideas intensified interrelationships among interstellar stardust, the origin of the solar system, and meteorites.

#### 2.2.4.3 $^{26}$Al: Fossil or Injected Fuel

During the mid 1970s it was discovered (Gray and Compston, 1974; Lee et al., 1977) that very refractory Al-rich minerals that apparently were among the first solids to form in the early solar system contained elevated isotopic ratios $^{26}$Mg/$^{27}$Mg. Furthermore, the number of excess $^{26}$Mg atoms within a mineral was shown to be proportional to the number of Al atoms. That correlation could be reproduced

(Lee et al., 1977) if the Mg had been initially isotopically normal and the initial Al contained a small component of radioactive $^{26}$Al, whose later decay produced the excess $^{26}$Mg atoms. The simplest way for this to have occurred is that the molecular cloud from which the sun had formed carried within it the large isotopic ratio $^{26}\mathrm{Al}/^{27}\mathrm{Al} = 5\times10^{-5}$ at the time the planetary system began to form. The reason for calling that ratio large is that it far exceeded the ratio that had been anticipated to exist within the ISM. Had that ratio been applicable to the ISM as a whole, it would have corresponded to 29 $M_\odot$ of radioactive $^{26}$Al nuclei – much more than was discovered in 1982 by Mahoney et al. by their detection of the 1.809 MeV gamma ray line that is emitted following its decay.

The immediate issue was whether these radioactive $^{26}$Al nuclei were actually alive in the early solar system or whether $^{26}$Al was alive only when Al-rich dust formed in earlier galactic supernovae. If the latter option were true, the excess $^{26}$Mg nuclei were fossils of radioactive $^{26}$Al decay within interstellar Al-rich dust. Existence of such $^{26}$Mg-rich and Al-rich interstellar dust had been predicted (Clayton, 1975a,b) in the first works to propose that dust containing evidence of extinct radioactivity should condense within supernova ejecta. This was an exciting new idea in the astronomy of radioactivity. Clayton (1977a,b) presented a solid-carrier model for excess $^{26}$Mg nuclei in Al-rich minerals based on the fossil picture. In the first case, that of live $^{26}$Al nuclei during the formation of the minerals, some explanation was required for why so much radioactive $^{26}$Al should have been present in the solar matter when it could not have been a general property of the ISM. In the second case, fossil excess $^{26}$Mg nuclei, why did interstellar Al dust containing abundant radioactive $^{26}$Al nuclei form within supernova ejecta and how did it participate in the chemical growth of the Al-rich minerals found in meteorites. In the first case, the heat of the radioactive decay of radioactive $^{26}$Al nuclei within those planetesimals that had been the parent bodies for the meteorites would have been sufficient to melt large parent bodies, allowing them to differentiate chemically (as the earth has done). In that case $^{26}$Al also played a role as fuel for melting of the meteorite parent bodies. In the second case, the excess $^{26}$Mg nuclei represented a fossil of interstellar decay. Tension between these two possibilities was reflected in the phrase "Fossil or Fuel" within the title of the paper by Lee et al., who argued in favor of the live $^{26}$Al nuclei in the solar gas. This debate raged for a few years but lost steam as chemical arguments for the growth of the mineral phases seemed increasingly likely to require live $^{26}$Al nuclei.

The struggle over the interpretation very short-lived extinct radioactivity was heightened later by the discovery of extinct $^{41}$Ca in meteorites. Its abundance relative to that of $^{41}$K is only $^{41}\mathrm{Ca}/^{41}\mathrm{K} = 1.5\times10^{-8}$ (Srinivasan et al., 1996); however, even that small abundance looms large because its mean lifetime is but $\tau_{41} = 0.144 \times 10^6$ year. Surely none can survive galactic nucleosynthesis to the time the solar system formed. So the supernova trigger injection would be needed to account for it as well as $^{26}$Al. But fossil effects are also definitely possible. Huge $^{41}$K excesses in Ca-rich supernova dust had been predicted two decades earlier (Clayton, 1975b).

The history of documentation of extinct radioactive nuclei that had been alive in the early solar system was led for decades by G.J. Wasserburg and his laboratory at Caltech (jokingly called "the lunatic asylum" owing to their research on lunar samples). His laboratory was of such renown for its measurements using traditional mass spectrometry with extreme sensitivity that Wasserburg was chosen recipient of the 1986 Crafoord Prize for geosciences. Similar mass spectrometric research of great consequence for this problem was conducted by Günter Lugmair in his laboratory at UCSD.

#### 2.2.4.4 Extinct Radioactivity and Mixing of Interstellar Phases

Xenology soon gave way to a host of newly discovered extinct radioactive nuclei. $^{129}$I and $^{244}$Pu could seen to be members of a larger class of now extinct parents. A very special class contains those nuclei having mean lifetimes long enough that some of that radioactive abundance might survive from galactic nucleosynthesis but short enough that their survival live to the time of the solar system depends on the timing of interstellar mixing. That class of nuclei also includes $^{129}$I and $^{244}$Pu. Their mean lifetimes are, in ascending order: $\tau(^{60}\mathrm{Fe}) = 2.16^{2} \times 10^{6}$ year; $\tau(^{53}\mathrm{Mn}) = 5.34 \times 10^{6}$ year; $\tau(^{107}\mathrm{Pd}) = 9.38 \times 10^{6}$ year; $\tau(^{182}\mathrm{Hf}) = 13.8 \times 10^{6}$ year; $\tau(^{129}\mathrm{I}) = 23.5 \times 10^{6}$ year; $\tau(^{244}\mathrm{Pu}) = 115 \times 10^{6}$ year; and $\tau(^{146}\mathrm{Sm}) = 149 \times 10^{6}$ year. The relative abundances that they possess and the three-isotope correlation plots by which they are measured are discussed in Chap. 6. What became abundantly clear is that simple exponential decay from a set of starting abundances is not consistent with the data. Something more complex is at work.

*Four layers of interpretation* The new data and new creative ideas led to a picture of cosmic radioactivity that is interpreted on four layers of complexity. Like the proverbial *Russian doll*, one opens each layer only to reveal a more complex version inside. What are these four layers?

1. *Mean Galactic Nucleosynthesis and the well-mixed ISM*: Using Eq. (A.12) for the mean interstellar ratio of a radioactive abundance to that of a reference isotope gives that mean abundance ratio to be

$$N(^{A}Z)/N(^{A*}Z) = P(A^{Z})/P(^{A*}Z)\ (k+1)\ \tau_Z/t_{\odot} \tag{2.22}$$

   If the halflife of species $^{A}Z$ is long enough that the ISM can be considered to be well-mixed, it may be interpretable by the mean expectation of Eq. (2.22). An example might be the uranium isotope ratio, since both have halflives of $10^{9}$ year or greater. The ISM probably mixes well during $10^{9}$ year.
2. *Mean Galactic Nucleosynthesis and the ISM Phases*: If the lifetime $\tau_Z$ is of order $10^{8}$ year or less, the abundances of that radioactive nucleus may differ in the different phases of the ISM. The three-phase mixing model (Clayton, 1983)

[2] The mean life of $^{60}$Fe is now known as 3.8 My.

that describes this in the mean is described below. The idea is that supernova ejecta appear initially in the hot ISM, and considerable mean time is required for it to work its way via mass exchange between phases into the molecular clouds where the sun might have formed. That delay reduces radioactive abundances, depending upon their lifetimes. This idea regards each phase of the ISM as having the mean concentration of radioactive nuclei everywhere in that phase; but the three phases have differing mean concentrations of the radioactivity, with the molecular-cloud abundance being the least. This picture divides the mean ratio Eq. (A.12) into differing steady-state ratios in each of the three phases. An example might be the iodine isotope ratio, since its halflife of 17 My is shorter than the expected phase change times (perhaps 50 Ma) for the ISM phases. Thus each phase will in the mean have different $^{129}I/^{127}I$ ratios.

3. *Mean Galactic Nucleosynthesis and Deviations from mean ISM mixing times*: All galactic samples of a given ISM phase will surely not be identical. Differences occur because nucleosynthesis does not follow the galactic mean rate at all locations, or because the interphase mixing times may vary from one place to another. Separate portions of a phase will scatter about that mean for that phase, perhaps by sizeable amounts. Astronomical observations of very old, very metal-poor stars reveal this phenomenon clearly (Burris et al., 2000). The iodine isotopes can illustrate how this works. Consider the molecular cloud from which the sun formed. It may have existed in a portion of the galaxy whose hot phase received less than the average amount of new nucleosynthesis during the past 24 My (the $^{129}I$ mean lifetime). Thus the $^{129}I$ concentration locally would be less than its galactic mean, whereas stable $^{127}I$ would have the same concentration everywhere. Furthermore, the time delay for the hot phase to be admixed into the solar molecular cloud may locally have exceeded the galactic mean delay, in which case the $^{129}I$ concentration in the solar molecular cloud would have decayed to a smaller value than typical of that phase. This layer of interpretation deals with specific local galactic workings that differ from the galactic mean.
4. *Nucleosynthesis from a single neighboring supernova*: The solar concentration of a short-lived radioactive nucleus may have been the result of a single galactic supernova. The shorter the halflife, the more likely it becomes that a neighboring dying star is needed to create it. In such a case the concept of mean galactic nucleosynthesis rate has no relevance. Short-lived $^{26}Al$ is historically the prime example of this situation. Variations on the supernova injection model then appear: When did that supernova occur? With what efficiency did its ejecta admix into the solar molecular cloud? How long was required for that mixing to occur? Such questions are so specific that they are frustrating. Almost any answer seems possible, depending upon hydrodynamic modeling for its credibility.

Finally, some radioactive nuclei may have multiple causes for their solar abundances. Each layer of interpretation may contribute. Take $^{244}Pu$ as a likely example ($\tau(^{244}Pu) = 115$ Ma). A portion of the solar $^{244}Pu$ abundance may survive at the level expected from mean galactic nucleosynthesis (layer 1). Layer 2 may have reduced that somewhat, depending on the mean interphase mixing times. Layer 3 describes a non-average $^{244}$Pu solar concentration owing to atypical (non-mean) local nucleosynthesis rate and mixing parameters. Finally, a portion of solar $^{244}$Pu abundance may have resulted from a single nearby supernova at the time of solar birth. It will be clear that scientific understanding must seek the best picture for simultaneously fitting all of the radioactivities. Each individual abundance will be inadequate by itself to determine the physics of its solar presence. But a model that fits them all is a serious contender for the truth. To advance more into this topic, consider the mean interphase mixing model.

*Mean mixing model for ISM radioactivity* Astrophysics provides this novel example of the layer-2 mathematics of radioactive decay. The distinct ISM phases have vastly differing temperatures and densities. These phases mix with one another on timescales that are longer than the shortest of the extinct radioactivities but shorter than the longest. As a consequence, the concentration $X$ of a radioactive nucleus will differ from phase to phase, with the differences greater for shorter halflife. Solar-system samples revealing the presence of once-live radioactive nuclei are obtained from meteorites and other planetary objects formed in the solar accretion disk. Measurements of solar samples therefore are of the concentrations in the cold molecular cloud (one of the phases) from which the sun was born. The key point is that radioactive concentrations in that molecular-cloud phase should be smaller than in the mean ISM, which includes all of its phases. This consequence of the mixing times among the phases, taking into account their levels of radioactivity, was devised by Clayton (1983).

*A simplified two-phase model* To focus on mathematical aspects of radioactivity rather than on aspects of astrophysics, consider first a simplified model that illustrates the essence of the problem. Suppose that the ISM consists of but two phases. Stars are born in phase 1, but the freshly synthesized radioactivity is ejected from supernovae into phase 2. Let those two phases have equal mass $M$ and exchange matter with the other phase on a timescale $T$. The mass exchange rate is then $(dM/dt)_{\mathrm{exch}} = M/T$. Matter from phase 1 joins phase 2 at that rate, and conversely matter from phase 2 joins phase 1 at that same rate. The masses of each phase then remain constant and equal.

In reality, spatial inhomogeneity also exists, depending on where in phase 2 the fresh radioactivity is created and how long is required for its homogenization. Roughly 30,000 supernovae occur per My in the Milky Way, and they have similar nucleosynthesis yields, so an almost homogeneous injection phase occurs faster than the output of a single supernova can be spread uniformly. Even so, concentration differences and isotopic differences will exist spatially in that injection phase. The best that can be done without a specific calculation defined for a specific configuration is to calculate the *average concentration* in the injection phase. To this end consider the injection phase to be uniformly mixed at all times (Layer 2). With that assumption the average difference in radioactive concentration between the phases can be calculated with a simple mathematical model. The answer can not be assumed to have applied exactly to the unique case of solar formation. Nonetheless, such a calculation reveals what is anticipated on average, without inhomogeneous effects.

Returning to the simplified two-phase model for those average concentrations, the mixing rate between them is $(dM/dt)_{\mathrm{exch}} = M/T$. Then with $X_1$ being the concentration (grams of $X$ per gram of ISM) of a radioactive nucleus in phase 1, the star-forming phase, its total radioactive mass $MX_1$ changes as follows: $d/dt(MX_1) = MdX_1/dt + X_1 dM/dt = MdX_1/dt$ because the second equality is a consequence of the unchanging mass $M$ of each phase. How shall one evaluate $dX_1/dt$? It changes owing to two effects: first, the radioactive decay of $X_1$; second, the mass exchange with phase 2 having concentration $X_2$. One writes for those two terms $MdX_1/dt = -MX_1/\tau + (X_2 - X_1)M/T$ where the first term is the radioactive decay rate with mean lifetime $\tau$. But the left-hand side must also vanish because $dX_1/dt = 0$ in the steady state. This reasoning yields at once $X_1 = \tau/(T+\tau)X_2$ This factor expressing the ratio of the concentrations can differ significantly from unity. Suppose the mixing time $T = 100$ Ma. Then for $^{129}$I, $X_1$ = (23.5 Ma/123.5 Ma)$X_2 = 0.19X_2$. The $^{129}$I concentration in star-forming clouds is then fivefold smaller than in the injection phase. This reduction is even greater for a nucleus having smaller mean lifetime. But for very long lifetimes, $\tau/(T+\tau)$ approaches unity, both phases having the same concentration.

Both $X_2$ and $X_1$ can be individually evaluated from the requirement that their average (since the two phases have the same mass) must equal the mean expected ISM concentration for that nucleus (Eqs. A.11 and A.12). Chapter 6 includes a table of the known extinct radioactive nuclei and their abundances in the early solar system.

The significance for differing ratios of extinct nuclei between the two phases is that those differences do not reflect the simple expectation of exponential decay. The ratio $\tau/(T+\tau)$ is very far from exponential in dependence on lifetime $\tau$. Clayton (1983) introduced that effect when comparing the solar abundance levels of differing radioactive nuclei. This simplified two-phase mean model illustrates that idea.

*Clayton three-phase model* The initial (Clayton, 1983) introduction of this physical idea for interpreting relative abundances of extinct radioactive nuclei in this manner actually suggested a three-phase ISM rather than only two. Those masses and phases were:

(1) $M_1$ is the molecular-cloud mass in which stars form. It exchanges mass with $M_2$.
(2) $M_2$ is the mass of large HI complexes that can not be disrupted by the supernova shock waves that frequently traverse the ISM. Phase 2 exchanges mass with $M_1$ with lifetime $T_1$ and also with $M_3$ with lifetime $T_2$.
(3) $M_3$ is the mass average HI clouds that are sufficiently small to be disrupted by passing supernova shock waves and which therefore are part of this warm neutral phase. Phase 3 exchanges mass with phase 2 with lifetime $T_2$.

It is not clear that this three-phase model is more realistic than the simpler two-phase model; but the need for fresh radioactivity to mix from its injection phase through a second phase before admixing with molecular clouds does amplify the distinction between radioactive lifetimes. Future progress with this multiphase idea, which definitely is significant for the extinct radioactivity problem, will require more modern astrophysical work on the differing phases, their masses, and how effectively they exchange mass with one another. Finally, it may be that time-dependence must be introduced. The growth of each phase, rather than static mean masses for each, needs evaluation. Over long times the global masses of each phase will be static; but their cyclic growth and disruption may be an essential part of the problem.

An insightful paper in regard to these issues has recently appeared (Huss et al., 2009). Those authors present current data for the observed abundances of each extinct radioactivity. They also review the sites of nucleosynthesis of each nucleus. Then they evaluate the mean abundance expected within the standard model of galactic chemical evolution moderated by the interphase mixing described above. Figure 2.2 (from Huss et al., 2009) displays the expectation as a function of the mean decay lifetime for a wide range of interphase-mixing timescales and compares those results to those for two differing free-decay intervals (the solid curves). Note that interphase mixing fits more extinct radioactivities than does either free-decay interval;but the four short-lived isotopes are too abundant for any galactic scheme,

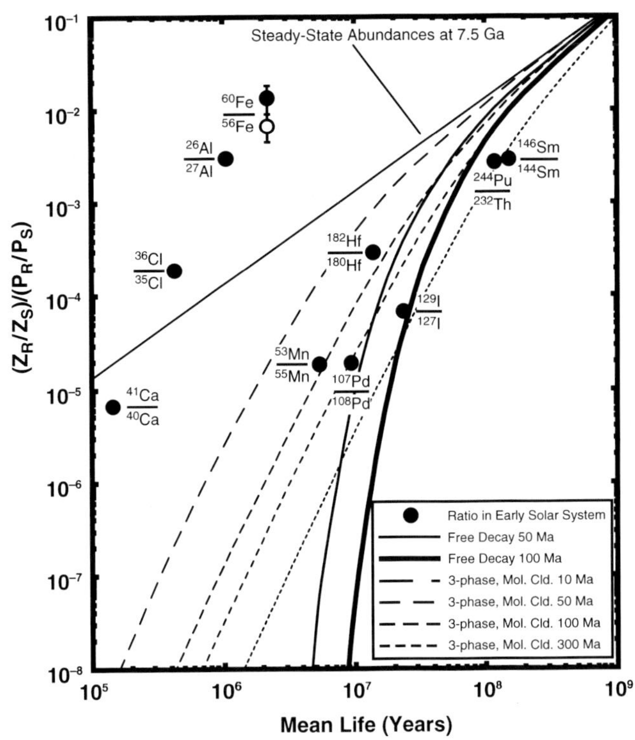


**Fig. 2.2** Calculated surviving levels of the extinct radioactive nuclei in the solar cloud at the time of solar birth are expressed as a ratio to a neighboring stable isotope of the same element. Galactic steady state modefied by the continuous interphase mixing is shown for several different mixing times. (from Huss et al., 2009)

apparently demanding injection from an adjacent supernova. Keep in mind that what is being examined by these mathematical procedures is a fundamental property of interstellar radioactivity, couched for convenience and transparency in terms of realistic analytic representations of the true galactic mixing processes.

### 2.2.5 Stardust: Radioactivity in Solid Samples of Presolar Stars

A presolar stardust grain is composed only of atoms from a single donor star. The word *stardust* applies to only those grains. It is used as a scientific word rather than a poetic word. It may be hard to accept that the isotopic composition of elements in an interstellar grain attests to it being a solid chunk of a single star. An average ISM grain has suffered a lengthy residence time within the interstellar medium before being incorporated into a new stellar accretion disk. Do not high energy interstellar ions implant within the grains as well as sputter them? Can interstellar chemistry add more atoms to mineralized grains? Would not grain ejection from a protostellar disk after chemically adding atoms to it within the hot dense disk structure also produce mineralized grains? Many questions flood the skeptical mind. Such doubts are reasonable considering the incomplete knowledge of these and other processes.

Stardust grains are not average interstellar grains. Those grains themselves provide the best answers to the doubts expressed above. Their well ordered crystal structures suggest thermal condensation at high temperatures, as would be expected within hot but slowly cooling gases leaving a star. Annealed crystals would not, in contrast, be expected from accreted interstellar atoms. The dramatic isotopic ratios measured within individual stardust grains, obtained from nearly homogeneous grains having large numbers of atoms ($10^6$–$10^{12}$, say), strongly suggests that the condensation was from a gas having the extremely non-solar isotopic composition of the stardust grain. The SiC grains, one of the most abundant of all presolar stardust grains, were assembled from C atoms having observed isotopic ratios in the grains between $^{12}C/^{13}C = 3$–$5{,}000$, whereas hot chemistry in an accretion disk would shift ratios toward interstellar norms, $^{12}C/^{13}C = 89$ in the case of the solar accretion disk. Similar evidence appears in isotopic compositions of N and Si, to name only the most studied of the elements. Those ratios attest to an extreme condensation environment wherein isotopic composition is not that found in the ISM. These isotopic studies of individual stardust grains have been made using SIMS (secondary-ion mass spectrometry) in terrestrial laboratories (Bernatowicz et al., 1997).

What followed from the discovery of stardust was nothing less than a revolution in astronomy. Solid pieces of stars are now characterized routinely in terrestrial laboratories, including isotopic analysis of many elements with a precision not attainable at telescopes. Distinguishing isotopes in stellar spectra is very difficult. Three scientists dominated the early development of this field experimentally. Robert M. Walker committed his laboratory at Washington University to development of new technology capable of better laboratory study of primitive solar system samples.

He foresaw the capability of the sputtering ion probe, and recruited Ernst Zinner to devote a decade to improving and studying that technique. Their laboratory at Washington University (St. Louis) was then ready to study the newly isolated meteoritic grains, which they documented to be stardust. Zinner and his many students led and dominated the new astronomy that blossomed from their isotopic analysis. An engaging history of this has been written in tribute to Walker and Zinner by McKeegan (2007). Thirdly, Edward Anders inspired his associates at University of Chicago to the detective-story pursuit that isolated stardust from the bulk meteorite rock. They termed their approach *burning down the haystack to find the needle*. It tracked the chemical carriers within the meteorites of isotopically anomalous neon and xenon through a sequential dissolving the bulk of the rock in strong acids (Amari et al., 1994) and finding that the undissolved residue became increasingly anomalous. The carriers of anomalous noble gas were, fortunately, carbonaceous, and therefore did not dissolve in acid. From these residues they were able to extract the individual stardust grains. This too was a profound adventure in scientific exploration. Finally it may be noted that predictions of the existence of isotopically anomalous stardust had existed for a decade prior to their experimental discovery.

Within oxide grains (Nittler et al., 1997) stunningly abnormal families of correlated O isotopes rule out any growth within mixed ISM. Equally stunning was the almost pure isolated $s$-process isotopic compositions that were predicted for noble xenon in red giant stardust (Clayton and Ward, 1978) and discovered (Srinivasan and Anders, 1978) in bulk carbon-rich residues of acid dissolution of meteorites. These $Xe_s$-rich residues are dominated by collections of SiC grains. By contrast, interstellar bombardment would implant normal Xe isotopic composition. Neon isotopes also revealed an exciting story. Individual mainstream SiC grains have been shown to carry almost pure $s$-process Mo, a quite reactive trace element rather than a noble gas, showing that the Mo-containing condensing gases had not mixed with the ISM prior to condensation. Isotopic evidence of stardust abounds. By careful consideration of the entirety of the known properties of stardust grains, researchers become confident that they are indeed solid samples of stars that died before the sun was born.

#### 2.2.5.1 Stardust Predicted

With the realization that computer models of massive-star supernovae generate shells of abundant intermediate-mass elements, (Hoyle and Wickramasinghe, 1970) argued that the adiabatic expansion and cooling of these newly synthesized nuclei within the supernova interior should be accompanied by condensation of a potentially large amount of silicate dust. The ejected gas cools to below 2,000 K after only several months when the density is still quite high. Their goal had been to account for the interstellar dust mass of silicates and of graphite. It did not occur to them that such grains could be identified by their unusual isotopic compositions. Nor did they suggest that they might be found within meteorites. They simply tried to explain the observed existence of dust in astronomy.

Clayton (1975a,b, 1978) advanced exciting observational considerations on the grains' condensed radioactive nuclei. It was suggested that grains from the supernova-condensation process would be identifiable by their extreme isotopic signatures, inherited from the isotopic compositions of those supernova shells. Radioactivity is prominent within supernova shells. Short-lived radioactivity was proposed to decay within each grain during its interstellar residence. So the ISM should contain everywhere numerous interstellar grains from the very large number of presolar supernovae. Also proposed was a chemical memory model (Clayton, 1977a,b, 1978, 1982) in which supernova grains were incorporated during later growth in the solar accretion disk of the larger solar-system solids. Many of those solids were incorporated into meteorites. That picture suggested potential explanations for several isotopic anomalies that had been discovered in meteorite solids; but it was rejected initially by meteoriticists, who focussed their doubts on the potential explanations of xenon anomalies from extinct radioactivity. Clayton and Wickramasinghe (1976), Clayton (1979) and Kozasa et al. (1989) demonstrated that a sufficient number of collisions of supernova-interior atoms will occur with any grain nucleation during expansion of the supernova interiors to grow grains greater than 1 $\mu$m in radius. None of these works was so optimistic as to predict that individual supernova grains would be found intact within meteorites; but they did establish the ideas by which those discoveries could be recognized. Radioactivity lay at the heart of these predictions.

In a paper outlining a system for the different types of precondensed matter in the early solar system, the supernova condensates were named SUNOCONs and the term STARDUST was restricted to hot thermal condensation during mass loss from other stars (Clayton, 1978), primarily red giants. NEBCON was suggested for grain mantles grown by nebular sticking of atoms and molecules to preexisting dust. These names have not found the favor of usage, despite describing a theoretical picture of what to expect in the early solar system. Only the term stardust is commonly used, and applied to all types of high-temperature thermal condensates from stars. The title *Precondensed matter: key to the early solar system* of Clayton (1978) explicitly contradicted that epoch's favored picture in which the initial solar system was initially hot and totally gaseous and condensed its solids from that solar gas.

The predictions of isotopically extreme stardust were, after some initial controversy, wonderfully confirmed by the discovery of supernova stardust and the later documentation of chemical memory within large solids, earning broad acceptance.

#### 2.2.5.2 Stardust Discovered: Fossil Extinct Radioactivity

Secondary-ion mass spectrometry (SIMS) of ions sputtered from isolated single meteoritic grains by a focused ion beam identified stardust in the late 1980s. A history may be read in Anders and Zinner (1993). First to be discovered (Bernatowicz et al., 1987; Zinner et al., 1989) was silicon carbide (SiC), with isotopic patterns that identified it as thermal condensation in matter flowing away from the photospheres of asymptotic-giant-branch red giants. This sensational identification of stardust is now beyond doubt, buttressed by a huge number of experimental and theoretical studies. These carbide grains can condense only after the composition changes caused by the third dredgeup of carbon, which eventually cause carbon to be more abundant than oxygen in the envelope. As long as O is more abundant than C, the CO molecule oxidizes (combusts) all carbon. Those SiC grains condensed after the carbon-star transition contained $^{26}Al/^{27}Al$ near $10^{-3}$ based on the excess $^{26}Mg$ atoms within their magnesium-bearing phases. When supernova SiC *X* grains were later identified as a subfamily of SiC stardust, many revealed $^{26}Al/^{27}Al$ near 0.1 when they condensed within the expanding supernova interior.

Oxide stardust too was identified and classified into isotopic families a few years later (Nittler et al., 1997). Its most common pattern is $^{17}O$-richness owing to secondary production in the parent star of $^{17}O$ from initial $^{16}O$ during hydrogen burning . At about the same time supernova stardust was identified unambiguously (Hoppe et al., 1996; Nittler et al., 1996) using the prediction (Clayton, 1975b) that excess $^{44}Ca$ would exist in the Ti-bearing phases within grains owing to the condensation of radioactive $^{44}Ti$ within a year after the explosion. Because the elemental ratio Ti/Ca is for crystal-lattice reasons larger than solar in SiC grains, the supernova SiC grains possess very large isotopic excess at $^{44}Ca$. These were found in the family of SiC grains called *X grains*, which were already suspected of being supernova grains on the basis of deficiencies in the heavy isotopes of Si and C and on excesses of $^{15}N$ (Alexander et al., 1990; Amari et al., 1992).

The extremely large isotopic excesses of $^{26}Mg$ and $^{44}Ca$ in these SiC X grains were the most dramatic discoveries of fossil extinct radioactivity in stardust. They demonstrated that radioactive $^{26}Al$ and $^{44}Ti$ had been quite abundant when the grains condensed during the supernova expansion. Predictions had also suggested that fossil abundances in interstellar grains might contribute to excesses of the daughter isotope in solids grown later from interstellar grains (*chemical memory*). But Lee et al. (1977) had emphasized grounds for believing that the excess $^{26}Mg$ within solids made in the solar system reflected instead live $^{26}Al$ at the time the solids were made. Both fossil and live $^{26}Al$ now appear to have existed. These discoveries dramatically spotlighted rich new roles for the astronomy of radioactivity in early solar system chemistry.

Many other fossil extinct radioactive nuclei have been added to the observed list. In supernova stardust these include excess $^{22}Ne$ owing to $^{22}Na$ decay, excess $^{41}K$ owing to $^{41}Ca$ decay, and excess $^{49}Ti$ owing to $^{49}V$ decay within the grains (see Clayton and Nittler, 2004). All had been predicted to exist in supernova dust, but their discovery surprised and delighted isotopic chemists and astronomers alike.

Following the exciting discovery of presolar stardust, its study has emerged as a new area of astronomy. In just 23 years its existence has evolved from a bewildering new discovery into several new techniques for measuring isotopic abundance ratios with high precision in stars. This has been especially important for astronomy with radioactivity, because in stardust the level of extinct radioactivity reflects its abundance during nucleosynthesis in stars, whereas in solar system samples it reflects the level of its survival. Precisely measured isotopic ratios for four to eight chemical elements endows each gemlike refractory mineral with significance for stellar structure and evolution, and for the chemical evolution of the Milky Way during the epoch 7-5

Gyr ago, and for new insight into nucleosynthesis (Clayton and Nittler, 2004). This rich harvest is compromised only by the fact that the donor stars can not be seen, because they died before our solar system began. Their stardust bears no label, save that of their measurable properties. The nature and evolutionary state of the donor stars must be ascertained from the detailed properties, primarily isotopic, of each grain and of the way each grain fits into the spectrum of the thousands of stardust grains that have been analyzed to date. The rapidly growing numbers of analyzed grains allows evolutionary trends within their distinct families to be identified. [3]

#### 2.2.5.3 Radioactivity and Chemistry of Condensation in Supernovae

An unanticipated aspect of astronomy with radioactivity lay in the role of radioactivity in the chemistry of the condensation process. The supernova interior offers a unique laboratory for condensation physics. It guarantees that chemistry begins with gaseous atoms, with no trace of previous molecules or grains. Gamma rays and their Compton-scattered electrons bathe the supernova core. By disrupting the CO molecule, they cause its abundance to be very much smaller than expected from states of chemical thermal equilibrium at the ambient temperature. The small CO molecular abundance enables other non-equilibrium paths to the condensation of carbon-bearing solids. This disequilibrium can be considered to be another aspect of radiogenic luminosity causing a higher degree of excitation in the gas than would be expected in the absence of radioactivity.

Because the supernova core is hydrogen-free, chemical condensation routes utilizing H are not relevant. Polycyclic aromatic hydrocarbons, for example, do not come into play save in the envelope; but 90–95% of the mass of ejected Mg, Al and Si emerges in the H-free core. And yet abundant dust condensation is observed to have occurred in SN1987A (Wooden et al., 1993; Colgan et al., 1994), in Cas A (Arendt et al., 1999; Dunne et al., 2003) and in the Kepler remnant (Dunne et al., 2003). Apparently a few solar masses condensed in Cas A and about one solar mass in Kepler, requiring high condensation efficiency for Mg, Al and Si and even for carbon and thereby establishing supernovae as a major contributor to the budget of thermally condensed interstellar grain cores.

The traditional guideline to condensation of solids had been to assume that chemical equilibrium applies during the expansion and temperature decline of the supernova interior. Formulae yield the equilibrium condensed masses of differing minerals (Lattimer et al., 1978; Ebel et al., 2000). This approach can not yield grain size but only the total condensed masses. The assumption of thermal equilibrium in the solid phaseexaggerates the ability of chemical reactions to maintain

[3] Readers wishing familiarity with these topics can best consult later chapters of this book, or Clayton and Nittler (2004), and *Astrophysical Implications of the Laboratory Study of Presolar Materials* (Bernatowicz and Zinner, 1997); the *Handbook of Isotopes in the Cosmos* by Clayton (2003) also presents many astrophysical consequences.

chemical equilibrium within the solids as the gas cools. Because of the rapid fall of density and temperature during the expansion of the supernova interior, thermal condensation must be accomplished within about two years, too fast to maintain equilibrium. A nonequilibrium theory of condensation based on nucleation theory followed by subsequent growth has been developed by Kozasa and collaborators (Kozasa et al., 1987, 1989, 1991; Todini and Ferrara, 2001). Their method identifies a *key molecular species* whose abundance controls the condensation. However, their questionable assumptions concerning gaseous mixing at the molecular level and on an outdated role for the CO molecule render their results of questionable value.

The nucleosynthesis problems posed by isotopic ratios within individual supernova grains can not be decoupled from physical questions about the chemistry of their condensation. Condensation chemistry is an essential aspect of astronomy with radioactivity. Because no single supernova mass zone is able to satisfy the isotopic ratios found in supernova grains,it has long been clear (Amari and Zinner, 1997) that some type of mixing before condensation is needed to produce their chemical and isotopic compositions; but it is not clear whether that mixing represents atom-scale gaseous mixing in the very young remnant or transport of a growing grain from one composition zone into another. But the biggest question is whether the requirement that the C abundance be greater than O abundance in order to condense SiC and graphite within supernovae is a valid requirement. For their discussion of supernova stardust, for example, Travaglio et al. (1999) took the view that mixing occurs at the atomic level, prior to condensation, and that only C>O portions of those mixtures support condensation of carbon. They arbitrarily and instantaneously mixed regions having C<O with the He shell having C>O to find mixing fractions that retained C>O overall. Kozasa et al. (1989) performed similar thought experiments. Such treatment is deeply flawed, however, because it seems certain that gases can not mix at the molecular level within a few years time (see Deneault et al., 2006; Fryxell et al., 1991). The so-called mixing calculated in hydrodynamic studies of supernovae, on the other hand, represents fluids of one composition exchanging places in the homologous expansion with overlying fluids of a different composition. This is a different use of the word *mixing*, and has introduced considerable confusion into published works. Mixing at the molecular level requires very much more time than the year that is available before condensation must occur. Later the expanding remnant is too dilute, its density too small, for condensation.

To circumvent the equilibrium CO trap for carbon, Clayton et al. (1999, 2001), and Deneault et al. (2006) called upon its dissociation by the fast compton electrons energized by supernova gamma rays. They advanced a kinetic theory of graphite growth and calculated its consequences in detail after advocating a specific nucleation model. The key idea is that small graphite test particles, if placed in a hot gas of C and O atoms, will associate with free C atoms faster than they can be oxidized by free O atoms. This is a property of reaction cross sections rather than of chemical equilibrium. Even though oxidation of carbon would surely be the ultimate end given adequate time, the expansion will terminate the chemistry after about two years with large graphite grains remaining. The graphite essentially is a metastable state of carbon. This theory is supported by observing that supernova 1987A ejected only $10^{-3}$ $M_\odot$ of CO molecules instead of the 0.1 $M_\odot$ of CO molecules that is first formed by association reactions in the hot gas before radioactive disruption reduces its abundance (Dalgarno 1994, 1995; Gearhart et al., 1999).

Similar issues probably surround the condensation of supernova SiC. It seems plausible that radioactive liberation of free C atoms from CO molecules will also facilitate the condensation of SiC in O-rich gas; but this is hotly debated. Although a kinetic route to SiC condensation has not been layed out, Deneault et al. (2003) have formulated a physical description of the ejecta enabling them to make several relevant conclusions derived from assuming that the radioactive CO-disruption mechanism is the correct key to SiC growth. They present animations of a hydrodynamic calculation showing that a reverse shock wave launched toward inner zones by the radially increasing value of the product $\rho r^3$ in the H envelope compresses a dense shell near radial mass coordinate $m = 3\,M_\odot$, where Si and O are the most abundant elements; but some C abundance remains there for possible carbon chemistry. They propose that SiC condenses in that zone, and they detailed several other physical processes that may produce the observed grain compositions. Mixing of a new type during condensation also occurs if the reverse shock generated by the radial expansion colliding with the presupernova wind arrives at the condensation zone at $m = 3\,M_\odot$ between six months to a year after the explosion, because that reverse shock slows the gas and forces the partially condensed SiC grains to propel forward through the decelerating gas into more $^{29,30}$Si-rich regions, giving perhaps a new interpretation for mixing during condensation.

Despite many uncertainties, it now appears certain that supernova grains studied by isotopic analysis will provide, through details of condensation chemistry, a new sampling spectrum of young supernova interiors, just as have gamma-ray lines and hard X-rays. If so, a very rich but complex discipline of astronomy with radioactivity will follow.

#### 2.2.5.4 Astronomy with Radioactivity Today

This introductory chapter has focussed on the history and the key physical ideas of astronomy with radioactivity. Effective research depends on a clear grasp of an interdisciplinary set of its central ideas. History itself often provides the best example by which the physical idea can be first grasped. But the remainder of this book looks forward, not backward. Every aspect of astronomy with radioactivity today involves grappling with a host of technical details. This book attempts to bring the reader to that point. Each aspect of astronomy with radioactivity today also involves grappling with the entire world of astronomy. Astronomy itself is many disciplines, and not even astronomers are always comfortable outside their own astronomical technique. How much harder it is to place the complex manifestations of radioactivity into a continuously changing astronomical fabric. But this is the direction of all scientific progress.

## References


Alexander, C.M.O'D., Swan, P., Walker, R.M.: Nature **348**, 715–717 (1990)

Amari, S., Zinner, E.: In Astrophysical Implications of the Laboratory Study of Presolar Materials, p. 287. Bernatowicz, T., Zinner, E. (eds.) American Institute of Physics. Woodbury, NY (1997)

Amari, S., Hoppe, P., Zinner, E., Lewis, R.S.: Astrophys. J. **394**, L43–46 (1992)

Amari, S., Lewis, R.S., Anders, E.: Geochim. Cosmochim. Acta **58**, 459–470 (1994)

Anders, Z.E.: Meteoritics **28**, 490–514 (1993)

Arendt, R.G., Dwek, E., Moseley, S.H.: Astrophys. J. **521**, 234 (1999)

Arnett, W.D.: In Eighth Texas Symposium on Relativistic Astrophysics. Papagiannis, M.D. (ed.) vol. 302, p. 90. New York (1977)

Arnett, W.D.: Supernovae and Mucleosynthesis: An Investigation of the History of Matter from the Big Bang to the Present. Princeton University Press, Princeton, NJ (1996)

Arnould, M., Beelen, W.: Astron. Astrophys. **33**, 215 (1974)

Arnould, M., Norgaard, H.: Astron. Astrophys. **64**, 195–213 (1978)

Arnould, M., Norgaard, H., Thielemann, F.K., Hillebrandt, W.: Astrophys. J. **237**, 931–950 (1980)

Arpesella, C., Back, H.O., Balata, M., et al.: Direct measurement of the Be7 solar neutrino flux with 192 days of Borexino data. Phys. Rev. Lett. **101**(9), 091, 302 (2008). doi:10.1103/PhysRevLett.101.091302, 0805.3843

Bahcall, J.N.: Astrophys. J. **139**, 318 (1964)

Beers, T.C., Christlieb, N.: Ann. Rev. Astron. Astrophys. **43**, 531 (2005)

Bernatowicz, T.J., Zinner, E.: Astrophysical Implications of the Laboratory Study of Presolar Materials. American Institute of Physics, Woodbury, NY (1997)

Bernatowicz, T.J., et al.: Nature **330**, 728–730 (1987)

Black, D.C.: Geochim. Cosmochim. Acta **36**, 377 (1972)

Bodansky, D., Clayton, D.D., Fowler, W.A.: Phys. Rev. Lett. **20**, 161 (1968a)

Bodansky, D., Clayton, D.D., Fowler, W.A.: Astrophys. Suppl. J. **16**, 299 (1968b)

Burbidge, E.M., Burbidge, G.R., Fowler, W.A., Hoyle, F.: Revs. Mod. Phys. **29**, 547–650 (1957)

Burris, D.L., Pilachowski, C.A., Armandroff, T.E., Sneden, C., Cowan, J.J., Roe, H.: Astrophys. J. **544**, 302 (2000)

Butcher, H.R.: Nature **328**, 127–131 (1987)

Cameron, A.G.W.: Astrophys. J. **121**, 144 (1955)

Cameron, A.G.W.: CRL 41, Chalk River, Ontario (1957)

Cameron, A.G.W.: Astrophys. J. **130**, 429 (1959)

Cameron, A.G.W., Truran, J.W.: Icarus **30**, 447 (1977)

Clayton, D.D.: Astrophys. J. **139**, 637 (1964)

Clayton, D.D.: Principles of Stellar Evolution and Nucleosynthesis. McGraw-Hill, New York (1968)

Clayton, D.D.: Nature **234**, 291 (1971)

Clayton, D.D.: Astrophys. J. **188**, 155 (1974)

Clayton, D.D.: Astrophys. J. **198**, 151 (1975a)

Clayton, D.D.: Astrophys. J. **199**, 765–769 (1975b)

Clayton, D.D.: Nature **257**, 36–37 (1975c)

Clayton, D.D.: Icarus **32**, 255–269 (1977a)

Clayton, D.D.: Icarus **35**, 398–410 (1977b)

Clayton, D.D.: Moon and Planets **19**, 109–137 (1978)

Clayton, D.D.: Astrophys. Spa. Sci. **65**, 179–189 (1979)

Clayton, D.D.: Astrphys. J. Lett. **244**, L97 (1981)

Clayton, D.D.: Q. J. R. Astron. Soc. **23**, 174–212 (1982)

Clayton, D.D.: Astrophys. J. **268**, 381–384 (1983)

Clayton, D.D.: Astrophys. J. **280**, 144 (1984)

Clayton, D.D.: Galactic chemical evolution and nucleocosmochronology: a standard model. In: Arnett, W.D., Truran, W. (eds.) Challenges and New Developments in Nucleosynthesis, p. 65. University of Chicago Press, Chicago (1985)

Clayton, D.D.: Nature **329**, 397 (1987)
Clayton, D.D.: Mon. Not. R. Astron. Soc. **234**, 1 (1988)
Clayton, D.D.: Meteoritics & Planetary Sci. **34**, A145–A160 (1999)
Clayton, D.D.: Handbook of Isotopes in the Cosmos. Cambridge University Press, Cambridge (2003)
Clayton, D.D.: Science **318**, 1876–1877 (2007)
Clayton, D.D., Fowler, W.A.: Ann. Phys. **16**, 51 (1961)
Clayton, D.D., Craddock, W.: Astrophys. J. **142**, 189 (1965)
Clayton, D.D., Silk, J.: Astrophys. J. Lett. **158**, L43 (1969)
Clayton, D.D., Hoyle, F.: Astrophys. J. Lett. **187**, L101 (1974)
Clayton, D.D., Hoyle, F.: Astrophys. J. **203**, 490 (1976)
Clayton, D.D., Wickramasinghe, N.C.: Astrophys. & Spa. Sci. **42**, 463–475 (1976)
Clayton, D.D., Ward, R.A.: Astrophys. J. **224**, 1000–1006 (1978)
Clayton, D.D., Leising, M.D., The, L.S., Johnson, W.N., Kurfess, J.D.: Astrophys. J. **399**, L141–144 (1992)
Clayton, D.D., Nittler, L.R.: Ann. Rev. Astron. Astrophys. **42**, 39 (2004)
Clayton, D.D., Fowler, W.A., Hull, T.E., Zimmerman, B.A.: Ann. Phys. **12**, 331 (1961)
Clayton, D.D., Colgate, S.A., Fishman, G.J.: Astrophys. J. **155**, 75 (1969)
Clayton, D.D., Hartmann, D.H., Leising, M.D.: Astrophys. J. **415**, L25 (1993)
Clayton, D.D., Liu, W., Dalgarno, A.: Science **283**, 1290–1292 (1999)
Clayton, D.D., Deneault, E., Meyer, B.S.: Astrophys. J. **56**, 480–493 (2001)
Colgan, S.W.J., et al.: Astrophys. J. **427**, 874 (1994)
Colgate, S.S.A., Fishman, G.J., Clayton, D.D.: Astrophys. J. **155**, 75 (1969)
Dalgarno, A.: Molecules and Grains in Space: An Overview. In I. Nenner (ed.) Molecules and Grains in Space. American Institute of Physics Conference Series vol. 312, p. 841 (1994)
Dalgarno, A.: Int. J. Mass Spectrom. Ion Processes **149**, 429–437 (1995)
Deneault, E.N., Clayton, D.D., Heger, A.: Astrophys. J. **594**, 312 (2003)
Deneault, E., Meyer, B.S., Clayton, D.D.: Astrophys. J. **638**, 234–240 (2006)
Diehl, R., et al.: Astron. & Astrophys. **298**, 445 (1995)
Dunne, L., Eales, S., Ivison, R., Morgan, H., Edmunds, M.: Nature **424**, 285–287 (2003)
Dwek, E., Moseley, S.H., Glaccum, W., Graham, J.R., Loewenstein, R.F., Silverberg, R.F., Smith, R.K.: Astrophys. J. **389**, L21–24 (1992)
Ebel, D., Grossman, L.: Geochim. Cosmochim. Acta **64**, 339–366 (2000)
Fowler, W.A., Wasserburg, G.J., Hoyle, F.: Phys. Rev. Lett. **4**, 112 (1960)
Fowler, W.A.: Nature **238**, 24–26 (1972). doi: 10.1038/238024a0
Fryxell, B., Müller, E., Arnett, D.: Astrophys. J. **367**, 619–634 (1991)
Gao, Y., Solomon, P.M.: Astrophys. J. **606**, 271 (2004)
Gearhart, R.A., Wheeler, J.C., Swartz, D.A.: Astrophys. J. **510**, 944–966 (1999)
Gilroy, K.K., Sneden, C., Pilachowski, C.A., Cowan, J.J.: Astrophys. J. **327**, 298 (1988)
Gray, C.M., Compston, W.: Nature **251**, 495 (1974)
Haxel, O., Jensen, J.H., Suess, H.E.: Phys. Rev. **75**, 1766 (1949)
Hoppe, P., et al.: Science **272**, 1314–1316 (1996)
Hoyle, F.: Mon. Not. R. Astron. Soc. **106**, 343–383 (1946)
Hoyle, F.: Astrophys. J. Suppl. **1**, 121–146 (1954)
Hoyle, F., Fowler, W.A.: Astrophys. J. **132**, 565 (1960)
Hoyle, F., Wickramasinghe, W.A.: Nature **226**, 62–63 (1970)
Hudson, G.B., Kennedy, B.M., Podosek, F.A., Hohenberg, C.M.: Proc. 19th Lunar Planet. Sci. Conf. **19**, 547–557 (1989)
Huss, G.R., Meyer, B.M., Srinivasan, G., Goswami, J.N., Sahijpal, S.: Geochim. Cosmochim. Acta, **73**, 4922 (2009)
Johnson, W.N., Harnden, F.R., Haymes, R.C.: Astrophys. J. **172**, L1 (1972)
Kaeppeler, F., et al.: Astrophys. J. **257**, 821 (1982)
Kozasa, T., Hasegawa, H.: Prog. Theor. Phys. **77**, 1402 (1987)
Kozasa, T., Hasegawa, H., Nomoto, K.: Astrophys. J. **344**, 325 (1989)

Kozasa, T., Hasegawa, H., Nomoto, K.: Astron. & Astrophys. **249**, 474 (1991)
Kurfess, J.D., et al.: Astrophys. J. Lett. **399**, L137 (1992)
Lattimer, J.M., Schramm, D.N., Grossman, L.: Astrophys. J. **219**, 230 (1978)
Lee, T., Papanastassious, D.A., Wasserburg, G.J.: Astrophys. J. **211**, L107 (1977)
Leising, M.D., Share, G.H.: Astrophys. J. **357**, 638 (1990)
MacCallum, C.J., Huters, J.F., Stang, P.D., Leventhal, M.: Astrophys. J. **317**, 877 (1987)
Mahoney, W.A., Ling, J.C., Jacobson, A.S., Lingenfelter, R.E.: Astrophys. J. **262**, 742 (1982)
Mahoney, W.A., Ling, J.C., Wheaton, A.S., Jacobson, A.S.: Astrophys. J. **286**, 578 (1984)
McKeegan, K.D.: Meteroritics & Planetary Sci. **42**, 1045–1054 (2007)
Nittler, L.R., et al.: Astrophys. J. **462**, L31–34 (1996)
Nittler, L.R., et al.: Astrophys. J. **483**, 475–495 (1997)
Ramaty, R., Lingenfelter, R.E.: Astrophys. J. **213**, L5 (1977)
Reynolds, J.H.: Phys. Rev. Lett. **4**, 8 (1960)
Rutherford, E.: Nature **123**, 313–314 (1929)
Seeger, P.A., Fowler, W.A., Clayton, D.D.: Astrophys. J. **11**, 121 (1965)
Share, G.H., et al.: Astrophys. J. **292**, L61 (1985)
Srinivasan, B., Anders, E.: Science **201**, 51 (1978)
Srinivasan, B., Sahijpal, S., Ulyanov, A.A., Goswami, J.N.: Geochim. Cosmochim. Acta **60**, 1823–1835 (1996)
Starrfield, S., Truran, J.W., Sparks, W.M., Kutter, G.S.: Astrophys. J. **176**, 169 (1972)
Suess, H.E., Urey, H.C.: Revs. Mod. Phys. **28**, 53 (1956)
Suntzeff, N.B., Phillips, M.M., Elias, J.H., Walker, A.R., Depoy, D.L.: Astrophys. J. **384**, L33–36 (1992)
Travaglio, C., et al.: Astrophys. J. **510**, 325 (1999)
The, L.S., Burrows, A., Bussard, R.: Astrophys. J. **352**, 731 (1990)
Timmes, F.X., Woosley, S.E., Weaver, T.A.: Astrophys. J. Suppl. **98**, 617 (1995)
Todini, P., Ferrara, A.: Mon. Not. R Astron. Soc. **325**, 726 (2001)
Truran, J.W.: Astron. Astrophys. **97**, 391–393 (1981)
Truran, J.W.: Nuclear theory of novae. In: Barnes, C.A., Clayton, D.D., Schramm, D.N. (eds.) Essays in Nuclear Astrophysics, p. 467. Cambridge University Press, Cambridge (1982)
Truran, J.W., Arnett, W.D., Cameron, A.G.W.: Can. J. Phys. **45**, 2315 (1967)
von Ballmoos, P., Diehl, R., Schonfelder, V.: Astrophys. J. **318,** 654 (1987)
Wasserburg, G.J., Papanastassiou, D.A.: Some short lived nucleides in the early solar system. In: Barnes, C.A., Clayton, D.D., Schramm, D.N. (eds.) Essays in Nuclear Astrophysics, p. 77. Cambridge University Press, Cambridge (1982)
Weaver, T.A., Woosley, S.E.: Phys. Rep., **227**, 65 (1993)
Wooden, D.H., et al.: Astrophys. J. Suppl. **88**, 477–508 (1993)
Woosley, S.E., Weaver, T.A.: Astrophys. J. **238**, 1017 (1980)
Woosley, S.E., Weaver, T.A.: Astrophys. J. Suppl. **101**, 181 (1995)
Woosley, S.E., Arnett, W.D., Clayton, D.D.: Astrophys. J. Suppl. **26**, 231 (1973)
Yokoi, K., Takahashi, K., Arnould, M.: Astron. & Astrophys. **117**, 65 (1983)
Zinner, E., Tang, M., Anders, E.: Geochim. Cosmochim. Acta **53**, 3273–3290 (1989)

# Part II
# Specific Sources of Cosmic Isotopes

The following three Chapters discuss various stellar sources of new isotopes. Here we focus on the processes which take place within an astrophysical object over the course of its evolution. For stars, this can be for millions to billions of years. For explosions, evolution may take place on characteristic time scales of hours or minutes, or even as short as seconds. In these three Chapters we present the material for later discussions of the evolution of larger objects ranging from stellar groups to galaxies. We grouped *stars* according to main themes related to *single stars* and *binary systems*. The section on single stars was further divided into Chapters each for the *two physically-different ranges of stellar masses* defined by separating *core-collapsing* from *white-dwarf-generating* stars. Binary systems can experience *mass transfer*, which adds complexity that leads to a rich variety of phenomena. We discuss sub-categories associated with different nuclear-reaction environments, spanning a broad range from *thermonuclear supernovae* through *novae* to *X-ray bursts*.

# Chapter 3
# Radioactivities in Low- and Intermediate-Mass Stars

**M. Lugaro and A. Chieffi**

Energy in stars is provided by nuclear reactions, which, in many cases, produce radioactive nuclei. When stable nuclei are irradiated by a flux of protons or neutrons, capture reactions push stable matter out of stability into the regime of unstable species. The ongoing production of radioactive nuclei in the deep interior of the Sun via proton-capture reactions is recorded by neutrinos emitted during radioactive decay. These neutrinos escape the inner region of the Sun and can be detected on Earth. Radioactive nuclei that have relatively long half lives may also be detected in stars via spectroscopic observations and in stardust recovered from primitive meteorites via laboratory analysis. The vast majority of these stardust grains originated from Asymptotic Giant Branch (AGB) stars. This is the final phase in the evolution of stars initially less massive than $\simeq$10 $M_{\odot}$, during which nuclear energy is produced by alternate hydrogen and helium burning in shells above the core. The long-lived radioactive nucleus $^{26}$Al is produced in AGB stars by proton captures at relatively high temperatures, above 60 MK. Efficient production of $^{26}$Al occurs in massive AGB stars ($>$4 : 5 $M_{\odot}$), where the base of the convective envelope reaches such temperatures. Several other long-lived radioactive nuclei, including $^{60}$Fe, $^{87}$Rb, and $^{99}$Tc, are produced in AGB stars when matter is exposed to a significant neutron flux leading to the synthesis of elements heavier than iron. Here, neutron captures occur on a timescale that is typically slower than $\beta$-decay timescales, resulting in a process known as *slow* neutron captures (the *s* process). However, when radioactive nuclei with half lives greater than a few days are produced, depending on the temperature and the neutron density, they may either decay or capture a neutron, thus branching up the path of neutron captures and defining the final *s*-process abundance distribution. The effect of these *branching points* is observable in the composition of AGB stars and stardust. This nucleosynthesis in AGB stars could produce some long-living radioactive nuclei in relative abundances that resemble those observed in the early solar system.

M. Lugaro (✉)
Monash University, Victoria 3800, Australia

A. Chieffi
I.N.A.F., 00133 Roma, Italy

Lugaro, M., Chieffi, A.: *Radioactivities in Low- and Intermediate-Mass Stars*. Lect. Notes Phys. **812**, 83–151 (2011)
DOI 10.1007/978-3-642-12698-7_3 

## 3.1 The Missing Element

The element with 43 protons in its nucleus, lying between molybdenum and ruthenium, was known for a long time as the *missing* element. Since the nineteenth century there had been many unsuccessful attempts at its discovery. Finally, in 1937 Italian physicist Emilio Segré and chemist Carlo Perrier found two isotopes of the missing element through measurements of radioactivity from discarded cyclotron parts: they observed several decay periods and proved they were occurring at Z = 43. Hence the missing element did not exist in nature because of its instability against nuclear decay. The discoverers named the missing element technetium (Tc), from $\tau\varepsilon\chi\nu\eta\tau\acute{o}\varsigma$, which means artificial in Greek, since it was the first element produced artificially. Fifteen years later, it was shown that Tc is not only made by men but also by stars. In 1952, astronomer Paul Merrill observed the absorption lines corresponding to the atomic structure of Tc in the spectra of several giant stars. Merrill was at first cautious about this result. To start with, the element he identified did not even exist on Earth. Second, up to then it was assumed, and not proved wrong, that all stars had the same chemical composition. This was in agreement with the accepted theory of the time that the elements were all produced during the Big Bang and their abundances in the Universe were not modified by any further process. Merrill's discovery in that respect was truly revolutionary: given the relatively short half lives of the Tc isotopes (a few million years at most), the Tc lines were the first indisputable demonstration that this radioactive element was made in situ in the stars where it was observed. This finding brought a radical change in the way we understand the origin of the elements, and the theory of stellar nucleosynthesis introduced in Chap. 2 began to take shape and garnered authority. In this chapter we discuss the life of those stars that, like our Sun, evolve twice through Red Giant stages. We describe how they produce long lived radioactive nuclei, like Tc, in their interiors, how the signature of such radioactivity is carried outside the star, and how it can be observed.

## 3.2 The Production of Radioactive Nuclei in Stellar Interiors

In Sect. 3.2.1 we first derive the four basic equations that control the quasi-equilibrium configuration of a self-gravitating gas sphere, namely, the hydrostatic equilibrium equation (that describes the balance between the pressure gradient and gravity) and the energy transport equation (due to photons and/or convection), plus the two associated continuity equations for mass and energy flux. Then, we show that energy losses, which occur mainly from the stellar surface in stars of mass less than $\sim$10 $M_\odot$, force the gas to contract and to heat, in accordance with the virial theorem. The progressive increase of the central temperature allows the activation of nuclear processes and we describe two sequences that convert protons into $^4$He nuclei ($\alpha$ particles): the PP chain and the CNO cycle. Since proton capture inevitably pushes matter out of the stability, both these sequences produce radioactive nuclei that decay by emitting neutrinos.

In Sect. 3.2.2 we briefly describe the quest for solar neutrinos and the various experiments that eventually allowed the demonstration that the lower than predicted neutrino flux from the Sun (the so-called Solar Neutrino Problem) is the consequence of neutrino oscillations among their three different flavors.

### 3.2.1 The Stellar Energy Source and Radioactive Isotopes

A star is, in first approximation, a spherically symmetric, gaseous cloud contracting under its own gravity and progressively heating up while losing energy from its surface in the form of photons. A strong temperature gradient, with the temperature decreasing from the centre to the surface, pushes the photon flux outward until the *mean free path*[1] of the photons eventually becomes larger than their distance from the surface, allowing their escape. The Virial theorem links the energy gained by the gravitational field $\Delta\Omega$ to that absorbed by the gas $\Delta U$: $\Delta U = -\Delta\Omega/3(\gamma - 1)$, where $\gamma$ is the ratio between two specific heats, that at constant pressure and that at constant volume, of the contracting gas. A stable quasi-equilibrium configuration exists for such a structure provided that $\gamma > 4/3$. In this case, a fraction of the energy gained by the gravitational field must be liberated from the structure before the gas cloud can contract further. In the case of a perfect gas ($\gamma = 5/3$) we obtain the classical result $\Delta U = -(1/2)\Delta\Omega$, stating that half of the gravitational energy liberated by each infinitesimal contraction is absorbed by the gas and half must be lost before an additional contraction can occur. The timescale over which the energy is lost from the system drives the timescale of contraction and keeps the structure in a quasi-equilibrium configuration. If, instead, $\gamma$ drops to 4/3, all the gravitational energy is absorbed by the gas and no time delay is required before a new contraction can occur. This is an unstable situation leading to collapse. In the evolutionary phases we discuss in this chapter $\gamma$ remains well above 4/3 and hence a stable quasi-equilibrium configuration is always assured.

The balancing forces required to maintain a stable stellar quasi-equilibrium configuration are due to pressure gradients and gravity. So, the first main equation of stellar structure describes the equilibrium between these two forces, at any given distance from the center of the star $r$:

$$dP/dr = -GM\rho/r^2 \tag{3.1}$$

where $P$ is the pressure, $G$ the gravitational constant, $M$ the cumulative mass inside $r$, and $\rho$ the density. Associated to this equation is a continuity equation for mass:

$$dM/dr = 4\pi r^2 \rho. \tag{3.2}$$

[1] The average distance a particle travels between collisions.

By assuming, to zero order, that $\rho$ is constant within the star, the integration of Eq. (3.2) implies that:

$$\rho \propto M/R^3. \tag{3.3}$$

Since the pressure at the surface of the star is much lower than where the radius approaches zero, the center of the star, the equation of hydrostatic equilibrium, (3.1), basically says that

$$P_c \propto M\rho/R_s \tag{3.4}$$

where $P_c$ is the central pressure and $R_s$ the stellar radius. By inserting the relation (3.3) into (3.4) one obtains:

$$P_c \propto M^2/{R_s}^4. \tag{3.5}$$

If the equation of state is that of a perfect gas, i.e. $P \propto \rho\ T/\mu$, relation (3.5) becomes:

$$T_c \propto \mu\ M/R \tag{3.6}$$

where $T$ is the temperature and $\mu$ is the mean molecular weight. This equation provides an important basic relationship among central temperature ($T_c$), mass, and radius of a star, which only relies on the assumption of hydrostatic equilibrium.

The second major equation describing the structure of the quasi-equilibrium configuration of a star determines the energy flux through the structure. In stationary situations the energy is transported by photons or electrons and application of the first Fick's law leads to the well known equation:

$$dT/dr = -3\kappa\rho L/(16ac\pi r^2 T^3) \tag{3.7}$$

where $\kappa$ is the opacity coefficient representing the impenetrability of a gas to light, $L$ the luminosity representing the amount of energy radiated per unit time, $a$ the radiation constant, and $c$ the speed of light. Furthermore, in this case a continuity equation

$$dL/dr = 4\pi r^2 \rho\varepsilon \tag{3.8}$$

controls the conservation of energy, where $\varepsilon$ represents the net local energy budget, i.e., the sum of the nuclear energy production rate $\varepsilon_{\text{nuc}}$, the neutrino energy loss rate $\varepsilon_\nu$, and the gravitational energy rate $\varepsilon_{\text{g}}$. Since the central temperature $T_c$ is much higher than the surface temperature, it is possible to obtain a basic relationship between central temperature, mass, luminosity and radius of a star, i.e.:

$${T_c}^4 \propto ML/R^4. \tag{3.9}$$

By combining this relation with the previous relation (3.6), derived from hydrostatic equilibrium (3.1), one eventually obtains the fundamental relation between mass and luminosity:

$$L \propto \mu^4 M^3. \tag{3.10}$$

Frequently, the energy produced locally cannot be transported quickly enough by radiation or conduction, and interior shells formally in an equilibrium condition can become unstable in the sense that a displacement from their equilibrium position is not fully counteracted by a restoring force. Instead, matter is accelerated even further from its original position and large scale motions of matter (*convection*) is established. Under these conditions energy is predominantly transported by buoyancy-driven motions of bulk material due to their much larger mean free path with respect to that of photons. A temperature gradient quite different from that described by Equation (3.7) must then be used in these regions. A direct consequence of these large scale motions is that matter is mixed throughout an unstable region.

The system formed by the two basic equations related to hydrostatic equilibrium and energy transport plus the two associated continuity equations and the equation of state (supplemented by an opacy coefficient $\kappa = f(\rho, T, \text{chem.comp.})$ and a total energy generation coefficient $\varepsilon = f'(\rho, T, \text{chem.comp.})$) constitutes the basic set of equations that describes the internal structure of a quasi-equilibrium (non rotating) stellar configuration at a given time. The temporal evolution of such a structure is determined by the rate at which energy is lost to the surroundings: the faster the energy is lost, the faster the structure evolves. Typical stellar luminosities range between $\sim 4 \times 10^{33}$ erg/s for a star of 1 $M_{\odot}$ and $\sim 4 \times 10^{36}$ erg/s for a star of 6 $M_{\odot}$ and the associated lifetimes can be estimated by dividing the total amount of available energy by the loss rate L, i.e. the stellar luminosity. If the only energy source was the gravitational field, the lifetime of a contracting gas cloud would be of the order of a few tens of millions of years (the Kelvin-Helmholtz timescale). Instead, as was known since the 1920s from radioactive dating of terrestrial rocks, that the age of the Earth is several Gyrs, much longer than the Kelvin-Helmholtz timescale. Thus, the Sun must be powered by different means. The lifetime of most stars is much larger than permitted by their gravitational energy reservoir alone. Instead of simple contraction, energy losses are replaced by the activation of nuclear fusion reactions among charged nuclear particles near the stellar core.

The efficiency of nuclear reactions, i.e., their *rate*, depends on the abundances of the reactant nuclei and the cross section of each reaction averaged over a Maxwellian distribution of relative velocities between the target and the projectile nuclei. For charged particle reactions the rate is mainly controlled by the Coulomb barrier generated by the number of protons in a nucleus. The nuclear reactions that activate at the lowest temperatures are those involving capture of protons, i.e., the nucleus of the lightest and most abundant element: hydrogen (H). Nature, however, does not allow the build up of stable nuclei made only of protons because in order to glue nucleons (i.e., proton and neutrons) together in a nucleus via the strong nuclear force, the repulsive electromagnetic force acting between protons needs to be diluted with a certain number of neutrons. The distribution of stable nuclei in the [N = number of neutrons, Z = number of protons] plane, the *chart of nuclides*, clearly shows that the region where nuclei are stable lies close to the N = Z line (the *valley of* $\beta$ *stability*) up to the element Ca, and then bends slightly towards the neutron-rich side as the repulsion between higher number of protons needs to be diluted with more and more neutrons. Nuclei outside this valley are radioactive, i.e., unstable, and decay towards their closest stable daughter through $\beta$-decay weak interaction reactions (as described in Chaps. 1, 2, 4, and 9). It follows that the build up of progressively heavier nuclei through the addition of protons naturally pushes the matter out of the stability valley, producing radioactive nuclei that decay back towards stability via $\beta^+$ decay.

A detailed analysis of the nuclear reactions involving the fusion of protons (H burning) foresees the existence of two processes. The PP chain activates at temperatures $\simeq$10 MK and operates through a sequence of proton captures and $\beta$ decays starting with a weak interaction p+p fusion. The processes involved in the PP chain are listed in Table 3.1 together with the mass defect ($Q_{tot}$ values) in MeV and the energy carried away by neutrinos $Q_\nu$. If the neutrino emission is described by an energy continuum, the maximum energy of this spectrum is reported. A direct build up of progressively heavier nuclei through successive proton captures stops very early, at $^3$He, because of the low cross section of the $^3$He+p reaction. Also, proton captures on the second most abundant isotope, $^4$He (or $\alpha$ particle, $N = Z = 2$), cannot even begin, because nuclei with atomic mass number A = N+Z = 5 do not exist.[2] In order to proceed with proton captures beyond $^3$He it is necessary to build up enough $^3$He nuclei to activate the capture of this nucleus by either another $^3$He nucleus, or $^4$He. The activation of $^3$He captures allows to overcome the non-existence of nuclei with A = 5, though it still does not allow the build up of an appreciable amount of nuclei heavier than He. In fact, the product of the $^3$He+$^3$He reaction is an $\alpha$ particle plus two protons, while the product of the $^3$He+$^4$He reaction is $^7$Be, whose fate, either proton or electron capture leads to the formation of $^8$Be, which very quickly decays in two $\alpha$ particles. In synthesis, the fusion of H mainly produces He, together with a number of radioactive nuclei that decay into

**Table 3.1** The PP chain

| Reaction | $Q_{tot}$ (MeV) | $Q_\nu$ (MeV) |
|---|---|---|
| $p + p \rightarrow d + e^+ + \nu$ | 1.442 | 0.42(spectrum) |
| $d + p \rightarrow ^3He$ | 5.494 | |
| $^3He + ^3He \rightarrow ^4He + 2p$ | 12.860 | |
| $^3He + ^4He \rightarrow ^7Be$ | 1.587 | |
| $^7Be + e^- \rightarrow ^7Li + e^+ + \nu$ | 0.862 | 0.861(90%)–0.383(10%) (lines) |
| $^7Be + p \rightarrow ^8B \rightarrow ^8Be + e^+ + \nu \rightarrow \alpha + \alpha$ | 18.209 | 14.060(spectrum) |
| $^7Li + p \rightarrow ^8Be \rightarrow \alpha + \alpha$ | 17.347 | |

[2] Their half lives are of the order of $10^{-22}$s.

their respective stable daughter nuclei emitting a neutrino. The energies and the number of neutrinos produced in these decays reflect the relative importance of the various PP-chain branches and the efficiency of the nuclear reactions in stars.

The second process converting protons to $\alpha$ particles is the CNO cycle. Given the high Coulomb barrier of the CNO nuclei, this cycle becomes efficient at temperatures (T $>$ 20 MK), significantly higher than those relevant to the PP chain. The main section of this sequence is characterized by the continuous conversion of C to N and viceversa. Let us start, e.g., with the capture of a proton by a $^{12}$C (Table 3.2). The outcome of this fusion is the radioactive nuclide $^{13}$N that quickly decays $\beta^+$ into $^{13}$C. Efficient proton captures by $^{13}$C lead to the synthesis of $^{14}$N. Proton captures by $^{14}$N produce $^{15}$O, a radioactive nuclide that quickly decays in $^{15}$N. The fusion of a proton and a $^{15}$N particle has, as the main outcome, a $^{12}$C nucleus plus an $\alpha$ particle. The sequence sketched above is called the CN cycle. If the temperature exceeds T $\sim$ 25 – 30 MK, also oxygen enters the game and the full CNO cycle activates: $^{16}$O begins to capture protons forming radioactive $^{17}$F that decays to $^{17}$O. The capture of a proton by this particle leads to a compound nucleus that preferentially splits into $^{14}$N and an $\alpha$ particle, and that partly turns into $^{18}$F, which quickly decays to $^{18}$O. Proton captures on $^{18}$O produce preferentially $^{15}$N plus an $\alpha$ particle. The activation of the channel $^{15}$N(p,$\gamma$)$^{16}$O closes the NO cycle, processing material back into oxygen. The total abundance by number of the CNO isotopes remains constant with time because the proton capture on any of them (and the subsequent $\beta^+$ decays) just produce another isotope in the same set.

For $T > 25 - 30$ MK, the full CNO cycle becomes efficient and quickly reaches a quasi-equilibrium in which the abundance of each nucleus settles on a steady state value determined by the balance between its production and destruction. For example, the equilibrium abundance of $^{13}$C (assuming that $^{13}$N decays instantaneously) is given by:

**Table 3.2** The CNO cycle

| Reaction | $Q_{\rm tot}$ (MeV) |
|---|---|
| $^{12}\mathrm{C} + \mathrm{p} \rightarrow {}^{13}\mathrm{N}$ | 1.944 |
| $^{13}\mathrm{N} \rightarrow {}^{13}\mathrm{C} + \mathrm{e}^+ + \nu$ | 2.220 |
| $^{13}\mathrm{C} + \mathrm{p} \rightarrow {}^{14}\mathrm{N}$ | 7.551 |
| $^{14}\mathrm{N} + \mathrm{p} \rightarrow {}^{15}\mathrm{O}$ | 7.297 |
| $^{15}\mathrm{O} \rightarrow {}^{15}\mathrm{N} + \mathrm{e}^+ + \nu$ | 2.754 |
| $^{15}\mathrm{N} + \mathrm{p} \rightarrow {}^{12}\mathrm{C} + \alpha$ | 4.966 |
| $^{15}\mathrm{N} + \mathrm{p} \rightarrow {}^{16}\mathrm{O}$ | 12.127 |
| $^{16}\mathrm{O} + \mathrm{p} \rightarrow {}^{17}\mathrm{F}$ | 0.600 |
| $^{17}\mathrm{F} \rightarrow {}^{17}\mathrm{O} + \mathrm{e}^+ + \nu$ | 2.761 |
| $^{17}\mathrm{O} + \mathrm{p} \rightarrow {}^{14}\mathrm{N} + \alpha$ | 1.192 |
| $^{17}\mathrm{O} + \mathrm{p} \rightarrow {}^{18}\mathrm{F}$ | 5.607 |
| $^{18}\mathrm{F} \rightarrow {}^{18}\mathrm{O} + \mathrm{e}^+ + \nu$ | 1.656 |
| $^{18}\mathrm{O} + \mathrm{p} \rightarrow {}^{15}\mathrm{N} + \alpha$ | 3.981 |
| $^{18}\mathrm{O} + \mathrm{p} \rightarrow {}^{19}\mathrm{F}$ | 7.994 |

$$\frac{dY_{^{13}C}}{dt} = Y_{^{12}C}\, Y_p\, \rho\, N_A\, <\sigma\, v>_{^{12}C+p} \;-\; Y_{^{13}C}\, Y_p\, \rho\, N_A\, <\sigma\, v>_{^{13}C+p} = 0$$

where $Y_i$ refers to the abundance by number of a given species $i$, $t$ is time, $\rho$ is the density, $\mathrm{N_A}$ is Avogadro's number and $<\sigma\ v>_{i+j}$ is the Maxwellian averaged product of the velocity $v$ times the nuclear cross section $\sigma$ for a given reaction between nuclei $i$ and $j$. The equilibrium condition immediately gives:

$$\frac{Y_{^{12}C}}{Y_{^{13}C}} = \frac{<\sigma\, v>_{^{13}C+p}}{<\sigma\, v>_{^{12}C+p}}$$

which means that the relative abundances between isotopes of the CNO cycle depend only on the ratio between the respective cross sections for proton capture. Typical isotopic and elemental ratios obtained in the temperature range $30 \leq T \leq 100$ MK are reported in Table 3.3.

The neutrinos emitted by the decay of radioactive nuclei synthesized by the CNO cycle have characteristic energies different from those emitted by the PP chain. Their detection would provide precious information about the relative efficiency of the various reactions involved in the CNO cycle.

In addition to the PP chain and the CNO cycle there is another sequence of proton captures that can become efficient in stars, although it does not play a role in the energy budget. In the temperature range 40–50 MK the proton captures listed in the upper part of Table 3.4 quickly bring to their equilibrium values the abundances of $^{20}$Ne, $^{21}$Ne, $^{22}$Ne, and $^{23}$Na, forming also in this case a NeNa cycle. For temperature in excess of 50 MK the $^{23}$Na(p,$\gamma$)$^{24}$Mg channel competes with the $^{23}$Na(p,$\alpha$)$^{20}$Ne so that matter from the NeNa cycle leaks towards more massive nuclei. At temperatures of order of 60 MK also the proton captures listed in the lower part of Table 3.4 fully activate so that all the nuclei between $^{20}$Ne and $^{27}$Al reach their equilibrium abundances. Is it worth noting that $^{26}$Al, a long-lived radioactive nucleus with half life $7.17 \times 10^5$ year, is included within this sequence. $^{26}$Al can be ejected into the interstellar medium by stellar outflows (winds) and its decay into $^{26}$Mg can be detected as diffuse $\gamma$-ray emission (Sect. 7.4) when the metastable $^{26}$Mg relaxes towards its ground state. Moreover, it can be included in dust grains that form around stars and decay within the already formed minerals. This nucleus is thoroughly discussed in Sects. 3.6.1 and 7.4, and in Chaps. 4 and 9. Typical

**Table 3.3** Typical isotopic ratios

| Isotopic ratio | Value | Solar value |
|---|---|---|
| $Y_{^{12}C}/Y_{^{13}C}$ | $\simeq 4$ | 89 |
| $Y_{^{14}N}/Y_{^{15}N}$ | $\simeq 4 \times 10^4 - 10^5$ | 272 |
| $Y_{^{17}O}/Y_{^{16}O}$ | $\simeq 10^{-2} - 10^{-3}$ | $3.8 \times 10^{-4}$ |
| $Y_C/Y_N$ | $\simeq 7 \times 10^{-3} - 2.5 \times 10^{-2}$ | 3.2 |
| $Y_N/Y_O$ | $\simeq 60 - 350$ | 0.13 |
| $Y_{^{18}O}/Y_{^{16}O}$ | $\simeq 2 \times 10^{-6}$ for $T < 50:60$ MK | $2 \times 10^{-3}$ |
| $Y_{^{18}O}/Y_{^{16}O}$ | declines to $\simeq 5 \times 10^{-8}$ at T$\simeq$ 100 MK | $2 \times 10^{-3}$ |

**Table 3.4** The NeNaMgAl sequence

| Reaction | $Q_{tot}$ (MeV) |
|---|---|
| $^{20}Ne + p \rightarrow ^{21}Na$ | 5.979 |
| $^{21}Na \rightarrow ^{21}Ne + e^{+} + \nu$ | 3.548 |
| $^{21}Ne + p \rightarrow ^{22}Na$ | 6.739 |
| $^{22}Na \rightarrow ^{22}Ne + e^{+} + \nu$ | 2.842 |
| $^{22}Ne + p \rightarrow ^{23}Na$ | 8.794 |
| $^{23}Na + p \rightarrow ^{20}Ne + \alpha$ | 2.377 |
| $^{23}Na + p \rightarrow ^{24}Mg$ | 11.693 |
| $^{24}Mg + p \rightarrow ^{25}Al$ | 6.548 |
| $^{25}Al \rightarrow ^{25}Mg + e^{+} + \nu$ | 4.277 |
| $^{25}Mg + p \rightarrow ^{26}Al$ | 6.307 |
| $^{26}Al \rightarrow ^{26}Mg + e^{+} + \nu$ | 4.004 |
| $^{26}Al + p \rightarrow ^{27}Si$ | 12.275 |
| $^{26}Mg + p \rightarrow ^{27}Al$ | 8.271 |
| $^{27}Si \rightarrow ^{27}Al + e^{+} + \nu$ | 4.812 |
| $^{27}Al + p \rightarrow ^{28}Si$ | 11.585 |
| $^{27}Al + p \rightarrow ^{24}Mg + \alpha$ | 1.601 |

$Y_{^{26}Al}/Y_{^{27}Al}$ equilibrium ratios produced by H burning range between $3 \times 10^{-2}$ at 60 MK and 0.8 at 100 MK. We refer the reader to the book by Cox and Giuli (1968) for a derivation of the basic stellar structure equations and detailed discussions of the physics involved in the study of stellar evolution.

Although the sequences of nuclear reactions that power stellar luminosity are now considered to be well understood, a wise Galilean approach suggests to verify experimentally (whenever possible) their occurrence in stars. Our Sun provides a unique opportunity to accomplish such verification via the detection of neutrinos produced by radioactive decay in its deep interior.

### 3.2.2 Solar Neutrinos – A Unique Opportunity

The long-term stability of characteristic solar properties, in particular its luminosity and surface temperature, can be explained only if the solar energy source is of nuclear origin and specifically involves proton captures, which are associated with abundant fuel and a long time scale. As discussed above, such an energy source inevitably results in the production of radioactive nuclei, which decay into their stable daughter nuclei through weak processes, hence emitting neutrinos. Modeling of the internal structure of the Sun predicts a central temperature at present of about 15 MK, and hence that the PP chain dominates (99.6%) over the CNO cycle (0.4%) converting H into $^4$He. The relative importance of the nuclear reactions in the PP chain in the Sun leads to the result that the majority (93%) of the neutrinos produced should come from $p(p, e^{+}\nu_e)d$ reactions (where d = deuterium, N = Z = 1) and be of relatively low energy ($E \leq 0.42$ MeV, see Table 3.1), while only a minor fraction of

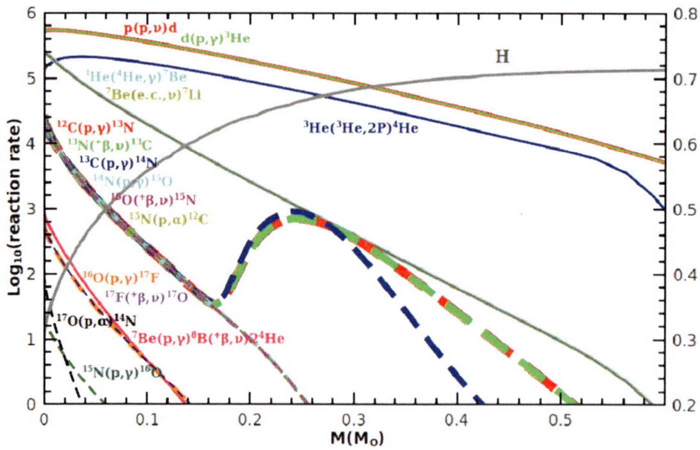


**Fig. 3.1** Rates of the reactions involved in the PP chain and the CNO cycle as a function of the mass coordinate in a solar-like stellar model having approximately the age of the Sun of 4.6 Gyr. The H abundance is also plotted and its range shown on the *right-side* y-axis

the total neutrino spectrum is expected to be contributed by the decay of $^{7}$Be ($\simeq$7%, E $\simeq$0.86 MeV) and $^{8}$B(0.0075%, E $<$ 15 MeV).

Figure 3.1 shows the rates of the nuclear reactions involved in the PP chain and CNO cycle as a function of the mass coordinate for a 1 $M_{\odot}$ stellar model of solar *metallicity* and an age close to the present age of the Sun, i.e., 4.6 Gyr. All the $\beta^{+}$ decays are concentrated near the center where the bulk of the synthesis of unstable nuclei takes place. The relative importance of the various PP reactions, and hence of the associated neutrino fluxes, depends on the rates of the nuclear reactions involved, which, in turn, are a function of density, temperature, and chemical composition. Since the internal structure of a model of the Sun depends on the adopted and somewhat uncertain input physics – e.g., the nuclear cross sections, the equation of state, the opacity, and the chemical composition of the gas from which the Sun formed – the detection of neutrinos from the Sun is fundamental not only to experimentally verify the nuclear origin of the solar luminosity, but also to confirm the overall reliability of the modeling of the internal structure of the Sun and the adopted input physics.

It is therefore understandable that the quest for the solar neutrinos started early, more than 40 years ago, with the Davis experiment (1967–1985) (Bahcall et al., 1985; Cleveland et al., 1998). This experiment, based on the interaction between an electron neutrino and $^{37}$Cl, has a threshold energy of roughly 0.8 MeV and hence could essentially only detect the $^{8}$B neutrinos, which constitute a minor fraction of the neutrino flux from the Sun. The Davis experiment provided two basic results, one very encouraging and another very stimulating. First, it detected solar neutrinos, demonstrating beyond any doubt that proton captures are occurring in the interior of the Sun. Second, the detected neutrino flux was roughly one third of the predicted value. Such a result stimulated much further work and a large number of papers on this puzzle were written over the decades. The discrepancy was considered by many physicists as a strong indication that the basic modeling of the Sun was wrong, however, it must be recalled that it was confined to a very minor branch of the PP chain. The discrepancy became more serious with the advent of the GALLEX experiment (Hampel et al., 1998) – a collaboration among France, Germany, Italy, Israel, Poland, and USA, led by MPIK Heidelberg, 1991–1997 – and the SAGE experiment (Abdurashitov et al., 1999) – a Russian-American collaboration, 1990–2000. These modern sophisticated experiments were designed to detect the bulk of the neutrinos produced in the Sun, i.e., the low energy neutrinos produced by the p+p reaction. They confirmed both the detection of a neutrino signature from the Sun and the existence of a discrepancy between theoretical and observed fluxes. This result plunged the basics of solar modeling into a crisis because these experiments were sensible to the total number of electron neutrinos emitted by the Sun. This number is theoretically robust, depending only on the basic assumption that the solar luminosity comes from the conversion of protons into $\alpha$ particles, and not on the details of the modeling of the internal structure of the Sun or of the cross sections of the nuclear reactions involved. If the solar luminosity is powered by the conversion of protons into $\alpha$ particles, the total number of electron neutrinos emitted per second by the Sun must be $2.38 \times 10^{39}$ ($L_{\odot}$ in MeV/s)/25 (energy provided per $\alpha$ nucleus in MeV) $\times$ 2 (number of $\nu_e$ produced per $\alpha$ nucleus) [$\nu_{\mathrm{e}}$/s], which corresponds, at a distance of one astronomical unit, to a flux of $6.78 \times 10^{10}$ $\nu_{\mathrm{e}}/\mathrm{s}$. Instead, the neutrino flux detected by GALLEX and SAGE was only half of this prediction.

The solution to this puzzling discrepancy arrived when it was shown that neutrinos oscillate among three different species: $\nu_e$, $\nu_\mu$ and $\nu_\tau$. None of the Davis, GALLEX, or SAGE experimental set-ups could detect the $\nu_\mu$ and $\nu_\tau$ neutrinos so that all these experiments simply missed the fraction of neutrinos that reach the Earth as $\nu_\mu$ and $\nu_\tau$. The Sudbury Solar Neutrino Observatory (SNO) experiment (a Canadian, USA, and UK collaboration that started in 1999, McDonald et al., 2002) was designed to detect all three neutrino flavors. In 2001 this experiment finally showed that there is agreement between the observed and predicted neutrino fluxes and hence demonstrated that our modeling of the interior of the Sun is basically correct. In 2008, the Borexino instrument finally also opened up the sub-MeV range for solar neutrinos and detected $^7$Be neutrinos (BOREXINO Collaboration et al., 2008). For comprehensive reviews on the solar neutrino problem and current status, see Bahcall and Peña-Garay (2004); Bahcall (2005); Oberauer (2010).

## 3.3 Evolution To and Through the First Giant Branch

In Sect. 3.3.1 we discuss the main evolutionary properties of stars once they leave the long lasting phase of central H burning described above, commonly referred to as the Main Sequence, and enter the phase known as the First, or Red, Giant Branch (RGB). At the end of central H burning a star is composed of a H-exhausted core made primarily of He and a H-rich envelope. Hydrogen burning shifts from the center to the base of the H-rich mantle while the envelope expands causing the surface of the star to reach radii 10–1,000 times the solar radius and to cool down to a few thousand K. This expansion triggers the formation of large scale convective motions extending from the surface down to deep regions in the star where partial H burning occurred during the Main Sequence. Some products of this H burning are thus brought to the surface in a process known as the 1st dredge-up. In this phase the He core grows in mass because the H-burning shell continuously converts H-rich matter into He-rich matter and deposits the ashes onto the He core. The temporal evolution of the He core depends on its initial size, i.e., the size it had just after the central H exhaustion, which is in turn mostly determined by the initial stellar mass. If the mass of the He core is less than $\simeq 0.35\,M_{\odot}$, which occurs for initial stellar masses less than $\sim 2\,M_{\odot}$, an *electron degenerate core* forms where matter reaches such extraordinarily high density, up to $\simeq 10^6$ g/cm$^3$, that the dominant contribution to its pressure arises from the Pauli exclusion principle, which prevents the electrons from occupying identical quantum states. This leads to an increase of the lifetime of this phase up to about 100 Myr, and forces the subsequent He ignition to occur quite far from the center. If the He core is instead more massive than $0.35\,M_{\odot}$, the electrons remain far from the degeneracy regime.

In Sect. 3.3.2 we discuss the conditions under which $^7$Li, the stable daughter of radioactive $^7$Be, may be produced, preserved, and brought to the stellar surface. This nucleus is typically destroyed in the PP chain (Table 3.1), because its destruction rate is efficient at temperatures lower than its production rate. A way to produce $^7$Li was proposed in 1971 by Cameron and Fowler (1971): if freshly synthesized $^7$Li is quickly brought to very low temperatures by mixing, then it can be preserved. If H burning occurs in a convective environment it is in principle possible to find a high Li abundance on the surface of a star, as observed in some stars belonging to the First Giant Branch. However, these observations are in fact difficult to explain because H burning occurs in a formally stable region well below the base of their convective envelopes. Additional mechanisms of mixing must be invoked to bring $^7$Li-rich material into the convective envelope.

### *3.3.1 The First Giant Branch*

During the Main Sequence phase of stellar evolution described in the previous section conversion of H into He via H burning in the centre of the star leads to a progressive increase of the mean molecular weight combined with a decrease of the amount of available fuel. The net result is a slight increase of the luminosity (because L scales with the fourth power of the molecular weight, Eq. (3.10) in Sect. 3.2.1), and a mild expansion of the surface of the star because of the formation of a molecular weight gradient (Stancliffe et al., 2009). Once H is exhausted in the central region of the star, the H-exhausted core, or He core, begins to contract on a gravitational timescale while the region of active nuclear burning smoothly shifts above the He core, where H is still abundant. Further evolution of the He core depends on its mass, which, in turn, depends on the initial total mass of the star. If the He-core is more massive than a threshold value of $\sim$0.35 M$_\odot$, which happens for an initial total mass of $\sim$2 M$_\odot$, its contraction induces strong heating of the He core itself, which quickly reaches a temperature of $\sim$100 MK at which fusion reaction of $\alpha$ particles (He burning) is activated. If, instead, the core is less massive than 0.35 M$_\odot$ the high densities reached render the electron gas degenerate, hence supporting the structure against gravity without the need for additional contraction.

This difference has a large impact on further evolution of the star because, in the latter case, the He core tends towards an almost isothermal configuration due to the large mean free path of degenerate electrons relative to that of photons. If the structure was isolated, as in the case of white dwarves, it would progressively cool down losing its stored energy through the surface. Instead, in the case discussed here, the degenerate structure heats up because it is surrounded by the H-burning shell, which continuously deposits freshly synthesized He onto the He core. The rate at which the maximum temperature increases with time in the degenerate He core is controlled by the growth rate of the He-core mass, which obviously coincides with the rate at which the H-burning shell converts H into He. Strong neutrino production (Itoh et al., 1989) in the center of the electron degenerate core, due to the interaction of photons with the plasma and/or to the scattering of photons on electrons, carries away energy from the core pushing the location of the maximum temperature outward in mass. The off-center location of the maximum temperature is the result of the balance between energy loss due to the neutrino emission, which scales directly with the density and pushes the maximum temperature outward, and the energy gain due to the compressional heating, which scales inversely with the density and pushes the temperature maximum back towards the center. The key stellar parameters that control the location of the maximum temperature are the CNO abundance and the initial mass of the star. The higher the CNO abundance, the faster the conversion of protons into $\alpha$ particles in the H-burning shell, the stronger the heating of the degenerate He core, and the closer the maximum temperature is to the center. The higher the initial mass of the star, the lower is the degree of electron degeneracy and the density in the He core, and hence the efficiency of neutrino production.

While the H-burning shell influences the evolution of the He core, the growth of the He core influences the evolution of the H-burning shell as well. In fact, the progressive heating of the core raises the temperature at the surface of the core, where H burning occurs. This results in a continuous positive feedback: the H burning shell deposits He onto the He core, which therefore heats up. Such a heating leads to an increase of the temperature and density in the H-burning shell, accelerating the H burning rate and increasing the conversion rate of H into He, and therefore the heating of the He core. As a consequence, the progressive increase of the H burning rate determined by the growth of the He core mass forces the H rich mantle of the star to expand and to cool. The cooling of the stellar envelope triggers a large increase of the opacity because atoms become partially ionised. The temperature gradient steepens, favoring the growth of convective instabilities that very quickly extend over a major part of the H-rich mantle from the surface down to near the outer border of the H-burning shell. A consequence of the growth of these convective motions within most of the H rich mantle is a large increase of the surface luminosity caused by the continuous increase of the H burning rate coupled to the fact that the convective envelope does not absorb or release energy but just transports it outward.

Since convective motions play a fundamental role in the physical and chemical evolution of any star, we briefly sketch the basic physical reason that leads to the growth of these large scale motions. The equilibrium condition provided by counterbalancing pressure gradients and gravity in stars does not necessarily imply stationary matter: a bubble of stellar matter may be considered stable against motion if a restoring force pushes it back towards its rest position when, for any reason, it is slightly displaced from its equilibrium location. Such a restoring force is simply given by Archimede's force, i.e., it depends on the density contrast between that of the environment and that of the bubble. If the density of an element of matter displaced towards a lower/higher density region turns out to be even lower/higher than that of its new surroundings, the element will continue to *raise/sink* and move away from its rest position, otherwise it will move back towards its equilibrium location.

Changes in the physical structure of the bubble during its motion play an important role in determining its density and thus its behavior. Mechanical equilibrium with the environment is certainly well verified so that it can be safely assumed that the internal pressure within the bubble istantaneously readjusts to that of the environment. More difficult is to determine the amount of heat that the bubble can exchange with the environment while moving. In the simplest case in which the bubble does not exchange any heat with the surroundings until it has covered a certain distance (adiabatic approximation), and assuming that the region is chemically homogeneous, the critical condition for the onset of large scale motions of the matter is that the temperature gradient of the environment must exceed the adiabatic gradient (Schwarzschild criterion). While the radiative temperature gradient remains less than the adiabatic temperature gradient, an element of matter will remain more/less dense than its surroundings if displaced towards less/more dense regions (within stars these displacements are typically connected to movements outward/inward in mass), and hence it will experience a restoring force that will keep it anchored to its rest location. On the contrary, when the radiative temperature gradient exceeds the adiabatic temperature gradient stochastic motion of the matter is not hampered by a restoring force, but it is amplified leading to the growth of large scale motions. Hence, convective regions are associated with steep temperature gradients, which typically occur either close to regions where energy production is strongly concentrated, or in regions where the mean free path of the photons, which scales with the inverse of the opacity, becomes so small that radiation energy transport becomes inefficient.

The determination of the temperature gradient in convective regions is quite complex. Here it suffices to say that while in the interior of a star the temperature gradient in a convective region remains very close to the adiabatic gradient, in a convective envelope the temperature gradient becomes much steeper (intermediate between the radiative and adiabatic case) because the low density in the outer envelope makes energy transport by convective eddies inefficient, so that both photons and eddies contribute to the outwards transport of thermal energy.

Since convective eddies have a very large mean free path with respect to that of photons, convection is a very efficient energy transport mechanism. In the specific case of the extended convective motions that form above the H-burning shell in Red Giant stars, energy transport is so efficient that virtually all the energy produced by the burning shell is transmitted to the surface without essentially any absorption by the convective layers. It follows that a star in the H-burning shell evolutionary phase is forced to increase in size to be able to get rid of the extra energy influx, while the drop of the surface temperature is limited by the presence of a maximum temperature gradient: the adiabatic temperature gradient, which cannot be overcome by much in the largest fraction of the envelope mass.

The mere existence of stars in the RGB phase constitutes evidence of (a) the presence of an active H-burning shell, demonstrated by the breaking of the mass-luminosity relation $L \propto M^3$ that holds during the Main Sequence phase, (b) the presence of a maximum temperature gradient, demonstrated by the only minor change of the surface temperature along the RGB, (c) the continuous increase of the energy production by the H-shell burning, demonstrated by the continuous increase of the surface luminosity, and d) the presence of an electron degenerate core (for stars with initial mass less than $\simeq$2 $M_\odot$), demonstrated by the existence of a relatively long lasting, $\sim$$10^8$ year, and thus observable RGB phase, which would be prevented if the He core was gravitationallly contracting.

Soon after the formation of the H-burning shell, the large scale motions that grow in the H-rich envelope and rapidly extend from the surface down to just above the top of the H-burning shell, bring to the stellar surface matter partially processed by proton-capture reactions during the Main Sequence phase. This mixing, referred to as the 1st dredge-up, modifies the stellar surface composition. The amplitude of these modifications depends on the initial stellar mass and metallicity, the general rule being that the amplitude of the changes of the surface composition scales with the initial stellar mass, directly up to 3 $M_\odot$ and then inversely for higher masses, and inversely with the metallicity. Figure 3.2 shows the abundance profiles of several nuclear species as a function of the mass location for a solar-like stellar model evolved to the RGB, just before the convective envelope deeply penetrates into the star. The solid vertical line shows the maximum inward peneration of the convective envelope. Since the convective motions reach layers in which the local chemical composition has previously been modified by nuclear burning, also the surface chemical composition is modified by the mixing induced by these large scale motions. In particular the surface He abundance is slightly increased by 5%, $^3$He increases by one order of magnitude, $^7$Li is destroyed, the $^{12}$C/ $^{13}$C ratio drops from the solar value of 89 to roughly 30, the $^{14}$N/ $^{15}$N ratio increases from the terrestrial value of 272 to $\simeq$ 500 while the oxygen isotopic ratios and those of heavier nuclei remain at their solar values. In stars more massive than the Sun the oxygen isotopic ratios are also modified with $^{16}$O/$^{17}$O decreasing by up to one order of magnitude,

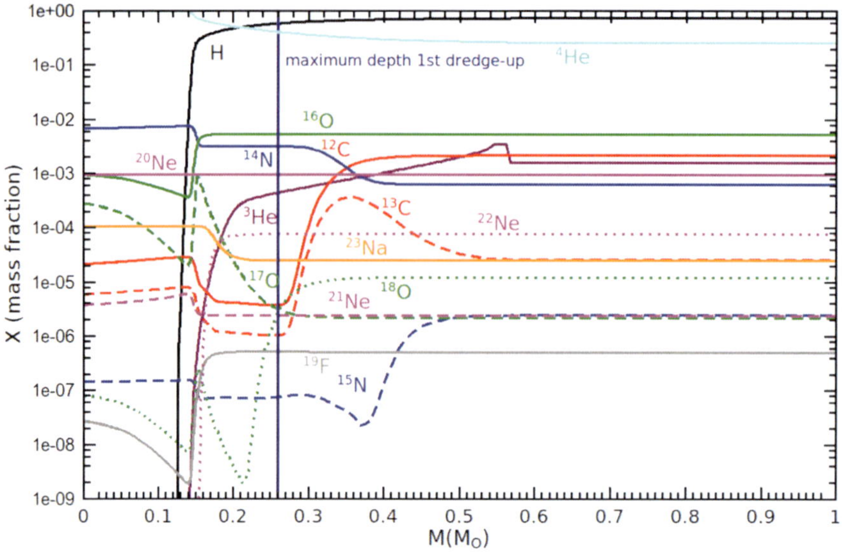


**Fig. 3.2** Snapshot of the abundances of several nuclear species in a solar-like stellar model just before the onset of the 1st dredge-up. The maximum inward penetration of the convective envelope during the 1st dredge-up is marked by the *vertical solid blue line*

from solar $\simeq$ 2700 to $\sim$ 250 and $^{18}O/^{16}O$ increasing mildly up to $\simeq$ 700 from the solar value of 500. A detailed quantitative determination of these changes depends on the specific stellar model considered.

The evolution of the star after the 1st dredge-up is characterized by the H-burning shell progressively converting H from the convective envelope into He, which is deposited onto the inert He core. The continuous mass transfer from the envelope to the core progressively reduces the mass of the envelope while its chemical composition does not change any more because the temperature within the convective envelope is too low to activate nuclear reactions.

The evolution along the RGB ends when the maximum temperature in the core is high enough, $\simeq$100 MK, to activate the burning of He via $3\alpha$ reactions, during which three $\alpha$ particles join into a $^{12}C$ nucleus. If the pressure is dominated by degenerate electrons the energy released by these reactions cannot be immediately balanced by an expansion of the core. Hence, He ignition occurs through a series of *flashes*, which progressively remove the degeneracy, shifting the burning towards the center. Once the electron degeneracy is fully removed, a quiescent central He-burning phase settles in.

All along the complex, and partly still mysterious, RGB evolutionary phase that links central H to central He burning, radioactive nuclei are produced by H-shell burning mainly via the CNO cycle. Most of them, however, have negligible lifetimes, so they could only be detected through the neutrinos they emit. Unfortunately, no Red Giant star is close enough to the Earth to allow the detection of neutrinos of nuclear origin produced in its interior. However, there are two unstable nuclei,

$^7$Be and $^{13}$N, whose half life may be comparable or even larger than some stellar timescales: for $^7$Be the half life is comparable to the envelope mixing turnover time, for $^{13}$N the half life is comparable to proton-capture timescale in extremely metal-poor stars, because stars of lower metallicity are more compact and hotter, due to their lower opacity.

In the next section we discuss specifically the abundance of $^7$Li, the stable daughter of $^7$Be, in giant stars, which could provide important clues about the presence of additional motions extending below the base of the convective envelope. This is important because, as we described above, the modeling of large scale motions within stars is still crude and their growth, timescale, and efficiency not yet well understood.

### 3.3.2 The Production of Li

Lithium (Li) isotopes[3] in stars are fragile as they are easily destroyed by proton-capture reactions once the temperature exceeds 3 MK. The destruction timescale drops from 100 million years at 3 MK to only 0.3 million years at 5 MK while their production through fusion reactions only occurs at much higher temperatures, between 10 MK and 25 MK. The lower limit is due to the fact that the synthesis of $^7$Li is initiated by the $^3$He($\alpha$,$\gamma$)$^7$Be reaction, which becomes efficient only at temperature of the order of 10 MK, while the upper value is due to activation of the $^7$Be(p,$\gamma$)$^8$B reaction, which overcomes the electron-capture reaction $^7$Be(e$^-$, $\nu$)$^7$Li above a temperature of the order of 25 MK. Hence, Li is efficiently produced in a temperature range where it is also efficiently destroyed and therefore there seems to be no room for Li production in a star. However, there are a number of Red Giant stars observed to be Li rich (Castilho et al., 2000; Balachandran, 2005; Uttenthaler et al., 2007).

A possible way out of such a puzzling situation was recognized by Cameron and Fowler (1971) and is based on the idea that instabilities, such as convection, rotation-induced instabilities, thermohaline mixing, etc., may bring freshly made $^7$Be from its production site to more external regions, where the temperature is low enough to inhibit proton captures on $^7$Li, on a timescale shorter than that of electron capture of $^7$Be. Note that the electron-capture rate of $^7$Be shows a mild increase as the temperature decreases.

A typical environment in which the Cameron-Fowler mechanism operates is during the Asymptotic Giant Phase (AGB) phase (Sect. 3.4.1), if the star is more massive than 4:5 M$_\odot$. These stars develop large scale motions in the H-rich mantle that extend from the surface down to regions where the temperature is high enough ($>$ 40 MK) for some nuclear burning to occur (Hot Bottom Burning), in particular via the $^3$He($\alpha$,$\gamma$)$^7$Be reaction. Figure 3.3 shows the cumulative turnover time

---

[3] Lithium has two stable isotopes $^6$Li and $^7$Li, of which $^7$Li is the more abundant representing 92% of solar Li.

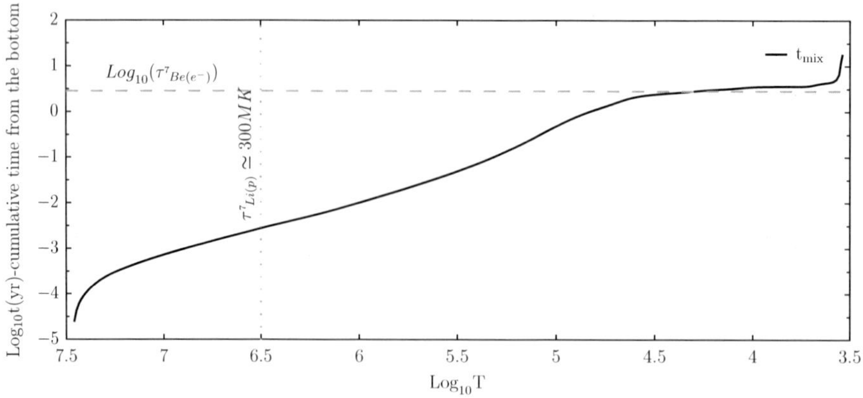


**Fig. 3.3** The cumulative turnover time of the convective eddies computed from the base of the convective envelope up to the surface as a function of temperature, T. The star is a 6M$_\odot$ star of solar metallicity some time after the beginning of the AGB phase. The *horizontal dashed grey line* marks the typical timescale of the $^7$Be(e$^-$, $\nu$)$^7$Li reaction while the *vertical dotted grey line* shows the threshold temperature below which the timescale of the proton capture on $^7$Li becomes larger than 300 million years

from the base of the convective envelope to the region of temperature T given in the abscissa for a 6 M$_\odot$ star of solar metallicity sometimes after the beginning of the AGB phase. The horizontal dashed grey line marks the typical timescale of the $^7$Be(e$^-$, $\nu$)$^7$Li reaction while the vertical dotted grey line shows the threshold temperature below which the timescale of the proton capture on $^7$Li becomes larger than 300 million years. In this environment $^7$Be produced above 10 MK is succesfully transfered before decaying to a region where its daughter $^7$Li can survive.

An increase of the Li abundance at the surface of RGB stars is more difficult to achieve. Though the turnover time scale within the convective envelope is also in this case rather short ($\simeq$1 year at Log$_{10}$(L/L$_\odot$) $\simeq$2), the temperature at the base of the convective envelope always remains well below 5 MK, too low to lead to an appreciable production of $^7$Be. Nonetheless, as mentioned above observations show the existence of a small number of Li-rich RGB stars. Figure 3.4 shows the internal structure of the region around the H-burning shell from a solar-like stellar model on the RGB. Here, $^7$Be is synthesized well below the region where large scale motions of the matter and hence mixing of the chemical composition occur. In this environment the Cameron-Fowler mechanism could operate only by assuming the presence of presently unidentified instabilities able to drive some mixing between the region rich in $^7$Be and the base of the convective envelope. The main constraint on this *extra* mixing is that it must get close enough to the active H-burning shell to reach the layers where the $^7$Be production occur, but it must not enter the region of the main nuclear burning. The reason is that the speed at which a star climbs along the RGB is regulated by the speed at which H is converted into He (see above) which, in turn, also depends on the amount of fuel that continuously enters

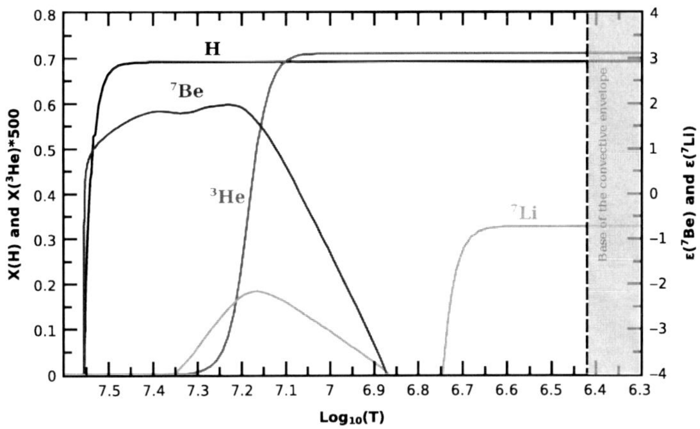


**Fig. 3.4** Abundance profiles for H, $^3$He, $^7$Be, and $^7$Li as function of the internal temperature in the region between the H-burning shell and the base of the convective envelope (*grey area*) for a solar-like stellar model on the RGB. The abundances of both $^7$Be and $^7$Li are given in the widely adopted logarithmic scale in which $\varepsilon(X) = \mathrm{Log}_{10}(N_X/N_H) + 12$, where $N_X$ and $N_H$ represent the abundances of element X and of hydrogen by number. In this scale the hydrogen abundance is equal to 12. X(H) and X($^3$He) represent the mass fraction of H and $^3$He

the burning region. If this extra mixing reached the active burning region, it would inevitably bring fresh H into the burning region, therefore altering the rate at which H is consumed by the H-burning shell and hence the timescale of evolution along the RGB. This evolutionary timescale is observationally well established from counting the number of stars on the RGB in many Galactic globular clusters, and already very well reproduced by current models of these stars without extra mixing.

There are other hints that point towards the presence of extra-mixing phenomena in RGB stars (and perhaps in AGB stars too, as discussed in Sects. 3.6.1 and 3.6.2). The observed surface $^{12}$C/$^{13}$C ratio and N abundance are, respectively, too low and too high with respect to the values predicted by the 1st dredge-up. Extra mixing would naturally lower the first ratio and raise the N abundance, this being the signature of H burning (see Table 3.3). A deeper mixing than predicted by current models would also reduce the abundance of $^3$He in the stellar envelope, which is increased by the 1st dredge-up, by bringing this nucleus down to regions where it is destroyed. This reduction is needed to avoid an increase of the $^3$He abundance in the interstellar medium, which is not observed, due to the material expelled by low-mass stars over the lifetime of the Galaxy.

Figures 3.5 and 3.6 show the isotopic abundances of several nuclei up to Al within a solar-like star while climbing the RGB (at $\mathrm{Log}(L/L_{\odot}) \simeq 3$). Note that we chose to use the temperature as the abscissa instead of mass to better clarify the temperature at which each nuclear species varies. The figures clearly shows that, for each given depth reached by an extra mixing process, a few nuclei are expected to be modified. For example, a drop of the oxygen abundance at the surface of an RGB

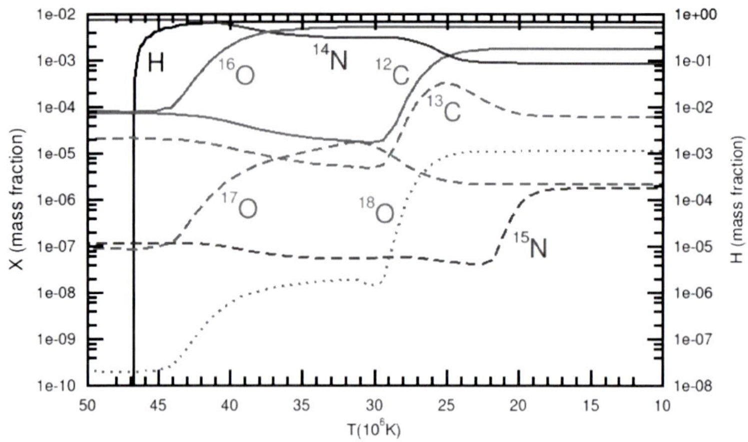


**Fig. 3.5** Abundance profiles of the CNO isotopes as function of the temperature on the RGB in a solar-like stellar model

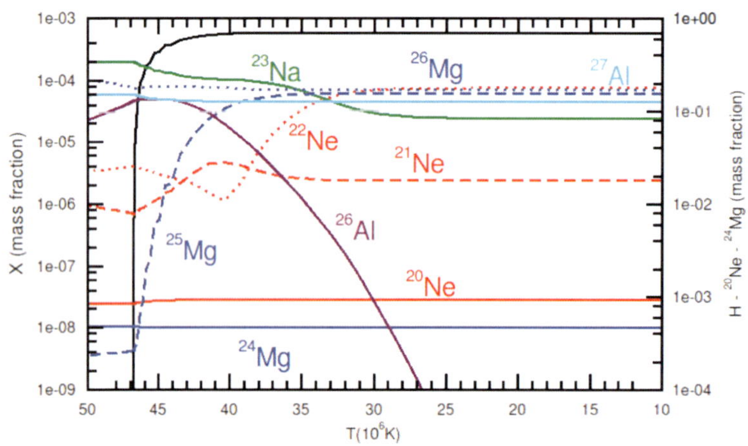


**Fig. 3.6** Abundance profiles of the Ne, Na, Mg and Al isotopes as function of the temperature on the RGB in a solar-like stellar model

star due to an extra mixing process (the depth reached by the extra mixing should extend down to at least 40 MK in this case), would also imply an increase of the surface abundances of both N and Na. Isotopes like $^{18}$O and $^{22}$Ne are expected to be fully destroyed, while the $^{12}$C/$^{13}$C ratio should drop and the $^{14}$N/$^{15}$N increase. Note that in any case it would be very difficult to obtain a surface change of the Ne and the Mg abundances because their most abundant isotopes, $^{20}$Ne and $^{24}$Mg, are not modified unless the mixing reaches down to the location of main H burning.

In summary, our modeling of mixing in stars is still oversimplified and unrealistic as it is based on a simple buoyancy model. Observational evidence of stellar abundances also involving radioactive nuclei and their daughters points out that mixing of matter outside the standard convective boundaries should occur in stars. These observations can be used to improve our description of mixing phenomena in stars.

## 3.4 Evolution in the Double Shell Burning Phase

We start Sect. 3.4.1 by describing the central He-burning phase and the direct scaling of the mass of the convective core resulting from He burning with the mass of the He core. The mass of the convective core determines the size of the initial He-exhausted core, an important parameter for subsequent evolutionary phases. As it happens previously when H burning shifts from the centre to a shell, also when He is exhausted in the core and He burning shifts from the centre to a shell, the envelope is forced to expand and convective motions extend from the external layers deeply inward into the star. In stars more massive than 4:5 $M_{\odot}$ the convective envelope even penetrates within the He core reducing its mass size and carrying to the stellar surface material processed by nuclear reactions (2nd dredge-up). If the He-exhausted core does not grow above $\sim$1.1 $M_{\odot}$, an electron degenerate core forms again, this time made of C and O, on top of which are located two burning shells: the He-burning and the H-burning shells. This marks the beginning of the double burning shell phase: the Thermally Pulsing Asymptotic Giant Branch (TP-AGB) phase.

The two key features of this phase are that (1) the two burning shells can not be simultaneously active, but they alternate within a cycle in producing the required energy and (2) He ignition within each cycle occurs through a thermal runaway (or thermal pulse, TP) that ends when enough energy is injected in the He-burning zone to convert its temperature and density profiles into a configuration that allows stable burning. The frequency of these thermal instabilities scales directly with the He-core mass. Such an abrupt injection of a quite large amount of energy ($\sim 10^{48}$ erg) induces first the growth of a convective shell within the (*intershell*) zone, located between the two shells, and second, soon after this convective region is extinguished, the expansion of the base of the H-rich envelope, which forces the convective envelope to penetrate well within the intershell zone (3rd dredge-up). The combination of these two successive convective episodes allows nuclei freshly synthetized by He burning to be carried up to the stellar surface. Moreover, the temperature at the base of the convective envelope scales directly with the He-core mass, and, in stars more massive that 4–5 $M_{\odot}$, reaches high enough values that H burning activates (Hot Bottom Burning, HBB).

In Sect. 3.4.2 we discuss Super-AGB stars, i.e., stars with initial mass that locates them in the interval between stars that develop an electron degenerate core after He is exhausted in the center and enter the AGB regime, and more massive stars that do not develop an electron degenerate core. Super-AGB stars ignite carbon out of center in semidegenerate conditions and go through a central C-burning phase. However, the C-exhausted core is not massive enough to heat up to the Ne burning, so an electron degenerate ONeMg core forms. These stars then go through a thermally pulsing phase. The final fate of these stars depends on the capability of their ONeMg core to reach the critical mass of $\sim$1.35 $M_\odot$ required to activate electron captures on $^{24}$Mg and $^{20}$Ne. Stars with a core that does not reach this crital mass lose all their H-rich envelope and end their life as ONeMg white dwarves, while stars with a core that reaches this critical mass explode as *electron-capture supernovae*.

We continue by briefly discussing mass loss during the AGB phase in Sect. 3.4.3. The strong increase in surface luminosity, coupled to luminosity variations and formation of dust grains in the atmospheres of AGB stars, strongly enhances the mass-loss rate in this phase with the consequence that all AGB stars lose their H-rich envelope, leaving behind the naked electron degenerate core as a cooling CO white dwarf. Finally, in Sect. 3.4.4, we discuss the different species of dust grains that form in the atmosphere of an AGB star. The key role here is played by the C/O number ratio in the atmosphere because the strong bond of the CO molecule results in trapping all of the atoms of the least abundant of the two elements. In an oxygen-rich gas (O>C) the species of dust are oxide grains, for example, $Al_3O_2$ (corundum) and many different types of silicates (SiO, $SiO_2$, etc.). In a carbon-rich gas (C>O), the species of dust are, for example, SiC (silicon carbide) and C itself (graphite). Some of this stellar AGB dust is now recovered from primitive meteorites, representing a real speck of an ancient AGB star that we can analyse in the laboratory.

### 3.4.1 Asymptotic Giant Branch (AGB) Stars

As anticipated at the end of Sect. 3.3.1, once the central temperature in a RGB star exceeds 100 MK, He in the core starts being converted into $^{12}$C via 3$\alpha$ reactions, and subsequently into $^{16}$O via $^{12}$C($\alpha, \gamma$) $^{16}$O reactions. The cross section of the 3$\alpha$ reaction has a tremendous dependence on the temperature: it scales roughly as $T^{23}$ in the range 100–300 MK, so that the energy produced by these reactions is very strongly concentrated towards the centre of the star where the temperature is at its maximum. The very large photon flux that forms in these conditions triggers the formation of large scale motions of the matter, which mix the material in the central part of the star (the convective core) in order to efficiently carry the energy outward. The mass of the convective core depends on the luminosity produced by the 3$\alpha$ reactions. This luminosity scales with the mass of the He core because the larger its mass, the larger is the amount of energy required to maintain the hydrostatic equilibrium (see Sect. 3.2). Hence, the size of the convective core scales directly with the mass of the He core. The mass of the He core, in turn, scales directly with the initial mass of the star, thus, in conclusion, the mass of the convective core scales directly with the initial mass of the star. Analogously to the Main Sequence central H-burning phase, the energy production in the core is dictated by the mass of the star (see Sect. 3.2.1). However, the role played by the total stellar mass in central H burning in the central He-burning phase is replaced by the He-core mass because the density contrast between the He core and the H-rich mantle is so large that the core does not feel the presence of the H rich mantle and evolves as if it was a naked He core.

In the meantime, the temperature at the He/H interface is high enough that also an efficient H-burning shell is active leading to continuous deposition of fresh He onto the He core. Moreover, large convective motions develop (in most cases) in the outer H-rich envelope. The actual extension and temporal variation of these convective regions depends on the initial mass and chemical composition of the star.

At variance with H burning, no radioactive nuclei are produced by the $3\alpha$ and the $^{12}C(\alpha,\gamma)\,^{16}O$ reactions because they convert matter along the valley of $\beta$ stability. Radioactivity during He burning is produced instead via the sequence of reactions that convert $^{14}N$ into $^{22}Ne$ via a double $\alpha$ capture and the radioactive decay of $^{18}F$: $^{14}N(\alpha,\gamma)\,^{18}F(e^+ + \nu)\,^{18}O(\alpha,\gamma)\,^{22}Ne$. In H-exhausted regions, $^{14}N$ is by far the most abundant nuclear species after He because a main effect of the CNO cycle, which operated in the previous H-burning phase (see Sect. 3.2.1) is to convert most of the initial C and O, the two most abundant elements beyond H and He, into N. Hence, during He burning $^{22}Ne$ becomes the most abundant isotope, after C and O, oncc $^{14}N$ is fully consumed by $\alpha$ captures.

When He is exhausted in the centre, He burning moves smoothly outward in mass leaving behind a CO core that begins to contract on a Kelvin-Helmholtz timescale. Similar to the H-burning shell, also the He-burning shell produces more energy than required to balance gravity because energy production is controlled by the size of the underlying core, and not by the mass of the star. The CO core increases progressively in mass because of the continuous deposition of CO-rich material made in the He-burning shell and the He-burning shell increases its energy production accordingly. As a consequence, the overlying He+H-rich mantle is forced to expand substantially and to cool down so much that the H-burning shell switches off. As during the RGB phase, this expansion progressively inhibits energy transport by radiation and large scale motions of the matter progressively extend inward from the outer envelope. In stars initially more massive than 4–5 $M_\odot$ the convective envelope penetrates even inside the He core (2nd dredge-up). The main consequences of this are a change of the surface abundances and a reduction of the He-core mass. Similar to what happens during the RGB, the formation of an extended convective envelope forces the star to expand at roughly constant surface temperature (because the onset of convective motions fixes a maximum value for the temperature gradient, see Sect. 3.2.1) and increasing luminosity. This phase is called Asymptotic Giant Branch (AGB). The specific phase when the He-burning shell advances in mass eroding the border of the He core from within is called Early Asymptotic Giant Branch (E-AGB).

The competition between the advancing He-burning shell and the sinking of the convective envelope during the 2nd dredge-up fixes the maximum mass that the CO core ($M_{CO}$) reaches in this phase. If $M_{CO}$ is larger than roughly 1.1 $M_\odot$, the core heats up to the C ignition temperature ($\sim 8 \times 10^8$ K), otherwise it turns into an electron degenerate CO core able to self-sustain against gravity without the need of additional contraction. The maximum initial stellar mass for which an electron degenerate CO core forms is of the order of 7–8 $M_\odot$, for solar metallicity stars. While the evolution of stars without an electron degenerate core is dictated by the self gravity of the core, the evolution of stars with an electron degenerate core is controlled by the burning shells. In the following we concentrate on the further evolution of the latter case, i.e., the AGB, while Chap. 4 describes the further evolution of the first case.

On the AGB three main regions may be identified: the electron degenerate CO core, a He-rich layer (also referred to as *intershell* since it is located between the He- and the H-burning shells), and a H-rich mantle, most of which forms an extended convective envelope. As the He-burning shell approaches the border of the He core, it quenches because of the steep temperature drop associated with the drastic reduction of the mean molecular weight caused by the change from a He-dominated to a H-dominated chemical composition. Being less and less supported by the extinguishing He burning shell, the mantle is forced to shrink, heat up, and progressively re-activate the H-burning shell at its base. The H-burning shell starts to deposit fresh He onto the He shell forcing the intershell to heat up again. At this point a fascinating evolutionary phase begins in which nuclear burning and instabilities coexist, realizing a unique *habitat* in which a large number of nuclear species may be synthesized: the Thermally Pulsing AGB (TP-AGB) phase.

Quite schematically, the TP-AGB phase consists of a sequence of cycles each of which may be divided in two main phases: a quiescent H-burning phase during which the He-burning shell is inactive, and a He-burning phase during which the H-burning shell is inactive. Though the two shells do not operate simultaneously, they process roughly the same amount of mass per cycle so that the intershell mass changes (shrinks) slowly in time. The transition from the active He-burning phase to the active H-burning phase occurs quiescently in the sense that the energy provided by the H-burning shell progressively replaces that provided by the dimming He-burning shell. Instead, the transition from the active H-burning phase to the active He-burning phase occurs in a traumatic way, which is responsible for the peculiar sequence of events that characterizes the TP-AGB phase.

The reason for such a traumatic He ignition is that the pileup of fresh He on top of an inert intershell leads to a T, $\rho$ profile in the intershell that is controlled by the compressional heating caused by the accretion of fresh He. This T, $\rho$ profile is quite different from the typical one determined by the presence of an active burning shell. The large amount of energy required to turn the T, $\rho$ profile from that determined by the accretion and that required by the steady He burning, coupled to the very steep dependence of the cross section of the $3\alpha$ nuclear reaction on the temperature, determines the growth of a thermal runaway (or *thermal pulse*, TP) in which a huge amount of energy is released over a very short timescale. This runaway comes to an end when enough energy has been deposited in the intershell to turn the T, $\rho$ profile into a profile suited for quiescent He burning. As an example, Fig. 3.7 shows as a gray line the typical Log(T), Log($\rho$) profile produced by the advancing H-burning shell just prior to $3\alpha$ ignition, while the black line shows the typical profile at the end of the thermal runaway during steady He burning. In this specific example roughly $\sim 10^{48}$ erg must be deposited in the intershell to perform the transition between the two configurations. This amount of energy is determined by the fact that the needed change of the $T$, $\rho$ structure in the intershell requires a reduction of the binding energy of the intershell.

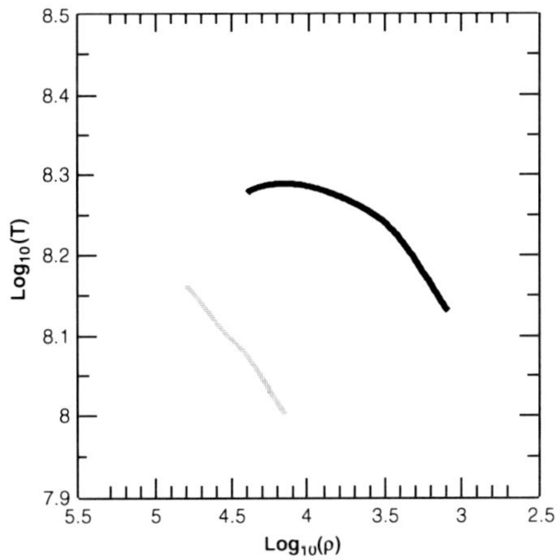


**Fig. 3.7** The *gray thick solid line* shows the typical Log(T) – Log($\rho$) profile in the intershell (in the range $10^{-3} < X_{He} < 0.9$) just prior the onset of a thermal pulse in a 3 $M_{\odot}$ of solar metallicity while the *black thick solid line* shows the typical profile in the same region at the end of the thermal runaway when the steady He burning occurs

The main effect of the rapid injection of energy into the intershell during the TP is the production of a very strong energy flux, which forces the growth of convective instabilities to efficiently carry the energy outward. This convective shell extends over most of the intershell region and plays a fundamental role in reshuffling the chemical composition within this region and hence influencing the detailed nucleosynthesis that occurs at this stage (see next sections). Once the TP comes to an end, the convective shell disappears and the quiescent He-burning shell phase begins. Another important side effect of the rapid energy injection caused by the TP is the expansion of the region above the He-burning shell, which forces a cooling of this region. The consequence is that the H-burning shell switches off, and the temperature gradient steepens. This favors the penetration of the convective envelope down into the intershell so that nuclei freshly synthetized in the deep interior of the star are efficiently brought up to the stellar surface (3rd dredge-up). Similar to what happens towards the end of the E-AGB phase, the He-burning shell progressively runs out of power as it approaches the border of the He-rich layer, where the temperature drops below the value necessary for the He burning. The overlying layers are forced to contract and heat so that a H-burning shell activates again and a new cycle starts.

To visually illustrate the sequence of events making up a full TP cycle and to make clear the pecularity of the TP-AGB phase, Fig. 3.8 shows the temporal evolution of the internal structure of a typical AGB star through three consecutive TPs.

The quantitative characteristics of the TPs depend on the core and envelope masses, the general rule being that larger CO core masses correspond to higher frequencies of thermal pulses, higher temperatures, and shorter lifetimes of the He convective shell. Typical TP frequencies (determined after the first 20 TPs or so) range between 2 and 3 TPs per $10^5$ year for a 3 $M_{\odot}$ star having $M_{CO}$ ~0.7 $M_{\odot}$ and

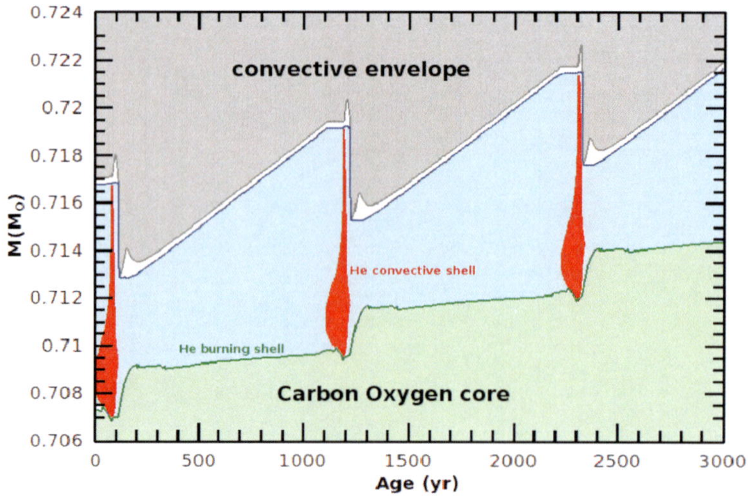


**Fig. 3.8** Temporal evolution of the internal structure of a typical AGB star through three consecutive TPs. The He convective shell is shown in red while the convective envelope is grey. The H and He burning shells are shown in *dark blue* and *dark green*, respectively, while the intershell is light cyan. The timescale between the end of the He convective shell episode and the beginning of the successive one has been rescaled in order to improve readibility. In particular it has been changed according to this formula: $t = t_{end\ conv\ sh} + (t - t_{end\ conv\ sh})/(\Delta t_{interpulse} \times 10^3)$. In this way the interpulse phase is rescaled to last $10^3$ yr

more than 50 TPs per $10^5$ year for a 6 $M_\odot$ star having $M_{CO}$ ~0.96 $M_\odot$, while the peak luminosity ranges between 1 and $10 \times 10^8\ L_\odot$.

Since the two burning shells process about the same amount of matter per cycle, the average growth rate of the CO core per cycle roughly equates that of the He core. This, coupled to the fact that He burning produces (per unit mass) roughly 10% of the energy produced by H burning, and that the luminosity of these stars does not change appreciably between the two burning phases, allows us to estimate the relative burning lifetimes ($t_{He}/t_H$). The amount of energy produced by He burning per cycle must balance the surface losses, i.e., $\varepsilon_{He} \times \Delta M_{He} = L_{surface} \times t_{He}$, where $\varepsilon_{He}$ represents the amount of energy liberated by the He burning per unit mass, $\Delta M_{He}$ the amount of mass processed by the He burning, $L_{surface}$ the luminosity of the star and $t_{He}$ the lifetime of the He burning phase. Similar for the H burning one may write that $\varepsilon_H \times \Delta M_H = L_{surface} \times t_H$. If the amount of mass processed is similar in the two cases (i.e., $\Delta M_{He} \simeq \Delta M_H$), the relative lifetimes scale roughly as the two nuclear burning rates, i.e., $t_{He}/t_H \sim 1/10$. The amount of He burnt during each He-burning episode is only partial, corresponding to about 25%–30% of the He present in the intershell. Of this, roughly 25% burns during the TP and the remainder during the quiescent He burning phase. The final nucleosynthetic result is that carbon is produced via the $3\alpha$ reaction, but it is only marginally converted into oxygen. The typical composition of the intershell after this partial He burning is represented by matter made by roughly 75% He and 23% $^{12}$C, while the remaining few percent are made up of $^{22}$Ne (from conversion of $^{14}$N as detailed above) and of $^{16}$O. The $^{22}$Ne nuclei are of interest as they act as a neutron source in the TPs when the temperature reaches 300 MK. The lifetime of the He convective shell varies between 100 year and 10 year for the 3 and 6 $M_\odot$ stellar models, respectively. Typically, the He burning shell is located between $7 \times 10^{-3}$ and $1.5 \times 10^{-2}$ $R_\odot$ from the center of the star, while the H-burning shell is located between $1 \times 10^{-2}$ and $2 \times 10^{-2}$ $R_\odot$. The intershell mass ranges roughly between $10^{-3}$ and $10^{-2}$ $M_\odot$. The surface radii of AGB stars vary between hundreds to thousands times the solar radius.

The final fate of AGB stars is to lose all their H-rich mantle before the electron degenerate core may grow to its most massive stable configuration (i.e., the Chandrasekhar mass). Such a destiny is due to the strong dependence of the mass-loss rate on the luminosity of the star and on its surface chemical composition (see Sect. 3.4.3). The maximum mass size reached by the CO core, which equates the mass of the newborn white dwarf, is determined by the competition between the speed at which the burning shells advance in mass and the efficiency of the mass loss that erodes the H-rich mantle from the surface.

The occurrence of the 3rd dredge-up significantly affects the evolutionary properties of an AGB star. First, it reduces the size of the He core anticipating the quenching of the quiescent He burning phase and hence its lifetime. Second, it slows down the overall growth rate of the CO core and the He-rich shell. Third, it carries to the stellar surface a fraction of the material freshly synthesized by partial He burning, i.e., C, $^{22}$Ne, and *slow* neutron-capture (*s* process) elements heavier than iron (see Sect. 3.5), drastically modifying the chemical composition of the star. In some cases, the star even changes from the usual oxygen-rich (O>C) composition to carbon-rich (C>O), with important consequences on the types of molecules and dust that can form and the ensuing mass loss (see Sect. 3.4.4). Unfortunately, the question of the maximum depth reached by the convective envelope during the 3rd dredge-up has always been highly debated and different results have been obtained over the years by different authors for AGB stars over the whole mass interval from 1.5 $M_\odot$ up to the more massive thermally pulsing stars. The reason is that, once the convective envelope enters the He core, a discontinuity in the opacity (H is much more opaque than He) determines the formation of a positive difference between the effective and adiabatic temperature gradients just at the border of the convective envelope. This is an unstable situation because the possible mixing of matter located just below the base of the convective envelope with the H-rich convective mantle is an irreversible process in the sense that these *additional* mixed layers become intrinsically convective because of the drastic increase of the opacity due to the mixing. It is therefore clear that even small different numerical techniques adopted by different authors may lead to quite different results.

Furthermore, the occurrence of the 3rd dredge-up is important because it creates a sharp discontinuity between the convective envelope and the radiative intershell.

Since a sharp discontinuity is not a realistic configuration in these conditions, the occurrence of the 3rd dredge-up allows the possibility that some kind of *diffusion* of protons occurs below the formal border of the convective envelope when it reaches its maximum inward extension at the end of the 3rd dredge-up smoothing out the discontinuity (though this is not obtained by applying the standard stability criteria for mixing). However, the shape, extent, and timescale over which the diffusion of protons in the He/C intershell may occur is unknown, its modeling is still artificial and not based on self-consistent computations.

This diffusion allows the formation of regions where a small amount of protons come in contact with matter that is predominantly composed of He and C, so that the ratio $Y(H)/Y(C) \ll 1$, but does not contain any $^{14}N$, since this nucleus has been fully converted into $^{22}Ne$ in the previous TP. When these proton-enriched layers begin to contract and to heat because of the quenching of the He burning, the CN cycle activates, but it can not go beyond the synthesis of $^{13}C$ due to the low proton concentration. As the temperature increases to roughly 90 MK, the $^{13}C(\alpha, n)$ $^{16}O$ reaction becomes efficient and a significant neutron flux is produced. Hence, this diffusion plays a pivotal role in the nucleosynthesis of species beyond the Fe peak via neutron captures. A detailed description of the properties of this neutron source and of its nucleosynthetic signature will be presented in Sect. 3.5.1. The lack of $^{14}N$ is crucial here, since this nucleus is a strong neutron poison and its presence would inhibit neutron captures by Fe and the elements heavier than Fe.

As already discussed in Sect. 3.3.2, typical temperatures at the base of the convective envelope do not exceed a few MK at most in the evolutionary phases prior to the AGB. Instead, another peculiarity of AGB stars is that during the H-burning phase the temperature at the base of the convective envelope may reach values in excess of several tens MK, and even exceed 100 MK, so that H-burning reactions activate within the convective envelope. The efficiency of this phenomenon, known as Hot Bottom Burning (HBB), scales directly with the temperature at the base of the envelope and hence with the CO-core mass, which in turn scales with the initial stellar mass. Hence, HBB is efficient in stars more massive than 4:5 $M_{\odot}$, depending on the metallicity. As the energy produced in the convective envelope sums to that produced by the H-burning shell, the core mass – luminosity relation changes (even strongly) in the presence of HBB. From a nucleosynthetic point of view the occurrence of an active H burning in a convective environment implies a redistribution of the processed material over all the convective zone, so the surface abundances turn towards the relative abundances typical of the H burning at high temperature. For example, an increase of the surface abundances of nuclei like $^{14}N$ and $^{26}Al$ (discussed in detail in Sect. 3.6.1), a temporaneous increase of $^{7}Li$, a reduction of $^{12}C$ and of the $^{12}C/^{13}C$ ratio and the other signatures of the CNO, NeNa and MgAl sequences.

We refer the reader to the review paper by Herwig (2005) and to the book chapter on the evolution of AGB stars by Lattanzio and Wood (2004) for a thorough presentation of the evolution of these cool giant stars.

### 3.4.2 Super-AGB Stars

In the previous section we identified stars that go through the double shell burning of the TP-AGB phase as those that develop an electron degenerate CO core where carbon burning fails to occur. There is, however, another class of stars that experience the double shell burning phase: those with initial total mass between the maximum mass that forms an electron degenerate CO core where C does not ignite ($M_{up}$) and the minimum mass that does not form an electron degenerate CO core ($M_{mas}$). Stars more massive than $M_{mas}$ evolve up to the final core collapse as described in Chap. 4. Depending on the initial chemical composition and the adopted physics, $M_{up}$ ranges between 6–8 $M_{\odot}$, and $M_{mas}$ ranges between 9–12 $M_{\odot}$. It is important at this point to recall that these limiting masses are somewhat uncertain because they depend on the size of the convective core, the carbon to oxygen ratio left by the He burning, the efficiency of the 2nd dredge-up and the cross section of the $^{12}C + ^{12}C$ nuclear reaction. Unfortunately, all these quantities are still subject to severe uncertainties.

Stars falling between these two limits form a partially electron degenerate core, but are massive enough to ignite C in the core, lift the degeneracy, and go through the C burning in the core. They are not massive enough, however, to avoid the electron degeneracy of the ONeMg core left by C burning. The evolution of stars in this relatively small mass interval, called Super-AGB stars, has not been studied extensively up to now because of the difficulty in computing the C-burning phase due to the removal of the degeneracy that occurs through a series of successive flashes, and the lack of massive computer power, which is needed to study this complex situation. This situation is rapidly changing and progress is currently under way on the computational modeling of Super-AGB stars (Siess, 2006, 2007).

Since these stars form an electron degenerate core after core C burning, they also go through a double shell burning phase similar to the AGB phase experienced by their less massive counterparts. Because their degenerate cores are more massive, following the trend shown by AGB stars, the frequency of the thermal pulses is higher (up to 500 TPs per $10^5$ year), the He peak luminosity is lower than in the normal AGB stars (up to $\sim 4 \times 10^6 L_{\odot}$), while the base of the convective envelope may reach temperatures as high as 110 MK, hence, H burning occurs within the convective envelope (Hot Bottom Burning). Similar to what happens in the more massive AGB stars, but quantitatively more pronounced, the luminosity produced in the convective envelope adds to that produced by the radiative H burning significantly altering the core mass - luminosity relation and the surface composition is modified by the signature of H burning. The possible occurrence of the 3rd dredge-up would also shuffle the surface chemical composition with the typical products of the partial He burning, i.e. C, and *s*-process elements. The efficiency of the 3rd dredge-up is very uncertain also for these stars. In principle, one could expect a lower efficiency of the 3rd dredge-up because the amount of energy released by a TP is lower and the overall temperature is much higher than in a standard AGB star, so that it could be more difficult to expand the base of the mantle and to steepen the temperature gradient up to a value that would allow the convective envelope to penetrate the He core. Quantitative estimates of the yields of the nuclei specifically produced by the TP Super-AGB stars are in progress (Doherty and Lattanzio, 2006).

The final fate of a Super-AGB star depends on the competition between the advancing burning shells, which increase the size of the ONeMg core, and the mass loss, which limits its growth. Also an efficent 3rd dredge-up would contribute to limiting the growth of the core. Stars more massive than a critical value reach the threshold electron degenerate core mass for the onset of electron captures on $^{24}$Mg and $^{20}$Ne after a certain number of TPs and eventually explode as *electron-capture supernovae*. Stars less massive than the critical value, instead, end their life as ONeMg white dwarfs. An estimate of the electron degenerate core mass above which electron captures become efficient in an ONeMg environment can be determined by considering that the threshold energy for electron capture is 6 MeV for $^{24}$Mg and 8 MeV for $^{20}$Ne and that the mass of a fully electron degenerate core having a Fermi energy of the order of 6 MeV is $\simeq 1.35\,M_{\odot}$. Thus, if the electron degenerate core grows to the threshold value of $\simeq 1.35\ M_{\odot}$ or so, electron captures are activated on $^{20}$Ne and $^{24}$Mg.

This process removes electrons and hence pressure from the center of the star, starting a runaway process that leads to the core collapse and final explosion as electron capture supernova. The explosion of these electron capture supernovae is similar to that of core collapse supernovae (see Chap. 4), with a few distinct features. During the initial collapse of the degenerate core, electron captures increase significantly the degree of neutronization of the matter, i.e., raise the global neutron over proton ratio because of the capture of the electrons by the protons. The nuclear species produced by explosive burning depend significantly on the neutron over proton ratio so that the higher the degree of neutronization of the matter the higher the production of neutron-rich nuclei: in particular $^{58}$Ni becomes favored with respect to $^{56}$Ni. Since the luminosity peak of a supernova correlates with the amount of $^{56}$Ni produced during the explosion, a natural feature of these electron captures supernovae is a lower luminosity with respect to typical core collapse supernovae. Also, the final kinetic energy of the ejecta is expected to be of the order of $0.1 \times 10^{51}$ erg, roughly one order of magnitude lower than in typical core collapse supernovae (see, e.g., Hoffman et al., 2008; Wanajo et al., 2009).

### *3.4.3 Winds from AGB Stars*

An observed peculiarity of AGB stars is that they show strong stellar winds, which carry material away from the surface of the star into its surroundings. Nuclei newly synthetised during the AGB phase and carried to the stellar surface by the 3rd dredge-up are shed into the interstellar medium so that AGB stars contribute to the chemical make-up of their environments and of new generations of stars. The mass loss rate due to winds in AGB star increases as the star evolves along the AGB and can reach values as high as $10^{-4}$ $M_{\odot}$/yr (to be compared, for example, to the solar mass loss rate of $10^{-11}$ $M_{\odot}$/yr) at the end of the AGB, which is known as the *superwind* (Iben and Renzini, 1983). This is a strong and dense but slow wind, with material leaving the star at relatively low speeds of 5–30 km/s.

The winds are caused by two main factors. First, large quantity of dust form around AGB stars and radiation pressure acting on this dust contributes to driving the winds. The extended envelopes of red giant and AGB stars, where the temperature drops down to ~1,000 K, are an ideal location for the formation of a large variety of molecules like CO, TiO, VO, as well as ZrO, when the gas has been enriched in heavy elements such as Zr by the *s* process and the 3rd dredge-up, and $C_2$, CN, and CH, when the gas has been enriched in carbon by the 3rd dredge-up. In the case of refractory elements, which have the property of condensing at high temperatures directly from gas into the solid state, the gas condenses into tiny particles, which then can grow into dust grains. Because of the large quantity of dust around them, AGB stars become obscured toward the end of their life and can only be seen as mid-infrared sources, since the dust absorbes the energy of the visual light coming from the star and reemits it as infrared light. Second, AGB stars are variable stars, meaning that their luminosity varies with time with changes occurring over relatively long periods >100 days. These luminosity variations are due to stellar pulsations, in the sense that the whole star expands and contracts. Pulsations produce changes in the stellar radius and temperature, which cause the variations in the stellar luminosity. When the pulsations attain a large amplitude they lead to strong stellar winds and a large mass-loss rate. Pulsation levitates matter above the photosphere and increases the wind density by about two orders of magnitude (Wood, 1979; Sedlmayr and Dominik, 1995; Dorfi et al., 2001).

The strong stellar winds driven by the combined effects of radiation pressure acting on dust and pulsation eventually erode the whole stellar envelope (Dupree, 1986; Willson, 2000). Hence, the winds govern the lifetime of AGB stars because when the envelope is almost completely lost the star moves away from the AGB phase into the hotter post-AGB phase. Toward the end of the post-AGB phase, the shell of material ejected by the AGB star may become illuminated by the radiation coming from the central star, and produce a planetary nebula. The former AGB stars is now referred to as a *planetary nebula nucleus* and finally turns into a cooling CO white dwarf.

### *3.4.4 Dust from Giant Stars and the Origin of Stardust*

The specific dust species that form in the atmosphere of AGB stars depends mainly on the C/O ratio. The difference in the type of dust that can form in a carbon-rich or oxygen-rich gas is due to the strong bond of the CO molecules: if O>C, all carbon atoms are locked into CO and only oxygen-rich dust can form, viceversa, if C>O, all oxygen atoms are locked into CO and only carbon-rich dust can form.[4]

[4] This general rule is debated in the case of dust formation in supernova ejecta, see Chap. 2, Sect. 2.2.5.3.

In an oxygen-rich gas (O>C) dust species are, for example, $Al_3O_2$ (corundum), $CaAl_{12}O_{19}$ (hibonite), $MgAl_2O_4$ (spinel), as well as many different types of silicates (SiO, $SiO_2$, etc.). In a carbon-rich gas (C>O), dust species are, for example, SiC (silicon carbide), TiC (titanium carbide), and C itself (graphite).

Formation of dust around AGB stars is well documented by spectroscopic observations in the infrared (e.g. Treffers and Cohen, 1974; Speck et al., 2000, 2009) and predicted to occur by theoretical models (e.g. Fleischer et al., 1992; Lodders and Fegley, 1995; Gail and Sedlmayr, 1999; Ferrarotti and Gail, 2002). It is now widely accepted that AGB stars are the most prolific source of dust in the Galaxy. When summing up the contribution of the different families of late red giant and AGB stars: i.e., spectroscopically, the M stars, the OH/IR stars,[5] and the carbon stars, it results that ~90% of all dust of stellar origin in the interstellar medium came from these sources (Whittet, 1992).

Thus, it is not surprising that the vast majority of stardust grains extracted from meteorites (Sect. 2.2.4. and Sect. 10.2) show the signature of an origin in AGB stars (Table 3.5. The main signatures of AGB nucleosynthesis imprinted in meteoritic stardust grains are: (1) the O isotopic composition of the majority of oxide and silicate grains showing excess in $^{17}O$ and deficits in $^{18}O$, and known as Population I and II of stardust oxide grains (Nittler et al., 1997), which match the O isotopic ratios observed around AGB stars via spectroscopic observations of CO molecular lines (e.g. Harris et al., 1987), and (2) the distribution of the $^{12}C/^{13}C$ ratios of >90% of SiC grains showing a peak between 50 and 60 (solar value is 89) and known as the *mainstream* SiC population, which match the distribution derived from spectroscopic observation of CO molecular lines in C-rich AGB stars (see Fig. 3 of Hoppe and Ott, 1997). The Ne composition measured in stardust SiC

**Table 3.5** Meteoritic stardust grain types, populations, and origins

| Type | Population | Origin |
|---|---|---|
| oxide and silicate grains | I | AGB stars |
| | II | AGB stars |
| | III | undetermined |
| | IV | supernovae |
| silicon carbide (SiC) | mainstream | AGB stars |
| | Y | AGB stars |
| | Z | AGB stars |
| | X | supernovae |
| | A+B | undetermined |
| | nova grains | novae |
| silicon nitride | | supernovae |
| graphite | low-density | supernovae |
| | high-density | undetermined |
| diamond | | undetermined |

Note: Table 5.2 gives complementary information to this table

[5] OH/IR stars are cool red giants with strong hydroxyl (OH) masers and infrared (IR) emissions.

grains - corresponding to the Ne-E(H) component rich in $^{22}$Ne is also a clear signature of material from the intershell of AGB stars, where $^{22}$Ne is abundant. Moreover, the elemental and isotopic abundances of the heavy elements Kr, Sr, Zr, Ru, Xe (the Xe—S component), Ba, Nd, Sm, W, and Pb present in trace amount and measured in SiC grains clearly show the imprint of the *s* process, which make inevitable their connection to AGB stars. Smaller SiC Populations Y and Z ($\simeq$1% each of the total recovered stardust SiC grains) are also attributed to AGB stars, but of metallicity down to 1/3–1/5 of the solar value (Hoppe et al., 1997; Amari et al., 2001b; Zinner et al., 2006).

With regards to the remaining types and populations of stardust grains, core-collapse supernovae have been invoked as the origin site of Population X of SiC grains ($\sim$1%) and the few recovered silicon nitride grains (Nittler et al., 1995), showing excesses in $^{28}$Si and evidence of the early presence of $^{44}$Ti, as well as low-density graphite grains and Population IV of oxide and silicate grains (with excess in $^{18}$O and $^{18}$Si Vollmer et al., 2008; Travaglio et al., 1999, see Chap. 4). Novae are invoked for a few SiC grains of unusal composition (excesses in $^{13}$C and $^{15}$N Amari et al., 2001a, see Chap. 5 Sect. 5.2), while the origin of SiC grains of Populations A+B ($\simeq$5% of all SiC grains, showing $^{13}$C/$^{12}$C $<$ 10) is still unclear (Amari et al., 2001c). Oxide and silicate grains with deficits in both $^{17}$O and $^{18}$O, known as Population III, have been attributed to stars of metallicity lower than solar, however, the Si isotopic composition of the silicate grains belonging to this population is very close to solar, which does not support this interpretation. The origin of this population remains to be ascertained, together with the origin of high-density graphite grains and of the very abundant and extremely tiny ($10^{-9}$ m) meteoritic diamond grains, the majority of which probably formed in the solar system. For more details in meteoritic stardust see, e.g., Clayton and Nittler (2004) and Lugaro (2005).

Given compelling evidence that most stardust came from AGB stars, the composition of these grains can be used as a stringent constraint for theoretical models of AGB stars and, viceversa, the models can be used to identify the mass and metallicity range of the parent stars of the grains. Data from the laboratory analysis of stardust are usually provided with high precision, down to a few percent errors, and for isotopic ratios. In comparison, data from spectroscopic observations of stellar atmospheres usually are provided with lower precision, errors typically >50%, and mostly for elemental abundances. Thus, the information from stardust grains represents a breakthrough in the study of AGB nucleosynthesis. Also, given that the abundances and isotopic compositions of elements heavier than Al and lighter than Fe, such as Si and Ti, are mostly unaltered by AGB nucleosynthesis, laboratory analysis of these elements in AGB stardust can be used to constrain in great detail the initial composition of the parent star of the grains, and in turn the chemical evolution of the Galaxy (e.g. Zinner et al., 2006).

Meteoritic stardust provides us with abundant and precise information on radioactive nuclei in stars because the initial abundance of radioactive nuclei at the time of the formation of the grains is recorded by the signature of their radioactive decay inside the grains, which is easily derived from measurements of the excesses in the abundances of their daughter nuclei. An important example is that of $^{26}$Al, where the initial $^{26}$Al abundance in a stardust grain is revealed by excesses in $^{26}$Mg. This will be discussed in detail in Sect. 3.6.1. In general, radioactive signatures in stardust have the potential to be used as clocks for the timescale of dust formation around stars and supernovae (Hoppe and Besmehn, 2002). Finally, stardust isotopic data provide a unique way to investigate the operation of the *s* process in AGB stars, as will be discussed in Sect. 3.5.5.

## 3.5 Neutron-Capture Nucleosynthesis in AGB Stars

In this section we show that:

- Free neutrons are produced in the TP-AGB phase by the $^{22}$Ne$(\alpha, n)^{25}$Mg reaction, which activates at ~300 MK and operates during He burning in the intershell convective region during thermal pulses, and the $^{13}$C$(\alpha, n)^{16}$O reaction, which activates at ~90 MK and operates in a radiative (and hence stable) region of the intershell during the H-burning phases. The free neutrons trigger the *s* process, which produces half of the cosmic abundances of the elements heavier than iron via neutron captures mostly occurring on stable and long-lived radioactive nuclei.
- Unstable isotopes with half lives higher than a few days can also suffer neutron captures during the *s* process, producing a wide variety of *branching points* on the *s*-process path, which define the details of the abundance distribution produced by the *s* process as a function of neutron density and temperature.
- The overall *s*-process abundance distribution is defined by stable nuclei with a magic number of neutrons at the three *s*-process peaks at Sr, Ba, and Pb, and by the total amount of free neutrons available.
- Several long-lived unstable isotopes are produced by the *s* process (details in Sect. 3.6.5). Among them are $^{93}$Zr and $^{99}$Tc. Observations of monoisotopic stable Nb (the daughter nucleus of $^{93}$Zr) and of Tc itself can be used as discriminant between intrinsic (on the AGB) and extrinsic (with a former AGB binary companion) *s*-process-enhanced stars.

### *3.5.1 Neutron Sources in AGB Stars*

In the double burning shell phase a nuclear reaction that may produce a copious neutron flux is $^{22}$Ne$(\alpha$,n$)^{25}$Mg. $^{22}$Ne is abundantly present in the intershell because it directly derives from the initial abundance of O (the most abundant nucleus after H and He) as a consequence of the operation of the CNO cycle first and of a double $\alpha$ capture on $^{14}$N later. This means that this neutron production channel is of *secondary* origin, i.e., its efficiency scales with the initial metallicity of the star. The relatively high Coulomb barrier of Ne (Z = 10) pushes the threshold temperature for $\alpha$ capture above 300 MK so that this process can activate only within

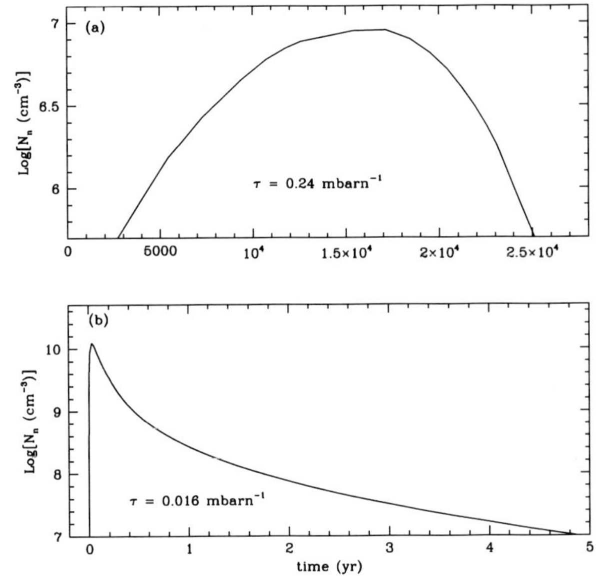


**Fig. 3.9** Neutron densities as functions of time corresponding to the activation of the two neutron sources in a 3 $M_\odot$ AGB star model of solar metallicity during the last interpulse-pulse cycle: (**a**) the $^{13}$C neutron source (the zero point in time represent the time from the start of the interpulse period, about 10,000 year, when the temperature reaches 79 MK); (**b**) the $^{22}$Ne neutron source (the zero in time corresponds to the time when the temperature in the TP reaches 250 MK)

a hot He-burning region. Since the temperature at the base of the He convective shell during thermal pulses scales directly with the mass of the H-exhausted core, only stars initially more massive than $\simeq$3 $M_\odot$ (Iben, 1975; Iben and Truran, 1978) can efficiently activate this nuclear reaction. Panel (b) in Fig. 3.9, shows a typical profile of the neutron density versus time associated with this neutron source. Its shape reflects the sharp rise of the temperature caused by the growth of the thermal instability and the following quite rapid decline due to the quenching of the instability. The high activation temperature and its very short duration (a few years) lead to a very high initial neutron density (reaching up to $N_n \simeq 10^{14}$ n/cm$^3$ in AGB stars of initial mass ~6 $M_\odot$) but to a small total amount of $^{22}$Ne burnt per cycle, so that the total number of neutrons released, i.e., the time-integrated neutron flux, or *neutron exposure* $\tau = \int_0^t N_n v_{th} dt$, remains quite small, of the order of a few hundredth of 1/mbarn (see Sect. 3.5.2). The signature of such an impulsive neutron flux on neutron-capture nucleosynthesis will be discussed in the next section. We only remark here an important difference between the neutron-capture nucleosynthesis occurring during AGB thermal pulses and that occurring in the He-convective shell of a massive star (other than the fact that in the AGB case the exposure to neutrons occurs recurrently): the mass of the He-convective shell in AGB stars is orders of magnitude smaller than that of a massive star so that the smaller dilution induced by the mixing allows, in the former case, many unstable nuclei to reach a higher equilibrium concentration. This occurrence favors the synthesis of stable nuclei on the neutron-rich side of the valley of $\beta$ stability.

The problem with the $^{22}$Ne neutron source is that AGB stars observed to be enriched in $s$-process elements have been identified as AGB stars of masses lower than $\sim$3 M$_\odot$ because (a) their relatively low luminosities (Frogel et al., 1990) match those of low-mass AGB models; (b) their surface is generally C enriched, an occurrence that rules out a significant HBB and hence an initial mass greater than 3 M$_\odot$; (c) excesses of $^{25}$Mg, predicted to be produced by $^{22}$Ne($\alpha$, $n$)$^{25}$Mg, and of $^{26}$Mg, predicted to be produced by the twin channel $^{22}$Ne($\alpha$, $\gamma$)$^{26}$Mg, with respect to $^{24}$Mg are not observed (Smith and Lambert, 1986; McWilliam and Lambert, 1988); (d) the high neutron density produced by the $^{22}$Ne channel, see Panel (b) in Fig. 3.9, would favor the synthesis of neutron-rich nuclei like $^{96}$Zr and elements as Rb, at odds with spectroscopic observations (Lambert et al., 1995; Abia et al., 2001) and the solar abundance distribution (Despain, 1980). Thus, for the vast majority of $s$-enhanced AGB stars, another nuclear fuel for the production of neutrons has to be invoked.

Nuclei of $^{13}$C are the best candidate for this role, given that the $^{13}$C($\alpha$, $n$)$^{16}$O reaction activates at temperatures from approximately 90 MK, which are easily reached in low-mass AGB stars. The achievement of the threshold temperature is, however, a necessary but not sufficient condition for a nuclear reaction to be effective: an additional requirement is the presence of a sufficient amount of reactants, in this case $^{13}$C. Models in which no mixing is allowed in the layers in radiative equilibrium do not naturally produce a significant concentration of $^{13}$C in the intershell region (see end of Sect. 3.4.1). In fact, the $^{13}$C available in the H-exhausted zone is that corresponding to the equilibrium value provided by the CNO cycle. As a neutron source for the $s$ process, this $^{13}$C suffers two major problems: its abundance is too low to power a significant neutron flux and its ratio with respect to $^{14}$N is too low ($^{13}$C/$^{14}$N$<<$ 1). The $^{14}$N($n$, $p$)$^{14}$C reaction[6] has a relatively high neutron-capture cross section of $\simeq$2 mbarn, with respect to typical cross section of the order of 0.1–0.01 mbarn for the light nuclei. Hence, it is a formidable poison that can even completely inhibit the $s$ process. Hence, the $^{13}$C neutron source represents a valid alternative to the $^{22}$Ne neutron source only if additional $^{13}$C is produced in an environment depleted in $^{14}$N. A way out of this problem is to assume that at the end of each 3rd dredge-up episode a small amount of protons penetrates the intershell region. The amount of protons engulfed in the He/C rich intershell must be small ($Y_p/Y_{^{12}C} << 1$) because they must allow the conversion of $^{12}$C into $^{13}$C, but not the conversion of $^{13}$C in $^{14}$N. (Note that the intershell is essentially free of $^{14}$N at the end of a thermal pulse because $^{14}$N nuclei have all been destroyed by $\alpha$ captures.) Once a small amount of protons has penetrated the intershell, the progressive heating caused by the deposition of fresh He synthetized by the H-burning shell induces the conversion of $^{12}$C in $^{13}$C. We can estimate the concentration of

[6] This reaction produces $^{14}$C, a radioactive nucleus with a half life of 5730 year. This nucleus is not carried to the stellar surface by the 3rd dredge-up because it is destroyed by $^{14}$C($\alpha$, $n$)$^{18}$O reactions during He-burning in the thermal pulse.

protons that allows the build up of $^{13}$C, but not of $^{14}$N, by considering that the production rate of $^{14}$N equates that of $^{13}$C when the concentration of $^{13}$C rises to a value of the order of 1/4 of that of $^{12}$C. Since the mass fraction of $^{12}$C in the intershell is about 0.2, the two rates equate each other for a $^{13}$C concentration of $X_{^{13}C} \simeq 0.20\ (13/12)/4 = 5 \times 10^{-2}$. If one requires the $^{13}$C production rate to dominate that of $^{14}$N, the $^{13}$C concentration must be reduced by at least a factor of 10, so that $X_{^{13}C} \simeq 5 \times 10^{-3}$. This abundance of $^{13}$C corresponds to a proton concentration of the order of $X_p = X_{^{13}C}/13 = 4 \times 10^{-4}$.

A self-consistent scenario able to produce this small amount of protons penetrating below the base of the convective envelope has not been found yet: several mechanisms have been proposed (e.g. Iben and Renzini, 1982; Herwig et al., 1997; Langer et al., 1999; Denissenkov and Tout, 2003) but none of them can presently be considered as widely accepted. A discussion of these alternative scenarios goes well beyond the purposes of the present discussion. What matters, and what modelers often pragmatically assume, is that a small amount of protons definitely penetrates in the intershell at the end of 3rd dredge-up. The detailed features of the $^{13}$C *pocket* obtained with such a procedure are subject to large uncertainties.

Nonetheless the basic properties of the neutron flux that is obtained in this way are considered relatively well understood (Gallino et al., 1998; Goriely and Mowlavi, 2000; Lugaro et al., 2003b). The activation of the $^{13}C(\alpha, n)^{16}O$ occurs well before the onset of the next thermal pulse and the *s*-process nucleosynthesis triggered by this neutron source occurs at low temperature in a radiative environment (see Sect. 3.5.2). Panel (a) in Fig. 3.9 shows the temporal evolution of this neutron flux. The rather long timescale over which this neutron flux remains active is determined by the speed at which the H-burning shell accretes matter on the He core, which means a typical timescale of the order of $10^4$ year. Given such a long timescale, $^{13}$C is totally consumed so that the total number of neutrons released is very large, with neutron exposures of the order of a tenth to a few mbarn$^{-1}$. The neutron density, instead, keeps to low values, up to $N_n \simeq 10^8$ n/cm$^3$. Let us finally remark that the neutron flux produced by the $^{13}$C neutron source is of *primary* origin, i.e., independent on the initial stellar metallicity, since the $^{13}$C is made from $^{12}$C synthetized starting from the initial H and He.

### 3.5.2 The s Process in AGB Stars

A fraction of the free neutrons produced in AGB stars by the $^{13}$C and $^{22}$Ne neutron sources described above is captured by Fe seed nuclei, leading to production of elements with large atomic mass numbers up to Pb (A = 208) and Bi (A = 209) via the *s* process. In general, a neutron flux that irradiates the surrounding matter reproduces a situation analogous to that occurring during H burning, where matter is irradiated by a flux of protons. While during a proton flux matter is pushed out of the valley of $\beta$ stability toward the proton-rich side, during a neutron flux matter is pushed out of the valley of $\beta$ stability valley toward the neutron-rich side. Thus, the presence of a neutron flux is inevitably associated to the synthesis of radioactive nuclei that, sooner or later, decay back towards the valley of $\beta$ stability.

During the *s* process, by definition, the timescale against $\beta$ decay of an unstable isotope is shorter that its timescale against neutron captures. Thus, neutron captures occur only along the valley of $\beta$ stability (Fig. 3.10). For this condition to hold neutron densities must be of the order of $N_n \sim 10^7$ n/cm$^3$. By comparison, during the *rapid* neutron-capture process (*r* process), neutron densities reach values as high as $10^{25}$ n/cm$^3$ so that neutron captures occur on a time scale less than a second (typically much shorter than that of radioactive decays) pushing matter towards very neutron-rich material. When the neutron flux is extinguished, the neutron-rich radioactive nuclei quickly decays back towards their stable isobars on the valley of $\beta$ stability. As presented in Chap. 4, the *r* process is believed to occur in explosive conditions in supernovae.

In stellar conditions, though, neutron densities during the *s* process can reach values orders of magnitude higher than $10^7$ n/cm$^3$. Depending on the peak neutron density, as well as on the temperature and density, which can affect $\beta$-decay rates, conditions may occur for the neutron-capture reaction rate of an unstable isotope to compete with its decay rate. These unstable isotopes are known as *branching points* on the *s*-process path. To calculate the fraction of the *s*-process flux branching off the main *s*-process path at a given branching point a *branching factor* is defined as:

$$f_{\text{branch}} = \frac{p_{\text{branch}}}{p_{\text{branch}} + p_{\text{main}}},$$

where $p_{\text{branch}}$ and $p_{\text{main}}$ are the probabilities per unit time associated to the nuclear reactions suffered by the branching point nucleus and leading onto the branch or onto the main path of the *s* process, respectively.

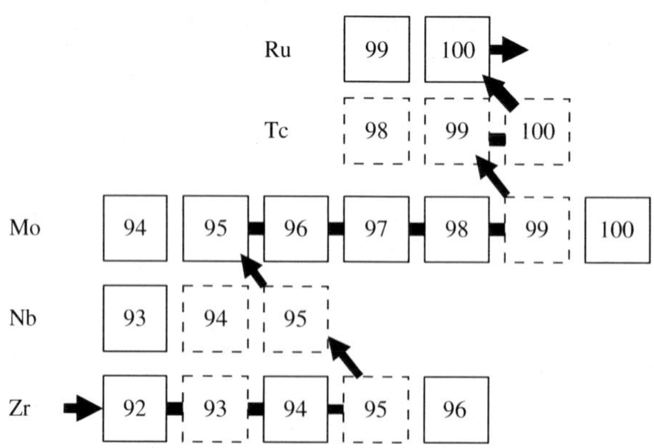


**Fig. 3.10** The main *s*-process path along the valley of $\beta$ stability from Zr to Ru is indicated by the *thick solid line* and *arrows*. *Solid* and *dashed boxes* represent stable and unstable nuclei, respectively. The radioactive nuclei $^{93}$Zr and $^{99}$Tc behave as stable during the *s* process as their half lifes (of the order of $10^5$–$10^6$ year) are longer than the timescale of the *s* process

There are several types of branching points: in the *classical* case $p_{\rm main}$ corresponds to $\lambda$, i.e., the probability per unit time of the unstable isotope to decay, and $p_{\rm branch}$ corresponds to $p_n$, i.e., the probability per unit time of the unstable isotope to capture a neutron $< \sigma v > N_n$, where $N_n$ is the neutron density and $< \sigma v >$ is the Maxwellian averaged product of the velocity $v$ and the neutron-capture cross section $\sigma$.[7] A typical example of this case is the isotope $^{95}$Zr in Fig. 3.10, which has a half life of 64 days, and can capture neutrons and produce the neutron-rich isotope $^{96}$Zr, classically a product of the *r* process, even during the *s* process. When the branching point is a long-lived, or even stable isotope, but its $\beta$-decay rate increases with temperature, the opposite applies: $p_{\rm branch} = \lambda$ and $p_{\rm main} = p_n$. In even more complex situations, a radioactive isotope may suffer both $\beta^+$ and $\beta^-$ decays, as well as neutron captures. In this case, three terms must be considered at denominators in the definition of the branching factor above: $p_n$, and $\lambda$ for both $\beta^+$ and $\beta^-$ decays.

Branching points have been fundamental in our understanding of the *s*-process conditions in AGB stars and will be discussed in more detail in Sect. 3.5.3. The low neutron density associated with the $^{13}$C neutron source does not typically allow the opening of branching points. On the other hand, the high neutron density associated with the $^{22}$Ne neutron source activate the operation of branching points on the *s*-process path, defining the details of the final abundance distribution.

It is possible to identify nuclei that can be produced only by the *s* process (*s-only* nuclei), which are shielded from *r*-process production by a stable isobar, or only by the *r* process (*r-only* nuclei), which are not reached by neutron captures during the *s* process as isotopes of the same element and same atomic mass number A − 1 are unstable. Examples of *s*-only nuclei are $^{96}$Mo and $^{100}$Ru shown in Fig. 3.10, which are shielded by the *r*-only nuclei $^{96}$Zr and $^{100}$Mo, respectively. These, in turn, are not typically produced by the *s* process as $^{95}$Zr and $^{99}$Mo are unstable. Proton-rich nuclei which cannot be reached by either the *s* or the *r* process must be produced via the *p process*, i.e., proton captures or photodisintegration of heavier nuclei, and are labelled as *p-only* nuclei (e.g., $^{94}$Mo in Fig. 3.10).

Models for the *s* process have historically been tested against the solar system abundances of the *s*-only isotopes, as these were the first precise available constraints. Once a satisfactory fit is found to these abundances, the selected theoretical distribution can be used to determine the contribution from the *s* process to each element and isotope. By subtracting this contribution to the total solar system abundance, an *r*-process contribution to each element can be obtained [8]

---

[7] Note that $\sigma$ is usually given in mbarn, corresponding to $10^{-27}$ cm$^2$, and that $< \sigma v >$ can be approximated to $\sigma \times v_{\rm thermal}$, where $v_{\rm thermal}$ is the thermal velocity. Neutron-capture cross sections for $(n, \gamma)$ reactions throughout this chapter are given at a temperature of 350 million degrees, corresponding to an energy of 30 keV, at which these rates are traditionally given. Values reported are from the Kadonis database (Karlsruhe Astrophysical Database of Nucleosynthesis in Stars, http://www.kadonis.org/) and the JINA reaclib database (http://groups.nscl.msu.edu/jina/reaclib/db/index.php), unless stated otherwise.

(e.g. Kaeppeler et al., 1982; Arlandini et al., 1999), which has been widely used to test $r$-process models, and to compare to spectroscopic observations of stars showing the signature of the $r$ process (Sneden et al., 2008). For example, it is found that $\simeq$80% of the solar abundance of Ba is due to the $s$ process, which is then classified as a typical $s$-process element, while $\simeq$5% of the solar abundance of Eu is due to the $s$ process, which is then classified as a typical $r$-process element.

Already $B^2FH$ had attributed to the operation of the $s$ process the three peaks in the solar abundance distribution at magic numbers of neutrons N = 50, the Sr, Y, and Zr peak, N = 82, the Ba and La peak, and N = 126, the Pb peak. This is because nuclei with a magic number of neutrons behave with respect to neutron-capture reactions in a similar way as atoms of noble gases do with respect to chemical reactions. Their energy levels, or shells, are fully populated by neutrons, in the case of magic nuclei, or by electrons, in the case of noble gases, and hence they are very stable and have a very low probability of capturing another neutron, in the case of magic nuclei, or of sharing electrons with another atom, in the case of noble gases. Nuclei with magic numbers of neutrons have small neutron-capture cross sections (of the order of a few to a few tens mbarn) with respect to other heavy nuclei, and they act as bottlenecks along the $s$-process path, leading to the observed abundance peaks. Nuclei located between the peaks, instead, have much higher neutron-capture cross sections (of the order of a few hundred to a few thousand mbarn). The neutron-capture chain in these local regions in-between magic nuclei quickly reaches equilibrium during the $s$ process. During a neutron-capture process the abundance $N_A$ of a stable isotope with atomic mass $A$ varies with time as:

$$\begin{aligned}\frac{dN_A}{dt} &= \text{production term} - \text{destruction term}\\ &= N_{A-1}N_n\sigma_{A-1} \times v_{\text{thermal}} - N_A N_n \sigma_A \times v_{\text{thermal}}.\end{aligned}$$

When replacing time with the neutron exposure $\tau$ one has:

$$\frac{dN_A}{d\tau} = N_{A-1}\sigma_{A-1} - N_A\sigma_A,$$

which, in steady-state conditions $\frac{dN_A}{d\tau} \to 0$ reached in between neutron magic nuclei, yields the simple rule to derive relative $s$-process abundances away from neutron magic numbers[9]:

$$N_A\sigma_A \simeq \text{constant}.$$

[8] The $p$-process contribution to elemental abundances is comparatively very small, $\simeq$1%, except in the case of Mo and Ru, which have magic and close-to-magic $p$-only isotopes, where it is up to $\simeq$25% and $\simeq$7%, respectively.

[9] For a detailed analytical description of the $s$ process refer to Chap. 7 of Clayton (1968).

It follows that the relative abundances of nuclei in-between the peaks are only constrained by their neutron-capture cross sections and do not provide information on the $s$-process neutron exposure. On the other hand, the relative abundances of the elements belonging to the three different peaks almost uniquely constrain the $s$-process neutron exposure. This is the reason behind the introduction and wide usage, both theoretically and observationally, of the $s$-process labels *light s* (*ls*) and *heavy s* (*hs*), corresponding to the average abundances of the $s$-process elements belonging to the first and second peak, respectively, as well as behind the importance of the determination of the abundance of Pb, representing the third $s$-process peak. In AGB stars, the high neutron exposure associated with the $^{13}$C neutron source drive the production of $s$-process elements even reaching up to the third $s$-process peak at Pb in low-metallicity AGB stars. On the other hand, the lower neutron exposure associated with the $^{22}$Ne neutron source typically produces $s$-elements only up to the first $s$-process peak at Sr.

It is now ascertained that the $s$ process is responsible for the production of about half the abundances of the elements between Sr and Bi in the Universe (see, e.g., Kaeppeler et al., 1982) and that it occurs in AGB stars.[10] The first direct evidence that the $s$ process occurs in AGB stars – and, more generally, that nucleosynthesis is happening inside stars – was the identification in the 1950s of the absorption lines of atoms of the radioactive element Tc in the atmospheres of some cool giant stars. The longest-living isotopes of Tc are $^{97}$Tc and $^{98}$Tc, with a half life of 4.0 and 4.2 million years, respectively. Since these stars would have taken billions of years to evolve to the giant phase, the observed Tc could have not been present in the star initially. It follows that the Tc must have been produced by the $s$ process inside the stars. Actually, neutron captures do not produce $^{97,98}$Tc, but the third longest-living isotope of this element: $^{99}$Tc (Fig. 3.10), which has a terrestrial half life of 0.21 million years. The presence of $^{99}$Tc in AGB stars has been confirmed by measurements of the Ru isotopic composition in stardust SiC grains, as will be discussed in Sect. 3.6.5.

The observation of Tc in giant stars has also been used to classify different types of $s$ process enhanced stars. If a given observed $s$ process enriched giant star shows the lines of Tc, then it must be on the AGB and have enriched itself of $s$-process elements. In this case it is classified as *intrisic* $s$-process enhanced star and typically belongs to one of the reddest and coolest subclasses of the spectroscopic class M: MS, S, SC, and C(N), where the different labels indicate specific spectral properties – S stars show zirconium oxide lines on top of the titanium oxide lines present in some M stars and C(N) stars have more carbon than oxygen in their atmospheres – or the transition cases between those properties – MS is the transition case between M stars and S stars and SC is the transition case between S and C(N) stars. On the other hand, if an $s$-process enriched giant star does not show the lines of Tc, it is classified as *extrinsic* $s$-process enhanced star. In this case its $s$-process

---

[10] Cosmic abundances of nuclei between Fe and Sr are also contributed by the $s$ process, but in this case by neutron captures occurring in massive stars during core He burning and shell C burning (Chap. 4 and, e.g., Raiteri et al., 1992).

enhancements have resulted from mass transfer from a binary companion, which was more massive and hence evolved first on the AGB phase. Stars belonging to the class of *extrinsic* $s$-process enhanced stars range from Ba stars in the Galactic disk, to the older halo populations of carbon-rich CH and Carbon-Enhanced Metal-Poor (CEMP) stars (e.g. Jorissen et al., 1998; Bond et al., 2000; Lucatello et al., 2005). Observations of Nb can also be used to discriminate intrisic from extrinsic $s$-process enhanced stars as Nb is destroyed during the $s$ process, but receives a radiogenic contribution over time due to the $\beta^-$ decay of $^{93}$Zr, with half life 1.5 million year, which is on the $s$-process path (see Fig. 3.10).

### 3.5.3 Branchings and the s Process in AGB Stars

Branching points at radioactive nuclei have provided for the past 50 years important tools to learn about conditions during the $s$ process in AGB stars. This is because branching factors depend on the neutron density and can also depend on the temperature and density of the stellar material. This happens in those cases when the decay rate of the branching nucleus is temperature and/or density dependent. These branching points are referred to as *thermometers* for the $s$ process. Traditionally, the solar abundances of isotopes affected by branching points were used to predict the neutron density and temperature at the $s$-process site using parametric models where parameters representing, e.g., the temperature and the neutron density were varied freely in order to match the observed abundances (e.g. Kaeppeler et al., 1982). Later, detailed information on branching points became available from spectroscopic observations of stellar atmospheres and from laboratory analyses of meteoritic stellar grains. At the same time, models for the $s$ process in AGB stars have evolved from parametric into stellar models, where the temperature and neutron density parameters governing the $s$ process are taken from detailed computation of the evolution of stellar structure (Gallino et al., 1998; Goriely and Mowlavi, 2000; Cristallo et al., 2009). For these models branching points are particularly useful to constrain neutron-capture nucleosynthesis and conditions inside the thermal pulse because, typically, they open at high neutron densities during the high-temperature conditions that allow the activation of the $^{22}$Ne neutron source in the convective intershell region.

As the temperature, density, and neutron density vary with time in the convective intershell region, branching factors also change over time. For example, a classical branching point, where the branching path corresponds to neutron capture, progressively opens while the neutron density reaches its maximum, and then closes again while the neutron density decreases and the main $s$-process path is restored. Of special interest is that toward the end of the thermal pulse the neutron density always decreases monotonically with the temperature and thus with time (Fig. 3.9) so that a *freeze out* time can be determined for a given nucleus, which represents the time after which the probability that the nucleus captures a neutron is smaller than unity and thus the abundances are *frozen* (Cosner et al., 1980). This can be calculated as the time when the neutron exposure $\tau$ left before the end of the neutron flux is $1/\sigma$, where $\sigma$ is the neutron-capture cross section of the nucleus. Hence, the abundances determined by branching points defined by unstable isotopes with higher $\sigma$ freeze out later during the neutron flux.

As a general rule of thumb, branching points that have the chance of being activated at the neutron densities reached in AGB stars are those corresponding to radioactive nuclei with half lives longer than at least a couple days. These correspond to similar half lives against capturing a neutron for neutron densities $\simeq 10^9 - 10^{11}$ n/cm$^3$, at AGB $s$-process temperatures. Isotopes with half lifes longer than approximately 10,000 years can be considered stable in this context as the $s$-process flux in AGB stars typically lasts less than this time. We refer to these isotopes as long-lived isotopes and we discuss their production in AGB stars in detail in Sect. 3.6.5. Very long-lived isotopes – half lives longer than $\sim$10 Myr – include for example $^{87}$Rb, and are considered stable in our context.

A list of unstable isotopes at which branching points that become relevant in the $s$-process reaction chain in AGB stars is presented in Appendix B of this book as a complete reference to be compared against observational information and as a tool for the building of $s$-process networks. Worth special mention are the branching points at $^{79}$Se, $^{85}$Kr, and $^{176}$Lu for the involvement of isomeric states of these nuclei, at $^{151}$Sm, one among a limited number of branching points for which an experimental estimate of the neutron-capture cross section is available, at $^{86}$Rb, responsible for the production of the very long-living $^{87}$Rb, and at $^{163}$Dy and $^{179}$Hf, which are stable nuclei in terrestrial conditions that become unstable in AGB stellar interiors.

Taken as a whole, the list of branching points that may be operating during the $s$ process in AGB stars sets a powerful group of constraints on our theoretical $s$-process scenarios. They are particularly effective when each of them is matched to the most detailed available observations of its effects. For example, some elemental abundance ratios and isotopic ratios that are affected by branching points can be measured from a stellar spectrum via identificaton and analysis of different emission or absorption lines. In these cases, model predictions can be compared directly to stellar observations of $s$-process-enhanced stars (Sect. 3.5.4). Isotopic ratios affected by branching points involving isotopes of refractory elements, but also of noble gases, have been or have the potential to be measured in meteoritic stardust SiC grains from AGB stars and provide unique constraints due to the large and expanding high-precision dataset available on the composition of stardust (Sect. 3.5.5). The values of the solar abundance ratios of $s$-only isotopes affected by branching points (e.g., $^{134}$Ba/$^{136}$Ba, $^{128}$Xe/$^{130}$Xe, and $^{176}$Hf/$^{176}$Lu) must be matched by any $s$-process model. When these involve nuclei with peculiar structure, such as $^{176}$Lu, combined investigation of nuclear properties and $s$-process models drives progress in our understanding of both.

One advantage of the computation of branching points in AGB stars is that the activation of one branching point is almost completely independent from the activation of all the other branching points because the overall neutron flux is only very marginally affected by the details of the $s$-process path. Thus, it is possible to include in a $s$-process nuclear network only the branching points of interest for a specific problem, or a specific element, hence keeping it simple and saving computational time.

One overall drawback of using branching points to understand the *s* process is that for the vast majority of the radioactive nuclei involved there exist only theoretical or phenomenological determinations of their neutron-capture cross sections and of the temperature and density dependence of their decay rates. This is due to the difficulty of producing experimental data for radioactive targets (see Chap. 9) and means that there are always some uncertainties associated to model predictions of the effect of branching points. These errors and their effect need to be carefully evaluated in every single case.

### 3.5.4 Signatures of s-Process Branching Points in Stars: Rb, Zr, Eu

The abundance of $^{87}$Rb, which can be produced in AGB stars via activation of the branching point at $^{86}$Rb, is a famous example of how detailed comparison of theoretical *s*-process abundances to the abundances observed in *s*-process-enhanced stars provide a stringent test to our understanding of the *s* process and AGB stars. The abundance of $^{87}$Rb is particularly interesting because the element Rb can be spectroscopically identified and its abundance determined in AGB stars. Overall, Rb is an *r*-process element – only 22% of its solar abundance can be ascribed to the *s* process (Arlandini et al., 1999) – made up of two isotopes: $^{85}$Rb and the very long-lived $^{87}$Rb, which is treated as a stable isotope in this context. Specifically, 92% of solar $^{85}$Rb is made by the *r* process because this nucleus has a relatively large neutron-capture cross section of 234 mbarn and thus it does not accumulate to high abundances during the *s* process. On the other hand, $^{87}$Rb, as described in Appendix B, has a magic number of neutrons $N = 50$, and thus a relatively small neutron-capture cross section of 15.7 mbarn. Hence, if it is reached by the *s*-process reaction chain via the activation of the branching points at $^{85}$Kr and $^{86}$Rb, it accumulates and is significantly produced. It follows that when these branching points are activated during the *s* process, the abundance $^{87}$Rb represents a fraction of the total abundance of *s*-process Rb larger than the initial solar fraction. This is illustrated in the top panel of Fig. 3.11. In the case of the massive AGB model, where the $^{22}$Ne neutron source is activated, the *s* process occurs at high neutron density, and branching points are open, almost half of the final total abundance of Rb is made by $^{87}$Rb. In the case of the low-mass AGB model, instead, where the $^{13}$C neutron source is activated, the *s* process occurs at low neutron density, and branching points are closed, only a quarter of the final total abundance of Rb is made by $^{87}$Rb.

The ratio of the abundance of Rb to that of a neighbouring *s*-process element, such as Sr, or Zr, whose overall abundance is instead not affected by the activation of branching points, can be determined in AGB stars and has been widely used as an indicator of the neutron density at which the *s* process occurs. Observations of Rb/Zr ratios lower than solar in MS, S, and C stars have strongly supported the theoretical scenario where the main neutron source in these low-mass AGB stars is the $^{13}C(\alpha, n)^{16}O$ reaction. This is because this neutron source produces neutron densities too low to increase the Rb/Zr ratio above the solar value (see lower panel of Fig. 3.11 and Lambert et al., 1995; Abia et al., 2001).

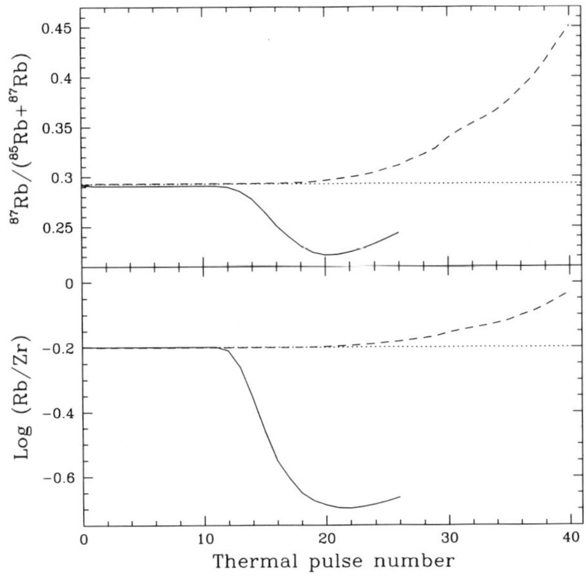


**Fig. 3.11** Ratio of $^{87}$Rb to the total abundance of Rb (*top panel*) and the Rb/Zr ratio (*lower panel*) computed in two solar metallicity AGB models (from van Raai et al., 2008). The *dotted lines* represent the initial solar ratios. The *evolution lines* represent a massive (6.5 $M_\odot$) AGB model experiencing the activation of the $^{22}$Ne neutron source only (*long-dashed lines*), and a low-mass (3 $M_\odot$) model experiencing the activation of the $^{13}$C neutron source only *(solid lines)* (except for a marginal activation of the $^{22}$Ne neutron source in the latest thermal pulses leading to the small final increase in the $^{87}$Rb and Rb abundances)

Massive AGB stars (>4:5 $M_\odot$) have only recently been identified in our Galaxy (García-Hernández et al., 2006, 2007). They belong to the group of OH/IR stars and they have been singled out as massive AGB stars on the basis of their location closer to the galactic plane, which indicates that they belong to a younger and thus more massive stellar population, and their longer pulsation periods ($\simeq$400 days). Rb/Zr ratios in these stars are observed to be well above the solar value, which has given ground to the theoretical scenario where the main neutron source in these massive AGB stars must be the $^{22}$Ne($\alpha, n$)$^{25}$Mg reaction, which produces neutron densities high enough to increase the Rb/Zr ratio above the solar value (see lower panel of Fig. 3.11).

Another indicator of the neutron density in AGB stars is the isotopic abundance of $^{96}$Zr, which is produced if the branching point at $^{95}$Zr is activated. Zr isotopic ratios were determined via observations of molecular lines of ZrO in a sample of S stars (Lambert et al., 1995). No evidence was found for the presence of $^{96}$Zr in these stars. This result provides further evidence that the neutron density in low-mass AGB stars stars must be low. A low $^{96}$Zr abundance has been confirmed by high-precision data of the Zr isotopic ratios in stardust SiC grains, which are discussed in the following Sect. 3.5.5.

Isotopic information from stellar spectra has also been derived for the typical $r$-process element Eu in old Main Sequence stars belonging to the halo of our Galaxy and enhanced in heavy neutron-capture elements (Sneden et al., 2008). This has been possible because the atomic lines of Eu differ significantly if the Eu atoms are made of $^{151}$Eu instead of $^{153}$Eu, the two stable isotopes of Eu. In old stars showing overall enhancements of $r$-process elements, the total Eu abundance is roughly equally divided between $^{151}$Eu and $^{153}$Eu. This fraction is consistent with the solar fraction, and it is expected by Eu production due to the $r$ process (Sneden et al., 2002; Aoki et al., 2003a). On the other hand, in two old stars showing overall enhancements in $s$- and $r$-process elements, roughly 60% of Eu is in the form of $^{151}$Eu (Aoki et al., 2003b). How these stars gained enhancements in the abundances of both $r$- and $s$-process elements is an unsolved puzzle of the study of the origin of the elements heavier than iron in the Galaxy (Jonsell et al., 2006). This is because according to our current knowledge, the $r$- and the $s$ process are completely independent of each other, and occur in very different types of stars, core-collpase supernovae and AGB stars, respectively. During the $s$ process, the Eu isotopic fraction is determined mostly by the activation of the branching point at $^{151}$Sm, and the observed 60% value is consistent with low neutron densities $\simeq 10^8$ $n$/cm$^3$ during the $s$ process. As more observations become available, the role of branching points during the $s$ process in AGB stars becomes more and more crucial to answering the questions on the origin of the heavy elements.

### 3.5.5 SiC Grains from AGB Stars and Branching Points

Stardust SiC grains from AGB stars represent a unique opportunity to study $s$-process conditions in the parent stars of the grains through the effect of the operation of branching points because SiC grains contain trace amounts of atoms of elements heavier than iron, which allow high-precision measurements of their isotopic ratios. Refractory heavy elements, such as Sr, Ba, Nd, and Sm, condensed from the stellar gas directly into the SiC grains while the grains were forming. Their isotopic composition have been determined from samples of meteoritic residual materials containing a large number of SiC grains using TIMS and SIMS (see Sect. 10.2 and Ott and Begemann, 1990; Zinner et al., 1991; Prombo et al., 1993; Podosek et al., 2004). High-resolution SIMS has also been applied to derive data in single stardust SiC grains for Ba with the NanoSIMS (see Sect. 10.2 and Marhas et al., 2007) and Eu with the SHRIMP (Sensitive High Resolution Ion Microprobe, Terada et al., 2006). Isotopic ratios in a sample containing a large number of SiC grains for many elements in the mass range from Ba to Hf were also measured by ICPMS (Sect. 10.2 and Yin et al., 2006).

A general drawback of these experimental methods is that they do not allow to separate ions of same mass but different elements. Hence, interferences by isotopes of the same mass (isobars) are present, which is especially problematic for the elements heavier than iron where a large number of stable isobars can be found.

Branching points, in particular, by definition affect the relative abundances of isobars, thus, with the methods above it is difficult to derive precise constraints on the effect of branching points on isotopic ratios. For example, the isobars $^{96}$Mo and $^{96}$Zr cannot be distinguished in these measurements, and thus it is not possible to derive information on the operation of the branching point at $^{95}$Zr.

Exceptions to this problem are the stable Eu isotopes, $^{151}$Eu and $^{153}$Eu, which do not have stable isobars and thus their ratio can be measured and used to constrain the neutron density and the temperature during the *s* process in the parent stars of the grains via the branching points at $^{151}$Sm and $^{152}$Eu (Terada et al., 2006), and the Ba isotopes, which are not affected by isobaric interferences because their isobars, the isotopes of the noble gas Xe isotopes, are present in very low amounts in the grains and are difficult to ionize and extract from the stardust (see specific discussion below in this section). The Ba isotopic ratios, in particular the $^{134}$Ba/$^{136}$Ba and the $^{137}$Ba/$^{136}$Ba ratios, can be affected by branching points at the Cs isotopes (see below and Prombo et al., 1993; Marhas et al., 2007).

The application of RIMS (Sect. 10.2) to the analysis of heavy elements in SiC grains has allowed to overcome the problem of isobaric interferences, at the same time providing an experimental method of very high sensitivity, which allows the measurements of trace elements in single stardust grains (Savina et al., 2003b). Since RIMS can select which element is ionized and extracted form the grains, mass interferences are automatically avoided. The Chicago-Argonne RIMS for Mass Analysis CHARISMA has been applied to date to the measurement of Zr (Nicolussi et al., 1997), Mo (Nicolussi et al., 1998a), Sr (Nicolussi et al., 1998b), Ba (Savina et al., 2003a) and Ru (Savina et al., 2004) in large single SiC grains (average size 3 $\mu$m), providing high-precision constraints on the operation of the *s*-process branching points that may affect the isotopic composition of these elements. A detailed comparison between data and models (Lugaro et al., 2003) shows that AGB stellar models of low mass and metallicity roughly solar, where the $^{13}$C$(\alpha,n)^{16}$O reaction is the main neutron source and the $^{22}$Ne$(\alpha,n)^{25}$Mg is only marginally activated, provide the best match to all measured isotopic ratios affected by branching points. This result is in agreement with the constraints mentioned above provided by the isotopic data for Ba (Marhas et al., 2007) and Eu (Terada et al., 2006) obtained via high-resolution SIMS.

For example, the $^{96}$Zr/$^{94}$Zr ratio is observed in all measured single SiC to be lower than solar by at least 50%. Low-mass AGB models can reproduce this constraint due to the low neutron density associated with the main $^{13}$C neutron source, in which conditions $^{96}$Zr behaves like a typical *r*-only nucleus and is destroyed during the neutron flux. Massive AGB stars ($>$4:5 M$_\odot$), on the other hand, experience high neutron densities and produce $^{96}$Zr/$^{94}$Zr ratios higher than solar. In more detail, the $^{96}$Zr/$^{94}$Zr ratio at the stellar surface of low-mass AGB stellar models reaches a minimum of $\simeq$90% lower than solar after roughly ten 3rd dredge-up episodes, and then may increase again, due to the marginal activation of the $^{22}$Ne neutron source in the latest thermal pulses. This predicted range allows to cover most of the $^{96}$Zr/$^{94}$Zr of single SiC grains (see Fig. 5 of Lugaro et al., 2003).

Another interesting example is the $^{134}$Ba/$^{136}$Ba ratio, where both isotopes are *s*-only nuclei. During the low-neutron density flux provided by the $^{13}$C neutron source the branching point at $^{134}$Cs is closed and the $^{134}$Ba/$^{136}$Ba ratio at the stellar surface reaches up to $\simeq$20% higher than the solar ratio after roughly ten 3rd dredge-up episodes. This value is too high to match the composition of single SiC grains. However, during the marginal activation of the $^{22}$Ne in the later thermal pulses, the branching point at $^{134}$Cs is activated, $^{134}$Ba is skipped during the *s*-process flux and the $^{134}$Ba/$^{136}$Ba ratio at the stellar surface is lowered down to the observed values roughly 10% higher than the solar value (see Fig. 14 of Lugaro et al., 2003).

The $^{137}$Ba/$^{136}$Ba ratio is another indicator of the neutron density because the activation of the chain of branching points along the Cs isotopes can produce $^{137}$Cs, which decays into $^{137}$Ba with a half life of 30 year. Grain data do not show any contribution of $^{137}$Cs to $^{137}$Ba, indicating that the Cs branching points beyond $^{134}$Cs are not activated in the parent stars of the grains (see Fig. 14 of Lugaro et al., 2003). This, again, excludes massive AGB stars, with an important neutron contribution from the $^{22}$Ne neutron source, as the parent stars of the grains.

Another example of the signature of the *s* process in meteorites is represented by very small variations, of the order of parts per ten thousand, observed in the osmium isotopic ratios of primitive chondritic meteorites. This fascinating anomaly looks like a *mirror s*-process signature, meaning that they show exactly the opposite behaviour expected if the meteorite had a component carrying an *s*-process signature. They are thus interpreted as a sign of incomplete assimilation of stardust SiC grains within the meteorite (Brandon et al., 2005). The branching points at $^{185}$W and $^{186}$Re make the $^{186}$Os/$^{188}$Os ratio a indicator of the neutron density for the *s* process and the value for this ratio observed in chondrites suggest a low neutron density of $N_n = 3 \times 10^8$ n/cm$^3$ (Humayun and Brandon, 2007), in agreement with other evidence discussed above.

Differently from refractory elements, the noble gases He, Ne, Ar, Kr, and Xe are not chemically reactive and can condense from gas into solid only at much lower temperatures than those around AGB stars. Still, they are found in SiC, even if in extremely low quantitites. Their atoms could have been ionized by the energy carried by the stellar winds during the AGB and post-AGB phase. These ions were then chemically reactive enough to be *implanted* into already formed dust grains.

It has been possible to extract noble gases from meteoritic samples by RIMS (see Chap. 11), laser gas extraction (Nichols et al., 1991) and stepped-heating combustion of the sample to high temperatures, up to $2,000^\circ$ (Lewis et al., 1994). In particular, for the heavy nobler gases Kr and Xe, since their abundances are very low in stardust and the stepped-heating experimental method does not provide high extraction efficiency, it has been possible to extract their ions only from a large amount of meteoritic residual material. The derived Kr and Xe isotopic data is thus the average over a large number – millions – of grains. Differential information as function of the grain sizes can still be obtained by preparing the meteoritical residual in a way that selects the size of the grains to be found in it.

The composition of Xe in SiC corresponds to the famous Xe-S component, one of the first signature of the presence of pure stellar material in primitives meteorites (see Chap. 2, Sect. 2.2.5), thus named because of its obvious *s*-process signature: excesses in the *s*-only isotopes $^{128,130}$Xe and deficits in the *r*-only and *p*-only isotopes $^{124,126,136}$Xe (all with respect to the solar composition). The $^{134}$Xe/$^{130}$Xe ratio may be affected by the operation of the branching point at $^{133}$Xe during the *s* process. This isotopic ratio in stardust SiC is very close to zero, indicating that the Xe trapped in SiC grains did not experience *s* process with high neutron density (Pignatari et al., 2004). This again allows the mass and metallicity of the parent stars of the grains to be constrained to low-mass AGB stars of roughly solar metallicity, in agreement with the conclusions drawn from the composition of the refractory elements.

The situation regarding the Kr isotopic ratios measured in SiC grains is much more complex. There are two branching points affecting the Kr isotopic composition: $^{79}$Se and $^{85}$Kr, changing the abundances of $^{80}$Kr and $^{86}$Kr, respectively, and both of them are tricky to model (see description in Appendix B). Moreover, the Kr atoms in stardust SiC appear to be consistent with implantation models of this gas into the grains only if these models consider two different components of implanted Kr (Verchovsky et al., 2004). One component was ionized and implanted in SiC at low energy, corresponding to a velocity of 5–30 km/s, typical of AGB stellar winds, the other component was ionized and implanted at high energy, corresponding to a velocity of a few thousands km/s, typical of the winds driven from the central star during the planetary nebular phase. In the second situation, which is the case also for all the He, Ar, and Ne atoms found in SiC, the isotopic composition of the noble gases indicate that they must have come directly from the deep He-rich and *s*-process-rich layers of the star, with very small dilution with the envelope material of initial solar composition. This is consistent with the fact that at this point in time the envelope is very thin as most of the initial envelope material has been peeled away by the stellar winds.

While the Kr AGB component is observed to be prominent in the small grains (of average size 0.4 $\mu$m) and shows low $^{86}$Kr/$^{82}$Kr and high $^{80}$Kr/$^{82}$Kr ratios, in agreement again with low neutron density *s* process AGB models, the Kr planetary-nebula component is observed to be prominent in the largest grains (average size 3 $\mu$m) and shows high $^{86}$Kr/$^{82}$Kr and low $^{80}$Kr/$^{82}$Kr ratios, as expected instead in pure He-rich intershell material due to the higher neutron density *s* process occurring in the final AGB thermal pulses (Pignatari et al., 2006). Actually, it is difficult to reproduce the $^{86}$Kr/$^{82}$Kr up to twice the solar value observed in the largest grains even using the final pure *s*-process intershell composition of low-mass and solar metallicity AGB stellar models. This high $^{86}$Kr/$^{82}$Kr ratios may be the signature of high-neutron density *s*-process nucleosynthesis occurring in *late* and *very late* thermal pulses during the post-AGB phase (see, e.g., Herwig et al., 1999), rather than during the AGB phase. Detailed *s*-process models are currently missing for this phase of stellar evolution.

In summary, the detailed information provided by stardust data on the isotopic ratios affected by branching points at radioactive nuclei on the *s*-process path has allowed us to pinpoint the characteristics of the neutron flux that the parent stars of stardust SiC grains must have experienced. The vast amount of information on the composition of light and heavy elements in SiC grains has allowed us to infer with a high degree of confidence that the vast majority of these grains came from C-rich AGB stars, i.e., C(N) stars, which have C>O in their envelope, the condition for SiC grain formation, of low mass and metallicity close to solar. In turn, the stardust data has been used to refine our theoretical ideas of the *s* process in these stars confirming that $^{13}$C nuclei must be the main neutron source, while the $^{22}$Ne neutron source is only marginally active.

## 3.6 Nucleosynthesis of Long-lived Isotopes in AGB Stars

### 3.6.1 $^{26}$Al

The famous long-lived radioactive nucleus $^{26}$Al (with half life of 0.7 Myr), of interest from the point of view of $\gamma$-ray observations, meteoritic stellar grains, and the composition of the early solar system, can be produced in AGB stars via proton captures on $^{25}$Mg, i.e., the $^{25}$Mg$(p,\gamma)^{26}$Al reaction, when the temperature is above $\simeq$60 MK (Mowlavi and Meynet, 2000; van Raai et al., 2008b). As detailed in Sect. 3.4.1, proton captures occur in AGB stars between thermal pulses in two different locations: (1) in the H-burning shell on top of the He-rich intershell, and (2) at the base of the convective envelope in massive AGB stars, above $\simeq 4\,M_\odot$ (in the process known as Hot Bottom Burning, HBB, Sect. 3.4.1).

In setting (1), the intershell material is progressively enriched in $^{26}$Al as proton captures in the H-burning shell convert 80% of $^{25}$Mg into $^{26}$Al. The efficiency of this conversion is determined by the fraction of 20% of $^{25}$Mg+$p$ reactions producing the isomeric, rather than the ground, state of $^{26}$Al, which quickly decays into $^{26}$Mg with a half life of $\simeq$6 s. Most intershell $^{26}$Al abundance is destroyed by neutron captures before having the chance of being dredged-up to the stellar surface via the 3rd dredge-up. This is because the neutron-capture cross sections of $^{26}$Al, in particular the $(n,p)$ and $(n,\alpha)$ channels, are very efficient: $\sigma$ $\simeq$250 and 180 mbarn, respectively.

More specifically, already during the interpulse period some $^{26}$Al is destroyed by neutron captures. This is because in the bottom layers of the ashes of H burning the temperature reaches 90 MK, high enough for the $^{13}$C$(\alpha$,n$)^{16}$O reaction to occur using as fuel the $^{13}$C nuclei in ashes of H burning produced by CNO cycling. Then, neutron captures in the following thermal pulse destroy most of the $^{26}$Al that was left over in the H-burning ashes. First, the $^{13}$C nuclei that had survived in the top layers of the H-burning ashes are engulfed in the convective pulse, where the temperature quickly reaches 200 MK and the $^{13}$C$(\alpha$,n$)^{16}$O reaction is very efficiently activated. Second, the neutrons that may be released by the $^{22}$Ne$(\alpha$,n$)^{25}$Mg reaction later on in the convective pulse, when the temperature is higher than roughly 250 MK, contribute to further destruction of $^{26}$Al. In this phase $^{26}$Al can be completely destroyed, depending on the temperature reached at the base of the convective pulse, which controls the efficiency of the $^{22}$Ne$(\alpha$,n$)^{25}$Mg reaction. If the temperature reaches up to 300 MK, the $^{26}$Al abundance is decreased by two orders of magnitude in the He-rich intershell at the end of the thermal pulse.

When the 3rd dredge-up occurs after the thermal pulse is extinguished, only a small mass of $^{26}$Al is carried from the intershell to the stellar surface, of the order of $10^{-8}$ $M_{\odot}$, mostly coming from a tiny region (roughly $10^{-4}$ $M_{\odot}$) at the top of the intershell, which was not ingested in the convective pulse and thus did not experience the availability of free neutrons. This small abundance of $^{26}$Al carried into the envelope translates into a small total contribution of the AGB winds to the abundance of $^{26}$Al in the interstellar medium (also defined as *yield*) of $10^{-7}$ $M_{\odot}$, for AGB stars of masses between 1 $M_{\odot}$ and 4 $M_{\odot}$, depending on the metallicity (upper panel of Fig. 3.12), though allowing a noticeable increase in the $^{26}$Al/$^{27}$Al ratio at the stellar surface, up to a typical value of $2 \times 10^{-3}$ (lower panel of Fig. 3.12).

The situation is very different for AGB stars of masses higher than approximately 4 $M_{\odot}$. Proton captures occurring in setting (2), i.e., HBB at the base of the convective envelope, combined with the 3rd dredged-up of $^{25}$Mg produced from efficient activation of the $^{22}$Ne($\alpha$,n)$^{25}$Mg reaction in the thermal pulse, produce large amounts of $^{26}$Al. These are directly mixed to the stellar surface via the envelope convection resulting in yields up to $10^{-4}$ $M_{\odot}$, and $^{26}$Al/$^{27}$Al ratios up to 0.5 (Fig. 3.12). During HBB the main channel for $^{26}$Al destruction is proton captures on

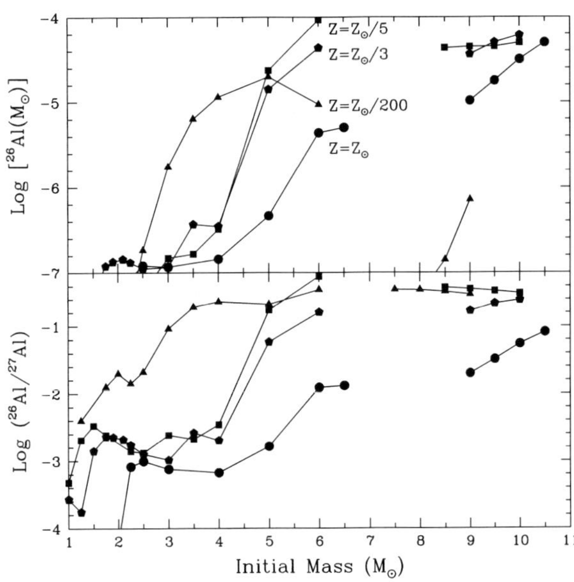


**Fig. 3.12** The yields of $^{26}$Al (*top panel*) for stellar models of different masses and metallicities (Z) from Karakas and Lattanzio, 2007 for AGB stars of masses up to 6 $M_{\odot}$, and from Siess and Arnould, 2008 for the more massive Super-AGB stars. Yields are defined as the total mass of $^{26}$Al (in $M_{\odot}$) lost in the wind during the whole evolution of the star (calculated as the average of the time-dependent envelope composition weighed on the mass lost at each time). The ratios of the yield of $^{26}$Al to the yield of $^{27}$Al are also shown in the *bottom panel*

$^{26}$Al itself, i.e., $^{26}$Al$(p, \gamma)^{27}$Si reactions. Also in Super-AGB stars HBB produces large quantitites of $^{26}$Al (Siess and Arnould, 2008).

Figure 3.12 shows the yields of $^{26}$Al and their ratio with the yield of $^{27}$Al for a variety of AGB stars and Super-AGB of different masses and metallicities. The plot shows how the efficiency of $^{26}$Al production increases with stellar mass and with decreasing metallicity of the stars. This is because the efficiency of the HBB depends on the temperature at the base of the convective envelope, which is higher for higher masses and lower metallicities. For example, a 3.5 $M_\odot$ star of metallicity 200 times lower than solar ejects the same amount of $^{26}$Al than a 6.5 $M_\odot$ star at solar metallicity. The reason is that the overall temperature is controlled by the mass of the CO core, which scales directly with the initial mass and inversely with the initial metallicity (see Sect. 3.4.1). In addition, the lower opacity in lower metallicity stars keeps the structure more compact and hence hotter.

In the models of Super-AGB presented in Fig. 3.12 the 3rd dredge-up is found to be negligible and HBB produces $^{26}$Al via proton captures on the $^{25}$Mg initially present in the envelope, without the contribution of $^{25}$Mg from the intershell. Still, these stars produce a large amount of $^{26}$Al since there is a large initial amount of $^{25}$Mg in the envelope due to the large envelope mass. The lower Super-AGB $^{26}$Al yield at metallicity solar/200 is due to very high HBB temperatures, at which the rate of the $^{26}$Al destruction reaction $^{26}$Al$(p, \gamma)^{27}$Si is significantly enhanced.

The yields predicted for the $^{26}$Al from AGB stars presented in Fig. 3.12 are quite uncertain since there are several stellar and nuclear uncertainties. First, there are uncertainties related to the modelling of HBB. In fact, the temperature reached at the base of the convective envelope, which governs the efficiency of the $^{25}$Mg$(p, \gamma)^{26}$Al reaction, depends on the modeling of the temperature gradient within the convectively unstable region. Hence, different treatments of the convective layers may lead to significantly different efficiencies of the HBB. Second, the uncertainty in the efficiency of the 3rd dredge-up already discussed in Section 3.4.1 also affects the $^{26}$Al yields: in the low-mass models it affects the dredge-up of $^{26}$Al itself, in the massive models it affects the dredge-up of $^{25}$Mg, which is then converted into $^{26}$Al via HBB. Third, the mass-loss rate is another major uncertain parameter in the modelling of AGB stars. The mass-loss rate determines the stellar lifetime and thus the time available to produce $^{26}$Al and the final $^{26}$Al yield.

Another model uncertainty is related to the possible occurrence of extra mixing at the base of the convective envelope in the low-mass AGB models that do not experience HBB. Such extra mixing in AGB stars would be qualitatively similar to the extra mixing in red giant stars described in Sect. 3.3.2. In the hypothesis of extra mixing, material travels from the base of the convective envelope inside the radiative region close to the H-burning shell, suffers proton captures, and is taken back up into the convective envelope. If the mixed material dips into the H-burning shell, down to temperatures higher than $\simeq$50 MK, then this mechanism could produce $^{26}$Al and contribute to some amount of this nucleus in the low-mass models (Nollett et al., 2003). Unfortunately, from a theoretical point of view, there is no agreement on which mechanism drives the extra mixing and on the features of the mixing. Some constraints on it, however, can be derived from the composition of MS, S, SC, and C stars as well as meteoritic stellar grains, as will be discussed in detail in Sect. 3.6.2.

As for nuclear uncertainties, the rate of the $^{26}$Al$(p, \gamma)^{27}$Si reaction is uncertain by three orders of magnitude in the temperature range of interest for AGB stars (Iliadis et al., 2001), with the consequence that $^{26}$Al yields from AGB stars suffer from uncertainties of up to two orders of magnitude (Izzard et al., 2007; van Raai et al., 2008b). New experiments and approaches to estimate this rate are needed to get a more precise determination of the production of $^{26}$Al in AGB stars.

In spite of all these important uncertainties, current models do indicate that at least some AGB models produce a significant amount of $^{26}$Al. These models cover a small range of stellar masses, only those suffering HBB on the AGB phase. When the yields presented in Fig. 3.12 are averaged over a Salpeter initial stellar mass function, the result is that AGB stars globally do not provide an important contribution to the present abundance of $^{26}$Al in the Galaxy. This contribution sums up to only 0.24% of the contribution from massive star winds and core-collapse supernovae (Limongi and Chieffi, 2006; Lugaro and Karakas, 2008). Adding up the contribution of Super-AGB stars only marginally increases the contribution of AGB stars to Galactic $^{26}$Al to 0.85% of the contribution coming from the more massive stars (see also Siess and Arnould, 2008).

### 3.6.2 Evidence of $^{26}$Al in AGB Stars

It may be possible to determine the abundance of $^{26}$Al in AGB stars using molecular lines of Al-bearing molecules. This was carried out by (Guelin et al., 1995) for the nearest carbon star, CW Leo, using rotational lines of AlF and AlCl molecules with different Al isotopic composition. One observed line was tentatively attributed to $^{26}$AlF, and from its observed strength an upper limit of 0.04 for the $^{26}$Al/$^{27}$Al ratio was inferred. No $^{26}$AlCl lines were detected, which led to an upper limit of 0.1. These values cannot be reached by solar metallicity AGB models (Fig. 3.12), however, this detection has not been confirmed so it is doubtful if it can represent a valid model constraint.

The main observational evidence of $^{26}$Al in AGB stars comes, instead, from stardust (see Fig. 3.13). Aluminium is one of the main component of most oxide stardust grains recovered to date and the initial amount of $^{26}$Al present in each grain can be derived from excesses in its daughter nucleus $^{26}$Mg. Magnesium is not a main component in corundum ($Al_3O_2$) and hibonite ($CaAl_{12}O_{19}$) grains, hence, in this cases, $^{26}$Mg excesses are all attributed to $^{26}$Al decay. In the case of spinel ($MgAl_2O_4$) grains, instead, Mg is a main component of the mineral and thus the contribution of $^{26}$Al to $^{26}$Mg needs to be more carefully evaluated by weighing the contribution of the two components. Specifically, there are two atoms of Al per each atom of Mg in spinel, which corresponds to a roughly 25 times higher ratio than in the average solar system material.

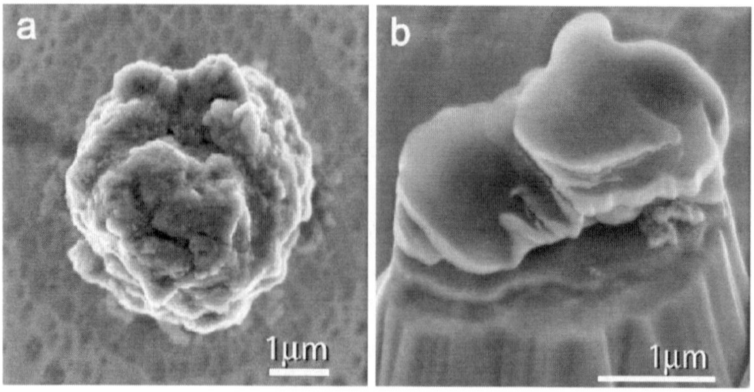


**Fig. 3.13** Scanning electron microscope images of dust grains from AGB stars. (**a**) 4-μm-sized silicon carbide (SiC) grain; ubiquitous excesses in $^{26}$Mg in such grains indicate prior presence of $^{26}$Al. (**b**) 2-μm-sized hibonite ($CaAl_{12}O_{19}$) grain KH15 (Nittler et al., 2008). The grain is sitting on a gold pedestal created by ion-probe sputtering during isotopic analysis. Excesses of $^{26}$Mg and $^{41}$K indicate that the grain originally condensed with live $^{26}$Al and $^{41}$Ca

The $^{26}$Al/$^{27}$Al ratios are observed to be different in the different populations of oxide and silicate grains (see, e.g., Fig. 8 of Nittler et al., 1997). Population I grains cover a wide range of $^{26}$Al initial abundance, from no detection to $^{26}$Al/$^{27}$Al $\simeq 0.02$. The presence of $^{26}$Al is used to discriminate Population I oxide grains coming from red giant or from AGB stars, since $^{26}$Al is expected to be present only in the winds of AGB stars. The $^{26}$Al/$^{27}$Al ratios of Population II grains lie at the upper end of the range covered by Population I grains, and reach up to $\simeq$0.1 (see also Fig. 6 of Zinner et al., 2007). This is qualitatively consistent with the strong $^{18}$O deficits observed in the Population II grains, since both signatures are produced by H burning. The mysterious Population III show low or no $^{26}$Al, which may indicate that these grains did not come from AGB stars. Finally, Population IV grains from supernovae show $^{26}$Al/$^{27}$Al ratios between 0.001 and 0.01 (see Chap. 4).

The $^{26}$Al/$^{27}$Al ratios together with the $^{18}$O/$^{16}$O ratios in Population I and II oxide and silicate grains provide an interesting puzzle to AGB modellers. Low-mass AGB models do not produce $^{26}$Al/$^{27}$Al ratios high enough and $^{18}$O/$^{16}$O ratios low enough to match the observations. Massive AGB models can produce $^{26}$Al/$^{27}$Al ratios high enough via HBB, however, in this case the $^{18}$O/$^{16}$O ratio is too low ($\sim 10^{-6}$) to match the observations (see Sect. 3.2.1). Grains with $^{18}$O/$^{16}$O $< 10^{-4}$ may have been polluted by solar material during the laboratory analysis, which would have shifted the $^{18}$O/$^{16}$O ratio to higher values with respect to the true ratio of the grain. This argument was invoked to attribute a massive AGB stars origin to a peculiar Population II spinel grain, named OC2 (Lugaro et al., 2007). However, also the $^{17}$O/$^{16}$O ratio presents a problem for this and similar grains because at the temperature of HBB this ratio is always much higher than observed (Boothroyd et al., 1995; Lugaro et al., 2007; Iliadis et al., 2008).

The extra-mixing phenomena mentioned in Sects. 3.3.2 and 3.6.1 have been hypothesized to operate in low-mass AGB stars below the base of the formal convective envelope to explain the composition of Population I and II grains with $^{26}$Al/$^{27}$Al ratios greater than $\simeq 10^{-3}$. This idea has been investigated in detail by (Nollett et al., 2003) using a parametric model where the temperature ($T_p$), determined by the depth at which material is carried, and the mass circulation rate ($M_{circ}$) in the radiative region between the base of the convective envelope and the H-burning shell are taken as two free and independent parameters. This model was originally proposed to explain observations of AGB stars and grains showing deficits in $^{18}$O (Wasserburg et al., 1995).

In the case of SiC grains, Al is present in the grains as a trace element in relatively large abundance, while Mg is almost absent. Again, this means that $^{26}$Mg excesses represent the abundance of $^{26}$Al at the time when the grains formed. Mainstream SiC grains from AGB stars show $^{26}$Al/$^{27}$Al ratios between $10^{-4}$ and $\simeq 2 \times 10^{-3}$. Models of C-rich AGB stars, i.e., the low-mass models in the lower panel of Fig. 3.12, which do not suffer HBB and hence can reach C/O>1 in their envelopes, match the observed upper value but do not cover the observed range down to the lower values (Zinner et al., 2007; van Raai et al., 2008b). It is difficult to interpret the lowest values as pollution of solar material as in this case one would expect a trend of decreasing $^{26}$Al/$^{27}$Al ratios with the total Al content in the grains, which is not observed. As discussed in Sect. 3.6.1, the main problem modelers have to consider in this context is the large uncertainty in the $^{26}$Al+$p$ reaction rate (van Raai et al., 2008b). The rate uncertainty results in an uncertainty of one order of magnitude in the $^{26}$Al/$^{27}$Al ratios predicted for C-rich AGB stars. If the current upper limit of the rate is employed in the models, then only the lower values observed in SiC can be matched and we may need to invoke extra-mixing phenomena also to explain the Al composition of SiC grains. On the other hand, if the lower or recommended values of the rates are used, then only the higher values observed in SiC can be matched.

In conclusion, observational constraints of $^{26}$Al in AGB stars provide the potential to investigate some of the most uncertain input physics in the modelling of AGB nucleosynthesis: mixing phenomena and reaction rates.

### *3.6.3 $^{60}$Fe*

The other famous long-lived radioactive nucleus $^{60}$Fe (with a recently revised half life of 2.6 My, Rugel et al., 2009), of interest from the point of view of $\gamma$-ray observations, meteoritic stellar grains, and the composition of the early solar system, can be produced in AGB stars (Wasserburg et al., 2006; Lugaro and Karakas, 2008) via the neutron-capture chain $^{58}$Fe$(n,\gamma)^{59}$Fe$(n,\gamma)^{60}$Fe, where $^{59}$Fe is a branching point, and destroyed via the $^{60}$Fe$(n,\gamma)^{61}$Fe reaction, whose rate has been measured experimentally by (Uberseder et al., 2009). This is the same chain of reactions responsible for the production of this nucleus in massive stars (see Chapter 4, Sect. 4.3.1.2). Given that $^{59}$Fe is an unstable nucleus with a relatively short half life of 44 days and with a neutron-capture cross section $\sigma \simeq 23$ mbarn (Rauscher and Thielemann, 2000), neutron densities of at least $10^{10}$ n/cm$^3$ are needed for this branching point to open at a level of 20%, allowing production of the long-living $^{60}$Fe. If the neutron density is higher than $10^{12}$ n/cm$^3$, then 100% of the neutron-capture flux goes through $^{60}$Fe.

From the description of the neutron sources in AGB stars (Sect. 3.5.2), it is clear that $^{60}$Fe can only be produced in the convective thermal pulses, where the neutron burst released by the $^{22}$Ne neutron source can reach the high neutron density required to open the branching point at $^{59}$Fe. Hence, the production of $^{60}$Fe in AGB stars is almost completely determined by the activation of the $^{22}$Ne neutron source. The $^{13}$C neutron source may instead destroy some $^{60}$Fe in the intershell (Wasserburg et al., 2006).

The AGB yields of $^{60}$Fe, and their ratios with the yields of $^{56}$Fe, are shown in Fig. 3.14. As the temperature at the base of the convective thermal pulses increases with increasing the stellar mass and decreasing the metallicity, the amount of $^{60}$Fe delivered to the interstellar medium increases, reaching up to $10^{-5}$ M$_\odot$, a value comparable to that delivered by a supernova of $\simeq$20 M$_\odot$ (Limongi and Chieffi, 2006). Ratios of the $^{60}$Fe and $^{56}$Fe abundances at the end of the AGB phase from the AGB neutron-capture models of (Wasserburg et al., 2006) also plotted in Fig. 3.14.

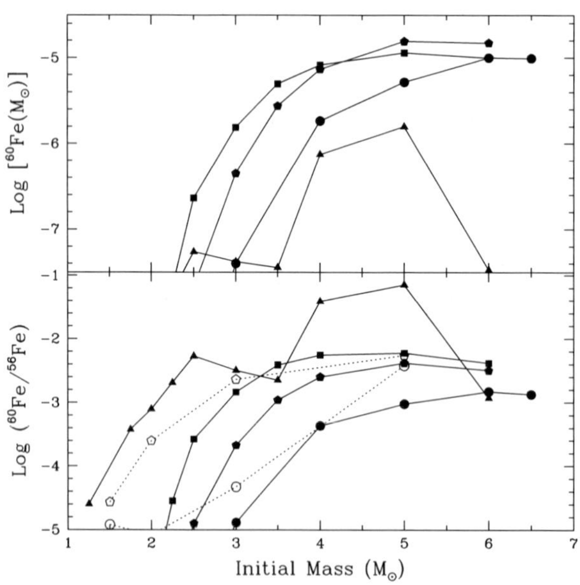


**Fig. 3.14** The yields of $^{60}$Fe (*top panel*) from (Karakas and Lattanzio, 2007) (see caption of Fig. 3.12 for definition of a yield) and the ratio of the yield of $^{60}$Fe to the yield of $^{56}$Fe (*bottom panel*) for stellar models of different masses and metallicities (*full symbols*) from Karakas and Lattanzio, 2007. The symbols representing the different metallicities are the same as in Fig. 3.12. For comparison, the ratios of the abundances of $^{60}$Fe and $^{56}$Fe at the end of the AGB evolution computed by Wasserburg et al., 2006 are also shown as *open symbols*

As for $^{26}$Al, also in the case of $^{60}$Fe stellar and nuclear uncertainties affect the results presented in Fig. 3.14 (and different choice in the model inputs are responsible for variations in the results obtained by different authors). First, the overall mass carried to the envelope via the 3rd dredged-up is essential to the determination of the envelope $^{60}$Fe abundance in AGB stars. This is because $^{60}$Fe is made only via neutron captures in the He-rich intershell and needs to be mixed into the envelope in order to show up at the stellar surface and to be carried to the interstellar medium by the winds. Hence, the $^{60}$Fe yield is directly related to the efficiency of the 3rd dredge-up. For example, models experiencing little or no 3rd dredge-up produce a null $^{60}$Fe yield. This important point applies to all long-living radioactive nuclei produced in AGB stars, except for the case of $^{26}$Al, which is made via HBB directly within the envelope. Second, the mass-loss rate affects the result as it determines the stellar lifetime and thus the number of thermal pulses and 3rd dredge-up episodes.

Nuclear physics inputs that contribute important uncertainties to the production of $^{60}$Fe are the rate of the neutron source reaction $^{22}$Ne($\alpha$,n)$^{25}$Mg, which determines how many neutrons are produced in the thermal pulses, and the neutron-capture cross section of $^{59}$Fe, which is estimated only theoretically (Rauscher and Thielemann, 2000) as the short half life of this nucleus hampers experimental determinations.

### *3.6.4 $^{36}$Cl and $^{41}$Ca*

Two more long-lived radioactive nuclei lighter than iron are of special interest because they are observed to be present in the early solar system and can be made by neutron captures in the intershell of AGB stars: $^{36}$Cl (with half life of 0.3 Myr) and $^{41}$Ca (with half life of 0.1 Myr). Differently from $^{60}$Fe, production of these nuclei does not require the activation of branching points, since $^{36}$Cl and $^{41}$Ca are made by neutron captures on $^{35}$Cl and $^{40}$Ca, respectively, which are stable nuclei with relatively high solar abundances. Neutron captures also destroy $^{36}$Cl and $^{41}$Ca via different channels, the predominants being $^{41}$Ca$(n,\alpha)^{38}$Ar, with $\sigma \simeq 360$ mbarn and $^{36}$Cl$(n,p)^{36}$S, with $\sigma \simeq 118$ mbarn.

Neutrons coming from the $^{22}$Ne neutron source are responsible for the production of $^{36}$Cl and $^{41}$Ca. As there are no branching points involved, this is not due to the high neutron density of this neutron flux, as it is for the production of $^{60}$Fe, but to the fact that neutrons released by the $^{22}$Ne in the thermal pulse affect the composition of the whole He-rich intershell material, where large initial quantities of the seed nuclei $^{35}$Cl and $^{40}$Ca are available. On the contrary, neutrons released by the $^{13}$C neutron source affect a small fraction of the intershell material, being the $^{13}$C pocket roughly 1/10th to 1/20th of the intershell (by mass) in the current models.

In general, to produce neutron-rich isotopes of elements lighter than iron by the $s$ process a small number of neutrons captured by seed nucleus are needed: only one in the cases of $^{36}$Cl and $^{41}$Ca. Hence, final abundances are determined to a higher level by the availability of seed nuclei, rather than that of free neutrons. For the light nuclei a production flux from the lighter to the heaviest elements does not occur (strictly speaking it is not correct to apply the *s*-process terminology in this case), instead, the nucleosynthetic process is very localized: neutron captures on the sulphur isotopes, for example, do not affect the abundances of the chlorine isotopes and so on. This is because neutron-capture cross section of nuclei lighter than iron are much smaller (by as much as 3 orders of magnitude) than those of typical nuclei heavier than iron. Hence, to produce nuclei heavier than iron by the *s* process, instead, including the relatively large number of long-living radioactive nuclei lying on the *s*-process path discussed in Sect. 3.6.5, a production flux from the lighter to the heavier elements occurs, where many neutrons are captured by the iron seeds and it is possible to reach up to the heaviest elements. Hence, the number of free neutrons plays a dominant role in this case.

The $^{36}$Cl/$^{35}$Cl and $^{41}$Ca/$^{40}$Ca abundance ratios at the end of the AGB evolution computed by Wasserburg et al. 2006 and by van Raai, Lugaro, and Karakas (which are based on the same codes and stellar models of Karakas and Lattanzio, 2007) are plotted in Fig. 3.15. As in the case of $^{60}$Fe, the main model uncertainties affecting these results is the efficiency of the 3rd dredge-up, the mass-loss rate, and the rate of the $^{22}$Ne($\alpha$,n)$^{25}$Mg reaction.

Moreover, while experimental estimates for the neutron-capture cross section of $^{36}$Cl and $^{41}$Ca are available (e.g. de Smet et al., 2006), a difficult problem is to provide a reliable set of values for the electron-capture rate of $^{41}$Ca, in particular as it is expected to vary significantly for different temperatures and densities

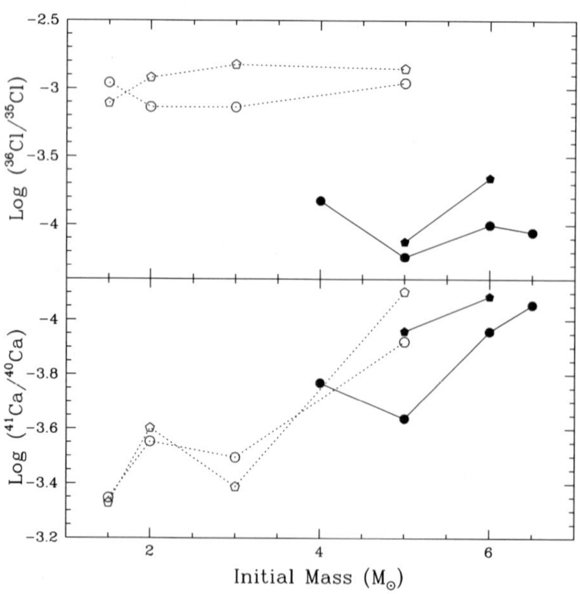


**Fig. 3.15** $^{36}$Cl/$^{35}$Cl and $^{41}$Ca/$^{40}$Ca abundance ratios at the end of the AGB evolution computed by (Wasserburg et al., 2006) (*open symbols*) and by van Raai, Lugaro, and Karakas (unpublished results, *full symbols*). The symbols representing the different metallicities are the same as in Fig. 3.12

relevant to stellar conditions (Chap. 9). As most electron captures in the $^{41}$Ca atom occurs on electrons belonging to the electron shell closest to the nucleus (the K shell), when the temperature increases to 100 MK and all electrons have escaped the atom leaving the nucleus bare, the half life of $^{41}$Ca increases by almost three orders of magnitude. However, if, still at a temperature of 100 MK, the density increases to $10^4$ g/cm$^3$, electrons are forced nearby $^{41}$Ca nuclei and the half life decreases back to its terrestrial value. The only set of theoretical data for this reaction are those provided by Fuller et al., 1982. Moreover, the temperature and density dependence of the electron-capture rate of $^{41}$Ca has never been properly implemented in AGB stellar models, in particular it has not yet been solved coupled to convective motions, both in the thermal pulses and in the stellar envelope, where material is constantly carried from hotter denser regions to cooler less dense regions and viceversa. Given these considerations, we are far from an accurate determination of the abundance of $^{41}$Ca made by AGB stars.

In summary, and in relevance to the early solar system composition of long-lived radioactive nuclei discussed in Chapter 6, AGB stars can produce some of the radioactive nuclei found to be present in the early solar system: $^{26}$Al via hot bottom burning and $^{41}$Ca and $^{60}$Fe via neutron captures in the thermal pulse and the 3rd dredge-up. In certain mixing conditions the abundances of these nuclei can be produced by AGB stars in the same proportions observed in the early solar system (Wasserburg et al., 2006; Trigo-Rodriguez et al., 2009). On the other hand, $^{36}$Cl cannot be produced in the observed amount. Uncertainties in the neutron-capture cross sections of $^{35}$Cl and $^{36}$Cl may play a role in this context.

Finally, a characteristic signature of the AGB stars inventory of long-living radioactive nuclei, is that, unlike supernovae (Chaps. 4 and Sect. 5.3), AGB stars cannot possibly produce $^{55}$Mn, another nucleus of relevance to early solar system composition. This is because $^{55}$Mn is a proton-rich nucleus, lying on the proton-rich side of the valley of $\beta$-stability, and thus it cannot be made by neutron captures.

## *3.6.5 Long-Lived Radioactive Isotopes Heavier than Fe*

### 3.6.5.1 Predicted Isotopic Abundances

The *s* process in AGB stars produces significant abundances of six long-lived radioactive nuclei heavier than iron: $^{81}$Kr, $^{93}$Zr, $^{99}$Tc, $^{107}$Pd, $^{135}$Cs, and $^{205}$Pb. The survival of $^{135}$Cs and $^{205}$Pb in stellar environments is however very uncertain and can even be prevented because of the strong and uncertain temperature and density dependence of their half lives, decreasing by orders of magnitudes in stellar conditions and determined only theoretically (as in the case of $^{41}$Ca. See detailed discussion by Mowlavi et al., 1998; Wasserburg et al., 2006, and also Appendix B). While $^{93}$Zr, $^{99}$Tc, $^{107}$Pd, and $^{205}$Pb are on the main *s*-process path and are produced by neutron captures on the stable isotopes $^{92}$Zr, $^{98}$Mo,[11] $^{106}$Pd, and $^{204}$Pb,

---

[11] Followed by fast decay of $^{99}$Mo, with a half life of 66 h.

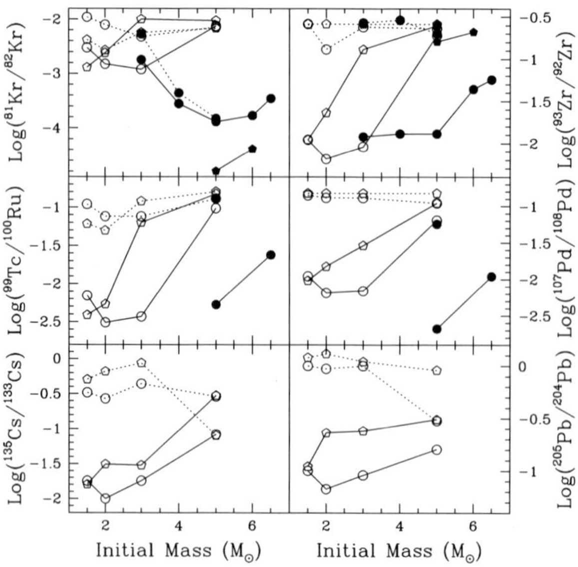


**Fig. 3.16** Abundance ratios of long-lived radioactive nuclei heavier than iron, with respect to one of their nearest stable isotope, at the end of the AGB evolution computed by (Wasserburg et al., 2006) (*open symbols*) and by van Raai, Lugaro, and Karakas (unpublished results, *full symbols*). The symbols representing the different metallicitics arc the same as in Fig. 3.12. Symbols connected by the *solid line* represent models computed without the inclusion of the $^{13}$C neutron source, symbols connected by the *dotted lines* represent models computed with the inclusion of the $^{13}$C neutron source

respectively, $^{81}$Kr and $^{135}$Cs are not on the main *s*-process path, but can be reached via the activation of branching points at $^{79}$Se and $^{80}$Br, and $^{134}$Cs, respectively (as described in Appendix B).

Figure 3.16 presents the abundance ratios of long-living radioactive isotopes heavier than iron produced during the *s* process in AGB stars to one of their nearest stable isotopes calculated by Wasserburg et al., 2006 and by van Raai et al. (unpublished results). For all ratios, except $^{81}$Kr/$^{82}$Kr, the inclusion of the $^{13}$C neutron source for models of masses lower than $\simeq$ 3–4 M$_\odot$, completely changes the results, since the $^{22}$Ne source is not significantly activated in these low-mass stellar models. It also makes an important difference in the absolute abundance of the all isotopes involved, with very low production factors with respect to the initial value if the $^{13}$C neutron source is not included (see Table 4 of Wasserburg et al., 2006). The case of $^{81}$Kr/$^{82}$Kr is different in that it does not feel the inclusion of the $^{13}$C neutron source as much as the other ratios because, even for the low-mass stars, the marginal activation of the $^{22}$Ne reaction in the latest thermal pulses affects the production of the *s*-process elements up to the first *s*-process peak, including Kr, and of $^{81}$Kr in particular via the branching point at $^{79}$Se.

For stellar models with initial masses higher than $\simeq$ 3–4 M$_\odot$, depending on the metallicity, the $^{22}$Ne neutron source is mainly responsible for the activation of the *s* process and thus the production of the heavy long-lived isotopes. Hence, in these models, the inclusion of a $^{13}$C neutron source typically does not make a significant difference in the final ratios, except in the case of $^{205}$Pb/$^{204}$Pb. This ratio is different in that it always feels the effect of the inclusion of the $^{13}$C neutron source because production of the element Pb, corresponding to the third and last *s*-process peak, is possible only if very large neutron exposures are available ($\sim$mbarn$^{-1}$), which can only be produced by the $^{13}$C neutron source.

It is interesting to discuss in detail the results for the 3 M$_\odot$ stellar model of 1/3 solar metallicity, because this model represents an example of the transition between the two regimes of the *s* process in AGB stars: when neutrons are provided by the $^{13}$C or by the $^{22}$Ne source. In this model the number of free neutrons produced by the $^{22}$Ne source is higher than in the solar metallicity model of the same mass partly because the temperature in the thermal pulses is slightly higher, but mostly because there is a smaller number of nuclei present to capture neutrons. Hence, the neutron flux coming from the $^{22}$Ne neutron source affects the production of the long-living isotopes up to $^{99}$Tc, but not that of the long-living isotopes of higher masses: for $^{107}$Pd, $^{133}$Cs, and $^{205}$Pb, $^{13}$C is still the main neutron source.

In addition to the main effect due to the shift from the $^{13}$C to the $^{22}$N regime with changing the initial mass and metallicity of the star, smaller variations due to the marginal effect of the $^{22}$Ne neutron source in the models of low-mass are always visible in the details of the production of the heavy long-living nuclei affected by the operation of the branching points activated in thermal pulses: $^{81}$Kr, and $^{135}$Cs. For example, restricting our view to the solar metallicity models of mass lower than 4 M$_\odot$ and computed with the inclusion of the $^{13}$C neutron source, the $^{81}$Kr/$^{82}$Kr ratio decreases with the stellar mass as $^{81}$Kr is progressively skipped by the branching point at $^{79}$Se at the higher neutron densities experienced by the higher mass models. The opposite happens for the $^{135}$Cs/$^{133}$Cs ratio, which increases with the stellar mass as the branching point at $^{134}$Cs becomes progressively more active.

When considering the effect of branching points on the production of heavy long-living radioactive nuclei by the *s* process in AGB stars it is worth noting that $^{129}$I and $^{182}$Hf – two long-lived radioactive isotopes of special interest for the composition of the early solar system – are not significantly produced in AGB stars. Production of $^{129}$I is not possible because the half life of $^{128}$I is only 25 minutes (see Appendix B), while $^{182}$Hf is produced,with $^{182}$Hf/$^{180}$Hf up to $\simeq$ 0.02 in the AGB envelope, via activation of the branching point at $^{181}$Hf (Appendix B). This results in $^{182}$Hf/$^{180}$Hf down to $10^{-6}$ (after dilution of AGB ejecta in the ISM, Wasserburg et al. 1994), which is too low to explain the early solar system value of $10^{-4}$ (see Chap. 6).

The main uncertainties affecting both sets of predictions shown in Fig. 3.16 are the detailed features of the proton diffusion leading to the production of the $^{13}$C neutron source (see end of Sect. 3.4.1). Ratios that depend on the activation of the $^{22}$Ne neutron source are also sensitive to the choice of the mass loss rate and of the $^{22}$Ne($\alpha$,n)$^{25}$Mg reaction rate. The treatment of branching points is also of importance in the determination of $^{81}$Kr and $^{135}$Cs. For example, in the case of $^{81}$Kr/$^{82}$Kr, the treatment of the temperature dependence of the decay rate of the branching point nucleus $^{79}$Se is fundamental to the final result, as demonstrated by the fact that including the temperature dependence of these decay rate (as carried out by Wasserburg et al., 2006) produce a $^{81}$Kr/$^{82}$Kr ratio two orders of magnitude larger than using the terrestrial value as constant (as done by van Raai, Lugaro, and Karakas, unpublished results).

In summary, due to the *s* 1process, AGB stars are a rich source of radioactive elements heavier than Fe. The signature of this production is confirmed, e.g., by the presence of Tc observed in AGB star and in meteoritic stardust grains (as discussed further in the next section) and possibly in primitive meteoritic solar-system materials (as discussed in detail in Chapter 6).

### Signatures in Stardust

The historical observation of Tc in late type giants (Sect. 3.5.2) was confirmed by the presence of $^{99}$Tc in single stardust SiC grains at the time of their formation discovered via laboratory analysis of the Ru isotopic composition of these grains (Savina et al., 2004). Since both Tc and Ru are refractory elements, they were included in SiC grains as trace elements during grain formation. To match the observational stardust data both the contribution of $^{99}$Ru and $^{99}$Tc predicted by AGB stellar models to the total nuclear abundance at mass 99 must be considered. Radiogenic decay of $^{99}$Tc occurs in the intershell in the absence of neutron fluxes, in the stellar envelope, and inside the grains.

On the other hand, there is no evidence for a contribution of $^{135}$Cs to $^{135}$Ba when comparing AGB model predictions to laboratory data of the $^{135}$Ba/$^{136}$Ba ratio in single SiC grains (see Fig. 16 of Lugaro et al., 2003). This is probably because Cs is not as refractory as Ba and thus was not included in the grains at the time of their formation.

## References

Abdurashitov, J.N., Bowles, T.J., Cherry, M.L., Cleveland, B.T., Davis, R., Elliott, S.R., Gavrin, V.N., Girin, S.V., Gorbachev, V.V., Ibragimova, T.V., Kalikhov, A.V., Khairnasov, N.G., Knodel, T.V., Lande, K., Mirmov, I.N., Nico, J.S., Shikhin, A.A., Teasdale, W.A., Veretenkin, E.P., Vermul, V.M., Wark, D.L., Wildenhain, P.S., Wilkerson, J.F., Yants, V.E., Zatsepin, G.T.: Measurement of the solar neutrino capture rate by SAGE and implications for neutrino oscillations in vacuum. Phys. Rev. Lett. **83**, 4686–4689 (1999). doi:10.1103/PhysRevLett.83.4686, arXiv:astro-ph/9907131

Abia, C., Busso, M., Gallino, R., Domínguez, I., Straniero, O., Isern, J.: The $^{85}$Kr s-process branching and the mass of carbon stars. ApJ **559**, 1117–1134 (2001). doi:10.1086/322383, arXiv:astroph/0105486

Amari, S., Gao, X., Nittler, L.R., Zinner, E., José, J., Hernanz, M., Lewis, R.S.: Presolar grains from Novae. ApJ **551**, 1065–1072 (2001a). doi:10.1086/320235, arXiv:astro-ph/0012465

Amari, S., Nittler, L.R., Zinner, E., Gallino, R., Lugaro, M., Lewis, R.S.: Presolar sic grains of type y: origin from low-metallicity asymptotic giant branch stars. ApJ **546**, 248–266 (2001b). doi:10.1086/318230

Amari, S., Nittler, L.R., Zinner, E., Lodders, K., Lewis, R.S.: Presolar SiC grains of type A and B: their isotopic compositions and stellar origins. ApJ **559**, 463–483 (2001c). doi:10.1086/322397

Aoki, W., Honda, S., Beers, T.C., Sneden, C.: Measurement of the Europium isotope ratio for the extremely metal poor, r-process-enhanced star CS 31082-001. ApJ **586**, 506–511 (2003a). doi:10.1086/367540, arXiv:astro-ph/0211617

Aoki, W., Ryan, S.G., Iwamoto, N., Beers, T.C., Norris, J.E., Ando, H., Kajino, T., Mathews, G.J., Fujimoto, M.Y.: Europium isotope ratios in s-process element-enhanced metal-poor stars: a new probe of the $^{151}$Sm branching. ApJ **592**, L67–L70 (2003b). doi:10.1086/377681, arXiv:astroph/0306544

Arlandini, C., Käppeler, F., Wisshak, K., Gallino, R., Lugaro, M., Busso, M., Straniero, O.: Neutron capture in low-mass asymptotic giant branch stars: cross sections and abundance signatures. ApJ **525**, 886–900 (1999). doi:10.1086/307938, arXiv:astro-ph/9906266

Bahcall, J.N.: Solar models and solar neutrinos. Physica Scripta Vol. T **121**, 46–50 (2005). doi:10.1088/0031-8949/2005/T121/006, arXiv:hep-ph/0412068

Bahcall, J.N., Peña-Garay, C.: Solar models and solar neutrino oscillations. New J. Phys. **6**, 63–89 (2004). doi:10.1088/1367-2630/6/1/063, arXiv:hep-ph/0404061

Bahcall, J.N., Cleveland, B.T., Davis R., Jr., Rowley, J.K.: Chlorine and gallium solar neutrino experiments. ApJ **292**, L79–L82 (1985). doi:10.1086/184477

Balachandran, S.C.: Anomalous abundances in red giants: the Li-rich stars. In: Barnes III, T.G., Bash, F.N. (eds.) Cosmic Abundances as Records of Stellar Evolution and Nucleosynthesis, vol. 336, pp. 113. Astronomical Society of the Pacific Conference Series (2005)

Bond, H., Sion, E., Murdin, P.: CH Stars and Barium Stars. Institute of Physics Publishing, Bristol (2000). doi:10.1888/0333750888/5413

Boothroyd, A.I., Sackmann, I.J., Wasserburg, G.J.: Hot bottom burning in asymptotic giant branch stars and its effect on oxygen isotopic abundances. ApJ **442**, L21–L24 (1995). doi:10.1086/187806

BOREXINO Collaboration, Arpesella, C., Bellini, G., Benziger, J., Bonetti, S., Caccianiga, B., Calaprice, F., Dalnoki-Veress, F., D'Angelo, D., de Kerret, H., Derbin, A., Deutsch, M., Etenko, A., Fomenko, K., Ford, R., Franco, D., Freudiger, B., Galbiati, C., Gazzana, S., Giammarchi, M., Goeger-Neff, M., Goretti, A., Grieb, C., Hardy, S., Heusser, G., Ianni, A., Ianni, A., Joyce, M., Korga, G., Kryn, D., Laubenstein, M., Leung, M., Litvinovich, E., Lombardi, P., Ludhova, L., Machulin, I., Manuzio, G., Martemianov, A., Masetti, F., McCarty, K., Meroni, E., Miramonti, L., Misiaszek, M., Montanari, D., Monzani, M.E., Muratova, V., Niedermeier, L., Oberauer, L., Obolensky, M., Ortica, F., Pallavicini, M., Papp, L., Perasso, L., Pocar, A., Raghavan, R.S., Ranucci, G., Razeto, A., Sabelnikov, A., Salvo, C., Schönert S, Simgen, H., Smirnov, O., Skorokhvatov, M., Sonnenschein, A., Sotnikov, A., Sukhotin, S., Suvorov, Y., Tarasenkov, V., Tartaglia, R., Testera, G., Vignaud, D., Vitale, S., Vogelaar, R.B., von Feilitzsch, F., Wojcik, M., Zaimidoroga, O., Zavatarelli, S., Zuzel, G.: First real time detection of 7Be solar neutrinos by Borexino. Phys. Lett. B **658**, 101–108 (2008). doi:10.1016/j.physletb.2007.09.054, 0708.2251

Brandon, A.D., Humayun, M., Puchtel, I.S., Leya, I., Zolensky, M.: Osmium isotope evidence for an s-process carrier in primitive chondrites. Science **309**, 1233–1236 (2005). doi:10.1126/science.1115053

Cameron, A.G.W., Fowler, W.A.: Lithium and the s-process in red-giant stars. ApJ **164**, 111–114 (1971). doi:10.1086/150821

Castilho, B.V., Gregorio-Hetem, J., Spite, F., Barbuy, B., Spite, M.: Detailed analysis of a sample of Li-rich giants. A&A **364**, 674–682 (2000)

Clayton, D.D.: Principles of Stellar Evolution and Nucleosynthesis. McGraw Hill, New York (1968)

Clayton, D.D., Nittler, L.R.: Astrophysics with presolar stardust. ARA&A **42**, 39–78 (2004). doi:10.1146/annurev.astro.42.053102.134022

Cleveland, B.T., Daily, T., Davis, R.J., Distel, J.R., Lande, K., Lee, C.K., Wildenhain, P.S., Ullman, J.: Measurement of the solar electron neutrino flux with the homestake chlorine detector. ApJ **496**, 505–526 (1998). doi:10.1086/305343

Cosner, K., Iben, I., Jr., Truran, J.W.: The effects of unthermalized isomeric states and of a time varying neutron flux on s-process branching ratios. ApJ **238**, L91–L96 (1980). doi:10.1086/183265

Cox, J.P., Giuli, R.T.: Principles of stellar structure. Gordon and Breach, New York (1968)

Cristallo, S., Straniero, O., Gallino, R., Piersanti, L., Domínguez, I., Lederer, M.T.: Evolution, nucleosynthesis, and yields of low-mass asymptotic giant branch stars at different metallicities. ApJ **696**, 797–820 (2009). doi:10.1088/0004-637X/696/1/797

de Smet, L., Wagemans, C., Heyse, J., Vermote, S., Van Gils, J.: Experimental determination of the $^{41}$Ca$(n, \alpha)^{38}$ Ar reaction cross section. In: Ninth International Symposium on Nuclei in the Cosmos, Proceedings of Science, PoS(NIC-IX)085 (2006)

Denissenkov, P.A., Tout, C.A.: Partial mixing and formation of the $^{13}$C pocket by internal gravity waves in asymptotic giant branch stars. MNRAS **340**, 722–732 (2003). doi:10.1046/j.1365-8711.2003.06284.x

Despain, K.H.: A difficulty with Ne-22 as the neutron source for the solar system s-process. ApJ **236**, L165–L168 (1980). doi:10.1086/183219

Doherty, C.L., Lattanzio, J.C.: Evolution and nucleosynthesis in super-AGB stars. Memorie della Societa Astronomica Italiana **77**, 828–832 (2006)

Dorfi, E.A., Höfner, S., Feuchtinger, M.U.: Pulsation and Mass Loss, pp. 137–154. Kluwer Academic Publisher, Dordrecht (2001)

Dupree, A.K.: Mass loss from cool stars. ARA&A **24**, 377–420 (1986). doi:10.1146/annurev.aa.24.090186.002113

Ferrarotti, A.S., Gail, H.P.: Mineral formation in stellar winds. III. Dust formation in S stars. A&A **382**, 256–281 (2002). doi:10.1051/0004-6361:20011580

Fleischer, A.J., Gauger, A., Sedlmayr, E.: Circumstellar dust shells around long-period variables. I – Dynamical models of C-stars including dust formation, growth and evaporation. A&A **266**, 321–339 (1992)

Frogel, J.A., Mould, J., Blanco, V.M.: The asymptotic giant branch of Magellanic cloud clusters. ApJ **352**, 96–122 (1990). doi:10.1086/168518

Fuller, G.M., Fowler, W.A., Newman, M.J.: Stellar weak interaction rates for intermediate mass nuclei. III – Rate tables for the free nucleons and nuclei with A = 21 to A = 60. ApJS **48**, 279–319 (1982). doi:10.1086/190779

Gail, H.P., Sedlmayr, E.: Mineral formation in stellar winds. I. Condensation sequence of silicate and iron grains in stationary oxygen rich outflows. A&A **347**, 594–616 (1999)

Gallino, R., Arlandini, C., Busso, M., Lugaro, M., Travaglio, C., Straniero, O., Chieffi, A., Limongi, M.: Evolution and nucleosynthesis in low-mass asymptotic giant branch stars. II. Neutron capture and the s-Process. ApJ **497**, 388–403 (1998). doi:10.1086/305437

García-Hernández, D.A., García-Lario, P., Plez, B., D'Antona, F., Manchado, A., Trigo-Rodríguez, J.M.: Rubidium-rich asymptotic giant branch stars. Science **314**, 1751–1754 (2006). doi:10.1126/science.1133706, arXiv:astro-ph/0611319

García-Hernández, D.A., García-Lario, P., Plez, B., Manchado, A., D'Antona, F., Lub, J., Habing, H.: Lithium and zirconium abundances in massive Galactic O-rich AGB stars. A&A **462**, 711–730 (2007). doi:10.1051/0004-6361:20065785, arXiv:astro-ph/0609106

Goriely, S., Mowlavi, N.: Neutron-capture nucleosynthesis in AGB stars. A&A **362**, 599–614 (2000)

Guelin, M., Forestini, M., Valiron, P., Ziurys, L.M., Anderson, M.A., Cernicharo, J., Kahane, C.: Nucleosynthesis in AGB stars: observation of Mg-25 and Mg-26 in IRC+10216 and possible detection of Al-26. A&A **297**, 183–196 (1995)

Hampel, W., Heusser, G., Kiko, J., Kirsten, T., Laubenstein, M., Pernicka, E., Rau, W., Roenn, U., Schlosser, C., Wojcik, M., von Ammon, R., Ebert, K.H., Fritsch, T., Heidt, D., Henrich, E., Stieglitz, L., Weirich, F., Balata, M., Hartmann, F.X., Sann, M., Bellotti, E., Cattadori, C., Cremonesi, O., Ferrari, N., Fiorini, E., Zanotti, L., Altmann, M., von Feilitzsch, F., Moessbauer, R., Berthomieu, G., Schatzman, E., Carmi, I., Dostrovsky, I., Bacci, C., Belli, P., Bernabei, R., D'Angelo, S., Paoluzi, L., Bevilacqua, A., Cribier, M., Gosset, L., Rich, J., Spiro, M., Tao, C., Vignaud, D., Boger, J., Hahn, R.L., Rowley, J.K., Stoenner, R.W., Weneser, J.: Final results of the $^{51}$Cr neutrino source experiments in GALLEX. Phys. Lett. B **420**, 114–126 (1998)

Harris, M.J., Lambert, D.L., Hinkle, K.H., Gustafsson, B., Eriksson, K.: Oxygen isotopic abundances in evolved stars. III – 26 carbon stars. ApJ **316**, 294–304 (1987). doi:10.1086/165201

Herwig, F.: Evolution of asymptotic giant branch stars. ARA&A **43**, 435–479 (2005). doi:10.1146/annurev.astro.43.072103.150600

Herwig, F., Bloecker, T., Schoenberner, D., El Eid, M.: Stellar evolution of low and intermediate-mass stars. IV. Hydrodynamically-based overshoot and nucleosynthesis in AGB stars. A&A **324**, L81–L84 (1997). arXiv:astro-ph/9706122

Herwig, F., Blöcker, T., Langer, N., Driebe, T.: On the formation of hydrogen-deficient post-AGB stars. A&A **349**, L5–L8 (1999). arXiv:astro-ph/9908108

Hoffman, R.D., Müller, B., Janka, H.: Nucleosynthesis in O − −Ne − −Mg supernovae. ApJ **676**, L127–L130 (2008). doi:10.1086/587621, 0712.4257

Hoppe, P., Besmehn, A.: Evidence for Extinct Vanadium-49 in Presolar Silicon Carbide Grains from Supernovae. **576**, L69–L72 (2002). doi: 10.1086/342785

Hoppe, P., Ott, U.: Mainstream silicon carbide grains from meteorites. In: Bernatowicz, T.J., Zinner, E. (eds.) American Institute of Physics Conference Series, American Institute of Physics Conference Series, vol. 402, pp. 27–58 (1997). doi:10.1063/1.53314

Hoppe, P., Annen, P., Strebel, R., Eberhardt, P., Gallino, R., Lugaro, M., Amari, S., Lewis, R.S.: Meteoritic silicon carbide grains with unusual Si isotopic compositions: evidence for an origin in low-mass, low-metallicity asymptotic giant branch stars. ApJ **487**, L101–L104 (1997). doi:10.1086/310869

Humayun, M., Brandon, A.D.: s-Process implications from osmium isotope anomalies in chondrites. ApJ **664**, L59–L62 (2007). doi:10.1086/520636

Iben, I., Jr.: Thermal pulses; p-capture, alpha-capture, s-process nucleosynthesis; and convective mixing in a star of intermediate mass. ApJ **196**, 525–547 (1975). doi:10.1086/153433

Iben, I., Jr., Renzini, A.: On the formation of carbon star characteristics and the production of neutron-rich isotopes in asymptotic giant branch stars of small core mass. ApJ **263**, L23–L27 (1982). doi:10.1086/183916

Iben, I., Jr., Renzini, A.: Asymptotic giant branch evolution and beyond. ARA&A **21**, 271–342 (1983). doi:10.1146/annurev.aa.21.090183.001415

Iben, I., Jr., Truran, J.W.: On the surface composition of thermally pulsing stars of high luminosity and on the contribution of such stars to the element enrichment of the interstellar medium. ApJ **220**, 980–995 (1978). doi:10.1086/155986

Iliadis, C., D'Auria, J.M., Starrfield, S., Thompson, W.J., Wiescher, M.: Proton-induced thermonuclear reaction rates for A = 20–40 nuclei. ApJS **134**, 151–171 (2001). doi:10.1086/320364

Iliadis, C., Angulo, C., Descouvemont, P., Lugaro, M., Mohr, P.: New reaction rate for O16($p$, $\gamma$) F17 and its influence on the oxygen isotopic ratios in massive AGB stars. Phys. Rev. C **77**(4), id. 045802 (2008). doi:10.1103/PhysRevC.77.045802, 0803.2757

Itoh, N., Adachi, T., Nakagawa, M., Kohyama, Y., Munakata, H.: Neutrino energy loss in stellar interiors. III – Pair, photo-, plasma, and bremsstrahlung processes. ApJ **339**, 354–364 (1989). doi:10.1086/167301

Izzard, R.G., Lugaro, M., Karakas, A.I., Iliadis, C., van Raai, M.: Reaction rate uncertainties and the operation of the NeNa and MgAl chains during HBB in intermediate-mass AGB stars. A&A **466**, 641–648 (2007). doi:10.1051/0004-6361:20066903, arXiv:astro-ph/0703078

Jonsell, K., Barklem, P.S., Gustafsson, B., Christlieb, N., Hill, V., Beers, T.C., Holmberg, J.: The Hamburg/ESO R-process enhanced star survey (HERES). III. HE 0338-3945 and the formation of the r + s stars. A&A **451**, 651–670 (2006). doi:10.1051/0004-6361:20054470, arXiv:astro-ph/0601476

Jorissen, A., Van Eck, S., Mayor, M., Udry, S.: Insights into the formation of barium and Tcpoor S stars from an extended sample of orbital elements. A&A **332**, 877–903 (1998). arXiv:astroph/9801272

Kaeppeler, F., Beer, H., Wisshak, K., Clayton, D.D., Macklin, R.L., Ward, R.A.: S-process studies in the light of new experimental cross sections – Distribution of neutron fluences and r-process residuals. ApJ **257**, 821–846 (1982). doi:10.1086/160033

Karakas, A., Lattanzio, J.C.: Stellar models and yields of asymptotic giant branch stars. Publications of the Astronomical Society of Australia **24**, 103–117 (2007). doi:10.1071/AS07021,0708.4385

Lambert, D.L., Smith, V.V., Busso, M., Gallino, R., Straniero, O.: The chemical composition of red giants. IV. The neutron density at the s-process site. ApJ **450**, 302–317 (1995). doi:10.1086/176141

Langer, N., Heger, A.,Wellstein, S., Herwig, F.: Mixing and nucleosynthesis in rotating TP-AGB stars. A&A **346**, L37–L40 (1999). arXiv:astro-ph/9904257

Lattanzio, J.C., Wood, P.R.: Evolution, nucleosynthesis and pulsation of AGB stars. In: Habing, H.J., Olofsson, H. (eds.) Asymptotic Giant Branch Stars, Astronomy and Astrophysics Library, vol. 145, pp. 23–104. Springer-Verlag, Berlin (2004)

Lewis, R.S., Amari, S., Anders, E.: Interstellar grains in meteorites: II. SiC and its noble gases. Geochim. Cosmochim. Acta **58**, 471–494 (1994). doi:10.1016/0016-7037(94)90478-2

Limongi, M., Chieffi, A.: The nucleosynthesis of $^{26}$Al and $^{60}$Fe in solar metallicity stars extending in mass from 11 to 120 solar masses: the hydrostatic and explosive contributions. ApJ **647**, 483–500 (2006). doi:10.1086/505164

Lodders, K., Fegley, B., Jr.: The origin of circumstellar silicon carbide grains found in meteorites. Meteoritics, **30**, 661–678 (1995)

Lucatello, S., Tsangarides, S., Beers, T.C., Carretta, E., Gratton, R.G., Ryan, S.G.: The binary frequency among carbon-enhanced, s-process-rich, metal-poor stars. ApJ **625**, 825–832 (2005). doi:10.1086/428104, arXiv:astro-ph/0412422

Lugaro, M.: Stardust from meteorites. An introduction to presolar grains. World Scientific, New York (2005)

Lugaro, M., Karakas, A.I.: $^{26}$Al and $^{60}$Fe yields from AGB stars. New Astron. Rev. **52**, 416–418 (2008). doi:10.1016/j.newar.2008.05.005

Lugaro, M., Davis, A.M., Gallino, R., Pellin, M.J., Straniero, O., Käppeler, F.: Isotopic compositions of Strontium, Zirconium, Molybdenum, and Barium in single presolar SiC grains and asymptotic giant branch stars. ApJ **593**, 486–508 (2003a). doi:10.1086/376442

Lugaro, M., Herwig, F., Lattanzio, J.C., Gallino, R., Straniero, O.: s-Process nucleosynthesis in asymptotic giant branch stars: a test for stellar evolution. ApJ **586**, 1305–1319 (2003b). doi:10.1086/367887, arXiv:astro-ph/0212364

Lugaro, M., Karakas, A.I., Nittler, L.R., Alexander, C.M.O'D., Hoppe, P., Iliadis, C., Lattanzio, J.C.: On the asymptotic giant branch star origin of peculiar spinel grain OC2. A&A **461**, 657–664 (2007). doi:10.1051/0004-6361:20065768, arXiv:astro-ph/0610464

Marhas, K.K., Hoppe, P., Ott, U.: NanoSIMS studies of Ba isotopic compositions in single presolar silicon carbide grains from AGB stars and supernovae. Meteorit. Planet. Sci. **42**, 1077–1101 (2007)

McDonald, A.B., Ahmad, Q.R., Allen, R.C., Andersen, T.C., Anglin, J.D., Barton, J.C., Beier, E.W., Bercovitch, M., Bigu, J., Biller, S.D., Black, R.A., Blevis, I., Boardman, R.J., Boger, J., Bonvin, E., Boulay, M.G., Bowler, M.G., Bowles, T.J., Brice, S.J., Browne, M.C., Bullard, T.V., Bühler, G., Cameron, J., Chan, Y.D., Chen, H.H., Chen, M., Chen, X., Cleveland, B.T., Clifford, E.T.H., Cowan, J.H.M., Cowen, D.F., Cox, G.A., Dai, X., Dalnoki-Veress, F., Davidson, W.F., Doe, P.J., Doucas, G., Dragowsky, M.R., Duba, C.A., Duncan, F.A., Dunford, M., Dunmore, J.A., Earle, E.D., Elliott, S.R., Evans, H.C., Ewan, G.T., Farine, J., Fergani, H., Ferraris, A.P., Ford, R.J., Formaggio, J.A., Fowler, M.M., Frame, K., Frank, E.D., Frati, W., Gagnon, N., Germani, J.V., Gil, S., Graham, K., Grant, D.R., Hahn, R.L., Hallin, A.L., Hallman, E.D., Hamer, A.S., Hamian, A.A., Handler, W.B., Haq, R.U., Hargrove, C.K., Harvey, P.J., Hazama, R., Heeger, K.M., Heintzelman, W.J., Heise, J., Helmer, R.L., Hepburn, J.D., Heron, H., Hewett, J., Hime, A., Howe, M., Hykawy, J.G., Isaac, M.C.P., Jagam, P., Jelley, N.A., Jillings, C., Jonkmans, G., Kazkaz, K., Keener, P.T., Klein, J.R., Knox, A.B., Komar, R.J., Kouzes, R., Kutter, T., Kyba, C.C.M., Law, J., Lawson, I.T., Lay, M., Lee, H.W., Lesko, K.T., Leslie, J.R., Levine, I., Locke, W., Luoma, S., Lyon, J., Majerus, S., Mak, H.B., Maneira, J., Manor, J., Marino, A.D., McCauley, N., McDonald, D.S., McFarlane, K., McGregor, G., Drees, R.M., Mifflin, C., Miller, G.G., Milton, G., Moffat, B.A., Moorhead, M., Nally, C.W., Neubauer, M.S., Newcomer, F.M., Ng, H.S., Noble, A.J., Norman, E.B., Novikov, V.M., O'Neill, M., Okada, C.E., Ollerhead, R.W., Omori, M., Orrell, J.L., Oser, S.M., Poon, A.W.P., Radcliffe,

T.J., Roberge, A., Robertson, B.C., Robertson, R.G.H., Rosendahl, S.S.E., Rowley, J.K., Rusu, V.L., Saettler, E., Schaffer, K.K., Schwendener, M.H., Schülke, A., Seifert, H., Shatkay, M., Simpson, J.J., Sims, C.J., Sinclair, D., Skensved, P., Smith, A.R., Smith, M.W.E., Spreitzer, T., Starinsky, N., Steiger, T.D., Stokstad, R.G., Stonehill, L.C., Storey, R.S., Sur, B., Tafirout, R., Tagg, N., Tanner, N.W., Taplin, R.K., Thorman, M., Thornewell, P.M., Trent, P.T., Tserkovnyak, Y.I., van Berg, R., van de Water, R.G., Virtue, C.J., Waltham, C.E., Wang, J., Wark, D.L., West, N., Wilhelmy, J.B., Wilkerson, J.F., Wilson, J.R., Wittich, P., Wouters, J.M., Yeh, M.: Direct evidence for neutrino flavor transformation from neutral-current interactions in SNO. In: Elias, V., Epp, R., Myers, R.C. (eds.) Theoretical Physics: MRST 2002, American Institute of Physics Conference Series, vol. 646, pp. 43–58 (2002). doi:10.1063/1.1524553

McWilliam, A., Lambert, D.L.: Isotopic magnesium abundances in stars. MNRAS **230**, 573–585 (1988)

Mowlavi, N., Meynet, G.: Aluminum 26 production in asymptotic giant branch stars. A&A **361**, 959–976 (2000)

Mowlavi, N., Goriely, S., Arnould, M.: The survival of $^{205}$Pb in intermediate-mass AGB stars. A&A **330**, 206–214 (1998). arXiv:astro-ph/9711025

Nichols, R.H., Jr., Hohenberg, C.M., Amari, S., Lewis, R.S.: $^{22}$Ne-E(H) and $^{4}$He measured in individual SiC grains using laser gas extraction. Meteoritics **26**, 377–378 (1991)

Nicolussi, G.K., Davis, A.M., Pellin, M.J., Lewis, R.S., Clayton, R.N., Amari, S.: s-Process zirconium in presolar silicon carbide grains. Science **277**, 1281–1283 (1997). doi:10.1126/science.277.5330.1281

Nicolussi, G.K., Pellin, M.J., Lewis, R.S., Davis, A.M., Amari, S., Clayton, R.N.: Molybdenum isotopic composition of individual presolar silicon carbide grains from the Murchison meteorite. Geochim. Cosmochim. Act **62**, 1093–1104 (1998a). doi:10.1016/S0016-7037(98)00038-6

Nicolussi, G.K., Pellin, M.J., Lewis, R.S., Davis, A.M., Clayton, R.N., Amari, S.: Strontium isotopic composition in individual circumstellar silicon carbide grains: a record of s-process nucleosynthesis. Phys. Rev. Lett. **81**, 3583–3586 (1998b). doi:10.1103/PhysRevLett.81.3583

Nittler, L.R., Hoppe, P., Alexander, C.M.O'D., Amari, S., Eberhardt, P., Gao, X., Lewis, R.S., Strebel, R., Walker, R.M., Zinner, E.: Silicon nitride from supernovae. ApJ **453**, L25–L28 (1995). doi:10.1086/309743

Nittler, L.R., Alexander, C.M.O'D., Gao, X.,Walker, R.M., Zinner, E.: Stellar sapphires: the properties and origins of presolar $AL_2O_3$ in meteorites. ApJ **483**, 475–495 (1997). doi:10.1086/304234

Nittler, L.R., Alexander, C.M.O'D., Gallino, R., Hoppe, P., Nguyen, A.N., Stadermann, F.J., Zinner, E.K.: Aluminum-, Calcium- and Titanium-rich Oxide stardust in ordinary chondrite meteorites. ApJ **682**, 1450–1478 (2008). doi:10.1086/589430, 0804.2866

Nollett, K.M., Busso, M., Wasserburg, G.J.: Cool bottom processes on the thermally pulsing asymptotic giant branch and the isotopic composition of circumstellar dust grains. ApJ **582**, 1036–1058 (2003). doi:10.1086/344817, arXiv:astro-ph/0211271

Oberauer, L.: Results from the solar neutrino experiment BOREXINO. J. Phys. Conf. Series **203**(1), id. 012081 (2010). doi:10.1088/1742-6596/203/1/012081

Ott, U., Begemann, F.: Discovery of s-process barium in the Murchison meteorite. ApJ **353**, L57–L60 (1990). doi:10.1086/185707

Pignatari, M., Gallino, R., Straniero, O., Reifarth, R., Käppeler, F., Davis, A.M.: Stellar origin of the meteoritic Xe-S anomalous component. Memorie della Societa Astronomica Italiana **75**, 182–185 (2004)

Pignatari, M., Gallino, R., Amari, S., Davis, A.M.: Krypton in presolar mainstream SiC grains from AGB stars. Memorie della Societa Astronomica Italiana **77**, 897–902 (2006)

Podosek, F.A., Prombo, C.A., Amari, S., Lewis, R.S.: s-Process Sr isotopic compositions in presolar SiC from the Murchison meteorite. ApJ **605**, 960–965 (2004). doi:10.1086/382650

Prombo, C.A., Podosek, F.A., Amari, S., Lewis, R.S.: S-process BA isotopic compositions in presolar SiC from the Murchison meteorite. ApJ **410**, 393–399 (1993). doi:10.1086/172756

Raiteri, C.M., Gallino, R., Busso, M.: S-processing in massive stars as a function of metallicity and interpretation of observational trends. ApJ **387**, 263–275 (1992). doi:10.1086/171078

Rauscher, T., Thielemann, F.K.: Astrophysical reaction rates from statistical model calculations. Atomic Data and Nuclear Data Tables **75**, 1–351 (2000). doi:10.1006/adnd.2000.0834, arXiv:astro-ph/0004059

Rugel, G., Faestermann, T., Knie, K., Korschinek, G., Poutivtsev, M., Schumann, D., Kivel, N., Günther-Leopold, I., Weinreich, R.,Wohlmuther, M.: New measurement of the Fe60 half-life. Phys. Rev. Lett. **103**(7), id. 072502 (2009). doi:10.1103/PhysRevLett.103.072502

Savina, M.R., Davis, A.M., Tripa, C.E., Pellin, M.J., Clayton, R.N., Lewis, R.S., Amari, S., Gallino, R., Lugaro, M.: Barium isotopes in individual presolar silicon carbide grains from the Murchison meteorite. Geochim. Cosmochim. Acta **67**, 3201–3214 (2003a). doi:10.1016/S0016-7037(03)00083-8

Savina, M.R., Pellin, M.J., Tripa, C.E., Veryovkin, I.V., Calaway, W.F., Davis, A.M.: Analyzing individual presolar grains with CHARISMA. Geochim. Cosmochim. Acta **67**, 3215–3225 (2003b). doi:10.1016/S0016-7037(03)00082-6

Savina, M.R., Davis, A.M., Tripa, C.E., Pellin, M.J., Gallino, R., Lewis, R.S., Amari, S.: Extinct technetium in silicon carbide stardust grains: implications for stellar nucleosynthesis. Science **303**, 649–652 (2004). doi:10.1126/science.3030649

Sedlmayr, E., Dominik, C.: Dust driven winds. Space Sci. Rev. **73**, 211–272 (1995). doi:10.1007/BF00751238

Siess, L.: Evolution of massive AGB stars. I. Carbon burning phase. A&A **448**, 717–729 (2006). doi:10.1051/0004-6361:20053043

Siess, L.: Evolution of massive AGB stars. II. model properties at non-solar metallicity and the fate of Super-AGB stars. A&A **476**, 893–909 (2007). doi:10.1051/0004-6361:20078132

Siess, L., Arnould, M.: Production of 26Al by super-AGB stars. A&A **489**, 395–402 (2008). doi:10.1051/0004-6361:200810147

Smith, V.V., Lambert, D.L.: The chemical composition of red giants. II – Helium burning and the s-process in the MS and S stars. ApJ **311**, 843–863 (1986). doi:10.1086/164823

Sneden, C., Cowan, J.J., Lawler, J.E., Burles, S., Beers, T.C., Fuller, G.M.: Europium isotopic abundances in very metal poor stars. ApJ **566**, L25–L28 (2002). doi:10.1086/339471, arXiv:astroph/0201456

Sneden, C., Cowan, J.J., Gallino, R.: Neutron-capture elements in the early galaxy. ARA&A **46**, 241–288 (2008). doi:10.1146/annurev.astro.46.060407.145207

Speck, A.K., Barlow, M.J., Sylvester, R.J., Hofmeister, A.M.: Dust features in the 10-mu m infrared spectra of oxygen-rich evolved stars. A&AS **146**, 437–464 (2000). doi:10.1051/aas:2000274

Speck, A.K., Corman, A.B., Wakeman, K., Wheeler, C.H., Thompson, G.: Silicon carbide absorption features: dust formation in the outflows of extreme carbon stars. ApJ **691**, 1202–1221 (2009). doi:10.1088/0004-637X/691/2/1202, 0810.2599

Stancliffe, R.J., Chieffi, A., Lattanzio, J.C., Church, R.P.: Why do low-mass stars become red giants? Publications of the Astronomical Society of Australia **26**, 203–208 (2009). doi:10.1071/AS08060, 0902.0406

Terada, K., Itoh, K., Hidaka, H., Yoshida, T., Iwamoto, N., Aoki, W., Williams, I.S.: Eu isotope measurements on single SiC grains from the Murchison meteorite: A new probe of s-process conditions in parent Asymptotic Giant Branch stars. New Astron. Rev. **50**, 582–586 (2006). doi:10.1016/j.newar.2006.06.006

Travaglio, C., Gallino, R., Amari, S., Zinner, E., Woosley, S., Lewis, R.S.: Low-density graphite grains and mixing in type II supernovae. ApJ **510**, 325–354 (1999). doi:10.1086/306551

Treffers, R., Cohen, M.: High-resolution spectra of cool stars in the 10- and 20-micron regions. ApJ **188**, 545–552 (1974). doi:10.1086/152746

Trigo-Rodriguez, J.M., Anibal Garcia-Hernandez, D., Lugaro, M., Karakas, A.I., van Raai, M., Garcia Lario, P., Manchado, A.: The role of massive Agb stars in the early solar system composition. Meteorit. Planet. Sci. **44**, 627–641 (2009)

Uberseder, E., Reifarth, R., Schumann, D., Dillmann, I., Pardo, C.D., Görres, J., Heil, M., Käppeler, F., Marganiec, J., Neuhausen, J., Pignatari, M., Voss, F., Walter, S., Wiescher, M.: Measurement of the $^{60}$Fe$(n, \gamma)$ $^{61}$Fe cross section at stellar temperatures. Phys. Rev. Lett. **102**(15), id. 151101 (2009). doi:10.1103/PhysRevLett.102.151101

Uttenthaler, S., Lebzelter, T., Palmerini, S., Busso, M., Aringer, B., Lederer, M.T.: Lowmass lithium-rich AGB stars in the Galactic bulge: evidence for cool bottom processing? A&A **471**, L41–L45 (2007). doi:10.1051/0004-6361:20077879, 0707.1380

van Raai, M.A., Lugaro, M., Karakas, A.I., García-Hernández, D.A.: Rubidium and zirconium production in massive AGB stars. In: Guandalini, R., Palmerini, S., Busso, M. (eds.) Evolution and Nucleosynthesis in AGB Stars, American Institute of Physics Conference Series, vol. 1001, pp. 146–153 (2008a). doi:10.1063/1.2916957

van Raai, M.A., Lugaro, M., Karakas, A.I., Iliadis, C.: Reaction rate uncertainties and $^{26}$Al in AGB silicon carbide stardust. A&A **478**, 521–526 (2008b). doi:10.1051/0004-6361:20078307, 0712.3702

Verchovsky, A.B., Wright, I.P., Pillinger, C.T.: Astrophysical significance of asymptotic giant branch stellar wind energies recorded in meteoritic SiC grains. ApJ **607**, 611–619 (2004). doi:10.1086/383230

Vollmer, C., Hoppe, P., Brenker, F.E.: Si isotopic compositions of presolar silicate grains from red giant stars and supernovae. ApJ **684**, 611–617 (2008). doi:10.1086/589913

Wanajo, S., Nomoto, K., Janka, H., Kitaura, F.S., Müller, B.: Nucleosynthesis in electron capture supernovae of asymptotic giant branch stars. ApJ **695**, 208–220 (2009). doi:10.1088/0004-637X/695/1/208, 0810.3999

Wasserburg, G.J., Busso, M., Gallino, R., Raiteri, C.M.: Asymptotic giant branch stars as a source of short-lived radioactive nuclei in the solar nebula. ApJ **424**, 412–428 (1994). doi:10.1086/173899

Wasserburg, G.J., Boothroyd, A.I., Sackmann, I.J.: Deep circulation in red giant stars: a solution to the carbon and oxygen isotope puzzles? ApJ **447**, L37–L40 (1995). doi:10.1086/309555

Wasserburg, G.J., Busso, M., Gallino, R., Nollett, K.M.: Short-lived nuclei in the early solar system: possible AGB sources. Nucl. Phys. A **777**, 5–69 (2006). doi:10.1016/j.nuclphysa.2005.07.015, arXiv:astro-ph/0602551

Whittet, D.C.B.: Dust in the galactic environment. Institute of Physics Publishing, New York (1992)

Willson, L.A.: Mass loss from cool stars: impact on the evolution of stars and stellar populations. ARA&A **38**, 573–611 (2000). doi:10.1146/annurev.astro.38.1.573

Wood, P.R.: Pulsation and mass loss in Mira variables. ApJ **227**, 220–231 (1979). doi:10.1086/156721

Yin, Q.Z., Lee, C.T.A., Ott, U.: Signatures of the s-process in presolar Silicon Carbide grains: Barium through Hafnium. ApJ **647**, 676–684 (2006). doi:10.1086/505188

Zinner, E., Amari, S., Lewis, R.S.: S-process Ba, Nd, and SM in presolar SiC from the Murchison meteorite. ApJ **382**, L47–L50 (1991). doi:10.1086/186210

Zinner, E., Nittler, L.R., Gallino, R., Karakas, A.I., Lugaro, M., Straniero, O., Lattanzio, J.C.: Silicon and carbon isotopic ratios in AGB stars: SiC grain data, models, and the galactic evolution of the Si isotopes. ApJ **650**, 350–373 (2006). doi:10.1086/506957

Zinner, E., Amari, S., Guinness, R., Jennings, C., Mertz, A.F., Nguyen, A.N., Gallino, R., Hoppe, P., Lugaro, M., Nittler, L.R., Lewis, R.S.: NanoSIMS isotopic analysis of small presolar grains: search for $Si_3$ $N_4$ grains from AGB stars and Al and Ti isotopic compositions of rare presolar SiC grains. Geochim. Cosmochim. Acta **71**, 4786–4813 (2007). doi:10.1016/j.gca.2007.07.012

# Chapter 4
# Massive Stars and Their Supernovae

**F.-K. Thielemann, R. Hirschi, M. Liebendörfer, and R. Diehl**

## 4.1 Cosmic Significance of Massive Stars

Our understanding of stellar evolution and the final explosive endpoints such as supernovae or hypernovae or gamma-ray bursts relies on the combination of

(a) (magneto-)hydrodynamics
(b) energy generation due to nuclear reactions accomanying composition changes
(c) radiation transport
(d) thermodynamic properties (such as the equation of state of stellar matter).

Hydrodynamics is essentially embedded within the numerical schemes which implement the physics of processes (b) to (d). In early phases of stellar evolution, hydrodynamical processes can be approximated by a hydro*static* treatment. Nuclear energy production (b) includes all nuclear reactions triggered during stellar evolution and explosive end stages, also among unstable isotopes produced on the way. Radiation transport (c) covers atomic physics (e.g. opacities) for photon transport, but also nuclear physics and neutrino nucleon/nucleus interactions in late phases and core collapse. The thermodynamical treatment (d) addresses the mixture of *ideal gases* of photons, electrons/positrons and nuclei/ions. These are fermions and bosons, in dilute media or at high temperatures their energies can often be approximated by Maxwell-Boltzmann distributions. At very high densities, the *nuclear* equation of state is required to relate pressure and density. It exhibits a complex behavior, with transitions from individual nuclei to clusters of nucleons

F.-K. Thielemann (✉)
University of Basel, 4056 Basel, Switzerland, f-k.thielemann@unibas.ch

R. Hirschi
University of Keele, Keele, ST5 5BG, UK

M. Liebendörfer
University of Basel, 4056 Basel, Switzerland

R. Diehl
Max Planck Institut für extraterrestrische Physik, 85748 Garching, Germany

Thielemann, F.-K. et al.: *Massive Stars and Their Supernovae*. Lect. Notes Phys. **812**, 153–231 (2011)
DOI 10.1007/978-3-642-12698-7_4 

with a background neutron bath, homogeneous phases of nucleons, the emergence of hyperons and pions up to a possible hadron-quark phase transition.

The detailed treatment of all these ingredients and their combined application is discussed in more depth in textbooks (Kippenhahn and Weigert, 1994; Maeder, 2009; Arnett, 1996; Iliadis, 2007), and/or the preceding Chap. 3, where the evolution of low and intermediate mass stars is addressed. That chapter also includes the stellar structure equations in spherical symmetry and a discussion of opacities for photon transport. Chapters 8 and 9 (tools for modeling objects and their processes) go into more detail with regard to modeling hydro*dynamics*, (convective) instabilities and energy transport as well as the energy generation due to nuclear reactions and the determination of the latter. Here we want to focus on the astrophysical aspects, i.e. a description of the evolution of massive stars and their endpoints with a special emphasis on the composition of their ejecta (in form of stellar winds during the evolution or of explosive ejecta). Low and intermediate mass stars end their evolution as AGB stars, finally blowing off a planetary nebula via wind losses and leaving a white dwarf with an unburned C and O composition. Massive stars evolve beyond this point and experience all stellar burning stages from H over He, C, Ne, O and Si-burning up to core collapse and explosive endstages. In this chapter we want to discuss the nucleosynthesis processes involved and the production of radioactive nuclei[1] in more detail. This includes all hydrostatic nuclear-burning stages experienced by massive stars, and explosive burning stages when a shock wave moves outward after a successful explosion was initiated, but also final wind ejecta from the hot proto-neutron star which emerged in the collapse and explosion phase. All these ejecta will enter the interstellar medium in galaxies, initially appearing as gas and dust in wind bubbles and supernova remnants, later determining the evolution of the larger-scale gas composition. The interstellar gas composition will evolve with time, and the composition of newly formed stars will witness this composition at the time of their formation.

Massive stars play an important role as contributors to the gas composition of the interstellar medium via wind losses or explosions. In astronomical terms they are the progenitors of blue supergiants (BSG), red supergiants (RSG), Wolf-Rayet (WR) and luminous blue variable (LBV) stars (Maeder and Meynet, 2010). At the end of their life, they explode as core collapse supernovae (cc-SNe), observed as SNe of type II or Ib,c (Woosley and Bloom, 2006) and also as long soft gamma-ray bursts GRBs (Piran, 2004). After collapse, their cores become neutron stars or black holes. They are one of the main sites for nucleosynthesis, which takes place during both pre-SN (hydrostatic) burning stages and during explosive burning. A weak *s* process occurs during core He- (and C-)burning (The et al., 2007; El Eid et al., 2009) and the *r* process probably occurs during the explosion (Qian and Woosley, 1996). These s(low) and r(apid) neutron capture processes are mainly responsible for the heavy nuclei beyong the Fe-group. Radioactive isotopes like $^{26}$Al and $^{60}$Fe

[1] We focus especially on long-lived radioactivities which can be observed with gamma-ray satellites, and refractory isotopes which can be observed in dust condensations included in meteorites.

detected by the INTEGRAL satellite are produced by massive stars, plus many more radioactivities from the final explosive ejecta (like e.g. $^{44}$Ti, $^{56}$Ni, $^{56}$Co etc., see Sects. 4.4.2 and 4.5). Chapters 2 and 3 discussed also many long-lived heavy nuclei beyond Fe with half-lives larger than $10^7$ and up to $10^{11}$ years. As massive stars are probably not the origin of heavy *s*-process nuclei (see Chap. 3), we will address here those nuclei which are clearly identified with the *r* process ($^{232}$Th, $1.4 \times 10^{10}$y, $^{235}$U, $7 \times 10^8$y, $^{236}$U, $2.3 \times 10^7$y, $^{238}$U, $4.5 \times 10^9$y, $^{244}$Pu, $8 \times 10^7$y, $^{247}$Cm, $1.6 \times 10^7$y) and where especially $^{232}$Th and $^{238}$U, with half-lives comparable to the age of the Galaxy/Universe, can also serve as chronometers.

Massive stars, even though they are much less numerous than low mass stars, contribute significantly (about two thirds) to the integrated luminosity of galaxies. At high redshifts $z$, or low metallicities $Z$, they are even more important drivers of characteristic phenomena and evolution. The first stars formed are thought to be all massive or even very massive, and to be the cause of the re ionisation of the universe. As discussed above, if the final core collapse leads to a black hole, the endpoint of this evolution can be the origin of the subset of (long, soft) gamma ray bursts (GRBs). GRBs are the new *standard candles* for cosmology at high redshifts. They are visible from higher redshifts than usual SNe (of type I or II) are, and thus will impose tighter constraints on cosmological models. Massive stars with their large energy output can be seen out to significant (cosmological) distances – either directly through their thermal photospheric emission, or indirectly through the impact on their surroundings (ionization, or heated dust). In their *collapsar* and GRB extremes, emission is beamed into a jet, which makes them visible even at greater distances. This can also give us information on the star formation history at a very early age of the universe ($z > 10$) beyond the reach of galaxy observations. Closer to home, recent surveys of metal poor halo stars provide a rich variety of constraints for the early chemical evolution of our Galaxy and thus the nucleosynthesis ejecta (*astro-archeology*).

## 4.2 Hydrostatic and Explosive Burning in Massive Stars

Following the motivation for studying massive stars in the previous section, we now discuss the ingredients for their modeling. Thermonuclear energy generation is one of the key aspects: It shapes the interior structure of the star, thus its evolutionary time scales, and the generation of new chemical elements and nuclei. Without understanding these, the *feedback* from massive stars *as it determines the evolution of galaxies* cannot be understood in astrophysical terms.[2] Thermonuclear burning, nuclear energy generation and resulting nuclear abundances are determined by thermonuclear reactions and weak interactions. The treatment of the required

---

[2] Empirical descriptions from observations of a multitude of galaxies are often utilized to substitute such astrophysical models in cosmological simulations.

nuclear/plasma physics, and a detailed technical description of reaction rates, their determination and the essential features of composition changes and reaction networks is presented in Chap. 9. Here we want to discuss which types of reactions are involved specifically in the evolution of massive stars and their catastrophic end stages. Nuclear burning can in general be classified into two categories:

(1) hydrostatic burning stages on timescales dictated by stellar energy loss
(2) explosive burning due to hydrodynamics of the specific event.

Massive stars (as opposed to low and intermediate mass stars) are the ones which experience explosive burning (2) as a natural outcome at the end of their evolution and they undergo more extended hydrostatic burning stages (1) than their low- and intermediate-mass cousins. Therefore, we want to address some of these features here in a general way, before describing the evolution and explosion in more detail.

The important ingredients for describing nuclear burning and the resulting composition changes (i.e. nucleosynthesis) are (i) strong-interaction cross sections and photodisintegrations, (ii) weak interactions related to decay half-lives, electron or positron captures, and finally (iii) neutrino-induced reactions. They will now be discussed[3].

### 4.2.1 Nuclear Burning During Hydrostatic Stellar Evolution

Hydrostatic burning stages are characterized by temperature thresholds, permitting thermal Maxwell-Boltzmann distributions of (charged) particles (nuclei) to penetrate increasingly larger Coulomb barriers of electrostatic repulsion. These are (two body) reactions as discussed in equations (9.6) and (9.9) of Chap. 9, representing terms of the type ${}_ir_j$ in the network equation (9.1). H-burning converts $^1$H into $^4$He via pp-chains or the CNO-cycles. The simplest pp-chain is initiated by $^1\mathrm{H}(\mathrm{p},e^+\nu)^2\mathrm{H}(\mathrm{p},\gamma)^3\mathrm{He}$ and completed by $^3\mathrm{He}(^3\mathrm{He},2\mathrm{p})^4\mathrm{He}$. The dominant CNO-cycle chain $^{12}\mathrm{C}(\mathrm{p},\gamma)^{13}\mathrm{N}(e^+\nu)^{13}\mathrm{C}(\mathrm{p},\gamma)^{14}\mathrm{N}(\mathrm{p},\gamma)^{15}\mathrm{O}(e^+\nu)^{15}\mathrm{N}(\mathrm{p},\alpha)^{12}\mathrm{C}$ is controlled by the slowest reaction $^{14}\mathrm{N}(\mathrm{p},\gamma)^{15}\mathrm{O}$. The major reactions in He-burning are the triple-alpha reaction $^4\mathrm{He}(2\alpha,\gamma)^{12}\mathrm{C}$ and $^{12}\mathrm{C}(\alpha,\gamma)^{16}\mathrm{O}$. The triple-alpha reaction, being essentially a sequence of two two-body reactions with an extremely short-lived intermediate nucleus $^8$Be, is an example for the term ${}_i\hat{r}_j$ in equation (9.1) in Chap. 9, which includes the product of three abundances. The H- and He-burning stages are also encountered in low and intermediate mass stars, leaving white dwarfs as central objects. They are discussed in much more detail with all minor reaction pathways in Chap. 3. Massive stars, the subject of the present chapter, undergo further burning stages up to those involving the production of Fe-group nuclei. Table 4.1 lists these burning stages and their typical central densities and temperatures, their duration and the typical luminosity in photons

[3] A review of the sources for this microphysics input is given for (i) in Chap. 9 and for (iii) in Chap. 8. We will review some of the required weak interaction rates (ii) in the subsections on late phases of stellar evolution/core collapse and the description of the explosion.

**Table 4.1** Burning stages of a 20 $M_\odot$ star

| Fuel | $\rho_c$ (g/cm$^3$) | $T_c$ ($10^9$ K) | $\tau$ (year) | $L_{phot}$ (erg/s) |
|---|---|---|---|---|
| Hydrogen | 5.6(0) | 0.04 | 1.0(7) | 2.7(38) |
| Helium | 9.4(2) | 0.19 | 9.5(5) | 5.3(38) |
| Carbon | 2.7(5) | 0.81 | 3.0(2) | 4.3(38) |
| Neon | 4.0(6) | 1.70 | 3.8(−1) | 4.4(38) |
| Oxygen | 6.0(6) | 2.10 | 5.0(−1) | 4.4(38) |
| Silicon | 4.9(7) | 3.70 | 2 days | 4.4(38) |

(from Woosley and Weaver 1995), which involve the reaction types given below. For further details see Sect. (4.3).

- *Heavy-ion fusion reactions:* In C-burning the reaction $^{12}$C($^{12}$C,$\alpha$)$^{20}$Ne dominates, in O-burning it is $^{16}$O($^{16}$O,$\alpha$)$^{28}$Si. The corresponding reaction rates $_i r_j$ (after integrating over a Maxwell-Boltzmann distribution of targets and projectiles) have the form given in equation (9.9) of Chap. 9 and contribute to the second term in equation (9.1). Reactions going beyond these key reactions are provided in Tables 4.2 and 4.3. Further features as well as the status of nuclear cross sections are discussed in recent reviews on hydrostatic burning stages (Haxton et al., 2006; Buchmann and Barnes, 2006; Costantini et al., 2009; Wiescher et al., 2010) and Chap. 9.
- *Photo-disintegrations:* The alternative to fusion reactions are photodisintegrations which start to play a role at sufficiently high temperatures $T$ when 30 $kT \approx Q$ (the Q-value or energy release of the inverse capture reaction). This ensures the existence of photons with energies $>Q$ in the Planck distribution and leads to Ne-Burning [$^{20}$Ne($\gamma,\alpha$)$^{16}$O, $^{20}$Ne($\alpha,\gamma$)$^{24}$Mg] at $T > 1.5\times10^9$ K (preceding O-burning) due to a small Q-value of $\approx$4 MeV and Si-burning at temperatures in excess of $3\times10^9$ K [initiated like Ne-burning by photodisintegrations]. Such photodisintegrations (after integrating over a thermal (Planck) distribution of photons at temperature $T$) have the form given in equation (9.4) of

**Table 4.2** Major reactions in carbon burning

| |
|---|
| (a) Basic energy generation |
| $^{12}$C($^{12}$C,$\alpha$)$^{20}$Ne  $^{12}$C($^{12}$C,p)$^{23}$Na |
| $^{23}$Na(p,$\alpha$)$^{20}$Ne  $^{23}$Na(p,$\gamma$)$^{24}$Mg  $^{12}$C($\alpha,\gamma$)$^{16}$O |
| (b) Fluxes $> 10^{-2}\times$(a) |
| $^{20}$Ne($\alpha,\gamma$)$^{24}$Mg  $^{23}$Na($\alpha$,p)$^{26}$Mg(p,$\gamma$)$^{27}$Al |
| $^{20}$Ne(n,$\gamma$)$^{21}$Ne(p,$\gamma$)$^{22}$Na ($e^+\nu$)$^{22}$Ne($\alpha$,n)$^{25}$Mg(n,$\gamma$)$^{26}$Mg |
| $^{21}$Ne($\alpha$,n)$^{24}$Mg  $^{22}$Ne(p,$\gamma$)$^{23}$Na  $^{25}$Mg(p,$\gamma$)$^{26}$Al($e^+\nu$)$^{26}$Mg |
| (c) Low temperature, high density burning |
| $^{12}$C(p,$\gamma$)$^{13}$N($e^+\nu$)$^{13}$C($\alpha$,n)$^{16}$O($\alpha,\gamma$)$^{20}$Ne |
| $^{24}$Mg(p,$\gamma$)$^{25}$Al($e^+\nu$)$^{25}$Mg |
| $^{21}$Ne(n,$\gamma$)$^{22}$Ne(n,$\gamma$)$^{23}$Ne($e^-\bar{\nu}$)$^{23}$Na(n,$\gamma$)$^{24}$Na($e^-\nu$)$^{24}$Mg + $s$ processing |

**Table 4.3** Major reactions in oxygen burning

| |
|---|
| (a) Basic energy generation |
| $^{16}$O($^{16}$O,$\alpha$)$^{28}$Si $^{16}$O($^{12}$O,p)$^{31}$P $^{16}$O($^{16}$O,n)$^{31}$S($e^+\nu$)$^{31}$P |
| $^{31}$P(p,$\alpha$)$^{28}$Si($\alpha,\gamma$)$^{32}$S |
| $^{28}$Si($\gamma,\alpha$)$^{24}$Mg($\alpha$,p)$^{27}$Al($\alpha$,p)$^{30}$Si |
| $^{32}$S(n,$\gamma$)$^{33}$S(n,$\alpha$)$^{30}$Si($\alpha,\gamma$)$^{34}$S |
| $^{28}$Si(n,$\gamma$)$^{29}$Si($\alpha$,n)$^{32}$S($\alpha$,p)$^{35}$Cl |
| $^{29}$Si(p,$\gamma$)$^{30}$P($e^+\nu$)$^{30}$Si |
| |
| Electron captures |
| $^{33}$S($e^-,\nu$)$^{33}$P(p,n)$^{33}$S |
| $^{35}$Cl($e^-,\nu$)$^{35}$S(p,n)$^{35}$Cl |
| |
| (b) High temperature burning |
| $^{32}$S($\alpha,\gamma$)$^{36}$Ar($\alpha$,p)$^{39}$K |
| $^{36}$Ar(n,$\gamma$)$^{37}$Ar($e^+\nu$)$^{37}$Cl |
| $^{35}$Cl($\gamma$,p)$^{34}$S($\alpha,\gamma$)$^{38}$Ar(p,$\gamma$)$^{39}$K(p,$\gamma$)$^{40}$Ca |
| $^{35}$Cl($e^-,\nu$)$^{35}$S($\gamma$,p)$^{34}$S |
| $^{38}$Ar($\alpha,\gamma$)$^{42}$Ca($\alpha,\gamma$)$^{46}$Ti |
| $^{42}$Ca($\alpha$,p)$^{45}$Sc(p,$\gamma$)$^{46}$Ti |
| |
| (c) Low temperature, high density burning |
| $^{31}$P($e^-\nu$)$^{31}$S $^{31}$P(n,$\gamma$)$^{32}$P |
| $^{32}$S($e^-,\nu$)$^{32}$P(p,n)$^{32}$S |
| $^{33}$P(p,$\alpha$)$^{30}$Si |

Chap. 9 and act similar to decays with a temperature-dependent decay constant, contributing (like decays) to the first term ${}_i\lambda_j$ in equation (9.1). In Table 4.4 we provide some of the main reactions of Ne-burning, which is initiated by the photodisintegration of Ne.

- *Electron capture reactions:* Massive stellar cores eventually lead to electron-gas degeneracy, i.e. the Pauli exclusion principle for fermions determines the population of energy states rather than the Boltzmann statistics, valid only for low densities/high temperatures. The Fermi energy of electrons is

$$E_F = \hbar^2/2m_e(3\pi^2)^{2/3}n_e^{2/3} \tag{4.1}$$

Here $n_e$ is the density of the electron gas $n_e = \rho N_A Y_e$, $\rho$ denotes the matter density and $N_A$ Avogadro's number. In late stages of O-burning, in Si-burning (and during the later collapse stage) this Fermi energy of (degenerate) electrons, increases to the level of nuclear energies (MeV). In a neutral, completely ionized plasma, the electron abundance $Y_e$ is equal to the total proton abundance $Y_e = \sum_i Z_i Y_i$ (summing over all abundances of nuclei, including protons/hydrogen) and limited by the extreme values 0 (only neutrons) and 1 (only protons) with typical values during stellar evolution close to 0.5 or slightly below. Such conditions permit electron captures on protons and nuclei, if the negative Q-value of the reaction can be overcome by the electron (Fermi) energy. The

**Table 4.4** Major reactions in neon burning

| |
|---|
| (a) Basic energy generation |
| $^{20}$Ne$(\gamma,\alpha)^{16}$O $\quad ^{20}$Ne$(\alpha,\gamma)^{24}$Mg$(\alpha,\gamma)^{28}$Si |
| |
| (b) Fluxes $> 10^{-2}\times$(a) |
| $^{23}$Na(p,$\alpha$)$^{20}$Ne $\quad ^{23}$Na($\alpha$,p)$^{26}$Mg($\alpha$,n)$^{29}$Si |
| $^{20}$Ne(n,$\gamma$)$^{21}$Ne($\alpha$,n)$^{24}$Mg(n,$\gamma$)$^{25}$Mg($\alpha$,n)$^{28}$Si |
| $^{28}$Si(n,$\gamma$)$^{29}$Si(n,$\gamma$)$^{30}$Si |
| $^{24}$Mg($\alpha$,p)$^{27}$Al($\alpha$,p)$^{30}$Si |
| $^{26}$Mg(p,$\gamma$)$^{27}$Al(n,$\gamma$)$^{28}$Al($e^-\bar{\nu}$)$^{28}$Si |
| |
| (c) Low temperature, high density burning |
| $^{22}$Ne($\alpha$,n)$^{25}$Mg(n,$\gamma$)$^{26}$Mg(n,$\gamma$)$^{27}$Mg($e^-\bar{\nu}$)$^{27}$Al |
| $^{22}$Ne left from prior neutron-rich carbon burning |

**Table 4.5** Electron capture

| |
|---|
| $p + e^- \rightarrow \nu_e + n$ or p($e^-$, $\nu_e$)n |
| $(A, Z) + e^- \rightarrow \nu_e + (A, Z-1)$ or $^{A}$Z($e^-$, $\nu_e$)$^{A}$Z-1 |
| |
| $E_F(\rho Y_e = 10^7\text{g/cm}^3) = 0.75\,\text{MeV}$ |
| $E_F(\rho Y_e = 10^9\text{g/cm}^3) = 4.70\,\text{MeV}$ |

general features for typical conditions are presented in Table 4.5, example reactions were already given in Table 4.3. Thus, at sufficiently high densities, electron captures – which are energetically prohibited – can become possible and lead to an enhanced *neutronization* of the astrophysical plasma, in addition to the role of beta-decays and electron captures with positive Q-values (Nomoto and Hashimoto, 1988). In degenerate Ne—O—Mg cores (after core C-burning of stars with $8 < M/\text{M}_\odot < 10$), electron captures on $^{20}$Ne and $^{24}$Mg cause a loss of degeneracy pressure support and introduce a collapse rather than only a contraction, which combines all further burning stages on a short collapse time scale (Nomoto, 1987). In Si-burning of more massive stars, electron capture on intermediate mass and Fe-group nuclei becomes highly important and determines the neutronization ($Y_e$) of the central core. As discussed in Chap. 9, such rates contribute to the one-body reaction terms $_i\lambda_j$ in Eq. (9.1) with the effective decay constants in Eq. (9.5) being a function of $T$ and $n_e = \rho N_A Y_e$, the electron number density.

- *Neutrino reactions* (see Table 4.6): Neutrino cross section on nucleons, nuclei and electrons are minute, by comparison to above reactions. It therefore requires high densities of the order $\rho > 10^{12}\text{g/cm}^3$ that also the inverse process to electron/positron capture (neutrino capture) can occur on relevant timescales. The same is true for other processes such as e.g. inelastic scattering, leaving a nucleus in an excited state which can emit nucleons and alpha particles. Such neutrino-induced reactions can be expressed in a similar way as photon and electron captures, integrating now over the corresponding neutrino distribution. The latter is,

**Table 4.6** Neutrino reactions

| | | |
|---|---|---|
| $\nu_e + n \leftrightarrow p + e^-$ | or | $\mathrm{n}(\nu_e, e^-)\mathrm{p}$ |
| $\bar{\nu}_e + p \leftrightarrow n + e^+$ | or | $\mathrm{p}((\bar{\nu}_e, e^+)\mathrm{n}$ |
| $\nu_e + (Z, A) \leftrightarrow (Z + 1, A) + e^-$ | or | ${}^A\mathrm{Z}(\nu_e, e^-){}^A\mathrm{Z}{+}1$ |
| $\bar{\nu}_e + (Z, A) \leftrightarrow (Z - 1, A) + e^+$ | or | ${}^A\mathrm{Z}(\bar{\nu}_e, e^+){}^A\mathrm{Z} - 1$ |
| $(Z, A) + \nu \leftrightarrow \nu + (Z, A)^*$ | | |

however, not necessarily in thermal equilibrium and not just a function of temperature and neutrino densities. Neutrino distributions are rather determined by (neutrino) radiation transport calculations (see Chap. 8, where also other neutrino scattering processes are discussed).

All the reactions presented above and occurring at different times in the sequence of burning stages, contribute to the three types of terms in the reaction network equation (9.1). If one is interested to show how nuclear abundances $Y_i$ enter in this set of equations, it can also be written in the form[4]

$$\frac{dY_i}{dt} = \sum_j P^i_j \lambda_j Y_j + \sum_{j,k} P^i_{j,k} \rho N_A < j,k > Y_j Y_k + \sum_{j,k,l} P^i_{j,k,l} \rho^2 N_A^2 < j,k,l > Y_j Y_k Y_l. \quad (4.2)$$

Core Si-burning, the final burning stage during stellar evolution, which is initiated by the photodisintegration ${}^{28}\mathrm{Si}(\gamma, \alpha){}^{24}\mathrm{Mg}$ close to $3 \times 10^9$ K – and followed by a large number of fusion and photodisintegration reactions – ends with nuclear reactions in a complete *chemical equilibrium*[5] (nuclear statistical equilibrium, NSE) and an abundance distribution centered around Fe (as discussed in Chap. 9 and Eqs. (9.14) and (9.15)). These temperatures permit photodisintegrations with typical Q-values of 8–10 MeV as well as the penetration of Coulomb barriers in capture reaction. In such an NSE the abundance of each nucleus $\mathrm{Y}_i$ is only dependent on temperature $T$, density $\rho$, its nuclear binding energy $B_i$, and via charge conservation on $\sum_i Z_i Y_i = Y_e$. $Y_e$ is altered by weak interactions on longer timescales. *A quasi-equilibrium* (QSE) can occur, if localized nuclear mass regions are in equilibrium with the background of free neutrons, protons and alphas, but offset from other regions of nuclei and thus their NSE values (Hix and Thielemann, 1996, 1999b; Hix

[4] The formal difference to Eq. (9.1) is that one does not sum here over the reactions but rather over all reaction partners (see also the equation following Table 3.2 in Chap. 3). However, in total, all the terms which appear are identical. Due to the different summation indices, the P's have a slightly different notation, $\lambda$'s stand for decay rates $L$ of Chap. 9 and $< j,k >$ for $< \sigma^* v >$ of reactions between nuclei $j$ and $k$, while $< j,k,l >$ includes a similar expression for three-body reactions (Nomoto et al., 1985). A survey of computational methods to solve nuclear networks is given in (Hix and Thielemann, 1999a; Timmes, 1999). (The abundances $Y_i$ occurring in Eq. (4.2) are related – like electron abundances $Y_e$ – to number densities $n_i = \rho N_A Y_i$ and mass fractions of the corresponding nuclei via $X_i = A_i Y_i$, where $A_i$ is the mass number of nucleus $i$ and $\sum X_i = 1$.)

[5] All strong (thermonuclear) and photodisintegration reactions are equilibrized, while weak interaction reactions, changing $Y_e$, may occur on longer timescales.

et al., 2007). Different quasi-equilibrium regions are usually separated from each other by slow reactions with typically small Q-values. Such boundaries between QSE groups, due to slow reactions, can be related to neutron or proton shell closures, like e.g. $Z = N = 20$, separating the Si- and Fe-groups in early phases of Si-burning.

All reactions discussed above, occurring during all stellar burning stages, are essentially related to nuclei from H to the Fe-group, and not much beyond.

- *Neutron capture processes*: Through neutron capture reactions, also during regular stellar evolution, there is a chance to produce heavier nuclei. During core and shell He-burning specific alpha-induced reactions can liberate neutrons which are responsible for the slow neutron capture process (*s* process). A major neutron source is the reaction $^{22}$Ne$(\alpha, n)^{25}$Mg, with $^{22}$Ne being produced via succesive $\alpha$-captures on the H burning CNO product $^{14}$N$(\alpha, \gamma)^{18}$F$(\beta^+)^{18}$O$(\alpha, \gamma)^{22}$Ne. If occurring, the mixing of $^{12}$C into H-burning shells can produce an even stronger neutron source $^{13}$C$(\alpha, n)^{16}$O via $^{12}$C$(p, \gamma)^{13}$N$(\beta^+)^{13}$C. In massive, rotating, low metallicity stars, mixing can lead to the production of "primary" $^{14}$N and $^{22}$Ne, i.e. a neutron source which does not reflect the initial metallicity of $^{14}$N in the CNO-cycle, and can thus be much stronger. The previous Chap. 3 has discussed in full detail the strong *s* process via a combination of $^{13}$C and $^{22}$Ne in He-shell flashes of low and intermediate mass stars. In a similar way mixing processes can also occur in massive stars due to rotation or convective instabilities. Without such mixing processes only secondary (metallicity-dependent) $^{22}$Ne is available for $^{22}$Ne$(\alpha, n)^{25}$Mg and core He-burning as well as shell C-burning lead to a weak *s* process (The et al., 2007). The *s* process can in principle form elements up to Pb and Bi through a series of neutron captures and $\beta^-$-decays, starting on existing heavy nuclei around Fe (Käppeler and Mengoni, 2006). Weak *s* processing, based on *secondary* $^{22}$Ne, does not go beyond mass numbers of $A = 80-90$. The production of heavier nuclei is possible in massive stars if *primary* $^{14}$N and $^{22}$Ne are available.

### 4.2.2 Explosive Burning

Many of the hydrostatic nuclear-burning processes occur also under explosive conditions at higher temperatures and on shorter timescales (see Fig. 4.1), when often the $\beta$-decay half-lives are longer than the explosive timescales, producing significant abundances of unstable isotopes as burning proceeds. This requires in general the additional knowledge of nuclear reactions for and among unstable nuclei. The fuels for explosive nucleosynthesis consist mainly of $N$=$Z$ nuclei like $^{12}$C, $^{16}$O, $^{20}$Ne, $^{24}$Mg, or $^{28}$Si (the ashes of prior hydrostatic burning), resulting in heavier nuclei, again with $N \approx Z$. At high densities also electron captures on nuclei $e^- +^A Z \rightarrow^A Z - 1 + \nu$ can occur at substantial rates due to energetic, degenerate electrons when Fermi energies are high, as already discussed for late hydrostatic burning stages.

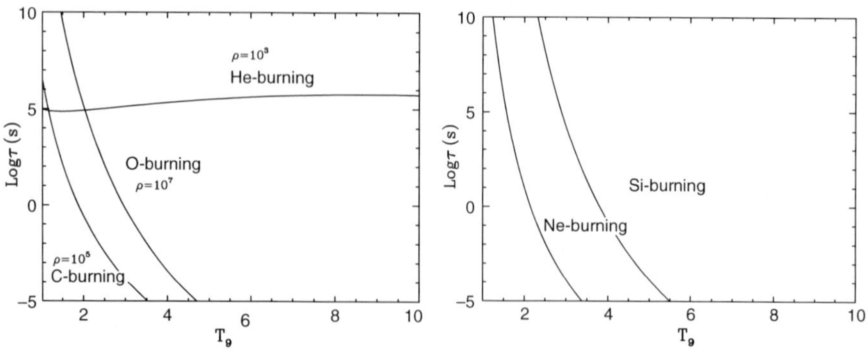


**Fig. 4.1** Burning timescales in ($\log_{10}$) seconds for fuel exhaustion of He-, C-, and O-burning (*left*) and Ne- and Si-burning (*right*), as a function of temperature. Density-dependent timescales are labeled with a chosen typical density (in g/cm$^3$). They scale with $1/\rho$ for C- and O-burning and $1/\rho^2$ for He-burning. Ne- and Si-burning, initiated by photodisintegrations, are not density-dependent. The almost constant He-burning timescale beyond $T_9$=$T/10^9$K=1 permits efficient destruction on explosive timescales only for high densities

Explosive Si-burning differs strongly from its hydrostatic counterpart and can be divided into three different regimes: (i) incomplete Si-burning and complete Si-burning with either (ii) a normal (high density, low entropy) or (iii) an $\alpha$-rich (low density, high entropy) freeze-out of charged-particle reactions during cooling from NSE. At initially high temperatures or during a *normal* freeze-out, the abundances remain in a full NSE. The full NSE can break up in smaller equilibrium clusters (quasi-equilibrium, QSE), for a detailed discussion see Hix and Thielemann (1996, 1999b); Hix et al. (2007). An example for such QSE-bevavior is an alpha-rich freeze-out, caused by the inability of the triple-alpha reaction $^4$He($2\alpha,\gamma$)$^{12}$C, and the $^4$He($\alpha$n,$\gamma$)$^9$Be reaction to keep light nuclei like n, p, and $^4$He, and nuclei beyond $A$=12 in an NSE during declining temperatures, when densities are low. This causes a large $\alpha$-particle abundance after freeze-out of nuclear reactions. This effect, most pronounced for core collapse supernovae, depends on the entropy of the reaction environment, being proportional to $T^3/\rho$ in a radiation dominated plasma (see Fig. 4.2).

$r$-process nucleosynthesis (*rapid* neutron capture) relates to environments of explosive Si-burning, either with low or high entropies, where matter experiences a normal or $\alpha$-rich freeze-out. The requirement of a neutron to seed-nuclei ratio of 10 to 150 after freeze-out of charged particle reactions[6] translates into $Y_e$ = 0.12–0.3 for a normal freeze-out. For a moderate $Y_e$>0.40 an extremely $\alpha$-rich freeze-out is needed (see the disussion in Sect. 4.4). Under these conditions the large mass fraction in $^4$He (with $N = Z$) permits ratios of remaining free neutrons to (small) abundances of heavier seed nuclei, which are sufficiently high to attain $r$-process

[6] Such neutron/seed ratio is required in order to produce all, including the heaviest, $r$-process nuclei via neutron capture from seed nuclei at their abundances before freeze-out.

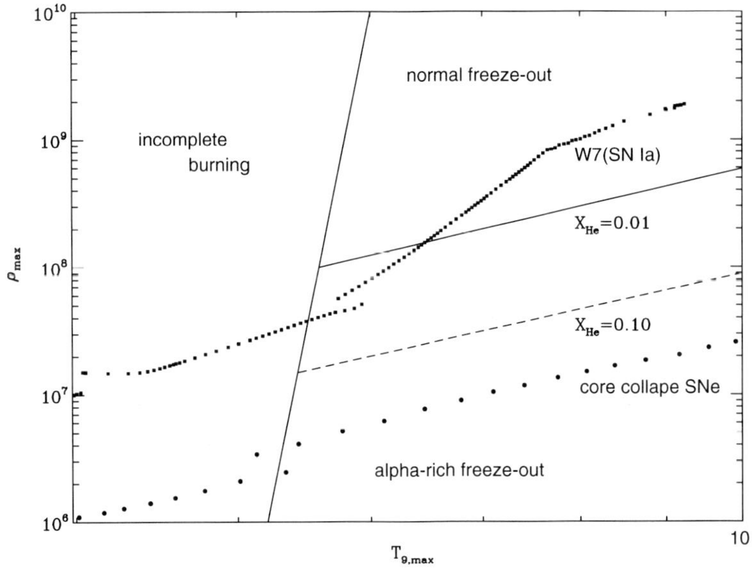


**Fig. 4.2** Final results of explosive Si-burning as a function of maximum temperatures and densities attained in explosions before adiabatic expansion. For temperatures in excess of $5 \times 10^9$ K any fuel previously existing is photodisintegrated into nucleons and alpha particles before re-assembling in the expansion. For high densities this is described by a full NSE with an Fe-group composition, favoring nuclei with maximum binding energies and proton/nucleon ratios equal to $Y_e$. For lower densities the NSE breaks into local equilibrium groups (quasi-equilibrium, QSE) with group boundaries determined by reactions with an insufficiently fast reaction stream. Alpha-rich freeze-out (insufficient conversion of alpha-particles into nuclei beyond carbon) is such a QSE behavior. Lines with 1 and 10% remaining alpha mass fraction are indicated as well as typical conditions for mass zones in type Ia and core collape supernovae

conditions. In many cases QSE-groups of neutron captures and photodisintegrations are formed in the isotopic chains of heavy elements during the operation of the *r* process.

## 4.3 Evolution of Massive Stars up to Core Collapse

In Sect. 4.2 we have discussed nuclear burning processes in detail, including also individual reactions which are of relevance during the evolution of massive stars. This relates to the main focus of this book, the production of (radioactive) nuclei in astrophysical environments. In the present section we will discuss the physics of stellar evolution and major related observational features; but we leave to review articles or textbooks for a technical description of how mass, energy, and momentum conservations equations as well as energy transport (via radiation or convective motions) are treated (Maeder, 2009; Maeder and Meynet, 2010; Heger et al., 2003; Limongi et al., 2000; Limongi and Chieffi, 2003, 2006; Ohkubo et al., 2008; El

Eid et al., 2009) (but see also the hydrostatic stellar struture/evolution equations in spherical symmetry, as presented in Chap. 3). Stellar-evolution calculations as discussed here are based on the *Geneva code* of A. Maeder and G. Meynet and their present and recent students (Maeder, 2009; Maeder and Meynet, 2010). This numerical implementation of stellar evolution includes (i) an adaptative reaction network for the advanced burning stages, which is capable to follow the detailed evolution of $Y_e$ and a large set of nuclei; (ii) a discretization of the stellar-structure equations, modified in order to damp instabilities occurring during the advanced stages of evolution; (iii) the treatment of dynamical shear in addition to the other mixing processes (such as, e.g., horizontal turbulence, secular shear and meridional circulation); and (iv) the treatment of convection as a diffusive process from O-burning onwards. This allows to follow the evolution of massive stars from their birth until the stage of Si-burning, including all nuclear burning stages discussed in Sect. 4.2, for a wide range of initial masses, metallicities and stellar rotation. Here the treatment of rotation and mixing effects is still utilizing methods based on spherical symmetry. Full multi-dimensional calculations of mixing processes during stellar evolution are on the scene (Meakin and Arnett, 2007; Arnett et al., 2009) and might open up a new era of understanding the evolution of stars.

### 4.3.1 Stellar Evolution with Rotation

The evolution of all stars (including massive stars discussed here) is initiated by core H-burning, during which the star is found on the so-called main sequence (MS) in the Hertzsprung-Russell (HR) diagram, which relates the stellar luminosity to the stellar surface temperature (color). The observational appearance of a star after the completion of core H-burning is affected by the fact that the H-burning region continues to move outward as a burning shell. The He-core contracts and ignites core He-burning in the center, which produces mainly C and O. The star's position in the HR diagram leaves the main sequence, and its radius increases due to the increased radiation pressure. Depending on the resulting surface temperature it becomes a blue or red supergiant (BSG or RSG). Radiation pressure can rise to such extreme values that stars (more massive than 20–30 $M_\odot$) blow off their outer parts through strong *stellar winds* of velocities up to 2,000 km/s, exposing the more-interior parts of the star, the helium (or in some cases, the carbon) shell. Such a Wolf-Rayet (WR) star loses between $10^{-6}$ and a few times $10^{-5}$ $M_\odot$ per year, in comparison to our Sun losing $10^{-14}$ of its $M_\odot$ per year through its solar wind. For non-rotating stars, the transition to the WR phase appears through the so-called Luminous Blue Variable stars (LBVs). LBVs are massive, intrinsically bright stars which display different scales of light and color variability, ranging from rapid microvariations to rare outbreaks of catastrophic mass loss. They represent a very short-lived (perhaps as little as 40,000 years) strongly mass-losing phase in the evolution of massive stars, during which they undergo deep erosion of the outer layers before they enter the Wolf-Rayet phase. For rotating stars, the WR phase may start before the star ends its main sequence, since rotation enhances mass loss and rotation-induced mixing further reduces the hydrogen content at the surface (Meynet and Maeder, 2003, 2005). In the following we discuss how these evolutionary phases depend on the initial properties of a star. Late burning phases progress much more rapidly than the H burning of the main sequence state. This is because the carriers of the star's energy loss, which drives the evolution of a star, change from photons to neutrinos, which escape immediately at the densities discussed here, while photons undergo a multitude of scattering processes until they finally escape at the photosphere.[7] The characteristics of late-burning stages are essentially identified by the size of a star's C+O-core after core He-burning.

The evolution of stars is governed mainly by three initial parameters: (1) its mass $M$, (2) its metallicity ($Z$, i.e. the mass fraction of pre-existing elements heavier than He from earlier stellar generations), and (3) the rotation rate or surface rotation velocity $v_{rot}$. Solar metallicity corresponds to[8] $Z = 0.02$. The evolution can also be influenced by interior magnetic fields, and by a close binary companion. Rotation significantly affects the pre-supernova state, through the impact it has on the H and He-burning evolution. Two mass groups are distinguishable: Either rotationally induced mixing dominates (for $M < 30\,M_\odot$), or rotationally increased mass loss dominates (for $M > 30\,M_\odot$). For massive stars around solar metallicity, mass loss plays a crucial role, in some cases removing more than half of the initial mass. Internal mixing, induced mainly by convection and rotation, also has a significant effect on the evolution of stars. An important result is the production of primary $^{14}N$ (via the CNO-cycle) and $^{22}Ne$ (via $\alpha$-captures in He-burning), due to mixing of burning products (such as $^{12}C$) with hydrogen or $\alpha$'s, respectively (see the discussion in Sect. 4.2).

The general impact of metallicity can be summarized in the following way: Lower metallicity implies a (slightly) lower luminosity due to the lack of CNO-cycling in hydrogen burning, which leads to slightly smaller convective cores. A lower metallicity also implies a lower opacity due to the lack of heavier elements with their many spectral lines, reducing therefore also radiation pressure and hence mass loss (as long as the chemical composition has not been changed by burning or mixing in the part of the star under consideration). This results in lower metallicity stars being more compact and experiencing less mass loss. Prescriptions for mass loss as used in the Geneva stellar evolution code are described in detail in Meynet and Maeder (2005). Mass loss rates depend on metallicity as $dM/dt \propto (Z/Z_\odot)^{0.5-0.86}$, where $Z$ is the mass fraction of heavy elements at the surface of the star. The effects can be seen in Fig. 4.3 which shows the interior structure of stars through so-called *Kippenhahn diagrams* of 20 $M_\odot$ models for different metallicities and rotation velocities of the stars. These diagrams indicate

---

[7] It takes a photon about $10^5$ years to reach the surface, after it has been launched in the hot core of, e.g., our Sun.

[8] The current value of solar metallicity is believed to be $Z = 0.014$, see Chap. 1; the value of $Z = 0.02$, which had been established before and was in common use till ~2005, remains a reference for comparisons, though.

regions (in radial mass coordinates) where matter is unstable against convection; here the energy transport is dominated by transporting hot matter rather than through the propagation of photons. The implications of such a behavior have already been described in the evolution of low and intermediate mass stars (Chap. 3), and the physical origin and treatment of these effects are addressed in Chap. 8.

With the exception of the outer convection zone, convective regions in most cases indicate *burning zones*, such as *core H-burning, core He-burning, core C-burning* etc. They testify also the ignition of outward moving burning shells of the same nuclear burning stages. When comparing models for decreasing metallicities (without rotation, left column of Fig. 4.3) one notices only minute reductions of the core sizes, but it is clearly seen that the outer (H-)burning shell moves further in towards smaller radial mass zones. In the third figure in this column we see merging of the H- and He-burning shells due to this effect, which leads to a largely-increased energy generation and extension of these combined burning zones.

How does rotation change this picture, and how do rotation-induced processes vary with metallicity? At all metallicities, rotation usually increases the core sizes via rotational mixing. The supply of more H-fuel leads to more energy generation and therefore a higher luminosity. The higher luminosity increases the radition pressure and stellar mass loss. The effect of increased core sizes (and smaller density gradients) can be viewed in all models with $v_{\rm rot} = 300$ km/s in the second column of Fig. 4.3. Clearly the convective core sizes are increased and the shell burning zones have moved outward. In the lowest metallicity case, the H/He-layers are separated again. In the intermediate metallicity case $Z = 10^{-5}$, the outer convection zone reaches the surface, and the star becomes a red supergiant. For metallicities $Z = 0.001$ (top row), the increased luminosity causes a sufficient increase in radiation pressure so that the mass loss is substantially enhanced (see the decrease of the stellar mass indicated by the top line). Mass loss becomes gradually unimportant for decreasing metallicities. For the rotating 20 $M_\odot$ models the stellar fraction lost is more than 50% for solar metallicities, 13% at $Z = 0.001$, less than 3% for $Z = 10^{-5}$, and less than 0.3% for $Z = 10^{-8}$.

This can be different for more massive stars (Meynet et al., 2006). In Fig. 4.4, we show results for low metallicity stars with $Z = 10^{-8}$ and fast rotation (500–800 km/s) from 9 to 85 $M_\odot$. The surface layers of massive stars usually accelerate due to internal transport of angular momentum from the core to the envelope. Since at low $Z$, stellar winds are weak, this angular momentum dredged up by meridional circulation remains inside the star, and the star reaches critical rotation more easily. At the critical limit, matter can be launched into a Keplerian disk which probably

**Fig. 4.3** Stellar structure (Kippenhahn) diagrams, which show the evolution of the structure as a function of the time left until the core collapses after the completion of core Si-burning. The initial parameters of the models are given on the *top* of each plot. Coloring (*shading*) marks convective zones, and the burning stages are denoted below the time axis. Non-rotating and moderately rotating 20 $M_\odot$ star models are shown, for different metallicities $Z$. $v_{\rm rot}$ indicates the rotation velocity at the surface of the star

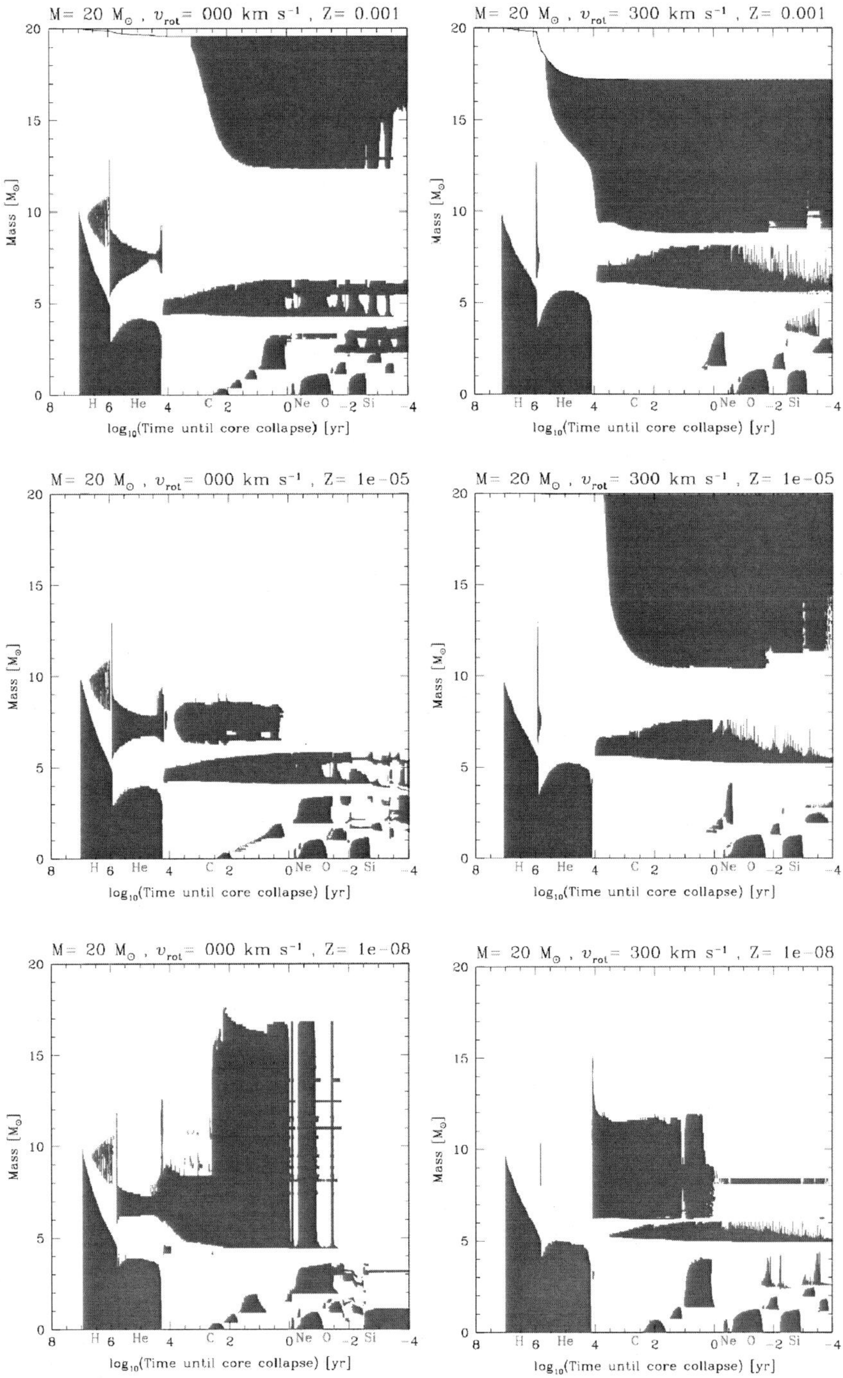


**Fig. 4.3** (continued)

dissipates under the action of the strong radiation pressure of the star. Such an effect can be seen for the 85 $M_{\odot}$ star, which loses in total more than 75% of its initial mass, and initially about 10% due to critical rotation. The remaining mass loss occurs during the red supergiant phase after rotation and convection have enriched the surface in primary CNO elements. We can also see that this effect becomes vanishingly small for stars with masses $M < 30\,M_{\odot}$. The two 20 $M_{\odot}$ models with varying metallicities and degrees of rotation again indicate the influence of metallicity and rotation on the compactness and mass loss of stars. In both cases the mass loss is negligible.

We have not shown here the evolution of extremely low metallicity stars. Below a metallicity of about $Z = 10^{-10}$, the CNO cycle cannot operate when H-burning starts after the star has been formed. The star therefore contracts until He-burning ignites, because the energy generation rate of H burning through the pp-chains cannot balance the effect of the gravitational force. Once enough C and O is produced, the CNO cycle can operate, and the star behaves like stars with $Z > 10^{-10}$ for the rest of the main sequence. Metal-free stellar evolution models are presented in Chieffi and Limongi (2004); Heger and Woosley (2002); Umeda and Nomoto (2005); Ekström et al. (2008).

Including the effects of both mass loss and rotation, massive star models better reproduce many observables of stars with metallicities around solar $Z$. For example, models with rotation allow chemical surface enrichments already on the main sequence of core hydrogen burning (MS), whereas without the inclusion of rotation, self-enrichment is only possible during advanced burning evolution such as the red supergiant RSG stage (Heger and Langer, 2000; Meynet and Maeder, 2000). Rotating star models also better reproduce the ratio of star types, for the ones which retain their hydrogen surface layer (O stars), which lose the hydrogen layer completely (WR stars), and which even lose their helium layer. The latter affects also the appearance of later core collapse supernova explosions of massive stars. Indeed, rotation changes the supernova type due to the mass loss of the hydrogen envelope (turning such an event in optical observations from a type II supernova with a strong plateau phase to a IIb event with a smaller plateau, or even a Ib event for the case of complete loss of the hydrogen envelope, and a Ic event with the additional loss of the He-envelope). This is discussed in more detail in Sect. 4.4. Both aspects, the chemical surface enrichment in MS stars as well as the ratio of type Ib+Ic to type II supernovae, as a function of metallicity, are drastically changed compared to non-rotating models, which underestimate these ratios (Georgy et al., 2009; Meynet and Maeder, 2005). The value of 300 km/s, used as the initial rotation velocity at

**Fig. 4.4** The same as Fig. 4.3. Stellar structure diagrams for rapidly rotating stars of metallicity $Z = 10^{-8}$, over a mass range from 9 to 85 $M_{\odot}$. We see the drastically increasing amount of mass loss with increasing mass (enhancing mixing of burning products to the surface, and increasing opacities, i.e. acting like increased metallicities, plus some mass loss from critical rotation for the most massive stars). The two metallicity cases shown for the 20 $M_{\odot}$ star show again that stars are less compact and show more and enhanced mass loss for higher metallicities (becoming a RSG which leads to the appearance of a large conductive envelope)

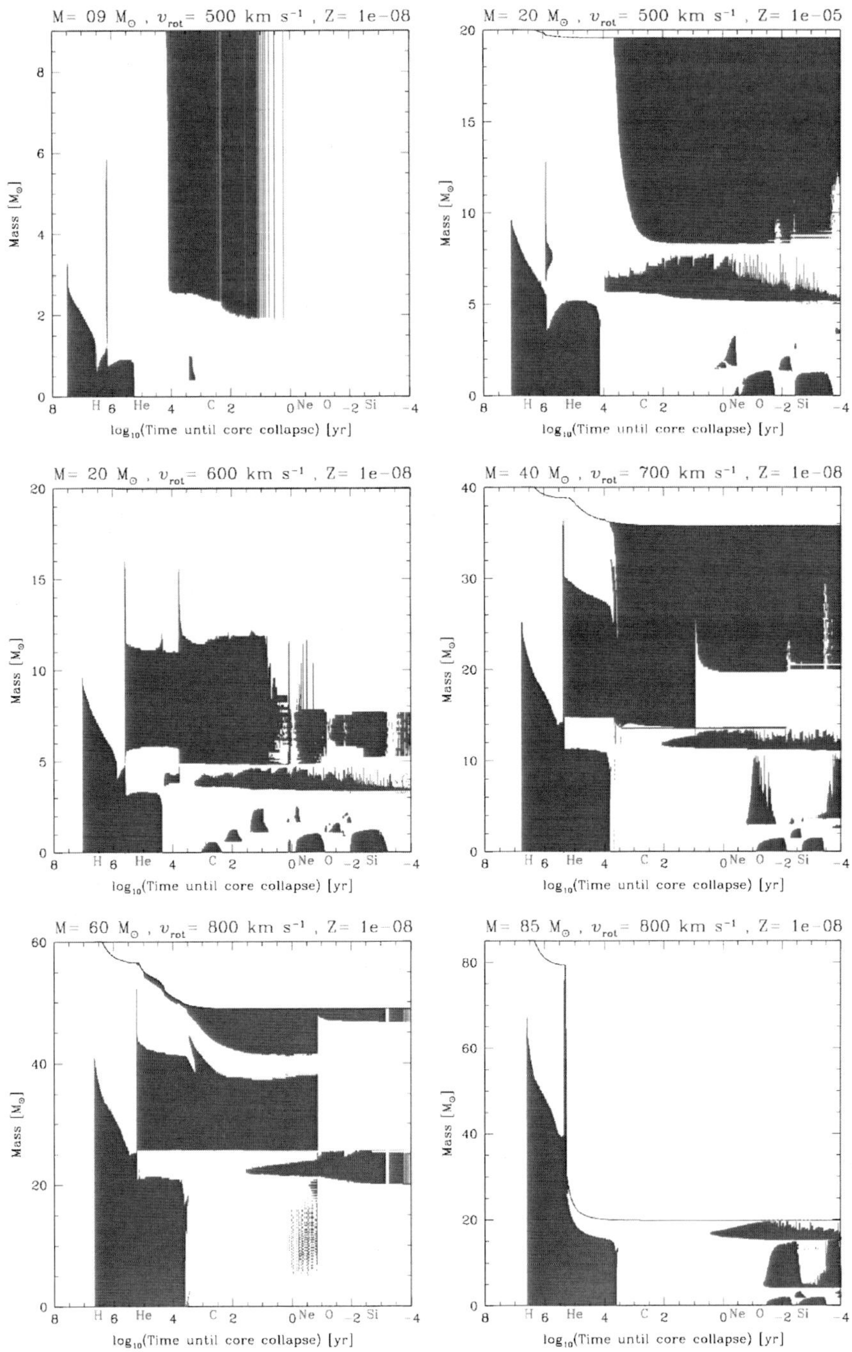


**Fig. 4.4** (continued)

solar metallicity, corresponds to an average velocity of about 220 km/s on the main sequence (MS), which is close to the average observed value (Fukuda, 1982; Meynet et al., 2008). Observed ratios of stars of different types in the Magellanic clouds, as compared to our Galaxy (Maeder et al., 1999; Martayan et al., 2007), point to stars rotating faster at lower metallicities. Fast initial rotation velocities in the range of 600–800 km/s (Hirschi et al., 2005) are supported by observations of very low-$Z$ stars (Chiappini et al., 2006).

Rotation affects all burning stages and the resulting Fe-core (we will discuss this issue further in the next subsection, see also Fig. 4.6). The size of the Fe-core in turn determines the final fate, whether a supernova explosion with neutron star formation or the collapse to a black hole occurs. The effects of rotation on pre-supernova models are most spectacular for stars between 15 and 25 $M_\odot$. It changes the total size/radius of progenitors (leading to blue instead of red supergiants) and the helium and CO core (bigger by a factor of $\sim$1.5 in rotating models). The history of convective zones (in particular the convective zones associated with shell H-burning and core He-burning) is strongly affected by rotation induced mixing (Hirschi et al., 2005). The most important rotation induced mixing takes place at low $Z$ while He is burning inside a convective core. Primary C and O are mixed from the convective core into the H-burning shell. Once the enrichment is strong enough, the H-burning shell is boosted (the CNO cycle depends strongly on the C and O mixing at such low initial metallicities). The shell becomes convective and leads to an important primary $^{14}$N production while the convective core mass decreases, leading to a less massive CO-core after He-burning than in non-rotating models. Convective and rotational mixing brings the primary CNO to the surface with interesting consequences for the stellar yields. The yield of $^{16}$O, being closely correlated with the mass of the CO-core, is reduced. At the same time the C yield is slightly increased (Hirschi et al., 2005), both due to the slightly lower temperatures in core He-burning. This is one possible explanation for the high [C/O] ratio observed in the most metal-poor halo stars (see Fig. 14 in Spite et al. (2005) and Fabbian et al. (2009)) and in damped Lyman alpha systems DLAs (Pettini et al., 2008).

The fate of rotating stars at very low $Z$ is therefore probably the following: $M < 30$–$40\ M_\odot$: Mass loss is insignificant and matter is only ejected into the ISM during the SN explosion. $30$–$40\ M_\odot < M < 60\ M_\odot$ : Mass loss (at critical rotation and in the RSG stage) removes 10–20% of the initial mass of the star. The star probably dies as a black hole without a SN explosion and therefore the feedback into the ISM is only due to stellar winds. $M > 60\ M_\odot$: A strong mass loss removes a significant amount of mass and the stars enter the WR phase. These stars therefore end as type Ib/c SNe and possibly as GRBs. This behavior is displayed in Fig. 4.5. At a metallicity $Z = 10^{-8}$, corresponding to an Fe/H ratio $\log_{10}[(Fe/H)/(Fe/H)_\odot]$ =[Fe/H]$\sim -6.6$, C and O are shown in models to be mixed into the H-burning shell during He-burning. This raises the importance of the shell, and leads to a reduction of the CO-core size. Later in the evolution, the H-shell deepens and produces large amounts of primary nitrogen. For the most massive stars ($M > 60\ M_\odot$), significant mass loss occurs during the red supergiant stage, caused by the surface enrichment in CNO elements from rotational and convective mixing.

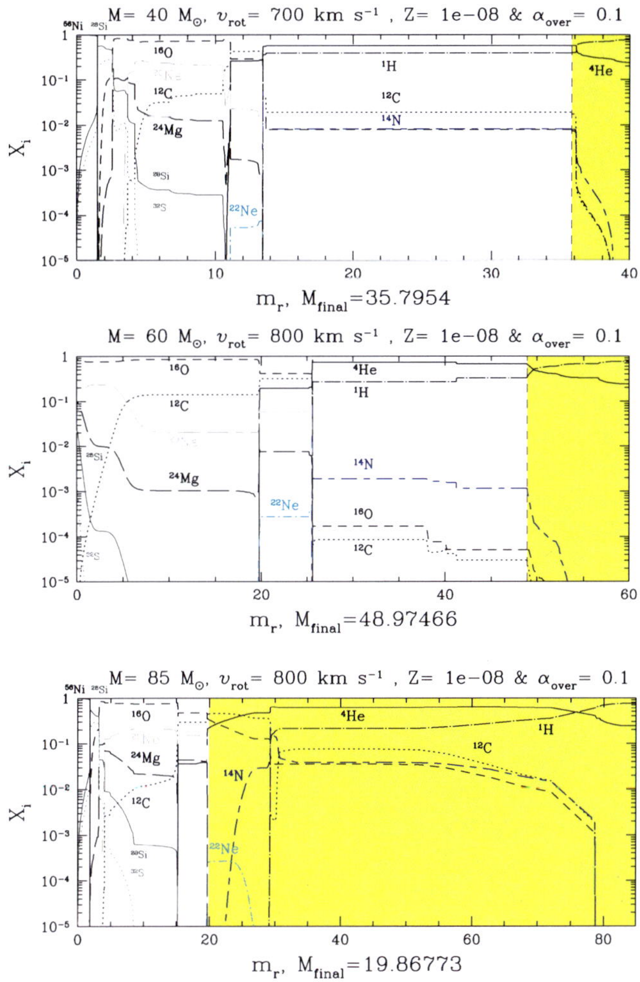


**Fig. 4.5** Abundance profiles for the 40 (*top*), 60 (*middle*) and 85 (*bottom*) $M_\odot$ models. The pre-SN and wind (*yellow shaded area*) chemical compositions are separated by a *vertical dashed line* located at the pre-SN total mass ($M_{\text{final}}$), given below each plot

The properties of non-rotating low-$Z$ stars are presented in Heger et al. (2003); Hirschi et al. (2008), and several groups have calculated their stellar yields (Heger and Woosley, 2002; Chieffi and Limongi, 2004; Tominaga et al., 2007). All results for the non-rotating stars (whether at solar metallicity or for low-$Z$ models) are consistent among these calculations, differences are understood from the treatments of convection and the rates used for $^{12}C(\alpha, \gamma)^{16}O$. The combined contributions to stellar yields by the wind and the later supernova explosion (see Sect. 4.4) will be provided separately. The results for stellar models with metallicities $Z$ close to solar can be described as follows: Rotating stars have larger yields in their stellar winds than the non-rotating ones, because of the extra mass loss and mixing due to rotation, for masses below ~30 $M_{\odot}$. The $^{12}C$ and $^{16}O$ yields are increased by a factor 1.5–2.5 by rotation. At high mass loss rates (above ~30 $M_{\odot}$), the rotating and non-rotating models show similar yield values. When the wind and explosive contributions are added, the total metal production of rotating stars is larger by a factor 1.5–2.5 (see Sect. 4.4). For very massive stars, the situation varies due to the extreme mass loss, as shown in Fig. 4.5.

In order to give a quantitative impression of the influence of initial mass, metallicity and rotation on the evolution of stars, we present in Tables 4.7 and 4.8 results for (a) non-rotating solar metallicity stars (Limongi and Chieffi, 2006) and (b) rotating stars for varying metallicities (Hirschi, 2007). Table 4.8 corresponds to the models shown in Figs. 4.3, 4.4, and 4.5. Given are the initial and final mass (in order to give an impression of the mass loss), as well as the core size after central H-burning (the He-core) and after central He-burning (the CO-core), and in Table 4.8 also the metallicity $Z$ and initial rotational surface velocity in km/s. As all burning stages after He-burning occur on significantly shorter timescales than the earlier burning phases, the CO-core size is the important quantity in order to determine the final outcome/fate of the star.

After this general discussion of stellar evolution, as it varies with initial mass, metallicity and rotation, we now focus on two long-lived isotopes $^{26}Al$ and $^{60}Fe$, which have important contributions from the earlier burning stages in explosion ejecta.

**Table 4.7** Stellar properties (Limongi and Chieffi, 2006)

| $M_{ini}/M_{\odot}$ | $M_{fin}/M_{\odot}$ | $M_{He}/M_{\odot}$ | $M_{CO}/M_{\odot}$ |
|---|---|---|---|
| 11 | 10.56 | 3.47 | 1.75 |
| 15 | 13.49 | 5.29 | 2.72 |
| 20 | 16.31 | 7.64 | 4.35 |
| 30 | 12.91 | 12.68 | 8.01 |
| 40 | 12.52 | 16.49 | 8.98 |
| 60 | 17.08 | 25.17 | 12.62 |
| 80 | 22.62 | 34.71 | 17.41 |

**Table 4.8** Stellar properties (Hirschi, 2007)

| $M_{ini}/M_{\odot}$ | $Z$ | $v_{rot}$ | $M_{fin}/M_{\odot}$ | $M_{He}/M_{\odot}$ | $M_{CO}/M_{\odot}$ |
|---|---|---|---|---|---|
| 9 | $1 \times 10^{-8}$ | 500 | 9.00 | 1.90 | 1.34 |
| 20 | $2 \times 10^{-2}$ | 300 | 8.76 | 8.66 | 6.59 |
| 20 | $1 \times 10^{-3}$ | 0 | 19.56 | 6.58 | 4.39 |
| 20 | $1 \times 10^{-3}$ | 300 | 17.19 | 8.32 | 6.24 |
| 20 | $1 \times 10^{-5}$ | 300 | 19.93 | 7.90 | 5.68 |
| 20 | $1 \times 10^{-5}$ | 500 | 19.57 | 7.85 | 5.91 |
| 20 | $1 \times 10^{-8}$ | 300 | 20.00 | 6.17 | 5.18 |
| 20 | $1 \times 10^{-8}$ | 600 | 19.59 | 4.83 | 4.36 |
| 40 | $1 \times 10^{-8}$ | 700 | 35.80 | 13.50 | 12.80 |
| 60 | $1 \times 10^{-8}$ | 800 | 48.97 | 25.60 | 24.00 |
| 85 | $1 \times 10^{-8}$ | 800 | 19.87 | 19.90 | 18.80 |

#### 4.3.1.1 $^{26}$Al

Long-lived $^{26}$Al is produced in core and shell H-burning via the NaMgAl-cycle (see Chap. 3) in the $^{25}$Mg$(p,\gamma)^{26}$Al reaction and will be eventually ejected in the stellar wind during the WR-phase. Gamma-ray observations of the 1.809 MeV decay line of $^{26}$Al in systems like the Wolf-Rayet binary system $\gamma^2$ Vel, being the closest known Wolf-Rayet (WR) star, serve as a constraint to nucleosynthesis in Wolf-Rayet stars. From observations of the $\gamma^2$ Vel binary system including a WR star, Oberlack et al. (2000) claimed that such WR stars must emit of the order $6\times10^{-5}\,M_\odot$ of $^{26}$Al by stellar winds. The amount of $^{26}$Al ejected into the interstellar medium is very sensitive to metallicity, initial stellar mass, rotation and mass loss rate, related to one or more of the physical effects discussed above. Results of detailed calculations can be found in Langer et al. (1995); Meynet et al. (1997); Palacios et al. (2005); Limongi and Chieffi (2006); Tur et al. (2009). Limongi and Chieffi (2006) provide an extended overview for the contribution from 11 to 120 $M_\odot$ stars. The dominant source for the $^{26}$Al production during stellar evolution is the $^{25}$Mg$(p,\gamma)^{26}$Al reaction. Therefore the resulting abundance depends (i) on this reaction rate converting $^{25}$Mg into $^{26}$Al, (ii) on the amount of $^{25}$Mg available, i.e. the total amount of matter in the NeNaMgAl-cycle (either in terms of the abundance/metallicity or in terms of the H-core size), and finally (iii) on the amount of $^{26}$Al distruction. In the part of the He-core (after H-burning) which undergoes He-burning, neutrons are produced via $(\alpha,n)$-reactions which destroy $^{26}$Al via $^{26}$Al$(n,p)^{26}$Mg and $^{26}$Al$(n,\alpha)^{23}$Na. A further question is related to the amount of matter being ejected in winds (i.e. mass loss) during stellar evolution before $^{26}$Al can decay inside the star via $\beta^+$-decay with a half-life of $7.17\times10^5$ year.

He-burning, with its neutrons released, is destructive for $^{26}$Al, but shell C-burning is again a source of $^{26}$Al, also via $^{25}$Mg$(p,\gamma)^{26}$Al, which is effective due to protons released in $^{12}$C($^{12}$C,$p$)$^{23}$Na (see Table 4.2 in Sect. 4.2). Convection in the C-burning shell brings in fresh $^{12}$C fuel and $^{25}$Mg which has been also produced in prior He-burning in the $^{22}$Ne$(\alpha,n)^{25}$Mg reaction. $^{26}$Al production may be effective also in Ne-burning, based on $^{25}$Mg left over from C-burning and protons released via $^{23}$Na$(\alpha,p)^{26}$Mg (see Table 4.4). This $^{26}$Al only survives if rapidly convected outwards to lower temperature environments ($^{26}$Al may decay rapidly in hot regions due to thermal population of its short-lived isomeric state; see Fig. 1.3 in Chap. 1).

A fraction of the $^{26}$Al produced during stellar evolution will again be destroyed, when a shock front is released in a supernova explosion and propagates through the stellar envelope; in particular, material from C and Ne-burning, being close to the Fe-core, will be affected. But there are also source processes for explosive $^{26}$Al production. The total yields, hydrostatic-evolution yields combined with the destruction and contribution from explosive burning, are given in Sect. 4.5.

#### 4.3.1.2 $^{60}$Fe

$^{60}$Fe is produced by neutron captures on $^{59}$Fe, and destroyed again via $^{60}$Fe$(n,\gamma)^{61}$Fe, i.e. during the *s* process. Generally, slow capture of neutrons released from the $^{22}$Ne$(\alpha, n)^{25}$Mg reaction in core He-burning leads to the so-called *weak s* process, producing nuclei up to nuclear mass numbers of around A=90. $^{59}$Fe is beta-unstable, thus in order for neutron capture to compete with this reaction branching (equating the neutron capture and beta-decay rates) requires a typcial neutron density of about $3\times10^{10}$cm$^{-3}$. These are relatively high neutron densities for an *s* process, which also ensure that the destruction of $^{60}$Fe via neutron captures dominates over its decay with its half-life of $2.6 \times 10^6$year (Fig. 7.22 in Chap. 7, Rugel et al. (2009)). Core He-burning will not provide sufficiently high-temperatures for the $^{22}$Ne$(\alpha, n)^{25}$Mg reaction to produce such high neutron densities. It requires the conditions in shell He-burning to do so. Apparently conditions are most favorable during shell He-burning at late evolutionary times when central O-burning has already active and a C-burning shell is existent as well (see Woosley and Weaver, 1995; Rauscher et al., 2002; Limongi and Chieffi, 2006; Tur et al., 2009). $^{60}$Fe yields are very sensitive to uncertainties in He-destruction reactions (such as the $3\alpha$-rate and $^{12}$C$(\alpha, \gamma)^{16}$O) which compete with the neutron source reaction $^{22}$Ne$(\alpha, n)^{25}$Mg and neutron(-capture) *poisons* which compete with the production and destruction rates of $^{60}$Fe via neutron captures (Rauscher et al., 2002; Tur et al., 2009; Giron et al., 2010; Uberseder et al., 2009). Such uncertainties amount to factors of up to 5 from present rate uncertainties. Another possible effect which has not really been looked into, yet, is the amount of $^{22}$Ne available in He-burning. An important effect in low metallicity stars is the production of primary $^{14}$N (not enherited from CNO of previous stellar generations, but produced inside the star due to mixing of He-burning products with H). This causes the production of $^{22}$Ne in He-burning and can at low metallicities (with small seed abundances of Fe) permit sizable *s* processing, affecting again the abundance of $^{60}$Fe.

### 4.3.2 Late Burning Stages and the Onset of Core Collapse

Stars more massive than about 8 $M_\odot$ will, after finishing core and shell H- and He-burning, lead to CO-cores which exceed the maximum stable mass of white dwarfs (the Chandrasekhar mass). For later burning stages, when the partial or full degeneracy of the electron gas is important, this critical limit $M_{\rm Ch}(\rho Y_e, T)$ decides upon further contraction and the central ignition of subsequent burning stages, i.e. C-, Ne-, O- and Si-burning. Dependent on the Fermi energy of the degenerate electron gas, electron capture on the C-burning products $^{20}$Ne and $^{24}$Mg can initiate a collapse, leading directly via nuclear statistical equilibrium to a central Fe-core. This evolution path occurs for stars in the range 8–10 $M_\odot$ (Nomoto, 1987). More massise stars will proceed through all burning stages until Si-burning will finally produce an Fe-core. All burning stages after core H- and He-burning proceed on timescales which are shorter by orders of magnitude. The reason is that the energy carried away by freely escaping neutrinos dominates over radiation losses by photons which undergo a cascade of scattering processes before their final escape. Most of these neutrinos are created when central

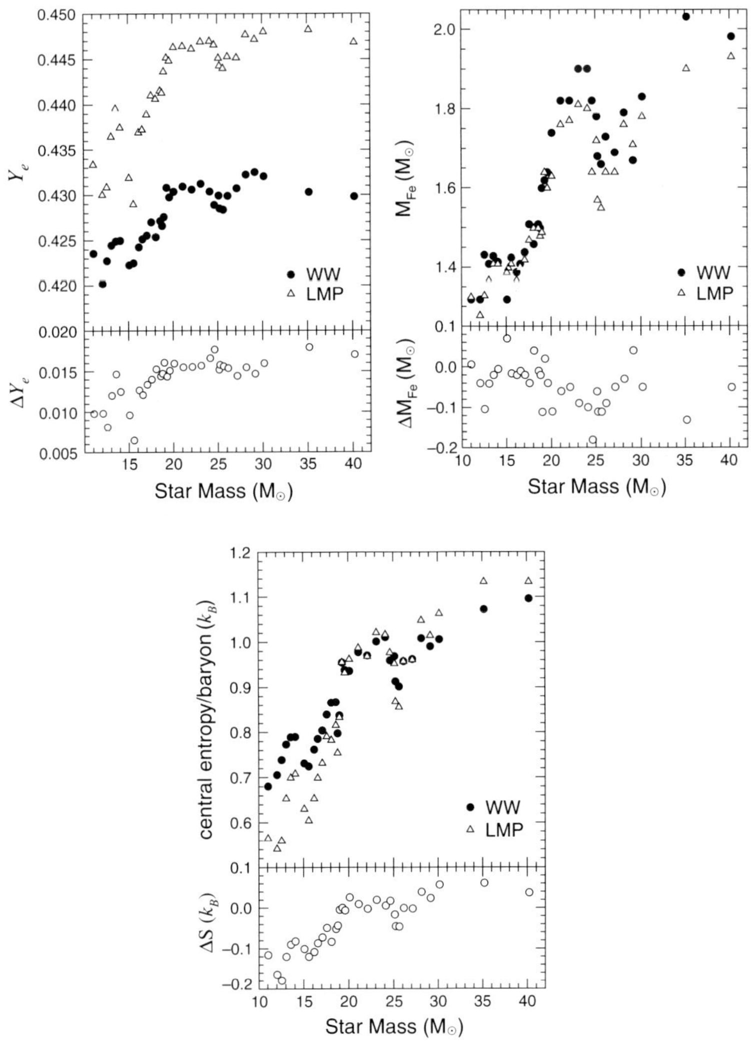


**Fig. 4.6** Comparison of the center values of $Y_e$ (*top left*), the iron core sizes (*top right*) and the central entropy (*bottom*) for 11–40 $M_\odot$ stars between the WW models and the ones using the shell model weak interaction rates (LMP) (Heger et al., 2001a). The lower parts define the changes in the 3 quantities between the LMP and WW models

densities and temperatures permit neutrino production via new particle reactions, different from beta-decay or electron capture on nuclei. Following neutrino production reactions are relevant: (i) $e^- + e^+$-pair annihilation (*pair neutrinos*), (ii) electron-photon scattering with neutrino-antineutrino pair creation (photo neutrinos), and (iii) neutrino-antineutrino pair creation from plasma oscillations (*plasmon neutrinos*). Neutrinos dominate the energy loss in stellar evolution from this point on, and lead to increasingly shorter burning timescales, although the photon radiation luminosity of the star remains roughly constant. The timescales for the individual burning stages are given in Table 4.1 in Sect. 4.2; these values refer to a 20 $M_\odot$ star with solar metallicity and no mass loss (Weaver and Woosley, 1993). Effects of mass loss, rotation and metallicity can change these timescales somewhat (up to 20%). Due to the large difference in evolution timescales, the dominant mass loss by stellar winds occurs during H- and He-burning, and the final outcome of stellar evolution is determined by the CO-core size after He-burning. Therefore, given all dependencies of stellar evolution via initial metallicities and rotation, the initial main sequence mass of a star is less indicative for the final outcome than the size of its CO-core.

In the late phases of O- and Si-burning (discussed in Sect. 4.2), electrons are moderately to strongly degenerate, dependent on the initial stellar mass, and will be characterized by increasing Fermi energies. This will allow for electron captures on burning products, and will make matter more neutron-rich, i.e decrease $Y_e$, the electron or proton to nucleon (neutrons plus protons) ratio. In high density O-burning ($\rho > 2 \times 10^7$ g/cm$^3$) two electron capture reactions become important and lead to a decrease in $Y_e$, $^{33}$S$(e^-, \nu)^{33}$P and $^{35}$Cl$((e^-, \nu)^{35}$S. Such effects become more extensive at even higher densities in Si-burning and a large range of nuclei has been identified to be of major importance $^{55-68}$Co, $^{56-69}$Ni, $^{53-62}$Fe, $^{53-63}$Mn, $^{64-74}$Cu, $^{49-54}$Sc, $^{50-58}$V, $^{52-59}$Cr, $^{49-54}$Ti, $^{74-80}$Ga, $^{77-80}$Ge, $^{83}$Se, $^{80-83}$As, $^{50-58}$V, and $^{75}$Zn (Aufderheide et al., 1994). The amount of electron capture and the resulting $Y_e$ has consequences for core sizes. (The core sizes of the late burning stages are shown in Figs. 4.3 and 4.4). The final size of the inner Fe-core represents the maximum mass which can be supported by the pressure of the degenerate electron gas. It is a function of $Y_e$, but also reflects temperature effects if the electron gas is not completely degenerate (Bethe, 1990), with $S_e$ being the entropy in electrons per baryon

$$M_{Ch}(Y_e, S_e) = 1.44(2Y_e)^2 \left[ 1 + \left( \frac{S_e}{\pi Y_e} \right)^2 \right] M_\odot. \tag{4.3}$$

Stars with masses exceeding roughly 10 $M_\odot$ reach a point in their evolution where their Si-burning core (which will turn eventually into their Fe-core) exceeds this critical mass. At this point they collapse and bounce, if not too massive, to explode in spectacular core collapse events known as type II or Ib/c supernovae. These explosions create a neutron star or black hole at the end of the life of a star. They play a preeminent role in the nucleosynthesis and chemical evolution of a galaxy.

The collapse is initiated by the capture of degenerate electrons on nuclei, which reduces the dominant contribution of the pressure (i.e. the one from the degenerate electron gas). Alternatively, for lower densities and higher temperatures (in more massive stars), the pressure supporting the core is reduced by endoergic photodisintegrations of nuclei, reducing the thermal energy. The evolution in the core is determined by the competition of gravity (that causes the collapse of the core) and weak interaction (that determines the rate at which electrons are captured and the rate at which neutrinos are trapped during the collapse).

The early phases of this final stage of stellar evolution are known as *presupernova evolution*. They follow the late-stage stellar evolution, and proceed until core densities of about $10^{10}$ g/cm$^3$ and temperatures between 5 and $10\times10^9$ K are reached. Until this point, modeling stellar evolution requires the consideration of extensive nuclear reaction networks, but is simplified by the fact that neutrinos need only be treated as a sink of energy and lepton number (due to their immediate escape). At later time and towards the collapse, this is no longer valid: As the weak interaction rates increase with the increasing density, the neutrino mean free paths shorten, so that the neutrinos eventually proceed from phases of free streaming, towards diffusion, and trapping. An adequate handling of the transitions between these transport regimes necessitates a detailed time- and space-dependent bookkeeping of the neutrino distributions in the core (see Chap. 8). During collapse, electron capture, accompanied by $\nu_e$ neutrino emission, dominates over electron antineutrino emission because the positron abundance is very low under electron-degenerate conditions. Later in the evolution the electron degeneracy is partially lifted, and in addition to the electron flavor neutrinos, also heavy neutrinos, $\nu_\mu$ and $\nu_\tau$ and their antiparticles, are usually included in numerical simulations of core collapse and postbounce evolution.

Advantageously, the temperature during the collapse and explosion are high enough that the matter composition is given by nuclear statistical equilibrium (NSE), i.e. without the need of reaction networks for the strong and electromagnetic interactions. The transition from a rather complex global nuclear reaction network, involving many neutron, proton and alpha fusion reactions and their inverses, to a quasi-statistical equilibrium, in which reactions are fast enough to bring constrained regions of the nuclear chart into equilibrium, to final and global nuclear statistical equilibrium is extensively discussed by Hix and Thielemann (1996, 1999b); Hix et al. (2007). In the late stages of Si-burning and the early collapse phase, weak interactions are dominated by electron captures on protons and nuclei. These are important equally in controling the neutronization of matter $Y_e$ and, in a large portion, also the stellar energy loss. Due to their strong energy dependence $\propto E_e^5$, the electron capture rates increase rapidly during the collapse while the density and the temperature increase (the electron Fermi energy $E_F$ scales with $\rho^{2/3}$, see Sect. 4.2).

The main weak interaction processes during the final evolution of a massive star are electron capture and $\beta$-decays. Their determination requires the calculation of Fermi and Gamow-Teller (GT) transitions. While the treatment of Fermi transitions (important only for $\beta$-decays) is straightforward, a correct description of the GT transitions is a difficult problem in nuclear structure physics. In astrophysical environments, nuclei are fully ionized. Therefore, electron capture occurs from the continuum of the degenerate electron plasma, and energies of the electrons are high enough to induce transitions to the Gamow-Teller resonance. Shortly after the discovery of this collective excitation, Bethe et al. (1979) recognized its importance for stellar electron capture. $\beta^-$-decay converts a neutron inside the nucleus into a proton and emits an electron. In a degenerate electron gas, with fully populated levels up to the Fermi energy $E_F$, all decays which would produce electrons with smaller energies than $E_F$ are not possible (*blocked*). Then, the decay rate of a given nuclear state is greatly reduced or even completely blocked at high densities. However, there is another pathway, as high temperatures populate a distribution of nuclear states: If an excited and thermally populated state of the decaying nucleus is connected by large GT transition probabilities to low-lying states in the daughter nucleus, producing electrons above the Fermi energy, such transition path can contribute significantly to the stellar $\beta$-decay rates. The importance of these states in the parent nucleus for $\beta$-decay in astrophysical environments was first recognized by Fuller et al. (1980, 1982, 1985).

Recent experimental data on GT distributions in iron group nuclei, measured in charge exchange reactions, show that the GT strength is strongly *quenched* (reduced), compared to the *independent-particle-model* value, and fragmented over many states in the daughter nucleus. An accurate understanding of these effects is essential for a reliable evaluation of the stellar weak-interaction rates, particularly for the stellar electron-capture rates (Fuller et al., 1980; Langanke and Martínez-Pinedo, 2000). The nuclear *shell-model* is the only known tool to reliably describe GT distributions in nuclei. When comparing the shell-model based rates (by Langanke and Martinez-Pinedo) with the those from Fuller et al., one finds that the shell-model based rates are almost always smaller at the relevant temperatures and densities, caused by the above mentioned quenching of the Gamow-Teller strength, and by a systematic misplacement of the energy of the Gamow-Teller resonance.

The influence of these shell-model rates on the late-stage evolution of massive stars has been investigated by Heger et al. (2001a,b), and compared to earlier calculations (Woosley and Weaver, 1995). Figure 4.6 illustrates the consequences of the shell model weak interaction rates for presupernova models in terms of the three decisive quantities: the central electron or proton to nucleon ratio $Y_e$, the entropy, and the iron core mass. The central values of $Y_e$ at the onset of core collapse increased by 0.01–0.015 for the new rates. This is a significant effect. For example, a change from $Y_e = 0.43$ in the Woosley and Weaver model for a 20 $M_\odot$ star to $Y_e = 0.445$ in the new models increases the respective Chandrasekhar mass by about 0.075 $M_\odot$ see Eq. (4.3). The new models also result in lower core entropies for stars with $M < 20\ M_\odot$, while for $M > 20\ M_\odot$, the new models actually have a slightly larger entropy. The Fe-core masses are generally smaller, where the effect is larger for more massive stars ($M > 20\ M_\odot$), while for the most common supernovae ($M < 20\ M_\odot$) the reduction is by about 0.05 $M_\odot$ (the Fe-core is here defined as the mass interior to the point where the composition is dominated by more than 50% of Fe-group elements with $A \geq 48$). This reduction seems opposite to the expected effect due to slower electron capture rates in the new models. It is, however, related to changes in the entropy profile during shell Si-burning which reduces the growth of the iron core just prior to collapse.

The evolution of $Y_e$ during the presupernova phase is plotted in Fig. 4.7. Weak processes become particularly important in reducing $Y_e$ below 0.5 after oxygen depletion ($\approx 10^7$ s and $10^6$ s before core collapse for the 15 and 25 M stars, respectively) and $Y_e$ begins a decline, which becomes precipitous during Si-burning. Initially electron captures occur much more rapidly than beta-decays. As the shell model rates are generally smaller, the initial reduction of $Y_e$ is smaller in the new models. The temperature in these models is correspondingly larger as less energy is radiated away by neutrino emission. An important feature of the new models is shown in the left panel of Fig. 4.7. For times between $10^4$ and $10^3$ s before core collapse, $Y_e$ increases due to the fact that $\beta$-decay becomes competitive with electron capture after Si-depletion in the core and during shell Si-burning. The presence of an important $\beta$-decay contribution has two effects (Aufderheide et al., 1994). Obviously it counteracts the reduction of $Y_e$ in the core, but also acts as an additional neutrino source, causing a stronger cooling of the core and a reduction in entropy. This cooling can be quite efficient, as often the average neutrino energy from the $\beta$-decays involved is larger than for the competing electron captures. As a consequence the new models have significantly lower core temperatures. At later stages of the collapse $\beta$-decay becomes unimportant again as an increased electron Fermi energy blocks/reduces its role. The shell model weak interaction rates predict the presupernova evolution to proceed along a temperature-density-$Y_e$ trajectory where the weak processes involve nuclei rather close to stability which will permit to test these effects in the next-generation radioactive ion-beam facilities.

Figure 4.7 identifies the two most important nuclei (the ones with the largest value for the product of abundance times rate) for the electron capture during various stages of the final evolution of 15 and 25 M$_\odot$ stars. An exhaustive list of the most important nuclei for both electron capture and beta-decay during the final stages of stellar evolution for stars of different masses is given in Heger et al. (2001b). In total, the weak interaction processes shift the matter composition to smaller $Y_e$ values

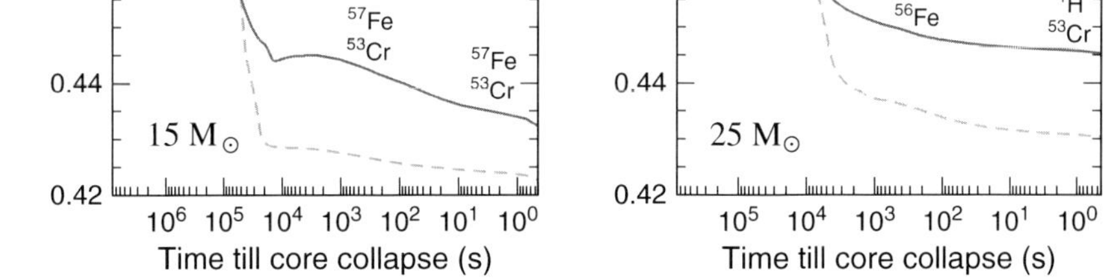


**Fig. 4.7** Evolution of the $Y_e$ value in the center of a 15 $M_\odot$ star (*left panel*) and a 25 $M_\odot$ star (*right panel*) as a function of time until bounce. The *dashed line* shows the evolution in the Woosley and Weaver models (WW) (Woosley and Weaver, 1995), while the *solid line* shows the results using the shell-model based weak-interaction rates of Langanke and Martínez-Pinedo (LMP). The two most important nuclei in the determination of the total electron-capture rate, for the calculations adopting the shell model rates, are displayed as a function of stellar evolution time

and hence more neutron-rich nuclei, subsequently affecting the nucleosynthesis. Its importance for the elemental abundance distribution, however, strongly depends on the location of the mass cut in the supernova explosion. It is currently assumed that the remnant will have a larger baryonic mass than the Fe-core, but smaller than the mass enclosed by the O-shell (Woosley et al., 2002). As the reduction of $Y_e$ occurs mainly during Si-burning, it is essential to determine how much of this material will be ejected.

## 4.4 Supernovae from Massive Stars and the Role of Radioactivity

### 4.4.1 The Explosion Mechanism

Supernova explosions are an application of numerical astrophysical modelling that has a long tradition. Continued improvements of the models are motivated by the following points: (i) open questions regarding the explosion mechanism; (ii) availability of observations for inidividual supernova explosions; (iii) interesting input physics that tests matter under conditions that are not accessible on earth; (iv) visibility in light and other photon wavelengths, cosmic rays, neutrino emission, decay gamma-rays of radioactive products, perhaps gravitational wave emission; (v) visibility on cosmological distances with improving statistical information on the events and (vi) their impact on the interstellar matter (e.g. abundances of metal-poor stars) and Galactic evolution.

As discussed in the previous sections, the death of massive stars $\approx 8 - 40\ M_\odot$ proceeds through several evolutionary and dynamical phases. At first, the modeling of a star must include the evolution through all nuclear burning stages until the resulting inner iron core grows beyond the maximum mass which can be supported by the dominant pressure of the degenerate electron gas. At this point, the inner stellar core enters a dynamical phase of gravitational collapse, during which it compactifies by $\sim$5 orders of magnitude. The nuclear saturation density (i.e. the density of stable nuclei $\approx 2 \times 10^{14}\,\mathrm{g/cm^3}$) is exceeded at the center of the collapse and a protoneutron star (PNS) is formed. The dynamical time scale reduces from a few hundreds of milliseconds at the onset of collapse to a few milliseconds after the core has bounced back at nuclear densities, see Fig. 4.8 from Liebendörfer et al. (2003).

The ensuing accretion phase onto the protoneutron star with fluid instabilities and radiative transfer phenomena, like the transport of neutrinos, is not well understood. It may last 0.5–10 s and can therefore be interpreted as a second evolutionary stage (much longer than the dynamical or transport time scale). Eventually it will lead to the observed vigorous supernova explosion, a dynamic phase where heavy elements are produced by explosive nucleosynthesis in an outward propagating shock wave. The processed matter is mixed by fluid instabilities and ejected into the interstellar medium, where it contributes to Galactic evolution. The remaining PNS at the center enters another evolutionary phase during which it cools by neutrino emission and contracts or even collapses to a black hole in a last dynamical phase.

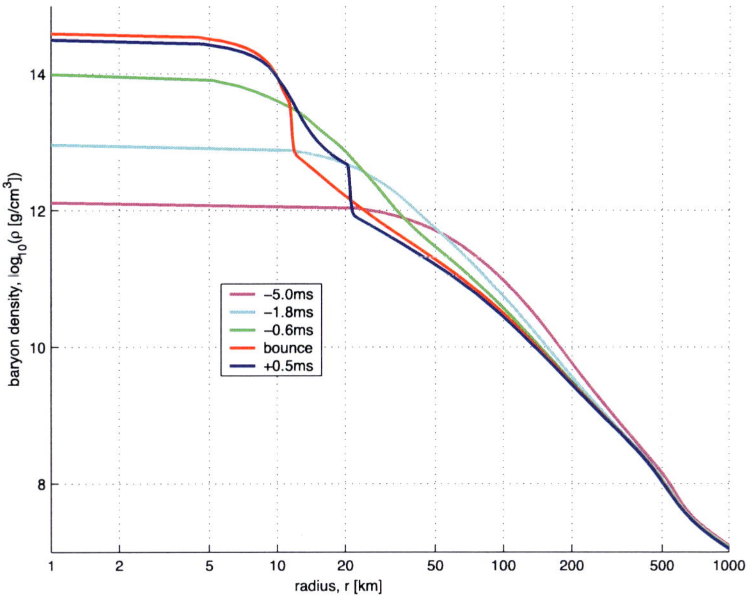


**Fig. 4.8** A sequence of density profiles of a 13 $M_\odot$ star before and after core bounce. For such a relatively low mass supernova with a small Fe-core the bounce occurs at a maximum density of less than twice nuclear matter density. At the bounce one recognizes the size of the homologous core (with roughly constant density). Thereafter the emergence of an outward moving density (shock) wave can be witnessed

While initially such calculations were performed in spherical symmetry and therefore lacked the consistent treatment of turbulent motion, presently performed research is done with multidimensional supernova models (Hix et al., 2003; Liebendörfer et al., 2005; Marek et al., 2005; Burrows et al., 2006a; Sumiyoshi et al., 2007; Langanke et al., 2008; Marek and Janka, 2009). The main ingredients are radiation (neutrino) transport, (relativistic) hydrodynamics, and the nuclear equation of state at such high densities. Recent progress has been made in the exploration of multidimensional hydrodynamics with idealized input physics. A refreshing view on the supernova mechanism has recently been suggested by pointing out that in certain axisymmetric simulations vibrational (so called PNS g-)modes are excited so that sound waves are emitted into the heating region. These sound waves are postulated to revive the stalled shock by dissipation of sound energy (Burrows et al., 2006a). Other efforts explore the role of magnetic fields and rotation in two-dimensional simulations with simplified input physics. One kind of proof-of-principle models is carried out in spherically symmetric approaches. The assumption of spherical symmetry is for many supernovae not compatible with observational constraints. However, one important advantage of spherically symmetric models is that sophisticated treatments of the neutrino-matter interactions can be included and that the neutrino spectra and transport are correctly treated in general relativistic space-time. Models of this kind try to address the question of how many neutrinos are emerging from the compactification of an inner stellar core, how is their emission distributed as a function of time and how do these neutrino fluxes generically affect the cooling, heating, or nucleosynthesis in the outer layers of the star without the complication of 3D dynamical fluid instabilities (Liebendörfer et al., 2003, 2004; Fischer et al., 2009a,b). The attempt to combine all these aspects with forefront methods is ongoing in order to achieve the final goal of understanding the multi-D explosion mechanism with up to date microphysics from the equation of state to all neutrino and nuclear interactions. For a review of the corresponding tools see Chap. 8.

The phase of stellar core collapse has intensively been studied in spherically symmetric simulations with neutrino transport. The crucial weak processes during the collapse and postbounce evolution are $\nu + (A, Z) \leftrightarrow \nu + (A, Z)$, $\nu + e^{\pm} \leftrightarrow \nu + e^{\pm}$, $p + e^- \leftrightarrow n + \nu_e$, $(A, Z) + e^- \leftrightarrow (A, Z-1) + \nu_e$ , $\nu + N \leftrightarrow \nu + N$, $n + e^+ \leftrightarrow p + \bar{\nu_e}$, $(A, Z) + e^+ \leftrightarrow (A, Z+1) + \bar{\nu_e}$, $\nu + (A, Z) \leftrightarrow \nu + (A, Z)^*$, $(A, Z)^* \leftrightarrow (A, Z) + \nu + \bar{\nu}$, $N + N \leftrightarrow N + N + \nu + \bar{\nu}$, $\nu_e + \bar{\nu_e} \leftrightarrow \nu_{\mu,\tau} + \bar{\nu}_{\mu,\tau}$, $e^+ + e^- \leftrightarrow \nu + \bar{\nu}$. Here, a nucleus is symbolized by its mass number $A$ and charge $Z$, $N$ denotes either a neutron or a proton and $\nu$ represents any neutrino or antineutrino. We note that, according to the generally accepted collapse picture (Bethe, 1990; Bethe et al., 1979), elastic scattering of neutrinos on nuclei is mainly responsible for the trapping, as it determines the diffusion time scale of the outwards streaming neutrinos. Shortly after trapping, the neutrinos are thermalized by energy downscattering, experienced mainly in inelastic scattering off electrons. The relevant cross sections for these processes are discussed in Martínez-Pinedo et al. (2006). The basic neutrino opacity in core collapse is provided by neutrino scattering off nucleons. Depending on the distribution of the nucleons in space and the wavelength of the neutrinos, various important coherence effects can occur: Most important during collapse is the binding of nucleons into nuclei with a density contrast of several orders of magnitude to the surrounding nucleon gas. Coherent scattering off nuclei dominates the scattering opacity of neutrinos (and scales with $A^2$). Moreover, these neutrino opacities should be corrected by an ion-ion correlation function, this occurs if the neutrino wavelength is comparable to the distances of scattering nuclei and quantum mechanical intererence effects appear (Sawyer, 2005; Burrows et al., 2006b). Even if current core collapse models include a full ensemble of nuclei in place of the traditional apprach with one representative heavy nucleus, it remains non-trivial to adequately determine correlation effects in the ion mixture. Depending on the Q-value of an electron-capturing nucleus, neutrinos are emitted with a high energy of the order of the electron chemical potential/Fermi energy. As the neutrino opacities scale with the squared neutrino energy, the initially trapped neutrinos will downscatter to lower energies until the diffusion time scale becomes comparable to the thermalization time scale. The thermalization in current collapse models occurs through neutrino-electron scattering because the energy transfer per collision with the light electron is more efficient than with the heavier nucleons. The contribution of inelastic scattering of neutrinos off heavy nuclei depends on the individual nuclei and affects only the high-energy tail of the neutrino spectrum.

Goldreich and Weber (1980) have shown that only the inner $M_{\rm Ch}(Y_e)$ (see the definition in Eq. (4.3)) undergo a homologous collapse ($v_{\rm collapse}(r) \propto r$), while at the edge of this core the velocity becomes supersonic and a fraction of the free-fall velocity. The inner core, falling at subsonic velocities where matter can communicate with sound speed, cannot communicate with the free-falling envelope. After the neutrinos are trapped, electron captures and neutrino captures are in equilibrium ($e^- + p \leftrightarrow n + \nu_e$) and the total lepton fraction $Y_L = Y_e + Y_\nu$ stays constant. $Y_e$ stops to decrease and $M_{Ch}$ stops shrinking. Typical values with the most recent electron capture rates (Langanke et al., 2003) of $Y_L \approx 0.3$ are found in numerical collapse calculations (Hix et al., 2003; Marek et al., 2005) which correspond to $M_{\rm Ch} \approx 0.5\,{\rm M}_\odot$. As soon as nuclear densities are reached at the center of the collapsing core, repulsive nuclear forces dominate the pressure in the equation of state. The collapse comes to a halt and matter bounces back to launch an outgoing pressure wave through the core. It travels through the subsonic inner core and steepens to a shock wave as soon as it faces supersonic infall velocities. Hence the matter in the PNS remains at low entropy $\sim 1.4$ k$_B$ per baryon while the supersonically accreting layers become shock-heated and dissociated at entropies larger than $\sim 6$ k$_B$ per baryon. Numerical simulations based on standard input physics and accurate neutrino transport exclude the possibility that the kinetic energy of the hydrodynamical bounce at nuclear densities drives a prompt supernova explosion because of dissociation and neutrino losses.

This can be seen in Fig. 4.8 presenting spherically symmetric calculations of a $13\,{\rm M}_\odot$ star. The inner core contains about $0.6\,{\rm M}_\odot$ of the initial Fe-core. The transition to free nucleons occurred only in this inner, homologous core and the outward moving shock runs through material consisting of Fe-group nuclei. The dissociation takes 8.7 MeV/nucleon or $8 \times 10^{18}$erg/g. Based on initial shock energies of $(4 - 8) \times 10^{51}$erg, this is sufficient for passing through 0.25–0.5 $\rm M_\odot$ and leads in essentially all cases to a stalling of the prompt shock. Only recently a possible exception was found (Sagert et al., 2009). If a hadron-quark phase transition occurs in the collapsed core at the appropriate time, releasing additional gravitational binding energy in the form of neutrinos from this second collapse, prompt explosions can be attained.

While the core collapse determines the state of the *cold* nuclear matter inside the PNS, the mass of the hot mantle around the PNS grows by continued accretion. The infalling matter is heated and dissociated by the impact at the accretion front and continues to drift inward. At first, it can still increase its entropy by the absorption of a small fraction of outstreaming neutrinos (heating region). Further in, where the matter settles on the surface of the PNS, neutrino emission dominates absorption and the electron fraction and entropy decrease significantly (cooling region). The tight non-local feedback between the accretion rate and the luminosity is well captured in computer simulations in spherical symmetry that accurately solve the Boltzmann neutrino transport equation for the three neutrino flavors. All progenitor stars between main sequence masses of 13 and 40 $\rm M_\odot$ showed no explosions in simulations of the postbounce evolution phase (Liebendörfer et al., 2003). This indicates that the neutrino flux from the PNS does not have the fundamental strength to blow off the surrounding layers for a vigorous explosion.

Improved electron capture rates on heavy nuclei overcame the idealized blocking of Gamow-Teller transitions in the traditionally applied single-particle model. In the single-particle picture of nuclei the so-called pf-shell is filled for $Z = 40$ or $N = 40$ for protons or neutrons respectively. Neutron numbers beyond $N = 40$ require a filling of the gd-orbits. If during core collapse nuclei ($Y_e$) become so neutron-rich that nuclei with $Z < 40$ and $N > 40$ dominate the NSE composition, electron capture would require the conversion of an $fp$ proton to a $gd$ neutron as all $pf$ neutron orbits are filled. This Pauli-blocked transition would lead to the dominance of electron capture on free protons rather than nuclei and under such conditions. The recent finding, that configuration mixing and finite temperature effects result in unfilled $pf$ neutron orbits, removes this Pauli-blocking and results in the fact that under these condition electron capture rates on nuclei dominate those on free protons (Langanke et al., 2003). Thus, there are two effects due to the new set of electron capture rates: (1) at low densities for less neutron-rich nuclei the total amount of electron capture is reduced with an improved description of Gamow-Teller transitions (see the discussion of the early collapse phase in Sect. 4.3), (2) at high densities in the late collapse phase the total amount of electron capture is enhanced, leading to smaller $Y_e$ and $Y_L$ values than before. Such changes caused a reduction of homologous core sizes down to $M_{\mathrm{Ch}} = 0.5\ \mathrm{M}_{\odot}$ (see discussion above and Hix et al. (2003)). This faster deleptonization in the collapse phase in comparison to captures on free protons alone thus resulted in a 20% smaller inner core at bounce.

Taking all this improved physics into acount leads in the entire simulations (i.e. all mass zones invoved) to conditions in densities $\rho$, electron abundance $Y_e$ and entropy $s$ per baryon, where properties like the equation of state or other

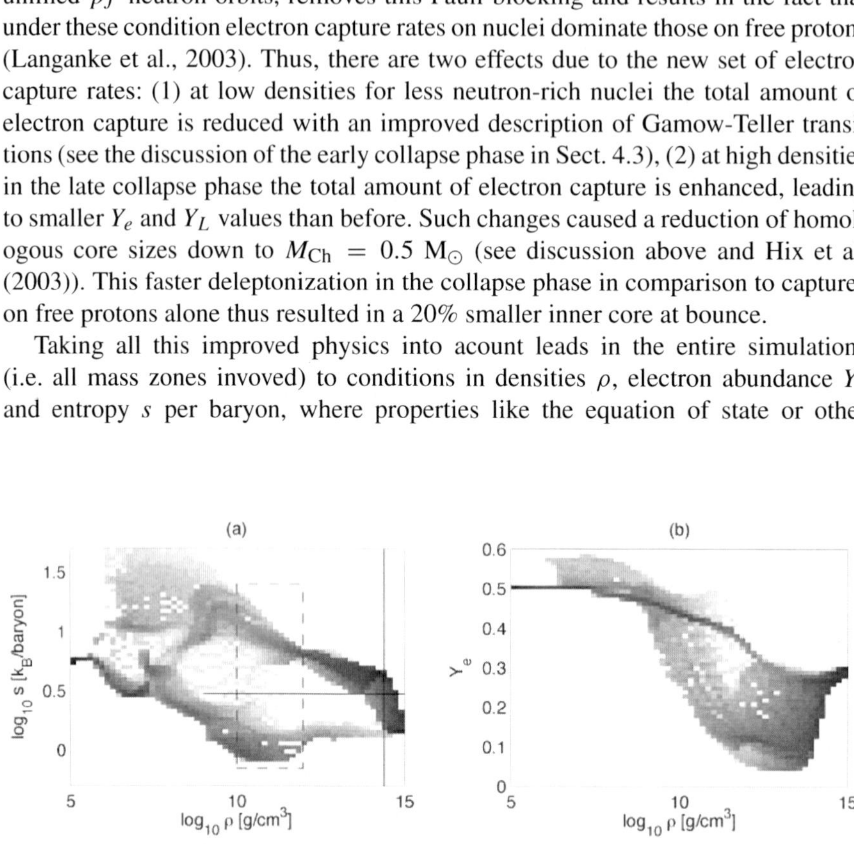


**Fig. 4.9** Overview of the conditions attained in a simulation of the collapse, bounce, and explosion (artificially induced) of a 20 $\mathrm{M}_{\odot}$ star. Shown are two histograms of the occurrence of conditions as a function of density $\rho$, specific entropy $s$ and electron fraction $Y_e$. The *shading* of a given bin corresponds to $\log_{10}(\int dmdt)$ in arbitrary units, where the integral over mass is performed over the mass $dm$ of matter whose thermodynamic state at a given time falls into the bin. The integral over time extends over the duration of a simulation. Hence, regions of dark shading correspond to states that are experienced by considerable mass for an extended time, while light or absent shading corresponds to conditions that are rarely assumed in the supernova simulation. The *vertical black line* indicates the nuclear density. The *horizontal black line* indicates an entropy of 3 $k_B$/baryon beyond which ions are dissociated. It clearly separates the conditions of cold infalling matter on the lower branch from the conditions of hot shocked matter on the upper branch

microscopic physics are needed in current supernova simulations. Figure 4.9 provides this information for a simulation of a 20 $M_\odot$ star (Liebendörfer et al., 2009).

Moreover, a comparison of the effects of the only two publicly available equations of state by Lattimer and Swesty (1991) and Shen et al. (1998a,b) is required. In simulations of massive progenitors that do not explode and exceed the maximum stable mass of the accreting neutron star in the postbounce phase, it was demonstrated that the neutrino signal changes dramatically when the PNS collapses to a black hole (Fischer et al., 2009b). Depending on the stiffness of the equation of state or the accretion rate from the external layers of the progenitor star, this can happen at very different time after bounce. Hence, the neutrino signal carries a clear imprint of the stiffness of the equation of state and the accretion rate to the observer of neutrinos.

The detailed treatment of the neutrino transport and interactions is of great importance for the nucleosynthesis. This has been shown in several recent studies (Fröhlich et al., 2006a,b; Pruet et al., 2005, 2006; Wanajo, 2006). This also opens an opportunity to investigate neutrino flavor oscillations among electron, muon and tau neutrinos. On the one hand, long term explosion runs achieve (low) density structures that allow for MSW (Mikheyev-Smirnov-Wolfenstein effect) neutrino flavor oscillations in the outer layers (Wolfenstein, 1978; Mikheyev and Smirnov, 1985). These may give additional hints on the expansion velocity and density distribution in case that the neutrinos can be observed from a near-by supernova. On the other hand, collective flavor transitions have recently been postulated in regions where the neutrino density exceeds the electron density (Duan et al., 2006, 2007; Fogli et al., 2007). This condition will be achieved in the evacuated zone that surrounds the PNS after the onset of an explosion. The impact of these collective neutrino flavor oscillations on the neutrino heating during the shock expansion, the neutrino wind, and the nucleosynthesis are important open points that require a detailed investigation under consideration of accurate neutrino transport and spectra.

The difficulty to reproduce explosions in spherically symmetric models of core-collapse and postbounce evolution stimulated the consideration of numerous modifications and alternatives to this basic scenario, mostly relying on multi-dimensional effects that could not be treated in spherical symmetry. It was discussed whether convection inside the PNS could accelerate the deleptonization and increase the neutrino luminosity (Wilson and Mayle, 1993). The convective overturn between the PNS and shock front was shown to increase the efficiency of neutrino energy deposition (Herant et al., 1994). Asymmetric instabilities of the standing accretion shock (Blondin et al, 2003; Foglizzo, 2009) may help to push the shock to larger radii and g-mode oscillations of the PNS may contribute to neutrino heating by the dissipation of sound waves between the PNS and the shock (Burrows et al., 2006a). Moreover, it has been suggested that magnetic fields have an impact on the explosion mechanism (Kotake et al., 2006). Most of the above-mentioned modifications of the explosion mechanism are essentially of a three-dimensional nature. In order to illustrate the complexity of the crucial accretion phase we show in Fig. 4.10 a slice through a three-dimensional simulation of core-collapse and postbounce evolution of a recent run (Liebendörfer et al., 2008). Its input physics uses the Lattimer-Swesty

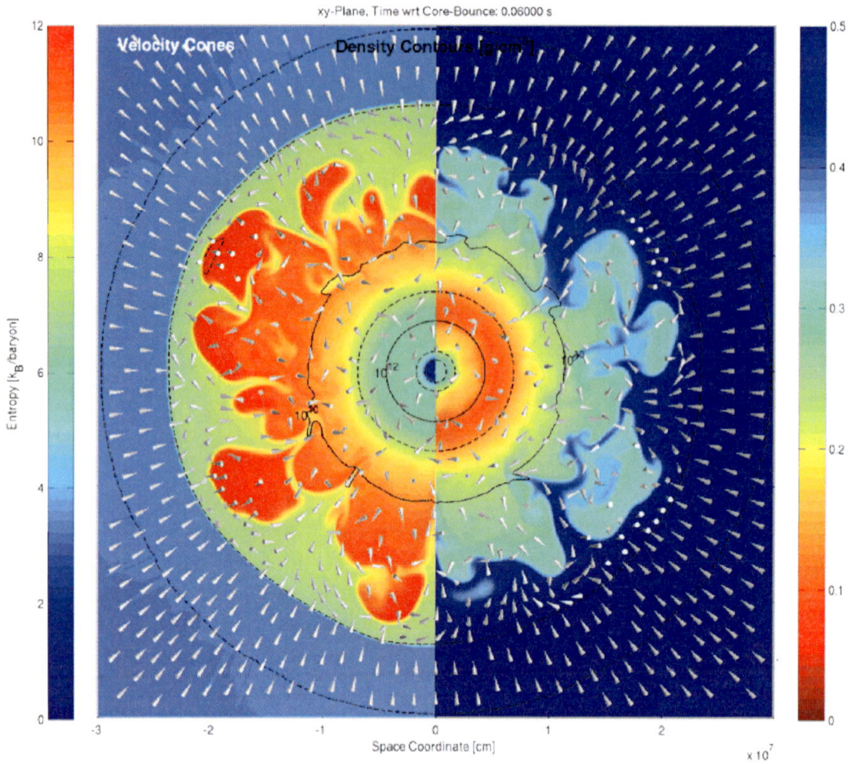


**Fig. 4.10** Illustration of the early accretion phase in a three-dimensional simulation with a resolution of $600^3$ zones and the isototropic diffusion source approximation for 3D neutrino transport (Liebendörfer et al., 2009). Shown are density contours as black lines for a $15\,M_\odot$ star from Woosley and Weaver (1995). *Left*: The color indicates the specific entropy and the cones the direction of the velocity. *Right*: The color refers to the magnetic field strength and the cones to its direction. The cool high-density interior of the PNS and the hot low-density accreted matter behind the standing accretion front are clearly distinguishable

equation of state (Lattimer and Swesty, 1991) and a parameterization of the neutrino physics for the collapse phase (Liebendörfer et al., 2005). The treatment of neutrino cooling and heating in the postbounce phase is based on multi-group diffusion (the isotropic diffusion source approximation of Liebendörfer et al. (2009)).

Initially, spherically symmetric supernova models were the most realistic among all feasible computer representations of the event. With increasing observational evidence for the complexity of the explosions (Hamuy, 2003) their primary purpose shifted from a realistic representation to the identification and understanding of the basic principles of the explosion mechanism. After the emergence of axisymmetric simulations with sophisticated and computationally intensive spectral neutrino transport (Buras et al., 2003; Walder et al., 2005) spherically symmetric models still have several assets. In the following subsection we will first describe purely phenomenological calculations based on artificially induced explosions via a *piston* or energy deposition in terms of a *thermal bomb*, purely in order to discuss nucleosynthesis effects. We will then also discuss still artificial explosions in spherical symmetry, however resulting from a *self-consistent* treatment including neutrino transport, which permits to analyse the effect of neutrinos on the nucleosynthesis of the innermost ejecta.

## 4.4.2 Nucleosynthesis in Explosions

### 4.4.2.1 Major Explosive Burning Processes

Despite considerable improvements of stellar models and numerical simulations in recent years, some fundamental problems remain in nucleosynthesis predictions. It has become evident that certain evolution aspects can only be followed in models going beyond one-dimensional simulations, such as convection, rotation, and the explosion mechanism. However, it is still not feasible to directly couple full reaction networks, containing several thousand nuclei, to multi-dimensional hydrodynamic calculations due to the lack of required computing power, even in modern computers. Thus, postprocessing after explosion models with parameterized networks still remains an important approach. One-dimensional models can directly accommodate increasingly larger networks but they cannot capture all of the necessary physics. As outlined in the previous subsection, it has become apparent that a self-consistent treatment of core collapse supernovae in 1D does not lead to successful explosions when using presently known input physics while 2D models show some promise. There are strong indications that the delayed neutrino mechanism works combined with a multi-D convection treatment for unstable layers (possibly with the aid of rotation, magnetic fields and/or still existent uncertainties in neutrino opacities). Therefore, hybrid approaches using certain parameterizations or approximations have been and are still necessary when predicting the nucleosynthetic yields required for the application described above.

Supernova nucleosynthesis predictions have a long tradition. All of these predictions relied on an artificially introduced explosion, either via a piston or a thermal bomb introduced into the progenitor star model. The mass cut between the ejecta and the remnant does not emerge from this kind of simulations but has to be determined from additional conditions. While the usage of artificially introduced explosions is justifiable for the outer stellar layers, provided we know the correct explosion energy to be dumped into the shock front (on the order of $10^{51}$ erg seen in observations), it clearly is incorrect for the innermost ejected layers which should be directly related to the physical processes causing the explosion. This affects the Fe-group composition, which has been recognized as a clear problem by many groups (Woosley and Weaver, 1995; Thielemann et al., 1990, 1996; Nakamura et al., 1999, 2001; Nomoto et al., 2006). The problem is also linked to the so-called neutrino wind, emitted seconds after the supernova explosion, and considered as a possible source of the *r* process to produce the heaviest elements via neutron captures (Qian and Woosley, 1996), as will be discussed below.

Given the above detailed discussion of the physics, problems and options regarding core collapse supernovae, we will adopt the following approach in order to predict the most reliable nucleosynthesis predictions for the ejecta in a 1D spherically symmetric treatment. The multiplication of neutrino capture cross sections on nucleons with a free parameter in 1D spherically symmetric calculations can mimic the enhanced energy deposition which multi-D models show. The free parameter is tuned to give correct explosion energies and $^{56}$Ni yields for a number of well known supernovae. This approach provides clear predictions for the mass cut between the remaining neutron star and the ejecta. It also includes the effect neutrinos can have on the correct $Y_e$ in the ejecta and the related nucleosynthesis. In the outer explosively burning layers, essentially only the energy in the shock front matters. The behavior of these zones can be easily understood from the maximum temperatures attained in the radiation bubble and for a first discussion we will just focus on these features, which can also be obtained with an artifically induced thermal bomb treatment.

For a given/known $Y_e$ and density $\rho$, the most significant parameter in explosive nucleosynthesis is the temperature, and a good prediction for the composition can already be made by only knowing $T_{\rm max}$, without having to perform complex nucleosynthesis calculations. Weaver and Woosley (1980) already recognized, that matter behind the shock front is strongly radiation dominated. Assuming an almost homogeneous density and temperature distribution behind the shock (which is approximately correct), one can equate the supernove energy with the radiation energy inside the radius $r$ of the shock front

$$E_{SN} = \frac{4\pi}{3} r^3 a T^4(r). \tag{4.4}$$

This equation can be solved for $r$. With $T = 5 \times 10^9$ K, the lower bound for explosive Si-burning with complete Si-exhaustion, and an induced thermal bomb energy of $E_{SN} = 10^{51}$erg, the result is $r \approx 3{,}700$ km. For the evolutionary model by Nomoto and Hashimoto (1988) of a 20 $\mathrm{M}_\odot$ star this radius corresponds to 1.7 $\mathrm{M}_\odot$, in excellent agreement with the exact hydrodynamic calculation. Temperatures which characterize the edge of the other explosive burning zones correspond to the following radii: incomplete Si-burning ($T_9 = 4$, $r = 4{,}980$ km), explosive O-burning (3.3, 6,430), and explosive Ne/C-burning (2.1, 11,750). This relates to masses of 1.75, 1.81, and 2.05 $\mathrm{M}_\odot$ in case of the 20 $\mathrm{M}_\odot$ star. The radii mentioned are model independent and vary only with the supernova energy. In the following we present a number of plots which show the different mass fractions $X_i = A_i Y_i$ as a function of radial mass $M(r)/\mathrm{M}_\odot$, passing outwards through a 20 $\mathrm{M}_\odot$ star through all explosive burning regions.

Matter between the mass cut $M(r)$=$M_{\rm cut}$ and the mass enclosed in the radius corresponding to explosive Si-burning with complete Si-exhaustion is indicated with $M$(ex Si-c). Then follows the zone of incomplete Si-burning until $M$(ex Si-i), explosive O-burning until $M$(ex O), explosive Ne/C-burning until $M$(ex Ne), and unprocessed matter from the C/Ne-core is ejected until $M$(C-core). The zones beyond explosive Ne/C-burning ($T_{max} < 2.1 \times 10^9$ K) are essentially unaltered and the composition is almost identical to the pre-explosive one. When performing such calculations for a variety of progenitors over a range of initial stellar masses, one can analyze the dependence of the mass involved in these different burning regimes as a function initial stellar mass (see Sect. 4.5).

Results for a 20 $M_\odot$ star (Nomoto and Hashimoto, 1988) are given as examples for the abundance behavior in a series of Figs. 4.11, 4.12, 4.13, and 4.14. It should be mentioned here that this still uses a simplified thermal bomb treatment for the pre-collapse model rather than the results from a 1D spherically symmetric simulation with enhanced neutrino capture rates, which insures an explosion also in 1D. The explosion energy used corresponds to a supernova energy of $10^{51}$erg. As mentioned before, this treatment cannot predict a self-consistent explosion and the position of the mass cut between neutron star and ejecta. Only the observation of $0.07 \pm 0.01$ $M_\odot$ of $^{56}$Ni in SN1987A (a 20 $M_\odot$ star) gives an important constraint, because $^{56}$Ni is produced in the innermost ejected zones. The explosive nucleosynthesis due to burning in the shock front is shown in Fig. 4.11 for a few major nuclei. Inside 1.7 $M_\odot$ all Fe-group nuclei are produced in *explosive* Si-burning during the SN II event. At 1.63 $M_\odot$ $Y_e$ changes from 0.494 to 0.499 and leads to a smaller $^{56}$Ni abundance further inside, where more neutron-rich Ni-isotopes share the abundance with $^{56}$Ni. This is an artifact of the $Y_e$ gradient in the pre-collapse model which can

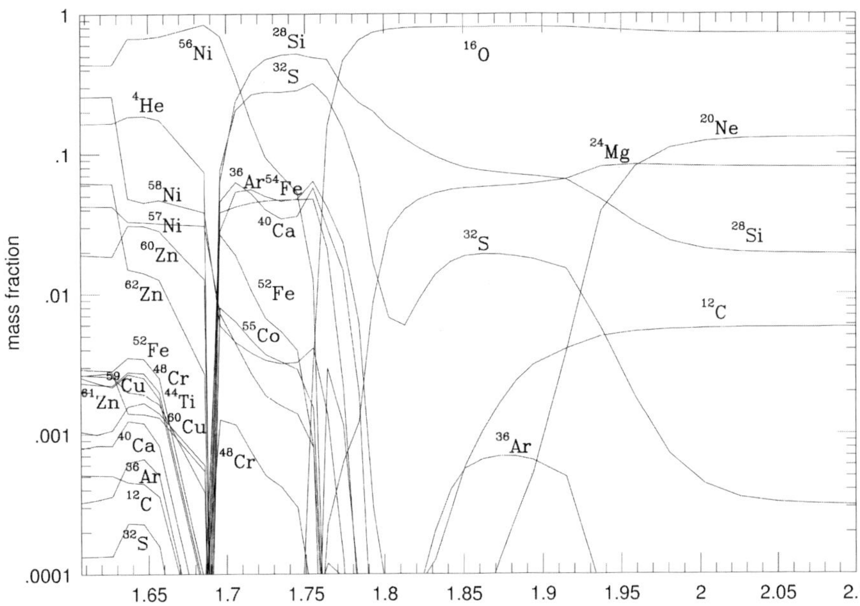


**Fig. 4.11** Mass fractions of a few major nuclei after passage of the supernova shockfront through a star with an intial mass of 20 $M_\odot$. Matter outside 2 $M_\odot$ is essentially unaltered. Mass zones further in experience explosive Si, O, Ne, and C-burning. For ejecting 0.07 $M_\odot$ of $^{56}$Ni the mass cut between neutron star and ejecta is required to be located at 1.6 $M_\odot$

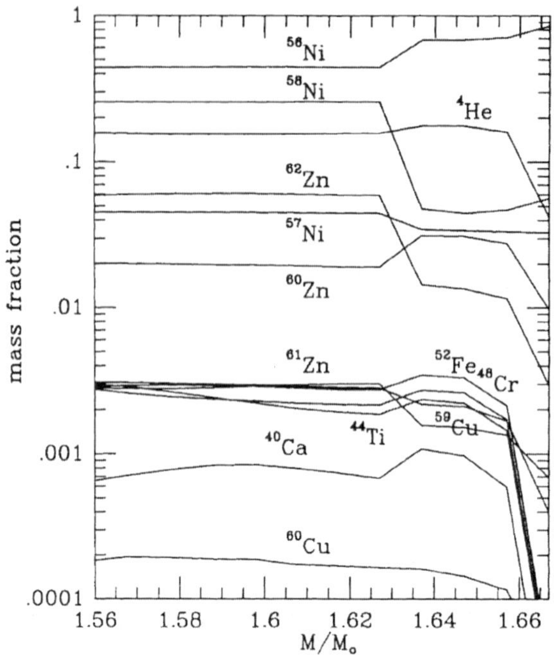


**Fig. 4.12** Mass fractions of the dominant nuclei in zones which experience alpha-rich freeze-out. Notice the relatively large amounts of Zn and Cu nuclei, which originate from alpha-captures on Ni and Co. One can recognize their strong decrease beyond 1.66 M$_\odot$, which goes parallel with the decrease of the $^4$He-abundance and other alpha-nuclei like $^{40}$Ca, $^{44}$Ti, $^{48}$Cr, and $^{52}$Fe. Nuclei which would dominate in a nuclear statistical equilibrium like $^{56,57,58}$Ni stay constant or increase even slightly. The increase of all nuclei with $N = Z$ at 1.63 M$_\odot$ and the decrease of nuclei with N>Z is due to the change in $Y_e$ in the original stellar model before collapse (see also Fig. 4.11)

be changed in a consistent explosion treatment via neutrino interactions with this matter.

In explosive Si-burning only alpha-rich freeze-out and incomplete Si-burning are encountered. Contrary to SNe Ia, densities in excess of $10^8$ g/cm$^3$, which would result in a normal freeze-out, are not attained in the ejecta of this 20 M$_\odot$ star (see also Fig. 4.2). The most abundant nucleus in the alpha-rich freeze-out is $^{56}$Ni. For the less abundant nuclei the final alpha-capture plays a dominant role transforming nuclei like $^{56}$Ni, $^{57}$Ni, and $^{58}$Ni into $^{60}$Zn, $^{61}$Zn, and $^{62}$Zn (see Fig. 4.12).

The region which experiences imcomplete Si-burning starts at 1.69 M$_\odot$ and extends out to 1.74 M$_\odot$. In the innermost zones with temperatures close to $4 \times 10^9$ K there exists still a contamination by the Fe-group nuclei $^{54}$Fe, $^{56}$Ni, $^{52}$Fe, $^{58}$Ni, $^{55}$Co, and $^{57}$Ni. Explosive O-burning occurs in the mass zones up to 1.8 M$_\odot$ (see Fig. 4.13). The main burning products are $^{28}$Si, $^{32}$S, $^{36}$Ar, $^{40}$Ca, $^{38}$Ar, and $^{34}$S. With mass fractions less than $10^{-2}$ also $^{33}$S, $^{39}$K, $^{35}$Cl, $^{42}$Ca, and $^{37}$Ar are produced. Explosive Ne-burning leads to an $^{16}$O-enhancement over its hydrostatic value in the mass zones up to 2 M$_\odot$ (see Fig. 4.14).

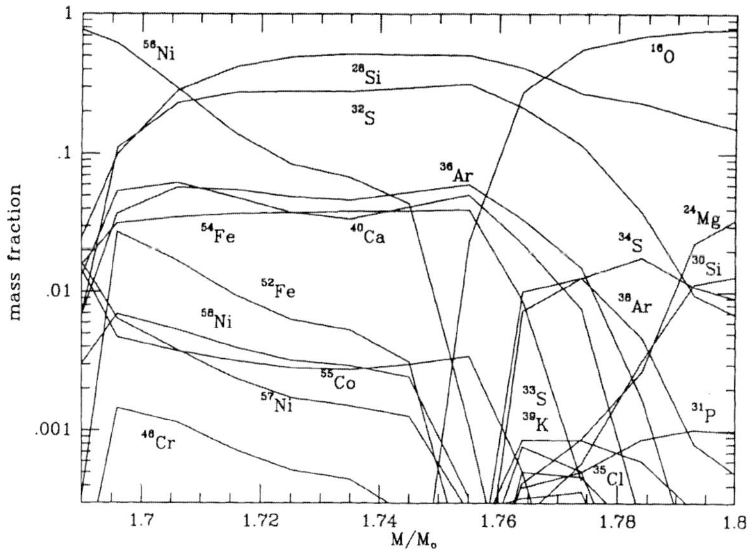


**Fig. 4.13** Mass fractions of nuclei in the zones of incomplete Si-burning M < 1.74 $M_{\odot}$ and explosive O-burning M < 1.8 $M_{\odot}$. The Si-burning zones are are characterized by important quantities of Fe-group nuclei besides $^{28}$Si, $^{32}$S, $^{36}$Ar, and $^{40}$Ca. Explosive O-burning produces mostly the latter, together with more neutron-rich nuclei like $^{30}$Si, $^{34}$S, $^{38}$Ar etc

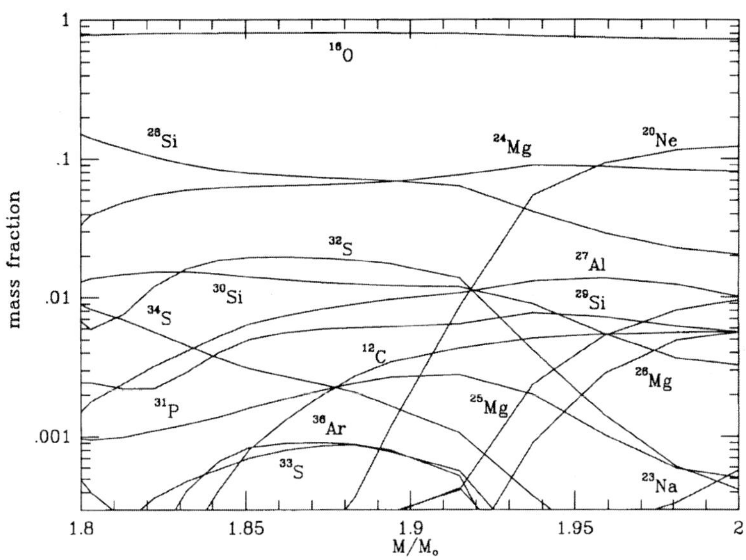


**Fig. 4.14** Composition in mass zones of explosive Ne and C-burning. The dominant products are $^{16}$O, $^{24}$Mg, and $^{28}$Si. Besides the major abundances, mentioned above, explosive Ne-burning supplies also substantial amounts of $^{27}$Al, $^{29}$Si, $^{32}$S, $^{30}$Si, and $^{31}$P. Explosive C-burning contributes in addition the nuclei $^{20}$Ne, $^{23}$Na, $^{24}$Mg, $^{25}$Mg, and $^{26}$Mg

#### 4.4.2.2 Explosive Burning Off Stability

The *p* Process

Up to now we discussed the production of heavy nuclei beyond the Fe-group only via slow neutron captures (the *s* process) in hydrostatic stellar evolution. A number of proton-rich (p-)isotopes of naturally occurring stable heavy nuclei cannot be produced by neutron captures along the line of stability. The currently most favored production mechanism for those 35 p-isotopes between Se and Hg is photodisintegration ($\gamma$ process) of intermediate and heavy elements at high temperatures in late (explosive) evolution stages of massive stars (Woosley and Howard, 1978; Rayet et al., 1990). However, not all p-nuclides can be produced satisfactorily, yet. A well-known deficiency in the model is the underproduction of the Mo-Ru region, but the region $151<A<167$ is also underproduced, even in recent calculations (Rauscher et al., 2002; Arnould and Goriely, 2003; Rapp et al., 2006; Dillmann et al., 2008). There exist deficiencies in astrophysical modeling and the employed nuclear physics. Recent investigations have shown that there are still considerable uncertainties in the description of nuclear properties governing the relevant photodisintegration rates. This has triggered a number of experimental efforts to directly or indirectly determine reaction rates and nuclear properties for the $p/\gamma$ process (Rauscher, 2006). Here it is important to investigate the sensitivity of the location of the $\gamma$-process path with respect to reaction rate uncertainties.

Concerning the astrophysical modeling, only a range of temperatures has to be considered which are related to the explosive Ne/O-burning zones of a supernova explosion (see Figs. 4.13 and 4.14), where partial (but not complete) photodisintegration of pre-existing nuclei occurs (from prior hydrostatic evolution or inherited metallicity), i.e. at $\approx 2-3\times10^9$ K. The $\gamma$ process starts with the photodisintegration of stable seed nuclei that are present in the stellar plasma. During the photodisintegration period, neutron, proton, and alpha-emission channels compete with each other and with beta-decays further away from stability. In general, the process, acting like "spallation" of pre-existing nucei, commences with a sequence of $(\gamma, n)$-reactions and moves the abundances to the proton-rich side. At some point in a chain of isotopes, $(\gamma, p)$ and/or $(\gamma, \alpha)$-reactions become faster than neutron emissions, and the flow branches and feeds other isotopic chains. At late times photodisintegrations become less effective, when decreasing temperatures shift the branching points and make beta-decays more important. Finally the remaining unstable nuclei decay back to stability. The branchings established by the dominance of proton and/or alpha-emission over neutron emission are crucial in determining the radioactive progenitors of the stable p-nuclei and depend on the ratios of the involved reaction rates. Numerous experimental and theoretical efforts have been undertaken to improve the reaction input, especially with respect to open questions in optical potentials for alpha particles and protons (Gyürky et al., 2006; Kiss et al., 2007, 2008; Yalçın et al., 2009).

Applications of *p*-process network calculations to the temperature profiles of initiated explosions have been performed by Rayet et al. (1995); Rapp et al. (2006); Dillmann et al. (2008). Here, in Fig. 4.15 we present the results of a 25 $M_\odot$ mass

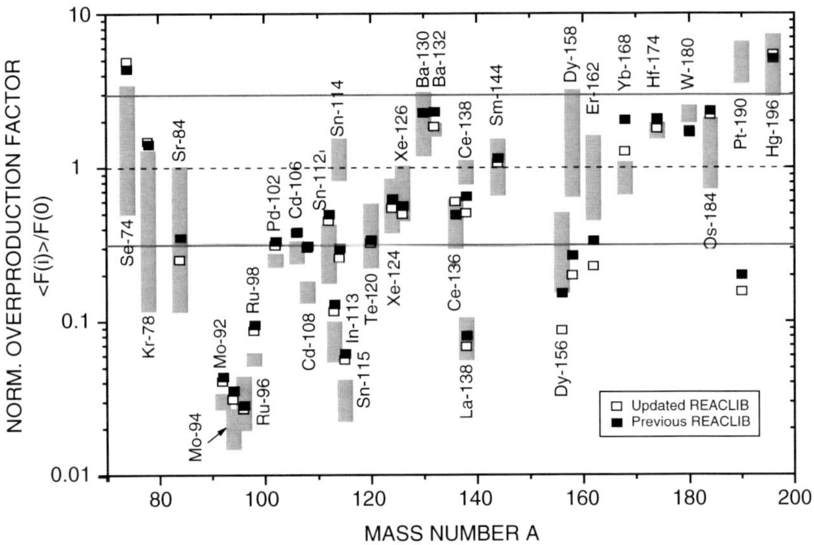


**Fig. 4.15** Normalized overproduction factors of $p$-process nuclei derived with the Rapp et al. (2006) (*open squares*) and Dillmann et al. (2008) (*full squares*) reaction library. In addition, the results from a range of stellar models (10–25 $M_{\odot}$) from Rayet et al. (1995) are given for comparison. A value equal to unity corresponds to relative solar abundances

model (Dillmann et al., 2008) with two reaction rate libraries without and with inclusion of all experimental improvements, existing at that point. It is noticed that the nuclear uncertainties cannot change the underproduction of especially the light p-nuclei. Another process seems to be required to supply these missing abundances.

### The $\nu p$ Process

Neutron-deficient nuclei can also be produced by two other astrophysical nucleosynthesis processes: the *rp* process in X-ray bursts (which, however, does not eject matter into the interstellar medium (Wallace and Woosley, 1981; Schatz et al., 1998; Fisker et al., 2008) and the recently discovered $\nu p$ process in core collapse supernovae (Fröhlich et al., 2006a,b; Pruet et al., 2006; Wanajo, 2006). The $\nu p$ process occurs in explosive environments when proton-rich matter is ejected under the influence of strong neutrino fluxes. This includes the innermost ejecta of core-collapse supernova (Buras et al., 2006; Thompson et al., 2005; Liebendörfer et al., 2008) and possible ejecta from black hole accretion disks in the collapsar model of gamma-ray bursts (Surman et al., 2006). The discussion of these innermost ejected mass zones has been skipped above, when discussing the results for explosive nucleosynthesis in a 20 $M_{\odot}$ star, utilizing a thermal bomb and the pre-collapse stellar conditions with the corresponding $Y_e$. Here, as discussed in the beginning of this subsection, we have boosted the energy deposition efficiencies by enhancing the neutrino and

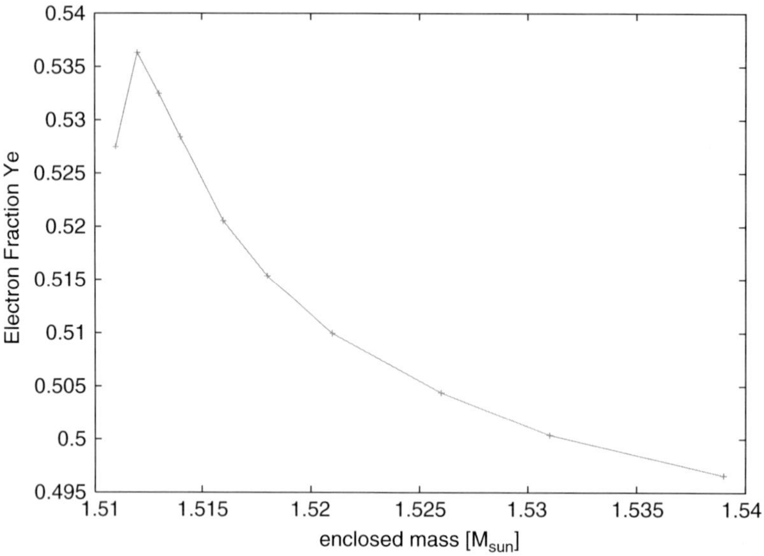


**Fig. 4.16** $Y_e$ of the innermost ejecta due to neutrino interactions with matter. At high temperatures electrons are not degenerate, thus the reduction of $Y_e$ due to electron captures is ineffective. For similar neutrino and antineutrino spectra the neutron-proton mass difference favors $\nu_e + n \leftrightarrow p + e^-$ over $\bar{\nu}_e + p \leftrightarrow n + e^+$

anti-neutrino captures on neutrons and protons in a 1D simulation. While this is not a fully self-consistant treatment, no external (artificial) energy is required to produce a successful explosion with a consistently emerging mass cut between neutron star and ejecta. Moreover, this treatment provides a $Y_e$ that is consistently determined by all weak interactions processes. The result is that explosions are obtained and the neutrino interaction with matter leads to a $Y_e$ enhanced beyond 0.5 (see Fig. 4.16) which overcomes nucleosynthesis problems for the Fe-group encountered previously.

The matter in these ejecta is heated to temperatures well above $10^{10}$ K and becomes fully dissociated into protons and neutrons. The ratio of protons to neutrons is mainly determined by neutrino and antineutrino absorptions on neutrons and protons, respectively. Similar neutrino and antineutrino energy spectra and fluxes produce proton-dominated matter in the reactions $\nu_e + n \leftrightarrow p + e^-$ and $\bar{\nu}_e + p \leftrightarrow n + e^+$, due to the n-p mass difference. When the matter expands and cools, the free neutrons and protons combine into $\alpha$-particles. Later, at temperatures around $5\times10^9$ K, alpha-particles assemble into heavier nuclei via unstable intermediate nuclei, e.g. the triple-$\alpha$ reaction via unstable $^8$Be, but – depending on the entropy and the expansion of matter – only a fraction of those form iron-group nuclei (alpha-rich freeze-out). In case of a proton-rich environment, there are also still free protons available at the time of the alpha freeze-out. Once the temperature drops to about $2\times10^9$ K, the composition of the ejecta consists mostly of $^4$He, protons, and iron group nuclei with N$\approx$Z (mainly $^{56}$Ni) in order of decreasing abundance.

Without neutrinos, synthesis of nuclei beyond the iron peak becomes very inefficient due to bottleneck (mainly even-even $N = Z$) nuclei with long beta-decay half-lives and small proton-capture cross sections. Such a nucleus is $^{64}$Ge. Thus, with the $Y_e$ determined by neutrino interactions with free neutrons and protons in the early very hot phase of dissociated nuclei, the nucleosynthesis leads to an alpha- and proton-rich freeze-out which does not stop at $^{56}$Ni but continues up to $^{64}$Ge (which later decays to $^{64}$Zn). This part of the story enables core collapse yields which produce Fe-group nuclei up to essentially $^{64}$Zn. The effect is seen in the bottom portion Fig. 4.17.

However, the matter is subject to a large neutrino/antineutrino flux from the proto-neutron star. While neutrons are bound in neutron-deficient $N = Z$ nuclei and neutrino captures on these nuclei are negligible due to energetics, antineutrinos are readily captured both on free protons and on heavy nuclei on a timescale of a few seconds. As protons are more abundant than heavy nuclei, antineutrino captures occur predominantly on protons, leading to residual neutron densities of $10^{14}$–$10^{15}$ cm$^{-3}$ for several seconds. These neutrons are easily captured by heavy neutron-deficient nuclei, for example $^{64}$Ge, inducing $(n, p)$ reactions with time scales much shorter than the beta-decay half-life. This permits further proton captures and allows the nucleosynthesis flow to continue to heavier nuclei (see upper part of Fig. 4.17). The $\nu p$ process (Fröhlich et al., 2006b) is this sequence of $(p, \gamma)$-reactions, followed

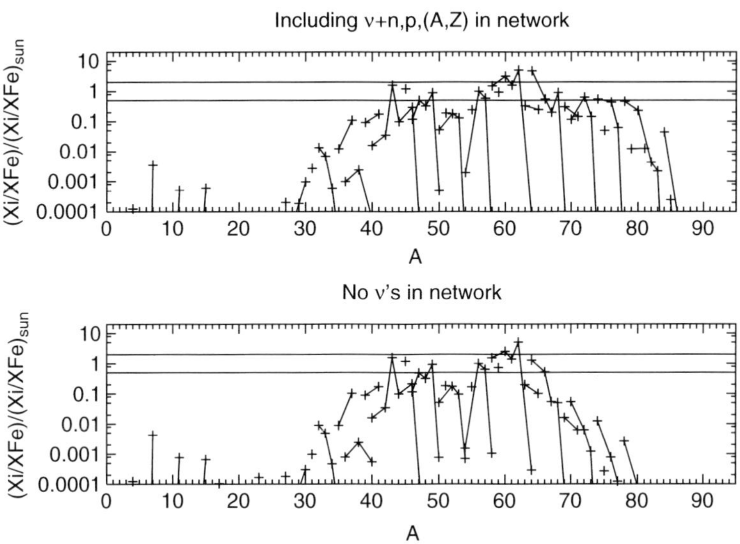


**Fig. 4.17** Final abundances in mass zones in the innermost ejecta which experienced neutrino irradiation, leading to proton-rich conditions ($Y_e > 0.5$). The *bottom part* of the figure shows the neucleosynthesis results in the innermost ejecta of explosive, after alpha-rich and proton-rich freeze-out from Si-burning, normalized to solar after decay. The *top part* of the figure also includes the interaction of anti-electron neutrinos with protons ($\bar{\nu}_e + p \rightarrow n + e^+$) which produces neutrons, permitting the late change of $^{64}$Ge via $^{64}$Ge$(n, p)^{64}$Ga. This feature permits further proton captures to produce havier nuclei (the so-called $\nu p$ process). Here matter up to $A = 85$ is produced

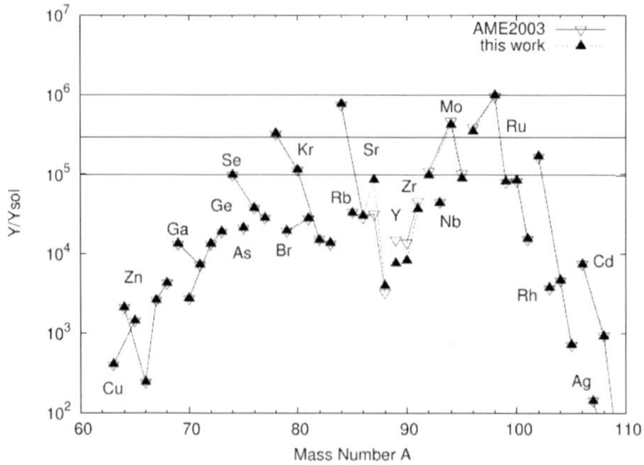


**Fig. 4.18** Final abundances in mass zones experiencing the $\nu p$ process, i.e. the innermost ejecta of explosive, alpha-rich freeze-out Si-burning, normalized to solar after decay for two sets of thermonuclear reaction rates/masses. Matter up to $A = 100$ can be produced easily

by $(n, p)$-reactions or beta-decays, where the neutrons are supplied by antineutrino captures on free protons.

In Fig. 4.18 we also show $\nu p$-process nucleosynthesis results from the innermost early neutrino wind ejecta evaporating from the neutron star surface after the onset of the explosion of a $15\,M_\odot$ star (Janka et al., 2003), also utilized in (Pruet et al., 2006; Fisker et al., 2009), which synthesizes efficiently nuclei even for $A > 90$. Two sets of astrophysical reaction rates were used in the reaction network, both based on theoretical rates from the NON-SMOKER code (Rauscher and Thielemann, 2000, 2004), but once with the latest excited state information and masses from the AME2003 compilation (Audi et al., 2003) and another set also with the latest mass measurements (Weber et al., 2008). Figure 4.18 shows the final abundances normalized to solar abundances after decay to stability for these two sets of thermonuclear reaction rates. Only nuclei produced in the p-rich ejecta are shown. As is clearly seen, there is no difference in the yields for the two different sets of rates except for a few nuclei in the mass range $85 < A < 95$, namely $^{87,88}$Sr, $^{89}$Y, and $^{90,91}$Zr. This can be directly traced back to the large change in the mass of $^{88}$Tc ($\Delta M = -1031\,keV$). This change in mass leads to an increase in the reaction rate for $^{88}$Tc$(\gamma, p)^{87}$Mo at the relevant temperatures and therefore a relative suppression of the opposite capture rate. These results show that the $\nu p$ process can easily produce the light p-nuclei of Mo and Ru, which are deficient in $p$-process calculations. Further processing depends on the expansion (speed) of matter and the overlying mass of ejecta.

## The *r* Process

A rapid neutron-capture process (*r* process) in an explosive environment is traditionally believed to be responsible for the nucleosynthesis of about half of the

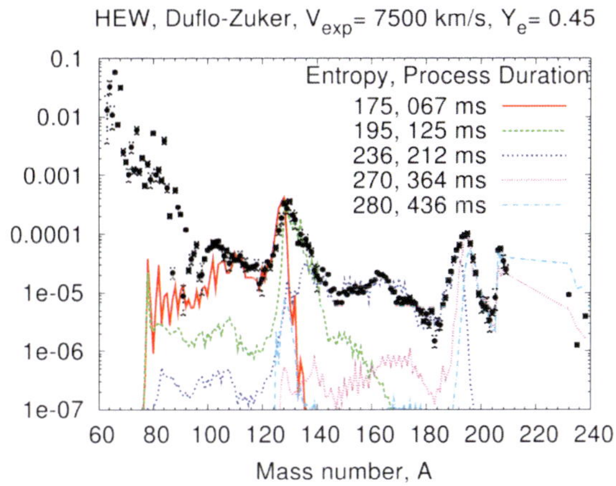


**Fig. 4.19** High entropy neutrino wind results for the mass model by Duflo and Zuker (1995), expansion parameters and proton/nucleon ratio $Y_e$ as given in the label, for a variation in extropies per baryon and $k_B$ (Farouqi et al., 2010)

heavy elements above Fe. While in recent years the (late) high entropy (neutrino) wind (HEW) of core-collapse supernovae has been considered to be one of the most promising sites, hydrodynamical simulations still encounter difficulties to reproduce the astrophysical conditions under which this process occurs. The classical *waiting-point* approximation, with the basic assumptions of an Fe-group seed, an $(n,\gamma)-(\gamma,n)$-equilibrium for constant neutron densities $n_n$ at a chosen temperature $T$, over a process duration $\tau$, and an instantaneous freezeout, has helped to gain improved insight into the systematics of an $r$ process in terms of its dependence on nuclear-physics input and astrophysical conditions (Cowan et al., 1991; Kratz et al., 1993, 2007) as well as the formation of long-lived actinides. This corresponds to a set of quasi-equilibria with each QSE group being represented by an isotopic chain. Taking a specific seed nucleus, the solar $r$-process pattern peaks can be reproduced by a variation/superposition of neutron number densities $n_n$ and durations $\tau$. Whether the solar $r$-process abundances are fully reproduced in each astrophysical event, i.e., whether each such event encounters the full superposition of conditions required, is a matter of debate (Wasserburg et al., 1996; Pfeiffer et al., 2001; Sneden et al., 2003; Honda et al., 2006; Qian and Wasserburg, 2007; Farouqi et al., 2009, 2010). In realistic astrophysical environments with time variations in $n_n$ and $T$, it has to be investigated whether at all and for which time duration $\tau$ the supposed $(n,\gamma)-(\gamma,n)$-equilibrium of the classical approach will hold and how freeze-out effects change this behavior. In general, late neutron captures may alter the final abundance distribution. In this case neutron capture reactions will be important. Also $\beta$-delayed neutrons can play a role in forming and displacing the peaks after freeze-out.

For many years since Woosley et al. (1994); Takahashi et al. (1994); Qian and Woosley (1996) the high entropy wind has been considered as the most promising (realistic?) environment, expelled from newly formed (hot) neutron stars in core-collapse supernovae, which continue to release neutrinos after the supernova shock wave is launched. These neutrinos interact with matter of the outermost proto-neutron star layers which are heated and ejected in a continuous wind. The late neutrino flux also leads to moderately neutron-rich matter (Qian and Woosley, 1996) via interactions with neutrons and protons and causes matter ejection with high entropies. (However, there are recent studies (Fischer et al., 2009a) from collapse calculations which predict a proton-rich wind composition for more than the first 10s after collapse.) Problems were encountered to attain entropies sufficiently high in order to obtain high neutron/seed ratios which can produce the heaviest $r$-process nuclei (Thompson et al., 2001; Wanajo et al., 2001; Terasawa et al., 2002). Recent hydrodynamic simulations for core-collapse supernovae support the idea that these entropy constraints can be fulfilled in the late phase (after the initial explosion) when a reverse shock is forming (Fryer et al., 2006; Arcones et al., 2007; Burrows et al., 2007; Janka et al., 2007; Panov and Janka, 2009).

The question is whether such high entropies occur at times with sufficiently high temperatures when an $r$ process is still underway (Kuroda et al., 2008). Exploratory calculations to obtain the necessary conditions for an $r$ process in expanding high-entropy matter have been undertaken by a number of groups (Hoffman et al., 1997; Meyer and Brown, 1997; Otsuki et al., 2000; Wanajo et al., 2001; Terasawa et al., 2002; Wanajo et al., 2004; Yoshida et al., 2004; Wanajo, 2007; Kuroda et al., 2008). Recent investigations (Farouqi et al., 2009, 2010) focussed (a) on the effects of varying nuclear physics input [mass models FRDM (Finite Range Droplett Model (Möller et al., 1995)), ETFSI-1 (Extended Thomas-Fermi with Strutinsky Integral) (Aboussir et al., 1995), ETFS-Q with quenching of shell closures far from stability (Pearson et al., 1996), the mass formula by Duflo & Zuker (DUFLO-ZUKER, Duflo and Zuker (1995)) and HFB-17 (a recent Hartree-Fock-Bogoliubov approach) (Goriely et al., 2009a)] and (b) the detailed understanding of the nuclear flow through the chart of nuclides, testing equilibria, freeze-out and delayed neutron capture. To investigate these effects we have applied a full network containing up to 6,500 nuclei and the corresponding nuclear masses, cross sections and $\beta$-decay properties.

The calculations presented here are based on trajectories for densities and temperatures originating from expansions with a complete parameter study in terms of entropy $S$, electron fraction $Y_e$ and expansion velocity $V_{\rm exp}$, the latter being related to the expansion timescale $\tau_{\rm exp}$ (Freiburghaus et al., 1999a; Farouqi et al., 2010). Here we only show the results utilizing the Duflo-Zuker mass model for a range of entropies. It is assumed that in the late phases of the neutrino wind of a deleptonized neutron star conditions with $Y_e < 0.5$ prevail (but see the discussion above).

Either higher entropies than obtained by the simulations discussed above or conditions with intrinsically high neutron densities (like expanding neutron star matter with $Y_e \approx 0.1 - 0.2$) can lead to neutron/seed ratios which are sufficiently high to reach fissionable nuclei in the $r$ process. The fission fragments can again capture neutrons and produce fissionable nuclei, leading to an $r$ process with fission recycling (Rauscher et al., 1994; Martínez-Pinedo et al., 2007). This requires reliable fission barriers (and fission fragment distributions) to test the possibility for the production of superheavy elements. It was shown recently that neutron-induced fission is more important in r-process nucleosynthesis than beta-delayed fission (Panov and Thielemann, 2003; Martínez-Pinedo et al., 2007). Thus, the need to provide a compilation of neutron-induced fission rates is obvious and has been performed recently (Panov et al., 2005, 2010; Goriely et al., 2009b). Comparison of rates obtained with different sets for mass and fission barrier predictions give a measure of the uncertainties involved.

### 4.4.3 Exotic SN Types: Hypernovae and GRBs

As was outlined in various parts of the preceeding sections, massive stars in the range of 8 to $\sim$130 $M_\odot$ undergo core-collapse at the end of their evolution and become Type II and Ib/c supernovae unless the entire star collapses into a black hole with no mass ejection (Heger et al., 2003). These Type II and Ib/c supernovae (as well as Type Ia supernovae, see Chap. 5) release large explosion energies and eject matter which underwent explosive nucleosynthesis, thus having strong dynamical, thermal, and chemical influences on the evolution of interstellar matter and galaxies. They have been the main focus of the present chapter up to now. The explosion energies of core-collapse supernovae are fundamentally important quantities, and an estimate of $E \sim 1 \times 10^{51}$ erg has often been used in calculating nucleosynthesis and the impact on the interstellar medium. (Here we use the explosion energy $E$ for the final kinetic energy of the explosion.) A good example is SN1987A in the Large Magellanic Cloud, whose energy is estimated to be $E = (1.0\text{–}1.5) \times 10^{51}$ ergs from its early light curve.

One of the most interesting recent developments in the study of supernovae (SNe) is the discovery of some very energetic supernovae (see e.g. Nomoto et al. (2006), whose kinetic energy (in spherically symmetric analysis, see also Piran (2004)) exceeds $10^{52}$ erg, about 10 times the kinetic energy of normal core-collapse SNe (hereafter $E_{51} = E/10^{51}$ erg). The most luminous and powerful of these objects, the Type Ic supernova (SN Ic) 1998bw, was probably linked to the gamma-ray burst GRB 980425, thus establishing for the first time a connection between gamma-ray bursts (GRBs) and the well-studied phenomenon of core-collapse SNe. However, SN 1998bw was exceptional for a SN Ic: it was as luminous at peak as a SN Ia, indicating that it synthesized $\sim 0.5\,M_\odot$ of $^{56}$Ni, and its kinetic energy was estimated at $E \sim 3 \times 10^{52}$ erg.

There is another class of supernovae which appears to be rather faint with apparently almost vanishing $^{56}$Ni ejection. Thus, the question emerges how these different objects are related, whether they correspond to different initial masses and how the explosion mechanism changes. The questions to be answered are the following:

- do 8–10 $M_\odot$ stars which produce an Fe-core in a collapse, initiated via electron capture after core He-burning (electron capture supernovae), have a different explosion mechanism after core collapse than more massive stars? Is here only a small amount of material involved outside the collapsing C-core and little Ni-ejection occurring?

- for which stellar progenitor masses do we have a transition from the formation of neutron stars to the formation of black holes after collapse?
- to which extent is this transition region shifted by the nuclear equation of state?
- for which transition region are initially neutron stars formed, causing a regular supernova explosion, and only fall back by the reverse shock swallows inner matter, leading to a small final Ni-ejection and faint light curves?
- for which progenitor masses are black holes formed directly during collapse and how can this be observed?
- what is the role of rotation and magnetic fields to cause gamma-ray bursts?
- can we give reliable nucleosynthesis yields for such events?

Before going into a too involved discussion of the causes of these events, let us first consider the possible effect which higher energy explosions have on the ejecta, i.e. nucleosynthesis products. Here we use the term *hypernova* to describe an extremely energetic supernova with $E \geq 10^{52}$ erg without specifying the explosion mechanism (Nomoto et al., 2001). Following SN 1998bw, other *hypernovae* of Type Ic have been discovered or recognised. Nucleosynthesis features in such hyper-energetic supernovae must show some important differences in comparison to normal supernova explosions. The higher explosion energies could lead to larger ejected $^{56}$Ni masses, as observed in such explosions. They also cause higher entropies in the innermost ejecta, which result in a more extreme alpha-rich freeze-out from explosive Si-burning. Such conditions permit the sizable production of Fe-group nuclei beyond $^{56}$Ni, up to $^{64}$Ge which decays to $^{64}$Zn (Nakamura et al., 2001). This feature could have an influence on abundance patterns observed in extremely metal-poor halo stars. In fact, the observational finding that Zn behaves like an Fe-group element in galactic evolution – and was underproduced in existing supernova models (which were not including the $\nu p$ process) – was used as a strong argument that a large fraction of massive stars explode as hypernovae (Nomoto et al., 2006; Kobayashi et al., 2006).

The observed frequency of type Ib supernovae is about 20% of SNe II at solar metallicity (Cappellaro et al., 1999). In four cases the typical spectrum of type Ic supernovae has been observed, associated with long soft gamma-ray bursts (Woosley and Bloom, 2006), indicating a link between SNe Ic and long soft GRBs. Prantzos and Boissier (2003) found an increase in the ratio of SNe Ibc/SNe II with metallicity. In order to understand this trend one has to understand stellar models as a function of metallicity, from the first stars in the Universe, i.e., metal-free, Population III (Pop III) stars which were born in a primordial hydrogen-helium gas cloud, to present metallicities. This is one of the important challenges of current astronomy and relates to Sect. 4.2, where we have discussed the evolution of massive stars from lowest metallicities to solar values, including rotation and metallicity-dependent winds. In fact Meynet and Maeder (2003) could reproduce this observed metallicity trend. On the other hand, different groups (Podsiadlowski et al., 1992; Vanbeveren et al., 2007; Eldridge et al., 2008) provide models from binary evolution with mass transfer (removing the H-rich envelope) which seems to reproduce this trend as well. Recent observations (Prieto et al., 2008) provide for the first time the individual ratios of SNe Ib/SNe II and SNe Ic/SNe II, rather than only the combined SNe Ibc/SNe II ratio. Georgy et al. (2009) have studied stellar evolution in detail as a function of initial mass, metallicity and rotation, based on the Geneva evolution models. As a full understanding of the GRB mechanism is pending, two options have been considered for the cases where the Fe-core is massive enough that the formation of a black hole in the collapse is expected: (i) nevertheless a supernova-type explosion is assumed, (ii) a black hole forms without a supernovae. They find that current models of stellar evolution can account for the observed number ratios of SNe Ib/SNe II and SNe Ic/SNe II and their variation with metallicity. In case (ii), i.e. when no supernova occurs after black holes are formed, single-star models can still account for more than one half of the combined SNe Ibc/SNe II ratio for metallicities above solar, however, low metallicity SNe Ic events have to come from binary evolution. If black hole formation is identified with the occurrance of GRBs, the resulting number is too large, indicating that only a fraction of such events, most probably very rapid rotators, result in GRBS after collapse (MacFadyen and Woosley, 1999).

As mentioned earlier in this subsection, the explanation of SN Ic SN1998bw is based on a very large progenitor mass $M$ and explosion energy $E$. The type Ic hypernovae 1998bw and 2003dh were clearly linked to the gamma-ray bursts GRB 980425 and GRB 030329, thus establishing the connection between long GRBs and core-collapse supernovae (SNe). SNe 1998bw and 2003dh were exceptional for SNe Ic: they were as luminous at peak as a SN Ia, indicating that they synthesized 0.3–0.5 $M_\odot$ of $^{56}$Ni, and their kinetic energies were estimated in the range $E_{51} = E/10^{51}$ erg $\sim$ 30–50. Other *hypernovae* have been recognized, such as SN 1997ef and SN 2002ap. These hypernovae span a wide range of properties, although they all appear to be highly energetic compared to normal core-collapse SNe. The mass estimates, obtained from fitting the optical light curves and spectra, place hypernovae at the high-mass end of SN progenitors.

In contrast, SNe II 1997D and 1999br were very faint SNe with very low kinetic energy. This leads to a diagram with the explosion energy $E$ or the ejected $^{56}$Ni mass $M(^{56}\mathrm{Ni})$ as a function of the main-sequence mass $M_{\mathrm{ms}}$ of the progenitor star which shows two branches. Therefore, one is led to the conclusion that SNe from stars with $M_{\mathrm{ms}} \gtrsim$ 20–25 $M_\odot$ have different $E$ and $M(^{56}\mathrm{Ni})$, show a bright, energetic *hypernova branch* at one extreme and a faint, low-energy SN branch at the other extreme. For the faint SNe, the explosion energy was so small that most $^{56}$Ni fell back onto the compact remnant. Thus the faint SN branch may become a *failed* SN branch at larger $M_{\mathrm{ms}}$. Between the two branches, there may be a variety of SNe.

This trend could be interpreted as follows. Stars more massive than ~25 $M_\odot$ form a black hole at the end of their evolution. Stars with non-rotating black holes are likely to collapse *quietly* ejecting a small amount of heavy elements (faint supernovae). A preceding stage could be the temporary formation of a neutron star and a supernova explosion, but fallback of matter leads to an increase of the neutron star mass beyond its maximum stable value. (The combination of mixing processes in the innermost ejecta and fallback can influence the ejecta composition.) In contrast, stars which formed rotating black holes are likely to give rise to hypernovae. Here disk and jet formation seems to be a necessary ingredient to understand the explosion. (An option is that hypernova progenitors might form from the rapidly rotating cores after spiraling-in of a companion star in a binary system.)

## 4.5 The Aftermath of Explosions

In the preceding sections we have given an overview of hydrostatic and explosive burning processes in massive stars, the individual phases of stellar evolution, the endstages like core collapse, explosive nucleosynthesis products from supernovae explosions and possible variations in outcome if core collapse ends in black hole formation, related possibly to hypernovae or gamma-ray bursts. What remains to be done is to (i) get a complete picture from stellar models and simulations how hydrostatic/wind and explosive contributions add up to the complete yields observed in such events, (ii) verify such models with individual observations, e.g. from lightcurves and from remnants, (iii) finally to integrate all these events/stellar yields over a mass distribution and metallicity evolution of galaxies, in order to make comparisons with overall galactic oberservations of very long-lived radioisotopes which average over several stellar generations.

### *4.5.1 Massive Stars and Their Complete Yields*

In Sect. 4.4 we have introduced in Eq. (4.4) a simplified rule which determined at which radius certain temperatures are attained in the explosion, assuming that the explosion energy is distributed at all times in a homogenous bubble within the radius of the present shock front position. If one knows the radial mass distribution $M(r)$ in pre-explosion models through which the shock front passes, one knows the amount of matter which encountered certain burning conditions. In Table 4.9 we provide this information for different initial stellar masses (still based on models from Nomoto and Hashimoto (1988)), at (up to) which radial mass position explosive (complete and incomplete) Si-burning, O-burning, Ne/C-burning are occurring (upper portion) and the size of these regions (in $M_{\odot}$) involved (lower portion). In addition, we give the size of the CO-core of prior He-burning in stellar evolution. To first order matter between explosive C/Ne-burning and the stellar surface is ejected unchanged. As this simplified treatment does not know anything of the explosion mechanism which produced this explosion energy, the position of the mass cut is not known and therefore also not the total amount of complete Si-burning material. The core sizes given (e.g. CO-core after He-burning) also make no difference whether this matter resulted from initial core burning of this burning stage or subsequent outward propagating shell burning (e.g. shell He-burning of shell C-burning) which produce specific isotopes of interest (e.g. $^{26}$Al, $^{60}$Fe as discussed in Sect. 4.3).

**Table 4.9** Masses and products in explosive and hydrostatic burning

| $M(r)$ | Burning site | $13\,M_\odot$ | $15\,M_\odot$ | $20\,M_\odot$ | $25\,M_\odot$ |
|---|---|---|---|---|---|
| Fe-core | hydr. Si-burning | 1.18 | 1.28 | 1.40 | 1.61 |
| mass cut | (expl. mechanism) | ? | ? | ? | ? |
| ex Si-c | expl. compl. Si-burning | 1.42 1.40 | 1.46 1.44 | 1.70 1.69 | 1.79 1.80 |
| ex Si-i | expl. incompl. Si-burning | 1.48 1.47 | 1.52 1.51 | 1.75 1.75 | 1.85 1.89 |
| ex O | expl. O-burning | 1.54 1.54 | 1.57 1.57 | 1.81 1.81 | 1.92 2.00 |
| ex C/Ne | expl. Ne-burning | 1.66 1.65 | 1.73 1.70 | 2.05 2.05 | 2.26 2.40 |
| CO-core | hydr. He-burning | 1.75 | 2.02 | 3.70 | 5.75 |
| $\Delta M$ | (Main) products, major radioactivities | | | | |
| ex. Si-c | "Fe", He; $^{56,57}$Ni, $^{61,62}$Zn, $^{59}$Cu, $^{52}$Fe, $^{48}$Cr | ? | ? | ? | ? |
| ex. Si-c | $^{44}$Ti, $\nu p$ process, *r* process? | ? | ? | ? | ? |
| ex. Si-i | Si, S, Fe, Ar, Ca; $^{55}$Co, $^{52}$Fe, $^{48}$Cr | 0.06 0.07 | 0.06 0.07 | 0.05 0.06 | 0.06 0.09 |
| ex. O | O, Si, S, Ar, Ca | 0.06 0.07 | 0.05 0.06 | 0.06 0.06 | 0.07 0.11 |
| ex. C/Ne | O, Mg, Si, Ne; $^{26}$Al, *p* process | 0.12 0.11 | 0.16 0.13 | 0.24 0.24 | 0.34 0.40 |
| hydr. He | O, Ne, Mg, Si, *s* process | 0.09 0.10 | 0.29 0.32 | 1.65 1.65 | 3.49 3.35 |

Note: All entries with ? are related to uncertainities in the explosion mechanism.

Initially we want to focus here on the explosive burning phases. We also want to add that Table 4.9 includes always two values for the radial masses involved, (a) from the simplified Eq. (4.4) applied to the appropriate stellar model ($M(r)$) and (b) resulting from an actual explosion calculation (initiated via a thermal bomb), as obtained in Thielemann et al. (1996). When comparing these numbers, we see a quite close agreement, except for the most massive star where non-negligible deviations are encountered.

Based on this information we want to discuss complete nucleosynthesis yields, including explosive processing (also the $\nu p$ process, affected by neutrinos in the innermost ejecta, as well as the $p$ process in explosive Ne/O-burning), hydrostatic yields from the outer layers (including $s$ process) which are ejected unaltered, and prior wind yields lost during stellar evolution. Then we concentrate on the long-lived radioactivities $^{26}$Al, $^{60}$Fe, $^{44}$Ti, other Fe-group and lightcurve-determining nuclei, including their origin which is e.g. important for $^{26}$Al and $^{60}$Fe, which have hydrostatic burning as well as explosive origins. The $r$ process in the neutrino wind or possibly polar jets has been presented qualitatively with entropy, $Y_e$ and expansion timescale as free parameters or expansion timescale of neutron star matter as a free parameter. Presently no realistic explosion models are available to discuss this matter in a real stellar context, but a short discussion of long-lived radioactive chronometers is presented.

Table 4.9 provides the following conclusions: The amount of ejected mass from the unaltered (essentially only hydrostatically processed) CO-core varies strongly over the progenitor mass range. The variation is still large for the matter from explosive Ne/C-burning, while the amount of mass from explosive O- and Si-burning is almost the same for all massive stars. Therefore, the amount of ejected mass from the unaltered (essentially only hydrostatically processed) CO-core and from explosive Ne/C-burning (C, O, Ne, Mg) varies strongly over the progenitor mass range, while the amount of mass from explosive O- and Si-burning (S, Ar, and Ca) is almost the same for all massive stars. Si has some contribution from hydrostatic burning and varies by a factor of 2–3. The amount of Fe-group nuclei ejected depends directly on the explosion mechanism which also affects the $Y_e$ in these inner zones. Thus, we have essentially three types of elements, which test different aspects of supernovae, when comparing with individual observations. The first set (C, O, Ne, Mg) tests the stellar progenitor models, the second (Si, S, Ar, Ca) the progenitor models and the explosion energy in the shock wave, while the Fe-group (beyond Ti) probes clearly in addition the actual supernova mechanism. Thus, we require that all three aspects of the predicted abundance yields are based on secure modeling (stellar evolution, explosion energy, and explosion mechanism) in order to be secure for their application in lightcurve modeling, radioactivities in remnants as well as the the in chemical evolution of galaxies.

#### 4.5.1.1 *r*-process ejecta

The biggest uncertainty exists for the absolutely innermost ejecta, i.e. the possible $r$-process ejection in the neutrino wind. This matter escapes after the supernova explosion shock wave was launched and the continuing neutrino escape from the remaining neutron star leads to its "surface erosion/evaporation", i.e. a neutron-rich wind which could trigger an *r* process (Qian and Woosley, 1996). Early modelling seemed to lead to a full *r*-process abundance distribution (Woosley et al., 1994), which was, however, already then questioned by other investigations (Takahashi et al., 1994), when utilizing the entropies obtained from their calculations. Results with present neutrino physics and detailed transport modeling seem to find the opposite behavior, i.e. proton-rich conditions for more than the first 10s after the explosion (Fischer et al., 2009a; Marek and Janka, 2009), as noticed first by Liebendörfer et al. (2003) and discussed in Sect. 4.4 together with the $\nu p$ process. A major question is if and how this turns to be neutron-rich in later phases, what physics causes this change (the nuclear EoS or neutrino properties?), and how very high entropies can be attained to produce also the heaviest nuclei. Present observations of low metallicity stars show huge variations in heavy *r*-process content and indicate that in most explosions the latter is not taking place, making the *full r* process a rare event. Then typical supernovae would only provide a weak *r*-process environment. Whether either high entropies are only attained in exceptional cases or other origins of low entropy, highly neutron-rich matter (neutron star mergers or neutron-rich jets from rotating core collapse supernovae, Freiburghaus et al. (1999b); Cameron (2001, 2003)) cause the main *r* process has to be explored, in parallel with the still remaining challenges of nuclear physics far from stability.

In the preceding sections we have shown that there exists a principle understanding about the nuclear working of the *r* process and that it is possible to reproduce solar system *r*-process abundances by superpositions of components with varying environment conditions. What seems not possible, yet, is to clearly identify, without doubt, the responsible astrophysical site. Taking, however, such *r*-process superposition fits as *zero-age* abundances with e.g. production ratios for $^{232}$Th/$^{238}$U or other actinide (chronometer) nuclei, one can predict such ratios also as a function of decay time (present-age abundances) and identify ages of very metal-poor stars which were born with a fresh pollution of an *r*-process pattern (see e.g. Cowan et al. (1991, 1999); Thielemann et al. (2002); Kratz et al. (2007) and references therein). The typical result is that these chronometers indicate an age of the oldest stars in our Galaxy in the range of 14–15 Gyr with an uncertainty of about 3–4 Gyr. A more complex utilization of long-lived *r* process radioactivities in terms of a continuing enrichment in galactic evolution is given in Chap. 2.

#### 4.5.1.2 Fe-group and beyond

The neutrino wind works via evaporation of the neutron star surface after the supernova shock wave emerged and caused the explosion. The innermost matter which experienced the shock is thus ejected earlier and a typical example for its composition is seen in Fig. 4.12, which shows the zones of complete Si-burning with alpha-rich freeze-out. The change in abundances at $M = 1.63\,\mathrm{M}_{\odot}$ is due to a change in $Y_e$ in the original stellar model, which was utilzed for explosive nucleosynthesis predictions just by introducing a thermal bomb of $10^{51}$erg. If one accounts correctly for the neutrino interactions during the collapse and explosion, this turns matter even slightly proton-rich ($Y_e > 0.5$), see the discussion of the $\nu p$ process in Sect. 4.4 and Figs. 4.16, 4.17 and 4.18 (Fröhlich et al., 2006a,b; Pruet et al., 2005, 2006; Wanajo, 2006). This results also in substantial fractions of $^{64}$Ge (decaying to $^{64}$Zn via $^{64}$Ga, both of short half-life in the minute to second regime and therefore not of interest in terms of radioactivities with long half-lifes), but within the $\nu p$ process also to the production of Sr and heavier nuclei. The isotopic ratios $^{58}$Ni, $^{60,61,62}$Zn change strongly. Alpha-nuclei like $^{40}$Ca, $^{44}$Ti, $^{48}$Cr, and $^{52}$Fe are affected as well. Higher entropies and $Y_e$-values close to 0.5 increase the fraction of these alpha-nuclei (and would in hypernovae also cause a substantial production of $^{64}$Ge as discussed in Sect. 4.4. It was also discussed there that probably only a small fraction of black-hole producing events actually lead to hypernovae (not 50% as assumed in some cases (Nomoto et al., 2006; Kobayashi et al., 2006)). However, this reduction in the $^{64}$Ge production can well be balanced by the larger $^{64}$Ge production in regular supernovae due to the correct inclusion of neutrino-interactions and their effect on increasing $Y_e$ to values larger than 0.5.

#### 4.5.1.3 Other explosive burning zones, $^{44}$Ti, $^{26}$Al

While such a discussion is of general interest for the composition of the Fe-group and filling in of lighter $p$-process nuclei, which are underproduced in the classical picture (see Sect. 4.4), we would like to concentrate here on long-lived radioactivities. The nucleus which is mainly produced in the complete Si-burning regime with alpha-rich freeze-out is $^{44}$Ti. As discussed above, its total amount depends on the matter experiencing this burning outside the mass cut (which is in principle unknown without successful explosion calculations). Several authors have made assumptions on its position, either based on total $^{56}$Ni ejecta in thermal bombs (Thielemann et al., 1996; Nomoto et al., 2006) or entropy jumps in the pre-collapse models in piston-induced explosions (Woosley and Weaver, 1995; Woosley et al., 2002). Based on these assumptions, different authors find somewhat different predictions, which are, however, a relatively flat function of the stellar progenitor mass. The values of Thielemann et al. (1996) range between $2 \times 10^{-5}$ and $1.5 \times 10^{-4}$ M$_\odot$. The interesting point here is that variations in the $Y_e$-structure can lead to changes up to a factor of 2. Here either the initial pre-collapse distribution was assumed (with possible $Y_e$-dips in the innermost parts from shell O-burning) or a smoothed flat $Y_e$-distributions closer to 0.5 (which also reproduces the solar $^{56}$Fe/$^{57}$Fe-ratio). This latter case leads to the higher values and will also be closer to reality when neutrino interactions are taken into account during the explosion (see discussion above on the $\nu p$ process and $Y_e$). Rauscher et al. (2002) and Tur et al. (2009) find smaller values (either $1.5 - 5 \times 10^{-5}$ and $3.5 - 6 \times 10^{-5}$ M$_\odot$). The latter is based on a readjustment of the triple-alpha and $^{12}$C$(\alpha, \gamma)^{16}$O-rates to most recent experimental values, which does not have a drastic influence, however. Indirectly, core sizes and other stellar properties, including explosion energies, can enter.

$^{44}$Ti is only made in the explosive phase of complete Si-burning with alpha-rich freeze-out from charged particle equilibrium. There have been investigations on the reactions producing ($^{40}$Ca($\alpha$, $\gamma$)$^{44}$Ti) and distroying ($^{44}$Ti($\alpha$, $p$)$^{47}$V) reactions as well as the half-life of $^{44}$Ti. Due to the fact, that this is a freeze-out from equilibrium, Hoffman et al. (1999) found that even rate changes by a factor of 6 change the $^{44}$Ti production only by a factor of 1.3 What is different between both approaches, leading to productions either larger or smaller than $5 \times 10^{-5}$ M$_\odot$, is the introduction of the explosion (i) a thermal bomb, (ii) a piston. This apparently leads to higher entropies in the first case and a more intense alpha-rich freeze-out. It should also be mentioned that non-spherical explosions can lead to larger $^{44}$Ti-production than spherical models (Nagataki et al., 1998; Nagataki, 2000). A comparison to observations from either SN 1987A or the SNR Cas A is discussed in the following subsection.

Complete explosive nucleosynthesis predictions for a range of progenitor stars with induced explosions have been given by a number of authors in recent years (Rauscher et al., 2002; Woosley et al., 2002; Nomoto et al., 2006; Limongi and Chieffi, 2006; Woosley and Heger, 2007; Umeda and Nomoto, 2008), updating some of the discussions made above, based on earlier models (Woosley and Weaver, 1995; Thielemann et al., 1996). Also specific investigations were undertaken for Pop III low metallicity stars (Umeda and Nomoto, 2005; Tominaga et al., 2007). We limit the present discussion to stars below 130 M$_\odot$ which still undergo core collapse and do not explode via explosive O-burning like the so-called pair-creation supernovae (Heger et al., 2003). While such explosions seem theoretically possible, provided that these massive cores can result from stellar evolution, the apparent absence of predicted abundance patterns in low metallicity stars plus our present understanding of massive stars with rotation (Maeder and Meynet, 2010), seem to exclude this outcome.

Then, the basic pattern given in Table 4.9 always applies. The abundances from incomplete Si-burning and explosive O-burning can explain Galactic chemical evolution. As mention in Sect. 4.4, the classical $p$ process takes place in explosive Ne-burning via photodisintegrations of pre-existing heavy nuclei, but even with the best nuclear input the underproduction of light p-nuclei cannot be solved. The solution can be obtained by adding a *light (heavy) element primary process* (LEPP, (Travaglio et al., 2004) where the best candidate is the $\nu p$ process. Thus, the classical $p$-process isotopes have to be explained by a superposition of the innermost proton-rich complete Si-burning ejecta with those of explosive Ne-burning in outer zones.

The scheme indicated in Table 4.9 is a bit too simplified when considering the ejecta of outer layers, whose composition was produced during stellar evolution and ejected essentially unaltered during the explosion. However, the *CO-core* scheme is not sufficient to describe massive-star yields. While it includes all matter which underwent He-burning, it does not differentiate between core He-burning and shell He-burning. The latter occurs at higher temperatures and has specific features different from core He-burning. In a similar way, the NeO-core contains all matter which underwent C-burning during stellar evolution, but also here, no difference is made between core C-burning and higher temperature shell C-burning. The same is true for Ne-burning.

$^{26}$Al production during stellar evolution was discussed in Sect. 4.3; now we include also the explosive production of $^{26}$Al. It occurs in the regions of explosive Ne/C-burning. Under these conditions $^{25}$Mg is produced via $^{24}$Mg$(n, \gamma)^{25}$Mg and the protons arise from $^{23}$Na$(\alpha, p)^{26}$Mg, similar to the reaction pattern shown in Table 4.4 for hydrostatic Ne-burning (and partially also C-burning). Under explosive conditions at temperatures of the order $2.3 \times 10^9$K, these burning stages act explosively in a combined way, and the temperatures are also suffuciently high to utilize the released protons for the $^{25}$Mg$(p, \gamma)^{26}$Al reaction. However, as also seen from Table 4.4, neutrons are abundantly produced. They act as the main destructive species via $(n, p)$ and $(n, \alpha)$ reactions. As can be seen from Table 4.9, the mass involved in explosive Ne/C burning is strongly dependent on the progenitor mass. Thus, we expect a dramatic increase with increasing initial stellar masses, which is exactly what we see. Limongi and Chieffi (2006) have analyzed in detail the contributions from (i) wind ejecta during stellar evolution, (ii) hydrostatic burning products ejected during the explosion, and (iii) explosive Ne/C-burning. The latter dominates up to about 60 $M_\odot$ and increases from initially about $2 \times 10^{-5}$ $M_\odot$ per event to $2-3 \times 10^{-4}$ $M_\odot$. Then wind ejecta start to take over and flatten out close to $10^{-3}$ $M_\odot$ at initial stellar masses of 120–140 $M_\odot$. The latter are subject to rotational effects (Langer et al., 1995; Meynet et al., 1997; Palacios et al., 2005) and increase in fact with higher rotation rates (see Sect. 4.3). Tur et al. (2009) have reanalyzed this behavior in the lower mass range from 15 to 25 $M_\odot$ and confirmed this trend. They also did not find a strong dependence of the result on the He-burning reactions triple-alpha and $^{12}$C$(\alpha, \gamma)^{16}$O. They show nicely how $^{26}$Al is produced starting in H-burning, but the final explosion produces close to a factor of 10 more of it. Yields from different studies have been assembled in Fig. 4.20.

#### 4.5.1.4 $^{60}$Fe

Massive-star yields for $^{60}$Fe are summarized in Fig. 4.21 and mentioned here for completeness, although not produced in explosive nucleosythesis. It is entirely produced in the s-proces during shell He-burning and thus a pure product of stellar evolution. The explosion only acts to eject the corresponding layers. As $^{60}$Fe is produced via neutron capture of beta-unstable $^{59}$Fe, a relatively high neutron density of about $3 \times 10^{10}$cm$^{-3}$ is required in order for its efficient production. This is only attained in shell He-burning during late evolution stages after core C-burning. The production ranges from $2 \times 10^{-6}$ to $8 \times 10^{-5}$ $M_\odot$ for initial stellar masses between 10 and 40 $M_\odot$. This result is dependent on the He-burning reactions triple-alpha and $^{12}$C$(\alpha, \gamma)^{16}$O, as they compete with the neutron producing reaction $^{22}$Ne$(\alpha, n)^{25}$Mg. There exist also uncertainties in $^{59}$Fe$(n, \gamma)^{60}$Fe and $^{60}$Fe$(n, \gamma)^{61}$Fe, which cause yield uncertainties by a factor of up to 5. If the star experiences strong mass loss, the He-burning shell does not encounter the higher density conditions required for the high neutron density of $3 \times 10^{10}$cm$^{-3}$. Thus

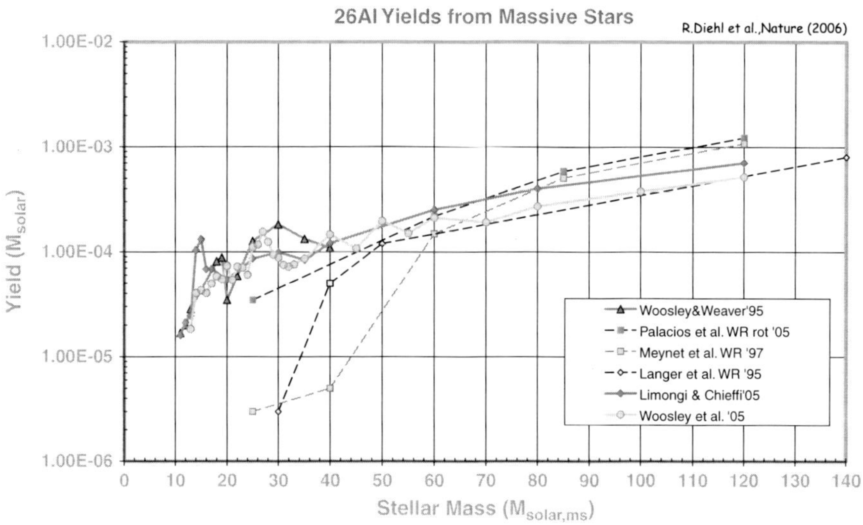


**Fig. 4.20** The $^{26}$Al yields from wind ejections and from the explosive release in the supernova, as a function of the initial mass of the star (as assembled by Diehl et al., 2006 (in their appendix)

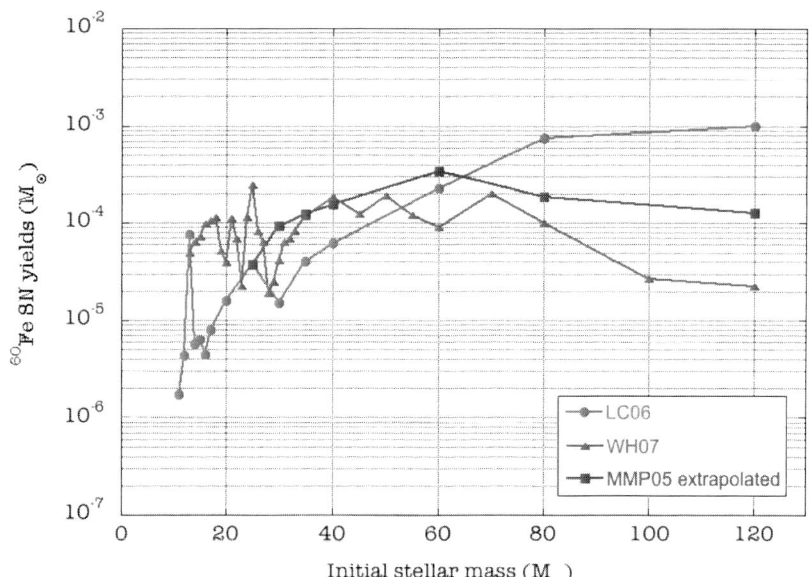


**Fig. 4.21** The $^{60}$Fe yields from the explosive release in the supernova, as a function of the initial mass of the star (as assembled by Martin et al., 2010)

for initial stellar masses in excess of 40 $M_\odot$, the mass loss treatment can also lead to variations in predicted yields of more than a factor of 10. Apparently a high mass loss rate is required to not overproduce $^{60}$Fe in high mass stars $M > 40\ M_\odot$ (Limongi and Chieffi, 2006) with respect to $\gamma$-ray line constraints (see Wang et al., 2007, and Chap. 7).

## *4.5.2 Observational Diagnostics: Lightcurves, Spectra and SNR*

### 4.5.2.1 Lightcurves

Supernova light curves are powered by radioactive decays. Very early interpretations of supernova lightcurves related them to the radioactive decay of $^{254}$Cf (Burbidge et al., 1957). In fact, a strong *r* process (with fission-cycling) would cause observable features based on the decay of heavy radioactive nuclei. This question was addressed very recently with respect to *r*-process ejecta from neutron star mergers (Metzger et al., 2010). Supernova lightcurves, however, are dominated by Fe-group ejecta. In addition to abundant $^{56}$Ni, there are a number of radioactive nuclei which will decay on time scales of ms up to $10^7$ year. Here we only want to concentrate on a few nuclei, which by a combination of their abundances and half-lives, can be of importance. These nuclei are $^{56}$Co ($^{56}$Ni), $^{57}$Co ($^{57}$Ni), $^{55}$Fe ($^{55}$Co), $^{44}$Ti, and $^{22}$Na (see Fig. 4.22). For a 20 $M_\odot$ star like SN 1987A they were predicted with total masses of 0.07, $3.12\times10^{-3}$, $3.03\times10^{-4}$, $1.53\times10^{-4}$, and $1.33\times10^{-7}\ M_\odot$ (Thielemann et al., 1990, 1996).

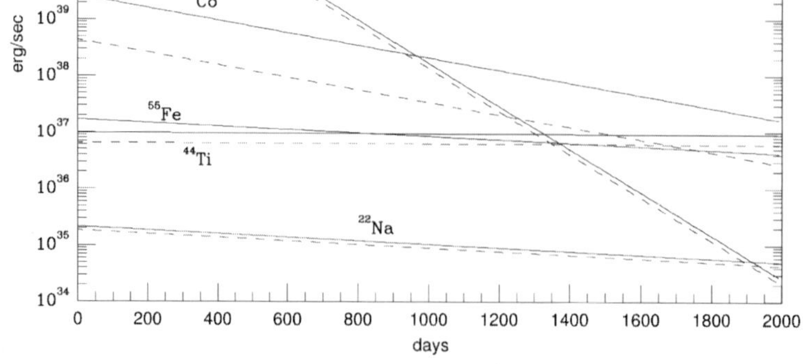


**Fig. 4.22** Total energy release due to the decay of long-lived radioactive species (*dashed lines*) and due to the release in terms of thermalized decay photons (*solid lines*). The ejected masses of radiactive species are takem from a 20 $M_\odot$ model for SN 1987A (Thielemann et al., 1990, 1996)

Observations of bolometric light curves, which reflects the thermalized energy of the radioactivities, constrained these values to $M(^{56}\mathrm{Ni}) \approx 0.071\ \mathrm{M}_\odot$ (e.g. Suntzeff and Bouchet (1990)) and $M(^{57}\mathrm{Ni}) \approx 3.3 \times 10^{-3}\ \mathrm{M}_\odot$ (Fransson and Kozma (1993) and references therein). Only more recently a very careful analysis confirmed an upper limit on $^{44}$Ti of the order $1.1 \times 10^{-4}\ \mathrm{M}_\odot$ (Lundqvist et al., 2001).

Generally, after beta-decay or electron capture, a daughter nucleus is produced in an excited state ( $^{55}$Fe is a notable exception, see below). The ground state is reached by one or several gamma transitions, observable by current gamma-ray detectors for nearby sources (see Sect. 10.1). Photons, positron-electron annihilations following $\beta^+$-decays, and the kinetic energy given to the decay products can contribute to the light curve at later times.

The number of photons released for each of the transitions, occuring in the daughter nucleus after beta-decay, is equal to the number of decays $N_d$, multiplied with the appropriate percentage of the occurrance (*branching ratio*) for the specific transition. The total energy released corresponds to the product of the number of decays with the decay Q-value:

$$N_d(t) = -\frac{dN}{dt}(t) = \lambda N_o \exp(-\lambda t) \quad \frac{dE}{dt}(t) = Q N_d(t) = Q \lambda N_o \exp(-\lambda t), \quad (4.5)$$

where $\lambda = \ln 2/t_{1/2}$ is the decay rate of the nucleus. The initial number of radioactive nuclei can be calculated from their total mass by $N_o = M/Am_u$, with $A$ being the nucleon number of the nucleus, $m_u$ the atomic mass unit, and $M$ the mass given above. When using the radioactivity half-lives of the relevant isotopes expected in supernova ejecta (i.e., 78.76d, 271.3d, 2.7y, 54.2y, and 2.602y, and atomic Q-values of 4.566, 0.835, 0.232, 3.919, and 2.842 MeV), we can estimate radioactive-energy generation rates in erg/s and the total number of decays per second. The Q-value used for $^{44}$Ti combines the subsequent decays of $^{44}$Ti and $^{44}$Sc. Q-values include all available energies, i.e. the kinetic energy of the decay products, the energy in photons, the annihilation energy of positron-electron pairs in $\beta^+$-decays, and the neutrino energy. At densities prevailing in the expanding remnant, neutrinos will escape freely and their energy has to be subtracted, which leaves corrected values for the appropriate energy deposits of 3.695, 0.136, 0.0, 2.966, and 2.444 MeV. Because the electron capture on $^{55}$Fe does only lead to an energetic neutrino, there is no local energy deposition from this isotope.[9] Gamma transitions following the decays of the other isotopes under consideration obtain candidate $\gamma$-rays at (rounded to full percent values): $^{56}$Co, 847 keV (100%), 1038 keV (14%), 1238 keV (68%), 1772 keV (16%), 2599 keV (17%); $^{57}$Co, 122 keV (86%), 136 keV (11%); $^{44}$Ti, 78 keV (93%),

[9] This situation was recently re-evaluated by Seitenzahl et al. (2009). The electron capture occurs from an electron in an atomic orbit, leaving a hole which can be filled by other electrons cascading down to fill this hole, thus emitting photons – X-rays – or depositing the energy in ejecting outer electrons – Auger electrons. Thus, in cases where only ground-state to ground-state electron capture occurs and the energy is emitted in an escaping neutrino, only Auger electrons or X-rays can contribute to local energy deposition.

68 keV (88%), 147 keV (9%), 1157 keV (100%); $^{22}$Na, 1275 keV (100%; branching ratios given as percentages per dcay). If positrons from $\beta^+$-decay slow down and annihilate with electrons locally within the supernova envelope, the full neutrino-loss corrected energy corresponding to the reaction Q-value will be deposited in the envelope. Observable signatures include high energy photons such as the ones from the gamma transitions, and their Compton scattered and completely thermalized descendants.[10]

Then the sum of all individual contributions discussed above would make up the bolometric lightcurve of the supernova (see Fig. 4.23). The *light curve*, i.e. the brightness as a function of time, will be dominated first by the decay of $^{56}$Co, and then $^{57}$Co and $^{44}$Ti, if one neglects possible radiation from a pulsar. $^{22}$Na never plays a dominant role for the lightcurve. At lower densities (and later times), escaping high energy photons or positrons lead to a reduction of the brightness of *bolometric* emission. This can be seen in late time observations as shown e.g. in Leibundgut and Suntzeff (2003) (see Fig. 4.23). An important consistency check would be

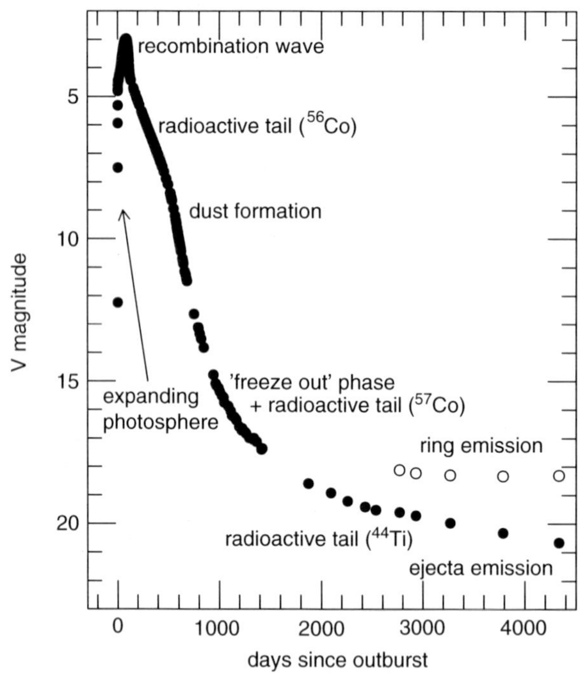


**Fig. 4.23** Observed (visual) lightcurve of SN1987A for the first 1,500 days from Leibundgut and Suntzeff (2003)

[10] Deposition of energy from radioactive decay involves absorption of high-energy photons, slowing down of ~MeV-type energy electrons and positrons, and proper treatment of temporary energy reservoirs such as ionization and inhibited radioactive decay from completely-ionized nuclei (see, e.g., Sim et al., 2009; Mochizuki et al., 1999; Woosley et al., 2007).

to compare this bolometric light curve (which includes only optical, UV and IR emission, hence thermalized gas and dust components) to the high-energy photons, more directly reflecting radioactive decays. At late times, those high energy photons escape freely.

The $\gamma$-ray detections of SN1987A were the first to identify $\gamma$-rays from $^{56}$Co decay. Later improved observations by CGRO for $^{56}$Co and $^{57}$Co were a direct proof of these unstable species in the right amounts. It turned out that Ni decay $\gamma$-ray lines were seen with the Gamma-Ray Spectrometer on the Solar Maximum Mission (Matz et al., 1988; Leising and Share, 1990) significantly earlier than expected from a spherically stratified distribution of elements, where the Fe-group nuclei are produced in the center. This is understood from deviations from spherical symmetry in the expanding remnant, bringing Ni-rich clumps to the surface earlier by convective instabilities, mixing $^{56}$Ni/$^{56}$Co to outer layers at early times. Gamma-ray line profiles measured with high spectral resolution indicated Doppler broadening of the lines from their ejecta motion (Tueller et al., 1990).

Many supernova remnants (such as the 300-year old Cas A remnant) show mixing in their ejecta (Vink, 2005). While there exist some theoretical indications that this is due to instabilities of the propagating shock wave, it is also associated with the expansion of the supernova into an inhomogeneous medium. SN1987A observations showed clearly that mixing is part of the supernova explosion itself. Several independent reasons lead to such a conclusion. The supernova produces large amounts of unstable long-lived nuclei, the dominant abundance is found in the doubly-magic nucleus $^{56}$Ni, which is produced in the innermost part of the ejecta. $^{56}$Ni decays with half-lives of 6.1 days to $^{56}$Co and 77.8 days to $^{56}$Fe. After the beta-transition, deexcitations to the ground state of the daughter nucleus lead to the emission of high energy gamma-rays. While these gamma-rays would escape freely at low densities, Compton scattering will reduce their energies into the X-ray and even thermal regime at higher densities. With decreasing densities during the expansion, initially only thermalized photons will escape, then X-rays and finally gamma-rays. For SN1987A X-ray observations with GINGA, HEXE, and balloons and gamma-ray observations with the SMM-satellite and balloons actually agreed with this general behavior (e.g. Sunyaev et al., 1990; Leising and Share, 1990). The main problem was that the predicted time scales did not agree with the observations, where X-rays and gamma-rays appeared earlier than predicted. An agreement could only be obtained when part of the $^{56}$Ni, being produced initially in the inner parts of the ejecta, was mixed out to larger distances (Fu and Arnett, 1989). Mixing is also required to explain the spread of expansion velocities seen in line widths of infrared observations for various elements and in the gamma ray lines of $^{56}$Co. The inferred velocities differ strongly from the much smaller ones, expected from an expanding remnant, which maintains the stratified composition from explosive and hydrostatic nuclear burning. Other indications came from the modeling of the optical light curve. The best agreement between calculated and observed light curves were obtained for a composition which mixed a small fraction of Ni all the way into the 10 $M_\odot$ hydrogen envelope and hydrogen into the deeper layers, containing mostly heavy elements (see e.g. Benz and Thielemann, 1990).

The lightcurve from SN1987A could be reproduced with theoretical modelling, including the effects of X-ray and $\gamma$-ray escape, as well as mixing of $^{56}$Ni. SNe Ib and Ic events, believed to be core collapse events without an overlying hydrogen envelope have to be treated accordingly. The combination of small masses involved (only He-cores or C-cores without H-envelope) and the assumption of mixing can reproduce the steeper decline than found in massive SNe II. A typical case of a type Ic supernova is SN 1998bw, associated with GRB 980425. The straight-forward modeling of the observed lightcurve (Sollerman et al., 2002), similar to the discussion in the beginning of this subsection (Nakamura et al., 2001), led to interpretations of a largely non-solar $^{56}$Ni/Fe to $^{56}$Ni/Fe ratio. The inclusion of internal conversion and Auger electrons, as suggested by Seitenzahl et al. (2009) could naturally explain the observed slowdown of the lightcurve without invoking such extreme abundance ratios.

We note that in recent years photon transport calculations have reached major improvements, and are now able to consistently reproduce both light curves and spectra from SNIa, and also from core collapse supernovae (e.g. Sim et al., 2009, for a description of the method). Presently, systematic uncertainties of the method are being investigated, and appear rather well understood, at least for SNe Ia (see, e.g., Woosley et al., 2007). As optical-to-IR light curves and spectra will be collected in abundance through large telescope survey programs for cosmological studies, it is likely that those (more indirect) measurements of core-collapse supernova nucleosynthesis will generate the tightest constraints to learn more about these events and their internal nuclear processes.

#### 4.5.2.2 Optical/IR Spectra and Dust Formation

In the preceding subsection we laid out the framework of understanding supernova lightcurves. What is missing here, is the evolution of supernova spectra. The receeding photosphere in terms of radial mass in an expanding, radiation-filled bubble can give clear indications of the element composition (as a function of time equivalent to declining Lagrangian mass). The problem of type II supernovae is that the huge H-envelope does not really contain much information in terms of nucleosynthesis. Type Ib and Ic supernovae, which lost their H- and possibly He-envelope, reveal much more information of the compact inner part, which experienced explosive processing (Matheson et al., 2001; Branch et al., 2002; Sauer et al., 2006). This is similar to type Ia supernovae, originating from exploding white dwarfs, which have been extensively utilized for abundance diagnostics.

Another issue, which has been discussed already in the subsection on supernova lightcurves is related to convective instabilities and mixing of nucleosynthesis products. The extent of mixing is responsible for the element mixture in the expanding and cooling remnant when chemical reactions and dust formation set in. For which compositions and conditions does this environment lead to presolar grains with high melting temperatures which can survive the interstellar medium and formation of the solar system in order to be detected today in meteoritic inclusions? When do we form oxides, diamonds, hybonites, carbides ... and how selectively do they

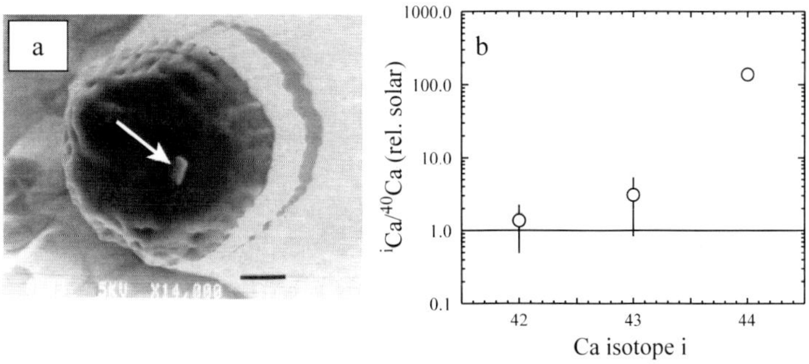


**Fig. 4.24** **(a)** Scanning electron microscope image of a presolar graphite grain following isotopic measurement with an ion microprobe (from Nittler et al., 1996). The *arrow* indicates TiC sub-grain originally enclosed in graphite but revealed by the ion-probe sputtering. **(b)** Calcium isotopic composition of graphite grain shown in **(a)**. The $^{44}$Ca/$^{40}$Ca ratio is 137 times the solar ratio, whereas the other Ca-isotopic ratios are normal within $2-\sigma$ errors (the error bar in the isotope 44 abundance is smaller than the symbol size). This is a clear signature of in situ decay of live $^{44}$Ti, originally condensed in the TiC sub-grain, and demonstrates that the grain formed in the ejecta of a supernova

include matter from the regions where they form? How can we relate $^{26}$Mg and $^{44}$Ca excesses in presolar grains to initially embedded $^{26}$Al and $^{44}$Ti? Figure 4.24 shows an example of a grain attributed to supernova condensation. A discussion of these questions can be found in Chap. 10.3.

#### 4.5.2.3 $^{26}$Al in the Vela Region

The Vela region appears prominent in several astronomical images of our Galaxy: It includes the Gum nebula and the Vela Supernova Remnant, both prominent agents to form nearby structures of the interstellar gas and bright in X- and radio emissions, and it includes the Vela pulsar where bright gamma-ray pulsations teach us about particle acceleration in neutron star magnetospheres, and furthermore with Vela X-1 a remarkable X-ray source and prototype of a binary system where a neutron star accretes wind material from a high-mass companion star. All those objects are relatively nearby, mostly in the foreground of the Vela molecular ridge which is one of the nearest star-forming regions and located in about 700 ($\pm$200) kpc distance (Massi et al., 2007).

Three prominent sources have been discussed in the context of measuring $^{26}$Al production for individual objects – all related to massive star and explosive nucleosynthesis, respectively ($^{26}$Al observations are discussed in Chap. 7 otherwise): The Vela supernova remnant, a recently-discovered supernova remnant called *Vela Junior*, and a Wolf-Rayet binary system $\gamma^2 Velorum$.

The Vela supernova remnant is relatively nearby at 250 pc, about 10,000 year old, and spans an area of about 8° diameter on the southern sky. It hosts the Vela pulsar and a plerionic pulsar nebula at its center. Such a nearby supernova explosion in the recent past seems a unique opportunity to calibrate the $^{26}$Al yield of a core-collapse supernova. With COMPTEL, diffuse and extended emission had been recognized from this direction (Diehl et al., 1995). But possibly-underlying extended $^{26}$Al emission limits the precision of this measurement, the $^{26}$Al gamma-ray flux attributed to this supernova remnant is 0.5–2.7 $10^{-5}$ph/cm$^2$/s. This is well within expectations (an $^{26}$Al yield of $10^{-4}$ $M_{\odot}$ would result in a gamma-ray flux of $\sim$2 $10^{-5}$ph/cm$^2$/s). INTEGRAL could not detect the supernova remnant, however, despite sufficient exposure. This may be due to the low surface brightness of this extended source and line broadening from the remaining ejecta motion of this young supernova remnant.

In refined X-ray imaging analysis of the Vela supernova remnant, a new supernova remnant was discovered being superimposed with a diameter of $\sim$2°and named RXJ0852.0-46.22 or *Vela Junior* for short (Aschenbach, 1998). Early speculations about its $^{26}$Al emission were stimulated from hints for $^{44}$Ti emission, which would have implied that this supernova remnant would be both young and nearby; these could not be substantiated by additional measurements and studies. Now absence of $^{26}$Al emission from Vela Jr. appears plausible, as the supernova remnant is probably older than 1,000 year and more distant than 740 pc (Katsuda et al., 2009).

The $\gamma^2 Velorum$ binary system is yet another tantalizing $^{26}$Al source: It consists of a Wolf Rayet star (*WR11*) and an O-star companion. If the binary interactions can be ignored for $^{26}$Al production of the WR11 star, this would be *the* opportunity to calibrate the $^{26}$Al ejection during the Wolf-Rayet phase of a massive star, as the distance to this object has been derived from Hipparcos measurements as only 258 pc. The Wolf-Rayet star currently has a mass of $\sim$9 $M_{\odot}$, with a 30 $M_{\odot}$ O star companion (De Marco et al., 2000). Modeling the Wolf-Rayet star evolution in detail, and also accounting for possibly rapid rotation as it delays the wind ejection phase, Mowlavi et al. (2005) show that the upper limit on $^{26}$Al emission from WR11 would be hard to understand (Fig. 4.25). But doubts have appeared on the Hipparcos distance measurement, and current belief is that the system is part of the Vela OB2 association and rather at a distance of 330–390 pc (see Eldridge, 2009, and references therein). Additionally, the system's age may also be somewhat higher than estimated earlier (beyond 5 My, rather than 3–5 My) (Eldridge, 2009), and part of the $^{26}$Al ejected in the earlier wind phase may now be spread over a shell extended by up to 6°(Mowlavi and Meynet, 2006), hence of lower surface brightness and still consistent with the non-detection by COMPTEL and INTEGRAL gamma-ray telescopes. Note, however, that binary mass transfer may have altered the evolution of the Wolf-Rayet star substantially; this mutual impact on stellar evolution in close binaries is still very uncertain, but could lead to orders of magnitude increases of $^{26}$Al production in rare cases (de Mink et al., 2009; Langer et al., 1998).

#### 4.5.2.4 Supernova Remnants

The very late phases of the expansion into the interstellar medium, when photons can escape freely from the expanding debree, permits the detection of long-lived radioactivities by detecting decay photons of specific energies. Such detections can then be identified with the amount of matter existing in radioactive isotopes.

The discovery of the 1157 keV $^{44}$Ti $\gamma$-ray line emission from the youngest Galactic SNR Cas A with COMPTEL (Iyudin et al., 1994) was the first direct

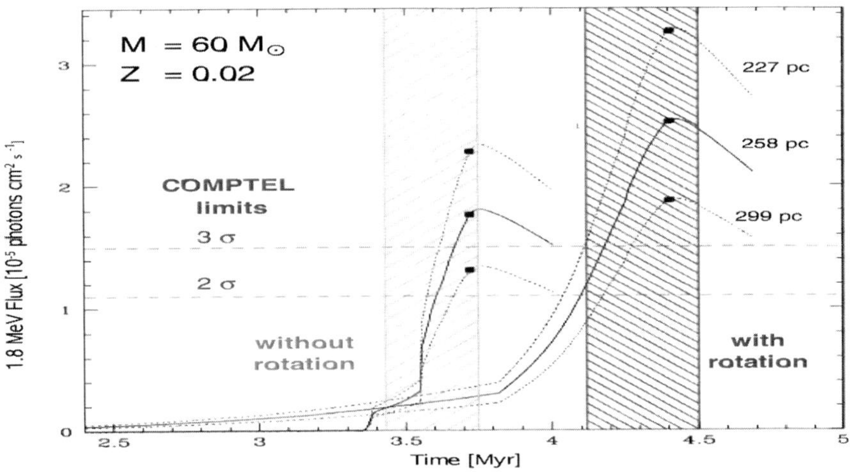


**Fig. 4.25** The $^{26}$Al gamma-ray flux estimated from ejection of $^{26}$Al-rich envelope material during the Wolf-Rayet phase (Mowlavi et al., 2005). Shown are two different stellar models, with and without stellar rotation. In addition to delaying the WR phase, rotation also increases the $^{26}$Al production. Shown are gamma-ray fluxes for different distances, not including the more recent and larger distance of 330–390 pc

proof that this isotope is indeed produced in SNe. This has been strengthened by the BeppoSAX/PDS detection of the two low-energy $^{44}$Ti lines (Vink et al., 2001). By combining both observations, Vink et al. (2001) deduced a $^{44}$Ti yield of $(1.5 \pm 1.0) \times 10^{-4}$ $M_{\odot}$. This value seemed higher than the predictions of most models (see the previous subsection), although it is not outside the error bars. Several aspects have been considered to explain this large value: a large energy of the explosion ($\approx 2 \times 10^{51}$ erg), asymmetries (Nagataki et al., 1998) currently observed in the ejecta expansion, and a strong mass loss of the progenitor consistent with the scenario of a Type Ib SN (Vink, 2004).

If $^{44}$Ti ejection, as seen in the Cas A event, was typical for core-collapse events, the $\gamma$-ray surveys made with COMPTEL(Dupraz et al., 1997; Iyudin et al., 1999) and with INTEGRAL (Renaud et al., 2006) should have seen several objects along the plane of the Galaxy through their $^{44}$Ti decay emission (see Sect. 7.6 for a detailed discussion of Galactic supernovae and $^{44}$Ti). From this, it had been concluded that $^{44}$Ti ejection is rather a characteristic of a rare subclass of core-collapse supernovae (The et al., 2006).

From the three different $\gamma$-ray lines resulting from the $^{44}$Ti decay chain, constraints for kinematic Doppler broadening can be derived: The Doppler broadening being a linear function of energy, it would broaden the 1,157 keV line to values in the few to tens of keV range, which can be measured with Ge spectrometers; the lower-energy lines at 68 and 78 keV would not show significant kinematic broadening. Martin et al. (2009) have exploited INTEGRAL/SPI spectrometer data to show that $^{44}$Ti ejecta as seen by above measurements need to be faster than 500 km/s, as the 1157 keV line is not found with SPI and assumed to be broadened such as to disappear in instrumental background.

#### 4.5.2.5 Ejecta from a Supernova Remnant on Earth

$^{60}$Fe has been discovered through accelerator-mass spectroscopy (AMS) analyses of ocean crust material (Knie et al., 2004) (Fig. 4.26). If taken from places on Earth which are remote from any antropogeneous contamination, such as in deep parts of the Pacific ocean, they provide a record of past composition of ocean water. Manganese crusts grow very slowly from sedimentation. Therefore, a rather small sample will cover tens of My of sedimentation history within a few cm of depth. $^{60}$Fe production from cosmic ray irradiation in the atmosphere is unlikely, other systematic contaminations also seem low. The age of each depth layer can be determined from Be isotopes produced by cosmic rays in the atmosphere of the Earth, also ingested into ocean water with other atmospheric gas and dust. The AMS method is one of the most-sensitive techniques to detect small amounts of specific isotopes, reaching a sensitivity of $10^{-16}$. Evaporization of chemically-prepared Fe-enriched crust samples and successive ion acceleration and mass spectrometry obtained the result shown in Fig. 4.26. This discovery was taken as evidence that debris from a very nearby supernova event must have been deposited on Earth about 3 million years ago. Unclear remain the deposition and crust uptake efficiencies, such that the quantitative estimation of interstellar $^{60}$Fe flux or supernova distance is uncertain; distances in the 10–30 pc range have been discussed. Studies of other sediment samples are underway to estimate these effects, and to confirm this exciting record

**Fig. 4.26** Ocean crust sample as analyzed by Knie et al. (2004) for $^{60}$Fe content. The insert graph shows their result, i.e. the number of $^{60}$Fe nuclei versus age as determined from cosmic-ray produced Be radioactivity

of nearby supernova activity. Supernova nucleosynthesis of radioactivities appears close to our lives, indeed.

## References

Aboussir, Y., Pearson, J.M., Dutta, A.K., Tondeur, F.: Nuclear mass formula via an approximation to the Hartree-Fock method. Atomic Data and Nuclear Data Tables **61**, 127 (1995). doi:10.1016/S0092-640X(95)90014-4

Arcones, A., Janka, H., Scheck, L.: Nucleosynthesis-relevant conditions in neutrino-driven supernova outflows. I. Spherically symmetric hydrodynamic simulations. A&A **467**, 1227–1248 (2007). doi:10.1051/0004-6361:20066983, arXiv:astro-ph/0612582

Arnett, D.: Supernovae and nucleosynthesis. an investigation of the history of matter, from the Big Bang to the present. Princeton University Press, Princeton (1996)

Arnett, D., Meakin, C., Young, P.A.: Turbulent Convection in Stellar Interiors. II. The Velocity Field, ApJ **690**, 1715–1729, (2009). doi:10.1088/0004-637X/690/2/1715, arXiv, 0809.1625

Arnould, M., Goriely, S.: The p-process of stellar nucleosynthesis: astrophysics and nuclear physics status. Phys. Rep. **384**, 1–84 (2003). doi:10.1016/S0370-1573(03)00242-4

Aschenbach, B.: Discovery of a young nearby supernova remnant. Nature **396**, 141–142 (1998). doi:10.1038/24103

Audi, G., Wapstra, A.H., Thibault, C.: The Ame2003 atomic mass evaluation (II). Tables, graphs and references. Nucl. Phys. A **729**, 337–676 (2003). doi:10.1016/S0375-9474(03)01809-8

Aufderheide, M.B., Fushiki, I., Woosley, S.E., Hartmann, D.H.: Search for important weak interaction nuclei in presupernova evolution. ApJS **91**, 389–417 (1994). doi:10.1086/191942

Benz, W., Thielemann, F.: Convective instabilities in SN 1987A. ApJ **348**, L17–L20 (1990). doi:10.1086/185620

Bethe, H.A.: Supernova mechanisms. Rev. Mod. Phys. **62**, 801–866 (1990). doi:10.1103/RevModPhys.62.801

Bethe, H.A., Brown, G.E., Applegate, J., Lattimer, J.M.: Equation of state in the gravitational collapse of stars. Nucl. Phys. A **324**, 487–533 (1979). doi:10.1016/0375-9474(79)90596-7

Blondin, J.M., Mezzacappa, A., DeMarino, C.: Stability of standing accretion shocks, with an eye toward core-collapse supernovae. ApJ **584**, 971–980 (2003). doi:10.1086/345812, arXiv:astroph/0210634

Branch, D., Benetti, S., Kasen, D., Baron, E., Jeffery, D.J., Hatano, K., Stathakis, R.A., Filippenko, A.V., Matheson, T., Pastorello, A., Altavilla, G., Cappellaro, E., Rizzi, L., Turatto, M., Li, W., Leonard, D.C., Shields, J.C.: Direct analysis of spectra of type Ib supernovae. ApJ **566**, 1005–1017 (2002). doi:10.1086/338127, arXiv:astro-ph/0106367

Buchmann, L.R., Barnes, C.A.: Nuclear reactions in stellar helium burning and later hydrostatic burning stages. Nucl. Phys. A **777**, 254–290 (2006). doi:10.1016/j.nuclphysa.2005.01.005

Buras, R., Rampp, M., Janka, H., Kifonidis, K.: Improved models of stellar core collapse and still no explosions: what is missing? Phys. Rev. Lett. **90**(24), 241,101 (2003). doi:10.1103/PhysRevLett.90.241101, arXiv:astro-ph/0303171

Buras, R., Janka, H., Rampp, M., Kifonidis, K.: Two-dimensional hydrodynamic core-collapse supernova simulations with spectral neutrino transport. II. Models for different progenitor stars. A&A **457**, 281–308 (2006). doi:10.1051/0004-6361:20054654, arXiv:astro-ph/0512189

Burbidge, E.M., Burbidge, G.R., Fowler, W.A., Hoyle, F.: Synthesis of the elements in stars. Rev. Mod. Phys. **29**, 547–650 (1957). doi:10.1103/RevModPhys.29.547

Burrows, A., Livne, E., Dessart, L., Ott, C.D., Murphy, J.: A new mechanism for core-collapse supernova explosions. ApJ **640**, 878–890 (2006a). doi:10.1086/500174, arXiv:astro-ph/0510687

Burrows, A., Reddy, S., Thompson, T.A.: Neutrino opacities in nuclear matter. Nucl. Phys. A **777**, 356–394 (2006b). doi:10.1016/j.nuclphysa.2004.06.012, arXiv:astro-ph/0404432

Burrows, A., Dessart, L., Ott, C.D., Livne, E.: Multi-dimensional explorations in supernova theory. Phys. Rep. **442**, 23–37 (2007). doi:10.1016/j.physrep.2007.02.001, arXiv:astro-ph/0612460

Cameron, A.G.W: Some properties of r-process accretion disks and jets. ApJ **562**, 456–469 (2001). doi:10.1086/323869

Cameron, A.G.W: Some nucleosynthesis effects associated with r-process jets. ApJ **587**, 327–340 (2003). doi:10.1086/368110

Cappellaro, E., Evans, R., Turatto, M.: A new determination of supernova rates and a comparison with indicators for galactic star formation. A&A **351**, 459–466 (1999). arXiv:astro-ph/9904225

Chiappini, C., Hirschi, R., Meynet, G., Ekström, S., Maeder, A., Matteucci, F.: A strong case for fast stellar rotation at very low metallicities. A&A **449**, L27–L30 (2006). doi:10.1051/0004-6361:20064866, arxi:v:astro-ph/0602459

Chieffi, A., Limongi, M.: Explosive yields of massive stars from $Z = 0$ to $Z = Zsolar$. ApJ **608**, 405–410 (2004). doi:10.1086/392523, arXiv:astro-ph/0402625

Costantini, H., Formicola, A., Imbriani, G., Junker, M., Rolfs, C., Strieder, F.: LUNA: a laboratory for underground nuclear astrophysics. Rep. Prog. Phys. **72**(8), 086,301 (2009). doi:10.1088/0034-4885/72/8/086301, 0906.1097

Cowan, J.J., Thielemann, F., Truran, J.W.: The R-process and nucleochronology. Phys. Rep. **208**, 267–394 (1991). doi:10.1016/0370-1573(91)90070-3

Cowan, J.J., Pfeiffer, B., Kratz, K., Thielemann, F., Sneden, C., Burles, S., Tytler, D., Beers, T.C.: R-process abundances and chronometers in metal-poor stars. ApJ **521**, 194–205 (1999). doi:10.1086/307512, arXiv:astro-ph/9808272

De Marco, O., Schmutz, W., Crowther, P.A., Hillier, D.J., Dessart, L., de Koter, A., Schweickhardt, J.: The gamma Velorum binary system. II. WR stellar parameters and the photon loss mechanism. A&A **358**, 187–200 (2000). arXiv:astro-ph/0004081

de Mink, S.E., Pols, O.R., Langer, N., Izzard, R.G.: Massive binaries as the source of abundance anomalies in globular clusters. A&A **507**, L1–L4 (2009). doi:10.1051/0004-6361/200913205, 0910.1086

Diehl, R., Bennett, K., Bloemen, H., Dupraz, C., Hermsen, W., Knoedlseder, J., Lichti, G., Morris, D., Oberlack, U., Ryan, J., Schoenfelder, V., Steinle, H., Varendorff, M., Winkler, C.: 1.809 MeV gammarays from the VELA region. A&A **298**, L25 (1995)

Diehl, R., Halloin, H., Kretschmer, K., Lichti, G.G., Schönfelder, V., Strong, A.W., von Kienlin, A., Wang, W., Jean, P., Knödlseder, J., Roques, J.P., Weidenspointner, G., Schanne, S., Hartmann, D.H., Winkler, C., Wunderer, C.: Radioactive $^{26}$Al from massive stars in the Galaxy. Nature **439**, 45–47 (2006). doi:10.1038/nature04364, arXiv:astro-ph/0601015

Dillmann, I., Rauscher, T., Heil, M., Käppeler, F., Rapp, W., Thielemann, F.: p-Process simulations with a modified reaction library. J. Phys. G Nucl. Phys. **35**(1), 014,029 (2008). doi:10.1088/0954-3899/35/1/014029, 0805.4756

Duan, H., Fuller, G.M., Qian, Y.: Collective neutrino flavor transformation in supernovae. Phys. Rev. D **74**(12), 123,004 (2006). doi:10.1103/PhysRevD.74.123004, arXiv:astro-ph/0511275

Duan, H., Fuller, G.M., Carlson, J., Qian, Y.: Neutrino mass hierarchy and stepwise spectral swapping of supernova neutrino flavors. Phys. Rev. Lett. **99**(24), 241,802 (2007). doi:10.1103/PhysRevLett.99.241802, 0707.0290

Duflo, J., Zuker, A.P.: Microscopic mass formulas. Phys. Rev. C **52**, 23 (1995). doi:10.1103/PhysRevC.52.R23, arXiv:nucl-th/9505011

Dupraz, C., Bloemen, H., Bennett, K., Diehl, R., Hermsen, W., Iyudin, A.F., Ryan, J., Schoenfelder, V.: COMPTEL three-year search for galactic sources of ^44 ^Ti gamma-ray line emission at 1.157MeV. A&A **324**, 683–689 (1997)

Ekström, S., Meynet, G., Chiappini, C., Hirschi, R., Maeder A: Effects of rotation on the evolution of primordial stars. A&A **489**, 685–698 (2008). doi:10.1051/0004-6361:200809633, 0807.0573

El Eid, M.F., The, L., Meyer, B.S.: Massive stars: input physics and stellar models. Space Sci. Rev. **147**, 1–29 (2009). doi:10.1007/s11214-009-9517-6

Eldridge, J.J.: A new-age determination for $\gamma^2$Velorum from binary stellar evolution models. MNRAS **400**, L20–L23 (2009). doi:10.1111/j.1745-3933.2009.00753.x, 0909.0504

Eldridge, J.J., Izzard, R.G., Tout, C.A.: The effect of massive binaries on stellar populations and supernova progenitors. MNRAS **384**, 1109–1118 (2008). doi:10.1111/j.1365-2966.2007.12738.x, 0711.3079

Fabbian, D., Nissen, P.E., Asplund, M., Pettini, M., Akerman, C.: The C/O ratio at low metallicity: constraints on early chemical evolution from observations of Galactic halo stars. A&A **500**, 1143–1155 (2009). doi:10.1051/0004-6361/200810095, 0810.0281

Farouqi, K., Kratz, K., Mashonkina, L.I., Pfeiffer, B., Cowan, J.J., Thielemann, F., Truran, J.W.: Nucleosynthesis modes in the high-entropywind of type ii supernovae: comparison of calculations with halo-star observations. ApJ **694**, L49–L53 (2009). doi:10.1088/0004-637X/694/1/L49, 0901.2541

Farouqi, K., Kratz, K., Pfeiffer, B., Rauscher, T., Thilemann, F., Truran, J.W.: Charged-particle and neutron-capture processes in the high-entropy wind of core-collapse supernovae (2010). ArXiv e-prints 1002.2346

Fischer, T., Whitehouse, S.C., Mezzacappa, A., Thielemann, F., Liebendörfer, M.: Protoneutron star evolution and the neutrino driven wind in general relativistic neutrino radiation hydrodynamics simulations (2009a). ArXiv e-prints 0908.1871

Fischer, T., Whitehouse, S.C., Mezzacappa, A., Thielemann, F., Liebendörfer, M.: The neutrino signal from protoneutron star accretion and black hole formation. A&A **499**, 1–15 (2009b). doi:10.1051/0004-6361/200811055, 0809.5129

Fisker, J.L., Schatz, H., Thielemann, F.: Explosive hydrogen burning during type I X-ray bursts. ApJS **174**, 261–276 (2008). doi:10.1086/521104

Fisker, J.L., Hoffman, R.D., Pruet, J.: On the origin of the lightest molybdenum isotopes. ApJ **690**, L135–L139 (2009). doi:10.1088/0004-637X/690/2/L135, 0711.1502

Fogli, G., Lisi, E., Marrone, A., Mirizzi, A.: Collective neutrino flavor transitions in supernovae and the role of trajectory averaging. J. Cosmol. Astro-Part. Phys. **12**, 10 (2007). doi:10.1088/1475-7516/2007/12/010, 0707.1998

Foglizzo, T.: A simple toy model of the advective-acoustic instability. I. Perturbative approach. ApJ **694**, 820–832 (2009). doi:10.1088/0004-637X/694/2/820, 0809.2302

Fransson, C., Kozma, C.: The freeze-out phase of SN 1987A – Implications for the light curve. ApJ **408**, L25–L28 (1993). doi:10.1086/186822

Freiburghaus, C., Rembges, J., Rauscher, T., Kolbe, E., Thielemann, F., Kratz, K., Pfeiffer, B., Cowan, J.J.: The astrophysical r-process: a comparison of calculations following adiabatic expansion with classical calculations based on neutron densities and temperatures. ApJ **516**, 381–398 (1999a). doi:10.1086/307072

Freiburghaus, C., Rosswog, S., Thielemann, F.: R-process in neutron star mergers. ApJ **525**, L121–L124 (1999b). doi:10.1086/312343

Fröhlich, C., Hauser, P., Liebendörfer, M., Martínez-Pinedo, G., Thielemann, F., Bravo, E., Zinner, N.T., Hix, W.R., Langanke, K., Mezzacappa, A., Nomoto, K.: Composition of the innermost core-collapse supernova ejecta. ApJ **637**, 415–426 (2006a). doi:10.1086/498224, arXiv:astro-ph/0410208

Fröhlich, C., Martínez-Pinedo, G., Liebendörfer, M., Thielemann, F., Bravo, E., Hix, W.R., Langanke, K., Zinner, N.T.: Neutrino-induced nucleosynthesis of a greater 64 nuclei: the np process. Phys. Rev. Lett. **96**(14), 142,502 (2006b). doi:10.1103/PhysRevLett.96.142502, arXiv:astro-ph/0511376

Fryer, C.L., Herwig, F., Hungerford, A., Timmes, F.X.: Supernova fallback: a possible site for the r-process. ApJ **646**, L131–L134 (2006). doi:10.1086/507071, arXiv:astro-ph/0606450

Fu, A., Arnett, W.D.: The late behavior of supernova 1987a. II. Gamma-ray transparency of the ejecta. ApJ **340**, 414 (1989). doi:10.1086/167403

Fukuda, I.: A statistical study of rotational velocities of the stars. PASP **94**, 271–284 (1982). doi:10.1086/130977

Fuller, G.M., Fowler, W.A., Newman, M.J.: Stellar weak-interaction rates for sd-shell nuclei. I – Nuclear matrix element systematics with application to Al-26 and selected nuclei of importance to the supernova problem. ApJS **42**, 447–473 (1980). doi:10.1086/190657

Fuller, G.M., Fowler, W.A., Newman, M.J.: Stellar weak interaction rates for intermediate-mass nuclei. II – A = 21 to A = 60. ApJ **252**, 715–740 (1982). doi:10.1086/159597

Fuller, G.M., Fowler, W.A., Newman, M.J.: Stellar weak interaction rates for intermediate-mass nuclei. IV – Interpolation procedures for rapidly varying lepton capture rates using effective log (ft)-values. ApJ **293**, 1–16 (1985). doi:10.1086/163208

Georgy, C., Meynet, G., Walder, R., Folini, D., Maeder, A.: The different progenitors of type Ib, Ic SNe, and of GRB. A&A **502**, 611–622 (2009). doi:10.1051/0004-6361/200811339, 0906.2284

Giron, S., Hammache, F., de Séréville, N., Beaumel, D., Burgunder, J., Caceres, L., Clement, E., Duchene, G., Flavigny, F., de France, G., Franchoo, S., Fernandez, B., Galaviz-Redondo, D., Gasques, L., Gibelin, J., Gillibert, A., Grevy, S., Guillot, J., Heil, M., Kiener, J., Lapoux, V., Maréchal, F., Matta, A., Matea, Y., Moukaddam, M., Nalpas, L., Obertelli, A., Perrot, L., Raabe, R., Scarpaci, J.A., Sorlin, O., Stefan, I., Stoedel, C., Takechi, M., Thomas, J.C., Togano, Y.: Study of $^{60}$Fe(n,$\gamma$) $^{61}$Fe reaction of astrophysical interest via d($^{60}$Fe,$p\gamma$) indirect reaction. In: Spitaleri, C., Rolfs, C., Pizzone, R.G. (eds.) American Institute of Physics Conference Series, vol. 1213, pp. 201–204 (2010). doi:10.1063/1.3362577

Goldreich, P., Weber, S.V.: Homologously collapsing stellar cores. ApJ **238**, 991–997 (1980). doi:10.1086/158065

Goriely, S., Hilaire, S., Girod, M., Péru, S.: First Gogny-Hartree-Fock-Bogoliubov nuclear mass model. Phys. Rev. Lett. **102**(24), 242,501 (2009a). doi:10.1103/PhysRevLett.102.242501

Goriely, S., Hilaire, S., Koning, A.J., Sin, M., Capote, R.: Towards a prediction of fission cross sections on the basis of microscopic nuclear inputs. Phys. Rev. C **79**(2), 024,612 (2009b). doi:10.1103/PhysRevC.79.024612

Gyürky, G., Kiss, G.G., Elekes, Z., Fülöp, Z., Somorjai, E., Palumbo, A., Görres, J., Lee, H.Y., Rapp, W., Wiescher, M., Özkan, N., Güray, R.T., Efe, G., Rauscher, T.: α-Induced cross sections of Cd106 for the astrophysical p process. Phys. Rev. C **74**(2), 025,805 (2006). doi:10.1103/PhysRevC.74.025805, arXiv:nucl-ex/0605034

Hamuy, M.: Observed and physical properties of core-collapse supernovae. ApJ **582**, 905–914 (2003). doi:10.1086/344689, arXiv:astro-ph/0209174

Haxton, W.C., Parker, P.D., Rolfs, C.E.: Solar hydrogen burning and neutrinos. Nucl. Phys. A **777**, 226–253 (2006). doi:10.1016/j.nuclphysa.2005.02.088, arXiv:nucl-th/0501020

Heger, A., Langer, N.: Presupernova evolution of rotating massive stars. II. Evolution of the surface properties. ApJ **544**, 1016–1035 (2000). doi:10.1086/317239, arXiv:astro-ph/0005110

Heger, A., Woosley, S.E.: The nucleosynthetic signature of population III. ApJ **567**, 532–543 (2002). doi:10.1086/338487, arXiv:astro-ph/0107037

Heger, A., Langanke, K., Martínez-Pinedo, G., Woosley, S.E.: Presupernova collapse models with improved weak-interaction rates. Phys. Rev. Lett. **86**, 1678–1681 (2001a). doi:10.1103/PhysRevLett.86.1678, arXiv:astro-ph/0007412

Heger, A., Woosley, S.E., Martínez-Pinedo, G., Langanke, K.: Presupernova evolution with improved rates for weak interactions. ApJ **560**, 307–325 (2001b). doi:10.1086/324092, arXiv:astroph/0011507

Heger, A., Fryer, C.L., Woosley, S.E., Langer, N., Hartmann, D.H.: How massive single stars end their life. ApJ **591**, 288–300 (2003). doi:10.1086/375341, arXiv:astro-ph/0212469

Herant, M., Benz, W., Hix, W.R., Fryer, C.L., Colgate, S.A.: Inside the supernova: A powerful convective engine. ApJ **435**, 339–361 (1994). doi:10.1086/174817, arXiv:astro-ph/9404024

Hirschi, R.: Very low-metallicity massive stars:. Pre-SN evolution models and primary nitrogen production, A&A **461**, 571–583 (2007). doi:10.1051/0004-6361:20065356, arXiv:astro-ph/0608170

Hirschi, R., Meynet, G., Maeder, A.: Stellar evolution with rotation. XIII. Predicted GRB rates at various Z. A&A **443**, 581–591 (2005). doi:10.1051/0004-6361:20053329, arXiv:astro-ph/0507343

Hirschi, R., Frischknecht, U., Thielemann, F., Pignatari, M., Chiappini, C., Ekström, S., Meynet, G., Maeder, A.: Stellar evolution in the early universe. In: Hunt, L.K., Madden, S.,

Schneider, R. (eds.) IAU Symposium, IAU Symposium, vol. 255, pp. 297–304 (2008). doi:10.1017/S1743921308024976

Hix, W.R., Thielemann, F.: Silicon burning. I. Neutronization and the physics of quasi-equilibrium. ApJ **460**, 869 (1996). doi:10.1086/177016, arXiv:astro-ph/9511088

Hix, W.R., Thielemann, F.: Computational methods for nucleosynthesis and nuclear energy generation. J. Comp. Appl. Math. **109**, 321–351 (1999a). arXiv:astroph/9906478

Hix, W.R., Thielemann, F.: Silicon burning. II. Quasi-equilibrium and explosive burning. ApJ **511**, 862–875 (1999b). doi:10.1086/306692, arXiv:astro-ph/9808203

Hix, W.R., Messer, O.E., Mezzacappa, A., Liebendörfer, M., Sampaio, J., Langanke, K., Dean, D.J., Martínez-Pinedo, G.: Consequences of nuclear electron capture in core collapse supernovae. Phys. Rev. Lett. **91**(20), 201,102 (2003). doi:10.1103/PhysRevLett.91.201102, arXiv:astro-ph/0310883

Hix, W.R., Parete-Koon, S.T., Freiburghaus, C., Thielemann, F.: The QSE-reduced nuclear reaction network for silicon burning. ApJ **667**, 476–488 (2007). doi:10.1086/520672

Hoffman, R.D., Woosley, S.E., Qian, Y.: Nucleosynthesis in neutrino-driven winds. II. Implications for heavy element synthesis. ApJ **482**, 951 (1997). doi:10.1086/304181, arXiv:astroph/9611097

Hoffman, R.D., Woosley, S.E., Weaver, T.A., Rauscher, T., Thielemann, F.: The reaction rate sensitivity of nucleosynthesis in Type II supernovae. ApJ **521**, 735–752 (1999). doi:10.1086/307568, arXiv:astro-ph/9809240

Honda, S., Aoki, W., Ishimaru, Y., Wanajo, S., Ryan, S.G.: Neutron-capture elements in the very metal poor star HD 122563. ApJ **643**, 1180–1189 (2006). doi:10.1086/503195, arXiv:astroph/0602107

Iliadis, C.: Nuclear Physics of Stars. Wiley-VCH Verlag (2007)

Iyudin, A.F., Diehl, R., Bloemen, H., Hermsen, W., Lichti, G.G., Morris, D., Ryan, J., Schoenfelder, V., Steinle, H., Varendorff, M., de Vries, C., Winkler, C.: COMPTEL observations of Ti-44 gamma-ray line emission from CAS A. A&A **284**, L1–L4 (1994)

Iyudin, A.F., Schönfelder, V., Bennett, K., Bloemen, H., Diehl, R., Hermsen, W., Knödlseder, J., Lichti, G.G., Oberlack, U., Ryan, J., Strong, A.W., Winkler, C.: COMPTEL all-sky survey in $^{44}$Ti line emission. Astrophys. Lett. Comm. **38**, 383 (1999)

Janka, H., Buras, R., Rampp, M.: The mechanism of core-collapse supernovae and the ejection of heavy elements. Nucl. Phys. A **718**, 269–276 (2003). doi:10.1016/S0375-9474(03)00727-9, arXiv:astro-ph/0212317

Janka, H., Langanke, K., Marek, A., Martínez-Pinedo, G., Müller, B.: Theory of core-collapse supernovae. Phys. Rep. **442**, 38–74 (2007). doi:10.1016/j.physrep.2007.02.002, arXiv:astro-ph/0612072

Käppeler, F., Mengoni, A.: Nuclear input for the s process: progress with experiments and theory. Nucl. Phys. A **777**, 291–310 (2006). doi:10.1016/j.nuclphysa.2005.01.028

Katsuda, S., Tsunemi, H., Mori, K.: Is Vela Jr. a young supernova remnant? Adv. Space Res. **43**, 895–899 (2009). doi:10.1016/j.asr.2009.01.004

Kippenhahn, R., Weigert, A.: Stellar Structure and Evolution. In: Kippenhahn, R., Weigert, A. (eds.) Stellar Structure and Evolution, XVI, **468** pp. 192 figs. Springer-Verlag Berlin Heidelberg New York. Also Astronomy and Astrophysics Library. Springer-Verlag Berlin Heidelberg New York. (1994)

Kiss, G.G., Gyürky, G., Elekes, Z., Fülöp, Z., Somorjai, E., Rauscher, T., Wiescher, M.: Ge70($p, \gamma$) As71 and Ge76(p,n)As76 cross sections for the astrophysical p process: sensitivity of the optical proton potential at low energies. Phys. Rev. C **76**(5), 055,807 (2007). doi:10.1103/PhysRevC.76.055807, 0711.1079

Kiss, G.G., Rauscher, T., Gyürky, G., Simon, A., Fülöp, Z., Somorjai, E.: Coulomb suppression of the stellar enhancement factor. Phys. Rev. Lett. **101**(19), 191,101 (2008). doi:10.1103/PhysRevLett.101.191101, 0809.2676

Knie, K., Korschinek, G., Faestermann, T., Dorfi, E.A., Rugel, G., Wallner, A.: $^{60}$Fe anomaly in a deep-sea manganese crust and implications for a nearby supernova source. Phys. Rev. Lett. **93**(17), 171,103 (2004). doi:10.1103/PhysRevLett.93.171103

Kobayashi, C., Umeda, H., Nomoto, K., Tominaga, N., Ohkubo, T.: Galactic chemical evolution: carbon through Zinc. ApJ **653**, 1145–1171 (2006). doi:10.1086/508914, arXiv:astro-ph/0608688

Kotake, K., Sato, K., Takahashi, K.: Explosion mechanism, neutrino burst and gravitational wave in core-collapse supernovae. Rep. Prog. Phys. **69**, 971–1143 (2006). doi:10.1088/0034-4885/69/4/R03, arXiv:astro-ph/0509456

Kratz, K., Bitouzet, J., Thielemann, F., Moeller, P., Pfeiffer, B.: Isotopic r-process abundances and nuclear structure far from stability – Implications for the r-process mechanism. ApJ **403**, 216–238 (1993). doi:10.1086/172196

Kratz, K., Farouqi, K., Pfeiffer, B., Truran, J.W., Sneden, C., Cowan, J.J.: Explorations of the r-processes: comparisons between calculations and observations of low-metallicity stars. ApJ **662**, 39–52 (2007). doi:10.1086/517495, arXiv:astro-ph/0703091

Kuroda, T., Wanajo, S., Nomoto, K.: The r-process in supersonic neutrino-driven winds: the role of the wind termination shock. ApJ **672**, 1068–1078 (2008). doi:10.1086/523795

Langanke, K., Martínez-Pinedo, G.: Shell-model calculations of stellar weak interaction rates: II. Weak rates for nuclei in the mass range/A=45-65 in supernovae environments. Nucl. Phys. A **673**, 481–508 (2000). doi:10.1016/S0375-9474(00)00131-7, arXiv:nucl-th/0001018

Langanke, K., Martínez-Pinedo, G., Sampaio, J.M., Dean, D.J., Hix, W.R., Messer, O.E., Mezzacappa, A., Liebendörfer, M., Janka, H., Rampp, M.: Electron capture rates on nuclei and implications for stellar core collapse. Phys. Rev. Lett. **90**(24), 241,102 (2003). doi:10.1103/PhysRevLett.90.241102, arXiv:astro-ph/0302459

Langanke, K., Martínez-Pinedo, G., Müller, B., Janka, H., Marek, A., Hix, W.R., Juodagalvis, A., Sampaio, J.M.: Effects of inelastic neutrino-nucleus scattering on supernova dynamics and radiated neutrino spectra. Phys. Rev. Lett. **100**(1), 011,101 (2008). doi:10.1103/PhysRevLett.100.011101, 0706.1687

Langer, N., Braun, H., Fliegner, J.: The production of circumstellar $^{26}$Al by massive stars. Ap&SS **224**, 275–278 (1995). doi:10.1007/BF00667858

Langer, N., Braun, H., Wellstein, S.: Massive Close Binaries as Source of Galactic $^{26}$AL. In: Hillebrandt, W., Muller, E. (eds.) Nuclear Astrophysics, vol. 18 (1998)

Lattimer, J.M., Douglas Swesty, F.: A generalized equation of state for hot, dense matter. Nucl. Phys. A **535**, 331–376 (1991). doi:10.1016/0375-9474(91)90452-C

Leibundgut, B., Suntzeff, N.B.: Optical light curves of supernovae. In: Weiler, K. (ed.) Supernovae and Gamma-Ray Bursters, Lecture Notes in Physics, Springer Verlag, Berlin, vol. 598, pp. 77–90 (2003)

Leising, M.D., Share, G.H.: The gamma-ray light curves of SN 1987A. ApJ **357**, 638–648 (1990). doi:10.1086/168952

Liebendörfer, M., Mezzacappa, A., Messer, O.E.B., Martinez-Pinedo, G., Hix, W.R., Thielemann, F.: The neutrino signal in stellar core collapse and postbounce evolution. Nucl. Phys. A **719**, 144 (2003). doi:10.1016/S0375-9474(03)00984-9, arXiv:astro-ph/0211329

Liebendörfer, M., Messer, O.E.B., Mezzacappa, A., Bruenn, S.W., Cardall, C.Y., Thielemann, F.: A finite difference representation of neutrino radiation hydrodynamics in spherically symmetric general relativistic spacetime. ApJS **150**, 263–316 (2004). doi:10.1086/380191, arXiv:astroph/0207036

Liebendörfer, M., Rampp, M., Janka, H., Mezzacappa, A.: Supernova simulations with Boltzmann Neutrino transport: a comparison of methods. ApJ **620**, 840–860 (2005). doi:10.1086/427203, arXiv:astro-ph/0310662

Liebendörfer, M., Fischer, T., Fröhlich, C., Thielemann, F., Whitehouse, S.: Nuclear physics with spherically symmetric supernova models. J. Phys. G Nucl. Phys. **35**(1), 014,056 (2008). doi:10.1088/0954-3899/35/1/014056, 0708.4296

Liebendörfer, M., Whitehouse, S.C., Fischer, T.: The isotropic diffusion source approximation for supernova neutrino transport. ApJ **698**, 1174–1190 (2009). doi:10.1088/0004-637X/698/2/1174, 0711.2929

Limongi, M., Chieffi, A.: Evolution, explosion, and nucleosynthesis of core-collapse supernovae. ApJ **592**, 404–433 (2003). doi:10.1086/375703, arXiv:astro-ph/0304185

Limongi, M., Chieffi, A.: The nucleosynthesis of $^{26}$Al and $^{60}$Fe in solar metallicity stars extending in mass from 11 to 120 $M_{solar}$: the hydrostatic and explosive contributions. ApJ **647**, 483–500 (2006). doi:10.1086/505164, arXiv:astro-ph/0604297

Limongi, M., Straniero, O., Chieffi, A.: Massive stars in the range 13–25 $M_{solar}$: evolution and nucleosynthesis. II. the solar metallicity models. ApJS **129**, 625–664 (2000). doi:10.1086/313424, arXiv:astro-ph/0003401

Lundqvist, P., Kozma, C., Sollerman, J., Fransson, C.: ISO/SWS observations of SN 1987A. II. A refined upper limit on the mass of $^{44}$Ti in the ejecta of SN 1987A. A&A **374**, 629–637 (2001). doi:10.1051/0004-6361:20010725, arXiv:astro-ph/0105402

MacFadyen, A.I., Woosley, S.E.: Collapsars: gamma-ray bursts and explosions in "Failed Supernovae". ApJ **524**, 262–289 (1999). doi:10.1086/307790, arXiv:astro-ph/9810274

Maeder, A.: Physics, Formation and Evolution of Rotating Stars. In: Maeder, A. (ed.) Physics, Formation and Evolution of Rotating Stars, Astronomy and Astrophysics Library, Volume ISBN 978-3-540-76948-4. Springer Berlin Heidelberg, (2009). doi:10.1007/978-3-540-76949-1

Maeder, A., Meynet, G.: Massive stars through the ages: evolution and wind composition. Rev. Mod. Phys. (2010). doi:10.1103/RevModPhys.62.801 (submitted)

Maeder, A., Grebel, E.K., Mermilliod, J.: Differences in the fractions of Be stars in galaxies. A&A **346**, 459–464 (1999). arXiv:astro-ph/9904008

Marek, A., Janka, H.: Delayed neutrino-driven supernova explosions aided by the standing accretion-shock instability. ApJ **694**, 664–696 (2009). doi:10.1088/0004-637X/694/1/664, 0708.3372

Marek, A., Janka, H., Buras, R., Liebendörfer, M., Rampp, M.: On ion-ion correlation effects during stellar core collapse. A&A **443**, 201–210 (2005). doi:10.1051/0004-6361:20053236, arXiv:astroph/0504291

Martayan, C., Frémat, Y., Hubert, A., Floquet, M., Zorec, J., Neiner, C.: Effects of metallicity, starformation conditions, and evolution in B and Be stars. II. Small magellanic cloud, field of NGC330. A&A **462**, 683–694 (2007). doi:10.1051/0004-6361:20065076, arXiv:astro-ph/0609677

Martin, P., Knödlseder, J., Vink, J., Decourchelle, A., Renaud, M.: Constraints on the kinematics of the $^{44}$Ti ejecta of Cassiopeia A from INTEGRAL/SPI. A&A **502**, 131–137 (2009). doi:10.1051/0004-6361/200809735

Martin, P., Knödlseder, J., Meynet, G., Diehl, R.: Predicted gamma-ray line emission from the Cygnus complex A&A **511**, 86 (2010). ArXiv e-prints 1001.1522

Martínez-Pinedo, G., Liebendörfer, M., Frekers, D.: Nuclear input for core-collapse models. Nucl. Phys. A **777**, 395–423 (2006). doi:10.1016/j.nuclphysa.2006.02.014, arXiv:astro-ph/0412091

Martínez-Pinedo, G., Mocelj, D., Zinner, N.T., Kelic, A., Langanke, K., Panov, I., Pfeiffer, B., Rauscher, T., Schmidt, K., Thielemann, F.: The role of fission in the r-process. Prog. Part. Nucl. Phys. **59**, 199–205 (2007). doi:10.1016/j.ppnp.2007.01.018

Massi, F., de Luca, M., Elia, D., Giannini, T., Lorenzetti, D., Nisini, B.: Star formation in the Vela molecular ridge. Large scale mapping of cloud D in the mm continuum. A&A **466**, 1013–1023 (2007). doi:10.1051/0004-6361:20066438, arXiv:astro-ph/0702687

Matheson, T., Filippenko, A.V., Li, W., Leonard, D.C., Shields, J.C.: Optical spectroscopy of type IB/C supernovae. AJ **121**, 1648–1675 (2001). doi:10.1086/319390, arXiv:astro-ph/0101119

Matz, S.M., Share, G.H., Leising, M.D., Chupp, E.L., Vestrand, W.T.: Gamma-ray line emission from SN1987A. Nature **331**, 416–418 (1988). doi:10.1038/331416a0

Meakin, C.A., Arnett, D.: Turbulent convection in stellar interiors. I. Hydrodynamic simulation. ApJ **667**, 448–475 (2007). doi:10.1086/520318, arXiv:astro-ph/0611315

Metzger, B.D., Martínez-Pinedo, G., Darbha, S., Quataert, E., Arcones, A., Kasen, D., Thomas, R., Nugent, P., Panov, I.V., Zinner, N.T.: Electromagnetic counterparts of compact object mergers powered by the radioactive decay of r-process nuclei, MNRAS, **840** (2010). doi: 10.1111/j.1365-2966.2010.16864.x, arXiv, 1001.5029

Meyer, B.S., Brown, J.S.: Survey of r-process models. ApJS **112**, 199 (1997). doi:10.1086/313032

Meynet, G., Maeder, A.: Stellar evolution with rotation. V. Changes in all the outputs of massive star models. A&A **361**, 101–120 (2000). arXiv:astro-ph/0006404

Meynet, G., Maeder, A.: Stellar evolution with rotation. X.Wolf-Rayet star populations at solar metallicity. A&A **404**, 975–990 (2003). doi:10.1051/0004-6361:20030512, arXiv:astro-ph/0304069

Meynet, G., Maeder, A.: Stellar evolution with rotation. XI. Wolf-Rayet star populations at different metallicities. A&A **429**, 581–598 (2005). doi:10.1051/0004-6361:20047106, arXiv:astroph/0408319

Meynet, G., Arnould, M., Prantzos, N., Paulus, G.: Contribution of Wolf-Rayet stars to the synthesis of $^{26}$Al. I. The $\gamma$-ray connection. A&A **320**, 460–468 (1997)

Meynet, G., Ekström, S., Maeder, A.: The early star generations: the dominant effect of rotation on the CNO yields. A&A **447**, 623–639 (2006). doi:10.1051/0004-6361:20053070, arXiv:astroph/0510560

Meynet, G., Ekström, S., Georgy, C., Maeder, A., Hirschi, R.: Massive star evolution: from the early to the present day Universe. In: Deng, L., Chan, K.L. (eds.) IAU Symposium, vol. 252, pp. 317–327 (2008). doi:10.1017/S1743921308023119

Mikheyev, S.P., Smirnov, A.Y.: Resonance enhancement of oscillations in matter and solar neutrino spectroscopy. Yadernaya Fizika, Sov J Nucl Phys **42**, 1441–1448, 913 (1985)

Mochizuki, Y., Takahashi, K., Janka, H.T., Hillebrandt, W., Diehl, R.: 44Ti: its effective decay rate in young supernova remnants, and its abundance in Cassiopeia A. A&A **346**, 831–842 (1999). arXiv:astro-ph/9904378

Mowlavi, N., Meynet, G.: $^{26}$Al production from $\gamma^2$-Velorum: Point source or diffuse emission at 1.8 MeV? New Astron. Rev. **50**, 484–486 (2006). doi:10.1016/j.newar.2006.06.060

Mowlavi, N., Knödlseder, J., Meynet, G., Dubath, P., Diehl, R., Lichti, G., Schanne, S., Winkler, C.: $^{26}$Al production in $\gamma^2$ Velorum. Nucl. Phys. A **758**, 320–323 (2005). doi:10.1016/j.nuclphysa.2005.05.177

Möller, P., Nix, J.R., Myers, W.D., Swiatecki, W.J.: Nuclear Ground-State Masses and Deformations, Atomic Data and Nuclear Data Tables **59**, 185 (1995). doi:10.1006/adnd.1995.1002, arXiv:nucl-th/9308022

Nagataki, S.: Effects of Jetlike explosion in SN 1987A. ApJS **127**, 141–157 (2000). doi:10.1086/313317, arXiv:astro-ph/9907109

Nagataki, S., Hashimoto, M., Sato, K., Yamada, S., Mochizuki, Y.S.: The high ratio of 44Ti/56Ni in Cassiopeia A and the axisymmetric collapse-driven supernova explosion. ApJ **492**, L45 (1998). doi:10.1086/311089, arXiv:astro-ph/9807015

Nakamura, T., Umeda, H., Nomoto, K., Thielemann, F., Burrows, A.: Nucleosynthesis in type II supernovae and the abundances in metal-poor stars. ApJ **517**, 193–208 (1999). doi:10.1086/307167, arXiv:astro-ph/9809307

Nakamura, T., Umeda, H., Iwamoto, K., Nomoto, K., Hashimoto, M., Hix, W.R., Thielemann, F.: Explosive nucleosynthesis in hypernovae. ApJ **555**, 880–899 (2001). doi:10.1086/321495, arXiv:astroph/0011184

Nittler, L.R., Amari, S., Zinner, E., Woosley, S.E., Lewis, R.S.: Extinct 44Ti in presolar graphite and SiC: proof of a supernova origin. ApJ **462**, L31 (1996). doi:10.1086/310021

Nomoto, K.: Evolution of 8–10 solar mass stars toward electron capture supernovae. II – Collapse of an O + NE + MG core. ApJ **322**, 206–214 (1987). doi:10.1086/165716

Nomoto, K., Hashimoto, M.: Presupernova evolution of massive stars. Phys. Rep.**163**, 13–36 (1988). doi:10.1016/0370-1573(88)90032-4

Nomoto, K., Thielemann, F., Miyaji, S.: The triple alpha reaction at low temperatures in accreting white dwarfs and neutron stars. A&A **149**, 239–245 (1985)

Nomoto, K., Mazzali, P.A., Nakamura, T., Iwamoto, K., Danziger, I.J., Patat, F.: The properties of hypernovae: SNe Ic 1998bw, 1997ef, and SN IIn 1997cy. In: Livio, M., Panagia, N., Sahu, K., (eds.) Supernovae and Gamma-Ray Bursts: the Greatest Explosions since the Big Bang, 144–170 (2001). arXiv:astro-ph/0003077

Nomoto, K., Tominaga, N., Umeda, H., Kobayashi, C., Maeda, K.: Nucleosynthesis yields of core-collapse supernovae and hypernovae, and galactic chemical evolution. Nucl. Phys. A **777**, 424–458 (2006). doi:10.1016/j.nuclphysa.2006.05.008, arXiv:astro-ph/0605725

Oberlack, U., Wessolowski, U., Diehl, R., Bennett, K., Bloemen, H., Hermsen, W., Knödlseder, J., Morris, D., Schönfelder, V., von Ballmoos, P.: COMPTEL limits on $^{26}$Al 1.809 MeV line emission from gamma$^2$ Velorum. A&A **353**, 715–721 (2000). arXiv:astro-ph/9910555

Ohkubo, T., Umeda, H., Maeda, K., Nomoto, K., Suzuki, T., Tsuruta, S., J Rees, M.: Evolution of core-collapse very massive population III stars. In: O'Shea, B.W., Heger, A. (eds.) First Stars III, American Institute of Physics Conference Series, vol. 990, pp. 244–246 (2008). doi:10.1063/1.2905553

Otsuki, K., Tagoshi, H., Kajino, T., Wanajo, S.: General relativistic effects on neutrino-driven winds from young, hot neutron stars and r-process nucleosynthesis. ApJ **533**, 424–439 (2000). doi:10.1086/308632, arXiv:astro-ph/9911164

Palacios, A., Meynet, G., Vuissoz, C., Knödlseder, J., Schaerer, D., Cerviño, M., Mowlavi, N.: New estimates of the contribution of Wolf-Rayet stellar winds to the Galactic $^{26}$Al. A&A **429**, 613–624 (2005). doi:10.1051/0004-6361:20041757, arXiv:astro-ph/0409580

Panov, I.V., Janka, H.: On the dynamics of proto-neutron star winds and r-process nucleosynthesis. A&A **494**, 829–844(2009). doi:10.1051/0004-6361:200810292, 0805.1848

Panov, I.V., Thielemann, F.: Final r-process yields and the influence of fission: The competition between neutron-induced and β-delayed fission. Nucl. Phys. A **718**, 647–649 (2003). doi:10.1016/S0375-9474(03)00875-3

Panov, I.V., Kolbe, E., Pfeiffer, B., Rauscher, T., Kratz, K., Thielemann, F.: Calculations of fission rates for r-process nucleosynthesis. Nucl. Phys. A **747**, 633–654 (2005). doi:10.1016/j.nuclphysa.2004.09.115, arXiv:astro-ph/0412654

Panov, I.V., Korneev, I.Y., Rauscher, T., Martínez-Pinedo, G., Kelić-Heil, A., Zinner, N.T., Thielemann, F.-K.: Neutron-induced astrophysical reaction rates for translead nuclei, A&A **513**, A61, (2010). doi: 10.1051/0004-6361/200911967, arXiv, 0911.2181

Pearson, J.M., Nayak, R.C., Goriely, S.: Nuclear mass formula with Bogolyubov-enhanced shell-quenching: application to r-process. Phys. Lett. B **387**, 455–459 (1996). doi:10.1016/0370-2693(96)01071-4

Pettini, M., Zych, B.J., Steidel, C.C., Chaffee, F.H.: C, N, O abundances in the most metal-poor damped Lyman alpha systems. MNRAS **385**, 2011–2024 (2008). doi:10.1111/j.1365-2966.2008.12951.x, 0712.1829

Pfeiffer, B., Kratz, K., Thielemann, F., Walters, W.B.: Nuclear structure studies for the astrophysical r-process. Nucl. Phys. A **693**, 282–324 (2001). doi:10.1016/S0375-9474(01)01141-1

Piran, T.: The physics of gamma-ray bursts. Rev. Mod. Phys. **76**, 1143–1210 (2004). doi:10.1103/RevModPhys.76.1143, arXiv:astro-ph/0405503

Podsiadlowski, P., Joss, P.C., Hsu, J.J.L.: Presupernova evolution in massive interacting binaries. ApJ **391**, 246–264 (1992). doi:10.1086/171341

Prantzos, N., Boissier, S.: On the relative frequencies of core-collapse supernovae sub-types: the role of progenitor metallicity. A&A **406**, 259–264 (2003). doi:10.1051/0004-6361:20030717, arXiv:astro-ph/0305376

Prieto, J.L., Stanek, K.Z., Beacom, J.F.: Characterizing supernova progenitors via the metallicities of their host galaxies, from poor dwarfs to rich spirals. ApJ **673**, 999–1008 (2008). doi:10.1086/524654, 0707.0690

Pruet, J., Woosley, S.E., Buras, R., Janka, H., Hoffman, R.D.: Nucleosynthesis in the hot convective bubble in core-collapse supernovae. ApJ **623**, 325–336 (2005). doi:10.1086/428281, arXiv:astroph/0409446

Pruet, J., Hoffman, R.D., Woosley, S.E., Janka, H., Buras, R.: Nucleosynthesis in early supernova winds. II. The role of neutrinos. ApJ **644**, 1028–1039 (2006). doi:10.1086/503891, arXiv:astroph/0511194

Qian, Y., Wasserburg, G.J.: Where, oh where has the r-process gone? Phys. Rep. **442**, 237–268 (2007). doi:10.1016/j.physrep.2007.02.006, 0708.1767

Qian, Y., Woosley, S.E.: Nucleosynthesis in neutrino-driven winds. I. The physical conditions. ApJ **471**, 331 (1996). doi:10.1086/177973, arXiv:astro-ph/9611094

Rapp, W., Görres, J., Wiescher, M., Schatz, H., Käppeler, F.: Sensitivity of p-process nucleosynthesis to nuclear reaction rates in a 25 $M_{solar}$ supernova model. ApJ **653**, 474–489 (2006). doi:10.1086/508402, arXiv:astro-ph/0608341

Rauscher, T.: Branchings in the $\gamma$ process path revisited. Phys. Rev. C **73**(1), 015,804 (2006). doi:10.1103/PhysRevC.73.015804, arXiv:astro-ph/0510710

Rauscher, T., Thielemann, F.: Astrophysical reaction rates from statistical model calculations. Atomic Data and Nuclear Data Tables **75**, 1–351 (2000). doi:10.1006/adnd.2000.0834, arXiv:astro-ph/0004059

Rauscher, T., Thielemann, F.: Predicted cross-sections for photon-induced particle emission. Atomic Data and Nuclear Data Tables **88**, 1–81 (2004). doi:10.1016/j.adt.2004.07.001, arXiv:nuclth/0403084

Rauscher, T., Applegate, J.H., Cowan, J.J., Thielemann, F., Wiescher, M.: Production of heavy elements in inhomogeneous cosmologies. ApJ **429**, 499–530 (1994). doi:10.1086/174339

Rauscher, T., Heger, A., Hoffman, R.D., Woosley, S.E.: Nucleosynthesis in Massive Stars with Improved Nuclear and Stellar Physics. ApJ **576**, 323–348 (2002). doi:10.1086/341728, arXiv:astroph/0112478

Rayet, M., Arnould, M., Prantzos, N.: The p-process revisited. A&A **227**, 271–281 (1990)

Rayet, M., Arnould, M., Hashimoto, M., Prantzos, N., Nomoto, K.: The p-process in Type II supernovae. A&A **298**, 517 (1995)

Renaud, M., Vink, J., Decourchelle, A., Lebrun, F., Terrier, R., Ballet, J.: An INTEGRAL/IBIS view of young Galactic SNRs through the $^{44}$Ti gamma-ray lines. New Astron. Rev. **50**, 540–543 (2006). doi:10.1016/j.newar.2006.06.061, arXiv:astro-ph/0602304

Rugel, G., Faestermann, T., Knie, K., Korschinek, G., Poutivtsev, M., Schumann, D., Kivel, N., Günther-Leopold, I., Weinreich, R., Wohlmuther, M.: New measurement of the Fe60 half-life. Phys. Rev. Lett. **103**(7), 072,502 (2009). doi:10.1103/PhysRevLett.103.072502

Sagert, I., Fischer, T., Hempel, M., Pagliara, G., Schaffner-Bielich, J., Mezzacappa, A., Thielemann, F., Liebendörfer, M.: Signals of the QCD phase transition in core-collapse supernovae. Phys. Rev. Lett. **102**(8), 081,101 (2009). doi:10.1103/PhysRevLett.102.081101, 0809.4225

Sauer, D.N., Mazzali, P.A., Deng, J., Valenti, S., Nomoto, K., Filippenko, A.V.: The properties of the 'standard' Type Ic supernova 1994I from spectral models. MNRAS **369**, 1939–1948 (2006). doi:10.1111/j.1365-2966.2006.10438.x, arXiv:astro-ph/0604293

Sawyer, R.F.: Effects of ion and electron correlations on neutrino scattering in the infall phase of a supernova. Phys. Lett. B **630**, 1–6 (2005). doi:10.1016/j.physletb.2005.09.032, arXiv:astroph/0505520

Schatz, H., Aprahamian, A., Goerres, J.,Wiescher, M., Rauscher, T., Rembges, J.F., Thielemann, F., Pfeiffer, B., Moeller, P., Kratz, K., Herndl, H., Brown, B.A., Rebel, H.: rp-Process nucleosynthesis at extreme temperature and density conditions. Phys. Rep. **294**, 167–264 (1998). doi:10.1016/S0370-1573(97)00048-3

Seitenzahl, I.R., Taubenberger, S., Sim, S.A.: Late-time supernova light curves: the effect of internal conversion and Auger electrons. MNRAS **400**, 531–535 (2009). doi:10.1111/j.1365-2966.2009.15478.x, 0908.0247

Shen, H., Toki, H., Oyamatsu, K., Sumiyoshi, K.: Relativistic equation of state of nuclear matter for supernova and neutron star. Nucl. Phys. A **637**, 435–450 (1998a). doi:10.1016/S0375-9474(98)00236-X, arXiv:nucl-th/9805035

Shen, H., Toki, H., Oyamatsu, K., Sumiyoshi, K.: Relativistic equation of state of nuclear matter for supernova explosion. Prog. Theoret. Phys. **100**, 1013–1031 (1998b). doi:10.1143/PTP.100.1013, arXiv:nucl-th/9806095

Sim, S.A., Kromer, M., Roepke, F.K., Sorokina, E.I., Blinnikov, S.I., Kasen, D., Hillebrandt, W.: Monte Carlo Radiative Transfer Simulations: Applications to Astrophysical Outflows and Explosions (2009). ArXiv e-prints 0911.1549

Sneden, C., Cowan, J.J., Lawler, J.E., Ivans, I.I., Burles, S., Beers, T.C., Primas, F., Hill, V., Truran, J.W., Fuller, G.M., Pfeiffer, B., Kratz, K.: The extremely metal-poor, neutron capture-rich star CS 22892-052: a comprehensive abundance analysis. ApJ **591**, 936–953 (2003). doi:10.1086/375491, arXiv:astro-ph/0303542

Sollerman, J., Holland, S.T., Challis, P., Fransson, C., Garnavich, P., Kirshner, R.P., Kozma, C., Leibundgut, B., Lundqvist, P., Patat, F., Filippenko, A.V., Panagia, N., Wheeler, J.C.: Supernova 1998bw – the final phases. A&A **386**, 944–956 (2002). doi:10.1051/0004-6361:20020326, arXiv:astro-ph/0204498

Spite, M., Cayrel, R., Plez, B., Hill, V., Spite, F., Depagne, E., François, P., Bonifacio, P., Barbuy, B., Beers, T., Andersen, J., Molaro, P., Nordström, B., Primas, F.: First stars VI – Abundances of C, N, O, Li, and mixing in extremely metal-poor giants. Galactic evolution of the light elements. A&A **430**, 655–668 (2005). doi:10.1051/0004-6361:20041274, arXiv:astro-ph/0409536

Sumiyoshi, K., Yamada, S., Suzuki, H.: Dynamics and neutrino signal of black hole formation in nonrotating failed supernovae. I. Equation of state dependence. ApJ **667**, 382–394 (2007). doi:10.1086/520876, 0706.3762

Suntzeff, N.B., Bouchet, P.: The bolometric light curve of SN 1987A. I – Results from ESO and CTIO U to Q0 photometry. AJ **99**, 650–663 (1990). doi:10.1086/115358

Sunyaev, R.A., Kaniovsky, A.S., Efremov, V.V., Grebenev, S.A., Kuznetsov, A.V., Englhauser, J., Doebereiner, S., PietschW, Reppin, C., Truemper, J., Kendziorra, E., Maisack, M., Mony, B., Staubert, R.: Hard X-rays from supernova 1987A – Results of observations performed with the Roentgen observatory onboard the KVANT module during 1987–1989. Pis ma Astronomicheskii Zhurnal **16**, 403–415 (1990)

Surman, R., McLaughlin, G.C., Hix, W.R.: Nucleosynthesis in the outflow from gamma-ray burst accretion disks. ApJ **643**, 1057–1064 (2006). doi:10.1086/501116, arXiv:astro-ph/0509365

Takahashi, K., Witti, J., Janka, H.: Nucleosynthesis in neutrino-driven winds from protoneutron stars II. The r-process. A&A **286**, 857–869 (1994)

Terasawa, M., Sumiyoshi, K., Yamada, S., Suzuki, H., Kajino, T.: r-Process nucleosynthesis in neutrino-driven winds from a typical neutron star with M=1.4 $M_{solar}$. ApJ **578**, L137–L140 (2002). doi:10.1086/344698, arXiv:astro-ph/0206322

The, L., El Eid, M.F., Meyer, B.S.: s-Process nucleosynthesis in advanced burning phases of massive stars. ApJ **655**, 1058–1078 (2007). doi:10.1086/509753, arXiv:astro-ph/0609788

The, L.S., Clayton, D.D., Diehl, R., Hartmann, D.H., Iyudin, A.F., Leising, M.D., Meyer, B.S., Motizuki, Y., Schönfelder, V.: Are $^{44}$Ti-producing supernovae exceptional? A&A **450**, 1037–1050 (2006). doi:10.1051/0004-6361:20054626, arXiv:astro-ph/0601039

Thielemann, F., Hashimoto, M., Nomoto, K.: Explosive nucleosynthesis in SN 1987A. II – Composition, radioactivities, and the neutron star mass. ApJ **349**, 222–240 (1990). doi:10.1086/168308

Thielemann, F., Nomoto, K., Hashimoto, M.: Core-collapse supernovae and their ejecta. ApJ **460**, 408 (1996). doi:10.1086/176980

Thielemann, F., Hauser, P., Kolbe, E., Martinez-Pinedo, G., Panov, I., Rauscher, T., Kratz, K., Pfeiffer, B., Rosswog, S., Liebendörfer, M., Mezzacappa, A.: Heavy elements and age determinations. Space Sci. Rev. **100**, 277–296 (2002)

Thompson, T.A., Burrows, A., Meyer, B.S.: The physics of proto-neutron star winds: implications for r-process nucleosynthesis. ApJ **562**, 887–908 (2001). doi:10.1086/323861, arXiv:astroph/0105004

Thompson, T.A., Quataert, E., Burrows, A.: Viscosity and rotation in core-collapse supernovae. ApJ **620**, 861–877 (2005). doi:10.1086/427177, arXiv:astro-ph/0403224

Timmes, F.X.: Integration of nuclear reaction networks for stellar hydrodynamics. ApJS **124**, 241–263 (1999). doi:10.1086/313257

Tominaga, N., Umeda, H., Nomoto, K.: Supernova nucleosynthesis in population III 13–50 $M_{solar}$ stars and abundance patterns of extremely metal-poor stars. ApJ **660**, 516–540 (2007). doi:10.1086/513063

Travaglio, C., Gallino, R., Arnone, E., Cowan, J., Jordan, F., Sneden, C.: Galactic evolution of Sr, Y, and Zr: a multiplicity of nucleosynthetic processes. ApJ **601**, 864–884 (2004). doi:10.1086/380507, arXiv:astro-ph/0310189

Tueller, J., Barthelmy, S., Gehrels, N., Teegarden, B.J., Leventhal, M., MacCallum, C.J.: Observations of gamma-ray line profiles from SN 1987A. ApJ **351**, L41–L44 (1990). doi:10.1086/185675

Tur, C., Heger, A., Austin, S.M.: Production of 26Al, 44Ti, and 60Fe in Core-Collapse Supernovae: Sensitivity to the Rates of the Triple Alpha and 12C(a,g)16O Reactions (2009). ArXiv e-prints 0908.4283

Uberseder, E., Reifarth, R., Schumann, D., Dillmann, I., Pardo, C.D., Görres, J., Heil, M., Käppeler, F., Marganiec, J., Neuhausen, J., Pignatari, M., Voss, F., Walter, S., Wiescher, M.: Measurement of the Fe60$(n, \gamma)$ $^{61}$Fe cross section at stellar temperatures. Phys. Rev. Lett. **102**(15), 151,101 (2009). doi:10.1103/PhysRevLett.102.151101

Umeda, H., Nomoto, K.: Variations in the abundance pattern of extremely metal-poor stars and nucleosynthesis in population III supernovae. ApJ **619**, 427–445 (2005). doi:10.1086/426097, arXiv:astro-ph/0308029

Umeda, H., Nomoto, K.: How much $^{56}$Ni can be produced in core-collapse supernovae? Evolution and Explosions of 30–100 $M_{solar}$ stars. ApJ **673**, 1014–1022 (2008). doi:10.1086/524767, 0707.2598

Vanbeveren, D., Van Bever, J., Belkus, H.: The Wolf-Rayet population predicted by massive single star and massive binary evolution. ApJ **662**, L107–L110 (2007). doi:10.1086/519454, arXiv:astro-ph/0703796

Vink, J.: Revealing the obscured supernova Remnant Kesteven 32 with Chandra. ApJ **604**, 693–699 (2004). doi:10.1086/381930, arXiv:astro-ph/0312328

Vink, J.: Gamma-ray observations of explosive nucleosynthesis products. Adv. Space Res. **35**, 976–986 (2005). doi:10.1016/j.asr.2005.01.097, arXiv:astro-ph/0501645

Vink, J., Laming, J.M., Kaastra, J.S., Bleeker, J.A.M., Bloemen, H., Oberlack, U.: Detection of the 67.9 and 78.4 keV lines associated with the radioactive decay of $^{44}$Ti in Cassiopeia A. ApJ **560**, L79–L82 (2001). doi:10.1086/324172, arXiv:astro-ph/0107468

Walder, R., Burrows, A., Ott, C.D., Livne, E., Lichtenstadt, I., Jarrah, M.: Anisotropies in the neutrino fluxes and heating profiles in two-dimensional, time-dependent, multigroup radiation hydrodynamics simulations of rotating core-collapse supernovae. ApJ **626**, 317–332 (2005). doi:10.1086/429816, arXiv:astro-ph/0412187

Wallace, R.K., Woosley, S.E.: Explosive hydrogen burning. ApJS **45**, 389–420 (1981). doi:10.1086/190717

Wanajo, S.: The rp-process in neutrino-driven winds. ApJ **647**, 1323–1340 (2006). doi:10.1086/505483, arXiv:astro-ph/0602488

Wanajo, S.: Cold r-process in neutrino-driven winds. ApJ **666**, L77–L80 (2007). doi:10.1086/521724, 0706.4360

Wanajo, S., Kajino, T., Mathews, G.J., Otsuki, K.: The r-process in neutrino-driven winds from Nascent, "Compact" neutron stars of core-collapse supernovae. ApJ **554**, 578–586 (2001). doi:10.1086/321339, arXiv:astro-ph/0102261

Wanajo, S., Goriely, S., Samyn, M., Itoh, N.: The r-process in supernovae: impact of new microscopic mass formulae. ApJ **606**, 1057–1069 (2004). doi:10.1086/383140, arXiv:astro-ph/0401412

Wang, W., Harris, M.J., Diehl, R., Halloin, H., Cordier, B., Strong, A.W., Kretschmer, K., Knödlseder, J., Jean, P., Lichti, G.G., Roques, J.P., Schanne, S., von Kienlin, A., Weidenspointner, G., Wunderer, C.: SPI observations of the diffuse $^{60}$Fe emission in the Galaxy. A&A **469**, 1005–1012 (2007). doi:10.1051/0004-6361:20066982, 0704.3895

Wasserburg, G.J., Busso, M., Gallino, R.: Abundances of actinides and short-lived nonactinides in the interstellar medium: diverse supernova sources for the r-processes. ApJ **466**, L109 (1996). doi:10.1086/310177

Weaver, T.A., Woosley, S.E.: Evolution and explosion of massive stars. In: Ehlers, J., Perry, J.J., Walker, M. (eds.) Ninth Texas Symposium on Relativistic Astrophysics, New York Academy Sciences Annals, vol. 336, pp. 335–357 (1980). doi:10.1111/j.1749-6632.1980.tb15942.x

Weaver, T.A., Woosley, S.E.: Nucleosynthesis in massive stars and the $^{12}C(\alpha, \gamma)$ $^{16}O$ reaction rate. Phys. Rep. **227**, 65–96 (1993). doi:10.1016/0370-1573(93)90058-L

Weber, C., Elomaa, V., Ferrer, R., Fröhlich, C., Ackermann, D., Äystö, J., Audi, G., Batist, L., Blaum, K., Block, M., Chaudhuri, A., Dworschak, M., Eliseev, S., Eronen, T., Hager, U., Hakala, J., Herfurth, F., Heßberger, F.P., Hofmann, S., Jokinen, A., Kankainen, A., Kluge, H., Langanke, K., Martín, A., Martínez-Pinedo, G., Mazzocco, M., Moore, I.D., Neumayr, J.B., Novikov, Y.N., Penttilä, H., Plaß, W.R., Popov, A.V., Rahaman, S., Rauscher, T., Rauth, C., Rissanen, J., Rodríguez D, Saastamoinen, A., Scheidenberger, C., Schweikhard, L., Seliverstov, D.M., Sonoda, T., Thielemann, F., Thirolf, P.G., Vorobjev, G.K.: Mass measurements in the vicinity of the rp-process and the np-process paths with the Penning trap facilities JYFLTRAP and SHIPTRAP. Phys. Rev. C **78**(5), 054,310 (2008). doi:10.1103/PhysRevC.78.054310, 0808.4065

Wiescher, M., Görres, J., Pignatari, M.: Experimental status of reactions in hand He-burning. Annu. Rev. Nucl. Part. Sci. **60**, 175–251 (2010). doi:10.1146/annurev.nucl.60.1.175

Wilson, J.R., Mayle, R.W.: Report on the progress of supernova research by the Livermore group. Phys. Rep. **227**, 97–111 (1993). doi:10.1016/0370-1573(93)90059-M

Wolfenstein, L.: Neutrino oscillations in matter. Phys. Rev. D **17**, 2369–2374 (1978). doi:10.1103/PhysRevD.17.2369

Woosley, S.E., Bloom, J.S.: The supernova gamma-ray burst connection. ARA&A **44**, 507–556 (2006). doi:10.1146/annurev.astro.43.072103.150558, arXiv:astro-ph/0609142

Woosley, S.E., Heger, A.: Nucleosynthesis and remnants in massive stars of solar metallicity. Phys. Rep. **442**, 269–283 (2007). doi:10.1016/j.physrep.2007.02.009, arXiv:astro-ph/0702176

Woosley, S.E., Howard, W.M.: The p-process in supernovae. ApJS **36**, 285–304 (1978). doi:10.1086/190501

Woosley, S.E., Weaver, T.A.: The evolution and explosion of massive stars. II. Explosive hydrodynamics and nucleosynthesis. ApJS **101**, 181 (1995). doi:10.1086/192237

Woosley, S.E., Wilson, J.R., Mathews, G.J., Hoffman, R.D., Meyer, B.S.: The r-process and neutrino heated supernova ejecta. ApJ **433**, 229–246 (1994). doi:10.1086/174638

Woosley, S.E., Heger, A., Weaver, T.A.: The evolution and explosion of massive stars. Rev. Mod. Phys. **74**, 1015–1071 (2002). doi:10.1103/RevModPhys.74.1015

Woosley, S.E., Kasen, D., Blinnikov, S., Sorokina, E.: Type Ia supernova light curves. ApJ **662**, 487–503 (2007). doi:10.1086/513732, arXiv:astro-ph/0609562

Yalçın, C., Güray, R.T., Özkan, N., Kutlu, S., Gyürky, G., Farkas, J., Kiss, G.G., Fülöp, Z., Simon, A., Somorjai, E., Rauscher, T.: Odd p isotope In113: measurement of α-induced reactions. Phys. Rev. C **79**(6), 065,801 (2009). doi:10.1103/PhysRevC.79.065801, 0906.4041

Yoshida, T., Terasawa, M., Kajino, T., Sumiyoshi, K.: Nucleosynthesis of light elements and heavy r-process elements through the n process in supernova explosions. ApJ **600**, 204–213 (2004). doi:10.1086/379766, arXiv:astro-ph/0305555

# Chapter 5
# Binary Systems and Their Nuclear Explosions

**J. Isern, M. Hernanz, and J. José**

## 5.1 Accretion onto Compact Objects and Thermonuclear Runaways

The nuclear energy supply of a typical star like the Sun would be $\sim 10^{52}$ erg if all the hydrogen could be incinerated into iron peak elements. Since the gravitational binding energy is $\sim 10^{49}$ erg, it is evident that the nuclear energy content is more than enough to blow up the Sun. However, stars are stable thanks to the fact that their matter obeys the equation of state of a classical ideal gas that acts as a thermostat: if some energy is released as a consequence of a thermal fluctuation, the gas expands, the temperature drops and the instability is quenched. The first researchers to discuss the scenario under which stars could explosively release their nuclear energy were Hoyle and Fowler (1960). They showed that this could occur under conditions of dynamic compression, as a consequence of collapse, or under electron degeneracy. They also pointed out in their seminal paper that hydrogen could only be responsible for mild explosions, like novae, as a consequence of the necessity to convert two protons into two neutrons, and that only the thermonuclear fusion of carbon could be energetic enough to feed a strong explosion. They did not consider helium because by this epoch the He-burning mechanism was not yet known.

Intermediate and low-mass stars ($M < 10$–$12\ M_{\odot}$) are able to get rid of their envelope and end their life as a white dwarf. On the contrary, massive stars form an iron core that grows until it reaches the Chandrasekhar mass and collapses to a neutron star or a black hole. The degenerate core of white dwarfs can have different

J. Isern (✉)
Institut de Ciències de l'Espai (CSIC) and Institut d'Estudis Espacials de Catalunya, Barcelona, Spain, isern@ieec.cat

M. Hernanz
Institut de Ciències de l'Espai (CSIC) and Institut d'Estudis Espacials de Catalunya, Barcelona, Spain

J. José
Universitat Politècnica de Catalunya and Institut d'Estudis Espacials de Catalunya, Barcelona, Spain

Isern, J. et al.: *Binary Systems and Their Nuclear Explosions*. Lect. Notes Phys. **812**, 233–305 (2011)
DOI 10.1007/978-3-642-12698-7_5 

chemical compositions, He, C—O or O—Ne, and different size depending on the mass and the single or binary nature of the progenitor. Single stars with masses in the range of 8–9 to 10–12 $M_\odot$ produce oxygen-neon cores, those in the mass range 0.5 to 8–9 $M_\odot$ produce carbon-oxygen cores, while stars with a mass in the range 0.08 to 0.5 $M_\odot$ produce helium cores, but the lifetime of such stars is so large that they cannot produce a white dwarf in a Hubble time. Members of a close binary system can be strongly perturbed by their companion and thereby produce different outcomes. For instance, stars with a mass of the order of 2.5 $M_\odot$ can end their life as He white dwarfs with a mass of the order of 0.4 $M_\odot$.

The destiny of isolated white dwarfs is to cool forever. However, if they are members of a close binary system, they can revive as a consequence of mass transfer from their companion. As the mass grows, the radius of the white dwarf shrinks and the density increases, as can be derived from simple dimensional arguments. The hydrostatic- and the degenerate, non-relativistic electron pressures have the functional form $P \sim M^2 R^{-4}$, and $P \sim M^{5/3} R^{-5}$, respectively. Thus it is always possible to find an equilibrium configuration defined by a mass – radius relation $R \sim M^{-1/3}$. However, as the density grows, the Fermi energy increases and electrons become relativistic. In the extreme case, the electron pressure takes the form $P \sim M^{4/3} R^{-4}$ and, since it has the same dependence on $R$ as hydrostatic pressure, there is no longer a definite lengthscale. Furthermore, according to the virial theorem, stars supported by relativistic particles are not gravitationally bound and the injection or removal of small amounts of energy can cause large expansion or contraction of the star.

The behavior of the different cores depends on the net rate at which energy is injected by the burning front or removed by electron captures on the ashes that were left from the previous burning cycle. Both quantities depend on the chemical composition of the stellar cores. Helium cores always experience a thermonuclear explosion because of the large energy content and the extreme flammability of He. Carbon-oxygen cores can explode or collapse, depending on the ignition density (Canal et al., 1990). If this density is larger than some critical value, $\sim (5 - 8) \times 10^9$ g/cm$^3$, the electron captures become dominant and they collapse to a neutron star (Bravo and García-Senz, 1999). ONe – cores ignite at such a density that they always tend to collapse (Nomoto and Kondo, 1991; Gutiérrez et al., 1996) and Fe – cores always collapse because of their inability to release nuclear energy.

### *5.1.1 Evolution of Degenerate Cores Before Ignition*

The behavior of the white dwarf interior during the accretion phase depends on the competition between the physical processes that increase the temperature of the material (compression, nuclear reactions in the inner core and possible burning of the freshly accreted matter) and those that cool the star (via neutrino and photon losses). Since the energy transport is dominated by electron conduction, one of the relevant timescales is the time taken by a thermal signal to cross the star, given by Henyey and L'Ecuyer (1969):

$$\tau_{\rm TH} = \frac{3\kappa\rho^2 c_{\rm P}}{64\sigma T^3} l^2 \tag{5.1}$$

where $\kappa$, $\rho$, T, $\sigma$ and $c_{\rm P}$ have their usual meanings and $l$ is the linear extent of the region considered, the radius of the white dwarf in this case.

The effects of the compression induced by the accreted mass can be separated into two terms (Nomoto, 1982). The first term is due to the increase in density at a fixed mass fraction as a consequence of the increase in mass, and its effects are quite uniform throughout the whole star. The second term corresponds to compression as matter moves inward in mass fraction space. It is negligible in the inner, strongly degenerate regions, where the major part of the compression work is invested in increasing the Fermi energy of electrons, but is very large in the external semi-degenerate layers. This means that a thermal wave generates in the outer layers and propagates towards the interior. A rough estimate of the compression-induced luminosity is: $L_{\rm c}/L_{\odot} = 1.4 \times 10^{-3} T_7 \dot{M}_{10}$, where $T_7$ is the temperature in units of $10^7$ K, and $\dot{M}_{10}$ is the mass accretion rate in units of $10^{-10}$ M$_{\odot}$/year.

The effects of this thermal wave on the physical state of the white dwarf interior depend on the time this wave takes to reach the center of the star, $\tau_{\rm TH}$, as compared with the time required for the star to reach the Chandrasekhar mass, $\tau_{\rm comp}$ (Hernanz et al., 1988). For low accretion rates, $\dot{M} \leq 3 \times 10^{-10}$ M$_{\odot}$/year, the thermal wave has time to reach the center, but normal cooling through the photosphere is dominant and the white dwarf evolves nearly isothermally with a temperature determined by the balance between compression and cooling, with the contribution of neutrinos and nuclear reactions. For high accretion rates, $5 \times 10^{-8} \leq \dot{M} \leq 10^{-6}$ M$_{\odot}$/year, compression heating dominates but, if the mass of the white dwarf is large enough, the thermal wave has no time to reach the center and, since there $\tau_{\rm TH} >> \tau_{\rm comp}$, these layers evolve with an adiabatic index:

$$\Gamma_3 - 1 = \frac{0.815 + 0.251\Gamma^{1/4}}{0.945 + 0.646\Gamma^{1/4}} \tag{5.2}$$

where $\Gamma$ is the Coulomb coupling constant. For typical values of $\Gamma \sim 100 - 200$, this index is $\sim 0.5$ and degenerate matter is heated gently. For intermediate accretion rates, the thermal wave has sufficient time to arrive at the central layers and they experience a sudden heating, followed by an evolution in the $\rho - $T diagram governed by the balance between the heating and cooling agents already mentioned (see Fig. 5.1).

As the temperature increases, fusion reactions start to become important. At low temperatures, neutrino emission is able to control them, but due to the different temperature dependences the energy production by nuclear reactions overwhelms neutrino losses and matter burning becomes unstable. This critical temperature, commonly called ignition temperature, $T_{\rm ig}$, is defined as $\varepsilon_{CC}(T_{\rm ig}) = \varepsilon_{\nu}(T_{ig})$. If ignition happens under degenerate conditions, a thermonuclear runaway occurs. The nature of this instability can be understood with the following argument. Assume that $P = P_{\rm e}(\rho) + P_{\rm i}(\rho, T)$, where $P_{\rm e}$ and $P_{\rm i}$ are the electron and ion pressure,

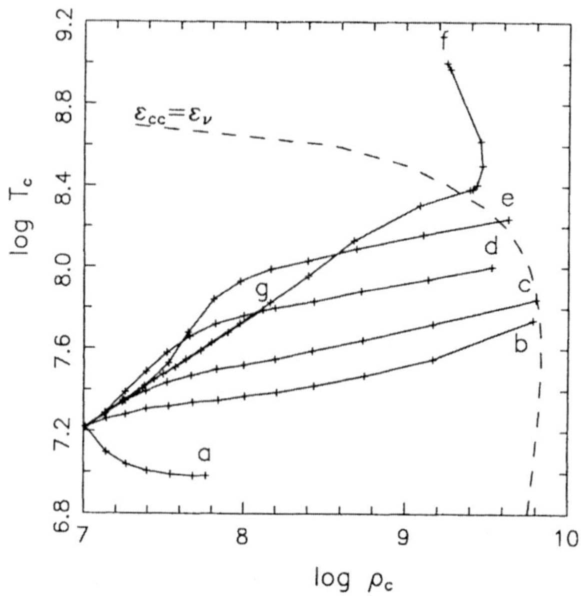


**Fig. 5.1** Evolution of the center of a C–O white dwarf in the $\log T - \log \rho$ plane for several accretion rates of pure C–O: (*a*) $10^{-10}\ M_\odot$/year; (*b*) $5 \times 10^{-10}\ M_\odot$/year; (*c*) $10^{-9}\ M_\odot$/year; (*d*) $5 \times 10^{-9}\ M_\odot$/year; (*e*) $5 \times 10^{-8}\ M_\odot$/year; (*f*) $5 \times 10^{-7}\ M_\odot$/year; (*g*) $5 \times 10^{-6}\ M_\odot$/year (Bravo et al., 1996). The *dashed line* represents the $^{12}$C ignition curve

respectively, and that ions behave as an ideal gas. Also assume that nuclear reactions release isochorically an amount of energy $\delta q$, and that matter expands adiabatically until pressure equilibrium is reached. The corresponding density change is:

$$\frac{\delta\rho}{\rho} = -\frac{2}{3\Gamma_1}\frac{P_\mathrm{i}}{P_\mathrm{e} + P_\mathrm{i}}\frac{Q}{kT}\frac{\delta N}{N} \tag{5.3}$$

where $\Gamma_1$ is the first adiabatic index, $Q$ is the energy released per fused nucleus and $\delta N$ is the number of nuclei that have fused. Since $Q \sim 1$ MeV and $kT \sim 1$ keV, if the ideal gas is dominant a small energy release will cause a large expansion with an associated cooling. On the contrary, if $P_\mathrm{e} >> P_\mathrm{i}$ adiabatic cooling is not efficient and matter will heat until $P_\mathrm{e} \sim P_\mathrm{i}$. At this point, if $\tau_\mathrm{nuc} << \tau_\mathrm{HD}$, nuclear reactions will continue until incineration of matter is complete. Here $\tau_\mathrm{nuc}^{-1} = d\ln(\varepsilon_\mathrm{nuc})/dt$, and $\tau_\mathrm{HD} = l/c_\mathrm{s}$ is the hydrodynamic time, $l$ the dimension of the burning region and $c_\mathrm{s}$ the sound velocity. Under hydrostatic equilibrium, $\tau_\mathrm{HD} \sim \tau_\mathrm{ff}$, where $\tau_\mathrm{ff} = (24\pi G\rho)^{-1/2} \sim 444\rho^{-1/2}$ is the free-fall time.

It is important to realize here that in the case of H-burning two protons have to be converted into two neutrons and that $\beta$-decays will control the total rate. At high temperatures, the longest $\beta$-decay timescale is that of $^{15}$O, with a mean lifetime of $\tau_{15O} = 178$ s (see Fig. 5.2) and the maximum energy production rate is:

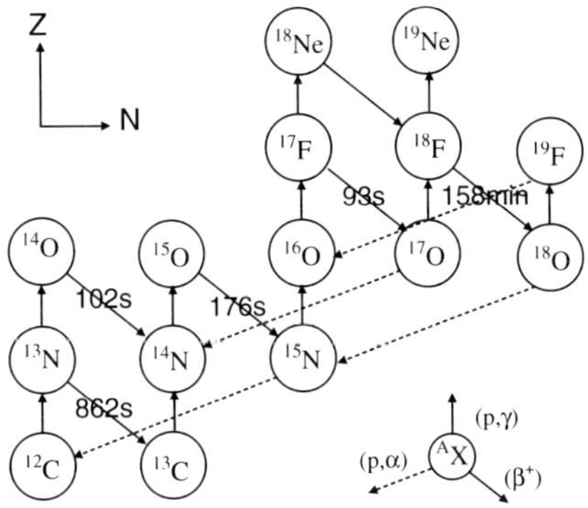


**Fig. 5.2** Scheme of the carbon-nitrogen-oxygen (CNO) cycle of hydrogen burning, which operates out of equilibrium in stellar explosions. The lifetimes of the $\beta^+$-unstable nuclei, which act as *bottlenecks* of the cycle, are displayed

$$\varepsilon_{CNO} \leq 1.3 \times 10^{14} \frac{X_{CNO}}{0.01} \text{ erg/g/s} \tag{5.4}$$

for an assumed energy release of 28 MeV per reaction (Mazurek and Wheeler, 1980). Therefore, complete burning cannot be achieved in a short time in comparison to the hydrodynamic time and H-driven explosions cannot involve all of the star.

### *5.1.2 The Thermonuclear Runaway*

When the central regions cross the ignition line, the temperature starts to rise and nuclear reactions accelerate. Conduction is rapidly overwhelmed by energy production and a convective core forms. This core grows very quickly as a consequence of the energy release enhancement and cannot prevent the continuous rise of the temperature. When the turnover timescale of the convective eddies is longer than the heating timescale, one or several bubbles enter into the dynamical regime (Nomoto et al., 1984; García-Senz and Bravo, 2005; Woosley et al., 2004), a thermonuclear runaway occurs, and a flame begins to propagate (Timmes and Woosley, 1992).

The igniting zone can be imagined as a highly turbulent region where the evolution of turbulent elements towards the thermonuclear runaway is governed by the balance between heating by nuclear burning and collision of turbulent eddies, and cooling by electron conduction and expansion pdV-work. In principle it is possible to assume a distribution of fluctuations characterized by their size, $\delta$, and their temperature excess, $\Delta T$. These fluctuations will be able to grow only if the conductive cooling is not able to evacuate the nuclear energy generated at the center of the bubble. Consequently their size has to be larger than:

$$\delta = \sqrt{\frac{2\sigma\Delta T}{\rho\varepsilon_{\rm nuc}}} \tag{5.5}$$

where $\sigma$ is the thermal conductivity and $\varepsilon_{\rm nuc}$ is the nuclear energy generation rate. For background temperatures in the range $(6-8)\times 10^8$ K, fluctuations must have a minimum size of 4 m–30 cm, respectively, to be able to grow. When this condition is satisfied, the temperature increases, the burning propagates by conduction (see next section) and the buoyancy accelerates the bubble to a substantial fraction of the sound speed (García-Senz and Bravo, 2005). During this time other bubbles can develop similar runaways, grow and float away when they reach a critical size of $\sim$1 km, such that the final outcome is an asynchronous ignition at multiple points.

The runaway of hydrogen is responsible for nova explosions. The mechanism for such explosions can be better understood after evaluating some relevant timescales (Starrfield, 1989): the accretion timescale, defined as $\tau_{\rm acc} \sim M_{\rm acc}/\dot{M}$ (which is of the order of $10^4$–$10^5$ year, depending on the accretion rate $\dot{M}$ and accreted mass $M_{\rm acc}$), the nuclear timescale $\tau_{\rm nuc} \sim c_{\rm p}T/\varepsilon_{\rm nuc}$ (which is as small as a few seconds at peak burning) and the hydrodynamic timescale ($\tau_{\rm HD} \sim H_{\rm p}/c_{\rm s} \sim (1/g)\sqrt{P/\rho}$; $H_{\rm p}$ is the pressure scale height). During the accretion phase, $\tau_{\rm acc} \leq \tau_{\rm nuc}$, accretion proceeds and the envelope mass increases. When degenerate ignition conditions are reached, degeneracy prevents envelope expansion and the thermonuclear runaway occurs. As temperature increases, the sudden release of energy would lift degeneracy in the envelope and ultimately halt the thermonuclear runaway, but this is not the case because $\tau_{\rm nuc} \ll \tau_{\rm HD}$. Therefore, the value of the nuclear timescale is crucial for the development of the thermonuclear runaway (TNR) and its final fate. In fact there are two main types of nuclear timescales: those related to $\beta^+$-decays, $\tau_{\beta+}$, and those related to proton capture reactions, $\tau_{(\rm p,\gamma)}$. In the early evolution towards the TNR, $\tau_{\beta+} < \tau_{(\rm p,\gamma)}$ and the CNO cycle operates at equilibrium. But as temperature increases up to $\sim 10^8$ K, the reverse situation occurs, ($\tau_{\beta+} > \tau_{(\rm p,\gamma)}$), and thus the CNO cycle is $\beta$-limited (see Fig. 5.2). In addition, since the large energetic output produced by nuclear reactions can not be evacuated by radiation only, convection sets in and transports the $\beta^+$-unstable nuclei to the outer cooler regions where they are preserved from destruction and where they will decay later on ($\tau_{\rm conv} < \tau_{\beta+}$), leading to envelope expansion, luminosity increase and mass ejection if the attained velocities are larger than escape velocity.

### *5.1.3 Physics of the Burning Front*

An explosion is the mechanical disruption of a system as a consequence of a rapid release of energy. In the case of an exploding white dwarf, enough mass, $\sim$0.3 $M_\odot$, has to be quickly embraced by the burning region to unbind the star. This can be accomplished either through detonation (Arnett, 1969) or through deflagration (Nomoto et al., 1976). A detonation is shock-induced burning propagating supersonically into an unburned medium, while a deflagration is a burning front that propagates by thermal conduction at subsonic velocities. Both, detonation and deflagration, are driven by a physical mechanism. However, there is a third possibility: spontaneous burning. This case occurs when the ignition conditions are reached nearly simultaneously in several points in such a way that burning spreads over a large region without any transport mechanism (Blinnikov and Khokhlov, 1986; Woosley and Weaver, 1986). The propagation velocity, a phase velocity in fact, can be estimated as $v_{sb} = (d\tau_{nuc}/dr)^{-1}$, where $\tau_{nuc}$ plays a critical role at the onset of burning. This velocity increases when the absolute values of the temperature and density gradients decrease. Thus, regions with $v_{sb} \geq c_s$ ignite spontaneously and the burning front propagates supersonically. Because of the strong dependence on $T$, the most important factor is the temperature profile.

In order to describe the properties of the burning front, either supersonic or subsonic, it is usually assumed (Landau and Lifshitz, 1959) that the unburned material is separated from the combustion products by a region of width $\delta$ where reactions take place. If $\delta << l$, where $l$ is the typical scale length of the system, it is possible to connect both sides of the front by means of conservation laws of mass, momentum and energy. In the frame associated with the front, these equations, known as the Rankine-Hugoniot jump conditions, can be written as (Landau and Lifshitz, 1959; Mazurek and Wheeler, 1980):

$$\rho_1 u_1 = \rho_0 u_0 \tag{5.6}$$

$$P_1 + \rho_1 u_1^2 = P_0 + \rho_0 u_0^2 \tag{5.7}$$

$$\varepsilon_1 + \frac{P_1}{\rho_1} + \frac{u_1^2}{2} = \varepsilon_0 + \frac{P_0}{\rho_0} + \frac{u_0^2}{2} + q \tag{5.8}$$

that are similar to those describing shock waves except for the presence of the term $q$ that represents the amount of energy released by reactions. The subscripts 0 and 1 denote fuel and ashes, respectively, $u$ is the matter velocity, $\varepsilon$ is the specific internal energy, and the remaining symbols have their usual meaning. The mass flux crossing the front is given by:

$$j = \rho_0 u_0 = \rho_1 u_1 \tag{5.9}$$

which can be written, using the mass and momentum conservation equations (Eqs. (5.6) and (5.7) respectively), as:

$$j^2 = -\frac{P_0 - P_1}{V_0 - V_1} \tag{5.10}$$

where $V = 1/\rho$, is the specific volume. The mass flux (and the velocity of the front with respect to the unburned material) is determined by the ratio between the difference of pressures and specific volumes at both sides of the burning front. Therefore, real solutions must satisfy: ($P_1 > P_0$, $V_1 < V_0$) or ($P_1 < P_0$, $V_1 > V_0$). The first solution corresponds to a detonation and the second one to a deflagration (Landau and Lifshitz, 1959).

The velocity at which a detonation propagates can be obtained from the energy conservation equation. Equation (5.8) can be written as:

$$\varepsilon_0 + q - \varepsilon_1 + \frac{1}{2}(P_0 + P_1)(V_0 - V_1) = 0 \qquad (5.11)$$

which is called the detonation adiabat (the case $q = 0$ is called the shock adiabat). The final state is obtained equating 5.10 and 5.11, once the properties of the front have been specified (see Fig. 5.3). The physical meaning of this solution is clear. A shock heats and compresses the material to a state ($P_{sh}$, $V_{sh}$) given by the intersection of Eq. (5.10) with the shock adiabat. Because of the temperature increment, material burns and reaches the state ($P_1$, $V_1$), defined by the intersection of Eq. (5.10) with the detonation adiabat. Since $q \geq 0$, then $P_1 < P_{sh}$, $V_1 > V_{sh}$ and the post – shock burning produces a rarefaction.

The family of solutions obtained from Eqs. (5.10) to (5.11) and $j$ as a free parameter has an extremum for which $j$ and the front velocity are minima. This solution, called the Chapman-Jouguet detonation, corresponds to the case where the Eq (5.10) is tangent to (5.11). This extremal solution has the following properties: (i) it is only determined by the thermodynamic properties of the material, including $q$, (ii) the entropy is maximum and (iii) the velocity with respect to the unburned material is minimal and equal to the sound velocity of the material behind the front. All the remaining solutions, called strong detonations, move supersonically with respect to the burned material and subsonically with respect to the unburned material. Therefore, if a detonation starts at the center of the white dwarf, all the material, from the center to very near the surface, will be incinerated to $^{56}$Ni.

Strong detonations are not allowed in stars. Since material must be at rest at the centre, the velocity has to decrease from a positive value behind the front to zero at

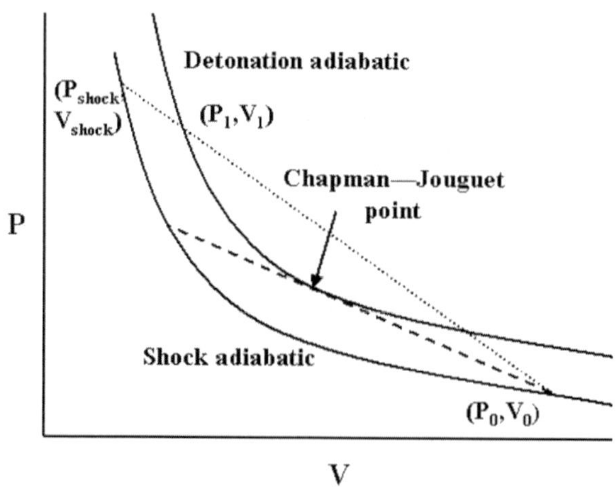


**Fig. 5.3** Adiabatic of a shock and a detonation in the P – V diagram. The dotted and dashed lines represent the condition of conservation of mass and momentum ($-j^2 =$ constant)

the centre. This means that a rarefaction wave has to follow the detonation. Since the velocity of a rarefaction wave is equal to the sound velocity of the material, it is necessary that the front moves at least with the sound velocity with respect to the burned material in order to not be overtaken by the rarefaction wave. Thus, due to the boundary conditions, the only acceptable detonations in stars are those of Chapman-Jouguet type.

In the case of deflagration solutions matter is subsonic on both sides of the front. Thus, any perturbation behind or ahead the front can affect it. As an example, consider a spherically symmetric burning front propagating outwards with a velocity $D$ and the unburned matter at rest. From Eqs. (5.6) and (5.7) it is possible to write:

$$V_0(P_1 - P_0) = v_1 D \tag{5.12}$$

where $v_1$ is the velocity measured in the frame fixed to the center of the star. Since $P_1 - P_0 < 0$ in deflagrations, and $D > 0$, the velocity of the burned matter must be negative, $v_1 < 0$, in contradiction to the boundary condition that requires matter to be at rest at the center. Thus a deflagration can only exist if it generates a shock precursor that boosts matter outwards (Mazurek and Wheeler, 1980).

The speed of the laminar flame can be estimated as follows (Landau and Lifshitz, 1959): the velocity of the burning front is $D \sim \delta/\tau_{\rm burn}$, where $\delta$ is the width of the front, and $\tau_{\rm burn}$ is the lasting time of the burning ($\tau_{\rm burn} \sim \varepsilon/\varepsilon_{\rm nuc}$, where $\varepsilon_{\rm nuc}$ has to be evaluated at the critical temperature). Since a stationary flame can only exist if $\tau_{\rm burn} \sim \tau_{\rm diff}$, where the diffusion time is given by $\tau_{\rm diff} \sim \delta^2/\chi$ and $\chi$ is the thermometric conductivity, the width of the front has to be

$$\delta \sim \sqrt{\frac{\varepsilon\chi}{\varepsilon_{\rm nuc}}} \tag{5.13}$$

and the laminar velocity

$$D \sim \sqrt{\frac{\varepsilon_{\rm nuc}\chi}{\varepsilon}} \tag{5.14}$$

In the case when a white dwarf is near the Chandrasekhar limit, $\delta \sim 10^{-4}$ cm, $D \sim 10^7$ cm/s $\sim 10^{-2} c_s$ and the density contrast between burned and unburned matter is $\Delta\rho/\rho \sim 0.2$. These values relax to 1 cm, $10^4$ cm/s and 0.5 respectively when $\rho \sim 10^7$ g/cm$^3$.

As a consequence of turbulence induced by instabilities like the Rayleigh-Taylor buoyancy-driven or the Kelvin-Helmholtz shear-driven instability, for instance, the flame surface is wrinkled and stretched in such a way that, despite the flame continuing to propagate at the laminar velocity, the effective burning rate is strongly enhanced. Buoyancy induces the formation of burning bubbles that rise into the fuel and generate turbulent motions. The turbulent motions decay downward to the smaller Kolmogorov scale and the eddies of this cascade interact with the flame, further wrinkling and stretching its surface, thereby further enhancing the burning rate. This effect acts down to the Gibson scale, defined as the size of the eddy that can turn over in a nuclear burning time. Below this scale, the laminar velocity is larger than the turbulent velocity and fuel is burned before the eddies are able to change the shape of the flame. If the Gibson length is large as compared with the width of the flame, the internal structure of the flame is not altered (the flamelet regime). In the opposite case, the turbulent motion is able to modify the internal structure of the flame and burning enters the so called distributed regime.

## *5.1.4 Scenarios Leading to a Thermonuclear Runaway*

Possible scenarios leading to a thermonuclear runaway can be classified according to the chemical composition of the donor (H, He,C+O, O+Ne) and the nature of the accretor, a white dwarf made of He, C+O, or O+Ne, or a neutron star. Some of the possible combinations are very rare, if not forbidden, and have not yet been associated with any observed astronomical event. Within the category of accreting white dwarfs it is possible to adopt the following scenarios (Iben and Tutukov, 1985; Nomoto, 1982; Webbink, 1984; Whelan and Iben, 1973):

### 5.1.4.1 Hydrogen Accretion

There are many astronomical objects containing a white dwarf that accretes hydrogen rich matter from a non-degenerate companion and that could suffer a thermonuclear runaway. The nature and the intensity of this instability depend on the accretion rate and the mass of the object. Typical examples are cataclysmic variables, classical novae, recurrent novae, symbiotic stars and supersoft X-ray sources.

If the accretion rate is smaller than $\sim\ 10^{-8-9}$ $M_{\odot}$/year, hydrogen accumulates on the surface of the white dwarf and becomes degenerate. When the accumulated mass reaches a critical value, $\Delta M_{\rm H} \sim 10^{-4} - 10^{-5}$ $M_{\odot}$, the exact values depending on the properties of the binary system, it experiences a strong flash that can be identified with the nova phenomenon (see Sect. 5.2). This flash expels almost all the accreted mass or even erodes the mass of the accreting object, for which reason the white dwarf is unable to reach the Chandrasekhar mass, except in the case when $M_{WD} > 1.3\,M_{\odot}$. However, since the chemical composition of such white dwarfs is a mixture of oxygen and neon, the fate of such scenario is collapse to a neutron star.

For intermediate rates, $10^{-8-9} \leq \dot{M}_{\rm H}(M_{\odot}/\text{year}) \leq 5 \times 10^{-7}$, hydrogen burns steadily or through mild flashes, and helium accumulates on the surface of the star. If the accretion rate is high enough, this helium is converted into carbon and oxygen through weak flashes or steady burning and the white dwarf approaches the Chandrasekhar mass. But, if the effective accretion rate of helium is in the range $10^{-9} \leq \dot{M}_{\rm H}(M_{\odot}/\text{year}) \leq 5 \times 10^{-8}$, the helium layer becomes degenerate and when it reaches a critical mass, $\Delta M_{\rm He} \sim 0.3\ M_{\odot}$, it ignites under degenerate conditions and experiences a thermonuclear runaway that can trigger the explosive destruction of the complete star. This scenario has been proposed for type Ia supernova progenitors (see Sect. 5.3.2).

If the accretion rate is larger than $\sim 5 \times 10^{-7}$ M$_\odot$/year a red giant-like envelope forms, a strong wind appears and the mass accumulates over the degenerate core at a rate (Hachisu et al., 1999):

$$\dot{M}_{\rm cr} \simeq 5.3 \times 10^{-7} \frac{1.7 - X}{X} (M_{\rm WD}/M_\odot - 0.4)\ {\rm M}_\odot/{\rm year} \qquad (5.15)$$

where $X$ is the mass fraction of H in the accreted matter. As before, hydrogen and helium burn peacefully and the white dwarf has the possibility to reach the Chandrasekhar mass.

#### 5.1.4.2 Helium Accretion

There are at least two scenarios in which a white dwarf can directly accrete helium from the companion. One consists of two degenerate objects, the primary made of carbon-oxygen and a secondary composed of helium, that merge as a consequence of the emission of gravitational waves. Since the mass of the secondary is small $\sim$ 0.3–0.4 M$_\odot$, the process of merging is self-regulated. The second scenario consists of a C+O white dwarf plus a non-degenerate helium star and the mass transfer is powered by helium burning in the secondary. As mentioned above, if $10^{-9} \leq \dot{M}_{\rm H}({\rm M}_\odot/{\rm year}) \leq 5\times10^{-8}$, helium ignites at the base of the freshly accreted mantle under degenerate conditions and can trigger the thermonuclear explosion of the accreting white dwarf despite the fact that its mass is smaller than the Chandrasekhar limit (see Sect. 5.3.2).

#### 5.1.4.3 Carbon-oxygen Accretion

Close enough binary systems formed by two intermediate mass stars can experience two episodes of common envelope evolution that result in the formation of two C/O white dwarfs with a separation that is smaller than the initial one. The emission of gravitational waves induces a loss of angular momentum that translates into additional reduction of the orbit at a rate:

$$\dot{r} = -\frac{64 G^3 m_1 m_2 (m_1 + m_2)}{5 c^5 r^3}$$

where $G$ is the gravitational constant, $c$ is the speed of light and $r$ is the separation of both stars. If the separation of the two white dwarfs is smaller than $\sim 3R_\odot$, nothing can prevent their merging in less than a Hubble time and the primary will start accreting a mixture of carbon and oxygen.

During the merging process, the secondary is destroyed in a few orbital periods after filling its Roche lobe (Benz et al., 1990) and forms a hot and thick accretion disk around the primary. The impact is not able to induce prompt ignition (Guerrero et al., 2004) and the final outcome depends on the subsequent evolution of the disk. If the accretion rate is spherically symmetric and larger than about $\dot{M} \geq 2.7 \times 10^{-6}$ M$_\odot$/year, carbon ignites off-center, the flame propagates conductively inwards and the white dwarf is converted into an O—Ne white dwarf before central carbon ignition. Upon further accretion the white dwarf collapses to a neutron star (Nomoto and Kondo, 1991). Recent calculations (Yoon and Langer, 2005) indicate that, at least in some cases, neutrino cooling is able to quench off-center carbon ignition. An open question is the effective rate at which matter is accreted (Piersanti et al., 2003a,b) since it also contains angular momentum that prevents the contraction of the primary unless it is dissipated. Therefore, the interplay between disk and star is crucial for understanding the outcome of such a scenario.

Van Horn and Hansen (1974), Hansen and van Horn (1975) were the first to point out that nuclear burning on the surface of neutron stars can also be unstable. The regimes of unstable burning have been extensively discussed elsewhere (Fujimoto et al., 1981). To summarize, for a chemical mixture with Z(CNO)$\sim$ 0.01: mixed H/He-burning is expected for $\dot{M} < 2 \times 10^{-10}\,M_{\odot}$/year, triggered by thermally unstable H-ignition; pure He-shell ignition for $2 \times 10^{-10} < \dot{M}\ (M_{\odot}/\text{year}) < (4.4-11.1) \times 10^{-10}$, following completion of H-burning; and mixed H/He-burning for $\dot{M} > (4.4-11.1) \times 10^{-10}\,M_{\odot}$/year, triggered by thermally unstable He ignition. A reduction of the CNO content lowers the critical accretion rates and substantially narrows the range for pure He bursts.

## 5.2 Classical Novae

The origin of the term *nova* comes from the latin *nova stella*, meaning that a *new star* appeared in the sky. But it has been known for a long time that the new star is in fact not new, and that a nova is more properly defined as an existing star that suddenly increases its luminosity by more than $\sim$10 magnitudes, i.e., by a factor larger than $10^4$ – and then returns to its previous faint state in a few months, or even years. In fact, already Newton in the seventeenth and eighteenth century talked about *temporary stars* shining suddenly and then vanishing. It was not until the twentieth century that novae and supernovae were distinguished from each other, once the distances to the *nebulae* where they had been discovered were better known, and thus some extragalactic objects turned out to be novae with much larger intrinsic brightness (*super-novae*). An interesting and complete historical perspective of novae can be found in Duerbeck (2008).

The discovery of the binarity of classical novae was made by Walker (1954), who observed DQ Her (a nova that exploded in 1934) and deduced that it was an eclipsing binary with a very short period. Later, Kraft showed that this was a common property of novae and of cataclysmic variables in general (Kraft, 1964). It is now well known that nova explosions occur on white dwarfs accreting hydrogen-rich matter from a main sequence star companion, in a close binary system of the cataclysmic variable type. Accumulation of matter on the white dwarf leads to hydrogen ignition in degenerate conditions, which prevents the self-adjustment of the envelope through expansion. Therefore, a thermonuclear runaway ensues (see Sect. 5.1.2) and the final consequence is mass ejection at large velocities (hundreds to thousands of km $s^{-1}$)

and a large increase of luminosity, even reaching the Eddington luminosity of the white dwarf ($10^{37}$–$10^{38}$ erg/s).

In contrast to type Ia supernovae, which also occur on white dwarfs in binary systems, novae do not experience a complete disruption of the white dwarf, because the outburst only affects the external hydrogen-rich layers, i.e. $10^{-4}$–$10^{-5}$ $M_{\odot}$. Therefore, the nova phenomenon is expected to recur, with periods of a few tens or hundreds thousand years, which is the typical accretion time required to build-up again a critical H-rich envelope ready to explode. Mass transfer onto the white dwarfs in cataclysmic variables is a long-lasting phase, so that many nova explosions on a given white dwarf must occur. However, for historical reasons the term *recurrent nova* is reserved for another type of eruptive phenomena, those that have more than one *recorded* nova outburst. These systems also correspond to white dwarfs experiencing a thermonuclear runaway of their H-rich envelope, but the companion star is in general a red giant, instead of a main sequence star. The binary system is not a cataclysmic variable anymore; both its period -and the related binary separation- and the mass transfer rate onto the white dwarf are larger, thus allowing for a faster build-up of the critical mass and thus a shorter recurrence period (decades rather than thousands of years).

It is worth mentioning that the term *recurrent novae* is also applied to white dwarf explosions, with similar outburst properties and recurrence periods than *genuine* recurrent novae, but with a non thermonuclear origin. A completely different case are the so-called *dwarf novae*, which have much smaller outburst amplitudes and which are produced by accretion disk instabilities in cataclysmic variables. Here we are only concerned with novae from thermonuclear explosions, i.e., *classical novae* and the sub-class of *recurrent novae* with thermonuclear origin.

The long term evolution of the white dwarfs in classical and recurrent novae is debated, since it is not clear if the mass of the white dwarf grows towards the Chandrasekhar mass or decreases after each explosion. Observations of nova ejecta often show overabundances with respect to solar of elements such as carbon, oxygen, neon, among others, indicating that some mixing between the core and the accreted envelope occurs. Then, some core mass is in principle ejected indicating that the white dwarf mass might in fact decrease. However, in *recurrent novae* no large overabundances are observed. On the other hand, *recurrent novae* should take place on very massive white dwarfs, which only need a very small amount of added mass to explode; this combined with a larger accretion rate leads to the very short recurrence period observed. All in all, recurrent novae are one of the possible scenarios of type Ia supernova progenitors, although their internal composition (likely ONe instead of CO) presents a problem for this scenario.

### *5.2.1 Observational Properties*

Most of the galactic classical novae have been discovered optically by amateur astronomers. In addition, some robotic telescopes, mainly devoted to search for optical counterparts of GRBs or to perform surveys are also finding novae and supernovae. Around 5 novae per year are being discovered in recent years in our galaxy, and several novae have been found as well in external galaxies (see Shafter (2008) and references therein). However, most of the galactic novae suffer from large optical extinction (reddening) by interstellar dust, and hence the real nova rate is expected to be much larger; it should be determined from extrapolations, either from extragalactic or from galactic data. In the first case, the dependence of the nova rate on the type of galaxy has been derived, indicating that early-type galaxies are more prolific nova producers; the derived nova rate is 15–24/year to $27 \pm 8$/year (Della Valle and Livio, 1994; Shafter et al., 2000). Larger rates are obtained when galactic data are extrapolated, taking into account the amount and distribution of galactic dust: $35 \pm 11$/year or $41 \pm 20$/year (Hatano et al., 1997; Shafter, 1997).

From optical light curves of classical novae one finds an increase in luminosity corresponding to a decrease of $m_V$ (apparent visual magnitude) of more than 9 magnitudes occurring in just a few days, and a pre-maximum halt 2 magnitudes before maximum, in some cases (Warner, 1995, and references therein). Nova light curves are classified according to their speed class, defined from either $t_2$ or $t_3$, i.e., the time needed to decay by 2 or 3 visual magnitudes after maximum. Speed classes range from very fast ($t_2 < 10$ days) and fast ($t_2 \sim 11$–25 days) to very slow ($t_2 \sim 151$–250 days) (Payne Gaposchkin, 1957). Some examples are the fast nova N Cyg 1992, which had $t_2 \sim 12$ days, the even faster nova N Her 1991 ($t_2 \sim 2$ days), and the slow nova N Cas 1993, which had $t_2 \sim 100$ days. An empirical relationship between the absolute magnitude at maximum $M_V$ and the speed class of novae shows that brighter novae have shorter decay times ($t_2$ or $t_3$). The theoretical explanation of this relationship (Livio, 1992) relies on novae reaching a maximum luminosity close to the Eddington limit and ejecting roughly all their envelope in a period similar to $t_3$. It was established that $L_{\text{max}}$ is an increasing function of $M_{\text{wd}}$ and that $t_3$ is a decreasing function of $M_{\text{wd}}$. From these two relationships an expression relating $M_V$ at maximum and $t_3$ is deduced. This empirical relation, valid both in the $V$ and $B$ photometric bands, is very often used to determine distances to novae, once visual extinction is known. Different calibrations of the maximum magnitude-rate of decline relationship (MMRD) exist, with that from Della Valle and Livio (1995) being the most commonly employed form.

It is important to point out that there are two distinct nova populations: disk novae, which are in general fast and bright ($M_V(\text{max}) \simeq -8$), and bulge novae, slower and dimmer ($M_V(\text{max}) \simeq -7$). This was first suggested by Della Valle et al. (1992) and later corroborated by their early post-outburst spectra (Williams, 1992) and based on the stronger group of emission lines they display (either FeII lines or He and N lines); FeII-type novae evolve more slowly and have a lower level of ionization, whereas He/N-type novae have larger expansion velocities and a higher level of ionization. It has been deduced that the faster and brighter He/N novae are concentrated closer to the galactic plane than those of the slower and dimmer FeII type, which would preferentially belong to the bulge population (Della Valle and Livio, 1998; Della Valle, 2002).

The optical light curves were extended with space-based observations to energy ranges not observable from the ground. An important step forward was the discovery of the luminosity increase in the ultraviolet when the optical started to decline, thanks to the IUE satellite (International Ultraviolet Explorer); the reason is that the spectral energy distribution shifts to higher energies when deeper and thus hotter regions of the expanding envelope are revealed (the photosphere recedes as a consequence of the decreasing opacity). On the other end of the spectrum, infrared observations (especially for novae which form dust) indicate an increase in luminosity once the ultraviolet luminosity starts to decline, which is interpreted as the resulting re-radiation (in the infrared) by dust grains of the ultraviolet energy they absorbed. Therefore, during optical decline the bolometric luminosity of classical novae remains constant, during a period of time which depends on the mass of the H-rich envelope remaining on the white dwarf after the nova explosion, which is expected to burn steadily. Evidence for residual H-burning came from observations in the soft X-ray range with ROSAT (Krautter et al., 1996; Balman et al., 1998; Orio et al., 2001), which revealed the related very hot photosphere. The bolometric luminosity deduced from observations is close to or even larger than the Eddington luminosity, and thus radiation pressure is probably responsible for ejection of nova envelopes (Kato and Hachisu, 1994).

A very important result deduced from nova observations in all spectral wavelengths is that their ejecta are often enriched in carbon, nitrogen and oxygen – as well as neon in many objects (around 1/3 of the total); the global metallicities in nova ejecta are well above solar metallicites (see Gehrz et al., 1998 for a review). This observational fact is one of the main drivers of theoretical models, which should be able to explain it. These metallicity enhancements are not likely to be produced in the TNR, because the temperatures achieved in nova explosions are not high enough. An alternative and more widely accepted explanation is that there is some mixing between accreted matter, assumed to be of solar composition, and the underlying CO or ONe core (Starrfield et al., 1978b; Prialnik et al., 1978, 1979). In fact, such enrichment is also required to power the nova explosion itself except for very slow novae.

### *5.2.2 Modeling Classical Novae*

The scenario of classical nova explosions consists of a white dwarf (either CO or ONe) accreting hydrogen-rich matter in a cataclysmic binary system, as a result of Roche lobe overflow from its main sequence companion. For accretion rates low enough, e.g. $\dot{M} \sim 10^{-9}$–$10^{-10}$ $M_\odot$/year, accreted hydrogen is compressed to degenerate conditions until ignition occurs, thus leading to a thermonuclear runaway (TNR, see Sect. 5.1.2). Explosive hydrogen burning synthesizes some $\beta^+$-unstable nuclei of short lifetimes (e.g. $^{13}$N, $^{14}$O, $^{15}$O, $^{17}$F, with $\tau$ = 862, 102, 176, and 93s respectively, see Fig. 5.2) which are transported by convection to the outer envelope, where they are saved from destruction. These decays lead to a large energy release in the outer shells which causes the nova outburst, i.e. a visual luminosity increase accompanied by mass ejection with typical velocities $10^2$–$10^3$ km s$^{-1}$. Another important effect of convection is that it transports unburned material to the burning shell.

As mentioned above, mixing at the core-envelope interface is essential both to power the TNR and to explain observed enhancements in metals in many novae. Several mechanisms have been suggested to explain this process, operating either prior or during the thermonuclear runaway, but none of them is completely satisfactory (see extensive review in Livio, 1994). Diffusion induced convection, first discussed by Prialnik and Kovetz (1984) and Kovetz and Prialnik (1985), can explain moderate enrichments but has difficulties to account for some of the largest observed enrichments (Kovetz and Prialnik, 1997). Other possibilities are shear mixing, convection induced shear mixing, and convective overshooting induced flame propagation. Recent multidimensional simulations have not yet succeeded completely in reproducing the necessary mixing needed to power the explosion (Glasner and Livne, 2002; Alexakis et al., 2004). Two approaches have been adopted to simulate mixing: parameterization (Starrfield et al., 1998; José and Hernanz, 1998) or follow-up of many successive eruptions, with inclusion of diffusion (Prialnik and Kovetz, 1995; Yaron et al., 2005). The latter is in principle self-consistent, but the treatment of mass-loss between successive outbursts is quite uncertain.

In the framework of one-dimensional models with parameterized mixing at the core-envelope interface, many observed nova properties can be replicated successfully, especially the ejecta abundances (Starrfield et al., 1998; José and Hernanz, 1998). However, multidimensional studies are needed to better understand both the accretion phase, the way in which the TNR initiates and propagates throughout the envelope, and the mass ejection phase (see José and Hernanz, 2007b for a recent review of multidimensional nova modeling).

### *5.2.3 Nucleosynthesis in Classical Novae*

Nova outbursts eject much less mass than supernova explosions, but novae are much more frequent events than supernovae in the Galaxy; this has raised the issue of the potential contribution of such stellar cataclysms to Galactic abundances. Although the mass injected into the ISM per novae is small, detailed numerical simulations have indicated novae as major players in the synthesis of some specific nuclear species, largely overabundant in their ejecta, such as $^{13}$C, $^{15}$N, and $^{17}$O, with a minor contribution to Galactic levels of other nuclei with $A < 40$, such as $^{7}$Li, $^{19}$F, or $^{26}$Al (Starrfield et al., 1998; José and Hernanz, 1998).

Radioactivities present in nova ejecta, previously synthesized during the explosion, also constitute a major source of positrons. Indeed, $^{13}$N and $^{18}$F, and to a lesser extent $^{22}$Na, are the major contributors. The synthesis of $^{13}$N and $^{18}$F naturally occurs during the operation of the CNO cycle. Actually, the triggering reaction that powers the onset of the thermonuclear runaway is $^{12}$C(p,$\gamma$)$^{13}$N, leading to $^{13}$N synthesis. The exact amount transported to the outer envelope and contributing to $\gamma$-ray emission once transparency allows for the escape of photons, depends on details of the evolution, specially on convection. Therefore, detection of positrons from $^{13}$N, through the associated electron-positron annihilation emission, would provide an important diagnostic of the dynamics of nova explosions.

The synthesis of $^{18}$F in novae proceeds through the hot CNO cycle. Regardless of the nature of the white dwarf hosting the explosion (CO or ONe), the initial abundance of $^{16}$O is large, and thus $^{16}$O is the main source for $^{18}$F formation, which can take place either through the reaction chain $^{16}$O(p,$\gamma$)$^{17}$F(p,$\gamma$)$^{18}$Ne($\beta^+$)$^{18}$F or via $^{16}$O(p,$\gamma$)$^{17}$F($\beta^+$)$^{17}$O(p,$\gamma$)$^{18}$F. The $^{18}$F yields are severely constrained by its destruction mode, whatever the production channel is. During the runaway, $^{18}$F destruction by beta decays can be neglected when compared to its destruction by proton captures mainly through $^{18}$F(p,$\alpha$)$^{15}$O, which is faster than $^{18}$F(p,$\gamma$)$^{19}$Ne (Hernanz et al., 1999). Other nuclear reactions affecting $^{18}$F synthesis are proton captures on $^{17}$O, i.e. $^{17}$O(p,$\gamma$)$^{18}$F and $^{17}$O(p,$\alpha$)$^{14}$N (Coc et al., 2000).

Another interesting isotope likely produced during nova outbursts is $^{7}$Li. Its synthesis is believed to proceed through the so-called *beryllium transport mechanism* (Cameron, 1955), in which the previously synthesized $^{7}$Be transforms into $^{7}$Li through electron capture ($\tau$=77 days, see Table 5.1) releasing a $\gamma$-ray photon of 478 keV. For this mechanism to be effective, $^{7}$Be has to be transported to the outer, cooler envelope layers, with a timescale shorter than its decay time, in order to preserve its fragile daughter $^{7}$Li from destruction. This mechanism requires a dynamic situation like the one encountered in novae.

The production of $^{7}$Li in novae is debated. The first studies based on parameterized one-zone models (Arnould and Norgaard, 1975), were followed by hydrodynamic computations (Starrfield et al., 1978b), which did not follow the accretion phase (i.e., they had an initial envelope already in place). These models indicated that the final amount of $^{7}$Li synthesized depends on the initial abundance of $^{3}$He and on the treatment of convection. Later works based on one or two-zone models (Boffin et al., 1993) pointed out the critical role played by the photodisintegration reaction $^{8}$B($\gamma$,p)$^{7}$Be. Finally, a complete hydrodynamic study, following both the accretion and the explosion phases, was performed (Hernanz et al., 1996). Formation of $^{7}$Be proceeds through $\alpha$-captures on $^{3}$He, $^{3}$He($\alpha$, $\gamma$)$^{7}$Be, since (p,$\gamma$) reactions can not bridge the A=5 gap; destruction occurs via $^{7}$Be(p,$\gamma$)$^{8}$B; however, at high temperatures ($T \approx 10^8$ K) this rate achieves quasi-equilibrium with the inverse photodisintegration reaction, $^{8}$B($\gamma$,p)$^{7}$Be. Indeed, it was shown that a critical issue is the amount of $^{7}$Be surviving the TNR, thanks to the efficient role played by $^{8}$B photodisintegration (Hernanz et al., 1996). $^{7}$Li formation is favored in CO novae with respect to ONe novae, because their faster evolution prior to the TNR (driven by the larger amount of $^{12}$C injected in the envelope) favors $^{7}$Be survival and thus final $^{7}$Li production.

A tentative detection of $^{7}$Li in the optical band, through the LiI doublet at 6,708 Å, was reported for Nova Vel 1999 (Della Valle et al., 2002), but it was later suggested that the spectral feature could instead correspond to a doublet from neutral nitrogen (Shore et al., 2003).

**Table 5.1** Radioactive isotopes synthesized in explosive events

| Isotope | Decay chain | Disintegration process | Lifetime | Line energy (keV) |
|---|---|---|---|---|
| $^{7}$Be | $^{7}$Be → $^{7}$Li | $e^-$-capture | 77 days | 478 |
| $^{22}$Na | $^{22}$Na → $^{22}$Ne | $\beta^+$ | 3.8 year | 1,275 |
| $^{26}$Al | $^{26}$Al → $^{26}$Mg | $\beta^+$ | $1.0 \times 10^6$ year | 1,809 |
| $^{44}$Ti | $^{44}$Ti → $^{44}$Sc → $^{44}$Ca | $e^-$-capture, $\beta^+$ | 89 year (5.4 h) | 78, 68, 1,157 |
| $^{56}$Ni | $^{56}$Ni → $^{56}$Co | $e^-$-capture | 8.8 days | 158, 812, 750, 480 |
| $^{56}$Co | $^{56}$Co → $^{56}$Fe | $\beta^+$ | 111 days | 847, 1,238 |
| $^{57}$Ni | $^{57}$Ni → $^{57}$Co → $^{57}$Fe | $e^-$-capture | (52 h) 390 days | 122, 136 |
| $^{60}$Fe | $^{60}$Fe → $^{60}$Co → $^{60}$Ni | $\beta^-$ | $2.0 \times 10^6$year (7.6 year) | 1,173, 1,332 |

Overproduction factors of $^{7}$Li with respect to solar values around 1,000 are predicted by CO nova models, meaning that novae can be important contributors to the Galactic $^{7}$Li (Hernanz et al., 1996) (up to 20% of the Galactic content) and may help to reproduce the steep rise of the observed lithium abundance between the formation of the Solar System and the present (Romano et al., 1999; Alibés et al., 2002).

Classical nova explosions are also sources of two important radioactive isotopes: $^{22}$Na and $^{26}$Al. In the pioneering work by Clayton and Hoyle (1974), it was mentioned that novae are potential emitters of 1,275 keV $\gamma$-rays resulting from $^{22}$Na decay. They assumed $^{22}$Na mass fractions in the ejecta of the order of $10^{-3}$, from the conversion of $^{20}$Ne to $^{22}$Na. In the last 15 years it has been shown that this conversion is not so efficient, but interestingly, the current accepted $^{22}$Na yields in the most prolific novae are not far from those historic predictions. The synthesis of $^{22}$Na in novae proceeds through $^{20}$Ne(p,$\gamma$)$^{21}$Na followed either by the decay of $^{21}$Na to $^{21}$Ne, i.e. $^{21}$Na($\beta^+$)$^{21}$Ne(p,$\gamma$)$^{22}$Na, or by a proton capture on $^{21}$Na, i.e. $^{21}$Na(p,$\gamma$)$^{22}$Mg($\beta^+$)$^{22}$Na (José et al., 1999).

The amount of $^{22}$Na synthesized during nova explosions has not yet been determined reliably. The first hydrodynamic models of nova outbursts did not include complete nuclear reaction networks covering the Ne-Na and Al-Mg regions. In the 1980s, the crucial role played by some uncertain nuclear reaction rates for the yields of $^{22}$Na (and $^{26}$Al, see below) was finally pointed out, and extensive nova nucleosynthesis models were computed, with parameterized models (i.e. through simplified one-zone models) with representative temperature-density temporal profiles taken from evolutionary nova models (Starrfield et al., 1978a). In the 1990s, new one-zone models for nova nucleosynthesis were developed, adopting various initial compositions which included the possibility of mixing with massive white dwarf cores. These models (Weiss and Truran, 1990; Nofar et al., 1991) investigated in detail the synthesis of $^{22}$Na and $^{26}$Al, in view of the then recent detection of galactic $^{26}$Al (and non detection of $^{22}$Na). Prompted by the recent discovery of large enrichments of neon in the spectra of some novae, these calculations explored the outcome of nova outbursts on massive, ONeMg white dwarfs. Interestingly, Weiss and Truran (1990) obtained $^{22}$Na yields as large as $10^{-4}$, which combined with envelope masses of $2 \times 10^{-5}$ $M_\odot$ gave $4 \times 10^{-9}$ $M_\odot$ of $^{22}$Na ejected into the interstellar medium. The most recent hydrodynamic models of ONe (or ONeMg) nova outbursts on masses larger than 1.0 $M_\odot$ provide $^{22}$Na yields in the range $10^{-4}$–$10^{-3}$ $M_\odot$ (José et al., 1999; Politano et al., 1995; Starrfield et al., 1998). It is worth mentioning that mixing can occur at various depths inside the stratified ONe white dwarf: inner ONe core, outer CO buffer or middle transition zone (José et al., 2003). If mixing occurs with the unburned CO buffer on top of the ONe core, no $^{22}$Na would be expected (José et al., 2003).

$^{26}$Al production is complicated by the presence of a short-lived (half-life 6.3s) isomer state, $^{26}$Al$^m$. In fact, when the temperature is smaller than $4 \times 10^8$ K (as is the case in novae), the ground ($^{26}$Al$^g$) and isomeric states must be treated as two separate isotopes, because they do not reach thermal equilibrium (Ward and Fowler, 1980).

The first calculations of $^{26}$Al synthesis during explosive hydrogen burning (Hillebrandt and Thielemann, 1982; Wiescher et al., 1986) suggested that novae are likely sites for the synthesis of this radioactive isotope, but not in very large amounts; these computations relied on solar or CNO-enhanced white dwarf envelopes. Later computations, again one-zone models, demonstrated the need for initial envelope enrichment in O, Ne and Mg, dredged-up from the white dwarf cores, to obtain larger amounts of $^{26}$Al (Weiss and Truran, 1990; Nofar et al., 1991).

The major seed nuclei for $^{26}$Al synthesis are $^{24,25}$Mg. At the early phases of the thermonuclear runaway (burning shell temperatures around $5\times10^7$ K), the dominant reaction is $^{25}$Mg(p,$\gamma$)$^{26}$Al$^{g,m}$; the subsequent reaction $^{26}$Al$^m$($\beta^+$)$^{26}$Mg(p,$\gamma$)$^{27}$Al produces the stable isotope $^{27}$Al. At larger temperatures ($\sim10^8$ K), the nuclear path $^{24}$Mg(p,$\gamma$)$^{25}$Al($\beta^+$)$^{25}$Mg dominates, with again $^{25}$Mg(p,$\gamma$)$^{26}$Al$^{g,m}$. When temperature reaches $2 \times 10^8$ K, (p,$\gamma$) reactions proceed very efficiently and reduce the amount of $^{25}$Al, leading to the formation of $^{26}$Si ($^{25}$Al(p,$\gamma$)$^{26}$Si) which decays into $^{26}$Al$^m$, thus by-passing $^{26}$Al$^g$ formation. Also $^{26}$Al itself (in both states) is destroyed to $^{27}$Si which decays into $^{27}$Al. In summary, the final amount of $^{26}$Al$^g$ and the ratio $^{26}$Al$^g$/$^{27}$Al mainly depend on the competition between the two nuclear paths $^{24}$Mg(p,$\gamma$)$^{25}$Al($\beta^+$)$^{25}$Mg(p,$\gamma$)$^{26}$Al$^{g,m}$ and $^{24}$Mg(p,$\gamma$)$^{25}$Al(p,$\gamma$)$^{26}$Si. The first channel is the only one producing $^{26}$Al$^g$, whereas both channels produce $^{27}$Al (through $^{26}$Al$^{g,m}$(p,$\gamma$)$^{27}$Si($\beta^+$)$^{27}$Al or $^{26}$Si($\beta^+$)$^{26}$Al$^m$($\beta^+$) $^{26}$Mg(p,$\gamma$)$^{27}$Al (José et al., 1999).

The final $^{26}$Al yields from novae sensitively depend on the initial mass of the white dwarf and on the degree of mixing between the accreted envelope and the core. Recent hydrodynamic models of ONe (or ONeMg) nova outbursts on masses larger than 1.0 $M_\odot$ suggest $^{26}$Al yields in the range $10^{-4}$–$10^{-2}$ (José et al., 1999; Politano et al., 1995; Starrfield et al., 1998). If mixing occurs with the CO buffer on top of the bare ONe nuclei (or in the transition zone), some $^{26}$Al would be expected (but no $^{22}$Na), since there is a non negligible amount of the seed nucleus $^{25}$Mg both in the CO buffer and in the transition zone (José et al., 2003).

## 5.3 SNIa Explosions

The pyrotechnical displays of the total explosion of stars are called supernovae. They are characterized by a sudden rise of the luminosity, followed by a steep decline lasting several weeks and eventually followed by a more gradual decline lasting many years. The total electromagnetic output, obtained by integrating the light curve, is $\sim10^{49}$ erg, while the luminosity at maximum can be as high as $\sim10^{10}$ $L_\odot$. The kinetic energy of supernovae can be estimated from the expansion velocity of the ejecta, $v_{\rm exp}$ $\sim$5,000–10,000 km s$^{-1}$, and turns out to be $\sim10^{51}$ erg. Such an amount of energy can only be obtained from the gravitational collapse of an electron degenerate core, forming a proto- neutron star or a black hole, or from its thermonuclear incineration to iron – peak isotopes. In the former case, the gravitational binding energy of a neutron star of $\sim$1.4 $M_\odot$ and $\sim$10 km radius is of the order of $\sim 10^{53}$ ergs and a weak coupling between the source of energy and matter is enough to fulfill the energetic requirements (Zwicky, 1938). In the second case, the nuclear specific energy of a carbon oxygen mixture is $q \sim 10^{18}$ erg/g, sufficient to obtain the required energy from burning $\sim 1\ M_\odot$ (Hoyle and Fowler, 1960).

Supernovae are classified according to their spectrum at maximum light. If hydrogen lines are absent, they are called Type I supernovae or SNI. If these lines are present, they are referred to as Type II or SNII (Minkowski, 1941). Also according to their spectra, SNI are further divided into three categories. If a prominent line of SiII is present, they are labeled SNIa. If this line is absent but there is a prominent HeI line, they are denoted SNIb. If both SiII and HeI lines are absent, the classification label is SNIc (Wheeler and Harkness, 1990).

The light curves of SNIa are characterized by a rapid rise in luminosity, up to an average maximum $M_V \approx M_B \approx -19.30 \pm 0.03 + 5\log(H_0/60)$ (Riess et al., 1999) in about 20 days, where $H_0$ is the Hubble constant, followed by a comparatively gentle decline divided into two different epochs. The first epoch after maximum light lasts $\sim$30 days and the luminosity typically drops by $\sim 3^{\mathrm{mag}}$, while the second phase is characterized by a slow decline with a characteristic time of $\sim$70 days. Infrared photometry shows that in the J, H and K bands there is a well defined minimum at $\sim$20 days after maximum and a secondary peak $\sim$30 days after maximum (Elias et al., 1985), although in some cases this secondary peak is absent.

This behavior can be understood in terms of the deposition of a large amount of energy, $\sim 10^{51}$ ergs, in the interior of a stellar envelope. If it is assumed that half of this energy is invested into the expansion of the debris and half into their internal energy, the average temperature is:

$$T = 6.3 \times 10^4 \left( \frac{E_{SN,51}}{R_{15}^3} \right)^{1/4} \tag{5.16}$$

where $E_{SN,51}$ is the energy released by the supernova in units of $10^{51}$ erg and $R_{15}$ the radius in units of $10^{15}$ cm. Consequently:

$$\frac{P_R}{P_G} = \frac{{}^{1}\!/_{3} a T^4}{\frac{\mathfrak{R}}{\mu} T \rho} \approx 1.6 \times 10^4 \frac{\left(R_{15} E_{SN,51}\right)^{3/4}}{M} \tag{5.17}$$

which means that a supernova is, in a certain sense, just a ball of light. Typical expansion velocities are:

$$v \approx 10^9 \sqrt{\left( \frac{E_{SN}}{10^{51}} \cdot \frac{M_\odot}{M} \right)} \mathrm{cm/s} \tag{5.18}$$

Since the internal energy is dominated by radiation and the expansion is nearly adiabatic, $T \propto R^{-1}$, the total energy will evolve as $E_{\mathrm{th}} \propto VT^4 \propto R^{-1}$. If the initial structure is compact, $\sim 10^8$ cm, the typical size of a relativistic degenerate stellar core, the energy decreases from $\sim 10^{51}$ erg to $\sim 10^{44}$ erg when the radius is $\sim 10^{15}$ cm, the typical radius at maximum light; i.e. the energy deposited by the shock is invested in the adiabatic expansion of material. On the contrary, if the initial structure is extended,

$$E_{\rm th} \approx E_{\rm th,0}\frac{R_0}{R_{\rm env}} \approx \frac{E_{SN}}{2}\frac{R_0}{R_{\rm env}} \tag{5.19}$$

the luminosity will be given by

$$L_P \approx \frac{E_{\rm th}}{\tau_{\rm diff}} \approx \frac{2\pi c}{9k_{Th}}\frac{E_{SN}}{M_{\rm env}}R_0 \tag{5.20}$$

where the diffusion time has been estimated as:

$$\tau_{\rm diff} \approx \frac{3R_{\rm env}^2}{\lambda c} \approx \frac{9\kappa M_{\rm env}}{4\pi c R_{\rm env}} \tag{5.21}$$

and the thermal energy will provide a luminosity plateau or a broad peak after maximum.

Therefore, the explosion of a compact object is able to account for the light curve of Type Ia supernovae, but an additional source of energy is necessary to explain the tail. Although in the past some other possibilities were considered, there is now broad consensus that this energy source is provided by the radioactive decay of $^{56}$Ni (Truran et al., 1967; Colgate and McKee, 1969):

$$^{56}\mathrm{Ni} \rightarrow ^{56}\mathrm{Co} \rightarrow ^{56}\mathrm{Fe}$$

with $q_{\rm Ni} \sim 7 \times 10^{49}$ erg/M$_\odot$, $q_{Co} \sim 1.5 \times 10^{50}$ erg/M$_\odot$ and $\tau_{1/2}(\mathrm{Ni}) = 6.1$ days, $\tau_{1/2}(Co) = 77.1$ days.

The majority of the energy released by the decay of $^{56}$Ni and $^{56}$Co is in the form of $\gamma$-rays of $\sim$1 MeV that are scattered and eventually thermalized via Compton scattering and photoelectric absorption. The resulting thermal photons diffuse and eventually escape. The observed light curve thus results from a competition of two time scales describing diffusive energy transport and dynamic expansion. As before, the diffusion time scale is dominated by Thomson scattering and by absorption from bound electrons. The contribution of true absorption to the total opacity is a complicated issue because of the departures from LTE, the Doppler effect introduced by the expansion velocity and the uncertainties introduced by the chemical composition and the energy levels of different ions. Nevertheless, the opacity seems to be confined within the range 0.2–0.03 cm$^2$/g. If it is assumed that the envelope expands with constant velocity, $R_{\rm env} \sim R_0 + v_{\rm exp}t$, the hydrodynamic time scale is $\tau_h \sim R_0/v_{\rm exp}$.

Initially the ejecta are opaque, $\tau_{\rm diff} \gg \tau_h$, and the luminosity is small. As time goes by, $\tau_{\rm diff} \sim \tau_h$, and photons begin to escape. Since the energy output decreases exponentially, a peak appears in the light curve that is equal to the instantaneous deposition of energy and therefore $L_{\max} \propto M_{\rm Ni}$. After the peak, the radiation trapped in the envelope diffuses outwards and the luminosity exceeds the instantaneous energy deposition rate. The width of the peak is determined by an effective diffusion time:

$$\tau_m = \sqrt{2\tau_{\rm diff}\tau_h} \propto \kappa^{1/2} M_{\rm env}^{3/4} M_{\rm Ni}^{-1/4} \tag{5.22}$$

Later, when the density is small enough, an increasing fraction of the $\gamma$-photons (and later also positrons) can escape and, consequently, the luminosity is smaller than the energy output of radioactive decays. Some radioactive energy may be stored in the form of ionization, and defer luminosity originating from radioactive energy. See Arnett (1996) and references therein for a complete discussion.

The observation of the bolometric light curves together with these simple relationships allow estimates of the mass expelled and the mass of the radioactive elements synthesized by SNIa: $M_{\rm env}$ = 0.4–1.4 $M_\odot$ and $M_{\rm Ni}$ = 0.1–1 $M_\odot$ (Stritzinger et al., 2006). One of the most striking properties of SNIa is their photometric homogeneity i.e. the light curve of the majority shows a very small dispersion at maximum, $\sigma_M \leq 0.3^{\rm mag}$ (Cadonau et al., 1985; Hamuy et al., 1996) when they are normalized to the peak. All these properties together immediately suggest that the most plausible scenario is the explosion of a CO white dwarf near the Chandrasekhar mass in a close binary system.

Spectroscopic observations at different epochs enable tomography of supernovae. During maximum light, the spectra of SNIa are characterized by various lines of neutral and singly ionized atoms of Si, Ca, Mg, S and O moving at high velocities ($v \sim$ 8,000–30,000 km/s) indicating that the outer layers are mainly composed of intermediate mass elements, i.e. that thermonuclear burning was not complete (Filippenko et al., 1992). Two weeks after maximum, permitted FeII lines are prominent, indicating that the photosphere has reached regions where the star was able to completely incinerate matter (Harkness, 1991). The nebular phase starts one month after the maximum, roughly when the tail of the lightcurve begins. During this epoch, the spectrum is dominated by forbidden FeII, FeIII and CoIII emission lines (Axelrod, 1980). The decrease of the Co lines, together with the relative intensity of CoIII with FeIII (Kuchner et al., 1994) provide support for the idea of a light curve tail powered by the decay of $^{56}$Co. In general, the lines of different elements have different expansion velocities, indicating a layered structure where the central regions are occupied by completely burned material, i.e. the iron peak elements. This property rules out the hypothesis of a prompt detonation since in this case the white dwarf would be completely converted to $^{56}$Ni.

Despite their remarkable photometric homogeneity, SNIa do exhibit some degree of heterogeneity. Already in 1973, it was proposed (Barbon et al., 1973) to subdivide SNIa into a fast and slow class according to the rate of decline of their light curve just after maximum, the transition from the peak to the tail, and the decline of the tail. The slow class is characterized by a broader and more luminous peak than the fast class. The most extreme cases are SN1991T, considered until recently the most energetic event with the broadest peak, and SN1991bg and SN1992K, which are the reddest, fastest and most subluminous Type Ia supernovae known to date (Phillips et al., 1992; Ruiz-Lapuente et al., 1992). The difference in brightness between the extreme cases is $\sim 2^{\mathrm{mag}}$. The large majority of SNIa also display a remarkable spectroscopic homogeneity (Branch et al., 1993) not only during the maximum but also during subsequent months. They are classified as *Branch-normal* and represent 85% of the total, although there are suggestions that this value should be reduced to 70%. The prototypes are SN1972E, 1981B, 1989B and 1994D. On the contrary, 91T-like events display FeIII lines before maximum while the 91bg-like supernovae lack the characteristic secondary maximum in the infrared.

This mildly inhomogeneous set of SNIa exhibits a correlation between peak magnitude, the width of the peak, and the expansion velocity, in the sense that the brightest SNIa show the largest expansion velocities (Pskovskii, 1977; Branch, 1981). The correlation between the brightness and the shape of the light curve was settled definitively when a clear relationship between the maximum of light and the magnitude decline during the first 15 days after maximum ($\Delta m_{15}$) was firmly established (Phillips, 1993). This correlation (which can be parameterized in terms of the decline rate (Phillips, 1993; Hamuy et al., 1996), the stretch parameter (Perlmutter et al., 1997), or via a multi-parameter fit of colors (Riess et al., 1996)) is used to renormalize the peak magnitudes and thereby substantially reduces the dispersion of the absolute magnitudes, making SNIa one of the most powerful tools for measuring cosmological distances.

In principle, these variations in the shape of the light curve can be understood in terms of the total amount of $^{56}$Ni synthesized, if the ejected mass is kept constant. Since the maximum of the luminosity is proportional to the $^{56}$Ni mass, the brightest events are those that have synthesized the largest amount of this material and consequently have larger expansion velocities and broader peaks since the opacity of iron peak elements is very large. In any case, this diversity of properties poses the question whether there are two explosion mechanisms, one for the *Branch-normal* supernovae and another for the peculiar events, or if there is a unique mechanism able to account for the broad range of behaviors.

The frequency and the impact on the chemical content of galaxies provide additional constraints on the different supernova mechanisms. The rate of supernovae in galaxies is usually normalized to the galaxy blue luminosity (Tammann, 1970) or to the mass assuming an average M/L-ratio for each galaxy type (Cappellaro et al., 2003). The most striking feature is that SNIb/c and SNII only appear in spiral- and irregular galaxies, and are associated with a young populations, while SNIa can appear in all galaxy types, indicating that they are related to the old stellar populations. Nevertheless, the SNIa rate per unit mass is almost 3 times larger in late spirals than in ellipticals, thus implying that at least a fraction of SNIa must be related to the young stellar population (Cappellaro et al., 2003). Furthermore, there is some evidence that, on average, SNIa in red or early type galaxies are dimmer, have faster light curves and slower expansion velocities than those in blue or late type galaxies (Hamuy et al., 1995, 1996; Branch et al., 1996). On the other hand, the frequency of supernovae in the Milky Way has been estimated (van den Bergh and Tammann, 1991) to be: $R_{\text{II}} = 3.32 \times 10^{-2}$/year, $R_{\text{Ib/c}} = 0.65 \times 10^{-2}$/year and $R_{\text{Ia}} = 0.41 \times 10^{-2}$/year. Taking into account that the mass of $^{56}$Ni ejected per event is roughly 0.07, 0.3 and 0.6 $M_{\odot}$ for SNII, SNIb/c and SNIa, respectively, it turns out that nearly half of the galactic iron was synthesized by Type Ia supernovae. This means that SNIa have to produce the right amount of iron peak isotopes to account for the observed isotopic Solar abundances.

## 5.3.1 *Chandrasekhar-Mass Models*

As discussed in Sect. 5.1.3, the outcome of carbon ignition under degenerate conditions in a white dwarf near the Chandrasekhar limit can be a detonation or a deflagration, depending on the particular structure at the moment of ignition, represented by density, temperature, chemical composition and velocity profiles. For instance, it is easier to generate the overpressure necessary to launch a detonation at low densities, $\sim 3 \times 10^7$ g/cm$^3$, than at high densities due to the degeneracy dependence on density and temperature.

### 5.3.1.1 The Prompt Detonation Model

Even though a pure detonation seems possible from a physical point of view (Blinnikov and Khokhlov, 1986), this kind of explosion cannot account for the observed SNIa spectra at maximum light. At densities above $\sim 10^7$ g/cm$^3$, the fuel is completely incinerated to Fe-peak elements and it leaves only a few hundredths of a solar mass of intermediate mass elements, which is not enough to produce the characteristic strong SiII line of SNIa. The rejection of a pure detonation as the SNIa mechanism is a consequence of the simplicity of this burning mode. As discussed in Sect. 5.1.3, in the absence of external perturbations like a piston, the Chapman-Jouget detonation is the only stable solution (other than a deflagration) of the Rankine-Hugoniot equations that define the burning front. Thus, there are no free parameters left, no time for modification of the fuel pre-combustion structure, no diversity, and pure detonations always produce the wrong result. Notice however that the presence of a shallow thermal gradient close to the ignition profile might induce the formation of shocks that could burn a large mass in a short time, starting the dynamical phase of the supernova (Blinnikov and Khokhlov, 1986; Bravo et al., 1996) and a mixture of deflagration and detonation regimes might be the result.

### 5.3.1.2 The Deflagration Model

Deflagrations are less constrained from a physical point of view, but their properties are strongly conditioned by the hydrodynamics of the explosion process itself. As described in Sect. 5.1.3, at the microscopic scale the speed of the flame only depends on the local physical conditions. Thus, the laminar velocity of the flame can be determined as a function of density (Timmes and Woosley, 1992): $v'_{\text{lam}} = \alpha \rho_9^{\beta}$ cm/s, where $\rho_9$ is the density in units of $10^9$ g/cm$^3$, and $\alpha$ and $\beta$ are fit parameters. For

$\rho_9 < 0.36$, $\alpha = 5.68 \times 10^6$ and $\beta = 1.46$, while for $2 \geq \rho_9 \geq 0.36$, $\alpha = 3.68 \times 10^6$ and $\beta = 1.03$. A further correction can be obtained taking into account the effect of Coulomb interactions: $v_{\rm lam} = K_{\rm cc} v'_{\rm lam}$, where $K_{\rm cc} = 0.894 - 0.0316 \log(\rho_9)$ (Bravo and García-Senz, 1999). The situation becomes extremely complicated when the flame is accelerated by the deformation induced by hydrodynamic instabilities. This acceleration is difficult to describe because it implies many length scales, from the global length scale, $\sim 10^7$–$10^8$ cm, to the microscopic width of the laminar flame, which strongly depends on the density as shown before.

One possibility (Woosley and Weaver, 1986) is to parameterize the velocity of the deflagration as a function of flame radius, $r$, as: $v_{\rm def,W} = A v_{\rm sound}\left(1 - {\rm e}^{-Br}\right)$. The parameters $A$ and $B$ are constrained by the condition that the flame should propagate at a small Mach number close to the center but should reach velocities as high as 0.1–0.5 Mach in the outer layers of the white dwarf. A second possibility (Khokhlov, 1995), assumes that the rate of surface creation by the turbulence is balanced by the rate of surface destruction due to flame propagation. The deflagration velocity should then be given by $v_{\rm def,K} = 0.5\sqrt{g_{\rm eff} L}$, where $L$ is the driving scale, $g_{\rm eff} = g{\rm At}$, $g$ is the gravitational acceleration and At is the Atwood number (Timmes and Woosley, 1992). Such kind of self-regulating regime has been implemented in several ways in many multidimensional simulations of SNIa (Gamezo et al., 2003; García-Senz and Bravo, 2005). A third possibility (Niemeyer and Woosley, 1997) is that the deflagration moves at the speed of the Rayleigh-Taylor bubbles in the nonlinear scale, the so-called Sharp-Wheeler model, in which the velocity increases linearly with time $t$, $v_{\rm def,NW} = 0.1 g_{\rm eff} t$. Finally, the concept of a subgrid-scale model that takes into account the dissipation of turbulent energy at microscopic scales has been adopted in many multidimensional simulations performed to date (Schmidt and Niemeyer, 2006; Röpke et al., 2006). In spite of the differences in the treatment of the flame, most three-dimensional simulations of SNIa produce quite homogeneous results.

The success or failure of a deflagration model depends on its ability to consume the fuel with the same speed as the front engulfs it, such that it does not leave unburned pockets of carbon and oxygen behind (Niemeyer and Woosley, 1997). High-resolution simulations aimed to explore the multipoint ignition scenario (Röpke et al., 2006) indicate that when the number of initial seeds increases, the ignition volume becomes saturated and the gross features of the explosion converge towards a unique solution. The optimal number of flame seeds is estimated to be in the range $\sim$ 100–400 per octant distributed in radius following a Gaussian up to $\sim$ 100–150 km. However, even in the most favorable case it is difficult to obtain more than 0.7 $M_\odot$ of $^{56}$Ni and a kinetic energy above $0.7 \times 10^{51}$ ergs, which is too small to account for the bulk of bright-normal SNIa. In addition, the deflagration always leaves a large mass of carbon and oxygen unburned, $M_{\rm ub} > 0.57\ M_\odot$ (Schmidt et al., 2006).

The present three-dimensional simulations of Type Ia supernovae based on a pure deflagration algorithm have to face the following problems when confronted with observations: (i) although the amount of Fe-group elements synthesized in the explosion is sufficient, the mass of $^{56}$Ni is not. (ii) the final kinetic energy is always smaller than the canonical value of $10^{51}$ ergs. (iii) the synthesis of intermediate-mass elements is scarce. (iv) the ejecta lack chemical stratification. (v) big clumps of radioactive $^{56}$Ni are present at the photosphere at the time of maximum luminosity.

Before discarding deflagrations as the main mode of Type Ia supernovae it is necessary to examine some still poorly understood aspects. For instance, it might be that the theoretical description of subsonic flames included in the hydrodynamical codes and used in the simulations is incomplete as is the case when they enter the distributed regime at densities lower than $\sim (1-3) \times 10^7$ g/cm$^3$. It is also important to notice that the influence of the initial conditions at the onset of the ignitions has not yet been clarified.

#### 5.3.1.3 The Delayed Detonation Model

In 1974, a burning regime was proposed (Ivanova et al., 1974) in which the initial flame was not able to unbind the star, leading to a pulsation and a delayed transformation of the deflagration into a detonation (deflagration-detonation transition or DDT). The DDT concept was later extended (Khokhlov, 1991; Khokhlov et al., 1993) to include the possibility of a transition to detonation without an intervening pulsation. The essential ingredient for the formation of a detonation is the existence of a non-uniformly preheated region with a level of fluctuations of temperature, density, and chemical composition such that a sufficiently large mass would be burnt before a sonic wave could cross it. The thermal gradient needed (Khokhlov, 1991) is:

$$\nabla T < \frac{\Theta T}{A v_{\rm sound} \tau_i} \tag{5.23}$$

where $A$ is a numerical coefficient, $A \sim 0.2$–$5.0$, $\tau_i = T/\dot{T}$ is the induction time at the temperature $T$, and $\Theta \sim 0.04$–$0.05$ is the Frank-Kameneetskii factor:

$$\Theta = -\frac{\partial \ln \tau_i}{\partial \ln T} \tag{5.24}$$

There are several mechanisms that can produce such fluctuations: adiabatic precompression in front of a deflagration wave, shock heating, mixing of hot ashes with fresh fuel (Khokhlov, 1991), accumulation of pressure waves due to a topologically complex geometrical structure of the flame front, or transition to the distributed burning regime (Niemeyer and Woosley, 1997). Among these possibilities the turbulence pre-conditioning has received more attention. In the case of a white dwarf expanding as a consequence of a deflagration, the turbulent velocity has to exceed the laminar flame velocity by a factor 1–8 at a length scale comparable to the detonation wave thickness (Khokhlov et al., 1997). This criterion is fulfilled for flame densities in the range $5 \times 10^6$ g/cm$^3$ $< \rho <$ $(2-5) \times 10^7$ g/cm$^3$ for reasonable assumptions. At densities above $10^8$ g/cm$^3$, a DDT transition is unlikely (Khokhlov et al., 1997) although the bubble fragmentation could increase the flame surface and facilitate a DDT at $\rho \sim 2 \times 10^8$ g/cm$^3$ (Zingale and Dursi, 2007). Despite the difficulties to justify the DDT models, the one-dimensional delayed detonation simulations are particularly successful in reproducing many key observational characteristics of SNIa (Hoeflich and Khokhlov, 1996), like the light curves and photospheric expansion velocities.

#### 5.3.1.4 The Pulsational Delayed Detonation Model

The 3D formulation of the pulsational delayed detonation model is the so-called Pulsational Reverse Detonation model or PRD. In this scenario the detonation is triggered by an accretion shock that forms above a quasi – hydrostatic core composed mainly of C—O with a mass of 0.8–1.15 $M_\odot$. Heating by the accretion shock ignites the fuel slightly below the core's surface. Because of the inertial confinement provided by the material falling through the accretion shock, the core cannot expand and cool efficiently. As a consequence, a detonation propagating inwards forms and burns most of the core. The resulting energetics as well as the nucleosynthesis are roughly in agreement with observations, specially concerning the observed stratification.

### *5.3.2 Sub-Chandrasekhar and Super-Chandrasekhar Models*

Sub-Chandrasekhar models assume a white dwarf with a mass $M_{\rm WD} \leq 1.1 - 1.2$ $M_\odot$ that accretes helium rich matter at a rate in the range $10^{-9} \leq \dot{M} \leq 5 \times 10^{-8}$ $M_\odot$/year. These rates allow the formation of a degenerate helium mantle around the initial CO core. When this mantle reaches a critical mass, $\sim$0.2–0.3 $M_\odot$, a thermonuclear runaway starts at its bottom and triggers the explosion of the star before reaching the Chandrasekhar limit. Notice that white dwarfs with an initial mass larger than 1.2 $M_\odot$ could reach the Chandrasekhar mass before exploding and experience central ignition.

One dimensional models indicate that before the thermonuclear runaway occurs, the base of the helium layers becomes convective and transports energy and part of the reactants away from the inner core boundary in such a way that He ignites above the interface. The high flammability of helium together with the low density of the envelope guarantees the formation of a detonation that incinerates the envelope and launches a shock wave inwards through the CO core. Because of the focusing effect of the spherical symmetry, this shock strengthens and induces the detonation of C in the central region that leads to a supernova explosion.

These explosions reproduce the gross features of SNIa explosions, specially sub-luminous ones like SN 1991bg and allow to explain with a single parameter, the initial mass of the white dwarf, their observed diversity. Despite such advantages, Sub-Chandrasekhar models are not the favorite to account for SNIa outbursts. The reason is that they predict the existence of a very fast moving layer composed of $^{56}$Ni and $^{4}$He that is not observed, as well resulting in light curves that rise too fast (Hoeflich and Khokhlov, 1996). This fact poses a fundamental problem. Since everything able to explode eventually does it, what prevents the explosion or explains why they have not yet been observed. A possible explanation could be a lack of spherical symmetry necessary to focus the inward shock wave, but 3D simulations assuming asynchronous ignitions in several points scattered through the envelope produce successful explosions (García-Senz and Bravo, 2005).

The recent analysis of SN2003fg (Howell et al., 2006) has shown this Type Ia supernova is about 2.2 times more luminous than the average. If the standard analysis is valid, the mass of $^{56}$Ni synthesized to account for such a luminosity is $\sim$1.3 $M_{\odot}$, and in order to account for the spectrum at maximum, the ejected mass should be 2.1 $M_{\odot}$ which implies that the exploding object should have a mass larger than the Chandrasekhar mass. The only way known to push the mass of a degenerate structure beyond the Chandrasekhar limit is through rotation. If it is assumed that the white dwarf rotates as a rigid body, it is possible to delay the ignition up to masses of the order of 1.4–1.5 $M_{\odot}$ (Piersanti et al., 2003a,b). If differential rotation is allowed, the ignition can be delayed up to masses $\sim$2 $M_{\odot}$ (Piersanti et al., 2003a). A follow-up of rigid rotation models has been calculated with a 1D hydrodynamic code modified to take into account the centrifugal force (Domínguez et al., 2006) and a weak dependence on the rotation period has been found. The problem is that these calculations assume that the transition from deflagration to detonation occurs at a fixed density and it is not known how rotation affects this change of regime of the burning front.

### 5.3.3 Nucleosynthesis in Thermonuclear Supernovae

The abundances of the elements synthesized in SNIa events depend on the peak temperature reached by the material and on the excess of neutrons versus protons. Roughly speaking, the SNIa material undergoes four burning regimes: (i) nuclear statistical equilibrium (NSE), (ii) incomplete Si burning, (iii) incomplete O burning and (iv) incomplete C—Ne burning (Woosley, 1986). The neutron excess depends on the initial abundance and distribution of the neutron rich isotopes like $^{22}$Ne, which depend on the metallicity and thermal history of the white dwarf, and on the extent of electron captures on the burned material, which mainly depends on the ignition density and on the burning regime. Another complication comes from the additional degrees of freedom introduced by 3D flames that open a variety of possible ignition modes as well as the possibility of leaving pockets of unburned material. Finally, the adopted nuclear reaction and electron capture rates are an important source of uncertainty. Despite all these factors it is possible to obtain some insight into the problem using parameterized 1D models with different propagation velocities of the burning front, different ignition densities and different initial metallicities. Three dimensional models still need some additional work.

The chemical composition of matter in nuclear statistical equilibrium (NSE) with equal number of protons and neutrons, i.e. with electron mole number $Y_e = 0.5$, peaks around $^{56}$Ni. When $Y_e$ takes values in the range 0.470–0.485, the peak moves towards $^{58}$Ni, and $^{54}$Fe; values in the interval 0.46–0.47 produce predominantly $^{56}$Fe; values in the range of 0.45–0.43 are responsible of the formation of $^{58}$Fe, $^{54}$Cr, $^{50}$Ti and $^{64}$Ni, while values below 0.43–0.42 are responsible for $^{48}$Ca. Parameterized models indicate that the amount of mass with $Y_{\mathrm{e}} < 0.45$ depends on the ignition density, while that with $0.47 < Y_{\mathrm{e}} < 0.485$ depends on the deflagration velocity. Therefore $^{58}$Fe, $^{54}$Cr, $^{50}$Ti, $^{64}$Ni and $^{48}$Ca are a measure of $\rho_{\mathrm{ig}}$ while $^{58}$Ni, $^{54}$Fe are a measure of $v_{\mathrm{def}}$ (Thielemann et al., 2004). It is important to realize that the change from using the Fuller et al. (1985) rates to the Langanke and Martínez-Pinedo (2000) rates strongly alleviates the chronic problem of producing an excess of neutronized species. In any case, to correctly evaluate the implications of the nucleosynthesis resulting from the different mechanisms and explosion scenarios it is necessary to integrate them into a galactic chemical evolution model that takes into account the contributions of all the iron – peak producers (Bravo et al., 1992; Matteucci et al., 1993). The nucleosynthesis yields also depend on a more subtle parameter, the abundance profiles of carbon and oxygen, which in turn are a function of the mass and initial metallicity of the progenitor of the white dwarf. In general, low mass progenitors produce white dwarfs with oxygen abundances in the center as large as $X_{\mathrm{O}} = 0.7$ (D'Antona and Mazzitelli, 1989; Salaris et al., 1997). This abundance can be enhanced as a consequence of the sedimentation induced by crystallization (Canal et al., 1990) if the white dwarf has had time to solidify before the start of the accretion phase. The differences in the energetic contents of carbon and oxygen nuclei translates into different $^{56}$Ni yields (Domínguez et al., 2001). A similar effect is produced by the abundance and distribution of $^{22}$Ne across the star (Bravo et al., 1992). These differences in the final production of $^{56}$Ni translate into a dispersion of the peak SNIa luminosities of $\sim$0.2 magnitudes. This value is smaller than the observed differences at high redshift and thus does not invalidate the use of SNIa for measuring distances but introduces some caution in the context of their use to determine the cosmological equation of state (Domínguez et al., 2001).

## 5.4 X-Ray Bursts and Superbursts

X-ray bursts were serendipitously discovered in 1975 by Grindlay et al. (1976), and independently, by Belian et al. (1976). In contrast to standard transient sources, characterized by lifetimes ranging from weeks to months, these new cosmic X-ray sources (a subset of the low-mass X-ray binary class, or LMXB) exhibit brief *bursts*, lasting from seconds to minutes (see Bildsten, 1998; Lewin et al., 1993, 1995; Strohmayer and Bildsten, 2006, for reviews).

The two bursting episodes reported by Grindlay et al. (1976). were based on observations performed with the Astronomical Netherlands Satellite (ANS) on a previously known X-ray source, 3U 1820-30, located in the globular cluster NGC

6624. Similar events were reported by Belian et al., from X-ray observations of sources in the Norma constellation, performed with two military *Vela-5* satellites, covering the 15-month period from May 1969 to August 1970.

One year later, three additional burst sources, one of them, the enigmatic *Rapid Burster* (XBT 1730-335), were identified within a few degrees of the Galactic center (Lewin et al., 1976). Within a year, 20 additional burst sources were discovered, mainly by SAS-3 and OSO-8 satellites. To date, more than 90 Galactic X-ray burst sources (hereafter, XRBs) have been identified (Liu et al., 2007), with burst durations of $\sim$10–100 s, and recurrence periods ranging from hours to days.

### 5.4.1 The Nature of Type I X-Ray Bursts

Maraschi and Cavaliere (1977), and independently, Woosley and Taam (1976), were the first to suggest the possibility that XRBs are powered by thermonuclear runaways on the surface of accreting neutron stars. However, it was soon realized that the quick succession of flashes exhibited by the Rapid Burster (with recurrent times as short as $\sim$10 s; see Fig. 5.4), didn't match the general pattern shown by these bursting sources. A major breakthrough in the understanding of the nature of these cataclysmic events was the discovery of two different kinds of bursts associated with the Rapid Burster (Hoffman et al., 1978): a classification of type I and type II bursts was then established, the former associated with thermonuclear flashes, the later linked to accretion instabilities. In fact, during many years, type II bursts were unequivocally linked with the Rapid Burster, the only object that showed both type I and type II bursts. More recently, a second member of the type II class, the bursting pulsar GRO J1744-28, has been identified. Hereafter, we will focus on type I X-ray bursts, the most frequent type of thermonuclear stellar explosion in the Galaxy (the third, in terms of total energy output after supernovae and classical novae).

The first evidence of the thermonuclear origin of type I XRBs came from lightcurve analysis, in particular the ratio between time-integrated persistent and burst fluxes, $\alpha$. It was soon realized that the ratio between the gravitational potential energy released by matter falling onto a neutron star (G $M_{NS}/R_{NS} \sim 200$ MeV/nucleon) during the accretion stage and the nuclear energy liberated during the burst ($\sim$ 5 MeV/nucleon, for a solar mixture transformed into Fe-group nuclei), match the values inferred for $\alpha$, in the range $\sim$ 40–100.

The spatial distribution of type I XRBs matches that of LMXBs, with a clear concentration towards the Galactic center. A significant fraction of XRBs is indeed found in globular clusters. This pattern suggests that they are associated with an old stellar population (Lewin et al., 1993). The donors transferring material onto the neutron stars in XRBs are faint, low – mass stars ($M < 1$ $M_\odot$), either Main Sequence or Red Giant stars. Recently, the first extragalactic XRBs were discovered in two globular cluster source candidates of the Andromeda galaxy (M31) (Pietsch and Haberl, 2005).

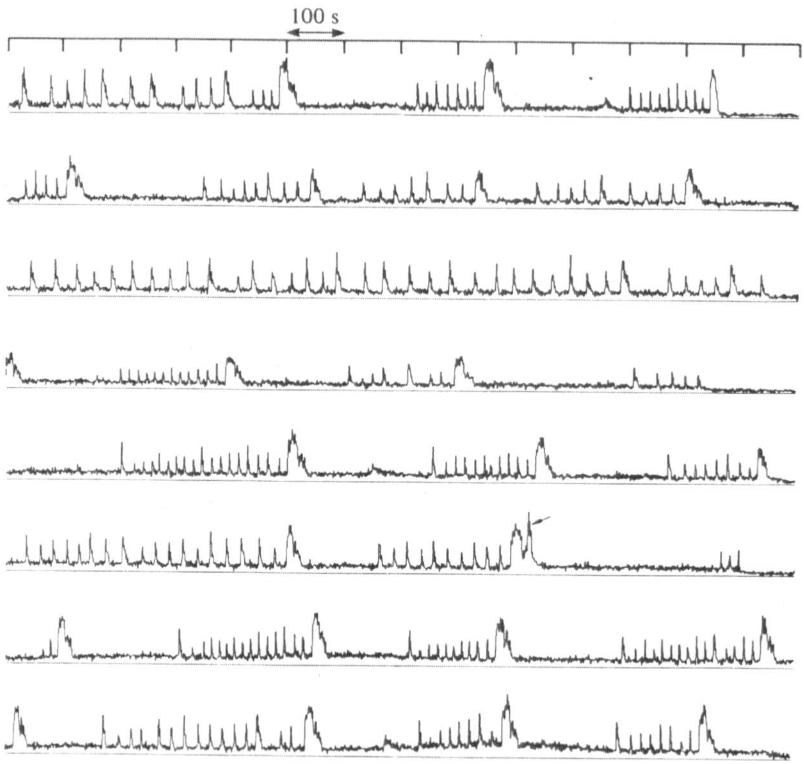


**Fig. 5.4** Type II bursts from the Rapid Burster, based on SAS-3 observations performed in march 1976. The burst pinpointed with an *arrow* is actually a type I burst. Image from Lewin (1977)

It is believed that mass transfer episodes are driven by Roche-lobe overflow, hence leading to the build-up of an accretion disk around the neutron star. The maximum mass-accretion rate is set by the Eddington limit ($\dot{M}_{Edd} \sim 2 \times 10^{-8}\ M_\odot$/year, for H-rich accretion onto a 1.4 $M_\odot$ neutron star). Typically, XRB sources have orbital periods ranging from 1 to 15 h (White et al., 1995).

The nature of the underlying primary star was initially a matter of debate. A model involving accretion onto massive black holes ($> 100\ M_\odot$) was proposed in the 1970s (Grindlay and Gursky, 1976). Nevertheless, XRB observations in globular clusters (from which reasonably accurate distance estimates can be obtained), performed with the OSO-8 satellite, best fitted with a blackbody spectrum with $kT \sim 0.87$–2.3 keV (Swank et al., 1977), suggested a source with much smaller dimensions than a super-massive black hole (either a neutron star or a stellar mass black hole). Other features, such as the harder X-ray spectra of XRB sources compared with most of the X-ray transients hosting black hole candidates, as well as the masses inferred from those systems, point towards a neutron star primary (van Paradijs and McClintock, 1995).

Indeed, the masses inferred for neutron stars in XRBs are quite uncertain. However, two lines observed in the XRB spectra of EXO 0748-676 (suggested to be H- and He-like Fe lines; see Cottam et al., 2002), plus the measurement of a 45 Hz neutron star spin frequency in the same source, allowed mass estimates in the range $1.5 < M_{NS}(M_\odot) < 2.3$, with a best fit for $1.8\,M_\odot$ (Villarreal and Strohmayer, 2004).

Light curves from X-ray bursts show a large variety of shapes (with single, double, or triple-peaked bursts; Fig. 5.5). Generally speaking, they are characterized by a fast rise (~1–10 s), a peak luminosity of $\sim 3 \times 10^{38}$ erg/s (Galloway et al., 2008; Lewin et al., 1993; Kuulkers et al., 2003), followed by a slower (sometimes exponential-like) decline (~10–100 s).

An interesting feature, observed in the spectra of many XRBs, is a 4.1 keV emission line (Waki et al., 1984), interpreted as Lyman $\alpha$ lines of helium-like Fe atoms,

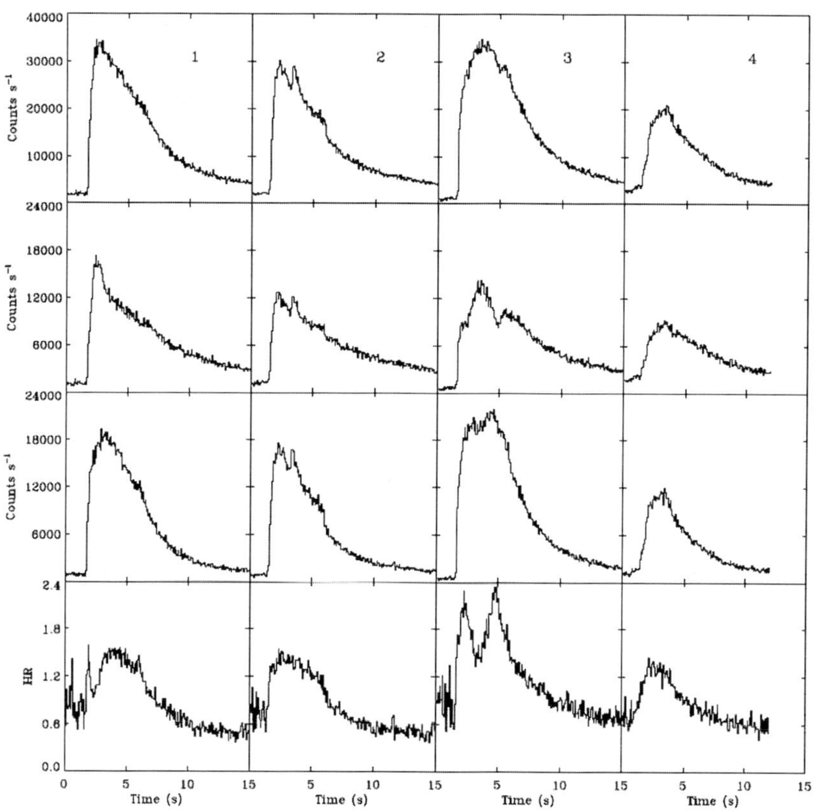


**Fig. 5.5** A suite of XRB lightcurves from the LMXB source 4U1728-34 as observed with the RXTE satellite. Each sequence (top to bottom), shows the overall count rates in the energy bands 2–60 keV, 2–6 kev, 6–30 keV, and the ratio (6–30 keV)/(2–6 keV). Figure from Strohmayer and Bildsten (2006)

broadened by Doppler and gravitational effects, likely originating at the inner edge of the accretion disk. Indeed, it has been suggested that time-resolved spectroscopy can in principle allow measurements of the surface gravitational redshift (Damen et al., 1990; Smale, 2001).

The fact that XRB sources do not exhibit X-ray pulsations suggest that the underlying neutron stars have weak magnetic fields ($< 10^{11}$ G). Indeed, pulsations are assumed to result from misalignment between the magnetic axis and the rotation axis of the neutron star. Moreover, it is unlikely that XRBs will originate from highly magnetized neutron stars, as a strong magnetic field would funnel the infalling charged plasma towards a small fraction of the neutron star surface, close to the magnetic caps; the effective accretion rate (per unit area) would be so high, that suppression of thermonuclear flashes would be expected (Joss, 1978; Taam and Picklum, 1978).

The understanding of the nature of XRBs requires also multiwavelength observations beyond the X-ray domain: in 1978, the first simultaneous optical/X-ray burst was detected from the source 1735-444 (Grindlay et al., 1978). The fluence in the optical burst was $\sim 2 \times 10^{-5}$ times that of the X-ray band, too large to be explained by the low-energy tail of the blackbody X-ray burst emission (Lewin et al., 1993). More important, the optical burst was delayed by ~3 s with respect to the X-ray peak (McClintock et al., 1979). A similar delay (~1.4 s) was also reported from Ser X-1 (Hackwell et al., 1979), and later, from many other sources (Lewin et al., 1993). These results suggest that optical emission observed from XRBs corresponds to reprocessing of X-rays in material within a few light-seconds from the source. Likely sites for this reprocessing include the accretion disk that surrounds the neutron star as well as the hemisphere of the secondary star directly illuminated by the X-ray source. Hence, the delay in the optical wavelengths results from travel-time differences between the X-rays leading directly to the observer and those that first intercept the disk, lose energy (becoming optical photons), and finally reach the observer.

The situation is less clear at other wavelengths: infrared emission has been suggested to accompany type I X-ray bursts. Indeed, detection of an infrared burst from the Rapid Burster has been claimed in the past (Kulkarni et al., 1979), although an unambiguous confirmation is lacking (Lewin et al., 1980). Also, although radio bursts have been reported from the Rapid Burster, no X-ray bursts were seen simultaneously (Hayakawa, 1981). More detailed observations at these wavelengths are required to disentangle the controversy.

### *5.4.2 Modeling X-Ray Bursts*

Modeling of type I XRBs and their associated nucleosynthesis has been extensively addressed by different groups. This reflects the astrophysical interest in determining the nuclear processes that power such explosions as well as in providing reliable estimates for the composition of the neutron star surface (Maraschi and Cavaliere,

1977; Woosley and Taam, 1976; Joss, 1977). Nonetheless, several thermal, radiative, electrical, and mechanical properties of the neutron star depend critically on the specific abundance pattern of its outer layers. Moreover, the diversity of shapes of XRB lightcurves is also probably due to different nuclear histories (see Heger et al., 2007, for a detailed analysis of the interplay between long bursts and the extension of the *rp* process), suggesting that the final chemical composition, at the end of the burst, is not unique.

The properties of the bursts are also affected by *compositional inertia*; that is, they are sensitive to the fact that accretion proceeds onto the ashes of previous bursts (Taam, 1980; Woosley et al., 2004). Indeed, this compositional inertia reduces the recurrence times between bursts, especially for scenarios involving accretion of metal-poor matter. Another critical quantity is the emerging heat flux from deeper layers of the neutron star (Ayasli and Joss, 1982; Fushiki and Lamb, 1987; Brown, 2000), which proved critical to the burst properties of pure He bursts (Bildsten, 1995).

The first studies of localized TNRs on neutron stars (Shara, 1982) suggested that heat transport was too inefficient to spread a local flame to the overall stellar envelope. Therefore, localized, volcanic-like explosions were predicted during X-ray bursts. However, it is worth noting that these studies relied only on radiative and conductive transport, ignoring the crucial role played by convection on the lateral thermalization of a TNR. The scenario was revisited by Fryxell and Woosley (1982), who suggested that the most likely outcome involves TNRs propagated by small-scale turbulences, in a deflagrative regime, leading to the horizontal spread of the front at typical velocities of $\sim 5 \times 10^6$ cm/s. Such speeds suggest that the time required for a flame to engulf the entire stellar surface is much longer than the characteristic spin periods of accreting neutron stars ($\sim$milliseconds). Hence, it was predicted that fast rotation of the neutron star could modulate localized burning regions, eventually allowing for a direct observation of the neutron star spin. Indeed, the discovery of high-frequency, burst oscillations in the X-ray source 4U1728-34 (360–600 Hz; see Strohmayer et al., 1996) provided first observational evidence for millisecond rotation periods in accreting neutron stars. Since then, burst oscillations have been claimed for many additional sources. Studies to constrain neutron star properties based on modeling of such oscillations are currently underway.

### *5.4.3 Nucleosynthesis in Type I X-Ray Bursts*

In contrast to classical nova outbursts, where the main nuclear activity is driven by proton-capture reactions in competition with $\beta^+$-decays, X-ray bursts are triggered by a combination of nuclear reactions, including H-burning (via *rp* process) and He-burning (that initiates with the triple $\alpha$-reaction, and is followed both by the breakout of the CNO cycle through $^{14,15}$O+$\alpha$, plus a competition of proton captures and ($\alpha$,p) reactions – the so-called $\alpha p$ process). Moreover, with a neutron star as the underlying compact object hosting the explosion, temperatures and densities in the accreted envelope reach quite high values: $T_{peak} > 10^9$ K (an order of magnitude higher than in nova outbursts), and $\rho \sim 10^6$ g/cm$^3$. As a result, detailed nucleosynthesis studies require the use of hundreds of isotopes, up to the SnSbTe mass region (Schatz et al., 2001), or even beyond (the nuclear activity in Koike et al., 2004 reaches $^{126}$Xe), and thousands of nuclear interactions extending to the proton drip line. In sharp contrast, the main nuclear activity for classical novae is limited to Ca, and runs close to the valley of stability.

Because of computational limitations, studies of XRB nucleosynthesis have usually been performed with limited nuclear reaction networks. More recently (Schatz et al., 1999, 2001), detailed nucleosynthesis calculations have been carried out with networks containing more than 600 isotopes (up to Xe, in Schatz et al., 2001), but using a one-zone approach, or also one-zone nucleosynthesis calculations with temperature and density profiles obtained with spherically symmetric evolutionary codes, linked to a 1,270-isotope network extending up to $^{198}$Bi (Koike et al., 2004). Other attempts (Parikh et al., 2008) include one-zone nucleosynthesis calculations, with temperature and density profiles obtained from the literature, and a large nuclear reaction network, containing 606 isotopes (up to $^{113}$Xe) and more than 3,500 nuclear processes. Note however that different numerical approaches and approximations have been adopted in all those works (hydrodynamic simulations with limited networks or one-zone calculations with detailed networks) and hence, the predicted nucleosynthesis in each case has to be taken with caution. Indeed, recent attempts have been made to couple hydrodynamic stellar calculations (in 1-D) and detailed networks with ~300 isotopes, up to $^{107}$Te (Fisker et al., 2008; Tan et al., 2007), with 1,392 nuclear processes and 325 isotopes, up to $^{107}$Te (José et al., 2010), or with networks containing up to 1,300 isotopes in an adaptive framework (Woosley et al., 2004).

To date, no multidimensional calculation for realistic XRB conditions has been performed (notice that the recent 2-D study by Zingale et al., 2001, relies on He-detonations on neutron stars, whereas TNRs are expected to propagate deflagratively in XRBs).

The relevant nuclear reaction path in XRBs (see Fig. 5.6) has been extensively discussed in the literature (José et al., 2010; Iliadis, 2007; Fisker et al., 2008): the most interesting nucleosynthesis is achieved for mixed H/He bursts, because of the complex nuclear reaction interplay (see details in Fisker et al., 2008; José et al., 2010). For illustrative purposes, we describe the main nuclear activity achieved for typical XRB conditions: a 1.4 $M_\odot$ neutron star, accreting solar-like material at a constant rate of $1.75 \times 10^{-9}$ $M_\odot$/year (José and Moreno, 2006; José et al., 2010). In general, such bursts are initiated by H-burning, specifically the cold mode of the CNO cycle (mainly through $^{12}$C(p,$\gamma$)$^{13}$N($\beta^+$)$^{13}$C(p,$\gamma$)$^{14}$N). At moderate temperatures, the main nuclear flow proceeds close to the valley of stability. When $T \sim 2 \times 10^8$ K, the nuclear activity already reaches $^{40}$Ca, with the most abundant species being H, $^4$He, and $^{14,15}$O. When $T$ approaches $\sim 3 \times 10^8$ K, the $3\alpha$ reaction dominates the nuclear flow, together with a combination of (p,$\gamma$) and (p,$\alpha$) reactions, and some $\beta^+$ decays (mainly $^{32,33}$Cl). At $T \sim 4 \times 10^8$ K, CNO-breakout ensues, initially led

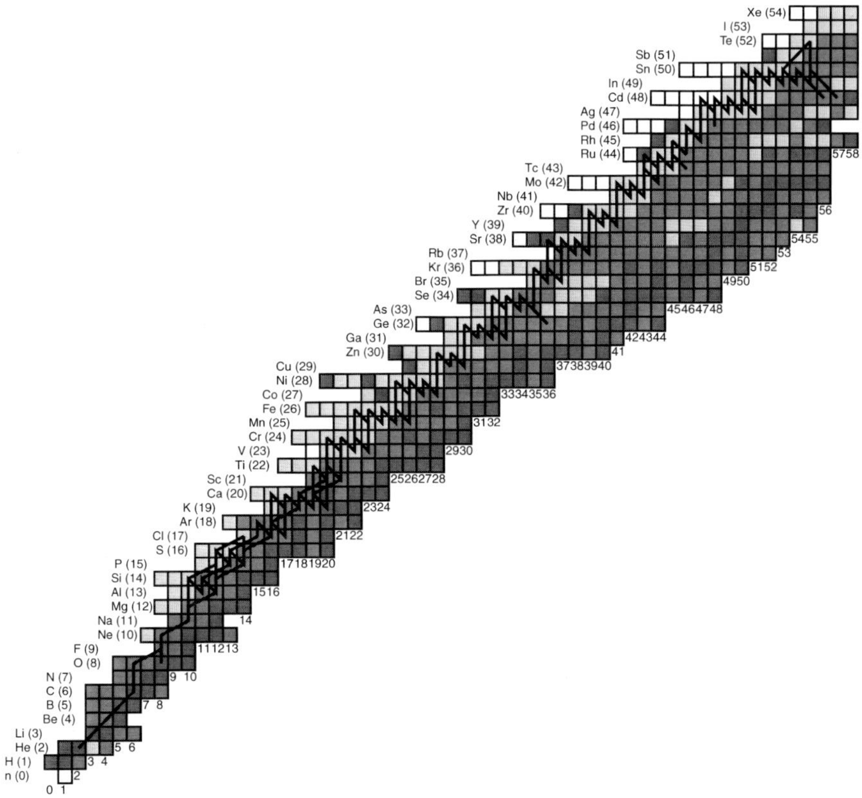


**Fig. 5.6** Main nuclear path during a typical X-ray burst. Figure from Schatz et al. (1999)

by $^{15}O(\alpha,\gamma)^{19}Ne$ (see Fisker et al., 2006, for a study of the impact of the $^{15}O(\alpha,\gamma)$ rate on the bursting behavior of an accreting neutron star), and followed by two consecutive proton-captures on $^{20}Na$ and $^{21}Mg$, where the flow recedes due to the strong photodisintegration reactions on $^{22}Al$. Following $^{21}Mg$-decay, the flow shifts to $^{21}Na(p,\gamma)^{22}Mg$, moving away from the valley of stability, towards the proton-drip line. As the rise of the temperature continues, and enough $^{14}O$ is build-up through the triple-$\alpha$ reaction, followed by $^{12}C(p,\gamma)^{13}N(p,\gamma)^{14}O$, the alternative path through $^{14}O(\alpha,p)^{17}F$ dominates the flow (Champagne and Wiescher, 1992; Woosley et al., 2004), by-passing the $^{15}O(\alpha,\gamma)^{19}Ne$ link to $^{21}Na$ through $^{17}F(p,\gamma)^{18}Ne(\alpha,p)^{21}Na$, with $^{18}Ne(\alpha,p)^{21}Na$ representing the main path towards heavier species.

When $T \sim 1.5 \times 10^9$ K, the most abundant species in the envelope become $^{18}Ne$, $^{21,22}Mg$ (from $^{18}Ne(\alpha,p)^{21}Na(p,\gamma)^{22}Mg$), $^{25}Si$, $^{28}S$-$^{28}P$, $^{33}Ar$-$^{33}Cl$, and $^{37}K$, the first isotope that achieves an abundance of 10% by mass. At this stage, the flow has reached $^{64}Ge$. Shortly after, the envelope achieves peak temperature, $T_{peak} \sim 1.7 \times 10^9$ K. The most abundant isotope (except for H) is $^{54}Ni$, and later, $^{64}Ge$ and $^{68}Se$. During the subsequent decline, the nuclear flow is dominated by a cascade of $\beta$-decays. The final composition of the envelope, which is not ejected by the TNR, is essentially composed of elements with A = 60–70, mainly $^{64}$Zn (originally as $^{64}$Ge, and $^{64}$Ga), and $^{68}$Zn ($^{68}$Se), with traces of other species. Explosions in lower metallicity envelopes are characterized by an extension of the main nuclear path by the *rp* process, much beyond $^{56}$Ni, up to the SnSbTe region (Schatz et al., 2001) or beyond, according to a recent reanalysis of the role of photodisintegration reactions in this mass region (Elomaa et al., 2009).

Most of the reaction rates required for these extensive nucleosynthesis calculations rely on theoretical estimates from statistical models, and may be affected by significant uncertainties. Efforts to quantify the impact of such nuclear uncertainties on the overall abundance pattern accompanying XRBs have been undertaken by different groups (Iliadis et al., 1999; Thielemann et al., 2001; Amthor et al., 2006), revealing a complex interplay between the nuclear activity and the shape of the light curve (Hanawa et al., 1983; Woosley et al., 2004). The most extensive work to date (Parikh et al., 2008), has helped to identify the most influential nuclear processes: $^{65}$As(p, $\gamma$)$^{66}$Se, $^{61}$Ga(p, $\gamma$)$^{62}$Ge, $^{12}$C($\alpha$, $\gamma$)$^{16}$O, $^{96}$Ag(p, $\gamma$)$^{97}$Cd, and in a lesser extent, $^{30}$S($\alpha$, p)$^{33}$Cl, $^{56}$Ni($\alpha$, p)$^{59}$Cu, $^{59}$Cu(p, $\gamma$)$^{60}$Zn, $^{86}$Mo(p, $\gamma$)$^{87}$Tc, $^{92}$Ru(p, $\gamma$)$^{93}$Rh, $^{102}$In(p, $\gamma$)$^{103}$Sn, and $^{103}$In(p, $\gamma$)$^{104}$Sn.

A major drawback in the modeling of X-ray bursts comes from the lack of observational nucleosynthetic constraints (beyond the obvious implications for the physics of the neutron star crusts, outlined at the beginning of this Section). The potential impact of XRB nucleosynthesis on Galactic abundances is still a matter of debate: although ejection from a neutron star is unlikely because of its large gravitational potential, radiation-driven winds during photospheric radius expansion may lead to ejection of a tiny fraction of the envelope, containing nuclear processed material (Weinberg et al., 2006; MacAlpine et al., 2007). Indeed, although it has been claimed that XRBs may help to explain the Galactic abundances of the problematic light *p-nuclei* (Schatz et al., 1998), new calculations have ruled out this possibility (Bazin et al., 2008; José et al., 2010). Finally, it has been proposed that a way to overcome the lack of observational constraints may come from the identification of gravitationally redshifted atomic absorption lines, which could be identified through high-resolution X-ray spectra (Bildsten et al., 2003; Chang et al., 2005, 2006; Weinberg et al., 2006). Indeed, although specific features have been reported in the spectra of 28 XRBs detected from EXO 0748-676 during a 335 ks observation with XMM-Newton (Cottam et al., 2002), interpreted as gravitationally redshifted absorption lines of Fe XXVI (during the early phase of the bursts), Fe XXV, and perhaps O VIII (during the late stages), no evidence for such spectral features was found neither after a 200 ks observation of GS 1826-24, from which 16 XRBs were detected (Kong et al., 2007), nor after a 600 ks observation of the original source EXO 0748-676 (Cottam et al., 2008; Rauch et al., 2008).

### 5.4.4 Superbursts

Whereas regular, type I XRBs are characterized by common features in terms of duration, energetics, and recurrence times, a few extremely energetic events have recently been detected thanks to better performances in monitoring achieved with

X-ray satellites (i.e., BeppoSAX, Chandra, or XMM-Newton). These rare and rather violent events are known as *superbursts* (see Kuulkers, 2004; Cumming, 2005, for reviews). The first observation of a superburst was reported by Cornelisse et al. (2000) in the framework of a "common" type I bursting source (c.f., the BeppoSAX source 4U1735-44).

To date, 15 superbursts from 10 different bursting sources have been discovered, including GX 17+2, for which 4 superbursts have been identified (in't Zand et al., 2004). Although the term *superburst* was first used by Wijnands (2001) to describe these very long X-ray bursts, historically the same name was applied to a relatively strong type I XRB reported from 4U1728-34 by Basinska et al. (1984), back in 1984.

Superbursts represent some sort of extreme X-ray bursts: they have long durations (see Fig. 5.7), with a typical (exponential) decay time ranging from 1 to 3 h (including an extreme case, KS 1731-260, that lasted for more than 10 h; see Kuulkers et al., 2002), extremely energetic (about ∼1,000 times more than a typical XRB, that is, $\sim 10^{42}$ erg), and with much longer recurrence periods (4.7 year for the system 4U 1636-53, for which two superbursts have been observed to date; see Wijnands, 2001). Although superburst sources also exhibit regular type I XRBs, their occurrence is quenched for about a month after each superburst.

The duration and energetics of superbursts suggest that they result from thermonuclear flashes occurring in deeper fuel layers than those from typical X-ray bursts (at densities exceeding $10^9$ g/cm$^3$; see Cumming and Bildsten, 2001), more likely, in the C-rich ashes resulting from type I X-ray bursts (first proposed by

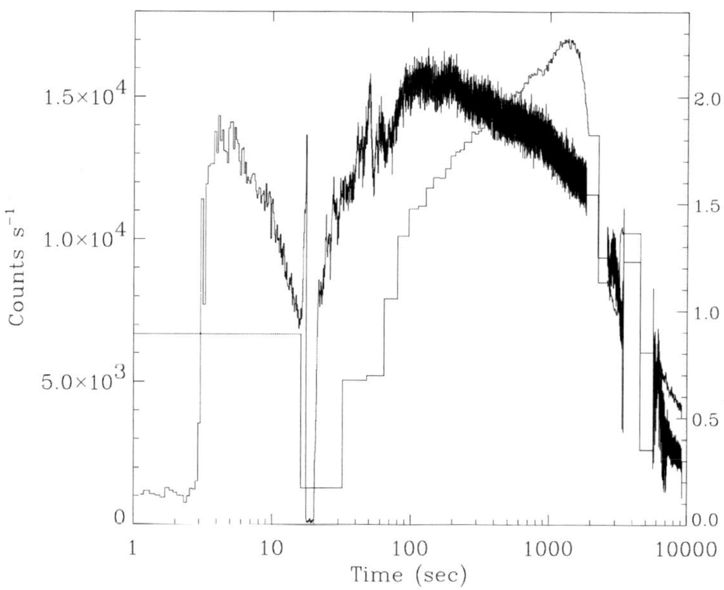


**Fig. 5.7** Time evolution of the X-ray luminosity of the 3-hr long superburst observed with the RXTE satellite in the source 4U 1820–30. Figure from Strohmayer and Brown (2002)

Woosley and Taam, 1976; see also Brown and Bildsten, 1998; Cumming and Bildsten, 2001; Weinberg and Bildsten, 2007).

Controversy remains over how much carbon is left after a type I burst: some studies (Schatz et al., 1999, 2001) have indeed shown than most of the C is burnt during the previous H/He burning episodes. However, other analyses (Cumming and Bildsten, 2001) led to the conclusion that even small amounts of carbon are enough to power a superburst (especially in neutron star oceans enriched from the heavy ashes driven by the *rp* process). Recent studies suggest that both stable and unstable burning of the accreted H/He mixture are required to power a superburst (in't Zand et al., 2003). Alternative models have also been proposed to account for the origin of such superbursts, including TNRs on strange quark matter stars (Page and Cumming, 2005).

## 5.5 Observational Diagnostics of Binary-Systems

### *5.5.1 Gamma-Rays from Radioactivity*

Novae and supernovae emit $\gamma$-rays because some of the nuclei they synthesize and eject into the interstellar medium are radioactive, either $\beta^+$-unstable (i.e., emitting a positron when decaying) or undergoing electron captures. Radioactive isotopes decay to excited states of their daughter nuclei, which de-excite to their ground states by emitting $\gamma$-ray photons with energies around 1 MeV, over a wide range of timescales. Table 5.1 shows the most relevant radioactive isotopes produced in novae and supernovae. Two additional isotopes, the $\beta^+$-unstable $^{13}$N and $^{18}$F ($\tau$=862 s and 158 min, respectively), are also important in the case of novae. The emitted $\gamma$-rays can be potentially detected, either in individual objects or as diffuse emission from the cumulative $\gamma$-ray output of many objects in the galaxy, whenever the lifetime of a given isotope is longer than the average period between two successive events producing it (see Sect. 5.5.1.3). In addition, the positrons emitted when $\beta^+$-unstable nuclei decay annihilate with electrons and produce emit $\gamma$-rays, powering a 511 keV line plus a continuum below this energy.

The shape and intensity of the $\gamma$-ray output of novae and supernovae, as well as its temporal evolution, depend not only on the number of $\gamma$-ray photons produced, but also on how they propagate through the expanding envelope and ejecta. The first step to compute the spectrum is to generate $\gamma$-rays according to the decay schemes of the corresponding radioactive isotopes. The number of photons generated in a particular object depends on the isotopic abundances and decay rates of the relevant nuclei. In addition to these *direct* $\gamma$-ray photons, positrons emitted as a consequence of the radioactive decays of the $\beta^+$-unstable nuclei (see Table 5.1) should be traced. Once photons are generated, their trip across the expanding ejecta should be simulated by taking into account the various interaction processes affecting their propagation, i.e., Compton scattering, e$^-$–e$^+$ pair production, and photoelectric absorption.

The treatment of positron annihilation deserves particular attention. The role of magnetic fields is crucial, but it is not well known how to handle it. Thus, some drastic approximations are often made. When a positron is emitted, it can either escape without interacting with the expanding envelope or annihilate with an ambient electron. In nova envelopes, it is safe to assume that positrons thermalize before annihilating. This approximation is wrong in less than 1% of cases in an electronic plasma (Leising and Clayton, 1987). In a neutral envelope, the excitation cross-section dominates any other interaction at energies above ~100 eV (Bussard et al., 1979), and thus positrons lose energy until they reach this value. In order to reproduce this braking effect, positrons should be propagated until they cross an equivalent column of ~0.2 g/cm$^2$, measured along a straight line (Chan and Lingenfelter, 1993). This is the mean range expected for a 0.6 MeV positron slowing to energies ~100 eV through elastic collisions with the surrounding medium, when the effect of magnetic fields on its propagation is neglected. Once thermalized, the positron covers a negligible distance and then annihilates.

For densities and temperatures typical of novae and SNIa envelopes, positrons form positronium (positron – electron system) in ~90% of annihilations (Leising and Clayton, 1987), while in the remaining 10% of cases they annihilate directly. Positronium is formed in the singlet state 25% of the time, leading to the emission of two 511 keV photons, and in the triplet state 75% of the time, leading to a three-photon annihilation continuum. The spectrum of photons produced from the triplet state was obtained by Ore and Powell (1949). Therefore, once a positron is produced, its trip should be followed until it escapes or covers the average energy-loss distance. In the latter case it produces positronium 90% of the time, resulting in triplet or singlet annihilations in a 3:1 ratio, while in 10% of the cases it annihilates directly. Monte Carlo codes, based for instance on the method described in Pozdnyakov et al. (1983) and Ambwani and Sutherland (1988), are well suited to compute the $\gamma$-ray output of novae and type Ia supernovae (Gómez-Gomar et al., 1998a,b).

#### 5.5.1.1 Gamma-Ray Emission from Individual Classical Novae

The potential of novae as $\gamma$-ray emitters was first pointed out by Clayton and Hoyle (1974), who stated that observable $\gamma$-rays from novae would come from electron-positron annihilation, with positrons from $^{13}$N, $^{14}$O, $^{15}$O and $^{22}$Na decays, as well as a result of the decay of $^{14}$O and $^{22}$Na to excited states of $^{14}$N and $^{22}$Ne nuclei, which de-excite by emitting photons at 2.312 and 1.274 MeV respectively. Some years later, Clayton (1981) noticed that another $\gamma$-ray line could be expected from novae when $^7$Be transforms (through an electron capture) to an excited state of $^7$Li, which de-excites by emitting a photon of 478 keV. The original idea came from Audouze and Reeves (1982), and both works were inspired by the contemporaneous papers mentioning the possibility of $^7$Li synthesis in novae (Arnould and Norgaard, 1975; Starrfield et al., 1978b). In fact, $^7$Li production in novae was, and continuous to be, a crucial topic (Hernanz et al., 1996), since Galactic $^7$Li is not well accounted for by other sources, either stellar (AGB stars), interstellar (spallation reactions by cosmic rays) or cosmological (Big Bang). The main ideas presented in these pioneering studies have remained unchanged; but some aspects have changed in the last 20 years, mainly related to new detailed nucleosynthesis studies of novae.

The $\gamma$-ray signatures of classical novae depend on their yields of radioactive nuclei (see the reviews by Leising (1991, 1993) and Hernanz (2002, 2008). CO and ONe novae differ in their production of $^7$Be, $^{22}$Na and $^{26}$Al, while they synthesize similar amounts of $^{13}$N and $^{18}$F. In both nova types, there should be line emission at 511 keV related to $e^-$–$e^+$ annihilation, and a continuum produced by Comptonized 511 keV emission and positronium decay.

The temporal evolution of the whole $\gamma$-ray spectrum of four representative nova models is shown in Fig. 5.8. The most prominent features of the spectra are the annihilation line at 511 keV and the continuum at energies between 20 and 30 keV and 511 keV (in both nova types), the $^7$Be line at 478 keV in CO novae, and the $^{22}$Na line at 1275 keV in ONe novae. Therefore, the main difference between spectra of CO and ONe novae are the long-lived lines, which directly reflect the different chemical composition of the expanding envelope ($^7$Be-rich in CO novae and $^{22}$Na-rich in ONe ones).

The early $\gamma$-ray emission, or *prompt* emission, of novae is related to the disintegration of the very short-lived radioisotopes $^{13}$N and $^{18}$F. The radiation is emitted as a line at 511 keV (direct annihilation of positrons and singlet state positronium), plus a continuum (Gómez-Gomar et al., 1998a; Hernanz et al., 2002). The continuum is related to both the triplet state positronium continuum and the Comptonization of the photons emitted in the line. There is a sharp cut-off at energies 20–30 keV (the exact value depending on the envelope composition) because of photoelectric absorption (see Fig. 5.8). The largest flux is emitted in the (20–250) keV range, since the continuum has its maximum at ~60 keV (ONe novae) and at ~45 keV (CO novae), followed by the flux in the (250–511) keV range (excluding the 511 keV line) and the flux in the 511 keV line (see Fig. 5.9). The two maxima in the light curves of the 511 keV line correspond to $^{13}$N and $^{18}$F decays, but the first maximum is difficult to resolve because its duration is extremely short; in addition, it is very model dependent: only $^{13}$N in the outermost zones of the envelope could be seen in $\gamma$-rays because of limited transparency at very early epochs and, therefore, the intensity of the first maximum depends on the efficiency of convection. This first maximum thus provides important insight into the dynamics of the envelope after the peak temperature is attained at its base.

The annihilation emission is the most intense $\gamma$-ray feature expected from novae, but unfortunately it has a very short duration, because of the short lifetime of the main positron producers ($^{13}$N and $^{18}$F). There are also positrons available from $^{22}$Na decay in ONe novae, but these contribute much less (they are responsible for the *plateau* at a low level, between $10^{-6}$ and $10^{-5}$ ph/cm$^2$/s, for $d = 1$ kpc; see Fig. 5.9). However, after roughly 1 week the envelope is so transparent that $^{22}$Na positrons escape freely without annihilating. In summary, annihilation radiation lasts only ~1 day at a high level, and 1–2 weeks at a lower level *plateau* (the latter only in ONe novae). Another fact preventing easy detection is the early (before the nova is discovered optically) appearance of $\gamma$-rays from electron-positron annihilation (see Fig. 5.9).

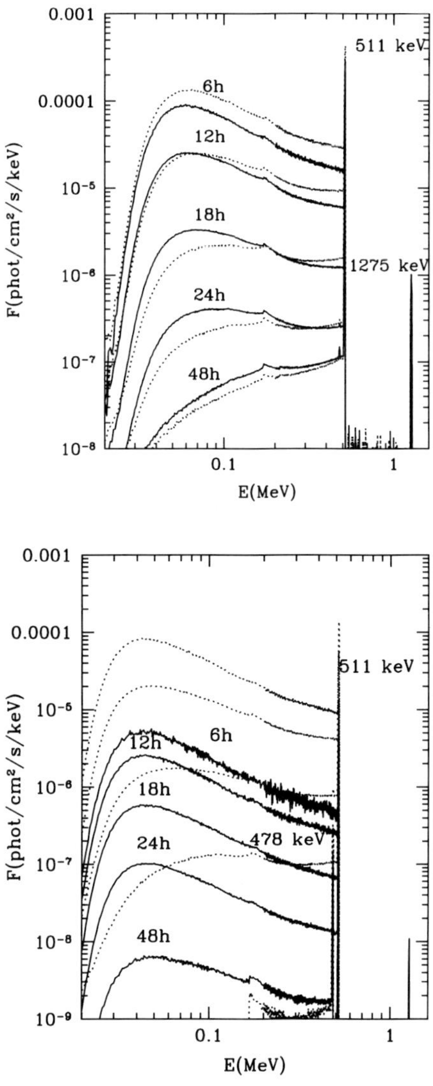


**Fig. 5.8** *Top panel*: spectra of ONe novae of masses 1.15 (*solid*) and 1.25 $M_\odot$ (*dotted*). at different epochs after $T_{max}$ (labels for *dotted lines* follow the same sequence as those for *solid lines*: from *top* to *bottom* 6, 12, 18, 14 and 48 h). *Bottom panel*: same for CO novae of masses 0.8 (*solid*) and 1.15 $M_\odot$ (*dotted*). Distance is 1 kpc

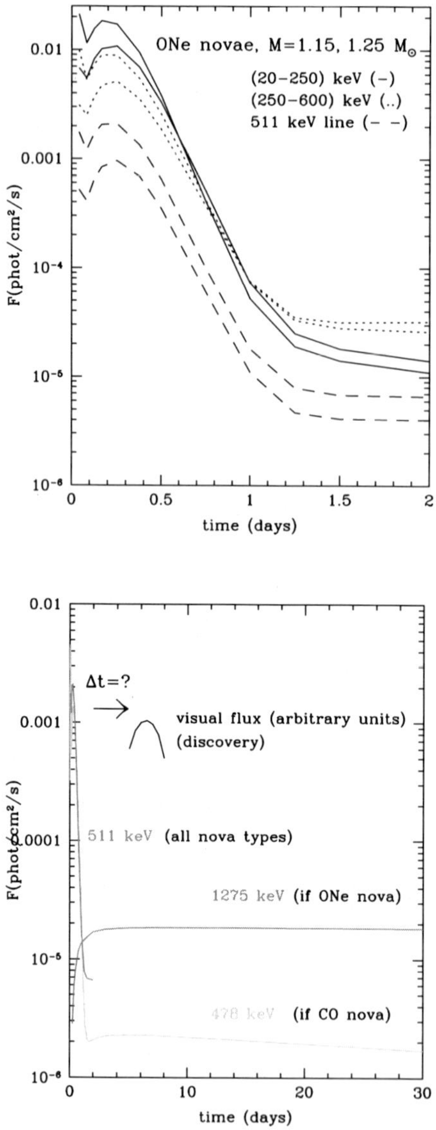


**Fig. 5.9** *Top panel*: *light curve* of two continuum bands below 511 keV for ONe novae. The *upper curves* correspond to the larger mass, at early times; but at later epochs the most massive nova emits a slightly smaller flux, except for the 511 keV *line*, because of larger transparency. The *light curve* of the 511 keV line is also shown. Distance is 1 kpc. *Bottom panel*: nova $\gamma$-ray light *curves*, as compared with visual ones. The *vertical scale* for the visual *light curve* is arbitrary

The most distinctive feature in the $\gamma$-ray spectra of CO novae is line emission at 478 keV, related to de-excitation of the $^{7}$Li which results from an electron capture on $^{7}$Be. The light curves of the 478 keV line are shown in Fig. 5.10: the flux reaches its maximum at day 13 and 5 in the more and less opaque models, with total masses 0.8 and 1.15 $M_{\odot}$, respectively. The width of the line is 3 and 8 keV for the 0.8 and 1.15 $M_{\odot}$ CO novae, respectively. The maximum flux is around $10^{-6}$ ph cm$^{-2}$ s$^{-1}$, for $d = 1$ kpc.

The $^{22}$Na line at 1,275 keV appears only in ONe novae, because CO novae do not synthesize this isotope. The rising phase of the 1,275 keV line light curves (see Fig. 5.10) lasts between 10 (1.25 $M_{\odot}$) and 20 days (1.15 $M_{\odot}$). Soon after the maximum, the line flux declines with the lifetime of $^{22}$Na, 3.75 years. The line intensities directly reflect the amount of $^{22}$Na ejected mass during this phase. The corresponding fluxes at maximum are typically around $10^{-5}$ ph/cm$^{2}$/s, at $d = 1$ kpc, and the width of the line is around 20 keV, which poses severe problems for its detectability with instruments having high energy resolution, like SPI onboard INTEGRAL.

There have been many unsuccessful attempts to detect $\gamma$-rays from novae. The main efforts have focused on the 1,275 keV line from $^{22}$Na in individual objects, but searches of the cumulative emission have also been performed. The annihilation line has also been searched for whenever wide field of view instruments were available, scanning zones of the sky where novae had exploded.

The most recent observational search for the 1,275 keV line from novae was performed with the COMPTEL instrument onboard the Compton Gamma-Ray Observatory (CGRO) (Iyudin et al., 1995). COMPTEL observed a number of recent novae during the period 1991–1993, five of which of the neon type (i.e. those expected to emit the 1,275 keV line). None was detected. The average $2\sigma$ upper limit for any nova of the ONe type in the galactic disk was around $3 \times 10^{-5}$ ph/cm$^{2}$/s, which translated into an upper limit of the ejected $^{22}$Na mass around $3.7 \times 10^{-8}$ $M_{\odot}$, for the adopted distances. This limit was constraining for models available at the time (Starrfield et al., 1992, 1993; Politano et al., 1995), but is not so for current models (José and Hernanz, 1998). The main reason for the discrepancy between models of different groups (José and Hernanz (1998) versus Politano et al. (1995) and Starrfield et al. (1998)) is the following: old models were based on the explosion on ONeMg white dwarfs, with some mixing between the accreted H-rich matter and the underlying material, whereas recent models adopt ONe white dwarfs as underlying cores, because more recent evolutionary calculations of stellar evolution predict much lower magnesium abundances (Ritossa et al., 1996; Dominguez et al., 1993). The smaller initial content of neon and magnesium makes $^{22}$Na synthesis much less favored. Different nuclear reaction rates also have an impact on the final yields obtained by different groups.

The first search for the 478 keV line from the galactic center and from some particular novae was performed with SMM/GRS (Harris et al., 1991), yielding upper limits around $10^{-3}$ ph/cm$^{2}$/s, corresponding to $^{7}$Be ejected masses around $10^{-7}$ $M_{\odot}$. These fluxes and masses are well above the current theoretical predictions and thus do not constrain the models. More recent analyses of novae during the period 1995–1997, have been possible thanks to the Transient Gamma-Ray Spectrometer (TGRS) onboard the Wind satellite. The flux limits from TGRS were a factor of

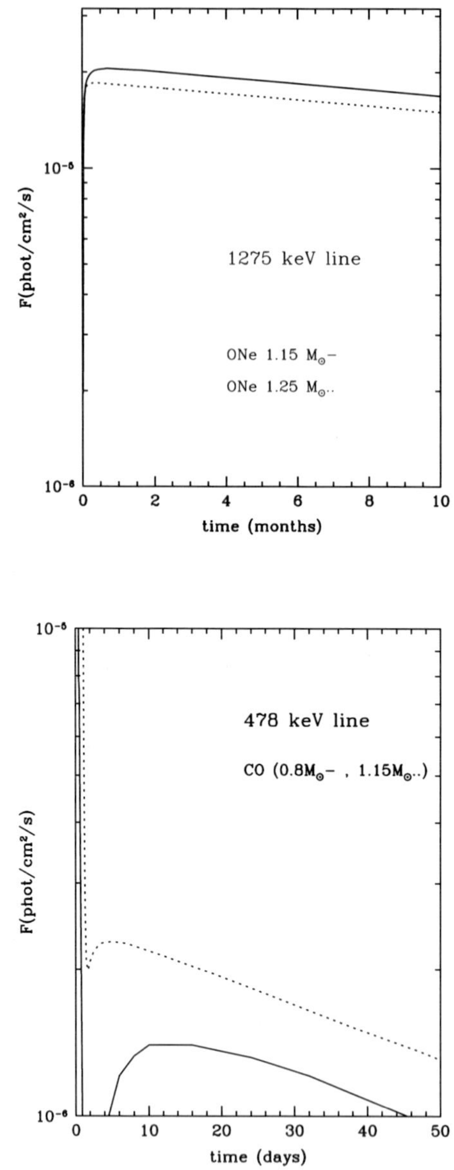


**Fig. 5.10** *Top panel*: *Light curve* of the 1,275 keV line for two ONe nova models. *Bottom panel*: *Light curve* of the 478 keV line for two CO nova models. Distance is 1 kpc

10 smaller than those from SMM observations, but the upper limits on $^7$Be ejected masses did not improve by the same factor, mainly because novae observed with TGRS were at larger distances than those observed with SMM (Harris et al., 2001).

As mentioned above, the emission resulting from $e^-$ to $e^+$ annihilation is the most intense $\gamma$-ray outcome of classical novae, but $\gamma$-rays are emitted well before the visual maximum of the nova, i.e. typically before the nova is discovered, and have a very short duration (see Fig. 5.9). Therefore, they can not be detected through observations pointing to a particular nova already discovered. Wide field of view instruments monitoring the sky in the appropriate energy range, like the Burst and Transient Source Experiment (BATSE) onboard CGRO, TGRS onboard Wind, and the Burst Alert Telescope (BAT) onboard Swift (Senziani et al., 2008), are best suited for the search of the 511 keV line and the continuum below it.

TGRS was very convenient to search for the 511 keV line, because of its large field of view, and also because its germanium detectors had enough spectral resolution to separate the cosmic 511 keV line from the nova line, provided that the latter is a bit blueshifted (this happens only at the beginning of the emission phase, when material is not completely transparent yet) (Harris et al., 1999). TGRS's field of view contained five new novae during the period 1995–1997; upper limits were obtained by Harris et al. (1999), who deduced that their method was sensitive enough to detect novae occurring out to about 0.8 kpc, for any nova type (CO and ONe).

Another instrument that was well suited for the detection of the prompt $\gamma$-ray emission from novae was BATSE onboard CGRO. Before the launch of CGRO in 1991, a prediction was made (Fishman et al., 1991) on the detectability of low-energy $\gamma$-rays from novae with the BATSE instrument, based on the models of $\gamma$-ray emission from Leising and Clayton (1987). BATSE had the advantage of continuously covering almost the whole sky, but on the other hand it was less sensitive and had poor energy resolution. More recently, a posteriori analyses of the background data at the explosion epoch of all classical novae discovered optically during the whole period of CGRO operation (1991–2000), searching for some signal, were performed (Hernanz et al., 2000). The 3-$\sigma$ sensitivity using the 511 keV data only is similar to that with WIND/TGRS (Harris et al., 1999), but TGRS's sensitivity required a particular line blueshift, whereas BATSE is independent of it. The 3-$\sigma$ sensitivity using the (250–511) keV data is a little more than a factor of 2 better than that from TGRS (Harris et al., 1999).

The 2002 launch of the ESA satellite International Gamma-Ray Laboratory, INTEGRAL, opened new perspectives for the detection of $\gamma$-rays from explosive events, with its two major instruments, the spectrometer SPI and the imager IBIS. SPI is made of 19 germanium detectors; its $3\sigma$ sensitivity at 1 MeV, for $10^6$ s observation time and narrow lines, is around $2.4 \times 10^{-5}$ ph/cm$^2$/s, with 2 keV energy resolution. However, this sensitivity degrades considerably for broad lines. Detection of $\gamma$-rays from novae with INTEGRAL is not too likely, because its detectability distance limits are small and, therefore, few novae are expected (Hernanz and José, 2004). This is due to both the small fluxes expected and the reduced (with respect to pre-launch estimates) inflight measured sensitivities at the relevant energies. Very small distances are needed to obtain a secure detection: around 0.2 kpc for the 478 keV line from $^7$Be and around 0.7 kpc for the 1,275 keV line from $^{22}$Na.

#### 5.5.1.2 Gamma-ray Emission from Individual Type Ia Supernovae

In Type Ia supernova ejecta, the dominant radioactive chains are $^{56}Ni \rightarrow ^{56} Co \rightarrow ^{56} Fe$ and $^{57}Ni \rightarrow ^{57} Co \rightarrow ^{57} Fe$ (see Table 5.1). The amount of radioactive material, its distribution within the ejecta as well as the density, velocity and chemical composition profiles are different for each model described in Sect. 5.3 and these differences affect the total intensity and the evolution of the different lines, as well as the importance and extension of the continuum component of the spectrum (Burrows and The, 1990; Kumagai and Nomoto, 1997; Gómez-Gomar et al., 1998b).

As shown in Fig. 5.11, 20 days after the explosion all models involving a prompt or a delayed detonation display strong lines because their high expansion rates induce a rapid decrease of the density. Lines are particularly intense for those models containing $^{56}Ni$ and $^{56}Co$ in the outer layers (pure detonation and sub – Chandrasekhar models). The maximum intensity of these lines is model dependent since it depends on the expansion rate and on the distribution of $^{56}Ni$. Pure deflagration models only display a continuum since they efficiently Comptonize high energy $\gamma$-rays. The shape of the continuum at low energies is limited in all models by the competing photoelectric absorption, which imposes a cut-off below 40–100 keV. The energy of the cut-off is determined by the chemical composition of the external

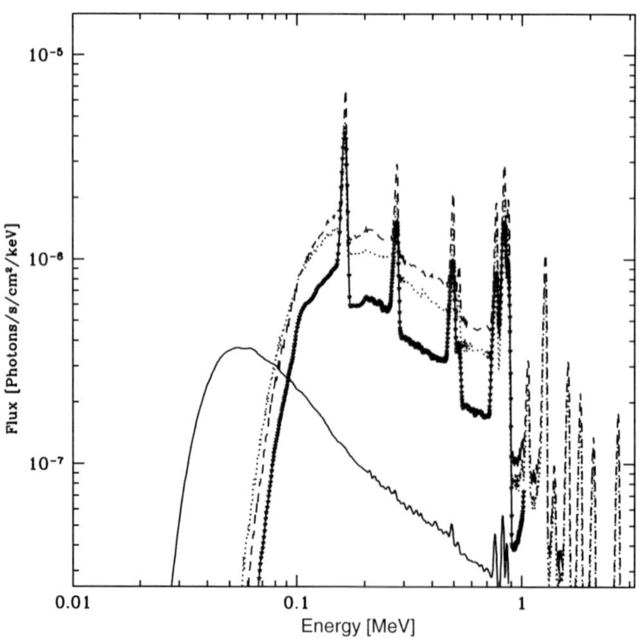


**Fig. 5.11** Gamma-ray spectrum for four models of SNIa explosion at 5 Mpc. Pure deflagration model (*solid line*), delayed detonation model (*long – dashed line*), detonation model (*dashed line*) and sub-Chandrasekhar model (*starred line*)

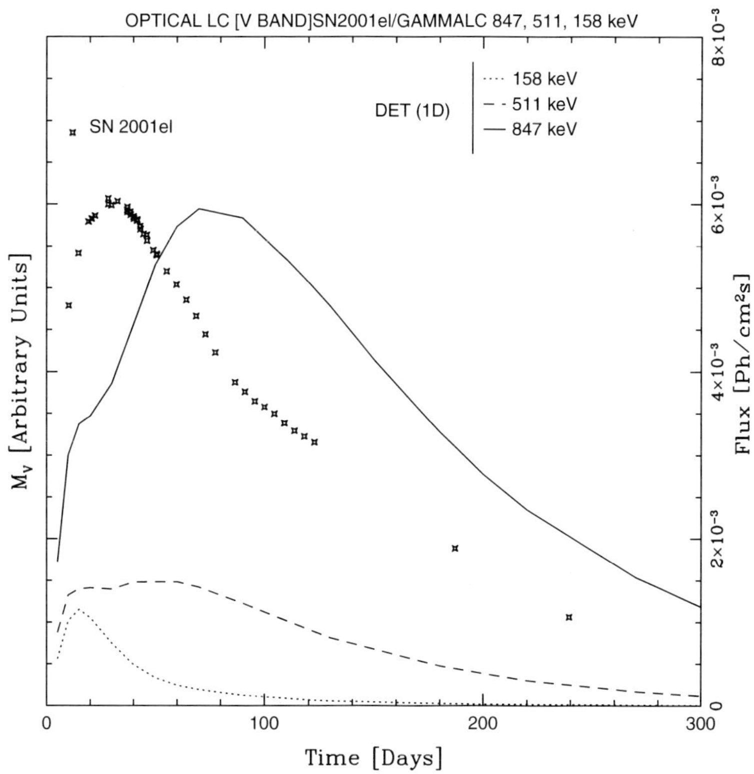


**Fig. 5.12** Evolution of the different lines as a function of time for a typical delayed detonation model. The distance is assumed to be 1 Mpc. The *optical light curve* in the visual has also been included in order to provide a time reference. Courtesy of A. Hirschmann

layers where most of the emergent continuum is formed at this epoch. Consequently, the continuum of those models containing low Z elements in the outer layers will extend to lower energies than that of those containing high Z elements. Therefore, it is possible to use these differences to discriminate among the different burning modes.

The 158, 812 keV $^{56}$Ni – lines peak very early, near the maximum of light and, because of absorption, they are much weaker than those of $^{56}$Co (see Fig. 5.12) Therefore, an early detection can provide information about the location of $^{56}$Ni in the debris. Nevertheless, because of its intensity, the most promising spectral feature is the 847 keV $^{56}$Co line, which reach maximum intensity roughly 2 months after the explosion in all models except for the pure deflagration case. At this moment, the intensity of the lines, except for the deflagration model, is determined by the total mass of the radioactive isotopes, while the effect of the differences in expansion velocities is secondary and the position of the low energy cut-off tends to converge towards 70 keV. Four months after the explosion, the ejecta are optically thin in all cases and the intensity of the lines is proportional to the total mass of the parent isotopes. The continuum is weak and dominated by the positronium annihilation component which shows a step below 170 keV, the energy of the backscattered 511 keV photons. At this epoch line profiles reveal the distribution in velocity of their parent isotopes in all layers of the ejecta. It is thus clear that there are noticeable differences between the spectra of these models, especially at early times, when the total amount of radioactive isotopes and their distribution, the expansion rate and the composition of the ejecta affect the emission.

Figure 5.12 also displays the evolution of one of the most prominent lines, the 511 keV annihilation one. Positrons emitted during the decay of $^{56}$Co thermalize because of ionization and excitation energy losses as well as other mechanisms, and eventually they annihilate either directly or through the formation of positronium. The degree of ionization and the structure of the magnetic field is crucial to determine the fraction of positrons that escape from SNIa. After 200 days almost all the high energy photons escape and the energy deposited by the annihilation of positrons is the only available source to power the light curve. Therefore, a careful determination of the 511 keV line is fundamental to understand the evolution of the supernova debris (Milne et al., 2001).

The early X – ray emission in the 6–8 keV region can provide an important diagnostic for discriminating between Chandrasekhar and sub-Chandrasekhar models. The $\gamma$-rays produced by disintegration of $^{56}$Ni and $^{56}$Co, together with the thermalized photons with energies above $\sim$7 keV, induce strong emission of the $K_\alpha$ – lines of Fe, Co and Ni. In the case of the Chandrasekhar models, $^{56}$Ni is so deeply placed that these photons are absorbed before escaping while in the sub-Chandrasekhar models they are produced in the outermost layers from where they freely escape producing a distinctive feature in the spectra. The total expected flux in the 5–10 keV band at 15 Mpc is $\sim 2 \times 10^{-7}$ ph/s/cm$^2$, which means that it could be detected from a reasonably close supernova (Pinto et al., 2001). Another feature that could also be used to distinguish among these two families of models is the 14.4 keV emission of $^{57}$Co, which is only expected in sub-Chandrasekhar models.

XMM and Chandra allow high spectroscopic and angular resolution studies of some galactic remnants of SNIa. In particular, for the X-rays from the Tycho supernova remnant the best fit is obtained with a one dimensional delayed detonation characterized by a quite high density transition placed in the range (2.2–2.5) $\times$ 10$^7$ g/cm$^3$ (Badenes et al., 2006), while in the case of G337.2–07 the best fit is obtained for a pulsational delayed detonation with a density transition at 7.7 $\times$ 10$^6$ g/cm$^3$ (Rakowski et al., 2006). In both cases, the X-ray spectrum strongly suggests a high degree of chemical stratification, a property that is lacking in most current three-dimensional models of SNIa.

Interestingly enough, DDT models also provide the best fit to the X-ray spectra of 22 clusters of galaxies (de Plaa et al., 2007) and is the only model able to match the observed Ar/Ca and the Ca/Fe ratios. It is important to remember that the chemical composition of the intracluster medium is representative of the average supernova yields, since it is the result of the contributions from many supernovae during the cluster?s life time.

#### 5.5.1.3 Contribution of Classical Novae to Diffuse Radioactivities

Some radioactive nuclei have lifetimes larger than the typical time elapsed between successive novae or type Ia supernova explosions in the Galaxy. For such cases, diffuse emission resulting from the cumulative effect of several sources is expected. This kind of emission should trace the galactic distribution of the corresponding sources of the given isotope. If detected, it would give a valuable information, not available from observations at other wavelengths because of interstellar extinction. The Galactic nova spatial distribution and the nova rate are in fact poorly known, since their determination relies on observations of novae in other galaxies or on extrapolations of observations in our Galaxy, taking into account the distribution of extinction related to interstellar dust (Della Valle and Livio, 1994; Shafter, 1997, 2002). $^{22}$Na and $^{26}$Al from novae are potential contributors to diffuse emission at 1,275 and 1,809 keV, respectively. For $^{22}$Na, there is the advantage that only novae are expected to contribute to its galactic content, whereas for $^{26}$Al massive stars and AGBs also clearly contribute. Therefore, the galactic 1,275 keV emission from $^{22}$Na should trace directly the spatial nova distribution; but unfortunately, as analyzed below, the predicted emission is too low for the performances of the current instruments. Concerning $^{26}$Al, since its emission has been detected in the Galaxy, an estimate of the nova contribution to the global line flux is needed (Diehl et al., 1995; Prantzos & Diehl, 1996).

The global flux at 1,275 keV depends on the amount of $^{22}$Na ejected per nova explosion and on the distribution and rate of ONe novae in the Galaxy (since only ONe novae produce $^{22}$Na). A detailed study of the diffuse galactic 1,275 keV line emission from novae showed that contributions from a few young and close novae dominate, yielding a very irregular distribution versus galactic longitude (Higdon and Fowler, 1987). A comparison with the upper limits from HEAO 3 observations (Mahoney et al., 1982) gave $5.6 \times 10^{-7}$ M$_\odot$ as upper limit to the mean $^{22}$Na yield per nova, for a disk nova population. It was clear from this work that the results were subject to many uncertainties, such as the galactic distribution, the bulge/disk ratio, the global rate and the fraction of ONe versus CO novae. A recent analysis of the cumulative emission at 1,275 keV from novae shows that the ejected $^{22}$Na masses needed for a detection of this emission with the SPI spectrometer, onboard the INTEGRAL satellite, are far above what current theoretical models predict ($\sim 10^{-7}$ versus a few $10^{-9}$ M$_\odot$) (Jean et al., 2000).

The production of $^{26}$Al by classical novae occurs again mainly in ONe novae, with low mass white dwarfs more prolific producers of $^{26}$Al than massive ones. A crude estimate of the global contribution of novae to the $^{26}$Al content in the Galaxy can be made, assuming that all novae contribute with the same amount of $^{26}$Al, M$_{\rm ejec}$($^{26}$Al), and that $^{26}$Al is active during a time equal to its lifetime $\tau$. Then the Galactic mass of $^{26}$Al coming from novae would be (Weiss and Truran, 1990; José et al., 1997)

$$M(^{26}\mathrm{Al})(M_\odot) = M_{\rm ejec}(^{26}Al)\,\tau\,R_{\rm nova}\,f_{\rm ONe} = 0.12\frac{M_{\rm ejec}}{10^{-8}\,M_\odot}\frac{R_{\rm nova}}{35/year}\frac{f_{\rm ONe}}{0.33} \tag{5.25}$$

where $R_{\text{nova}}$ is the total galactic nova rate and $f_{\text{ONe}}$ is the fraction of ONe novae. Adopting typical $^{26}$Al ejected masses (i.e., $2 \times 10^{-8}$ M$_\odot$), the contribution of novae to galactic $^{26}$Al would be ~0.2 M$_\odot$, more than a factor of 10 below the observed mass, in agreement with the current idea (deduced from the observed 1.809 MeV line sky map) that galactic $^{26}$Al comes mainly from massive stars (Knödlseder, 1999) A complete analysis of the global contribution of novae to the $^{26}$Al in the Galaxy was carried out by Kolb and Politano (1997), applying galactic nova population models, adopting the $^{26}$Al yields from Politano et al. (1995) and taking very large ejected masses (larger than those from typical hydrodynamic models). The authors concluded that the nova contribution could range between 0.15 and 3 M$_\odot$, but this number largely depended on the unknown degree of mixing in novae, which largely influences their $^{26}$Al yield, in addition to other parameters of the population synthesis code, like for instance the mass ratio (primary versus secondary star masses) distribution in zero-age main sequence binaries.

## 5.5.2 Dust from Novae and Thermonuclear Supernovae

Astrophysics has basically relied on electromagnetic radiation (collected by ground-based telescopes as well as by space-borne observatories) as the basic tool to determine stellar properties. But since the mid-1980s, new methods that rely on matter rather than on radiation, have become available as well.

### 5.5.2.1 Stardust Mineralogy

Back in 1973, A.G.W. Cameron speculated in a seminal paper (Cameron, 1973) that primitive carbonaceous chondrites may host *presolar grains*, tiny spherules of *stardust* condensed in the outflows of stars in advanced stages or in the ejecta of stellar explosions, containing a record of the nuclear history of their stellar parent bodies. Indeed, presolar grains have been isolated from meteorites, suggesting that the chemical processes that affected some meteoritic bodies were apparently mild and non-destructive to the grains.

The stellar paternity of these grains can be assessed by their anomalous isotopic composition, significantly different from that of the Solar System, and attributed to a suite of nucleosynthetic processes that took place in their parent stellar sources. In turn, the discovery of isotopically anomalous grains embedded in meteorites provided evidence of the chemical heterogeneity of the solar nebula (Cameron, 1962). Moreover, although grains are difficult to date because of their low content in radioactive species, their large isotopic anomalies, including $^{14}$N/$^{15}$N, $^{12}$C/$^{13}$C, or silicon ratios far beyond the values reported from any other Solar System sample, suggest an ancient origin, with an age older than the Solar System itself (thus the label *presolar*).

Diamonds were the first presolar grains isolated from meteorites (Lewis et al., 1987). This was followed by the isolation of SiC grains (Bernatowicz et al., 1987; Tang and Anders, 1988), and graphite (Amari et al., 1990). These three carbona-

**Table 5.2** Inventory of known presolar grain types (adapted from Zinner, 2005; Lodders, 2005)

| Grain type | Characteristic size | Potential stellar sources[a] | Discovery papers |
|---|---|---|---|
| Nanodiamond | 2 nm | AGB SN | Lewis et al. (1987) |
| SiC | 0.1–20 μm | AGB, SN, J-stars, CN | Bernatowicz et al. (1987) and Tang and Anders (1988) |
| Graphite | 1–20 μm | SN, AGB, CN | Amari et al. (1990) |
| Corundum | 0.2–3 μm | RGB, AGB, SN | Hutcheon et al. (1994); Nittler et al. (1994) |
| Spinel | 0.2–3 μm | RGB, AGB, SN | Nittler et al. (1997) and Choi et al. (1998) |
| Hibonite | 0.2–3 μm | RGB, AGB, SN | Choi et al. (1999) |
| $Si_3N_4$ | 0.3–1 μm | AGB, SN | Nittler et al. (1995) |
| Silicates (olivine, pyroxene) | 0.1–0.3 μm | RGB, AGB, SN | Messenger et al. (2003) and Nguyen and Zinner (2004) |

[a] Acronyms: Asymptotic Giant Branch Stars (AGB), Supernovae (SN), Classical Novae (CN), Red Giant Branch Stars (RGB).

ceous phases were identified because of their isotopically anomalous noble gas (Xe, Ne) components. So far, silicon carbide (SiC), graphite (C), diamond (C), silicon nitride ($Si_3N_4$), silicates (Messenger et al., 2003; Nguyen and Zinner, 2004; Mostefaoui and Hoppe, 2004), and oxides, such as corundum ($Al_2O_3$), or spinel ($MgAl_2O_4$), have been identified as presolar grains (see Table 5.2). In fact, all SiC grains extracted from meteorites are of presolar origin; approximately half of the graphite grains are presolar; only $\sim 2\%$ of the spinel grains, and scarcely 0.001–0.02% of the silicates, are presolar.

Those grains, identified and extracted from meteorites, are systematically analyzed in the laboratory with ever improving precision. Such laboratory analyses revealed a variety of isotopic signatures that point towards several stellar progenitors, such as asymptotic giant branch stars and supernovae (see Clayton and Nittler, 2004; Lodders, 2005; Meyer and Zinner, 2006, for recent reviews).

Several meteoritic bodies have been used to study presolar grains, basically very primitive, mildly metamorphosed, carbonaceous chondrites, such as Murchison, or Allende. Indeed, the anomalous size of Murchison's grains (Lodders, 2005; Zinner, 2005), much larger than those isolated from other meteorites (for reasons not yet understood), as well as the large number of samples available, made Murchison one of the favorite targets for studies of presolar grains.

#### 5.5.2.2 Silicon Carbide Grains

SiC grains have been most extensively studied (Fig. 5.13). They are often classified into different populations (presumably reflecting different stellar birthplaces) on the basis of their C, N, and Si isotopic ratios (Hoppe and Ott, 1997).

It is widely accepted that about ~93% of all SiC grains, the so-called *mainstream population*, are formed in the winds accompanying solar-metallicity AGB stars (Gallino et al., 1993; Lugaro et al., 2003; Ott and Begemann, 1990). About ~1% correspond to *X grains*, which are characterized by moderate excesses of $^{12}C$ and $^{15}N$, large $^{26}Al/^{27}Al$ ratios, and excesses of $^{28}Si$, features pointing towards a supernova origin (Amari et al., 1992; Hoppe et al., 2000; see also Sect. 5.3.3). In addition, a variety of carbon-rich J-type stars are expected to account for ~4–5% of the overall SiC grains, the so-called *A* and *B grains* (with born-again AGB stars, such as the Sakurai's object V4334 Sgr, or other C-rich stellar types, like R- or CH-stars, not being totally excluded; see Amari et al., 2001c). Other populations include *Y* (~1%) and *Z grains* (~1%), whose origin is probably linked to low-metallicity AGB stars (Amari et al., 2001b; Hoppe et al., 1997). A rare variety of SiC grains (<1%), together with a couple of graphite grains, that exhibit a suite of isotopic signatures characteristic of classical nova outbursts, have been reported in recent years (Amari et al., 2001a; Amari, 2002).

#### 5.5.2.3 Supernova Grains

SiC grains of type X, low-density graphites, and the very rare silicon nitrates are believed to originate in ejecta accompanying supernovae. Many of the iso-

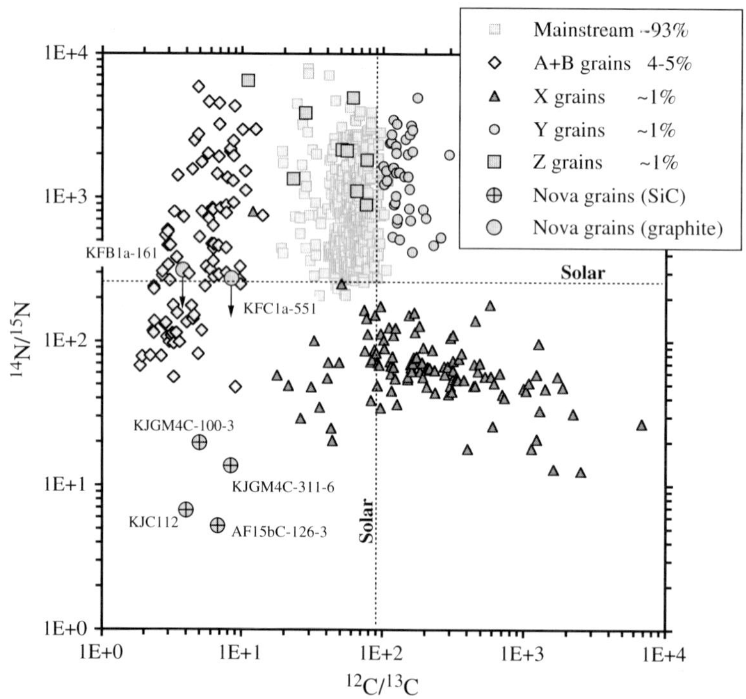


**Fig. 5.13** Carbon and nitrogen isotopic ratios for the different SiC grain populations. Error bars are smaller than the symbols

topic signatures of these grains (namely, moderate excesses of $^{12}$C, and $^{15}$N, large $^{26}$Al/$^{27}$Al ratios, and excesses of $^{28}$Si) are qualitatively consistent with supernova models, although some of these features can also be produced by other stellar sources.

Both thermonuclear and core-collapse supernovae have been proposed as potential sources for these grains, although type II supernova models seem to be preferred (so far, there is little evidence of grains condensed in a thermonuclear supernova).

A clear fingerprint of their supernova origin is the excess of $^{44}$Ca (attributed to in situ decay of $^{44}$Ti), present in $\sim$10–20% of the X grains (Amari et al., 1992). Moreover, there is also evidence for extinct $^{49}$V in some of these grains (Amari et al., 1992). Other isotopic anomalies, measured in individual X grains, include Mo, Zr, Sr, Ba, and Fe (Pellin et al., 1999), in puzzling proportions, neither matching those reported from AGB grains, nor the theoretical estimates from r and *p*-process in supernova simulations. It is also worth noting that many of the isotopic signatures measured in these grains are less extreme than those predicted for pure supernova ejecta. A possible explanation (Clayton et al., 2002) suggests that grains crossing the SN ejecta at high velocities may implant atoms from the environment, hence diluting the original signatures.

A major problem to quantitatively match the grain data with supernova models (type II, in particular) is the need for selective mixing between different stellar layers (Lodders, 2005; Zinner, 2005), at a much larger scale than that suggested by observations and/or simulations. In particular, the high abundances of $^{15}$N measured in the grains cannot be accounted for with current supernova models.

Among the oxide grain propulation, two corundum grains (T48, with an almost pure $^{16}$O composition; and S-C122 with 3 times the solar $^{18}$O/$^{16}$O ratio), have also been linked to type II supernovae (Nittler et al., 1998; Messenger et al., 2005; Nittler et al., 2008). However, the lack of additional supernova signatures and the problems to match the specific $^{17}$O/$^{16}$O ratios measured in these grains make it hard to reconcile with a supernova origin.

#### 5.5.2.4 Nova Grains

Infrared and ultraviolet observations have revealed dust forming episodes in the shells ejected during classical nova outbursts (Gehrz et al., 1998; Gehrz, 2002). Their relatively high frequency (about $\sim$30–35 nova explosions per year, just in our Galaxy (Shafter, 2002)), has raised the issue of the potential contribution of novae to the different grain populations.

Since the pioneering studies of dust formation in novae by D.D. Clayton and F. Hoyle (Clayton and Hoyle, 1976) (a concept already suggested by Cameron in 1973), all efforts devoted to the identification of potential nova grains relied mainly on the search for low $^{20}$Ne/$^{22}$Ne ratios (since noble gases, such as Ne, do not condense into grains, $^{22}$Ne was attributed to in situ $^{22}$Na decay, a clear imprint of a classical nova explosion). Indeed, Clayton and Hoyle pointed out several isotopic signatures (large overproduction of $^{13,14}$C, $^{18}$O, $^{22}$Na, $^{26}$Al or $^{30}$Si), that may help in the identification of such nova candidate grains. Twenty-five years later, most of these signatures still hold, in view of our current understanding of nova explosions (see José et al., 2006; José and Shore, 2008; Starrfield et al., 2008, for recent reviews), except $^{14}$C, bypassed by the main nuclear path in novae, and $^{18}$O, slightly overproduced by novae although grains nucleated in this environment are expected to be much more anomalous in $^{17}$O (Kovetz and Prialnik, 1997; José and Hernanz, 1998; Starrfield et al., 1998, 2009).

A major step forward in the discovery of presolar nova candidate grains was achieved by Amari et al. (2001a) and Amari (2002), who reported on several SiC and graphite grains, isolated from the Murchison and Acfer 094 meteorites (Fig. 5.14), with an abundance pattern qualitatively similar to nova model predictions: low $^{12}$C/$^{13}$C and $^{14}$N/$^{15}$N ratios, high $^{30}$Si/$^{28}$Si, and close-to-solar $^{29}$Si/$^{28}$Si; and high $^{26}$Al/$^{27}$Al and $^{22}$Ne/$^{20}$Ne ratios for some of the grains (José et al., 2004). But in order to quantitatively match the grain data, one had to assume a mixing process between material newly synthesized in the nova outburst and more than ten times as much unprocessed, isotopically close-to-solar, material before grain formation.

Concerns about the likely nova paternity of these grains have recently been raised (Nittler and Hoppe, 2005), after three additional micron-sized SiC grains were also isolated from the Murchison meteorite with similar trends (in particular, low $^{12}$C/$^{13}$C and $^{14}$N/$^{15}$N ratios), but with additional imprints (mainly non-solar Ti features), from which a supernova origin cannot be excluded.

It is not clear if both samples (hereafter, A01 and NH05, respectively) correspond to the same progenitor. After all, their isotopic signatures are not identical: for instance, grains from the NH05 sample have much larger $^{26}$Al/$^{27}$Al ratios and are more heavily depleted in $^{29}$Si than grains from the A01 sample. Furthermore, grain M11-334-2 (NH05 sample) is deficient in $^{30}$Si with respect to solar (whereas $^{30}$Si excesses, characteristic of the A01 sample, are expected in the ejecta from ONe novae). In turn, spectroscopic analyses of the nova ejecta support the main conclusions achieved by different studies of nova outbursts, revealing a nucleosynthetic

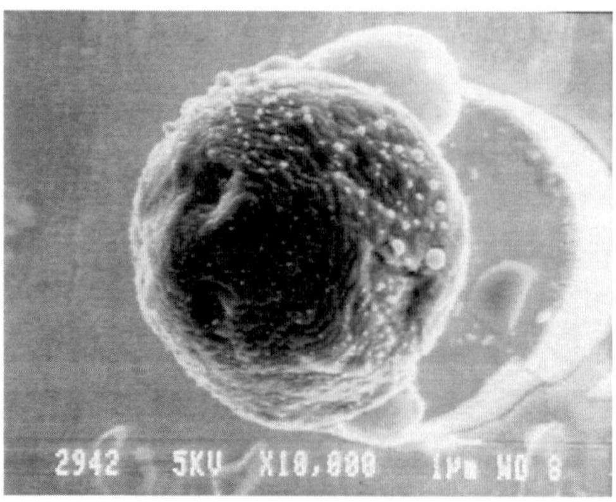


**Fig. 5.14** The nova candidate graphite grain KFC1a-511 after Secondary Ion Mass Spectrometry (SIMS). Image courtesy of S. Amari

endpoint around calcium, and a *normal* (solar) nucleosynthetic pattern for Ti or heavier species. These evidences suggest that the presence of anomalous Ti features in the A01 sample cannot be inferred from measurements performed in the NH05 sample.

Moreover, it must be stressed that the presence of anomalous Ti does not necessarily rule out a possible nova paternity (with the exception of $^{44}$Ti, attributed to in situ decay of $^{44}$Ca, an isotope clearly linked to a supernova explosion): titanium is, indeed, very close to the nucleosynthetic endpoint for novae (calcium), and hence, it could easily be reached by a sligthly more violent outburst. This could be driven by explosions in cooler white dwarfs, or following lower mass-accretion rate episodes (José and Hernanz, 2007a; Glasner and Truran, 2009). Furthermore, explosions in metal-deficient envelopes, such as those expected for primordial nova systems, could have a similar effect (José et al., 2007; José and Hernanz, 2007b).

Other nova candidate grains have been proposed: indeed, the SiC grain 240-1 (Nittler et al., 2006), also isolated from Murchison, exhibits both $^{12}$C/$^{13}$C and $^{14}$N/$^{15}$N ratios lower than for any other presolar grain reported so far. These isotopic features are consistent with pure nova ejecta from a white dwarf with a mass ranging between 1.0 and 1.2 $M_\odot$ (a supernova origin is unlikely because of the extremely low $^{12}$C/$^{13}$C ratios). However, the $^{29}$Si excesses measured in this grain do not match the usual predictions from nova models (which usually reflect $^{29}$Si deficits, with respect to solar). Also a putative nova origin has been attributed to the oxide grain T54 (Nittler, 1997), with $^{16}$O/$^{17}$O $\sim$ 71 and $^{16}$O/$^{18}$O $\sim$ 2000, likely condensed in the shells ejected from a nova outburst on a 0.8 $M_\odot$ CO white dwarf. Unfortunately, no additional isotopic determinations were carried out on this grain. Finally, it should be stressed that novae may contribute to the oxide grain population as well as to the *A*+*B* subclass of SiC grains (those characterized by low $^{12}$C/$^{13}$C ratios that do not show large $^{29}$Si excesses).

Future nova candidate grains will reveal more clues on the mechanisms powering nova explosions. To achieve this, cosmochemists will have to rely on a much wider range of isotopic determinations for proper identification of the stellar source (to disentangle, for instance, which grains are formed in supernova blasts and which in nova explosions). Novae hosting very massive white dwarfs (around 1.35 $M_\odot$) likely imprint additional signatures in the grains condensing in their ejecta (in particular, a suite of sulfur anomalies as well as severe $^{31}$P overproduction). New techniques for laboratory analysis need to be developed to unambiguously identify such signatures, avoiding potential contamination of the samples by sulfuric acid, one of the standard methods used during the separation process.

## 5.6 Accretion in Binaries: Special Cases

It was seen in the case of the merging of two white dwarfs that if the total mass was larger than the Chandrasekhar mass, the final outcome could either be a SNIa or a collapse to a neutron star. However in the large majority of cases, the total mass is smaller than the critical value and the final result is a white dwarf "born again".

The fraction of the secondary that is expelled is not yet known and, consequently, the influence of such systems on the chemical evolution of the Galaxy has not been yet elucidated. The total amount of freshly synthesized elements during the impact is small (Guerrero et al., 2004), except in the case of a secondary made of helium that shows an enhancement of Ca, Mg, Si, S and Fe, and confined to a corona around the primary.

In the collapse case, since the primary is rapidly rotating, as expected from the transfer of angular momentum from the disk to the star, a centrifugally supported disk made of heavily neutronized species, $Y_e \sim 0.1$, will form around the proto-neutron star. As a consequence of the neutrino irradiation, electron neutrino captures will increase the electron mole number to a value $Y_e \sim 0.5$ and $\alpha$-particles will form inducing a wind that will blow away the disk. During this process it is expected that $\sim 10^{-2}$ $M_\odot$ of $^{56}$Ni will be synthesized and that the event will look as a dim SNIa – like transient (Metzger et al., 2009).

Merging of close binaries as well as close encounters in densely populated stellar systems, like globular clusters or galactic nuclei, can also provide violent scenarios able to trigger a nucleosynthetic activity other than the conventional thermonuclear explosions described up to now. The center of the Milky Way, for instance, contains a massive black hole surrounded by a swarm of stars, many of them white dwarfs. Close encounters are very common and tidal torques can produce extreme deformations of the stars or even trigger an explosion.

The tidal interaction between a white dwarf and a black hole is characterized by three length scales (Carter and Luminet, 1983; Rosswog et al., 2009): (i) the stellar radius, $R_{\rm WD}$, (ii) the gravitational radius of the black hole $R_{rmg} = 2GM_{\rm BH}/c^2 \simeq 3 \times 10^{11} M_{\rm BH,6}$, where $M_{\rm BH,6}$ is the mass of the black hole in units of $10^6$ $M_\odot$, and (iii) the tidal radius, $R_\tau \simeq 1.2 \times 10^{11} M_{\rm BH,6}^{1/3} (R_{\rm WD}/10^9\,{\rm cm})(M_{\rm WD}/0.6\,{\rm M}_\odot)^{-1/3}$ cm, that is the distance from the black hole at which $M_{\rm BH}/R_\tau^3$ equals the mean density of the passing star.

The strength of the tidal encounter can be estimated from the dimensionless parameter $\beta = R_\tau/R_{\rm P}$, where $R_{\rm P}$ is the pericenter distance, assuming a parabolic orbit. When $\beta \geq 1$, the star is disrupted in a single flyby. The energy to tear apart the star (the binding energy of the star) is supplied by the orbital energy. In the case of white dwarfs, the ratio between the total disruption radius and the gravitational radius is

$$\beta_{\rm g} = \frac{R_\tau}{R_{\rm g}} \simeq 0.4 M_{\rm BH,6}^{-2/3} \left(\frac{R_{\rm WD}}{10^9\,{\rm cm}}\right) \left(\frac{M_{\rm WD}}{0.6 M_\odot}\right)^{-1/3} \quad (5.26)$$

therefore, if the mass of the black hole is high enough, the tidal radius is inside the gravitational radius and the white dwarf is swallowed without being disrupted. This critical mass is

$$M_{\rm BH,lim} \simeq 2.5 \times 10^5 \left(\frac{R_{\rm WD}}{10^9\,{\rm cm}}\right)^{3/2} \left(\frac{M_{\rm WD}}{0.6\,{\rm M}_\odot}\right)^{-1/2} \quad (5.27)$$

The dynamics of the encounter can be described as follows (Carter and Luminet, 1983). When the star is far from the black hole, the tidal interaction is negligible and the white dwarf is in hydrostatic equilibrium. As soon as it enters the Roche lobe, tidal interaction quickly grows in strength. As a consequence, matter is compressed by the flattening of the star. When the white dwarf is flat enough, the internal energy becomes dominant and the star experiences a bounce that reduces the pressure and makes the tidal interaction dominant once more. Depending on the parameters of the encounter ($M_{\rm BH}$, $M_{\rm WD}$, $\beta$) and on the chemical composition of the white dwarf, a vigorous thermonuclear burning can occur during the compression phase that can even produce a substantial amount of iron peak elements (Rosswog et al., 2009). Finally, when the star is far enough, self-gravitation recovers control and, depending on the balance between internal and gravitational energies, matter is partially swallowed by the black hole, partially ejected to the interstellar medium and partially remains bound. During the expansion phase the radioactive debris, the $\beta$-decay of $^{56}$Ni, mainly, can emit light and produce some kind of peculiar, sub-luminous SNIa (Rosswog et al., 2009).

## References

Alexakis, A., Calder, A.C., Heger, A., Brown, E.F., Dursi, L.J., Truran, J.W., Rosner, R., Lamb, D.Q., Timmes, F. X., Fryxell, B., Zingale, M., Ricker, P.M., Olson, K.: On Heavy element enrichment in classical novae. ApJ **602**, 931–937 (2004). doi:10.1086/381086, arXiv:astro-ph/0307126

Alibés, A., Labay, J., Canal, R.: Galactic cosmic rays from superbubbles and the abundances of lithium, beryllium, and boron. ApJ **571**, 326–333 (2002). doi:10.1086/339937, arXiv:astroph/0202097

Amari, S.: Presolar grains from novae: their isotopic ratios and radioactivities. New Astron. Rev. **46**, 519–524 (2002). doi:10.1016/S1387-6473(02)00194-X

Amari, S., Anders, A., Virag, A., Zinner, E.: Interstellar graphite in meteorites. Nature **345**, 238–240 (1990). doi:10.1038/345238a0

Amari, S., Hoppe, P., Zinner, E., Lewis, R.S.: Interstellar SiC with unusual isotopic compositions – Grains from a supernova? ApJ **394**, L43–L46 (1992). doi:10.1086/186468

Amari, S., Gao, X., Nittler, L.R., Zinner, E., José, J., Hernanz, M., Lewis, R.S.: Presolar grains from novae. ApJ **551**, 1065–1072 (2001a). doi:10.1086/320235, arXiv:astro-ph/0012465

Amari, S., Nittler, L.R., Zinner, E., Gallino, R., Lugaro, M., Lewis, R.S.: Presolar SiC grains of type Y: origin from low-metallicity asymptotic giant branch stars. ApJ **546**, 248–266 (2001b). doi:10.1086/318230

Amari, S., Nittler, L.R., Zinner, E., Lodders, K., Lewis, R.S.: Presolar SiC grains of type A and B: their isotopic compositions and stellar origins. ApJ **559**, 463–483 (2001c). doi:10.1086/322397

Ambwani, K., Sutherland, P.: Gamma-ray spectra and energy deposition for type IA supernovae. ApJ **325**, 820–827 (1988). doi:10.1086/166052

Amthor, M.A., Galaviz, D., Heger, A., Sakharuk, A., Schatz, H., Smith, K.: Sensitivity of type I X-ray bursts to rp-process reaction rate. In: International Symposium on Nuclear Astrophysics – Nuclei in the Cosmos, Geneva (2006)

Arnett, D.: Supernovae and Mucleosynthesis: An Investigation of the History of Matter from the Big Bang to the Present. Princeton University Press, Princeton (1996)

Arnett, W.D.: A possible model of supernovae: detonation of 12C. Ap&SS **5**, 180–212 (1969). doi:10.1007/BF00650291

Arnould, M., Norgaard, H.: The explosive thermonuclear formation of 7Li and 11B. A&A **42**, 55–70 (1975)

Audouze, J., Reeves, H.: The origin of the light elements. In: Essays in Nuclear Astrophysics, pp. 355–375. Cambridge University Press, Cambridge (1982)

Axelrod, T.S.: Late time optical spectra from the Ni-56 model for Type 1 supernovae. PhD thesis, University of California, Santa Cruz. (1980)

Ayasli, S., Joss, P.C.: Thermonuclear processes on accreting neutron stars – a systematic study. ApJ **256**, 637–665 (1982). doi:10.1086/159940

Badenes, C., Borkowski, K.J., Hughes, J.P., Hwang, U., Bravo, E.: Constraints on the physics of type Ia supernovae from the x-ray spectrum of the tycho supernova remnant. ApJ **645**, 1373–1391 (2006). doi:10.1086/504399, arXiv:astro-ph/0511140

Balman, S., Krautter, J., Oegelman, H.: The x-ray spectral evolution of classical nova V1974 Cygni 1992: a reanalysis of the ROSAT data. ApJ **499**, 395–406 (1998)

Barbon, R., Ciatti, F., Rosino, L.: Light curves and characteristics of recent supernovae. A&A **29**, 57–67 (1973)

Basinska, E.M., Lewin, W.H.G., Sztajno, M., Cominsky, L.R., Marshall, F.J.: X-ray observations of the burst source MXB 1728-34. ApJ **281**, 337–353 (1984). doi:10.1086/162103

Bazin, D., Montes, F., Becerril, A., Lorusso, G., Amthor, A., Baumann, T., Crawford, H., Estrade, A., Gade, A., Ginter, T., Guess, C.J., Hausmann, M., Hitt, G.W., Mantica, P., Matos, M., Meharchand, R., Minamisono, K., Perdikakis, G., Pereira, J., Pinter, J., Portillo, M., Schatz, H., Smith, K., Stoker, J., Stolz, A., Zegers, R.G.T.: Production and β Decay of rp-Process Nuclei Cd96, In98, and Sn100. Phys. Rev. Lett. **101**(25), 252501/1-25201/4 (2008). doi:10.1103/PhysRevLett.101.252501, 0810.3597

Belian, R.D., Conner, J.P., Evans, W.D.: The discovery of X-ray bursts from a region in the constellation Norma. ApJ **206**, L135–L138 (1976). doi:10.1086/182151

Benz, W., Cameron, A.G.W., Press, W.H., Bowers, R.L.: Dynamic mass exchange in doubly degenerate binaries. I – 0.9 and 1.2 solar mass stars. ApJ **348**, 647–667 (1990). doi:10.1086/168273

Bernatowicz, T., Fraundorf, G., Ming, T., Anders, E., Wopenka, B., Zinner, E., Fraundorf, P.: Evidence for interstellar SiC in the Murray carbonaceous meteorite. Nature **330**, 728–730 (1987). doi:10.1038/330728a0

Bildsten, L.: Propagation of nuclear burning fronts on accreting neutron stars: x-ray bursts and sub-hertz noise. ApJ **438**, 852–875 (1995). doi:10.1086/175128

Bildsten, L.: Thermonuclear burning on rapidly accreting neutron stars. In: Buccheri, R., van Paradijs, J., Alpar, A. (eds.) NATO ASIC Proc. 515: The Many Faces of Neutron Stars, p. 419. Kluwer Academic Publishers, Dordrecht (1998)

Bildsten, L., Chang, P., Paerels, F.: Atomic spectral features during thermonuclear flashes on neutron stars. ApJ **591**, L29–L32 (2003). doi:10.1086/377066, arXiv:astro-ph/0303147

Blinnikov, S.I., Khokhlov, A.M.: Development of detonations in degenerate stars. Sov. Astron. Lett. **12**, 131–132 (1986)

Boffin, H.M.J., Paulus, G., Arnould, M., Mowlavi, N.: The explosive thermonuclear formation of Li-7 revisited. A&A **279**, 173–178 (1993)

Branch, D.: Some statistical properties of type I supernovae. ApJ **248**, 1076–1080 (1981). doi:10.1086/159237

Branch, D., Fisher, A., Nugent, P.: On the relative frequencies of spectroscopically normal and peculiar type IA supernovae. AJ **106**, 2383–2391 (1993). doi:10.1086/116810

Branch, D., Romanishin, W., Baron, E.: Statistical connections between the properties of type IA supernovae and the B-V colors of their parent galaxies, and the value of H 0. ApJ **465**, 73–78 (1996). doi:10.1086/177402, arXiv:astro-ph/9510071

Bravo, E., García-Senz, D.: Coulomb corrections to the equation of state of nuclear statistical equilibrium matter: implications for SNIa nucleosynthesis and the accretion-induced collapse of white dwarfs. MNRAS **307**, 984–992 (1999). doi:10.1046/j.1365-8711.1999.02694.x

Bravo, E., Isern, J., Canal, R., Labay, J.: On the contribution of Ne-22 to the synthesis of Fe-54 and Ni-58 in thermonuclear supernovae. A&A **257**, 534–538 (1992)

Bravo, E., Tornambe, A., Domínguez, I., Isern, J.: Clues to Type IA SN progenitors from degenerate carbon ignition models. A&A **306**, 811–822 (1996)

Brown, E.F.: Nuclear heating and melted layers in the inner crust of an accreting neutron star. ApJ **531**, 988–1002 (2000). doi:10.1086/308487, arXiv:astro-ph/9910215

Brown, E.F., Bildsten, L.: The ocean and crust of a rapidly accreting neutron star: implications for magnetic field evolution and thermonuclear flashes. ApJ **496**, 915–933 (1998). doi:10.1086/305419, arXiv:astro-ph/9710261

Burrows, A., The, L.: X- and gamma-ray signatures of type IA supernovae. ApJ **360**, 626–638 (1990). doi:10.1086/169150

Bussard, R.W., Ramaty, R., Drachman, R.J.: The annihilation of galactic positrons. ApJ **228**, 928–934 (1979). doi:10.1086/156920

Cadonau, R., Tammann, G.A., Sandage, A.: Type I supernovae as standard candles. In: Bartel, N. (ed.) Supernovae as Distance Indicators. Lecture Notes in Physics, vol. 224, pp. 151–165. Springer Verlag, Berlin (1985). doi:10.1007/3-540-15206-7-56

Cameron, A.G.W.: Origin of anomalous abundances of the elements in giant stars. ApJ **121**, 144–160 (1955). doi:10.1086/145970

Cameron, A.G.W.: The formation of the sun and planets. Icarus **1**, 13–69 (1962). doi:10.1016/0019-1035(62)90005-2

Cameron, A.G.W.: Interstellar grains in museums? In: Greenberg, J.M., van de Hulst, H.C. (eds.) Interstellar Dust and Related Topics, IAU Symposium, vol. 52, pp. 545–547. Reidel, Dordrecht (1973)

Canal, R., Isern, J., Labay, J.: The origin of neutron stars in binary systems. ARA&A **28**, 183–214 (1990). doi:10.1146/annurev.aa.28.090190.001151

Cappellaro, E., Barbon, R., Turatto, M.: Supernova statistics. ArXiv Astrophysics e-prints arXiv:astro-ph/0310859 (2003)

Carter, B., Luminet, J.: Tidal compression of a star by a large black hole. I Mechanical evolution and nuclear energy release by proton capture. A&A **121**, 97–113 (1983)

Champagne, A.E., Wiescher, M.: Explosive hydrogen burning. Ann. Rev. Nucl. Part Sci **42**, 39–76 (1992). doi:10.1146/annurev.ns.42.120192.000351

Chan, K., Lingenfelter, R.E.: Positrons from supernovae. ApJ **405**, 614–636 (1993). doi:10.1086/172393

Chang, P., Bildsten, L., Wasserman, I.: Formation of resonant atomic lines during thermonuclear flashes on neutron stars. ApJ **629**, 998–1007 (2005). doi:10.1086/431730, arXiv:astroph/0505062

Chang, P., Morsink, S., Bildsten, L., Wasserman, I.: Rotational broadening of atomic spectral features from neutron stars. ApJ **636**, L117–L120 (2006). doi:10.1086/499428, arXiv:astroph/0511246

Choi, B., Huss, G.R., Wasserburg, G.J., Gallino, R.: Presolar corundum and spinel in ordinary chondrites: origins from AGB stars and a supernova. Science **282**, 1284–1289 (1998). doi:10.1126/science.282.5392.1284

Choi, B., Wasserburg, G.J., Huss, G.R.: Circumstellar hibonite and corundum and nucleosynthesis in asymptotic giant branch stars. ApJ **522**, L133–L136 (1999). doi:10.1086/312239

Clayton, D.D.: Li-7 gamma-ray lines from novae. ApJ **244**, L97–L98 (1981). doi:10.1086/183488

Clayton, D.D., Hoyle, F.: Gamma-ray lines from novae. ApJ **187**, L101–L103 (1974). doi:10.1086/181406

Clayton, D.D., Hoyle, F.: Grains of anomalous isotopic composition from novae. ApJ **203**, 490–496 (1976). doi:10.1086/154104

Clayton, D.D., Nittler, L.R.: Astrophysics with presolar stardust. ARA&A **42**, 39–78 (2004). doi:10.1146/annurev.astro.42.053102.134022

Clayton, D.D., Meyer, B.S., The, L., El Eid, M.F.: Iron implantation in presolar supernova grains. ApJ **578**, L83–L86 (2002). doi:10.1086/344434

Coc, A., Hernanz, M., José, J, Thibaud, J.: Influence of new reaction rates on 18F production in novae. A&A **357**, 561–571 (2000). arXiv:astro-ph/0003166

Colgate, S.A., McKee, C.: Early supernova luminosity. ApJ **157**, 623–643 (1969). doi:10.1086/150102

Cornelisse, R., Heise, J., Kuulkers, E., Verbunt, F., in't Zand, J.J.M. The longest thermonuclear X-ray burst ever observed? A BeppoSAX Wide Field Camera observation of 4U 1735–1744. A&A **357**, L21–L24 (2000). arXiv:astro-ph/0003454

Cottam, J., Paerels, F., Mendez, M.: Gravitationally redshifted absorption lines in the X-ray burst spectra of a neutron star. Nature **420**, 51–54 (2002). doi:10.1038/nature01159, arXiv:astroph/0211126

Cottam, J., Paerels, F., Méndez, M, Boirin, L., Lewin, W.H.G., Kuulkers, E., Miller, J.M.: The burst spectra of EXO 0748-676 during a Long 2003 XMM-Newton observation. ApJ **672**, 504–509 (2008). doi:10.1086/524186, 0709.4062

Cumming, A.: Superbursts: a new probe of the rp-process. Nucl. Phys. A **758**, 439–446 (2005). doi:10.1016/j.nuclphysa.2005.05.081

Cumming, A., Bildsten, L.: Carbon flashes in the heavy-element ocean on accreting neutron stars. ApJ **559**, L127–L130 (2001). doi:10.1086/323937, arXiv:astro-ph/0107213

Damen, E., Magnier, E., Lewin, W.H.G., Tan, J., Penninx, W., van Paradijs, J.: X-ray bursts with photospheric radius expansion and the gravitational redshift of neutron stars. A&A **237**, 103–109 (1990)

D'Antona, F., Mazzitelli, I.: The fastest evolving white dwarfs. ApJ **347**, 934–949 (1989). doi:10.1086/168185

Della Valle, M.: Nova populations. In: Hernanz, M., José, J. (ed.) Classical Nova Explosions, American Institute of Physics Conference Series, vol. 637, pp. 443–456. American Institute of Physics, New York (2002). doi:10.1063/1.1518244

Della Valle, M., Bianchini, A., Livio, M., Orio, M.: On the possible existence of two classes of progenitors for classical novae. A&A **266**, 232–236 (1992)

Della Valle, M., Livio, M.: On the nova rate in the Galaxy. A&A **286**, 786–788 (1994)

Della Valle, M., Livio, M.: The calibration of novae as distance indicators. ApJ **452**, 704–709 (1995). doi:10.1086/176342

Della Valle, M., Livio, M.: The spectroscopic differences between disk and thick-disk/bulge novae. ApJ **506**, 818–823 (1998). doi:10.1086/306275

Della Valle, M., Pasquini, L., Daou, D., Williams, R.E.: The evolution of Nova V382 Velorum 1999. A&A **390**, 155–166 (2002). doi:10.1051/0004-6361:20020611, arXiv:astro-ph/0205135

de Plaa, J., Werner, N., Bleeker, J.A.M., Vink, J., Kaastra, J.S., Méndez M.: Constraining supernova models using the hot gas in clusters of galaxies. A&A **465**, 345–355 (2007). doi:10.1051/0004-6361:20066382, arXiv:astro-ph/0701553

Diehl, R., Dupraz, C., Bennett, K., Bloemen, H., Hermsen, W., Knoedlseder, J., Lichti, G., Morris, D., Ryan, J., Schoenfelder, V., Steinle, H., Strong, A., Swanenburg, B., Varendorff, M., Winkler, C.: COMPTEL observations of galactic 26Al emission. A&A **298**, 445–460 (1995)

Domínguez, I., Tornambe, A., Isern, J.: On the formation of o-ne white dwarfs in metal-rich close binary systems. ApJ **419**, 268–275 (1993). doi:10.1086/173480

Domínguez, I., Höflich, P., Straniero, O.: Constraints on the progenitors of type Ia supernovae and implications for the cosmological equation of state. ApJ **557**, 279–291 (2001). doi:10.1086/321661, arXiv:astro-ph/0104257

Domínguez, I., Piersanti, L., Bravo, E., Tornambé, A., Straniero, O., Gagliardi, S.: rotating type Ia SN progenitors: explosion and light curves. ApJ **644**, 21–29 (2006). doi:10.1086/503534

Duerbeck, H.W.: Novae: an historical perspective. In: Bode, M.F., Evans, A. (eds.) Classical Novae, Second Edition, pp. 1–15. Cambridge University Press, Cambridge (2008) 245

Elias, J.H., Matthews, K., Neugebauer, G., Persson, S.E.: Type I supernovae in the infrared and their use as distance indicators. ApJ **296**, 379–389 (1985). doi:10.1086/163456

Elomaa, V., Vorobjev, G.K., Kankainen, A., Batist, L., Eliseev, S., Eronen, T., Hakala, J., Jokinen, A., Moore, I.D., Novikov, Y.N., Penttilä, H., Popov, A., Rahaman, S., Rissanen, J., Saastamoinen, A., Schatz, H., Seliverstov, D.M., Weber, C., Äystö, J.: Quenching of the SnSbTe cycle in the rp process. Phys. Rev. Lett. **102**(25), 252 501/1-252501/4 (2009). doi:10.1103/PhysRevLett.102.252501

Filippenko, A.V., Richmond, M.W., Branch, D., Gaskell, M., Herbst, W., Ford, C.H., Treffers, R.R., Matheson, T., Ho, L.C., Dey, A., Sargent, W.L.W., Small, T.A., van Breugel, W.J.M.: The subluminous, spectroscopically peculiar type IA supernova 1991bg in the elliptical galaxy NGC 4374. AJ **104**, 1543–1556 (1992). doi:10.1086/116339

Fishman, G.J., Wilson, R.B., Meegan, C.A., Brock, M.N., Horack, J.M., Paciesas, W.S., Pendleton, G.N., Harmon, B.A., Leising, M.: Detectability of early low-energy gamma rays from nearby novae by BATSE/GRO. In: Durouchoux, P., Prantzos, N. (eds.) Gamma-Ray Line Astrophysics, American Institute of Physics Conference Series, vol. 232, pp. 190–192. American Institute of Physics, New York (1991). doi:10.1063/1.40956

Fisker, J.L., Görres, J., Wiescher, M., Davids, B.: The importance of 15O($\alpha$, $\gamma$) 19Ne to x-ray bursts and superbursts. ApJ **650**, 332–337 (2006). doi:10.1086/507083, arXiv:astro-ph/0410561

Fisker, J.L., Schatz, H., Thielemann, F.: Explosive hydrogen burning during type I x-ray bursts. ApJS **174**, 261–276 (2008). doi:10.1086/521104

Fryxell, B.A., Woosley, S.E.: Finite propagation time in multidimensional thermonuclear runaways. ApJ **261**, 332–336 (1982). doi:10.1086/160344

Fujimoto, M.Y., Hanawa, T., Miyaji, S.: Shell flashes on accreting neutron stars and X-ray bursts. ApJ **247**, 267–278 (1981). doi:10.1086/159034

Fuller, G.M., Fowler, W.A., Newman, M.J.: Stellar weak interaction rates for intermediate-mass nuclei. IV – Interpolation procedures for rapidly varying lepton capture rates using effective log (ft)-values. ApJ **293**, 1–16 (1985). doi:10.1086/163208

Fushiki, I., Lamb, D.Q.: New insights from a global view of X-ray bursts. ApJ **323**, L55–L60 (1987). doi:10.1086/185056

Gallino, R., Raiteri, C.M., Busso, M.: Carbon stars and isotopic BA anomalies in meteoritic SiC grains. ApJ **410**, 400–411 (1993). doi:10.1086/172757

Galloway, D.K., Muno, M.P., Hartman, J.M., Psaltis, D., Chakrabarty, D.: Thermonuclear (type I) X-ray bursts observed by the Rossi X-ray timing explorer. ApJS **179**, 360–422 (2008). doi:10.1086/592044, arXiv:astro-ph/0608259

Gamezo, V.N., Khokhlov, A.M., Oran, E.S., Chtchelkanova, A.Y., Rosenberg, R.O.: Thermonuclear supernovae: simulations of the deflagration stage and their implications. Science **299**, 77–81 (2003). doi:10.1126/science.1078129, arXiv:astro-ph/0212054

García-Senz, D., Bravo, E.: Type Ia Supernova models arising from different distributions of igniting points. A&A **430**, 585–602 (2005). doi:10.1051/0004-6361:20041628, arXiv:astro-ph/0409480

Gehrz, R.D.: Infrared and radio observations of classical novae: physical parameters and abundances in the ejecta. In: Hernanz, M., José, J. (eds.) Classical Nova Explosions, American Institute of Physics Conference Series, vol. 637, pp. 198–207. American Institute of Physics, New York (2002). doi:10.1063/1.1518200

Gehrz, R.D., Truran, J.W., Williams, R.E., Starrfield, S.: Nucleosynthesis in classical novae and its contribution to the interstellar medium. PASP **110**, 3–26 (1998). doi:10.1086/316107

Glasner, A., Livne, E.: Multidimensional nova simulations. In: Hernanz, M., José, J. (eds.) Classical Nova Explosions, American Institute of Physics Conference Series, vol. 637, pp. 124–133. American Institute of Physics, New York (2002). doi:10.1063/1.1518189

Glasner, S.A., Truran, J.W.: Carbon-nitrogen-oxygen "breakout" and nucleosynthesis in classical novae. ApJ **692**, L58–L61 (2009). doi:10.1088/0004-637X/692/1/L58, 0812.3984

Gómez-Gomar, J., Hernanz, M., José, J., Isern, J.: Gamma-ray emission from individual classical novae. MNRAS **296**, 913–920 (1998a). doi:10.1046/j.1365-8711.1998.01421.x, arXiv:astroph/9711322

Gómez-Gomar, J., Isern, J., Jean, P.: Prospects for Type IA supernova explosion mechanism identification with gamma rays. MNRAS **295**, 1 9 (1998b). doi:10.1046/j.1365-8711.1998.29511115.x, arXiv:astro-ph/9709048

Grindlay, J., Gursky, H.: Scattering model for X-ray bursts – Massive black holes in globular clusters. ApJ **205**, L131–L133 (1976). doi:10.1086/182106

Grindlay, J., Gursky, H., Schnopper, H., Parsignault, D.R., Heise, J., Brinkman, A.C., Schrijver, J.: Discovery of intense X-ray bursts from the globular cluster NGC 6624. ApJ **205**, L127–L130 (1976). doi: 10.1086/182105 263

Grindlay, J.E., McClintock, J.E., Canizares, C.R., Cominsky, L., Li, F.K., Lewin, W.H.G., van Paradijs, J.: Discovery of optical bursts from an X-ray burst source, MXB1735-44. Nature **274**, 567–568 (1978). doi:10.1038/274567a0

Guerrero, J., García-Berro, E., Isern, J.: Smoothed particle hydrodynamics simulations of merging white dwarfs. A&A **413**, 257–272 (2004). doi:10.1051/0004-6361:20031504

Gutiérrez, J., García-Berro, E., Iben, I. Jr., Isern, J., Labay, J., Canal, R.: The final evolution of ONeMg electron-degenerate cores. ApJ **459**, 701–705 (1996). doi:10.1086/176934

Hachisu, I., Kato, M., Nomoto, K.: A wide symbiotic channel to type IA supernovae. ApJ **522**, 487–503 (1999). doi:10.1086/307608, arXiv:astro-ph/9902304

Hackwell, J.A., Grasdalen, G.L., Gehrz, R.D., Cominsky, L., Lewin, W.H.G., van Paradijs, J.: The detection of an optical burst coincident with an X-ray burst from MXB 1837 + 05/Ser X-1/. ApJ **233**, L115–L119 (1979). doi:10.1086/183088

Hamuy, M., Phillips, M.M., Maza, J., Suntzeff, N.B., Schommer, R.A., Aviles, R.: A Hubble diagram of distant type IA supernovae. AJ **109**, 1–13 (1995). doi:10.1086/117251

Hamuy, M., Phillips, M.M., Suntzeff, N.B., Schommer, R.A., Maza, J., Antezan, A.R., Wischnjewsky, M., Valladares, G., Muena, C., Gonzales, L.E., Aviles, R., Wells, L.A., Smith, R.C., Navarrete, M., Covarrubias, R., Williger, G.M., Walker, A.R., Layden, A.C., Elias, J.H., Baldwin, J.A., Hernandez, M., Tirado, H., Ugarte, P., Elston, R., Saavedra, N., Barrientos, F., Costa, E., Lira, P., Ruiz, M.T., Anguita, C., Gomez, X., Ortiz, P., Della Valle, M., Danziger, J., Storm, J., Kim, Y., Bailyn, C., Rubenstein, E.P., Tucker, D., Cersosimo, S., Mendez, R.A., Siciliano, L., Sherry, W., Chaboyer, B., Koopmann, R.A., Geisler, D., Sarajedini, A., Dey, A., Tyson, N., Rich, R.M., Gal, R., Lamontagne, R., Caldwell, N., Guhathakurta, P., Phillips, A.C., Szkody, P., Prosser, C., Ho, L.C., McMahan, R., Baggley, G., Cheng, K., Havlen, R., Wakamatsu, K., Janes, K., Malkan, M., Baganoff, F., Seitzer, P., Shara, M., Sturch, C., Hesser, J., Hartig, A.N.P., Hughes, J., Welch, D., Williams, T.B., Ferguson, H., Francis, P.J., French, L., Bolte, M., Roth, J., Odewahn, S., Howell, S., Krzeminski, W.: BVRI light curves for 29 Type IA supernovae. AJ **112**, 2408–2437 (1996). doi:10.1086/118192, arXiv:astro-ph/9609064

Hanawa, T., Sugimoto, D., Hashimoto, M.: Nucleosynthesis in explosive hydrogen burning and its implications in ten-minute interval of X-ray bursts. PASJ **35**, 491–506 (1983)

Hansen, C.J., van Horn, H.M.: Steady-state nuclear fusion in accreting neutron-star envelopes. ApJ **195**, 735–741 (1975). doi:10.1086/153375

Harkness, R.: Type ia Supernovae. In: Danziger, I.J., Kjaer, K. (eds.) Supernova 1987A and other supernovae, European Southern Observatory Astrophysics Symposia, vol. 37, pp. 447–456. European Southern Observatory, Garching (1991)

Harris, M.J., Leising, M.D., Share, G.H.: A search for the 478 keV line from the decay of nucleosynthetic Be-7. ApJ **375**, 216–220 (1991). doi:10.1086/170183

Harris, M.J., Naya, J.E., Teegarden, B.J., Cline, T.L., Gehrels, N., Palmer, D.M., Ramaty, R., Seifert, H.: Transient gamma ray spectrometer observations of gamma-ray lines from novae. I. Limits on the positron annihilation line in five individual novae. ApJ **522**, 424–432 (1999). doi:10.1086/307625, arXiv:astro-ph/0004164

Harris, M.J., Teegarden, B.J., Weidenspointner, G., Palmer, D.M., Cline, T.L., Gehrels, N., Ramaty, R.: Transient gamma-ray spectrometer observations of gamma-ray lines from novae. III. The 478 keV line from 7Be decay. ApJ **563**, 950–957 (2001). doi:10.1086/323951

Hatano, K., Branch, D., Fisher, A., Starrfield, S.: On the spatial distribution and occurrence rate of Galactic classical novae. MNRAS **290**, 113–118 (1997)

Hayakawa, S.: Galactic X-rays observed with X-ray astronomy satellite 'Hakucho'. Space Sci. Rev. **29**, 221–290 (1981). doi:10.1007/BF00229297

Heger, A., Cumming, A., Galloway, D.K., Woosley, S.E.: Models of type I x-ray bursts from GS 1826-24: a probe of rp-process hydrogen burning. ApJ **671**, L141–L144 (2007). doi:10.1086/525522, 0711.1195

Henyey, L., L'Ecuyer, J.: Studies in stellar evolution. VIII. The time scale for the diffusion of energy in the stellar interior. ApJ **156**, 549–558 (1969). doi:10.1086/149988

Hernanz, M.: Gamma-ray emission from classical novae. In: Hernanz, M., José, J. (eds.) Classical Nova Explosions, American Institute of Physics Conference Series, vol. 637, pp. 399–408. American Institute of Physics, New York (2002). doi:10.1063/1.1518237

Hernanz, M.: Gamma-rays from classical novae. In: Bode, M.F., Evans, A. (eds.) Classical Novae, Second Edition, pp. 252–284. Cambridge University Press, Cambridge (2008) 272

Hernanz, M., José, J.: $\gamma$-rays from classical novae: expectations from present and future missions. New Astron. Rev. **48**, 35–39 (2004). doi:10.1016/j.newar.2003.11.005

Hernanz, M., Isern, J., Canal, R., Labay, J., Mochkovitch, R.: The final stages of evolution of cold, mass-accreting white dwarfs. ApJ **324**, 331–344 (1988). doi:10.1086/165898

Hernanz, M., José, J., Coc, A., Isern, J.: On the synthesis of 7Li and 7Be in novae. ApJ **465**, L27–L30 (1996). doi:10.1086/310122

Hernanz, M., José, J., Coc, A., Gómez-Gomar, J., Isern, J.: Gamma-ray emission from novae related to positron annihilation: constraints on its observability posed by new experimental nuclear data. ApJ **526**, L97–L100 (1999). doi:10.1086/312372, arXiv:astro-ph/9910111

Hernanz, M., Smith, D.M., Fishman, J., Harmon, A., Gómez-Gomar, J., José, J., Isern, J., Jean, P.: BATSE observations of classical novae. In: McConnell, M.L., Ryan, J.M. (eds.) Proceedings of the fourth COMPTON Symposium, American Institute of Physics Conference Series, vol. 510, pp. 82–86. American Institute of Physics New York (2000). doi:10.1063/1.1303179

Hernanz, M., Gómez-Gomar, J., José, J.: The prompt gamma-ray emission of novae. New Astron. Rev. **46**, 559–563 (2002). doi:10.1016/S1387-6473(02)00201-4, arXiv:astro-ph/0110203

Higdon, J.C., Fowler, W.A.: Gamma-ray constraints on Na-22 yields in nova explosions. ApJ **317**, 710–716 (1987). doi:10.1086/165317

Hillebrandt, W., Thielemann, F.: Nucleosynthesis in novae – a source of Ne-E and Al-26. ApJ **255**, 617–623 (1982). doi:10.1086/159864

Hoeflich, P., Khokhlov, A.: Explosion models for type IA supernovae: a comparison with observed light curves, distances, H 0, and Q 0. ApJ **457**, 500–528 (1996). doi:10.1086/176748, arXiv:astro-ph/9602025

Hoffman, J.A., Marshall, H.L., Lewin WHG Dual character of the rapid burster and a classification of X-ray bursts. Nature **271**, 630–633 (1978). doi:10.1038/271630a0

Hoppe, P., Annen, P., Strebel, R., Eberhardt, P., Gallino, R., Lugaro, M., Amari, S., Lewis, R.S.: Meteoritic silicon carbide grains with unusual Si isotopic compositions: evidence for an origin in low-mass, low-metallicity asymptotic giant branch stars. ApJ **487**, L101–L104 (1997). doi:10.1086/310869

Hoppe, P., Ott, U.: Mainstream silicon carbide grains from meteorites. In: Bernatowicz, T.J., Zinner, E. (eds.) Astrophysical implications of the laboratory study of presolar materials, American Institute of Physics Conference Series, vol. 402, pp. 27–58. American Institute of Physics, New York (1997). doi:10.1063/1.53314

Hoppe, P., Strebel, R., Eberhardt, P., Amari, S., Lewis, R.S.: Isotopic properties of silicon carbide X grains from the Murchison meteorite in the size range 0.5–1.5 $\mu$m. Meteorit Planet Sci **35**, 1157–1176 (2000)

Howell, D.A., Sullivan, M., Nugent, P.E., Ellis, R.S., Conley, A.J., Le Borgne, D., Carlberg, R.G., Guy, J., Balam, D., Basa, S., Fouchez, D., Hook, I.M., Hsiao, E.Y., Neill, J.D., Pain, R., Perrett, K.M., Pritchet, C.J.: The type Ia supernova SNLS-03D3bb from a super-Chandrasekhar-mass white dwarf star. Nature **443**, 308–311 (2006). doi:10.1038/nature05103, arXiv:astro-ph/0609616

Hoyle, F., Fowler, W.A.: Nucleosynthesis in supernovae. ApJ **132**, 565–590 (1960). doi:10.1086/146963

Hutcheon, I.D., Huss, G.R., Fahey, A.J., Wasserburg, G.J.: Extreme Mg-26 and O-17 enrichments in an Orgueil corundum: identification of a presolar oxide grain. ApJ **425**, L97–L100 (1994). doi:10.1086/187319

Iben, I. Jr., Tutukov, A.V.: On the evolution of close binaries with components of initial mass between 3 solar masses and 12 solar masses. ApJS **58**, 661–710 (1985). doi:10.1086/191054

Iliadis, C., Endt, P.M., Prantzos, N., Thompson, W.J.: Explosive hydrogen burning of 27Si, 31S, 35Ar, and 39Ca in novae and x-ray bursts. ApJ **524**, 434–453 (1999). doi:10.1086/307778

Iliadis, C.: Nuclear Physics of Stars. Wiley-VCH Verlag, Weinheim (2007) in't Zand, J.J.M., Cornelisse, R., Cumming, A.: Superbursts at near-Eddington mass accretion rates. A&A **426**, 257–265 (2004). doi:10.1051/0004-6361:20040522, arXiv:astro-ph/0407083

in't Zand, J.J.M., Kuulkers, E., Verbunt, F., Heise, J., Cornelisse, R.: A superburst from 4U 1254-69. A&A **411**, L487–L491 (2003). doi:10.1051/0004-6361:20031586, arXiv:astro-ph/0310364

Ivanova, L.N., Imshennik, V.S., Chechetkin, V.M.: Pulsation regime of the thermonuclear explosion of a star's dense carbon core. Ap&SS **31**, 497–514 (1974). doi:10.1007/BF00644102

Iyudin, A.F., Bennett, K., Bloemen, H., Diehl, R., Hermsen, W., Lichti, G.G., Morris, D., Ryan, J., Schoenfelder, V., Steinle, H., Strong, A., Varendorff, M., Winkler, C.: COMPTEL search for 22Na line emission from recent novae. A&A **300**, 422–428 (1995)

Jean, P., Hernanz, M., Gómez-Gomar, J., José, J.: Galactic 1.275-MeV emission from ONe novae and its detectability by INTEGRAL/SPI. MNRAS **319**, 350–364 (2000). doi:10.1046/j.1365-8711.2000.03587.x, arXiv:astro-ph/0004126

José, J., Coc, A., Hernanz, M.: Nuclear Uncertainties in the NENA-MGAL cycles and production of 22Na and 26Al during nova outbursts. ApJ **520**, 347–360 (1999). doi:10.1086/307445, arXiv:astro-ph/9902357

José, J., García-Berro, E., Hernanz, M., Gil-Pons, P.: The first nova explosions. ApJ **662**, L103–L106 (2007). doi:10.1086/519521

José, J., Hernanz, M.: Nucleosynthesis in classical novae: CO versus ONe White Dwarfs. ApJ **494**, 680–690 (1998). doi:10.1086/305244, arXiv:astro-ph/9709153

José, J., Hernanz, M.: The origin of presolar nova grains. Meteorit Planet Sci **42**, 1135–1143 (2007a)

José, J., Hernanz, M.: TOPICAL REVIEW: nucleosynthesis in classical nova explosions. J Phys G Nucl Phys **34**, R431–R458 (2007b). doi:10.1088/0954-3899/34/12/R01

José, J., Hernanz, M., Amari, S., Lodders, K., Zinner, E.: the imprint of nova nucleosynthesis in presolar grains. ApJ **612**, 414–428 (2004). doi:10.1086/422569, arXiv:astro-ph/0405332

José, J., Hernanz, M., Coc, A.: New results on 26Al production in classical novae. ApJ **479**, L55–L58 (1997). doi:10.1086/310575, arXiv:astro-ph/9701181

José, J., Hernanz, M., García-Berro, E., Gil-Pons, P.: The impact of the chemical stratification of white dwarfs on the classification of classical novae. ApJ **597**, L41–L44 (2003). doi:10.1086/379782, arXiv:astro-ph/0309451

José, J., Hernanz, M., Iliadis, C.: Nucleosynthesis in classical novae. Nucl. Phys. A **777**, 550–578 (2006). doi:10.1016/j.nuclphysa.2005.02.121

José, J., Moreno, F.: Hydrodynamic models of type I x-ray bursts. In: International Symposium on Nuclear Astrophysics – Nuclei in the Cosmos (2006)

José, J., Moreno, F., Parikh, A., Iliadis, C.: Hydrodynamic models of type I x-ray bursts: metallicity effects. ApJS American Institute of Physics, New York (2010)

José, J., Shore, S.N.: Observational mysteries and theoretical challenges for abundance studies. In: Bode, M.F., Evans, A. (eds.) Classical Novae, Second Edition, pp. 121–137. Cambridge University Press, Cambridge (2008) 286

Joss, P.C.: X-ray bursts and neutron-star thermonuclear flashes. Nature **270**, 310–314 (1977). doi:10.1038/270310a0

Joss, P.C.: Helium-burning flashes on an accreting neutron star – a model for X-ray burst sources. ApJ **225**, L123–L127 (1978). doi:10.1086/182808

Kato, M., Hachisu, I.: Optically thick winds in nova outbursts. ApJ **437**, 802–826 (1994). doi:10.1086/175041

Khokhlov, A.M.: Delayed detonation model for type IA supernovae. A&A **245**, 114–128 (1991)

Khokhlov, A.M.: Propagation of turbulent flames in supernovae. ApJ **449**, 695–713 (1995). doi:10.1086/176091

Khokhlov, A., Mueller, E., Hoeflich, P.: Light curves of Type IA supernova models with different explosion mechanisms. A&A **270**, 223–248 (1993)

Khokhlov, A.M., Oran, E.S., Wheeler, J.C.: Deflagration-to-detonation transition in thermonuclear supernovae. ApJ **478**, 678–688 (1997). doi:10.1086/303815, arXiv:astro-ph/9612226

Knödlseder, J.: Implications of 1.8 MEV gamma-ray observations for the origin of 26Al. ApJ **510**, 915–929 (1999). doi:10.1086/306601

Koike, O., Hashimoto, M., Kuromizu, R., Fujimoto, S.: Final products of the rp-process on accreting neutron stars. ApJ **603**, 242–251 (2004). doi:10.1086/381354

Kolb, U., Politano, M.: The contribution of O-Ne-Mg novae to the 26Al production in the Galaxy. A&A **319**, 909–922 (1997)

Kong, A.K.H., Miller, J.M.,Méndez, M., Cottam, J., Lewin, W.H.G., Paerels, F., Kuulkers, E., Wijnands, R., van der Klis, M.: Nondetection of gravitationally redshifted absorption lines in the x-ray burst spectra of GS 1826-24. ApJ **670**, L17–L20 (2007). doi:10.1086/524137, 0708.0413

Kovetz, A., Prialnik, D.: CNO abundances resulting from diffusion in accreting nova progenitors. ApJ **291**, 812–821 (1985). doi:10.1086/163117

Kovetz, A., Prialnik, D.: The composition of nova ejecta from multicycle evolution models. ApJ **477**, 356–367 (1997). doi:10.1086/303675

Kraft, R.P.: Are all novae binary stars? Leaflet Astron. Soc. Pac. **9**, 137–144 (1964)

Krautter, J., Oegelman, H., Starrfield, S., Wichmann, R., Pfeffermann, E.: ROSAT x-ray observations of nova V1974 Cygni: the rise and fall of the brightest supersoft x-ray source. ApJ **456**, 788–797 (1996). doi:10.1086/176697

Kuchner, M.J., Kirshner, R.P., Pinto, P.A., Leibundgut, B.: Evidence for Ni-56 yields Co-56 yields Fe-56 decay in type IA supernovae. ApJ **426**, L89–L92 (1994). doi:10.1086/187347

Kulkarni, P.V., Ashok, N.M., Apparao, K.M.V., Chitre, S.M.: Discovery of IR bursts from Liller I/MXB1730-333. Nature **280**, 819–820 (1979). doi:10.1038/280819a0

Kumagai, S., Nomoto, K.: Gamma-rays and X-rays from Type Ia supernovae. In: Ruiz-Lapuente, P., Canal, R., Isern, J. (eds.) NATO ASIC Proc. 486: Thermonuclear Supernovae, pp. 515–540 Kluwer Academic Publishers, Dordrecht (1997)

Kuulkers, E.: The observers' view of (very) long X-ray bursts: they are super! Nucl. Phys. B Proc. Suppl. **132**, 466–475 (2004). doi:10.1016/j.nuclphysbps.2004.04.081, arXiv:astro-ph/0310402

Kuulkers, E., den Hartog, P.R., in't Zand, J.J.M., Verbunt, F.W.M., Harris, W.E., Cocchi, M.: Photospheric radius expansion X-ray bursts as standard candles. A&A **399**, 663–680 (2003). doi:10.1051/0004-6361:20021781, arXiv:astro-ph/0212028

Kuulkers, E., in't Zand, J.J.M., van Kerkwijk, M.H., Cornelisse, R., Smith, D.A., Heise, J., Bazzano, A., Cocchi, M., Natalucci, L., Ubertini, P.: A half-a-day long thermonuclear X-ray burst from KS 1731-260. A&A **382**, 503–512 (2002). doi:10.1051/0004-6361:20011654, arXiv:astro-ph/0111261

Landau, L.D., Lifshitz, E.M.: Fluid mechanics. Pergamon Press, Oxford (1959)

Langanke, K., Martínez-Pinedo, G.: Shell-model calculations of stellar weak interaction rates: II. Weak rates for nuclei in the mass range/A=45–65 in supernovae environments. Nucl. Phys. A **673**, 481–508 (2000). doi:10.1016/S0375-9474(00)00131-7, arXiv:nucl-th/0001018

Leising, M.D., Clayton, D.D.: Positron annihilation gamma rays from novae. ApJ **323**, 159–169 (1987). doi:10.1086/165816

Leising, M.D.: Gamma-ray lines from classical novae. In: Durouchoux, P., Prantzos, N. (eds.) Gamma-Ray Line Astrophysics, American Institute of Physics Conference Series, vol. 232, pp. 173–182 American Institute of Physics, New York (1991). doi:10.1063/1.40932

Leising, M.D.: Hard emission from classical novae. A&AS **97**, 299–301 (1993)

Lewin, W.H.G.: X-ray burst sources. In: Papagiannis M.D. (ed.) Eighth Texas Symposium on Relativistic Astrophysics, New York Academy Sciences Annals, vol. 302, pp. 210–227 (1977). doi:10.1111/j.1749-6632.1977.tb37050.x

Lewin, W.H.G., Clark, G., Doty, J.: X-Ray Bursts. IAU Circ. **2922** (1976)

Lewin, W.H.G., Cominsky, L.R., Walker, A.R., Robertson BSC simultaneous IR and X-ray burst observation of SER X-1. Nature **287**, 27–28 (1980). doi:10.1038/287027a0

Lewin, W.H.G., van Paradijs, J., Taam, R.E.: X-Ray bursts. Space Sci. Rev. **62**, 223–389 (1993). doi:10.1007/BF00196124

Lewin, W.H.G., van Paradijs, J., Taam, R.E.: X-ray bursts. In: Lewin, W.H.G., van Paradijs, J., van den Heuvel, E.P.J. (eds.) X-ray binaries, pp. 175–232. Cambridge University Press, Cambridge (1995)

Lewis, R.S., Ming, T., Wacker, J.F., Anders, E., Steel, E.: Interstellar diamonds in meteorites. Nature **326**, 160–162 (1987). doi:10.1038/326160a0

Liu, Q.Z., van Paradijs, J., van den Heuvel, E.P.J.: A catalogue of low-mass X-ray binaries in the Galaxy, LMC, and SMC (Fourth edition). A&A **469**, 807–810 (2007). doi:10.1051/0004-6361:20077303, 0707.0544

Livio, M.: Classical novae and the extragalactic distance scale. ApJ **393**, 516–522 (1992). doi:10.1086/171524

Livio, M.: Topics in the theory of cataclysmic variables and X-ray binaries. In: Shore, S.N., Livio, M., van den Heuvel, E.P.J. (eds.) Saas-Fee Advanced Course 22: Interacting Binaries, pp. 135–262. Springer-Verlag, Berlin (1994)

Lodders, K.: Presolar grains from meteorites: Remnants from the early times of the solar system. Chemie der Erde/Geochemistry **65**, 93–166 (2005). doi:10.1016/j.chemer.2005.01.001, arXiv:astro-ph/0501430

Lugaro, M., Davis, A.M., Gallino, R., Pellin, M.J., Straniero, O., Käppeler, F.: Isotopic compositions of strontium, zirconium, molybdenum, and barium in single presolar SiC grains and asymptotic giant branch stars. ApJ **593**, 486–508 (2003). doi:10.1086/376442

MacAlpine, G.M., Ecklund, T.C., Lester, W.R., Vanderveer, S.J., Strolger, L.: A spectroscopic study of nuclear processing and the production of anomalously strong lines in the crab nebula. AJ **133**, 81–88 (2007). doi:10.1086/509504, arXiv:astro-ph/0609803

Mahoney, W.A., Ling, J.C., Jacobson, A.S., Lingenfelter, R.E.: Diffuse galactic gamma ray line emission from nucleosynthetic Fe-60, Al-26, and Na-22 - Preliminary limits from HEAO 3. ApJ **262**, 742–748 (1982). doi:10.1086/160469

Maraschi, L., Cavaliere, A.: X-ray bursts of nuclear origin? In: X-ray Binaries and Compact Objects, pp. 127–128 (1977)

Matteucci, F., Raiteri, C.M., Busson, M., Gallino, R., Gratton, R.: Constraints on the nucleosynthesis of CU and Zn from models of chemical evolutino of the Galaxy. A&A **272**, 421–429 (1993)

Mazurek, T.J., Wheeler, J.C.: Thermonuclear explosions in stars. Fundam. Cosmic Phys. **5**, 193–286 (1980)

McClintock, J.E., Canizares, C.R., Cominsky, L., Li, F.K., Lewin, W.H.G., van Paradijs, J., Grindlay, J.E.: A 3-s delay in an optical burst from X-ray burst source MXB1735-44. Nature **279**, 47–49 (1979). doi:10.1038/279047a0

Messenger, S., Keller, L.P., Lauretta, D.S.: Supernova olivine from cometary dust. Science **309**, 737–741 (2005). doi:10.1126/science.1109602

Messenger, S., Keller, L.P., Stadermann, F.J., Walker, R.M., Zinner, E.: Samples of stars beyond the solar system: silicate grains in interplanetary dust. Science **300**, 105–108 (2003). doi:10.1126/science.1080576

Metzger, B.D., Piro, A.L., Quataert, E.: Nickel-rich outflows from accretion discs formed by the accretion-induced collapse of white dwarfs. MNRAS **396**, 1659–1664 (2009). doi:10.1111/j.1365-2966.2009.14909.x, 0812.3656

Meyer, B.S., Zinner, E.: Nucleosynthesis, pp. 69–108. The University of Arizona Space Science Series, Tucson (2006)

Milne, P.A., The, L., Leising, M.D.: Late Light Curves of Type Ia Supernovae. ApJ **559**, 1019–1031 (2001). doi:10.1086/322352, arXiv:astro-ph/0104185

Minkowski, R.: Spectra of supernovae. PASP **53**, 224–225 (1941). doi:10.1086/125315

Mostefaoui, S., Hoppe, P.: Discovery of abundant in situ silicate and spinel grains from red giant stars in a primtive meteorite. ApJ **613**, L149–L152 (2004). doi:10.1086/424842

Nguyen, A.N., Zinner, E.: Discovery of ancient silicate stardust in a meteorite. Science **303**, 1496–1499 (2004). doi:10.1126/science.1094389

Niemeyer, J.C., Woosley, S.E.: The thermonuclear explosion of chandrasekhar mass white dwarfs. ApJ **475**, 740–753 (1997). doi:10.1086/303544, arXiv:astro-ph/9607032

Nittler, L.R.: Presolar oxide grains in meteorites. In: Bernatowicz, T.J., Zinner, E. (eds.) Astrophysical implications of the laboratory study of presolar materials, American Institute of Physics Conference Series, vol. 402, pp. 59–82. American Institute of Physics, New York (1997). doi:10.1063/1.53320

Nittler, L.R., Alexander, C.M.O., Gallino, R., Hoppe, P., Nguyen, A.N., Stadermann, F.J., Zinner, E.K.: Aluminum-, calcium- and titanium-rich oxide stardust in ordinary chondrite meteorites. ApJ **682**, 1450–1478 (2008). doi:10.1086/589430, 0804.2866

Nittler, L.R., Alexander, C.M.O., Gao, X., Walker, R.M., Zinner, E.: Stellar sapphires: the properties and origins of presolar AL 2O 3 in meteorites. ApJ **483**, 475–495 (1997). doi:10.1086/304234

Nittler, L.R., Alexander, C.M.O., Nguyen, A.N.: Extreme 13C and 15N enrichments in a murchison presolar SiC grain. Meteorit. Planet. Sci. Suppl. **41**, 5316–5316 (2006)

Nittler, L.R., Alexander, C.M.O., Wang, J., Gao, X.: Meteoritic oxide grain from supernova found. Nature **393**, 222–222 (1998). doi:10.1038/30377

Nittler, L.R., Hoppe, P.: Are presolar silicon carbide grains from novae actually from supernovae? ApJ **631**, L89–L92 (2005). doi:10.1086/497029

Nittler, L.R., Hoppe, P., Alexander, C.M.O., Amari, S., Eberhardt, P., Gao, X., Lewis, R.S., Strebel, R., Walker, R.M., Zinner, E.: Silicon nitride from supernovae. ApJ **453**, L25–L28 (1995). doi:10.1086/309743

Nittler, L.R., O'D Alexander, C.M., Gao, X., Walker, R.M., Zinner, E.K.: Interstellar oxide grains from the Tieschitz ordinary chondrite. Nature **370**, 443–446 (1994). doi:10.1038/370443a0

Nofar, I., Shaviv, G., Starrfield, S.: The formation of Al-26 in nova explosions. ApJ **369**, 440–450 (1991). doi:10.1086/169772

Nomoto, K.: Accreting white dwarf models for type 1 supernovae. II – Off-center detonation supernovae. ApJ **257**, 780–792 (1982). doi:10.1086/160031

Nomoto, K., Kondo, Y.: Conditions for accretion-induced collapse of white dwarfs. ApJ **367**, L19–L22 (1991). doi:10.1086/185922

Nomoto, K., Sugimoto, D., Neo, S.: Carbon deflagration supernova, an alternative to carbon detonation. Ap&SS **39**, L37–L42 (1976). doi:10.1007/BF00648354

Nomoto, K., Thielemann, F., Yokoi, K.: Accreting white dwarf models of type I supernovae. III – Carbon deflagration supernovae. ApJ **286**, 644–658 (1984). doi:10.1086/162639

Ore, A., Powell, J.L.: Three-photon annihilation of an electron-positron pair. Phys. Rev. **75**, 1696–1699 (1949). doi:10.1103/PhysRev.75.1696

Orio, M., Covington, J., Ögelman, H.: X-ray emission from classical and recurrent novae observed with ROSAT. A&A **373**, 542–554 (2001). doi:10.1051/0004-6361:20010537, arXiv:astroph/0104219

Ott, U., Begemann, F.: Discovery of s-process barium in the Murchison meteorite. ApJ **353**, L57–L60 (1990). doi:10.1086/185707

Page, D., Cumming, A.: Superbursts from strange stars. ApJ **635**, L157–L160 (2005). doi:10.1086/499520, arXiv:astro-ph/0508444

Parikh, A., José, J., Moreno, F., Iliadis, C.: The effects of variations in nuclear processes on type I x-ray burst nucleosynthesis. ApJS **178**, 110–136 (2008). doi:10.1086/589879, 0802.2819

Payne Gaposchkin, C.H.: The galactic novae. Interscience Publishers, New York (1957)

Pellin, M.J., Davis, A.M., Lewis, R.S., Amari, S., Clayton, R.N.: Molybdenum isotopic composition of single silicon carbide grains from supernovae. In: Lunar and Planetary Institute Science Conference Abstracts, Lunar and Planetary Institute Science Conference Abstracts, vol. 30, abstract 1969 (1999)

Perlmutter, S., Gabi, S., Goldhaber, G., Goobar, A., Groom, D.E., Hook, I.M., Kim, A.G., Kim, M.Y., Lee, J.C., Pain, R., Pennypacker, C.R., Small, I.A., Ellis, R.S., McMahon, R.G., Boyle, B.J., Bunclark, P.S., Carter, D., Irwin, M.J., Glazebrook, K., Newberg, H.J.M., Filippenko, A.V.,

Matheson, T., Dopita, M., Couch, W.J., The supernova cosmology project measurements of the cosmological parameters omega and lambda from the first seven supernovae at Z _ 0.35. ApJ **483**, 565–581 (1997). doi:10.1086/304265, arXiv:astro-ph/9608192

Phillips, M.M.: The absolute magnitudes of Type IA supernovae. ApJ **413**, L105–L108 (1993). doi:10.1086/186970

Phillips, M.M., Wells, L.A., Suntzeff, N.B., Hamuy, M., Leibundgut, B., Kirshner, R.P., Foltz, C.B.: SN 1991T – Further evidence of the heterogeneous nature of type IA supernovae. AJ **103**, 1632–1637 (1992). doi:10.1086/116177

Piersanti, L., Gagliardi, S., Iben, I., Jr., Tornambé, A.: Carbon-oxygen white dwarf accreting co-rich matter. II. Self-regulating accretion process up to the explosive stage. ApJ **598**, 1229–1238 (2003a). doi:10.1086/378952

Piersanti, L, Gagliardi, S., Iben, I. Jr., Tornambé, A.: Carbon-oxygen white dwarfs accreting co-rich matter. I. A comparison between rotating and nonrotating models. ApJ **583**, 885–901 (2003b). doi:10.1086/345444, arXiv:astro-ph/0210624

Pietsch, W., Haberl, F.: XMM-newton detection of type I x-ray bursts inM31. A&A **430**, L45–L48 (2005). doi:10.1051/0004-6361:200400128, arXiv:astro-ph/0412373

Pinto, P.A., Eastman, R.G., Rogers, T.: A test for the nature of the type IA supernova explosion mechanism. ApJ **551**, 231–243 (2001). doi:10.1086/320059, arXiv:astro-ph/0008330

Politano, M., Starrfield, S., Truran, J.W., Weiss, A., Sparks, W.M.: Hydrodynamic studies of accretion onto massive white dwarfs: ONeMg-enriched nova outbursts. I. Dependence on white dwarf mass. ApJ **448**, 807–821 (1995). doi:10.1086/176009

Pozdnyakov, L.A., Sobol, I.M., Syunyaev, R.A.: Comptonization and the shaping of X-ray source spectra – Monte Carlo calculations. Astrophys. Space Phys. Rev. **2**, 189–331 (1983)

Prantzos, N., Diehl, R.: Radioactive 26Al in the galaxy: observations versus theory. Phys. Rep. **267**, 1–69 (1996). doi:10.1016/0370-1573(95)00055-0

Prialnik, D., Kovetz, A.: The effect of diffusion on prenova evolution – CNO-enriched envelopes. ApJ **281**, 367–374 (1984). doi:10.1086/162107

Prialnik, D., Kovetz, A.: An extended grid of multicycle nova evolution models. ApJ **445**, 789–810 (1995). doi:10.1086/175741

Prialnik, D., Shara, M.M., Shaviv, G.: The evolution of a slow nova model with a Z = 0.03 envelope from pre-explosion to extinction. A&A **62**, 339–348 (1978)

Prialnik, D., Shara, M.M., Shaviv, G.: The evolution of a fast nova model with a Z = 0.03 envelope from pre-explosion to decline. A&A **72**, 192–203 (1979)

Pskovskii, I.P.: Light curves, color curves, and expansion velocity of type I supernovae as functions of the rate of brightness decline. Sov. Astron. **21**, 675–682 (1977)

Rakowski, C.E., Badenes, C., Gaensler, B.M., Gelfand, J.D., Hughes, J.P., Slane, P.O.: Can ejecta-dominated supernova remnants be typed from their x-ray spectra? The case of G337.2-0.7. ApJ **646**, 982–1000 (2006). doi:10.1086/505018, arXiv:astro-ph/0604246

Rauch, T., Suleimanov, V., Werner, K.: Absorption features in the spectra of X-ray bursting neutron stars. A&A **490**, 1127–1134 (2008). doi:10.1051/0004-6361:200810129, 0809.2170

Riess, A.G., Press, W.H., Kirshner, R.P.: A precise distance indicator: type IA supernova multicolor light-curve shapes. ApJ **473**, 88–109 (1996). doi:10.1086/178129, arXiv:astro-ph/9604143

Riess, A.G., Filippenko, A.V., Li, W., Treffers, R.R., Schmidt, B.P., Qiu, Y., Hu, J., Armstrong, M., Faranda, C., Thouvenot, E., Buil, C.: The rise time of nearby type IA supernovae. AJ **118**, 2675–2688 (1999). doi:10.1086/301143, arXiv:astro-ph/9907037

Ritossa, C., García-Berro, E., Iben, I. Jr.: On the evolution of stars that form electrondegenerate cores processed by carbon burning. II. Isotope abundances and thermal pulses in a 10 Msun model with an ONe core and applications to long-period variables, classical novae, and accretion-induced collapse. ApJ **460**, 489–505 (1996). doi:10.1086/176987

Romano, D., Matteucci, F., Molaro, P., Bonifacio, P.: The galactic lithium evolution revisited. A&A **352**, 117–128 (1999) arXiv:astro-ph/9910151

Röpke, F.K., Hillebrandt, W., Niemeyer, J.C., Woosley, S.E.: Multi-spot ignition in type Ia supernova models. A&A **448**, 1–14 (2006). doi:10.1051/0004-6361:20053926, arXiv:astro-ph/0510474

Rosswog, S., Ramirez-Ruiz, E., Hix, W.R.: Tidal disruption and Ignition of white dwarfs by moderately massive black holes. ApJ **695**, 404–419 (2009). doi:10.1088/0004-637X/695/1/404, 0808.2143

Ruiz-Lapuente, P., Cappellaro, E., Turatto, M., Gouiffes, C., Danziger, I.J., Della Valle, M., Lucy, L.B.: Modeling the iron-dominated spectra of the type IA supernova SN 1991T at premaximum. ApJ **387**, L33–L36 (1992). doi:10.1086/186299

Salaris, M., Domínguez, I., García-Berro, E., Hernanz, M., Isern, J., Mochkovitch, R.: The cooling of CO white dwarfs: influence of the internal chemical distribution. ApJ **486**, 413–419 (1997). doi:10.1086/304483, arXiv:astro-ph/9704038

Schatz, H., Aprahamian, A., Barnard, V., Bildsten, L., Cumming, A., Ouellette, M., Rauscher, T., Thielemann, F., Wiescher, M.: End point of the rp process on accreting neutron stars. Phys. Rev. Lett. **86**, 3471–3474 (2001). doi:10.1103/PhysRevLett.86.3471, arXiv:astro-ph/0102418

Schatz, H., Aprahamian, A., Goerres, J., Wiescher, M., Rauscher, T., Rembges, J.F., Thielemann, F., Pfeiffer, B., Moeller, P., Kratz, K., Herndl, H., Brown, B.A., Rebel, H.: rp-process nucleosynthesis at extreme temperature and density conditions. Phys. Rep. **294**, 167–264 (1998). doi:10.1016/S0370-1573(97)00048-3

Schatz, H., Bildsten, L., Cumming, A., Wiescher, M.: The rapid proton process ashes from stable nuclear burning on an accreting neutron star. ApJ **524**, 1014–1029 (1999). doi:10.1086/307837, arXiv:astro-ph/9905274

Schmidt, W., Niemeyer, J.C.: Thermonuclear supernova simulations with stochastic ignition. A&A **446**, 627–633 (2006). doi:10.1051/0004-6361:20054145, arXiv:astro-ph/0510427

Schmidt, W., Niemeyer, J.C., Hillebrandt, W., Röpke, F.K.: A localised subgrid scale model for fluid dynamical simulations in astrophysics. II. Application to type Ia supernovae. A&A **450**, 283–294 (2006). doi:10.1051/0004-6361:20053618, arXiv:astro-ph/0601500

Senziani, F., Skinner, G.K., Jean, P., Hernanz, M.: Detectability of gamma-ray emission from classical novae with Swift/BAT. A&A 485, 223–231 (2008). doi: 10.1051/0004-6361:200809863 277

Shafter, A.W.: On the nova rate in the galaxy. ApJ **487**, 226–236 (1997). doi:10.1086/304609

Shafter, A.W.: The galactic nova rate. In: Hernanz, M., José, J. (eds.) Classical Nova Explosions, American Institute of Physics Conference Series, vol. 637, pp. 462–471. American Institute of Physics, New York (2002). doi:10.1063/1.1518246

Shafter, A.W.: Extragalactic novae. In: Bode, M.F., Evans, A. (eds.) Classical Novae, Second Edition, pp. 335–359. Cambridge University Press, Cambridge (2008) 246

Shafter, A.W., Ciardullo, R., Pritchet, C.J.: Novae in external galaxies: M51, M87, and M101. ApJ **530**, 193–206 (2000). doi:10.1086/308349

Shara, M.M.: Localized thermonuclear runaways and volcanoes on degenerate dwarf stars. ApJ **261**, 649–660 (1982). doi:10.1086/160376

Shore, S.N., Schwarz, G., Bond, H.E., Downes, R.A., Starrfield, S., Evans, A., Gehrz, R.D., Hauschildt, P.H., Krautter, J., Woodward, C.E.: The early ultraviolet evolution of the ONeMg nova V382 velorum 1999. AJ **125**, 1507–1518 (2003). doi:10.1086/367803, arXiv:astro-ph/0301415

Smale, A.P.: A second intense burst with photospheric radius expansion from X2127+119 in M15. ApJ **562**, 957–962 (2001). doi:10.1086/323517, arXiv:astro-ph/0107593

Starrfield, S.: Thermonuclear processes and the classical nova outburst. In: Classical Novae, pp. 39–60. John Wiley & Sons Ltd, New York (1989)

Starrfield, S., Iliadis, C., Hix, W.R., Timmes, F.X., Sparks, W.M.: The effects of the pep nuclear reaction and other improvements in the nuclear reaction rate library on simulations of the classical nova outburst. ApJ **692**, 1532–1542 (2009). doi:10.1088/0004-637X/692/2/1532, 0811.0197

Starrfield, S., Iliadis, C., Hix, W.R.: Thermonuclear processes. In: Bode, M.F., Evans, A. (eds.) Classical Novae, Second Edition, pp. 77–101. Cambridge University Press, Cambridge (2008) 286

Starrfield, S., Shore, S.N., Sparks, W.M., Sonneborn, G., Truran, J.W., Politano, M.: A prediction of the gamma-ray flux from Nova Herculis 1991. ApJ **391**, L71–L74 (1992). doi:10.1086/186401

Starrfield, S., Truran, J.W., Sparks, W.M. CNO abundances and hydrodynamic studies of the Nova outburst. V – 1.00-solar-mass models with small mass envelopes. ApJ **226**, 186–202 (1978a). doi:10.1086/156598

Starrfield, S., Truran, J.W., Politano, M., Sparks, W.M., Nofar, I., Shaviv, G. 22Na and 26Al production in nova outbursts. Phys. Rep. **227**, 223–234 (1993). doi:10.1016/0370-1573(93)90067-N

Starrfield, S., Truran, J.W., Sparks, W.M., Arnould, M. On Li-7 production in nova explosions. ApJ **222**, 600–603 (1978b). doi:10.1086/156175

Starrfield, S., Truran, J.W., Wiescher, M.C., Sparks, W.M.: Evolutionary sequences for Nova V1974 Cygni using new nuclear reaction rates and opacities. MNRAS **296**, 502–522 (1998). doi:10.1046/j.1365-8711.1998.01312.x

Stritzinger, M., Leibundgut, B., Walch, S., Contardo, G.: Constraints on the progenitor systems of type Ia supernovae. A&A **450**, 241–251 (2006). doi:10.1051/0004-6361:20053652, arXiv:astroph/0506415

Strohmayer, T., Bildsten, L.: New views of thermonuclear bursts, pp. 113–156. Cambridge University Press, Cambridge (2006)

Strohmayer, T.E., Brown, E.F.: A Remarkable 3 Hour Thermonuclear Burst from 4U 1820-30. ApJ 566, 1045–1059 (2002). doi:10.1086/338337 242

Strohmayer, T.E., Zhang, W., Swank, J.H., Smale, A., Titarchuk, L., Day, C., Lee, U.: Millisecond x-ray variability from an accreting neutron star system. ApJ **469**, L9–L12 (1996). doi:10.1086/310261

Swank, J.H., Becker, R.H., Boldt, E.A., Holt, S.S., Pravdo, S.H., Serlemitsos, P.J.: Spectral evolution of a long X-ray burst. ApJ **212**, L73–L76 (1977). doi:10.1086/182378

Taam, R.E.: X-ray bursts from thermonuclear runaways on accreting neutron stars. ApJ **241**, 358–366 (1980). doi:10.1086/158348

Taam, R.E., Picklum, R.E.: Nuclear fusion and carbon flashes on neutron stars. ApJ **224**, 210–216 (1978). doi:10.1086/156367

Tammann, G.A.: On the frequency of supernovae as a function of the integral properties of inter mediate and late type spiral galaxies. A&A **8**, 458–475 (1970)

Tan, W.P., Fisker, J.L., Görres, J., Couder, M., Wiescher, M.: O15(α, γ)Ne19 breakout reaction and impact on x-ray bursts. Phys. Rev. Lett. **98**(24), 242503/1-242503/4 (2007). doi:10.1103/PhysRevLett.98.242503, 0706.2349

Tang, M., Anders, E.: Interstellar silicon carbide - How much older than the solar system? ApJ **335**, L31–L34 (1988). doi:10.1086/185333

Thielemann, F., Brachwitz, F., Freiburghaus, C., Kolbe, E., Martinez-Pinedo, G., Rauscher, T., Rembges, F., Hix, W.R., Liebendörfer, M., Mezzacappa, A., Kratz, K., Pfeiffer, B., Langanke, K., Nomoto, K., Rosswog, S., Schatz, H., Wiescher, W.: Element synthesis in stars. Prog. Part. Nucl. Phys. **46**, 5–22 (2001). doi:10.1016/S0146-6410(01)00103-X, arXiv:astro-ph/0101476

Thielemann, F., Brachwitz, F., Höflich, P., Martinez-Pinedo, G., Nomoto, K.: The physics of type Ia supernovae. New Astron. Rev. **48**, 605–610 (2004). doi:10.1016/j.newar.2003.12.038

Timmes, F.X., Woosley, S.E.: The conductive propagation of nuclear flames. I – degenerate C + O and O + NE + MG white dwarfs. ApJ **396**, 649–667 (1992). doi:10.1086/171746

Truran, J.W., Arnett, W.D., Cameron, A.G.W.: Nucleosynthesis in supernova shock waves. Can. J. Phys. **45**, 2315–2319 (1967)

van den Bergh, S., Tammann, G.A.: Galactic and extragalactic supernova rates. ARA&A **29**, 363–407 (1991). doi:10.1146/annurev.aa.29.090191.002051

van Horn, H.M., Hansen, C.J.: A model for the transient x-ray sources. ApJ **191**, 479–482 (1974). doi:10.1086/152987

van Paradijs, J., McClintock, J.E.: Optical and ultraviolet observations of x-ray binaries. In: Lewin, W.H.G., van Paradijs, J., van den Heuvel, E.P.J. (eds.) X-ray Binaries, pp. 58–125. Cambridge University Press, Cambridge (1995)

Villarreal, A.R., Strohmayer, T.E.: Discovery of the neutron star spin frequency in EXO 0748-676. ApJ **614**, L121–L124 (2004). doi:10.1086/425737, arXiv:astro-ph/0409384

Waki, I., Inoue, H., Koyama, K., Matsuoka, M., Murakami, T., Ogawara, Y., Ohashi, T., Tanaka, Y., Hayakawa, S., Tawara, Y., Miyamoto, S., Tsunemi, H., Kondo, I.: Discovery of absorption lines in X-ray burst spectra from X1636-536. PASJ **36**, 819–830 (1984)

Walker, M.F.: Nova DQ Herculis (1934): an eclipsing binary with very short period. PASP **66**, 230–232 (1954). doi:10.1086/126703

Ward, R.A., Fowler, W.A.: Thermalization of long-lived nuclear isomeric states under stellar conditions. ApJ **238**, 266–286 (1980). doi:10.1086/157983

Warner, B.: Cataclysmic Variable Stars. Cambridge University Press, Cambridge (1995)

Webbink, R.F.: Double white dwarfs as progenitors of R Coronae Borealis stars and Type I supernovae. ApJ **277**, 355–360 (1984). doi:10.1086/161701

Weinberg, N.N., Bildsten, L.: Carbon detonation and shock-triggered helium burning in neutron star superbursts. ApJ **670**, 1291–1300 (2007). doi:10.1086/522111, 0706.3062

Weinberg, N.N., Bildsten, L., Schatz, H.: Exposing the nuclear burning ashes of radius expansion type I x-ray bursts. ApJ **639**, 1018–1032 (2006). doi:10.1086/499426, arXiv:astro-ph/0511247

Weiss, A., Truran, J.W.: Na-22 and Al-26 production and nucleosynthesis in novae explosions. A&A **238**, 178–186 (1990)

Wheeler, J.C., Harkness, R.P.: Type I supernovae. Rep. Prog. Phys. **53**, 1467–1557 (1990). doi:10.1088/0034-4885/53/12/001

Whelan, J., Iben, I. Jr.: Binaries and supernovae of type I. ApJ **186**, 1007–1014 (1973). doi:10.1086/152565

White, N.E., Nagase, F., Parmar, A.N.: The properties of X-ray binaries. In: Lewin, W.H.G., van Paradijs, J., van den Heuvel, E.P.J. (eds.) X-Ray Binaries, pp. 1–57. Cambridge University Press, Cambridge (1995)

Wiescher, M., Gorres, J., Thielemann, F., Ritter, H.: Explosive hydrogen burning in novae. A&A **160**, 56–72 (1986)

Wijnands, R.: Recurrent very long type I x-ray bursts in the low-mass x-ray binary 4U 1636-53. ApJ **554**, L59–L62 (2001). doi:10.1086/320922, arXiv:astro-ph/0103125

Williams, R.E.: The formation of novae spectra. AJ **104**, 725–733 (1992). doi:10.1086/116268

Woosley, S.E.: Nucleosynthesis and stellar evolution. In: Audouze, J., Chiosi, C., Woosley, S.E. (eds.) Saas-Fee Advanced Course 16: Nucleosynthesis and Chemical Evolution, pp. 1–109. Geneva Observatory, Sauverny (1986)

Woosley, S.E., Heger, A., Cumming, A., Hoffman, R.D., Pruet, J., Rauscher, T., Fisker, J.L., Schatz, H., Brown, B.A., Wiescher, M.: Models for type I x-ray bursts with improved nuclear physics. ApJS **151**, 75–102 (2004). doi:10.1086/381533, arXiv:astro-ph/0307425

Woosley, S.E., Taam, R.E.: Gamma-ray bursts from thermonuclear explosions on neutron stars. Nature **263**, 101–103 (1976). doi:10.1038/263101a0

Woosley, S.E., Weaver, T.A.: The physics of supernova explosions. ARA&A **24**, 205–253 (1986). doi:10.1146/annurev.aa.24.090186.001225

Yaron, O., Prialnik, D., Shara, M.M., Kovetz, A.: an extended grid of nova models. II. The parameter space of nova outbursts. ApJ **623**, 398–410 (2005). doi:10.1086/428435, arXiv:astroph/0503143

Yoon, S., Langer, N.: On the evolution of rapidly rotating massive white dwarfs towards supernovae or collapses. A&A **435**, 967–985 (2005). doi:10.1051/0004-6361:20042542, arXiv:astroph/0502133

Zingale, M., Dursi, L.J.: Propagation of the first flames in type Ia supernovae. ApJ **656**, 333–346 (2007). doi:10.1086/510306, arXiv:astro-ph/0610297

Zingale, M., Timmes, F.X., Fryxell, B., Lamb, D.Q., Olson, K., Calder, A.C., Dursi, L.J., Ricker, P., Rosner, R., MacNeice, P., Tufo, H.M.: Helium detonations on neutron stars. ApJS **133**, 195–220 (2001). doi:10.1086/319182

Zinner, E.K.: Presolar Grains, pp 17–40. Elsevier B, Amsterdam (2005)

Zwicky, F.: On collapsed neutron stars. ApJ **88**, 522–525 (1938). doi:10.1086/144003

# Part III
# Special Places to Observe Cosmic Isotopes

The preceding Chapters presented what happens in individual source objects. We started from a theoretical understanding of these objects, and pointed out where observational data, in particular those related to radioactive decays, contributed significantly. Now we make the connection to the more complex astrophysical sites for which we have measurements of radioactivities. Here theory has a more difficult task, as observations cover a very broad range, and the interplay of different objects and of processes acting on different time scales adds complexity. – The Sun and the solar system are unique places for which we actually hold material samples and posses detailed historic knowledge. We would like to link the origins of life on Earth to our cosmic roots through broader and more general astrophysical concepts, but we must bear in mind that special circumstances can distort our vantage point. Interstellar gas, on the other hand, is the reservoir where the specific sources discussed in preceding Chapters contribute their nucleosynthesis ejecta to the medium out of which they were formed originally, and continue being formed. We observe this gas in its current state, but the sources also bear a record of its past evolution. Therefore, one Chapter addresses the complex astrophysics of chemical evolution, emphasizing the theme of interstellar gas as an observable, yet very special *object* for studies of cosmic nucleosynthesis. Different themes of modern astrophysics define the substructure of this Chapter, discussing the various components of our Galaxy, mixing and feedback processes in interstellar gas, and the role and propagation of cosmic rays. This Chapter concludes with a discussion of the formation rates of stellar objects discussed in the preceding part of the book.

# Chapter 6
# The Early Solar System

**M. Busso**

This chapter presents a (partial) review of the information we can derive on the early history of the Solar System from radioactive nuclei of very different half-life, which were recognized to have been present alive in pristine solids. In fact, radioactivities open for us a unique window on the evolution of the solar nebula and provide tools for understanding the crucial events that determined and accompanied the formation of the Sun. Discussing these topics will require consideration of (at least) the following issues. (i) The determination of an age for solar system bodies, as it emerged especially from the application of radioactive dating. (ii) A synthetic account of the measurements that proved the presence of radioactive nuclei (especially those of half-life lower than about 100 Million years) in the Early Solar System (hereafter ESS). (iii) An explanation of their existence in terms of galactic nucleosynthesis, and/or of local processes (either exotic or in-situ) preceding and accompanying the formation of the Sun. This will also need some reference to the present scenarios for star formation, as applied to the ESS.

## 6.1 The Age of the Solar System

In the second half of the XIXth century, Helmholtz and Kelvin independently noticed that, evolving under the opposite effects of gravity and thermodynamics, the Sun would survive only for about 20 Million years (the *Kelvin-Helmholtz time scale*). Hints on the fact that this estimate was too short first came from Charles Darwin and his disciples, as this datum appeared to be by far insufficient to permit the biological and geological evolution they were discovering.

The first role played by radioactivities in the history of our understanding of the solar system was then the solution of this dilemma concerning its age. This solution became available rather soon, less than two decades after the same discovery of radioactivity as a physical phenomenon.

---

M. Busso (✉)
Università di Perugia, 06123 Perugia, Italy

Busso, M.: *The Early Solar System*. Lect. Notes Phys. **812**, 309–344 (2011)
DOI 10.1007/978-3-642-12698-7_6 

### 6.1.1 The Beginnings

In 1895 Wilhelm Conrad Röntgen showed the amazing imaging properties of the energetic radiation emanating (sometimes) from matter; the X-rays. Soon after Becquerel made the fortuitous discovery that uranium produced X-rays leaving traces on photographic plates. Further researches on these *beams* were carried out by Pierre and Marie Curie. In 1898 they showed that pitchblende (the major uranium ore, mainly made of uranium oxide $UO_2$), emitted X-rays 300 times stronger than those expected from U-rich compounds. This required the presence of another X-ray emitter, a nucleus that was called polonium (from Marie's native land). Then they discovered radium, inventing the term *radio-activity*, and showed that the X-ray emission could be quantitatively predicted as a function of time.

It was then Rutherford who unambiguously showed how radioactivity were an intrinsic property of certain atoms, linked to their intimate structure; and Arthur Holmes, in 1911, presented the first systematic attempt at dating rocks, based on their content of U and Pb. The ages of the rocks derived from radioactivity were scattered from 0.64 to more than 1.4 Gigayears (Gy), in sharp contrast with all non-nuclear estimates. The decades to follow would prove that Holmes was right: the age of the solar system has now been suggested to be $4.566 \pm 0.001$ Gy, see e.g. Allègre et al. (1995).

Rutherford (1929) also demonstrated that the ratio $^{235}U/^{238}U$ (0.007 presently) was probably as high as 0.3 at the Sun's formation, and concluded that an extrapolation backward would yield the production ratio in the extra-solar environment where nuclei had been produced. So it was Rutherford who first suggested to use radioactivity for dating not only planetary rocks and the Sun, but even the environment from where the Solar System was formed; and he correctly estimated that U production was already going on 4–5 Gyrs ago, although at his epoch the age of the Sun was still largely debated. Most of the tools for dating samples with radioactive nuclei had been, at this point, made available.

Rutherford's and Holmes' suggestion of using radioactivities as clocks capable of dating ancient samples remains as one of the most important scientific results of the twentieth century. Depending on the isotope used and its lifetime, the technique is good for measuring very different ages: from those of historical and archaeological objects to those of materials in the ESS and in old stars of the Galaxy. These tools are generally known under the name of *radio-chronology* or, in geological and astronomical contexts, *nuclear cosmo-chronology* (Clayton, 1988). For a detailed history of the Earth's datation see Dalrymple (1991); for an account of more modern efforts, after Holmes' work, see Chapter 2 in this volume.

### 6.1.2 Long-Lived Nuclei for Solar System Dating

Estimates of the solar-system age are now based on the measured abundances of long-lived radioactive (parent) nuclei and of their stable daughters in pristine meteorites. This kind of radiometric dating can be performed through modern upgrades

**Table 6.1** Possible decay modes at atomic mass $A$ and charge $Z$

| Decay mode | Particles involved | Daughter nucleus |
|---|---|---|
| $\beta^-$-decay | Emission of an $e^-$ and $\bar{\nu}$ | $(A,\ Z+1)$ |
| $\beta^+$-decay | Emission of an $e^+$ and a $\nu$ | $(A,\ Z-1)$ |
| $e^-$-capture | $e^-$ capture, $\nu$ emission | $(A,\ Z-1)$ |
| Double $\beta^-$-decay | Emission of two $e^-$ and two $\bar{\nu}$s | $(A,\ Z+2)$ |
| Double $e^-$-capture | Capture of two $e^-$, two $\nu$ emissions | $(A,\ Z-2)$ |
| $e^-$-capture, $e^+$-emission | Capture of an $e^-$, emission of an $e^+$ and two $\nu$s | $(A,\ Z-2)$ |
| Double $e^+$ emission | Emission of two $e^+$ and two $\nu$s | $(A,\ Z-2)$ |
| $\alpha$-decay | An $\alpha$ particle ($A=4,\ Z=2$) is emitted | $(A-4,\ Z-2)$ |
| Proton emission | A proton is ejected | $(A-1,\ Z-1)$ |
| Neutron emission | A neutron is emitted | $(A-1,\ Z)$ |
| Spontaneous fission | Two or more smaller nuclei emitted | |
| Cluster decay | Emission of a nucleus of $A1,\ Z1$ | $(A-A1, Z-Z1)$ $+(A1,\ Z1)$ |
| $\gamma$-decay | Photon emission from excited states | $(A,\ Z)$ |

of the mass spectrometer, from tiny samples of ancient solid materials, making use of unstable isotopes preserving today a residual abundance. This implies that their lifetime must be very long; and, actually, the shortest-lived nucleus fulfilling this requirement is $^{235}$U ($\tau = 1.015 \cdot 10^9$ year).

Various parent-daughter couples can be used, exploiting the existing channels for natural decay (schematically summarized in Table 6.1). General references where these techniques are discussed in an astrophysical environment are numerous: see e.g. Chap. 2, and Clayton (1988), Pagel (1997), Cowley (1995) and Dalrymple (1991).

Age calculation takes into account the presence of the radioactive parent ($P$) and also the original amount of the stable daughter ($D$) at the beginning of the time

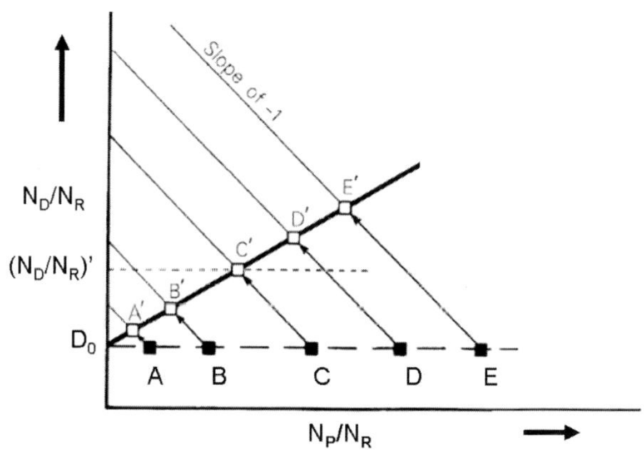


**Fig. 6.1** The scheme for the construction of an isochron in the decay of an unstable parent to its daughter. See text for comments

interval (see Fig. 6.1). In general, when there is a non-radiogenic isotope of the daughter element in the mineral, it can be used as a reference to compute abundance ratios (in this case this reference nucleus is indicated as $R$).

Suppose a series of rocks are characterized, at formation, by the values $A$, $B$, ... $E$ of the ratio $N_P/N_R$; they all obviously have the same *initial* $(N_D/N_R)_0 = D_0$. After a time interval $t$, the content of $P$ will be diminished in each sample, so that, for example:

$$(N_P/N_R)_t^A = (N_P/N_R)_0^A \exp(-\lambda t)$$

where $\lambda$ is the decay constant ($\lambda = 0.6931/t_{1/2}$). As the product $\lambda t$ is the same for all rocks, the new contents of $P$ (abscissas of the measured points $A^{'}$, $B^{'}$, ...) are proportional to the original ones and the measurement results fall on a straight line (*isochrone*). The value of $N_D/N_R$ is the sum of the *radiogenic contribution* and of the *initial value* $D_0$. It is straightforward that:

$$(N_D/N_R)_t = D_0 + (N_P/N_R)_t[\exp(\lambda t) - 1] \tag{6.1}$$

When measuring rocks with the same age $\Delta t$, the isochrone passing through the points $A^{'}$, $B^{'}$, ... will intersect the ordinate axis at a value providing the initial isotopic ratio of the daughter, $D_0$. Least square minimization then allows to determine both the age and the initial $D_0$ ratio. For a radiogenic nucleus $R$ decaying to a stable isotope $D$, with a half-life $t_{1/2}$, we derive that:

$$\Delta t = \frac{1}{\lambda} \times \ln \left[ 1 + \frac{\Delta D}{\Delta R} \right]. \tag{6.2}$$

A recent example of such age determination, involving $^{87}$Rb ($P$), $^{87}$Sr ($D$) and $^{86}$Sr ($R$), can be found in Anderson et al (2007).

Several such parent-daughter couples are available for isotope geology. Examples are the $^{40}$K – $^{40}$Ar pair ($^{40}$K producing $^{40}$Ar through $\beta^+$-decay and $e^-$-captures, with a half-life of 1.25 Gy), and the $^{147}$Sm – $^{143}$Nd pair (the first $\alpha$-decays to the second with a half life of 10.6 Gy).

Alternatively, the series of $\alpha$ and $\beta$ decays of uranium (and sometimes transuranic nuclei) can be exploited. Perhaps the most commonly-used pair is in this case U-Pb, actually made of two separate chains, one leading from $^{238}$U to $^{206}$Pb, with a half-life of 4.5 Gy, the second leading from $^{235}$U to $^{207}$Pb, with a half-life of 0.7 Gy. The two processes are shown in Fig. 6.2. Very recently, cautions have been advanced on the accuracy of the U-Pb datation, due to the discovery of live $^{247}$Cm in the ESS. This nucleus decays to $^{235}$U, with a half-life of 15.6 Million years, so that the initial inventory of U must be corrected for this effect in order to avoid inaccuracies in the solar system age, which can be as high as 5 Myr (Brennecka, 2010). (See also Sect. 2.2.4. for early solar system radioactivity studies.)

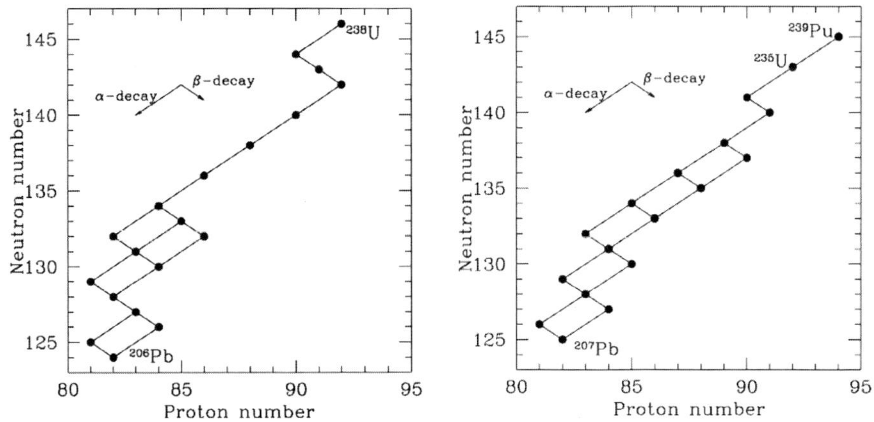


**Fig. 6.2** The decay chains of $^{238}$U and $^{235}$U, commonly used for dating rocks. The U-Pb method is applied to the mineral zircon ($ZrSiO_4$), containing some U (because it can chemically substitute Zr) and strongly rejecting lead

## 6.2 Short-Lived Radioactive Nuclei in the Early Solar System

Measurements revealing the presence of radioactive species in the ESS, with mean lifetimes lower than about 100 Million years (generally *in this context* referred to as *short-lived*) were the source of enormous progress in our knowledge of the solar origins, and have by now been recognized as one of the most important scientific achievements in the second half of the twentieth century. Identification of the decay product allows us to draw a plot similar to Fig. 6.1, if we replace the parent nucleus P (that no longer exists) in the abscissa by another stable isotope of the same element (sometimes called the *substitute*, or *index* nucleus), which supposedly entered into the sample together with P (Busso et al., 1999; Wasserburg et al., 2006). In this way, after a careful demonstration that the excess in the stable daughter's abundance is quantitatively correlated with the chemical properties of the parent element (using the index isotope), one can reconstruct isochrones and derive the initial concentration of P in the sample. In a diagram similar to Fig. 6.1, and using the pair $^{26}$Al – $^{26}$Mg as an example, we use as abscissa the ratio $^{27}$Al/$^{24}$Mg ($^{27}$Al is then the index nucleus) and on the ordinate we plot $^{26}$Mg/$^{24}$Mg. This can be done, for example, for a mineral intrinsically poor in Mg, where the presence of $^{26}$Mg can be ascribed largely to the $^{26}$Al-decay. A description of the procedure can be found in Lee et al. (1976). Fig. 6.3 shows the technique applied to $^{26}$Al and $^{41}$Ca, as obtained by Sahijpal et al. (1998). These authors showed that $^{26}$Al and $^{41}$Ca were either both present and correlated, or both absent from the samples, thus demonstrating their likely common origin.

The above studies revealed an unexpected complexity of the physical processes that affected the early phases of the solar system. It became clear that stellar nucleosynthesis, Galactic chemical evolution and the energetic processes in the winds of the early Sun are all contributing; any single explanation alone leads to more or less

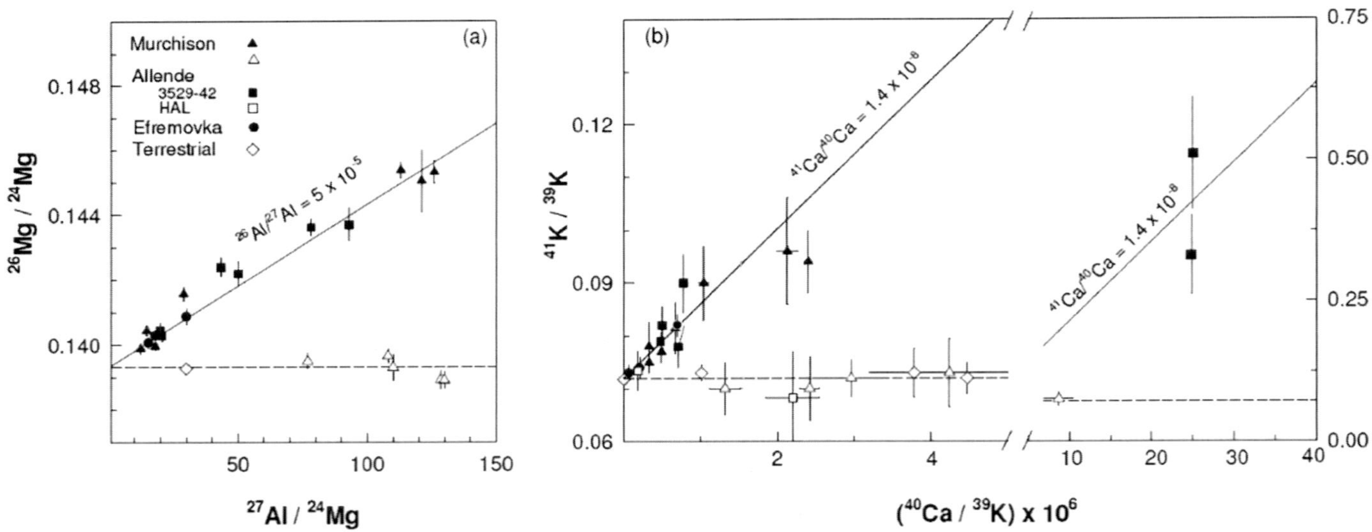


**Fig. 6.3** **Panel a** shows the Al-Mg data from CAIs in different meteorites; filled symbols demonstrate the presence in ESS of $^{26}$Al at the level $^{26}Al/^{27}Al = 5 \cdot 10^{-5}$. **Panel b** displays in a similar way the Ca – K data from the same samples. The first published data, and the *line* showing $[^{41}Ca/^{40}Ca]_0 = 1.4 \cdot 10^{-8}$ were due to Srinivasan et al. (1994, 1996). Other data and the demonstration of the correlation with Al are from Sahijpal et al. (1998). This figure is reproduced from these last authors and from Wasserburg et al. (2006). Copyright Nature Publishing Group

obvious conflicts with ESS radio-isotope measurements (see also Chaps. 2 and 7). These insights began from the work of Reynolds (1960), who was the first to show that a now extinct nucleus, $^{129}$I ($\tau$ = 23 Million years), had existed alive in the ESS. This was demonstrated analyzing the concentration of its decay product, $^{129}$Xe, in meteorites (Jeffery and Reynolds, 1961). It was then suggested by Wasserburg et al. (1960) and by Cameron (1960) that the abundance of $^{129}$I could be attributed to the continuous production of $r$-process nuclei in the Galaxy, provided the solar nebula had been isolated for about $10^8$ years.

Shortly after the discovery of iodine, the presence of another relatively long-lived nucleus, $^{244}$Pu ($\tau$ = 115 Million years), was inferred from excesses in planetary differentiates (PD) of neutron-rich Xe isotopes (Rowe and Kuroda, 1965). It was then demonstrated that the enrichment in Xe isotopes correlated with excess fission tracks in meteoritic materials rich in both U, Th, and rare earth elements, showing that fission of heavy nuclei had occurred *in situ* (Wasserburg et al., 1969; Stoerzer and Pellas, 1977)

More systematic measurements, extending to nuclei of shorter lifetime and to very early ESS constituents, became possible after 1969, thanks to the fall of two extremely primitive meteorites: one in the Mexican village Allende and the second in the Australian village Murchison. They are rich in very refractory aggregates, called *Aluminium-Calcium Inclusions*, or CAIs (MacPherson et al., 2003), which are believed to be among the first-condensing solids in the solar nebula.

Measurements of Xe isotopes on CAIs in Allende (Podosek and Lewis, 1972) soon confirmed both the presence of $^{129}$I and the in situ fission of actinides. Even

**Table 6.2** Short-lived nuclei in the ESS[a]

| Rad. | Ref. | Mean life (Million years) | $(N_P/N_I)_{ESS}$ |
|---|---|---|---|
| $^{10}$Be | $^{9}$Be | 2.2 | $5.2 \cdot 10^{-4}$ |
| $^{26}$Al | $^{27}$Al | 1.05 | $5 \cdot 10^{-5}$ |
| $^{36}$Cl | $^{35}$Cl | 0.43 | $10^{-4}$ |
| $^{41}$Ca | $^{40}$Ca | 0.15 | $\geq 1.5 \cdot 10^{-8}$ |
| $^{53}$Mn | $^{55}$Mn | 5.3 | $6.7 \cdot 10^{-5}$ |
| $^{60}$Fe | $^{56}$Fe | 2.2[b] | $10^{-7}$–$10^{-6}$ |
| $^{107}$Pd | $^{108}$Pd | 9.4 | $2.0 \cdot 10^{-5}$ |
| $^{129}$I | $^{127}$I | 23 | $1.0 \cdot 10^{-4}$ |
| $^{146}$Sm | $^{144}$Sm | 148 | $1.0 \cdot 10^{-2}$ |
| $^{182}$Hf | $^{180}$Hf | 13 | $2.0 \cdot 10^{-4}$ |
| $^{244}$Pu | $^{232}$Th | 115 | $3 \cdot 10^{-3}$ |
| $^{247}$Cm | $^{232}$Th | 23 | $1\text{-}2.4 \cdot 10^{-4}$ |
| $^{135}$Cs | $^{133}$Cs | 2.9 | $1.6 \cdot 10^{-4}$ (?) |
| $^{205}$Pb | $^{204}$Pb | 22 | $1$–$2 \cdot 10^{-4}$ (?) |

[a] In the lower panel we include isotopes for which only hints or upper bounds are available. [For references see Wasserburg et al. (2006) and, for $^{247}$Cm, Brennecka et al. (2010).]

[b] This is the value used in most works. Recent measurements by Rugel et al. (2009) suggest intead 3.8 Million years.

more important was the discovery of $^{26}$Mg excesses that, thanks to the technique illustrated above in Figs. 6.1 and 6.3, were unambiguously proven to be the relics of the earlier presence of $^{26}$Al (Lee et al., 1976, 1977). The meanlife of $^{26}$Al ($\tau = 1.05$ Million years) is much shorter than for iodine, and its presence seemed to confirm older suggestions (Urey, 1955; Urey and Donn, 1956), where it had been anticipated that $^{26}$Al would be a very likely source for the early heating of planetesimals, as other internal energy sources appeared insufficient to melt them. Searches for $^{26}$Al had been done before the fall of Allende (Schramm and Wasserburg, 1970), but they could not find the nucleus in solid solar-system samples. After the works by Lee et al. (1976, 1977), several studies then confirmed the homogeneous presence of $^{26}$Al at the high (often called *canonical*) ratio $^{26}\mathrm{Al}/^{27}\mathrm{Al} = 5\times10^{-5}$. The existence of values higher than the standard one (sometimes referred to as *super canonical*) was suggested a few years ago (Young et al., 2005; Liu et al., 2005), but not confirmed afterwards.

The above discoveries were soon followed by others: the $p$-nucleus $^{146}$Sm ($\tau =$ 148 Million years) was identified in meteoritic materials more evolved than CAIs, showing the signs of ongoing planetary differentiation: it was identified from its decay product, $^{142}$Nd (Lugmair and Marti, 1977). Then the abundance of $^{107}$Pd ($\tau = 9.4$ Million years) was established, again in planetary differentiates (Kelley and Wasserburg, 1978). The fact that planetary cores could contain a relatively short-lived nucleus like $^{107}$Pd was a proof that they formed very quickly in the history of the solar system.

Excesses of $^{205}$Tl were also inferred by Chen and Wasserburg (1987). They would indicate the presence of $^{205}$Pb ($\tau = 22$ Million years). Recently Nielsen et al. (2006) found evidence of a correlation between the $^{205}$Tl excess in iron meteorites and $^{204}$Pb. This would imply the in-situ decay of $^{205}$Pb, at the level $^{205}\mathrm{Pb}/^{204}\mathrm{Pb} =$ 1-2$\times10^{-4}$. We notice that $^{205}$Pb is a shielded (*s-process only*) n-capture nucleus, whose presence would certify that the ESS was somehow influenced by recent $s$-process nucleosynthesis events, either from high mass stars (Busso and Gallino, 1985) or from low-intermediate mass red giants (Gallino et al., 1998); see also later, in Sect. 6.4.3.

The presence of $^{53}$Mn ($\tau = 5.3$ Million years) in CAIs, through a correlation of $^{53}$Cr with Mn, was suggested by Birck and Allègre (1985, 1988). This was later confirmed by measurements of abundant $^{53}$Mn in PDs, made first by Lugmair et al. (1992) and subsequently by Lugmair and Shukolyukov (1998) and Hutcheon et al. (1998). $^{53}$Mn is very effectively produced in supernovae (SNe II or SNIa), and is also a possible outcome of spallation processes.

Tungsten isotopic anomalies discovered in early planetary materials, with deficiencies in $^{182}$W, and their correlation with hafnium in chondritic meteorites demonstrated the presence of $^{182}$Hf ($\tau = 13$ Million years), see Lee and Halliday (1995, 1996) and Harper and Jacobsen (1996).

The presence of $^{60}$Fe ($\tau = 2.2$ Million years) was early discovered by Shukolyukov and Lugmair (1993a,b). Subsequently $^{60}$Ni excesses were shown to be correlated with Fe/Ni ratios in chondrites, and the $^{60}$Fe concentration was found to be rather high in the ESS (Mostefaoui et al., 2003, 2004; Tachibana and Huss,

2003). This nucleus is a product of neutron captures at a relatively high neutron density, and can be synthesized both in massive stars exploding as supernovae and in intermediate-mass stars in their final evolutionary stages, but not in spallation processes (Lee et al., 1998). Recent estimates of the precise initial $^{60}$Fe content of the ESS have oscillated noticeably, ranging from a minimum of $^{60}$Fe/$^{56}$Fe = $10^{-7}$ to a maximum some 20 times higher (Wadhwa et al., 2007). Despite many discussions and claimed revisions, the situation still remains essentially as defined by Tachibana et al. (2006), where a recommended range of $(0.5\text{–}1)\times10^{-6}$ was indicated; see also discussion in Gounelle and Meibom (2010) and Huss et al. (2009). A remarkable suggestion by Bizzarro et al. (2007), according to which deficits in $^{60}$Ni would exist in early meteorites as compared to the planets, and would indicate a contamination in $^{60}$Fe that occurred shortly after, not before, the solar nebula formation, seem now to be the result of experimental uncertainties and have not since been confirmed (Dauphas et al., 2008a,b).

$^{10}$Be ($\tau = 2.3$ Million years) was shown to be present in the ESS by McKeegan et al. (2000); as it is not produced by stellar nucleosynthesis, its presence certifies either the in situ proton bombardment of small solids, or some contributions from galactic cosmic rays. Recent measurements fix the initial $^{10}$Be/$^{9}$Be ratio at $(0.4\text{–}1)\times10^{-3}$ (Liu et al., 2005).

Also very-short lived nuclei, with lifetimes below 1 Million year, were discovered in the last decade of the twentieth century. The presence of $^{41}$Ca ($\tau = 0.15$ Million year) was ascertained by Srinivasan et al. (1994, 1996). The initial ratio derived was $^{41}$Ca/$^{40}$Ca = $1.5\times10^{-8}$, a datum that was later suggested to be a lower limit (Gounelle et al., 2006). $^{41}$Ca is abundantly produced from neutron captures in stellar environments, but the short time scale for its decay poses hard constraints on any astrophysical scenario for its formation. At the abundances observed, it might also be produced by proton bombardment, but its proven correlation with $^{26}$Al, already shown in Fig. 6.3 (Sahijpal et al., 1998), and the recent evidence that $^{26}$Al must be exotic (Fitoussi et al., 2008), indicate that also $^{41}$Ca should come from stellar nucleosynthesis.

Hints on the presence of alive $^{36}$Cl ($\tau = 0.43$ Million years) in the ESS were early found by Goebel et al. (1982), through measurements of $^{36}$Ar abundances in Allende samples (including CAIs). The large shifts found in $^{36}$Ar/$^{38}$Ar correlated with Cl abundances, and would imply an initial ratio $^{36}$Cl/$^{35}$Cl $\simeq 2\times10^{-8}$ (U. Ott, personal communication). A more recent study by Murty et al. (1997) also found $^{36}$Ar shifts that were attributed to $^{36}$Cl. Subsequent works (Lin et al., 2004; Leshin et al., 2004) clearly demonstrated a correlation of $^{36}$S/$^{34}$S with $^{35}$Cl/$^{34}$S in late-formed halogen-rich phases in CAIs ($^{36}$S is another decay product of $^{36}$Cl). Sulphur anomalies were shown to be uncorrelated with $^{26}$Al. These results demonstrate the presence of $^{36}$Cl in the ESS with a high concentration ($^{36}$Cl/$^{35}$Cl $\simeq 10^{-4}$). Stellar sources would not be capable of explaining $^{36}$Cl at such a high abundances, and it would require to be attributed to bombardment of solids by the early solar wind, while $^{26}$Al would in this case come from stars. These results were subsequently strengthened by the work by Hsu et al. (2006), where the decoupling of chlorine and aluminium was unambiguously and clearly demonstrated. Further evidence have then been added (Jacobsen et al., 2009; Matzel et al., 2010), confirming that the use of $^{36}$Cl as a chronometer for ESS events is unfeasible and that chlorine was added by phenomena internal to the system, occurred well after the injection of $^{26}$Al.

The wealth of new measurements on ESS samples has now become impressive. The above account is certainly incomplete, but should at least focus our attention on the fact that no simple explanation for the origin of the complex nucleosynthesis pattern revealed by short-lived radioactivities can be invoked. It is evident that, although solid materials formed in a very short lapse of time (from a fraction of a Million year to a few Million years), they maintain the records of several phenomena, from a blend of different stellar nucleosynthesis processes, to solar wind bombardment and possibly also to galactic cosmic ray spallation.

Table 6.2 summarizes the status of ESS radioactivities today; for notations, see later Eqs. (6.2) and (6.3). A lively scientific debate is now considering the various possible sources for the production of short lived nuclei, including the continuous synthesis in the Galaxy, presented e.g. in Schramm et al. (1970) and in Chap. 7; the local pollution by a nearby star, early introduced by Cameron and Truran (1977) for a massive supernova and recently addressed by Wasserburg et al. (2006) and by Trigo-Rodríguez et al. (2009) for a lower-mass star; and, for several species, the production by spallation processes in the winds of the forming Sun, for which see Shu et al. (2001). Uncertainties in the interpretations still remain, and are due to the poorly known details of stellar nucleosynthesis and of magnetic winds in star-forming regions.

The isochrones constructed thanks to short-lived radio-nuclei thus became a precious tool for estimating Million year-scale age differences among samples that are, on average, 4.5 Gy old. This offers a clock, with an accuracy that might be as high as 0.1 Million years, for dating the fast events that drove planetary formation (McKeegan and Davis, 2005). For a more detailed account of the methods discussed in this section and of the science built in the last 50 years with short-lived nuclei I refer the reader to dedicated reviews – see e.g. Arnould and Prantzos (1999), Busso et al. (1999), Kratz et al. (2004), McKeegan and Davis (2005), Goswami et al. (2005), Wasserburg et al. (2006), and Wadhwa et al. (2007). At the moment of this report, perhaps the most complete and recent update can be found in Huss et al. (2009).

## 6.3 Expected Conditions in the Early Solar System

The presence, in the ESS, of short-lived radioactive nuclei, some of which exclusively produced in stars, calls for a detailed study of the mutual relations between the forming Sun and the sources of galactic nucleosynthesis. In order to do this one needs to know the general scheme of low-mass star formation, then inquiring whether some special condition was necessary, for the Sun to carry live nuclei of short mean-life. The conditions prevailing in star-forming regions have been ascertained in the last 30 years, mainly thanks to the advances in the techniques for infrared-, millimeter- and radio-astronomy. Recent accounts can be found in Lada and Lada (2003) and in Zuckerman and Song (2004). Some 70–90% of stars are born in clusters or multiple systems, while the remaining part undergoes more isolated processes of slow accretion. In the first phases not only gravity, but also galactic magnetic fields, velocity fields of the interstellar medium (ISM), and possibly the extra-pressure exerted by supernovae and other triggering phenomena can affect the proto-stellar clouds (Boss, 2005). In most cases, and especially when stars are born through the formation of isolated cloud cores or globules, the initial phases are characterized by a slow accretion of cool materials thanks to plasma processes, namely the diffusion of positive and negative particles at virtually the same rate, due to interactions via the electric fields. This mechanism, called *ambipolar diffusion* allows weakly ionized or neutral gas to separate from the galactic plasma and to accumulate until the minimum mass necessary for gravitational contraction (the so-called *Jeans mass*) is reached. Then the collapse starts from inside out and a protostar and a disk are formed: see e.g. Shu et al. (1987a,b) and references therein.

### 6.3.1 Processes in Star-Forming Clouds

The various phases that lead a forming star to reach its final structure and burn hydrogen on the Main Sequence correspond to different classes of objects seen in molecular clouds at long wavelengths. Starting from a proto-stellar cloud (often indicated as being of class –I), one has first the formation of a condensed core with a surrounding envelope (class 0). This core then evolves, with the growth of the central condensation, through classes I, II and III, characterized by a varying spectral energy distribution at mid-infrared wavelengths, informative of the amount of cold circumstellar dust. Class I objects have large mid-infrared excesses and are optically invisible, Class II sources have infrared fluxes decreasing for increasing wavelength, as the percentage of dust is reduced to some $\sim 0.01\ M_\odot$. In the total emission, a flattened disk becomes gradually more important than the outside fading envelope and the central star becomes visible in the form of a variable object, of the T Tauri group (a star approaching the Main Sequence). Class III sources, then, have essentially no remaining mid-infrared excess from the original envelope, being *naked* T Tauri stars, with disks whose masses decrease in time, although they are partially preserved up to the Main Sequence phase. The whole duration of the above processes is highly uncertain theoretically and also strongly dependent on the total mass. For solar-like stars the whole pre-Main Sequence evolution may last for about 20 Million years, half of which spent in the form of a pre-collapse cloud core. Then the formation of the central protostar is fast (virtually in free-fall conditions), and the T Tauri star may need another 10 Million years to approach the main Sequence and start the fusion of hydrogen into helium.

The above mentioned time scales suggest that the proto-Sun must have been isolated from galactic nucleosynthesis processes for a time of one to several $\times 10^7$ years. Such a long quiescent time would imply large dilution factors and long free-decay periods for any radioactive nucleus created in a galactic evolution scenario. Roughly speaking, one might expect that nuclei with meanlives longer than about

10 Million years ($^{244}$Pu, $^{129}$I, $^{182}$Hf etc.) might have survived the pre-stellar phases. On the contrary, other short-lived nuclei, with mean-lives up to a few Million years, would require a *local* production. With this term one may mean either that some nearby stellar source contaminated the ESS during its initial evolutionary stages, or that radioactive nuclei were produced in the solar nebula itself. Very short-lived nuclei will have to be attributed to either of these local processes; an exception to this rule is $^{53}$Mn. In fact, despite its rather short lifetime ($\tau = 5.3$ Million years), its production in supernovae is so huge that the requirement, for it, of a late addition has been questioned (Wasserburg et al., 2006).

Concerning the local production of radioactive nuclei, the stellar origin was early proposed by Cameron and Truran (1977), who suggested that a nearby supernova might have been responsible both for introducing unstable nuclei in the ESS and for triggering the same contraction of the Sun. An alternative to this assumed stellar event would require that, in the ESS, a combination of magnetized winds and particle bombardments might have induced spallation processes on early solids, breaking heavy nuclei into lighter unstable isotopes.

This last idea descends from the observation of strong bipolar emissions from star forming clouds, showing that matter is lost along the rotational axis through magnetic winds (Shu et al., 1987a,b; Boss, 2005). The physical origins of these winds have been identified in three main phenomena, namely: (i) a coronal wind, originating directly from the star; (ii) a disk wind, starting from the surface of the accretion disk over a wide range of distances from the central star (from less than one to more than a hundred astronomical units), which is probably the main source of mass loss in T Tauri variables (Hartmann and Kenyon, 1996); and (iii) a wind driven by the interaction of the stellar corona with the inner edge of the accretion disk, launched at a distance of a few stellar radii from the center. This last is the so-called *X wind* (Shu et al., 1997), illustrated in Fig. 6.4, and is the type of magnetic interaction that might be important also for the explanation of extinct radioactivities in the solar nebula. In such a case, its role would be that of lifting early solids from the symmetry plane of the disk, exposing them to high temperatures and to fluxes of energetic particles from the Sun. The irradiated dust would then fall back on the disk itself at large distances from the internal regions where it had been originally

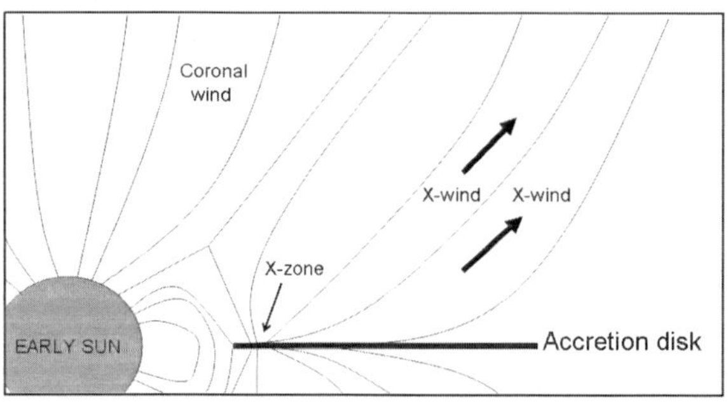


**Fig. 6.4** A schematic view of the X-wind model

created, preserving the record of new radioactive species produced by solar spallation (Shu et al., 1997; Boss, 2005). Confirmations of these lifting and transporting processes recently came from the first results of the STARDUST mission (Brownlee et al., 2006).

### *6.3.2 General Constraints for a Stellar Encounter*

Since the original proposal by Cameron and Truran (1977), the idea of a close stellar encounter for explaining the presence in the ESS of radioactive nuclei with short mean-life has evolved into different branches, and both massive and low/intermediate-mass stars have been considered as possible sources (Podosek and Nichols, 1997; Wasserburg et al., 1998; Busso et al., 1999; Meyer and Clayton, 2000; Wasserburg et al., 2006; Gounelle et al., 2006; Takigawa et al., 2008; Huss et al., 2009).

Various objections to the trigger model have however been advanced, on the basis of the growing knowledge of star formation (Evans, 1999). In the violent scenarios of cluster formation, events of triggering are known (Kothes et al., 2001; Palla and Stahler, 2000; Zinnecker, 2002), induced either by SN ejecta, or by massive star winds from OB associations. In such an environment, the required distance of a SN explosion from the solar nebula can be estimated from the dilution needed to obtain the ESS abundances of short-lived nuclei, starting with their yields in SN ejecta. Useful distances are from a fraction of a pc to a few pc, depending on the SN model; see also Cameron et al. (1995). However, both the models for the evolution SN remnants and the observational evidence indicate that star formation triggering occurs when the shock fronts have slowed down to a few km/sec or less (Preibisch and Zinnecker, 1999). Such slow motions may require that the expanding SN shells have traveled much larger distances, of the order of 50 pc: see e.g. Scheffler and Elsaesser (1987), Chap. 6. This would imply a dilution in excess by a factor of about 10 as compared to the requirements by Cameron et al. (1995). A SN trigger like the one originally suggested (Cameron and Truran, 1977), if it ever occurred, must therefore be a very rare or unique event.

Recent studies have considered in some detail the probability that the Sun be formed in a cluster massive enough to contain from one to several Supernovae, and that the timely explosion of these supernovae be at the origin of ESS radioactivities (Adams and Laughlin, 2001; Hester and Desch, 2005; Williams and Gaidos, 2007; Smith and Brooks, 2007). In such a case the contaminated nebula may have already evolved forming a protostellar disc (Adams and Laughlin, 2001; Ouellette et al., 2005, 2007). A recent analysis in the above framework is due to Gounelle and Meibom (2010). For the contamination of a contracting nebula having already developed the protostar/disc structure these authors estimated the useful distance of the SN source to be very small (corresponding to less than 1 Million year in time). It was also argued that massive-star evolution timescales are too long for them to explode as supernovae with the due timeliness to contaminate forming stars in a same cluster. As a consequence, in these works the probability of contamination from a supernova belonging to the same cluster (the so-called Orion Nebula Cluster-like scenario) turns out to be very small (of the order of $10^{-3}$). Similarly low probabilities were found by the above authors in the case of a more distant supernova, for a cloud still in its initial evolutionary stages, those of a molecular cloud core, a situation that would mimic the original idea by Cameron and Truran (1977). If *instead* the Sun condensed alone, then a nearby supernova is even less likely then otherwise estimated. One has further to remember that all studies involving massive star evolution suffer for the large uncertainties involved in crucial cross sections, especially the one for the $^{12}C(\alpha, \gamma)^{16}O$ reaction. A recent careful analysis of SN yields for $^{26}$Al and $^{60}$Fe by Tur et al. (2009) concludes with warnings about the still non-quantitative nature of such predictions.

For some radioactive nuclei an alternative was suggested by Wasserburg et al. (1994, 2006) and by Busso et al. (1999, 2003). This idea foresees the close-by passage of an AGB star in its final evolutionary stages, producing $^{26}$Al, $^{41}$Ca, $^{60}$Fe, $^{107}$Pd and *s*-process radioactivities like $^{135}$Cs and $^{205}$Pb. For updates on the production of some radioactive nuclei in AGB stars of intermediate mass see Trigo-Rodríguez et al. (2009). However, also the close encounter with an AGB star is hard to motivate on statistical grounds. In this case we have not to face the risk of cloud disruption (as for the fast supernova winds), but we require the simultaneous occurrence of two rare phenomena (an isolated, low-mass, cloud-core collapse and a planetary nebula ejection by an old star passing there just by chance). Critical remarks on this idea, pointing out that such an event must be extremely rare, were advanced by Kastner and Myers (1994). One has however to warn that most remarks have been recently considered as being no longer valid (Trigo-Rodríguez et al., 2009). This is due to recent upgrades in our knowledge of the Initial Mass Function and also to the very poor statistics available at the time of the analysis by Kastner and Myers (1994).

In order to overcome the difficulties related to the low probability of suitable encounters, and of supernovae in particular, very recently the hypothesis was advanced that the presence of $^{26}$Al in the oldest Solar System solids and its uniform distribution could be due to the inheritance of $^{26}$Al itself from the parent giant molecular cloud, where $^{26}$Al would have been contributed by very massive stars evolving to the Wolf-Rayet stage (Gaidos et al., 2009b). Although this idea is interesting and has now good reasons for being further analyzed by other authors, in its first formulation it again leads to the conclusion that the presence of $^{26}$Al is not a common event, and that the probability of this occurrence is rather low. Moreover, ad hoc independent hypotheses had to be advanced to avoid overproduction (by a factor of 10) of $^{60}$Fe from the supernovae expected in the cloud. The *combined* probability of these special occurrences for $^{60}$Fe and of the production of $^{26}$Al by WR stars would certainly be extremely small. I guess that, needing two series of different (rare) events, the joint explanation of both $^{26}$Al and $^{60}$Fe in this scenario would turn out to be even less likely than for a single encounter (either a Supernova or an AGB star).

The situation depicted above is embarrassing. A few radioactive isotopes alive in the ESS seem to require necessarily a stellar production very close in time and space to the formation of the Sun. However, realistic (although highly uncertain) models of how such a production might have occurred, always lead to unlikely conditions. It is not comfortable, for science, to call for special or unique conditions, occurred only for our Sun. Somehow, this seems to shed doubts on the accepted principle, according to which our position in the Universe has nothing special. However, a more satisfactory view of the injection of radioactive nuclei into the ESS is not available at the present moment.

## 6.4 The Galactic Inheritance

As mentioned, for radioactive nuclei with sufficiently long meanlife ($\tau \geq 10$ Million years) the ESS abundance was probably inherited from the equilibrium established in the ISM by galactic evolution. This topic is dealt with extensively in Chap. 7; see also Chap. 2 for the effects on the ESS. For illustration purposes, we shall limit the discussion to the simplest possible scheme, representing the Galaxy as a *closed box* with *instantaneous recycling*, evolving for a time interval $T$. For this case the equilibrium ratio $N_P/N_R$ of a radioactive nucleus $P$ with respect to a stable reference isotope $R$ produced in the same process was derived by Schramm et al. (1970):

$$N_P(T)/N_R(T) \simeq P_P \cdot p(T)\tau_{\mathrm{P}}/P_R < p > T. \tag{6.3}$$

Here $P_R < p >$ is the stellar production rate of the stable reference nucleus, expressed as the product of an invariant stellar production factor $P_R$ and of the average over $T$ of a variable scaling factor $p(t)$. $P_P$ is the stellar production factor for the parent isotope and $p(T)$ is the scaling factor for it at time $T$. If the ESS is separated from the ISM for a time interval $\Delta_1$ before forming the first solid condensates, then the abundance ratio in the oldest meteoritic material will be decreased with respect to the ISM equilibrium by $\exp(-\tau_P/\Delta_1)$. This old material is usually identified with CAIs. Any younger solid body, formed at a time $t = \Delta_2$ after the CAI's condensation, will have an abundance ratio further reduced by $\exp(-\tau_P/\Delta_2)$. In this framework, Wasserburg et al. (1996) demonstrated that $^{244}$Pu (an actinide nucleus produced by the $r$ process and with a relatively long mean life) and several other nuclei have ESS abundances compatible with uniform production in the Galaxy over about $10^{10}$ years. On this subject, see also Cameron (1993), Cameron et al. (1995) and Meyer and Clayton (2000).

We already stated that the solar nebula must have been isolated from the last events of galactic nucleosynthesis for a time $\Delta_1 > 10^7$ years (see Sect. 6.3.1), i.e. the time required by a cloud core to evolve. For reproducing the ESS amounts of $^{182}$Hf this delay was found by Wasserburg et al. (1996) to be adequate. That scheme, however, would overestimate the concentration of $^{129}$I (and also of $^{107}$Pd) by a large amount, so that the above authors assumed that these isotopes were synthesized by a different supernova type, presenting an $r$-process distribution substantially different from the typical producers of heavy $r$-nuclei and whose last explosion would have occurred long before the one accounting for these last.

More specifically, assuming $r$-process production factors that were standard at that moment, the above authors confirmed that reproducing $^{129}$I would require a delay of about 70 Million years. This suggestion stimulated a series of studies, both theoretical and observational, on the ensuing multi-modal nature of the $r$-process. The constraints on the $r$ process coming from ESS radioactivites were complemented from studies of early stars in the galaxy, where the distribution of nuclei beyond Fe attributed to the $r$ process is best studied. Both phenomenological models (Qian and Wasserburg, 2007, 2008) and full $r$-process calculations in the so-called High Entropy Wind (HEW) scenario (Farouqi et al., 2006, 2008; Kratz et al., 2007) contributed to a lively discussion. At the moment of this report, it appears that the original proposition of (only) two distinct $r$-process components, one producing $^{129}$I and the other producing actinides and $^{182}$Hf (Wasserburg et al., 1996; Qian et al., 1998), should rather be evolved into a more sophisticated picture of fast neutron capture products coming from a superposition of many nucleosynthesis environments, all possibly produced in the high-entropy wind region, but each characterized by its own specific entropy range. Then the formation of the solar system would retain the signature of a special blend of distributions, the one characterizing the last $r$-process-producing episodes. In this view ESS radioactivities would in fact suggest a multi-modal $r$ process, as noted by Wasserburg et al. (1996); but this would no longer lead to the possibility of inferring universal rules from the particular sample of supernova products experienced by the early Sun; see also, on this, Kratz et al. (2008).

The complexity of the situation is underlined by recent evidence on the $^{247}$Cm abundance in early solids. Until a few years ago, only a broad upper bound existed: $(^{247}\mathrm{Cm}/^{235}\mathrm{U})_{ESS} << 2 \times 10^{-3}$, as established by Chen and Wasserburg (1981a,b). Subsequently, Stirling et al. (2006) reported a more constraining bound for this ratio, close to $10^{-4}$. Finally, unambiguous evidence of live $^{247}$Cm in ESS, at an abundance ratio $(^{247}\mathrm{Cm}/^{235}\mathrm{U})_{ESS} = 1{-}2.4{\cdot}10^{-4}$ has been found by Brennecka et al. (2010). This has two main implications. First, it shows that the ancient inventory of $^{235}$U might have been modified by actinide decays and should be corrected accordingly, to avoid errors in the U-Pb datation system. Second, the low inferred $^{247}$Cm abundance confirms that fast neutron-capture processes must be more complicated than assumed in the simple uniform production models, adopting *standard* $r$-process production factors. If such models were calibrated to reproduce the ESS concentration of $^{182}$Hf, then not only the predicted abundance of $^{129}$I, but also that of $^{247}$Cm (and, to a lesser extent, of $^{244}$Pu) would be higher than observed. This would then require a significant time interval ($>$100 Million years) between the termination of actinide nucleosynthesis and the formation of the solar nebula; such scenario would not produce $^{182}$Hf.

Fixing the isolation time of the solar nebula to 70 Million years, so that the abundance of $^{129}$I predicted by uniform production models would fit the observed ESS concentration, the inherited quantity of $^{107}$Pd would turn out to be insufficient to explain its measured meteoritic abundance (Wasserburg et al., 1996, 2006). As $^{107}$Pd can be produced also by the $s$ process in AGB stars, it was then natural to conclude that a late contamination event might have occurred, induced by the close-by passage of an AGB star in its final stages of evolution, producing Pd but also $^{26}$Al, $^{41}$Ca, $^{60}$Fe, and possibly $^{205}$Pb.

This scenario was recently reconsidered by Gounelle and Meibom (2010), who noticed that, reducing the isolation time to 43 Million years, the discrepancies of the predictions for iodine and palladium abundances with respect to the measured requirements would be reduced to a factor of 3 (iodine being overabundant, palladium deficient, so that the discrepancy *between* the two is actually a factor of 10). According to those authors discrepancies at such a level might still be acceptable, in view of the large uncertainties affecting both models and observations, so that both $^{129}$I and $^{107}$Pd might be considered as products of the same continuous nucleosynthesis in the Galaxy.[1] The recent indications on the complexity of the *r*-process and the measurement of the $^{247}$Cm abundance suggest that this interpretation is not correct.

We then return to our still insufficient picture of the *r* process. Recent research in this field by Farouqi et al. (2008) and Kratz et al. (2008) suggests that a fast process of neutron captures is only responsible for nuclei above $A \geq 120$, so that attributing $^{107}$Pd to the same process that produced $^{129}$I would be inappropriate. Below $A = 120$ a blend of different nucleosynthesis mechanisms should be active, possibly induced by fast charged-particle processes, not by a "weak" *r* process, as previously assumed (Farouqi et al., 2008). Should this be confirmed, both the schemes by Wasserburg et al. (2006) and by Gounelle and Meibom (2010) would require revisions. Unfortunately, no general prescription accounting completely for solar abundances in the mass range $A = 100 - 120$ is still available. Contributions from a single neutron-burst, like the one ensuing from the passage of a SN shock through the He-rich layers of the stellar mantle (Meyer and Adams, 2006), are also possible, with yields for some heavy radioactive nuclei hardly distinguishable from those of the *r* process (Huss et al., 2009). In such case, a late production of $^{107}$Pd can neither be securely affirmed nor securely denied.

## 6.5 Local Production of Radioactive Nuclei

We now consider those nuclei that cannot be ascribed to normal galactic evolution, as their very short meanlife would make them decay during the delay time required by a molecular-cloud core to evolve towards a new star. For some of them (e.g. $^{60}$Fe) spallation processes cannot be at play, so referring to a nearby nucleosynthesis episode seems unavoidable, despite the already mentioned low probability of such an event.

For a prompt contamination, the relations that short-lived nuclei should obey at $t = \Delta_1$ are (Wasserburg et al., 1994, 1998):

[1] One might however remember the old joke on the interpretation of experimental results, saying that: "agreement within a factor of ten" actually means "disagreement".

$$\alpha_{R,S}^{\Delta_1} = \left(\frac{N_R}{N_S}\right)_{\Delta_1} \simeq q_R(w) \cdot \frac{N_R(w)}{N_S(w)} f_0 \cdot e^{-\frac{\Delta_1}{\tau_R}} \tag{6.4}$$

where $\alpha_{R,S}^{\Delta_1}$ is the abundance ratio (radioactive nucleus to stable reference) established in the solar nebula at $t = \Delta_1$, $q_S(w)$ is the production factor of the stable nucleus $S$ in the stellar wind, $N_R(w)/N_S(w)$ is the abundance ratio (radioactive to stable) in the wind, $f_0$ is the dilution factor at $t = 0$ and $\tau_R$ is the mean life of the nucleus $R$. Abundances in PDs ($t = \Delta_2$) can then be found by considering an exponential decay for the nuclei of interest. According to recent results from absolute dating of ESS samples (Zinner and Göpel, 2002; Amelin et al., 2005) the values of $\Delta_2$ should not exceed few Million years.

### 6.5.1 Short-Lived Nuclei: A Nearby Supernova Origin?

More than a decade ago Wasserburg et al. (1998), from the assessment of SN yields by Woosley and Weaver (1995), indicated a fundamental constraint that must be satisfied by a SN explosion (at a time $t = \Delta_1$ before the formation of CAIs), providing short-lived nuclei in the ESS. The admixture of SN ejecta with the material of the forming solar nebula should be compatible with the general knowledge of isotopic anomalies on stable elements in meteorites. The main doubt that was raised in that paper is that a close SN would introduce several anomalies over the distribution of stable isotopes and mainly on typical SN products like oxygen and $\alpha$-rich nuclei, at levels that should be within the present experimental possibilities. These problems were underlined and repeated in great detail by Wasserburg et al. (2006), considering also models by Rauscher et al. (2002) and by Limongi and Chieffi (2003). A remarkable anomaly, known since several years, concerns $^{16}$O (Clayton, 1973; Clayton and Nittler, 2004). However, it is not associated to shifts in other oxygen isotopes (Thiemens and Heidenreich, 1983; Mauersberger et al., 1999) and it is now attributed to chemical processes which do not involve any nucleosynthesis input. In this context, Young et al. (2009) recently noted that the $^{18}$O/$^{17}$O ratio is anomalous in the solar system as compared to the ISM. This anomaly was interpreted as due to the addition of SN-polluted material to the solar nebula, from various recent SNe in a picture of sequential star formation. Notice however that this solar anomaly with respect to the present ISM, if confirmed, can certainly be a signature of nucleosynthesis, but not necessarily of a solar contamination by a SN. Addition of material richer in $^{17}$O than in $^{18}$O in the recent chemical evolution of the nearby Galaxy, from low mass stars experiencing extra-mixing (Nollett et al., 2003; Palmerini et al., 2009), would induce the same difference, which should in this case be interpreted as an ongoing production of $^{17}$O (and destruction of $^{18}$O) in the 4.56 Gy after the Sun's formation. This would underline the relevance of low mass star nucleosynthesis in changing the chemical composition of objects younger than the Sun. This is a characteristic of recent galactic chemical evolution that is now known from other contexts (D'Orazi et al., 2009). Similar conclusions (although not involving extra mixing) were presented by Gaidos et al. (2009a), who also pointed out how this scenario could explain the gradual decrease in the $^{12}C/^{13}C$ ratio of the Galaxy.

Supernova nucleosynthesis models have been significantly improved since the time of the analysis by Wasserburg et al. (1998), but the problem of introduction of unobserved anomalies still remains. Any SN injection model proposed to account for short-lived nuclei of the ESS also must be required to also be compatible with the isotopic abundances of stable elements measured in meteorites. In view of current supernova-nucleosynthesis uncertainties as presented e.g. by Tur et al. (2009) this demonstration of overall consistency seems unfeasible at present seems hardly possible today.

With this cautionary note in mind, we still face a wealth of interesting new result, both in modeling massive star explosion and nucleosynthesis and in analyzing the possible pollution of the solar nebula from a supernova (Chieffi and Limongi, 2004; Meyer, 2005; Limongi and Chieffi, 2006; Nomoto et al., 2006; Huss et al., 2007; Woosley and Heger, 2007; Takigawa et al., 2008; Kuroda et al., 2008; Huss et al., 2009). These studies and the general constraints from ESS abundances (e.g. avoiding the overproduction of $^{53}$Mn and of $^{60}$Fe as compared to $^{26}$Al, guaranteeing enough production for very short-lived nuclei like $^{41}$Ca, etc) focused the attention on SN models including fallback of a consistent amount of matter and including internal mixing among a previously-layered composition (Joggerst et al., 2009; Takigawa et al., 2008). In particular, Takigawa et al. (2008) show that virtually all SN models not modified by these effects would overproduce $^{53}$Mn by large amounts, and most models would also overproduce $^{60}$Fe. In constrast, a *faint supernova*, including fallback and mixing, more likely is able to reproduce the observed distribution of short-lived radioactivities. The mass cut below which material has to fall back and the level of mixing among the stellar layers may then be varied as free parameters to reproduce the measurements in the ESS. We recall that the first idea of a modified supernova model with a high mass cut (to avoid $^{53}$Mn overproduction) was suggested by the Clemson group (Meyer and Clayton, 2000). The models by Takigawa et al. (2008) show now an impressively good accord between the ejecta of a fall-back SN around 30 $M_{\odot}$ and the ESS record of $^{26}$Al, $^{41}$Ca, $^{53}$Mn and $^{60}$Fe. A slight deficit in $^{53}$Mn from this model is not a real problem, in view of the abundant production of this isotope in the continuous galactic nucleosynthesis processes.

Comparisons between the recent studies by Limongi and Chieffi (2006) and by Woosley and Heger (2007) also clarify that accounting for the galactic inventory of $^{26}$Al and $^{60}$Fe (for any reasonable value of their abundance ratio in the range indicated by measurements) is possible, provided the proper mass loss description is chosen. In this respect the stronger mass loss rates by Langer et al. (1989) seem to be preferable to the more recent choice by Nugis and Lamers (2000). The reproduction of the observed $^{26}$Al/$^{60}$Fe in the Galaxy remains an important consideration, and if a massive star could also explain the ESS radioactivities without introducing anomalies incompatible with known meteoritic abundances, this would certainly offer a convincing global picture, coming from a unique astrophysical site.

Summing up, the scenario for a late supernova production of short-lived nuclei is still the subject of an intense and controversial debate. Among the reasons that make SNe appealing candidates are: (i) SNe are in general the site where most nucleosynthesis processes occur. (ii) A close-by SN might explain the abundances of ESS radioactivities in the same framework in which the equilibrium abundances of radioactive nuclei in the Galaxy are explained. (iii) Despite the difficulties we mentioned, star formation in the vicinity of SNe and possibly triggered by them is known to occur in various galactic environments. (iv) Faint supernovae, with internal mixing and limited or no contribution to iron do exist, justifying the speculative part of the models required by ESS radioactivities.

However, serious concerns arise by considering other relevant issues, such as: (i) The fact that the probability of a close-by SN encounter has been recently shown to be extremely small (Gounelle and Meibom, 2010), possibly smaller than for an AGB star (Kastner and Myers, 1994; Trigo-Rodríguez et al., 2009). (ii) The need for ad-hoc choices of parameters, like e.g. the mixing extension and the mass cut, in any SN model suitable to explain the ESS radioactivities and the fact that results strongly depend on uncertain choices for cross sections (Tur et al., 2009). (iii) The lack of a convincing answer to the risk of introducing unwanted anomalies on stable isotopes. As an example, the best model by Takigawa et al. (2008), from a 30 $M_\odot$ star, ejects almost 7 $M_\odot$ of processed material, where at least C, O, Ne, and Mg should be highly enhanced (maybe by factors close to 100, as other SN nucleosynthesis calculations suggest). At the adopted dilution factor of $4\times10^{-4}$ this should induce anomalies at percent level on major elements, which are not observed. This inconsistency should be disproved before any model involving a SN can be considered as a real, quantitative possibility.

### *6.5.2 Contributions from a Nearby AGB Star?*

The scenario of a close-by contamination by an AGB star has been extensively explored (Wasserburg et al., 1994, 1995, 2006). A recent detailed analysis of the contributions from an intermediate mass star experiencing H burning at the base of the envelope (Hot Bottom Burning, or HBB) was presented by Trigo-Rodríguez et al. (2009); these authors also demonstrated that previous criticisms to the AGB scenario might be inconsistent, or at least should be looked at with caution.

The most appealing property of the above AGB models is that they do not include free parameters (i.e. parameters that are not constrained otherwise). As an example, neutron capture nucleosynthesis at the relatively low efficiency required to explain the ESS radioactivities does not need neutrons from the $^{13}$C neutron source, which still lacks modeling from first principles. Hence the only process still needing parameters not provided by the main stellar evolution history is the efficiency of extra-mixing processes above the H-burning shell, which are necessary to produce $^{26}$Al if the mass of the star is too small to host HBB (i.e. lower than 4 $M_\odot$). The existence of thermally-pulsing AGB stars with minimum $s$-process production, so that a $^{13}$C-pocket becomes unnecessary, was known already for low mass stars (the Mira protoptype, *o Cet* being an example) and has been recently demonstrated also for intermediate masses (García-Hernández et al., 2007).

The effectiveness of extra mixing can be independently established, for low mass stars, from consideration of abundances in presolar grains of AGB origin (see Sects. 3.5 and 10.2), whose reproduction offers a way to constrain in a rather clear way the occurrence of any non-convective circulation of matter. Although a unique paradigm for extra mixing is still lacking, recent studies address this old problem: Eggleton et al. (2006, 2008) underlined the role of instabilities (thermohaline mixing) induced by the molecular weight decrease associated with the burning of two $^{3}$He nuclei, producing $^{4}$He and two protons. Alternatively, a similar effect can be provided by magnetic fields; in magnetic flux tubes the gas density is smaller than in the outside environment, so that the buoyancy of magnetized blobs, processed by proton captures, can occur (Busso et al., 2007b; Nordhaus et al., 2008; Denissenkov et al., 2009). In any case, as shown by Nollett et al. (2003) and by Palmerini et al. (2009), the actual extra-mixing products do not in general depend on the specific physical model (a most remarkable exception being offered by lithium).

The minimization of free parameters might be even better for an intermediate mass star, where H-burning at the base of the convective envelope provides the high-temperature environment to produce $^{26}$Al and non-convective mixing mechanisms are not required (and actually not expected to occur). In general, both low and intermediate mass star models will need a free choice for mass loss rates, which only now begin to be quantitatively determined, thanks to extensive sky surveys in the infrared by space-borne instruments (Guandalini et al., 2006; Busso et al., 2007a; Guandalini and Busso, 2008).

A crucial point for AGB models is the ESS abundance of $^{60}$Fe. A low mass star ($M$ = 1.5–2 $M_{\odot}$) would be incapable of producing enough $^{60}$Fe to account for the initial ESS $^{60}$Fe/$^{56}$Fe ratio (at least a few$\times 10^{-7}$). Hence the calculations by both Wasserburg et al. (2006) and Trigo-Rodríguez et al. (2009) pointed toward stellar masses of 3–7 $M_{\odot}$, higher than in the original suggestions by Wasserburg et al. (1994). These recent studies show that both a 3 $M_{\odot}$ and a 6.5 $M_{\odot}$ star can reproduce the ESS measurements at least for $^{26}$Al, $^{41}$Ca, $^{60}$Fe, and $^{107}$Pd, for dilution factors of a few $\times 10^{-3}$. This is shown in Table 6.3, where the best cases from Wasserburg et al. (2006) and Trigo-Rodríguez et al. (2009) are summarized; note that the estimates for $^{135}$Cs and $^{205}$Pb are available only for the results of Wasserburg et al. (2006). In the 3 $M_{\odot}$ model $^{26}$Al would come from extra-mixing, in the 6.5 $M_{\odot}$ star from HBB. We must also underline that AGB stars of intermediate mass would imprint in the solar nebula only marginal anomalies in stable nuclei. The most critical effect is probably a 1% shift in $^{17}$O for a star of 6.5 $M_{\odot}$, see Trigo-Rodríguez et al. (2009). A lower-mass star would have essentially no effect on stable isotopes, except carbon. This is a special merit of AGB stars. If one also considers that heavy $s$ process nuclei like $^{205}$Pb are typical products of AGB nucleosynthesis, some cautions should be used before excluding an intermediate-mass star as a polluting source, at least until massive stars cannot be shown to provide the same consistency. Very recently a remarkable nuclear physics result has added new arguments in this direction. The half-life of $^{60}$Fe was remeasured by Rugel et al. (2009) and the new value is about a factor of two larger than the one previously adopted. If this will be confirmed, the main challenge for AGB models (that of producing enough $^{60}$Fe) would be considerably alleviated and AGB nucleosynthesis

**Table 6.3** Recent predictions for short lived nuclei from an AGB star

| Parent $P$ | Index $I$ | 3.0 $M_\odot$, $Z = Z_\odot/3$ $f_0 = 4.0 \times 10^{-3}$ $(N_P/N_I)_{\Delta_1}$ $\Delta_1 = 0.53$ Million years | (W+06) $(N_P/N_I)_{\Delta_2}$ $\Delta_2 = 6.5$ Million years | 6.7$M_\odot$, $Z = Z_\odot$, $f_0 = 3.3 \times 10^{-3}$ $(N_P/N_I)_{\Delta_1}$ $\Delta_1 = 0.53$ Million years | (TR+09) $(N_P/N_I)_{\Delta_2}$ $(\Delta_2) = 6.0$ Million years | Measured or extrapolated at $t = \Delta_i$ |
|---|---|---|---|---|---|---|
| $^{26}$Al | $^{27}$Al | $5.0 \cdot 10^{-5}$ | $8.5 \cdot 10^{-8}$ | $3.2 \cdot 10^{-5}$ | $9.8 \cdot 10^{-8}$ | $5.0 \cdot 10^{-5}$ $(\Delta_1)$ |
| $^{41}$Ca | $^{40}$Ca | $1.5 \cdot 10^{-8}$ | – | $1.5 \cdot 10^{-8}$ | – | $\geq 1.5 \cdot 10^{-8}$ $(\Delta_1)$ |
| $^{60}$Fe | $^{56}$Fe | $2.1 \cdot 10^{-6}$ | $1.0 \cdot 10^{-7}$ | $2.6 \cdot 10^{-6}$ | $1.7 \cdot 10^{-7}$ | $0.5-1 \cdot 10^{-6}$ $(\Delta_1)$ |
| $^{107}$Pd | $^{108}$Pd | $3.8 \cdot 10^{-5}$ | $2.0 \cdot 10^{-5}$ | $3.8 \cdot 10^{-5}$ | $2.0 \cdot 10^{-5}$ | $2.0 \cdot 10^{-5}$ $(\Delta_2)$ |
| $^{93}$Zr | $^{92}$Zr | $2.5 \cdot 10^{-4}$ | $1.2 \cdot 10^{-5}$ | $1.6 \cdot 10^{-4}$ | $8.6 \cdot 10^{-6}$ | (?) |
| $^{99}$Tc | $^{100}$Ru | $1.9 \cdot 10^{-5}$ | – | $1.4 \cdot 10^{-5}$ | – | (?) |
| $^{135}$Cs | $^{133}$Cs | $3.6 \cdot 10^{-5}$ | $3.5 \cdot 10^{-6}$ | n.a. | n.a. | $1.6 \times 10^{-4}$ (?) |
| $^{205}$Pb | $^{204}$Pb | $\leq 3.4 \cdot 10^{-4}$ | $\leq 2.5 \cdot 10^{-4}$ | n.a. | n.a. | $1-2 \times 10^{-4}$ (?) |

would offer an even more promising scenario than so far suspected. This scenario does not suffer for the same nuclear uncertainties discussed for massive stars and, despite all contrary arguments, it remains the only really quantitative model on the market, thanks to its relatively simple physics.

Doubts on an AGB pollution of the ESS were raised from the low probability of a close encounter (a mere 1%, see Kastner and Myers, 1994) (see also Sect. 6.3.2). Recently Trigo-Rodríguez et al. (2009) presented several reasons to revise this estimate, and moreover Gounelle and Meibom (2010) indicated that encounter probabilities for massive stars are at least as small as for AGBs. If a close encounter with a star producing radioactive nuclei ever occurred, this must have been necessarily an unusual event, maybe a unique occurrence. Relative statistics may then not help judging between these scenarios.

Deciding between an AGB star and a massive star as a source for ESS pollution would benefit from a better knowledge of the initial concentration of $^{205}$Pb. Among heavy, $n$-capture nuclei, $^{135}$Cs can be produced by both the $s$ and $r$ processes, while $^{205}$Pb is a purely $s$-process nucleus. At the low neutron exposure required, AGB stars do not produce it much more efficiently than a massive star (we are in a situation very different from the main component of the $s$ process, which is instead the exclusive realm of AGB stars). However, if $^{205}$Pb was really present in the ESS at the level suggested by the measurements of Nielsen et al. (2006), then a massive star would be essentially excluded for a question of dilution factors. For a dilution of some $10^{-3}$, Wasserburg et al. (2006) obtained an ESS concentration of a few $\times 10^{-4}$ for $^{205}$Pb, in the case of AGB production. This has to be interpreted as an upper limit, given the uncertainties affecting the decay rate of $^{205}$Pb in the stellar interior. Very recently Baker et al. (2010) found evidence of live $^{205}$Pb in the ESS from small variations of the $^{205}$Tl abundance in carbonaceous chondrites, which correlate well with Pb/Tl ratios. These authors inferred an initial $^{205}$Pb abundance very close to the predictions by Wasserburg et al. (2006), obtained assuming an AGB star as the nucleosynthesis source. If confirmed, this result would considerably help in disentangling among the available pollution scenarios. Looking at recent results from Woosley and Heger (2007), $^{205}$Pb appears to be produced by a 25 $M_{\odot}$ star more or less at the same efficiency as for an AGB star of intermediate mass, but the material would then need to be diluted 10 times more, to avoid overproduction of lighter species. This is still a very difficult matter to put in focus, because of the large uncertainties affecting both the measurements and the predictions. We underline this point here because it deserves further attention. It would in particular be very important if theoretical nuclear physics could firmly establish the complex behavior of the weak interactions regulating the pair $^{205}$Pb-$^{205}$Tl as a function of temperature, in the range between, say, $10^6$ and $10^8$ K (pertinent to the stellar layers that any $^{205}$Pb produced by neutron captures must cross while being dredged-up to the surface nucleus of an AGB star, or ejected in a SN explosion).

In summary, also the models of an AGB contamination contain merits and problems. Among the merits are: (i) The AGB phase is a experienced by all stars below about 7–8 $M_{\odot}$, i.e. by the large majority of galactic objects evolving within a cosmological time scale. Possible candidates are abundant in any galactic epoch. (ii) AGB models are simpler, involve a much smaller number of free parameters than for massive stars and their main features are not critically dependent on uncertain reaction rates. Most parameters can be fixed independently, from the abundant observations, and the predictions can therefore be rather quantitative. (iii) A polluting AGB star of about 3 $M_{\odot}$(or even less, accepting the revision of the $^{60}$Fe lifetime by Rugel et al., 2009) would not introduce any unwanted anomaly on stable nuclei. (iv) An AGB star would be able to produce also heavy nuclei like $^{135}$Cs and $^{205}$Pb, at concentrations close to those suggested by the present, very limited, data. Firm experimental constraints would however be needed. Problems for an AGB contamination model may be: (i) An AGB encounter is difficult to imagine for a cloud core; apart from the low probability derived by statistical calculations, a proximity of forming stars to AGBs or planetary nebulae is certainly not a common fact (if it can occur at all), while it is demonstrated for SNe. (ii) A small dilution is required, due to the small amount of mass ejected: this requires very tight constraints on the timing. (iii) As nuclei deriving by the continuous galactic production are accounted for by SN nucleosynthesis, a late event of different nature appears rather ad-hoc.

## 6.6 Short-Lived Nuclei: In-Situ Production

For a few shorter-lived species, especially $^{10}$Be, $^{26}$Al, $^{36}$Cl, $^{41}$Ca, $^{53}$Mn, alternative models for the formation in spallation reactions, from the bombardment of fast particles coming from the magnetically-active early Sun were proposed; see e.g. Shu et al. (1997); Gounelle et al. (2001). The same magnetic phenomena occurring in most stars, at least during their Main Sequence but probably also in their subsequent Red Giant phases (Andrews et al., 1988), provide a site where nucleosynthesis of radioactive isotopes occurs: this has been shown for very short-lived nuclei (Tatischeff et al., 2006) but might be true also for longer-lived species of interest for the ESS (Palmerini and Busso, 2008).

### 6.6.1 Radioactivities from the Bombardment of Early Solids

This mechanism takes advantage of the already described X-wind scenario (Lee et al., 1998), lifting the recently formed CAIs from the disk plane and transporting them to large distances. In such models the hypothesis was also advanced that CAIs and chondrules might be produced in those same winds. If this were the case, then the high abundance measured in CAIs for $^{26}$Al ($^{26}$Al/$^{27}$Al = $5{\cdot}10^{-5}$) would not be indicative of a uniform situation, and many properties attributed to $^{26}$Al would have to be reconsidered, especially its use as a precise chronometer. A thorough discussion of the spallation mechanisms in the ESS was presented by Gounelle et al. (2006), on the basis of nuclear parameters for spallation processes that were standard at that epoch. On the basis of models for the structure of CAIs and for the flux of irradiating particles, the authors first determined the conditions for the production of $^{7}$Be, whose very short mean life makes it a discriminating nucleus. They suggested that, for a flux of particles from the Sun of the order of $F \simeq 2 \times 10^{10}$ cm$^{-2}$ s$^{-1}$, both $^{7}$Be and $^{10}$Be could be produced at the observed levels within the uncertainties. A byproduct of the adopted fluxes was also a noticeable contribution to $^{26}$Al, $^{36}$Cl, and $^{53}$Mn, whose abundances in the ESS were therefore suggested to come form proton bombardment, leaving space to a nearby supernova only to produce $^{60}$Fe. Further analysis lead Gounelle et al. (2009) to suggest that $^{60}$Fe itself (at a level close to the lower limit of the uncertainty range for its abundance) might derive not from an isolated, rare event, but rather from the superposition of various SN contributions, much like for longer-lived species.

On the above subject one can however notice that SN nucleosynthesis has a complex pattern: $^{60}$Fe cannot be produced alone. Other radioactive species would come out, and their predicted abundance in the ESS should be verified. The effects of several recent supernovae would also be detectable from many abundance anomalies in stable elements at levels within the reach of present measurements. This issue is the same previously raised by Wasserburg et al. (1998, 2006) for one isolated SN.

After the work by Gounelle et al. (2006), it was shown that the cross sections for some crucial spallation processes, in particular $^{24}$Mg($^{3}$He,p)$^{26}$Al, needed strong revisions (Fitoussi et al., 2008). This leads to the conclusion that $^{26}$Al cannot be produced at sufficient levels by spallation processes in the early solar nebula, so that its synthesis seems now to require unambiguously nucleosynthesis processes in stars. The clear decoupling between $^{26}$Al and $^{36}$Cl (Hsu et al., 2006) in early meteorites then suggests that $^{36}$Cl, instead, was formed in the the X-wind scenario, together with Be-isotopes. The already mentioned correlation of $^{41}$Ca with $^{26}$Al (see

Fig. 6.3) would instead suggest a stellar origin for $^{41}$Ca. Further limits to the level of production of radioactive nuclei from solid particle bombardment were put, after the first results of the STARDUST mission (Brownlee et al., 2006), by new models using improved cross sections for the reactions leading to $^{7}$Be, $^{10}$Be, $^{26}$Al, $^{36}$Cl, $^{41}$Ca, and $^{53}$Mn (Duprat and Tatischeff, 2008). According to these new findings, the role of non-thermal nucleosynthesis in the ESS might actually be limited to the production of $^{10}$Be and $^{41}$Ca.

All these difficulties of the solar-wind models then call again for a last-minute stellar contamination, at least for $^{26}$Al, $^{41}$Ca, $^{60}$Fe, $^{135}$Cs, $^{205}$Pb. At this moment the need for an exceptional nucleosynthesis event of some kind, close in time and space to the forming Sun, cannot be avoided. Although this event looks very unusual in the light of our understanding of star formation, we might be forced to accept that an unlikely event occurred at least once, if the alternative is no explanation at all.

### 6.6.2 Solar Activity and the Production of $^{7}$Be

Nuclear interactions occurring in solar explosions are revealed by their induced prompt emission of $\gamma$-ray lines; these are due to the de-excitation of nuclei excited by reactions occurring between flare-accelerated particles and the solar atmospheric material. The first observations of these solar $\gamma$-ray lines were obtained by the experiment GRS (Gamma Ray Spectrometer), on board of the OSO-7 space-borne observatory (Chupp et al., 1973). Subsequently, the evidence for spallation-induced reactions in flares grew thanks to the measurements of various instruments: the Solar Maximum Mission (Share and Murphy, 1995), the Compton Gamma-Ray Observatory (Share et al., 1997) and the Ramaty High Energy Solar Spectroscopic Imager (RHESSI), see e.g. Lin et al. (2003).

In recent years, it was suggested that also delayed X- and $\gamma$-ray emission might occur from solar flares, thanks to the production of short-lived radioactive nuclei, whose subsequent decay would be accompanied by emission lines (Ramaty and Mandzhavidze, 2000; Kozlovsky et al., 2002). In particular, detailed predictions of line emissions form such decays in solar active regions, after the occurrence of intense flares, were presented by Tatischeff et al. (2006, 2007). These authors estimated the cross sections for the formation of several radio-nuclei from interactions of protons and $^{3,4}$He particles with isotopes of elements up to Ni. Then, on the basis of a thick-target model, they provided expected yields for 25 radio-nuclei with half-life in the range from 10 min ($^{13}$N) to 77 days ($^{56}$Co). Fluxes in $\gamma$-ray lines interesting for a possible detection from future experiments were found for $^{52}$Mn (1,434 keV), $^{60}$Cu (1,332 and 1,792 keV), $^{34}$Cl (2,127 keV), $^{24}$Na (1,369 and 2,754 keV) and $^{55}$Co (931.1 keV).

The above findings are important for explaining the $\gamma$-ray fluxes of solar activity phenomena and for providing information on details of the solar plasma physics. On the contrary, spallation processes on solar and stellar atmospheres are not expected to be relevant for galactic nucleosynthesis, as production of nuclei by similar processes occurring in Galactic Cosmic Rays would dominate by several orders of magnitude (Tatischeff and Thibaud, 2008). Concerning the ESS radioactivities discussed in this Chapter, the works mentioned in this subsection might be of interest for the early inventory of very short-lived nuclei, like $^{7}$Be. In fact, it has been found that this isotope might have been present alive in CAIs (Chaussidon et al., 2003, 2006). Its production must be essentially contemporaneous to the same CAI formation, due to its very short half-life (53 days), and its (uncertain) ESS abundance was originally indicated as being roughly compatible with X-wind models (Gounelle et al., 2003; Chaussidon et al., 2006). Recent revisions suggest, instead, that its production in that environment should be largely insufficient (Duprat and Tatischeff, 2008). The formation of $^{7}$Be in solar flares might therefore be considered as a promising alternative possibility to the irradiation of solids. It has also been noticed that production of $^{7}$Be (hence of its daughter $^{7}$Li) might explain the unexpected detection of Li in several M-type dwarfs, although it would probably be insufficient to account for the trends of the Li abundance in open clusters (Tatischeff et al., 2008).

The above studies represent a remarkable bridge linking the models for the in-situ production of ESS radioactivities to those for their stellar synthesis. Indeed, some recent models for the deep-mixing phenomena occurring in evolved red giants and accounting, among other things, for the production of abundant $^{26}$Al and Li (Palmerini and Busso, 2008; Palmerini et al., 2009; Guandalini et al., 2009) consider stellar magnetic fields as the main engine for the transport of proton-capture ashes to the stellar envelope, following ideas by Busso et al. (2007b). Those models account for the production and destruction of Li in red giants through mechanisms of magnetic buoyancy occurring at rates respectively faster or slower than that for $^{7}$Be decay. Such models, however, did not include, so far, the possible contributions from spallation processes in the transported material itself, which is moving relativistically along toroidal flux tubes and their $\Omega$-shaped instabilities. A check of the relevance of stellar spallation processes for explaining the formation of super-Li rich stars, and for the evolution of the Li abundance in the Sun seems to be really necessary now.

## 6.7 Early Solar System Lessons

The general scenarios explored so far in order to account for the presence of short-lived radioactive nuclei in the ESS is far from satisfactory and still quite confused. We can summarize the indications emerged from the many efforts dedicated to this field in the past years in the following points:

- Decay of radioactive isotopes of long lifetime (longer than 1 Gy) offers us a number of tools for estimating the global age of the Earth and of the other solid bodies orbiting around our star. They tell us that the solar system was formed when the Galaxy was already quite mature, having spent 2/3 of its present age. The resulting age of the solar system is close to 4.566 Gy.

- The ESS abundances of shorter-lived isotopes, with half lives from 10 to a few 100 Million years, can be used for reconstructing the history of nucleosynthesis in the solar neighborhood, as their initial concentration in the solar nebula is probably compatible with the equilibrium abundances established by SN explosions and nuclear decay in the local interstellar medium, during several cycles of molecular cloud aggregation and destruction. From these nuclei it was inferred that the *r* process has a multi-modal nature. This suggestion, which has produced considerable insight on both solar system formation and the abundance patterns observed in old stars, has stimulated further studies that are now producing a very complex scenario. First of all, it has been shown that some of the radioactive nuclei commonly ascribed to the *r* process might be produced also in a single neutron burst in evolved massive stars. Secondly, the classical *r* process itself has undergone important revisions, and seems now to be limited to the production of very heavy nuclei ($A > 120$).
- The solar nebula must have undergone a process of autonomous evolution in which it was isolated from SN nucleosynthesis contaminations for a rather long time. One can reconstruct, from the initial abundance ratio of heavy n-rich species, that the last episodes of SN nucleosynthesis contributing to these nuclei affected the solar material from $10^7$ to $10^8$ years ago. More stringent constraints require a better knowledge of fast nucleosynthesis processes beyond Fe.
- There are also signs that the forming Sun was affected by a last episode of stellar nucleosynthesis (much closer in time than those mentioned above) producing some short lived radioactivities (like $^{26}$Al, $^{41}$Ca, $^{60}$Fe, and possibly $^{135}$Cs, $^{205}$Pb), although this event has a low probability of occurrence from a purely statistical point of view. The stellar contamination had to occur immediately before the contraction of the solar parent cloud (with a time separation from it of $\Delta_1 \leq 1$–2 Million years). The importance of this last event was crucial for several reasons. In particular, decays from some radioactive nuclei generated in this event, like $^{26}$Al and $^{60}$Fe, produced daughter isotopes in excited states. Their de-excitation is now considered as the main heating source for differentiating the early bodies (Urey, 1955; Schramm et al., 1970).
- Attribution of this last event to a massive Supernova of some peculiar type or to an AGB star is still a matter of lively debates, as both scenarios face unsolved problems and have considerable drawbacks.
- Subsequently, the forming Sun itself, in its fully convective, pre-Main Sequence phase, must have added new nuclei ($^7$Be, $^{10}$Be, $^{36}$Cl) through spallation processed occurring either in coronal flares or in the interactions of the solar wind with early solids that were forming in the inner regions of an accretion disk.

Despite the many uncertainties, the wealth of the short-lived or very-short-lived radioactivities discussed in this Chapter is now an invaluable source of information on the timing of the first events occurred in the solar nebula.

## References


Adams, F.C., Laughlin, G.: Constraints on the birth aggregate of the solar system. Icarus **150**, 151–162 (2001)

Allègre, C.J., Manhès, G., Göpel, C.: The age of the earth. GeoChim. Cosmochim. Acta **59**, 2445–1456 (1995)

Amelin, Y., Ghosh, A., Rotenberg, E.: Unraveling the evolution of chondrite parent asteroids by precise U-Pb dating and thermal modeling. GeoChim. Cosmochim. Acta **69**, 505–518 (2005). doi: 10.1016/j.gca.2004.05.047

Anderson, F.S., Whitaker, T.J., Young, D., Peterson, B.: Rb-Sr dating using LDRIMS. In: Chronology of Meteorites and the Early Solar System, Kauai, Hawaii, pp. 22–23 (2007)

Andrews, A.D., Rodono, M., Linsky, J.L., Brown, A., Butler, C.J., Catalano, S., Scaltriti, F., Busso, M., Nha, I., Oh, J.Y., Henry, M.C.D., Hopkins, J.L., Landis, H.J., Engelbrektson, S.: Rotational modulation and flares on RS CVn and BY DRA stars. A&A **204**, 177–192 (1988)

Arnould, M., Prantzos, N.: Cosmic radioactivities. New Astron. **4**, 283–301 (1999). doi: 10.1016/S1384-1076(99)00016-0, arXiv:astro-ph/9907275

Baker, R.G.A., Schönbächler, M., Rehkämper, M., Williams, H.M., Halliday, A.N.: Earth & Planet. Sci. Lett. **291**, 39–47 (2010)

Birck, J., Allègre, C.J.: Evidence for the presence of Mn-53 in the early solar system. Geophs. Res. Lett. **12**, 745–748 (1985). doi: 10.1029/GL012i011p00745

Birck, J., Allègre, C.J.: Manganese-chromium isotope systematics and the development of the early solar system. Nature **331**, 579–584 (1988). doi: 10.1038/331579a0

Bizzarro, M., Ulfbeck, D., Trinquier, A., Thrane, K., Connelly, J.N., Meyer, B.S.: Evidence for a Late Supernova Injection of 60Fe into the Protoplanetary Disk. Science **316**, 1178–1181 (2007). doi: 10.1126/science.1141040

Boss, A.P.: The Solar Nebula, p. 63. Elsevier, B., Oxford (2005)

Brennecka, G.A.: A complication in determining the precise age of the solar system. In: Planetary Science Research Discoveries (on-line publication): http://www.psrd.hawaii.edu. Tech. rep. (2010)

Brennecka, G.A., Weyer, S., Wadhwa, M., Janney, P.E., Zipfel, J., Anbar, A.D.: 238U/235U variations in meteorites: extant 247Cm and implications for Pb-Pb dating. Science **327**, 449–451 (2010). doi: 10.1126/science.1180871

Brownlee, D., Tsou, P., Aléon, J., Alexander, C.M.O'D. Araki, T., et al.: Comet 81P/Wild 2 under a microscope. Science **314**, 1711 (2006). doi: 10.1126/science.1135840

Busso, M., Gallino, R.: The production of neutron-rich isotopes during He burning in massive stars. A&A **151**, 205–214 (1985)

Busso, M., Gallino, R., Wasserburg, G.J.: Nucleosynthesis in asymptotic giant branch stars: relevance for galactic enrichment and solar system formation. ARA&A **37**, 239–309 (1999). doi: 10.1146/annurev.astro.37.1.239

Busso, M., Gallino, R., Wasserburg, G.J.: Short-lived nuclei in the early solar system: a low mass stellar source? Publ. Astron. Soc. Aust. **20**, 356–370 (2003). doi: 10.1071/AS03035

Busso, M., Guandalini, R., Persi, P., Corcione, L., Ferrari-Toniolo, M.: Mid-infrared photometry of mass-losing asymptotic giant branch stars. AJ **133**, 2310–2319 (2007a). doi 10.1086/512612, arXiv:astro-ph/0701501

Busso, M., Wasserburg, G.J., Nollett, K.M., Calandra, A.: Can extra mixing in RGB and AGB stars be attributed to magnetic mechanisms? ApJ **671**, 802–810 (2007b). doi: 10.1086/522616, 0708.2949

Cameron, A.G.W.: New neutron sources of possible astrophysical importance. AJ **65**, 485 (1960). doi: 10.1086/108085

Cameron, A.G.W.: Nucleosynthesis and star formation. In: Levy, E.H., Lunine, J.I. (eds.) Protostars and Planets III, pp. 47–73. Arizona: University of Arizona Press (1993)

Cameron, A.G.W., Hoeflich, P., Myers, P.C., Clayton, D.D.: Massive supernovae, orion gamma rays, and the formation of the solar system. ApJ **447**, L53 (1995). doi: 10.1086/309554

Cameron, A.G.W., Truran, J.W.: The supernova trigger for formation of the solar system. Icarus **30**, 447–461 (1977). doi: 10.1016/0019-1035(77)90101-4

Chaussidon et al.: EGS - AGU - EUG Joint Assembly, Abstracts from the meeting held in Nice, France, 6–11 April 2003, abstract #9292 (Publication Date:04/200)

Chaussidon, M., Robert, F., Russel, S.S., Gounelle, M., Ash, R.D.: B and Mg isotopic variations in Leoville mrs-06 type B1 cai:origin of 10Be and 26Al. In: EGS – AGU – EUG Joint Assembly, p. 9292 (2003)

Chen, J.H., Wasserburg, G.J.: Precise isotopic analysis of uranium in picomole and subpicomole quantities. Anal. Chem. **53**, 2060–2067 (1981a). doi: 10.1021/ac00236a027

Chen, J.H., Wasserburg, G.J.: The isotopic composition of uranium and lead in Allende inclusions and meteoritic phosphates. Earth Planet. Sci. Lett. **52**, 1–15 (1981b). doi: 10.1016/0012-821X(81)90202-8

Chen, J.H., Wasserburg, G.J.: A search for evidence of extinct lead 205 in iron meteorites. In: Lunar and Planetary Institute Science Conference Abstracts (Houston: LPI), vol. 18, p. 165 (1987)

Chieffi, A., Limongi, M.: Explosive yields of massive stars from Z = 0 to Z = Zsolar. ApJ **608**, 405–410 (2004). doi: 10.1086/392523, arXiv:astro-ph/0402625

Chupp, E.L., Forrest, D.J., Higbie, P.R., Suri, A.N., Tsai, C., Dunphy, P.P.: Solar gamma ray lines observed during the solar activity of August 2 to August 11, 1972. Nature **241**, 333–335 (1973). doi: 10.1038/241333a0

Clayton, D.D.: Nuclear cosmochronology within analytic models of the chemical evolution of the solar neighbourhood. MNRAS **234**, 1–36 (1988)

Clayton, R.N.: Oxygen isotopic composition of the Luna 20 soil. Geochim. Cosmochim. Acta **37**, 811–813 (1973). doi: 10.1016/0016-7037(73)90177-4

Clayton, D.D., Nittler, L.R.: Astrophysics with presolar stardust. ARA&A **42**, 39–78 (2004). doi: 10.1146/annurev.astro.42.053102.134022

Cowley, C.R.: An Introduction to Cosmochemistry. Cambridge University Press, Cambridge (1995)

Dalrymple, G.B.: The Age of the Earth. Stanford University Press, Stanford (1991)

Dauphas, N., Cook, D.L., Sacarabany, A., Fröhlich, C., Davis, A.M., Wadhwa, M., Pourmand, A., Rauscher, T., Gallino, R.: $^{60}$Fe in the cosmic blender. Geochim. Cosmochim. Acta (Suppl.) **72**, 200 (2008a)

Dauphas, N., Cook, D.L., Sacarabany, A., Fröhlich, C., Davis, A.M., Wadhwa, M., Pourmand, A., Rauscher, T., Gallino, R.: Iron 60 evidence for early injection and efficient mixing of stellar debris in the protosolar nebula. ApJ **686**, 560–569 (2008b). doi: 10.1086/589959, 0805.2607

Denissenkov, P.A., Pinsonneault, M., MacGregor, K.B.: Magneto-thermohaline mixing in red giants. ApJ **696**, 1823–1833 (2009). doi: 10.1088/0004-637X/696/2/1823, 0806.4346

D'Orazi, V., Magrini, L., Randich, S., Galli, D., Busso, M., Sestito, P.: Enhanced production of barium in low-mass stars: evidence from open clusters. ApJ **693**, L31–L34 (2009). doi: 10.1088/0004-637X/693/1/L31, 0901.2743

Duprat, J., Tatischeff, V.: On non-thermal nucleosynthesis of Short-Lived Radionuclei in the early solar system. New Astron. Rev. **52**, 463–466 (2008). doi: 10.1016/j.newar.2008.06.016

Eggleton, P.P., Dearborn, D.S.P., Lattanzio, J.C.: Deep mixing of 3He: reconciling big bang and stellar nucleosynthesis. Science **314**, 1580–1583 (2006). doi: 10.1126/science.1133065, arXiv:astroph/0611039

Eggleton, P.P., Dearborn, D.S.P., Lattanzio, J.C.: Compulsory deep mixing of 3He and CNO isotopes in the envelopes of low-mass red giants. ApJ **677**, 581–592 (2008). doi: 10.1086/529024

Evans, N.J., II: Physical conditions in regions of star formation. ARA&A **37**, 311–362 (1999). doi: 10.1146/annurev.astro.37.1.311, arXiv:astro-ph/9905050

Farouqi, K., Kratz, K., Mashonkina, L.I., Pfeiffer, B., Thielemann, F.: News from r-process nucleosynthesis: consequences from the ESS to early stars. In: Guandalini, R., Palmerini, S., Busso, M. (eds.) Evolution and Nucleosynthesis in AGB Stars. American Institute of Physics Conference Series (New York: A.I.P.), vol. 1001, pp. 245–253 (2008). doi: 10.1063/1.2916971

Farouqi, K., Kratz, K., Pfeiffer, B., Rauscher, T., Thielemann, F.: Neutron captures and the r-process. In: Woehr, A., Aprahamian, A. (eds.) Capture Gamma-Ray Spectroscopy and Related Topics. American Institute of Physics Conference Series (New York: A.I.P.), vol. 819, pp. 419–422 (2006). doi: 10.1063/1.2187894

Fitoussi, C., Duprat, J., Tatischeff, V., Kiener, J., Naulin, F., Raisbeck, G., Assunção, M., Bourgeois, C., Chabot, M., Coc, A., Engrand, C., Gounelle, M., Hammache, F., Lefebvre, A., Porquet, M., Scarpaci, J., de Séréville, N., Thibaud, J., Yiou, F.: Measurement of the 24Mg(3He,p)26Al cross section: Implication for Al26 production in the early solar system. Phys. Rev. C **78**(4), 044613 (2008). doi: 10.1103/PhysRevC.78.044613

Gaidos, E., Krot, A.N., Huss, G.R.: On the oxygen isotopic composition of the solar system. ApJ **705**, L163–L167 (2009a). doi: 10.1088/0004-637X/705/2/L163, 0909.3589

Gaidos, E., Krot, A.N., Williams, J.P., Raymond, S.N.: 26Al and the formation of the solar system from a molecular cloud contaminated by Wolf-Rayet winds. ApJ **696**, 1854–1863 (2009b). doi: 10.1088/0004-637X/696/2/1854, 0901.3364

Gallino, R., Arlandini, C., Busso, M., Lugaro, M., Travaglio, C., Straniero, O., Chieffi, A., Limongi, M.: Evolution and nucleosynthesis in low-mass asymptotic giant branch stars. II. Neutron capture and the s-process. ApJ **497**, 388 (1998). doi: 10.1086/305437

García-Hernández, D.A., García-Lario, P., Plez, B., Manchado, A., D'Antona, F., Lub, J., Habing, H.: Lithium and zirconium abundances in massive Galactic O-rich AGB stars. A&A **462**, 711–730 (2007). doi: 10.1051/0004-6361:20065785, arXiv:astro-ph/0609106

Goebel, R., Begemann, F., Ott, U.: On neutron-induced and other noble gases in Allende inclusions. Geochim. Cosmochim. Acta **46**, 1777–1792 (1982). doi: 10.1016/0016-7037(82)90117-X

Goswami, J.N., Marhas, K.K., Chaussidon, M., Gounelle, M., Meyer, B.S.: Origin of short-lived radionuclides in the early solar system. In: Krot, A.N., Scott, E.R.D., Reipurth, B. (ed.) Chondrites and the protoplanetary disk. Astronomical Society of the Pacific Conference Series (San Francisco: ASPC), vol. 341, p. 485 (2005)

Gounelle, M., Meibom, A.: The origin of 60Fe and other short-lived radionuclides in the early solar system. In: Montmerle, T., Ehrenreich, D., Lagrange, A.-M. (eds.) EAS Publications Series, EAS Publications Series (Les Ulis: EDP Science Pub.), vol. 41, pp. 301–311 (2010). doi: 10.1051/eas/1041021

Gounelle, M., Meibom, A., Hennebelle, P., Inutsuka, S.: Supernova propagation and cloud enrichment: a new model for the origin of 60Fe in the early solar system. ApJ **694**, L1–L5 (2009). doi: 10.1088/0004-637X/694/1/L1, 0904.1661

Gounelle, M., Shang, S., Glassgold, A.E., Shu, F.H., Rehm, E.K., Lee, T.: Early solar system irradiation and beryllium-7 synthesis. In: Mackwell, S., Stansbery, E. (eds.) Lunar and Planetary Institute Science Conference Abstracts, Lunar and Planetary Institute Science Conference Abstracts (Houston: LPI), vol. 34, p. 1833 (2003).

Gounelle, M., Shu, F.H., Shang, H., Glassgold, A.E., Rehm, K.E., Lee, T.: Extinct radioactivities and protosolar cosmic rays: self-shielding and light elements. ApJ **548**, 1051–1070 (2001). doi: 10.1086/319019

Gounelle, M., Shu, F.H., Shang, H., Glassgold, A.E., Rehm, K.E., Lee, T.: The irradiation origin of beryllium radioisotopes and other short-lived radionuclides. ApJ **640**, 1163–1170 (2006). doi: 10.1086/500309, arXiv:astro-ph/0512517

Guandalini, R., Busso, M.: Infrared photometry and evolution of mass-losing AGB stars. II. Luminosity and colors of MS and S stars. A&A **488**, 675–684 (2008). doi: 10.1051/0004-6361:200809932, 0806.4591

Guandalini, R., Busso, M., Ciprini, S., Silvestro, G., Persi, P.: Infrared photometry and evolution of mass-losing AGB stars. I. Carbon stars revisited. A&A **445**, 1069–1080 (2006). doi: 10.1051/0004-6361:20053208, arXiv:astro-ph/0509739

Guandalini, R., Palmerini, S., Busso, M., Uttenthaler, S.: Extra-mixing in luminous cool red giants: hints from evolved stars with and without Li. Publ. Astron. Soc. Aust. **26**, 168–175 (2009). doi: 10.1071/AS08063, 0905.4458

Harper, C.L., Jacobsen, S.B.: Evidence for 182Hf in the early Solar System and constraints on the timescale of terrestrial accretion and core formation. Geochim. Cosmochim. Acta **60**, 1131–1153 (1996). doi: 10.1016/0016-7037(96)00027-0

Hartmann, L., Kenyon, S.J.: The FU orionis phenomenon. ARA&A **34**, 207–240 (1996). doi: 10.1146/annurev.astro.34.1.207
Hester, J.J., Desch, S.J.: Understanding our origins: star formation in hii region environments. In: Krot, A.N., Scott, E.R.D., Reipurth, B. (eds.) Chondrites and the Protoplanetary Disk. Astronomical Society of the Pacific Conference Series (San Francisco: ASPC), vol. 341, p. 107 (2005)
Hsu, W., Guan, Y., Leshin, L.A., Ushikubo, T., Wasserburg, G.J.: A late episode of irradiation in the early solar system: evidence from extinct 36Cl and 26Al in meteorites. ApJ **640**, 525–529 (2006). doi: 10.1086/500043
Huss, G.R., Goswami, J.N., Meyer, B.S., Sahijpal, S., Wasserburg, G.J.: Stellar sources of short-lived radionuclides. In: Chronology of Meteorites and the Early Solar System (Houston: LPI), pp. 71–72 (2007)
Huss, G.R., Meyer, B.S., Srinivasan, G., Goswami, J.N., Sahijpal, S.: Stellar sources of the short-lived radionuclides in the early solar system. Geochim. Cosmochim. Acta **73**, 4922–4945 (2009)
Hutcheon, I.D., Krot, A.N., Keil, K., Phinney, D.L., Scott ERD 53Mn-53Cr dating of fayalite formation in the CV3 chondrite mokoia: evidence for asteroidal alteration. Science **282**, 1865 (1998). doi: 10.1126/science.282.5395.1865
Jacobsen, B., Matzel, J.E.P., Hutcheon, I.D., Ramon, E., Krot, A.N., Ishii, H.A., Nagashima, K., Yin, Q.: The 36Cl-36S systematics of wadalite from the Allende meteorite. In: Lunar and Planetary Institute Science Conference Abstracts (Houston: LPI), vol. 40, p. 2553 (2009)
Jeffery, P.M., Reynolds, J.H.: Origin of excess Xe129 in stone meteorites. J. Geoph. Res. **66**, 3582–3583 (1961). doi: 10.1029/JZ066i010p03582
Joggerst, C.C., Woosley, S.E., Heger, A.: Mixing in zero- and solar-metallicity supernovae. ApJ **693**, 1780–1802 (2009). doi: 10.1088/0004-637X/693/2/1780, 0810.5142
Kastner, J.H., Myers, P.C.: An observational estimate of the probability of encounters between mass-losing evolved stars and molecular clouds. ApJ **421**, 605–615 (1994). doi: 10.1086/173676
Kelley, W.R., Wasserburg, G.J.: Evidence for the existence of Pd-107 in the early solar system. Geophys. Res. Lett. **5**, 1079–1082 (1978). doi: 10.1029/GL005i012p01079
Kothes, R., Uyaniker, B., Pineault, S.: The Supernova Remnant G106.3+2.7 and its Pulsar-Wind Nebula: relics of triggered star formation in a complex environment. ApJ **560**, 236–243 (2001). doi: 10.1086/322511, arXiv:astro-ph/0106270
Kozlovsky, B., Murphy, R.J., Ramaty, R.: Nuclear deexcitation gamma-ray lines from accelerated particle interactions. ApJS **141**, 523–541 (2002). doi: 10.1086/340545
Kratz, K., Farouqi, K., Mashonkina, L.I., Pfeiffer, B.: Nucleosynthesis modes in the high-entropy-wind of type II supernovae. New Astron. Rev. **52**, 390–395 (2008). doi: 10.1016/j.newar.2008.06.015
Kratz, K., Farouqi, K., Pfeiffer, B., Truran, J.W., Sneden, C., Cowan, J.J.: Explorations of the r-processes: comparisons between calculations and observations of low-metallicity stars. ApJ **662**, 39–52 (2007). doi: 10.1086/517495, arXiv:astro-ph/0703091
Kratz, K., Pfeiffer, B., Cowan, J.J., Sneden, C.: r-process chronometers. New Astron. Rev. **48**, 105–108 (2004). doi: 10.1016/j.newar.2003.11.014
Kuroda, T., Wanajo, S., Nomoto, K.: The r-process in supersonic neutrino-driven winds: the role of the wind termination shock. ApJ **672**, 1068–1078 (2008). doi: 10.1086/523795
Lada, C.J., Lada, E.A.: Embedded clusters in molecular clouds. ARA&A **41**, 57–115 (2003). doi: 10.1146/annurev.astro.41.011802.094844, arXiv:astro-ph/0301540
Langer, N., El Eid, M.F., Baraffe, I.: Blue supergiant supernova progenitors. A&A **224**, L17–L20 (1989)
Lee, D., Halliday, A.N.: Hafnium-tungsten chronometry and the timing of terrestrial core formation. Nature **378**, 771–774 (1995). doi: 10.1038/378771a0
Lee, D., Halliday, A.N.: Hf-W isotopic evidence for rapid accretion and differentiation in the early solar system. Science **274**, 1876–1879 (1996). doi: 10.1126/science.274.5294.1876
Lee, T., Papanastassiou, D.A., Wasserburg, G.J.: Demonstration of 26Mg excess in Allende and evidence for 26Al. Geophys. Res. Lett. **3**, 109–109 (1976). doi: 10.1029/GL003i002p00109
Lee, T., Papanastassiou, D.A., Wasserburg, G.J.: Aluminum-26 in the early solar system - Fossil or fuel. ApJ **211**, L107–L110 (1977). doi: 10.1086/182351

Lee, T., Shu, F.H., Shang, H., Glassgold, A.E., Rehm, K.E.: Protostellar cosmic rays and extinct radioactivities in meteorites. ApJ **506**, 898–912 (1998). doi: 10.1086/306284

Leshin, L.A., Guan, Y., Lin, Y.: Implications of meteoritic 36Cl abundance for the origin of short-lived radionuclides in the early solar system. In: Krot, A., Scott, E., Keil, K., Reipurth, B. (eds.) Workshop on chondrites and the protoplanetary disk (Houston: LPI), p. 9084 (2004)

Limongi, M., Chieffi, A.: Evolution, explosion, and nucleosynthesis of core-collapse supernovae. ApJ **592**, 404–433 (2003). doi: 10.1086/375703, arXiv:astro-ph/0304185

Limongi, M., Chieffi, A.: The nucleosynthesis of 26Al and 60Fe in Solar metallicity stars extending in mass from 11 to 120 Msolar: the hydrostatic and explosive contributions. ApJ **647**, 483–500 (2006). doi: 10.1086/505164, arXiv:astro-ph/0604297

Lin, R.P., Krucker, S., Hurford, G.J., Smith, D.M., Hudson, H.S., Holman, G.D., Schwartz, R.A., Dennis, B.R., Share, G.H., Murphy, R.J., Emslie, A.G., Johns-Krull, C., Vilmer, N.: RHESSI observations of particle acceleration and energy release in an intense solar gamma-ray line flare. ApJ **595**, L69–L76 (2003). doi: 10.1086/378932

Lin, Y., Guan, Y., Leshin, L.A., Ouyang, Z.,Wang, D.: Evidence for live 36Cl in Ca-Al-rich inclusions from the Ningqiang Carbonaceous Chondrite. In: Mackwell, S., Stansbery, E. (eds.) Lunar and Planetary Institute Science Conference Abstracts (Houston: LPI), vol. 35, p. 2084 (2004)

Liu, M., Iizuka, Y., McKeegan, K.D., Tonui, E.K., Young, E.D.: Supra-canonical 26Al/27Al ratios in an unaltered allende CAI. In: Mackwell, S., Stansbery, E. (eds.) Lunar and Planetary Institute Science Conference Abstracts (Houston: LPI), vol. 36, p. 2079 (2005)

Lugmair, G.W., Marti, K.: Sm-Nd-Pu timepieces in the Angra DOS Reis meteorite. Earth Planet. Sci. Lett. **35**, 273–284 (1977). doi: 10.1016/0012-821X(77)90131-5

Lugmair, G.W., MacIsaac, C., Shukolyukov, A.: The 53Mn-53Cr isotope system and early planetary evolution. In: Lunar and Planetary Institute Science Conference Abstracts (Houston: LPI), vol. 23, p. 823 (1992)

Lugmair, G.W., Shukolyukov, A.: Early solar system timescales according to 53Mn-53Cr systematics. Geochim. Cosmochim. Acta **62**, 2863–2886 (1998). doi: 10.1016/S0016-7037(98) 00189-6

MacPherson, G.J., Huss, G.R., Davis, A.M.: Extinct ^1^0Be in Type A calcium-aluminum-rich inclusions from CV chondrites. Geochim. Cosmochim. Acta **67**, 3165–3179 (2003). doi: 10.1016/S0016-7037(02)01298-X

Matzel, J.E.P., Jacobsen, B., Hutcheon, I.D., Krot, A.N., Nagashima, K., Yin, Q., Ramon, E., Weber, P.K., Wasserburg, G.J.: Distribution and origin of 36Cl in Allende CAIs. In: Lunar and Planetary Institute Science Conference Abstracts (Houston: LPI), vol. 41, p. 2631 (2010)

Mauersberger, K., Erbacher, B., Krankowsky, D., Gunther, J., Nickel, R.: Ozone isotope enrichment: isotopomer-specific rate coefficients. Science **283**, 370 (1999). doi: 10.1126/science.283.5400.370

McKeegan, K.D., Chaussidon, M., Robert, F.: Incorporation of short-lived 10Be in a calcium-aluminum-rich inclusion from the Allende meteorite. Science **289**, 1334–1337 (2000). doi: 10.1126/science.289.5483.1334

McKeegan, K.D., Davis, A.M.: Early Solar System Chronology, p. 431. Elsevier, B., Amsterdam (2005)

Meyer, B.S.: Synthesis of short-lived radioactivities in a massive star. In: Krot, A.N., Scott, E.R.D., Reipurth, B. (eds.) Chondrites and the Protoplanetary Disk, Astronomical Society of the Pacific Conference Series (San Francisco: ASPC), vol. 341, p. 515 (2005)

Meyer, B.S., Adams, D.C.: Neutron burst production of 60Fe necessarily implies production of 182Hf. In: Mackwell, S., Stansbery, E. (eds.) Lunar and Planetary Institute Science Conference Abstracts (Houston: LPI), vol. 37, p. 1403 (2006)

Meyer, B.S., Clayton, D.D.: Short-lived Radioactivities and the birth of the sun. Space Sci. Rev. **92**, 133–152 (2000). doi: 10.1023/A:1005282825778

Mostefaoui, S., Lugmair, G.W., Hoppe, P., El Goresy, A.: Evidence for live iron-60 in Semarkona and Chervony Kut: a NanoSIMS study. In: Mackwell, S., Stansbery, E. (eds.) Lunar and Planetary Institute Science Conference Abstracts (Houston: LPI), vol. 34, p. 1585 (2003)

Mostefaoui, S., Lugmair, G.W., Hoppe, P., El Goresy, A.: Evidence for live 60Fe in meteorites. New Astron. Rev. **48**, 155–159 (2004). doi: 10.1016/j.newar.2003.11.022

Murty, S.V.S., Goswami, J.N., Shukolyukov, Y.A.: Excess 36Ar in the Efremovka meteorite: a strong hint for the presence of 36Cl in the early solar system. ApJ **475**, L65–L68 (1997). doi: 10.1086/310449

Nielsen, S.G., Rehkämper, M., Halliday, A.N.: Large thallium isotopic variations in iron meteorites and evidence for lead-205 in the early solar system. Geochim. Cosmochim. Acta **70**, 2643–2657 (2006). doi: 10.1016/j.gca.2006.02.012

Nollett, K.M., Busso, M., Wasserburg, G.J.: Cool bottom processes on the thermally pulsing asymptotic giant branch and the isotopic composition of circumstellar dust grains. ApJ **582**, 1036–1058 (2003). doi: 10.1086/344817, arXiv:astro-ph/0211271

Nomoto, K., Tominaga, N., Umeda, H., Kobayashi, C., Maeda, K.: Nucleosynthesis yields of core-collapse supernovae and hypernovae, and galactic chemical evolution. Nucl. Phys. A **777**, 424–458 (2006). doi: 10.1016/j.nuclphysa.2006.05.008, arXiv:astro-ph/0605725

Nordhaus, J., Busso, M.,Wasserburg, G.J., Blackman, E.G., Palmerini, S.: Magnetic mixing in red giant and asymptotic giant branch stars. ApJ **684**, L29–L32 (2008). doi: 10.1086/591963, 0806.3933

Nugis, T., Lamers, H.J.G.L.M. Mass-loss rates of Wolf-Rayet stars as a function of stellar parameters. A&A **360**, 227–244 (2000)

Ouellette, N., Desch, S.J., Hester, J.J., Leshin, L.A.: A Nearby supernova injected short-lived radionuclides into our protoplanetary disk. In: Krot, A.N., Scott, E.R.D., Reipurth, B. (eds.) Chondrites and the Protoplanetary Disk, Astronomical Society of the Pacific Conference Series (San Francisco: ASPC), vol. 341, p. 527 (2005)

Ouellette, N., Desch, S.J., Hester, J.J.: Interaction of supernova ejecta with nearby protoplanetary disks. ApJ **662**, 1268–1281 (2007). doi: 10.1086/518102, 0704.1652

Pagel, B.E.J.: Nucleosynthesis and Chemical Evolution of Galaxies. Cambridge University Press, Cambridge, UK (1997)

Palla, F., Stahler, S.W.: Accelerating star formation in clusters and associations. ApJ **540**, 255–270 (2000). doi: 10.1086/309312

Palmerini, S., Busso, M.: 26Al production from magnetically induced extramixing in AGB stars. New Astron. Rev. **52**, 412–415 (2008). doi: 10.1016/j.newar.2008.05.003, 0806.2733

Palmerini, S., Busso, M., Maiorca, E., Guandalini, R.: Nucleosynthesis of light-element isotopes in evolved stars experiencing extended mixing. Publ. Astron. Soc. Aust. **26**, 161–167 (2009). doi: 10.1071/AS08040, 0905.4365

Podosek, F.A., Lewis, R.S.: 129I and 244Pu abundances in white inclusions of the Allende meteorite. Earth Planet. Sci. Lett. **15**, 101 (1972). doi: 10.1016/0012-821X(72)90048-9

Podosek, F.A., Nichols, R.H.: Short-lived radionuclides in the solar nebula. In: Bernatowicz, T.J., Zinner, E. (eds.) American Institute of Physics Conference Series (San Francisco: ASPC), vol. 402, pp. 617–647 (1997). doi: 10.1063/1.53321

Preibisch, T., Zinnecker, H.: The history of low-mass star formation in the Upper Scorpius OB Association. AJ **117**, 2381–2397 (1999). doi: 10.1086/300842

Qian, Y., Vogel, P., Wasserburg, G.J.: Supernovae as the site of the r-process: implications for gamma-ray astronomy. ApJ **506**, 868–873 (1998). doi: 10.1086/306285, arXiv:astro-ph/9803300

Qian, Y., Wasserburg, G.J.: Where, oh where has the r-process gone? Phys. Rep. **442**, 237–268 (2007). doi: 10.1016/j.physrep.2007.02.006, 0708.1767

Qian, Y., Wasserburg, G.J.: Abundances of Sr, Y, and Zr in metal-poor stars and implications for chemical evolution in the early galaxy. ApJ **687**, 272–286 (2008). doi: 10.1086/591545, 0807.0809

Ramaty, R., Mandzhavidze, N.: Gamma-rays from Solar Flares. In: Martens, P.C.H., Tsuruta, S., Weber, M.A. (ed.) Highly Energetic Physical Processes and Mechanisms for Emission from Astrophysical Plasmas, IAU Symposium, vol. 195, (San Francisco: ASPC) p. 123 (2000)

Rauscher, T., Heger, A., Hoffman, R.D., Woosley, S.E.: Nucleosynthesis in massive stars with improved nuclear and stellar physics. ApJ **576**, 323–348 (2002). doi: 10.1086/341728, arXiv:astroph/0112478

Reynolds, J.H.: Isotopic composition of primordial xenon. Phys. Rev. Lett. **4**, 351–354 (1960). doi: 10.1103/PhysRevLett.4.351

Rowe, M.W., Kuroda, P.K.: Fissiogenic xenon from the pasamonte meteorite. J. Geoph. Res. **70**, 709–714 (1965). doi: 10.1029/JZ070i003p00709

Rugel, G., Faestermann, T., Knie, K., Korschinek, G., Poutivtsev, M., Schumann, D., Kivel, N., Günther-Leopold, I., Weinreich, R., Wohlmuther, M.: New measurement of the Fe60 half-life. Phys. Rev. Lett. **103**(7), 072502 (2009). doi: 10.1103/PhysRevLett.103.072502

Rutherford, E.: Origin of actinium and age of the earth. Nature **123**, 313–314 (1929)

Sahijpal, S., Goswami, J.N., Davis, A.M., Lewis, R.S., Grossman, L.: A stellar origin for the short-lived nuclides in the early solar system. Nature **391**, 559 (1998). doi: 10.1038/35325

Scheffler, H., Elsaesser, H.: Physics of the Galaxy and Interstellar Matter. Springer: Berlin (1987)

Schramm, D.N., Tera, F., Wasserburg, G.J.: The isotopic abundance of ^26Mg and limits on ^26Al in the early Solar System. Earth Planet. Sci. Lett. **10**, 44–59 (1970). doi: 10.1016/0012-821X(70)90063-4

Schramm, D.N., Wasserburg, G.J.: Nucleochronologies and the mean age of the elements. ApJ **162**, 57 (1970). doi: 10.1086/150634

Share, G.H., Murphy, R.J.: Gamma-ray measurements of flare-to-flare variations in ambient solar abundances. ApJ **452**, 933 (1995). doi: 10.1086/176360

Share, G.H., Murphy, R.J., Ryan, J.: Solar and stellar Gamma Ray observations with COMPTON. In: Dermer, C.D., Strickman, M.S., Kurfess, J.D. (eds.) Proceedings of the Fourth Compton Symposium, American Institute of Physics Conference Series, vol. 410, (San Francisco: ASPC) pp. 17–36 (1997). doi: 10.1063/1.54117

Shu, F.H., Adams, F.C., Lizano, S.: Star formation in molecular clouds – Observation and theory. ARA&A **25**, 23–81 (1987a). doi: 10.1146/annurev.aa.25.090187.000323

Shu, F.H., Lizano, S., Adams, F.C.: Star formation in molecular cloud cores. In: Peimbert, M., Jugaku, J. (eds.) Star Forming Regions, IAU Symposium, vol. 115, (Dordrecht: Reidel) pp. 417–433 (1987b)

Shu, F.H., Shang, H., Glassgold, A.E., Lee, T.: X-rays and fluctuating X-winds from protostars. Science **277**, 1475–1479 (1997). doi: 10.1126/science.277.5331.1475

Shu, F.H., Shang, H., Gounelle, M., Glassgold, A.E., Lee, T.: The origin of chondrules and refractory inclusions in Chondritic meteorites. ApJ **548**, 1029–1050 (2001). doi: 10.1086/319018

Shukolyukov, A., Lugmair, G.W.: Fe-60 in eucrites. Earth Planet. Sci. Lett. **119**, 159–166 (1993a). doi: 10.1016/0012-821X(93)90013-Y

Shukolyukov, A., Lugmair, G.W.: Live Iron-60 in the early solar system. Science **259**, 1138–1142 (1993b). doi: 10.1126/science.259.5098.1138

Smith, N., Brooks, K.J.: A census of the Carina Nebula – II. Energy budget and global properties of the nebulosity. MNRAS **379**, 1279–1292 (2007). doi: 10.1111/j.1365-2966.2007.12021.x, 0705.3053

Srinivasan, G., Sahijpal, S., Ulyanov, A.A., Goswami, J.N.: Ion microprobe studies of Efremovka CAIs: II. Potassium isotope composition and 41Ca in the early Solar System. Geochim. Cosmochim. Acta **60**, 1823–1835 (1996). doi: 10.1016/0016-7037(96)00054-3

Srinivasan, G., Ulyanov, A.A., Goswami, J.N.: Ca-41 in the early solar system. ApJ **431**, L67–L70 (1994). doi: 10.1086/187474

Stirling, C.H., Halliday, A.N., Potter, E., Andersen, M.B., Zanda, B.: A low initial abundance of 247Cm in the early solar system and implications for r-process nucleosynthesis. Earth Planet. Sci. Lett. **251**, 386–397 (2006). doi: 10.1016/j.epsl.2006.09.023

Stoerzer, D., Pellas, P.: Angra DOS Reis – Plutonium distribution and cooling history. Earth Planet. Sci. Lett. **35**, 285–293 (1977). doi: 10.1016/0012-821X(77)90132-7

Tachibana, S., Huss, G.R.: The initial abundance of $^{60}$Fe in the solar system. ApJ **588**, L41–L44 (2003). doi: 10.1086/375362

Tachibana, S., Huss, G.R., Kita, N.T., Shimoda, G., Morishita, Y.: $^{60}$Fe in Chondrites: debris from a nearby supernova in the early solar system? ApJ **639**, L87–L90 (2006). doi: 10.1086/503201

Takigawa, A., Miki, J., Tachibana, S., Huss, G.R., Tominaga, N., Umeda, H., Nomoto, K.: Injection of short-lived radionuclides into the early solar system from a faint supernova with mixing fallback. ApJ **688**, 1382–1387 (2008). doi: 10.1086/592184, 0808.1441

Tatischeff, V., Kozlovsky, B., Kiener, J., Murphy, R.J.: Delayed X- and gamma-ray line emission from solar flare radioactivity. ApJS **165**, 606–617 (2006). doi: 10.1086/505112, arXiv:astroph/0604325

Tatischeff, V., Kozlovsky, B., Kiener, J., Murphy, R.J.: Radioactive line emission from solar flares. In: ESA Special Publication (Nordwijk: ESA Publ.), vol. 622, p. 135 (2007)

Tatischeff, V., Thibaud, J.: Li production by stellar flares in young open clusters. New Astron. Rev. **52**, 423–426 (2008). doi: 10.1016/j.newar.2008.06.013

Tatischeff, V., Thibaud, J., Ribas, I.: Nucleosynthesis in stellar flares. ArXiv e-prints 0801.1777 (2008)

Thiemens, M.H., Heidenreich, J.E., III: The mass-independent fractionation of oxygen – a novel isotope effect and its possible cosmochemical implications. Science **219**, 1073–1075 (1983). doi: 10.1126/science.219.4588.1073

Trigo-Rodríguez, J.M., García-Hernández, D.A., Lugaro, M., Karakas, A.I., van Raai, M., Garcáa Lario, P., Manchado, A.: The role of massive AGB stars in the early solar system composition. Meteorit. Planet. Sci. **44**, 627–641, 0812.4358 (2009)

Tur, C., Heger, A., Austin, S.M.: Production of 26Al, 44Ti, and 60Fe in core-collapse supernovae: sensitivity to the rates of the triple alpha and 12C(a,g)16O reactions. ArXiv e-prints 0908.4283 (2009)

Urey, H.C.: The cosmic abundances of potassium, uranium, and thorium and the heat balances of the earth, the moon, and mars. Proc. Natl. Acad. Sci. **41**, 127–144 (1955). doi: 10.1073/pnas.41.3.127

Urey, H.C., Donn, B.: Chemical heating for meteorites. ApJ **124**, 307 (1956). doi: 10.1086/146223

Wadhwa, M., Amelin, Y., Davis, A.M., Lugmair, G.W., Meyer, B., Gounelle, M., Desch, S.J.: From dust to planetesimals: implications for the solar protoplanetary disk from short-lived radionuclides. In: Reipurth, B., Jewitt, D., Keil, K. (eds.) Protostars and Planets V, pp. 835–848. Tucson: University of Arizona Press (2007)

Wasserburg, G.J., Fowler, W.A., Hoyle, F.: Duration of nucleosynthesis. Phys. Rev. Lett. **4**, 112–114 (1960). doi: 10.1103/PhysRevLett.4.112

Wasserburg, G.J., Huneke, J.C., Burnett, D.S.: Correlations between fission tracks and fission type xenon in meteoritic whitlockite. J. Geoph. Res. **74**, 4221–4232 (1969). doi: 10.1029/JB074i017p04221

Wasserburg, G.J., Busso, M., Gallino, R.: Abundances of actinides and short-lived nonactinides in the interstellar medium: diverse supernova sources for the r-processes. ApJ **466**, L109–L113 (1996). doi: 10.1086/310177

Wasserburg, G.J., Busso, M., Gallino, R., Nollett, K.M.: Short-lived nuclei in the early Solar System: possible AGB sources. Nucl. Phys. A **777**, 5–69 (2006). doi: 10.1016/j.nuclphysa.2005.07.015, arXiv:astro-ph/0602551

Wasserburg, G.J., Busso, M., Gallino, R., Raiteri, C.M.: Asymptotic Giant Branch stars as a source of short-lived radioactive nuclei in the solar nebula. ApJ **424**, 412–428 (1994). doi: 10.1086/173899

Wasserburg, G.J., Gallino, R., Busso, M.A.: Test of the supernova trigger hypothesis with 60Fe and 26Al. ApJ **500**, L189–L193 (1998). doi: 10.1086/311414

Wasserburg, G.J., Gallino, R., Busso, M., Goswami, J.N., Raiteri, C.M.: Injection of freshly synthesized Ca-41 in the early solar nebula by an asymptotic giant branch star. ApJ **440**, L101–L104 (1995). doi: 10.1086/187771

Williams, J.P., Gaidos, E.: On the likelihood of supernova enrichment of protoplanetary disks. ApJ **663**, L33–L36 (2007). doi: 10.1086/519972, 0705.3459

Woosley, S.E., Heger, A.: Nucleosynthesis and remnants in massive stars of solar metallicity. Phys. Rep.**442**, 269–283 (2007). doi: 10.1016/j.physrep.2007.02.009, arXiv:astro-ph/0702176

Woosley, S.E., Weaver, T.A.: The evolution and explosion of massive stars. II. Explosive hydrodynamics and nucleosynthesis. ApJS **101**, 181 (1995). doi: 10.1086/192237

Young, E.D., Gounelle, M., Smith, R.L., Morris, M.R., Pontoppidan, K.M.: The oxygen isotopic composition of the solar system in a galactic context: new results for CO in young stellar objects and implications for the birth environment of the solar system. In: Lunar and Planetary Institute Science Conference Abstracts (Houston: LPI), vol. 40, p. 1967 (2009)

Young, E.D., Simon, J.I., Galy, A., Russell, S.S., Tonui, E., Lovera, O.: Supra-canonical 26Al/27Al and the residence time of CAIs in the solar protoplanetary disk. Science **308**, 223–227 (2005). doi: 10.1126/science.1108140

Zinnecker, H.: From local star formation to global star formation. Ap&SS **281**, 147–157 (2002). doi: 10.1023/A:1019532503571

Zinner, E., Göpel, C.: Aluminum-26 in H4 chondrites: implications for its production and its usefulness as a fine-scale chronometer for early solar system events. Meteorit. Planet. Sci. **37**, 1001–1013 (2002)

Zuckerman, B., Song, I.: Young stars near the sun. ARA&A **42**, 685–721 (2004). doi: 10.1146/annurev.astro.42.053102.134111

# Chapter 7
# Distributed Radioactivities

R. Diehl, D.H. Hartmann, and N. Prantzos

## 7.1 The Role of Radioactivities

Radioactive isotopes are ejected into the surroundings of their sources, and become observable through their gamma-ray line emission once having left dense production sites from where not even gamma-rays may escape. Nuclear gamma-rays can penetrate material layers of integrated thickness of a few grams/cm$^2$. A typical interstellar cloud would have $\sim$0.1 g/cm$^2$, SNIa envelopes are transparent to gamma-rays after 30–100 days, depending on explosion dynamics. In this chapter, we discuss such radioactivities in interstellar space. Depending on radioactive lifetime, we thus address isotopes which originate from single sources (the short-lived isotopes) or up to thousands of sources as accumulated in interstellar space over the radioactive lifetime of long-lived isotopes (see Table 7.1).

Radioactive isotopes provide us with a *clock*, which carries information on the time scale of processes between their (nucleo-)synthesis and the time an actual isotopic-abundance determination is obtained through measurement. Different processes are investigated with different isotopes, making use of their characteristic decay times. Key isotopes and their specific applications are summarized:

- $^{44}$Ti with its decay time of 89 years can provide diagnostics on mixing of ejecta in core-collapse supernova explosions. Being produced in the inner regions of the supernova close to the proto-neutron star, its distribution within the supernova remnant is influenced by the Rayleigh-Taylor instabilities which are believed to develop in this inner region, from the unstable layering of hot interiors and cooler stellar envelope. With a telescope of sufficient spatial resolution[1] the nearby

R. Diehl (✉)
Max Planck Institut für extraterrestrische Physik, 85748 Garching, Germany

D.H. Hartmann
Clemson University, Clemson SC 29634-0978, USA

N. Prantzos
Institut d'Astrophysique, 75104 Paris, France

[1] The NuSTAR mission planned for 2012 is equipped with such imaging capability up to 80 keV.

Diehl, R. et al.: *Distributed Radioactivities*. Lect. Notes Phys. **812**, 345–436 (2011)
DOI 10.1007/978-3-642-12698-7_7 

**Table 7.1** Radioactivities with gamma-ray line emission, sorted by ascending radioactive mean lifetime (updated from Diehl et al., 2006c)

| Decay chain | Lifetime(y) | $\gamma$-ray energy (keV) (branching ratio [%]) | Site (detections) | Process type |
|---|---|---|---|---|
| $^{7}Be \rightarrow {}^{7}Li$ | 0.21 | 478 (100) | Novae | Explosive H burning |
| $^{56}Ni \longrightarrow {}^{56}Co \longrightarrow {}^{56}Fe$ | 0.31 | 847 (100), 1,238 (68) 2,598 (17), 1,771 (15) and 511 from $e^+$ | SNe (SN1987A, SN1991T) | NSE burning |
| $^{57}Co \longrightarrow {}^{57}Fe$ | 1.1 | 122 (86), 136 (11) | SNe (SN1987A) | NSE burning |
| $^{22}Na \longrightarrow {}^{22}Ne$ | 3.8 | 1275 (100) and 511 from $e^+$ | Novae | Explos. H burning |
| $^{44}Ti \longrightarrow {}^{44}Sc \longrightarrow {}^{44}Ca$ | 89 | 68 (95), 78 (96) 1,156 (100) and 511 from $e^+$ | SNe (Cas A) | NSE $\alpha$ freeze-out |
| $^{26}Al \longrightarrow {}^{26}Mg$ | $1.04\ 10^6$ | 1,809 (100) and 511 from $e^+$ | ccSNe, WR Novae, AGB (Galaxy) (Cygnus; Sco-Cen; Orion; Vela) | H burning ($\nu$-proc.) |
| $^{60}Fe \longrightarrow {}^{60}Co \longrightarrow {}^{60}Ni$ | $3.8\ 10^6$ | 1,173 (100), 1,332 (100) 59 (2) | SNe (Galaxy) | He,C shell burning |
| $e^+ \longrightarrow Ps,.. \longrightarrow \gamma\gamma(\gamma)$ | $\sim 10^7$ | 2·511 ($\sim$100), cont $<$510 | Radioactivities Pulsars, $\mu$QSOs, ... (Galactic bulge; disk) | $\beta^+$ decay rel. plasma |

Cas A supernova remnant at 3.4 kpc distance could be resolved (diameter about 4 arcmin), and the morphology of the image in $^{44}$Ti decay gamma-rays then reflects the effects of this inner mixing agent, as we assume a homologous expansion of the supernova remnant. If sufficient spatial resolution cannot be obtained, modern X- and gamma-ray detectors can provide spectral resolution which can resolve Doppler shifts or broadening of the $^{44}$Ti decay lines, thus mapping out a velocity distribution of $^{44}$Ti ejecta in their current state. The smoothness (or raggedness) of such a distribution would encode clumping of ejecta in space, and thus can be related to mixing agents in core-collapse supernova interiors.

- $^{26}$Al and $^{60}$Fe have radioactive decay times of 1.04 and 3.8 My, respectively. Therefore, dispersal of nucleosynthesis ejecta from a single source can be traced over a time scale of millions of years (My). This is approximately the time scale expected to characterize the return of stellar gas through an explosive event into the ambient interstellar gas in the source surroundings. Thus, one of the questions addressed would be how far out in space away from the source ejecta may travel within a few My. Another question would be if the dispersal shows any preferred directions, such as possibly escape flows from the galactic-plane (*chimneys*). Yet another aspect is the state of turbulence in hot interstellar cavities of age 1 My or younger, yet older than the 0.1 My scale which can be addressed through observations of gas in supernova remnants by observations of atomic transitions in partly-ionized plasma. Measuring the width of radioactive-decay gamma-ray lines encodes turbulence through Doppler broadening. Note that such interstellar cavities are filled with tenuous and hot gas, and current observations of such objects are dominated by the signals from the denser (but also cooler, and presumably less-turbulent) cavity walls along the line-of-sight of any such observation.
- $^{59}$Ni is an example of a radioactive isotope, which can decay only through electron capture. Therefore, its decay implies an environment of incomplete if any ionization, i.e., an environment below $10^5$ K. On the other extreme, cosmic ray nuclei are relativistic and always fully ionized. Therefore, the observed abundance of $^{59}$Ni in cosmic ray nuclei directly encodes the delay time between nucleosynthesis of this $^{59}$Ni ($\tau_{\rm decay} \sim 10^5$ years) and its acceleration to cosmic-ray energies (and hence full ionization) (see Sect. 8 and e.g. Mewaldt et al. (2001)). $^{57}$Co and $^{56}$Ni are more-shortlived electron-capture only radio-isotopes, similarly useful to constrain delays before ionization. More sophisticated analysis is required to understand the cosmic-ray abundances of unstable isotopes which include electron capture as a main decay mechanism, such as $^{7}$Be, $^{49}$V, $^{51}$Cr, and $^{55}$Fe: Their abundances result from how much of their trajectory had involved low energies with chances to attach electrons. Hence they address the topic of cosmic-ray reacceleration and/or successive cosmic-ray acceleration by different supernova shock regions (Mewaldt et al., 2001).
- Radioactive species abundances in cosmic rays can be combined with known production cross sections for spallation reactions to determine the effectively-traversed path length of cosmic rays. Isotopes $^{10}$Be, $^{26}$Al, $^{36}$Cl, and $^{54}$Mn have been used to constrain the time spent in the Galaxy to $\sim 10^7$ years (Yanasak et al., 2001a).

- Early-solar system astrophysics involving homogeneity and effectiveness of mixing before first meteorites were formed can be addressed by 'shortlived' (0.1-few My) radioactivities, such as $^{10}$Be, $^{26}$Al, $^{36}$Cl, $^{41}$Ca, $^{53}$Mn, and $^{60}$Fe. Probes on a somewhat longer time scale are $^{129}$I, $^{146}$Sm, $^{182}$Hf, and $^{244}$Pu. Their abundance measurements and implications are dealt with in detail in Chap. 6.
- Long-lived ($>$100 My) radioactivities ($^{187}$Re, $^{235}$U, $^{238}$U, $^{232}$Th) provide age dating of nucleosynthesis at cosmic time scales, and also dating of the age of the solar system (see Chaps. 2 and 6).
- Positrons produced in various radioactive decays add interesting aspects, from their propagation properties as charged particles (Jean et al., 2009). They will be directed by the morphology of the magnetic field, and will lose energy in collisions with ambient gas. Their annihilation is favored in a rather narrow energy range between ~7–200 eV (corresponding to temperatures in the 1,000 K regime) (Guessoum et al., 2005). Positrons may deposit radioactive energy efficiently in expanding envelopes from stellar explosions, or may reveal their annihilation site through characteristic 511 keV gamma-rays. This yields information which complements other observables of the same radioactive decays (e.g. bolometric broad-band emission, or $^{26}$Al-decay gamma-rays).

Each of above is discussed in more detail throughout this book, at their appropriate places, addressing their nucleosynthesis models (Chaps. 3, 4 and 5), observational techniques (Chap. 10), and astrophysical lessons (Chaps. 2–7).

When the lifetime of a radioactive nucleus is not very large w.r.t. the timescale between two nucleosynthetic events in the Galaxy, those events are expected to be seen as point-sources in the light of that radioactivity. In the opposite case a diffuse emission along the Galaxy is expected from the cumulated emission of hundreds or thousands of sources[2]. Characteristic timescales between two explosions are ~1–2 weeks for novae, ~50–100 year for SNII+SNIb and ~200–400 year for SNIa (see Sect. 7.2.3). Comparing those timescales to the decay lifetimes of Table 7.1 one sees that in the case of the long-lived $^{26}$Al and $^{60}$Fe a diffuse emission is expected; the spatial profile of that emission should reflect the Galactic distribution of the underlying sources. All the other radioactivities of Table 7.1 should be seen as point sources in the Galaxy except, perhaps, $^{22}$Na from Galactic novae.

## 7.2 The Milky Way

The Galactic distribution of any kind of stellar source of radioactivities is somewhat related to the distribution of stars in the Milky Way. Similarly, the birthrate of any kind of radioactivity source is somewhat related to the Galactic star formation rate.

---

[2] Notice that the distinction between a point and a diffuse source depends also on the resolving power of the observing instrument, i.e. diffuse emission is seen if typical distances – projected on the sky – between point sources are larger than the instrument's resolution. But even detectors with good spatial resolution would see diffuse emission if the emitting particles/nuclei travel far away from their sources during their lifetimes; this may be the case of Galactic positrons (see Sect. 7.7).

In this section we present a summary of current knowledge about the stellar populations of the Milky Way and their spatial distribution and we discuss the birthrates of stars and supernovae. Long lived radioactivities, like $^{26}$Al and $^{60}$Fe are expected to be thermalised in the ISM; some of their properties should reflect then the ISM properties. Moreover, positrons produced by various processes (including radioactivity), slow down and finally annihilate in the ISM and the resulting electromagnetic emission also reflects the ISM properties. We present then in this section a brief overview of the various phases of the ISM in the bulge and the disk (including the spiral arms) of the Miky Way.

### *7.2.1 Stellar Populations*

The Milky Way is a typical spiral galaxy, with a total baryonic mass of $\sim 5 \times 10^{10}$ $M_\odot$, of which more than 80% is in the form of stars. Stars are found in three main components: the central bulge, the disk and the halo, while the gas is found essentially in the plane of the disk. Because of its low mass, estimated to $4 \times 10^8$ $M_\odot$ i.e. less than 1% of the total (Bell et al., 2007), the Galactic halo plays no significant role in the production of distributed radioactivities. The bulge contains ~1/3 of the total mass and an old stellar population (age>10 Gigayears (Gy)). The dominant component of the Milky Way is the so-called *thin disk*, a rotationally supported structure composed of stars of all ages (0–10 Gy). A non negligible, contribution is brought by the *thick disk*, an old (>10 Gy) and kinematically distinct entity identified by Gilmore and Reid (1983).

To a first approximation, and by analogy with external galaxies, the bulge of the Milky Way can be considered as spherical, with a density profile either exponential or of Sercic-type $\rho r \propto r^{1/n}$ with $n > 1$. Measurements in the near infrared (NIR), concerning either integrated starlight observations or star counts revealed that the bulge is not spherical, but elongated. Recent models suggest a tri-axial ellipsoid, but its exact shape is difficult to determine (Rattenbury et al., 2007) because of the presence of a Galactic bar. The mass of the bulge lies in the range 1–2 $\times 10^{10}$ $M_\odot$ (Robin et al., 2003). By comparing colour-magnitude diagrams of stars in the bulge and in metal-rich globular clusters, Zoccali et al. (2003) find that the populations of the two systems are co-eval, with an age of ~10 Gy.

The innermost regions of the bulge, within a few hundred pc, are dominated by a distinct, disk-like component, called the Nuclear Bulge which contains about 10% of the bulge stellar population ($\sim 1.5 \times 10^9$ $M_\odot$) within a flattened region of radius $230 \pm 20$ pc and scaleheight $45 \pm 5$ pc (Launhardt et al., 2002). It is dominated by three massive stellar clusters (Nuclear Stellar Cluster or NSC in the innermost 5 pc, Arches and Quintuplex), which have a mass distribution substantially flatter than the classical Salpeter IMF [3](Figer, 2008). Finally, in the center of the Milky Way,

[3] Stars are born with a mass distribution called Initial mass function (IMF). Observed IMFs of young stellar clusters in the Milky Way and other galaxies have similar IMFs, with the upper part ($M > 1$ $M_\odot$) described by a power-law ($dN/dM \propto M^{-(1+X)}$, where $X$ is the slope of the IMF; in most cases, $X$=1.35 (as determined by Salpeter (1955), for the local IMF.

at the position of SgrA* source, lies the super-massive Galactic black hole (SMBH) with a total mass of $\sim 4 \times 10^6$ $M_\odot$ (Gillessen et al., 2008).

The Sun is located in the thin disk of the Milky way, at a distance of $R_\odot \sim 8$ kpc from the Galactic center (Groenewegen et al., 2008) and references therein). Furthermore, the Sun is not located exactly on the plane, but at a distance from it $z_\odot \sim 25$ pc, as evaluated from the recent analysis of the Sloan Digital Sky Survey (SDSS) data (Jurić et al., 2008).

In studies of the Milky Way the solar neighborhood plays a pivotal role, since local properties can, in general, be measured with greater accuracy than global ones. The total baryonic surface density of the solar cylinder (which is defined as a cylinder of radius 500 pc centered on Sun's position and extending perpendicularly to the Galactic plane up to several kpc) is estimated to $\Sigma_T = 48.8$ $M_\odot$ pc$^{-2}$ (Flynn et al., 2006), with $\sim 13$ $M_\odot$ pc$^{-2}$ belonging to the gas (see Sect. 7.2.2). This falls on the lower end of the dynamical mass surface density estimates from kinematics of stars perpendicularly to the plane) which amount to $\Sigma_D$=50–62 $M_\odot$ pc$^{-2}$ (Holmberg and Flynn, 2004) or 57–66 $M_\odot$ pc$^{-2}$ (Bienaymé et al., 2006). Thus, the values for the baryon content of the solar cylinder, summarized in Table 7.2, should be considered rather as lower limits (Flynn et al., 2006): the total stellar surface density could be as high as 40 $M_\odot$ pc$^{-2}$.

The density profiles of the stellar thin and thick disks can be satisfactorily fit with exponential functions, both in the radial direction and perpendicularly to the Galactic plane. The recent SDSS data analysis of star counts, with no a priori assumptions as to the functional form of the density profiles finds exponential disks with scalelengths as displayed in Table 7.2 (from Jurić et al. (2008)). The thin and thick disks cannot extend all the way to the Galactic center, since dynamical arguments constrain the spatial co-existence of such rotationally supported structures with the pressure-supported bulge. The exact shape of the *central hole* of the disks is poorly known (see, e.g. Freudenreich (1998); Robin et al. (2003), for parametrizations), but for most practical purposes (i.e. estimate of the total disk mass) the hole can be considered as trully void of disk stars for disk radius $R < 2$ kpc.

The data presented in this section (as summarized in Table 7.2), allow one to estimate the total mass of the thin and thick disks as $M_{D,\mathrm{thin}} = 2.3 \times 10^{10}$ $M_\odot$

**Table 7.2** Properties of the stellar populations of the thin and thick disk[a] (from Prantzos et al., 2010)

| | | | Thin | Thick |
|---|---|---|---|---|
| Mass density | $\rho_{0,\odot}$ | ($M_\odot$ pc$^{-3}$) | $4.5 \times 10^{-2}$ | $5.3 \times 10^{-3}$ |
| Surface density | $\Sigma_\odot$ | ($M_\odot$ pc$^{-2}$) | 28.5 | 7 |
| Scaleheight | $H_\odot$ | (pc) | 300 | 900 |
| Scalelength | $L$ | (pc) | 2,600 | 3,600 |
| Star mass | $M_D$ | ($10^{10}$ $M_\odot$) | 2.3 | 0.53 |
| $\langle$Age$\rangle_\odot$ | $\langle$A$\rangle_\odot$ | (Gyr) | 5 | 10 |
| $\langle$Metallicity$\rangle_\odot$ | $\langle$[Fe/H]$\rangle_\odot$ | (dex) | −0.1 | −0.7 |

[a] The indice $\odot$ here denotes quantities measured at Galactocentric distance $R_\odot = 8$ kpc. Average quantites are given within $\langle\ \rangle$.

and $M_{D,\mathrm{thick}} = 0.5 \times 10^{10}$ $M_\odot$, respectively, in the galactocentric distance range 2–15 kpc. Overall, the disk of the Galaxy is twice as massive as the bulge.

### 7.2.2 Interstellar Matter

Interstellar gas is primarily composed of hydrogen, but it also contains helium ($\simeq$10% by number or 28% by mass) and heavier elements, called *metals* ($\simeq$0.12% by number or 1.5% by mass in the solar neighborhood). All the hydrogen, all the helium, and approximately half the metals exist in the form of gas; the other half of the metals is locked up in small solid grains of dust. Overall, gas and dust appear to be spatially well correlated (Boulanger and Perault, 1988; Boulanger et al., 1996).

Interstellar gas can be found in molecular, atomic (cold or warm) and ionized (warm or hot) forms. The physical properties of the different gas components in the Galactic disk were reviewed by Ferrière (2001) and are summarized in Table 7.3. The gas properties in the Galactic bulge are less well established, but on the whole, all gas components appear to be hotter and denser in the bulge than in the disk (Ferrière et al., 2007).

Spatially, the molecular gas is confined to discrete clouds, which are roundish, gravitationally bound, and organized hierarchically from large complexes (size $\sim$20–80 pc, mass $\sim 10^5$–$2 \times 10^6$ $M_\odot$) down to small clumps (size $\lesssim 0.5$ pc, mass $\lesssim 10^3$ $M_\odot$) (Goldsmith, 1987). The cold atomic gas is confined to more diffuse clouds, which often appear sheet-like or filamentary, cover a wide range of sizes (from a few pc up to $\sim$2 kpc), and have random motions with typical velocities of a few km/s[1] (Kulkarni and Heiles, 1987). The warm and hot components are more widespread and they form the intercloud medium.

The different gas components also differ by their spatial distributions at large scales. The observational situation was reviewed by Ferrière (2001) for the Galactic disk and by Ferrière et al. (2007) for the Galactic bulge. Figure 7.1 gives the radial variation of the azimuthally-averaged surface densities of $H_2$, HI, HII and the total gas (accounting for a 28% contribution from He). The distributions of those ISM phases have also different scaleheights, which increase with galactocentric radius (*flaring*), as can be seen in Fig. 7.3; the HII layer (not appearing in that figure) has an even larger scaleheight, of $>1$ kpc.

**Table 7.3** Physical properties (typical temperatures, hydrogen densities and ionization fractions) of the different ISM phases in the Galactic disk (from Prantzos et al., 2010)

| Phase | | $T$ (K) | $n_\mathrm{H}$ (cm$^{-3}$) | $x_\mathrm{ion}$ |
|---|---|---|---|---|
| Molecular | (MM) | 10–20 | $10^2$–$10^6$ | $\lesssim 10^{-4}$ |
| Cold neutral | (CNM) | 20–100 | 20–100 | $4 \times 10^{-4}$–$10^{-3}$ |
| Warm neutral | (WNM) | $10^3$–$10^4$ | 0.2–2 | 0.007–0.05 |
| Warm ionized | (WIM) | $\sim$8,000 | 0.1–0.3 | 0.6–0.9 |
| Hot ionized | (HIM) | $\sim 10^6$ | 0.003–0.01 | 1 |

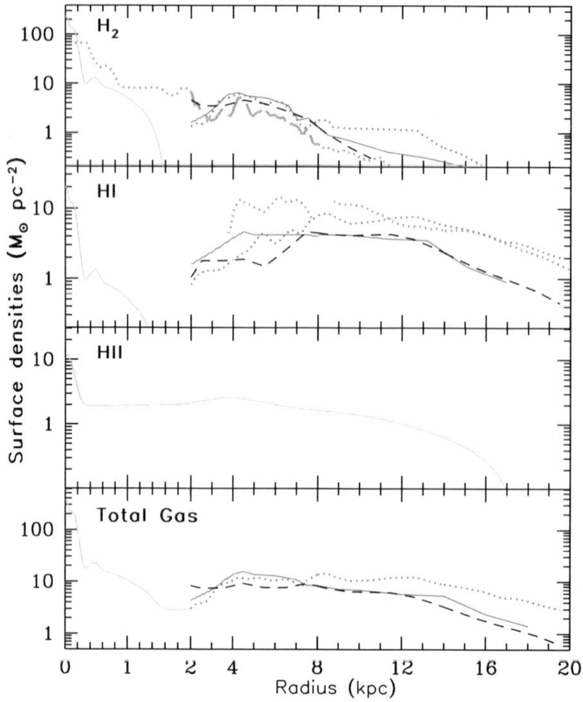


**Fig. 7.1** Azimuthally-averaged surface densities of interstellar atomic, molecular and ionized hydrogen as functions of Galactic radius. The total gas (*bottom*) includes a 40% contribution by helium. Notice the change of scale at $R = 2$ kpc. For $R < 2$ kpc (bulge) data are derived by Ferrière et al. (2007), based on a compilation of earlier works: Sawada et al. (2004) for the molecular gas in the Central Molecular Zone, Liszt and Burton (1980) for the neutral gas in the tilted disk, and Cordes and Lazio (2002) for the ionized gas. In all panels, disk data ($R$ >2 kpc) are from: Dame (1993) (*solid*); Olling and Merrifield (2001) (*dotted*) ; Nakanishi and Sofue (2006) for HI and Nakanishi and Sofue (2003) for $H_2$, respectively (*dashed*); and Kalberla and Dedes (2008) for HI and and Pohl et al. (2008) for $H_2$, respectively (*dot-dashed*). The curve in the HII panel is from the NE2001 free-electron density model of Cordes and Lazio (2002) (for simplicity, we identified the hydrogen density with the free-electron density, i.e., we neglected the contribution of free electrons originating from helium in the HIM) (From Prantzos et al., 2010)

The total interstellar masses (including helium and metals) of the three gas components in the Galactic disk are highly uncertain. The masses given below are estimated from references in Fig. 7.1, inside 20 kpc and they are in the range: $\sim$(0.9–2.5) $\times$ $10^9$ $M_\odot$ for the molecular component, $\sim$(0.65–1.1) $\times$ $10^9$ $M_\odot$ for the atomic component, and $\sim$1.5 $\times$ $10^9$ $M_\odot$ for the ionized component. The total interstellar mass in the Galaxy is probably comprised between $\sim$0.9 $\times$ $10^{10}$ $M_\odot$ and $\sim$1.5 $\times$ $10^{10}$ $M_\odot$, representing $\sim$15–25% of the baryonic Galaxy mass or $\sim$25–35% of the total mass of the Galactic disk.

The gas distributions of Fig. 7.1 are azimuthally averaged, whereas the Milky Way disk displays a characteristic spiral pattern, concerning mostly molecular gas and young stars. Despite several decades of study, neither the number (2 or 4) nor the

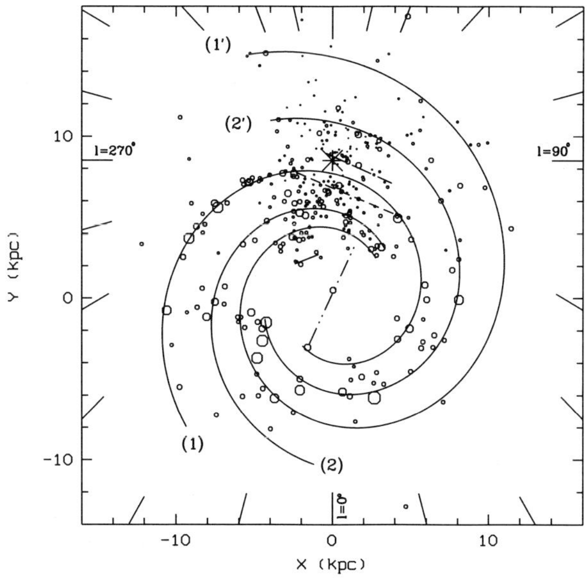


**Fig. 7.2** Spiral structure of Milky Way. The Sun position is given by the large star symbol. *1*: Sagittarius-Carina arm, *2*: Scutum-Crux arm, *1*: Norma-Cygnus arm and *2*: Perseus arm. The local arm feature (*long dashed line*), the bar (*dashed-dot-dot line*), the expected departure from the logarithmic spiral observed for the Sagittarius-Carina arm (*short dashed line*) and a feature probably linked to the 3-kpc arm (*solid line*) are sketched. From Russeil et al. (2007)

precise form of the spiral arms is well established yet. A possible description of the spiral arm pattern is provided in Fig. 7.2, which shows deviations from a logarithmic spiral. Recent work of Pohl et al. (2008) suggests a more complex picture: the Milky Way has 4 spiral arms, two of which end inside the corotation radius, while the other two start at the end point of the Galactic bar (~4 kpc), continue through corotation and branch at 7 kpc into 4 arms, which continue up to 20 kpc.

The spiral pattern of the Milky Way is of key importance for the case of distributed radioactivites produced in massive stars (like $^{26}$Al and $^{60}$Fe), since massive stars are born and die inside spiral arms. Because of their small ejection velocities (a few hundreds of km/s) and ~Million year lifetimes, radioactive nuclei decay also inside spiral arms, and it is expected that their characteristic $\gamma$-ray emission would trace the spiral pattern. Indeed, enhanced emission towards the direction of the spiral arms is detected in the case of $^{26}$Al, albeit at a low significance level (see Sect. 7.4).

The gas distribution in the Galactic bulge is of particular interest in the case of the Galactic positrons, since most of the detected $e^+$ annihiation emission originates from that region (see Sect. 7.7). According to Ferrière et al. (2007), the molecular gas tends to concentrate in the so-called central molecular zone (CMZ), a thin sheet parallel to the Galactic plane, which, on the plane of the sky, extends out to $R \sim 250$ pc at longitudes $l > 0°$ and $R \sim 150$ pc at $l < 0°$ and has a FWHM thickness ~30 pc. Outside the CMZ, the molecular gas is contained in a significantly tilted disk, extending (on the plane of the sky) out to $R \sim 1.3$ kpc on each side of the GC and having a FWHM thickness $\sim$70 pc. The spatial distribution of the atomic gas is arguably similar to that of the molecular gas, with the atomic layer being about three times thicker than the molecular layer. The ionized gas is not confined to either the CMZ or the tilted disk; it appears to fill the entire bulge and to connect with the ionized gas present in the disk.

The dramatic density and temperature contrasts between the different ISM phases as well as the supersonic random motions observed in all of them suggest a highly turbulent state, attributed to the powerful winds and the supernova explosions of massive stars. Interstellar turbulence manifests itself over a huge range of spatial scales, from $\lesssim 10^{10}$ cm up to $\gtrsim 10^{20}$ cm; throughout this range, the power spectrum of the free-electron density in the local ISM is consistent with a Kolmogorov-like power law (Armstrong et al., 1995).

## 7.2.3 Star Formation and Supernova Rates

### 7.2.3.1 Star Formation

Determination of absolute values of star formation rates (SFR, in $M_\odot$/year) constitutes one of the most challenging tasks in modern astrophysics. Each one of the tracers used is sensitive to only some part of the stellar initial mass function (IMF) and has its own advantages and drawbacks (e.g. Kennicutt, 1998a). In the case of the Milky Way, methods based on counts of various short-lived objects (with lifetimes less than a few Million year) are used; these include e.g. counts of supernova remnants (detected through their synchrotron radio-emission, resulting from electrons accelerated by their shock waves); pulsars (detected through synchrotron emission from electrons, accelerated in the pulsar magnetospheres, e.g. Lyne et al. (1985) and Case and Bhattacharya (1998)); or OB associations (McKee and Williams, 1997). Distance determinations to those objects are crucial for an accurate evaluation of the Galactic profile of their surface density. Those methods establish in fact the *relative* star formation rate across the Galactic disk, i.e. with respect to its value at some Galactocenric distance (for instance, the solar neighborhood). The derivation of an *absolute* value (i.e. the *calibration* of the SFR profile) remains a delicate enterprise.

Surface density profiles of various stellar tracers appear in Fig. 7.3. They all increase towards the inner disk and pass through a broad maximum at the distance of 4 kpc (the *molecular ring*). Those profiles can be compared to the star and gas profiles ($\Sigma_*(R)$ and $\Sigma_G(R)$, respectively in panels a and b of the same figure), through the so-called Schmidt-Kennicutt law $\left(\text{SFR} \propto \Sigma_G^n\right)$, modified Schmidt-Kennicutt law (SFR$\propto \Sigma_G^n \Omega$, with $\Omega$ being the rotation period), or Dopita-Ryder law (SFR $\propto \Sigma_G^n \Sigma_*^m$). As discussed in Boissier et al. (2003), all those empirically and/or theoritically motivated laws provide satisfactory fits to the data of the MW disk (taking into account the large uncertainties in the data) and have been extensively used in models of the chemical evolution of the Galaxy. For the calibration of the SFR profile one needs to know either the total SFR of the disk or

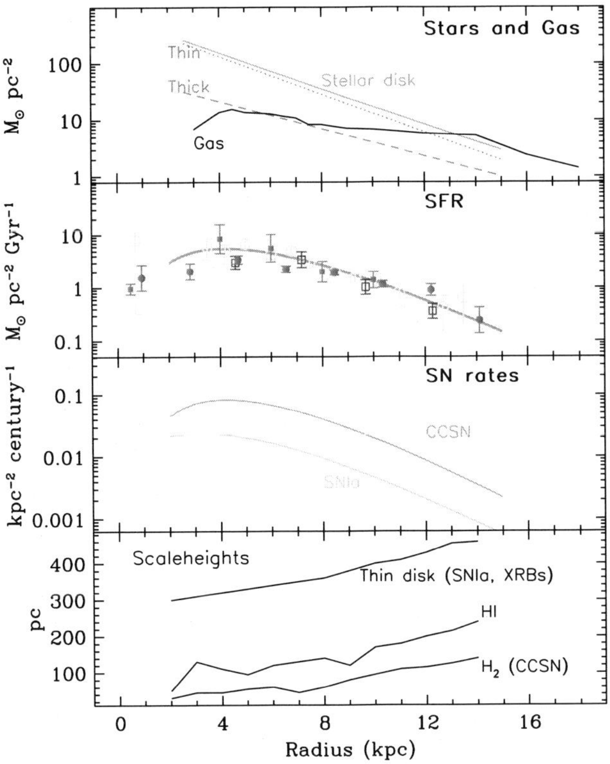


**Fig. 7.3** Surface densities of stars+gas, SFR, SN rates and scaleheights of gas and stars as a function of Galactocentric distance. Star profiles are from data of Table 7.2 and gas profile is the one of Dame (1993) (bottom pane of Fig. 7.2). Data for SFR are from : Lyne et al. (1985) (*open cicles*); Case and Bhattacharya (1998) (*filled circles*); McKee and Williams (1997) (*open squares*); Guibert et al. (1978) (*filled squares*). The *solid curve* is an approximate fit, normalized to 2 M$_\odot$/year for the whole Galaxy. The same curve is used for the CCSN rate profile (third panel), normalized to 2 CCSN/century; the SNIa rate profile is calculated by Eq. (7.1) and normalized to 0.5 SNIa/century (Table 7.4). From Prantzos et al. (2010)

the local one in the solar neighborhood. A *ball-park* estimate of the former value is obtained by noting that the late spectral type (Sbc) of the Galaxy suggests a slow formation at a relatively steady rate $\langle SFR \rangle$ over the past $\Delta T$ ~10 Gy, leading to $\langle SFR \rangle$=M$_{D,\rm thin}/\Delta T$ ~2.3 M$_\odot$/year. Most empirical estimates of the present-day total SFR, based on the aforementioned tracers, produce values within a factor of two of the $\langle SFR \rangle$ (e.g. Robitaille and Whitney, 2010, and references therein).

In the context of Galactic positrons, special attention should be paid to the star formation activity in the central regions of the bulge. The massive star population of the three major star clusters inside the Nuclear Bulge clearly indicate important recent star formation, obviously fed from the gas of the Central Molecular Zone (CMZ). Deep field observations of late-type stars with the *NICMOS/HUBBLE* (Figer et al., 2004) and with *SINFONI/VLT* (Maness et al., 2007) suggest that the star forming activity in that region has proceeded at a relatively steady rate, of the order of a few $10^{-2}$ $M_{\odot}$/year, over the past ~10 Gy.

### 7.2.3.2 Supernova Rates

From the theoretical point of view, SN are now classified mainly in *thermonuclear* supernovae (the explosion energy being due to the thermonuclear disruption of a white dwarf accreting matter in a binary system) and *core collapse* supernovae (CCSN, where the energy originates from the gravitational collapse of the iron core of a massive star having exhausted all its nuclear fuel). Thermonuclear supernovae are identified with SNIa (lacking hydrogen in their spectra) and are observed in all types of galaxies: old ellipticals with no current star formation, but also young, star forming, spiral and irregular galaxies. All other supernova types (SNII, SNIb, SNIc) are exclusively observed in the star forming regions of spirals (i.e. inside spiral arms) and irregulars. The degree of mass loss suffered by the massive star prior to the explosion determines the appearance of the core collapse supernova as SNII (little H lost), SNIb (all H and little He lost) or SNIc (all H and most He lost).

No supernovae have been observed in the Galaxy in the past four centuries, and the handfull of so called *historical supernovae* offers a very biased estimate of the Galactic SN frequency. Some methods used to determine the Galactic SN rate are based exclusively on Galactic data (e.g. counts of SN remnants, the present-day SFR, the present-day mass of $^{26}$Al) and are, in principle, able to evaluate solely the core-collapse SN rate; all those methods suffer from various systematic uncertainties (e.g. shape of the adopted massive star IMF, yields of $^{26}$Al etc.) and converge to a value of $R_{\mathrm{CCSN}}$= a few per century (Diehl et al. (2006a), and references therein).

The most accurate way to evaluate the Galactic SN rate is, probably, through statistics of SN rates in external galaxies. The work of Mannucci et al. (2005), corrected for various observational biases, offers a valuable database for such an estimate and can be used, along with the stellar masses of the various Galactic components, to derive the Galactic rate of the main SN types (Table 7.4). Note that only SNIa are (perhaps) important producers of positrons, as discussed in Sect. 7.7.3. The spatial distribution of core collapse SN in the Galaxy should obviously follow the one of the SFR (Fig. 7.3). Such an azimuthally averaged surface density masks the fact that CCSN are exclusively concentrated inside spiral arms. The scaleheight of core collapse SN should be comparable to the scaleheight of the molecular gas, i.e. less than 100 pc, and little varying with Galactocentric distance.

More difficult is the evaluation of the radial profile of SNIa, since the progenitor white dwarfs may originate from stars of a wide variety of stellar masses (1–8 $M_{\odot}$) and corresponding lifetimes (10–0.05 Gy). Various models have been developped in order to calculate the SNIa rate (e.g. Greggio, 2005). A useful empirical (and, therefore, realistic) approach is the one adopted in Scannapieco and Bildsten (2005), where the SNIa rate is calculated as the sum of two terms: one depending on the stellar mass $M_*$ and one on the SFR of the system, i.e.

**Table 7.4** Supernova rates in the Milky Way (From Prantzos et al. 2010)

| | | | SNIa | | Core collapse SN | |
|---|---|---|---|---|---|---|
| | Stellar mass[a] $10^{10}$ $M_\odot$ | Spectral type | Specific rate[b] SNuM | Rate/ century | Specific rate[b] SNuM | Rate/ century |
| Bulge | 1.4 | E0 | 0.044 | 0.062 | – | – |
| Nuclear bulge | 0.15 | Sbc/d-Irr[c] | 0.17–0.77 | 0.025–0.115 | 0.86–2.24 | 0.13–0.33 |
| Thin disk | 2.3 | Sbc | 0.17 | 0.4 | 0.86 | 2 |
| Thick disk | 0.5 | E0 | 0.044 | 0.022 | – | |
| Total bulge | 1.5 | | | 0.087–0.18 | | 0.13–0.33 |
| Total disk | 2.8 | | | 0.42 | | 2 |
| Total Milky Way | 4.3 | | | 0.5–0.6 | | 2.13–2.33 |
| Bulge/Disk ratio | $<0.5$ | | | 0.21–0.43 | | 0.06–0.15 |

SNuM: = 1 SN per $10^{10}$ $M_\odot$/century
[a] See Sect. 7.2.1 for references
[b] Mannucci et al. (2005)
[c] Very uncertain, in view of uncertainties in star formation efficiency and slope of IMF (see text).

$$\frac{R_{\mathrm{SNIa}}}{\mathrm{century}} = A\,\frac{M_*}{10^{10}\,\mathrm{M}_\odot} + B\,\frac{SFR}{\mathrm{M}_\odot/\mathrm{year}} \tag{7.1}$$

Recent studies converge on $A$ ~0.04–0.05 (Scannapieco and Bildsten, 2005; Sullivan et al., 2006), but give discrepant values for $B$. Adopting $A = 0.05$ and using the data of Table 7.4 leads to $B = 0.14$. The parametrized SNIa profile in the Milky Way disk appears in the third panel of Fig. 7.3. Taking into account the nature of the SNIa progenitors, it is expected that the distribution of SNIa vertically to the disk plane will follow the corresponding distribution of the thin disk, i.e. with a scaleheight of 300 pc (an insignificant contribution from the thick disk is also expected).

## 7.3 Processes in the Interstellar Medium: Mixing and Feedback

In this Section, we discuss the processes which occur around (and are stimulated from) sources of new isotopes, and address how radioactivities can help to enlighten their treatment towards a most realistic view. We will first describe the astrophysical issues on different scales, and then the approach to model successive enrichment of cosmic matter from nucleosynthesis. This will present the framework called *galactic chemical evolution*, which makes use of the star-formation and stellar-evolution physics discussed in the previous Section of this Chapter and also in Chap. 3 and 4, in order to analyze and study how measurements of radioactivities can advance our understanding of stars and galaxies.

### *7.3.1 Scales of Interstellar-Medium Processing*

The interstellar medium is a key mediator for the outputs of nucleosynthesis sources, i.e. ejected matter, ionizing radiation, and kinetic energy from winds and explosions. These process the interstellar medium into phases and states which determine further star formation; this is called *feedback*, and determines the evolution of normal disk galaxies.[4] Turbulence generated by stellar winds and explosions will drive how interstellar gas eventually forms stars, or ceases to form new stars, thus driving galactic evolution on a more fundamental level.[5] Feedback from nucleosynthesis sources occurs throughout a galaxy, and influences its embedded objects. Exactly how matter spreads from nucleosynthesis sites into next-generation stars will determine chemical enrichment over a galaxy's evolution (*mixing*). Major other drivers

[4] In *active galaxies* the central supermassive black hole also plays a role, and even dominates over the impact from massive stars for central regions, and for entire galaxies in late (largely-processed) evolution such as at low redshifts.

[5] In this book we only address the scales and processes within a galaxy, as it is driven by massive stars and can be traced by radioactive material; in this and further Sections of this Chapter, the broader (cosmic evolution) context is also relevant.

of galactic evolution are material inflows from extragalactic space through clouds, streams, or mergers, but also a supermassive black hole in a galaxy's center will incur major effects in a galaxy's evolution.

Chemical evolution of the universe at large involves mixing of material at different scales: The early phase of forming a star (before/until planets are being formed), stellar winds and explosions, clusters of co-evolving stars, the disks of typical galaxies, and intergalactic space. We trace matter in its different appearances as plasma (ionized atoms and their electrons), atoms and molecules, and dust particles. We briefly discuss the different spatial scales in more detail:

(a) At the smallest scale, a stellar/planetary formation site evolves from decoupling of its parental interstellar cloud (i.e. no further material exchange with nucleosynthesis events in the vicinity) until the star and its planets have settled and overcome the disk accretion phase with its asteroid collision and jet phases. This phase may have a typical duration of ~My. Issues here are how inhomogeneities in composition across the early solar nebula are smoothed out over the time scales at which chondrites, planetesimals, and planets form. (Chondrites are *early* meteorites, and the most-common meteorites falling on Earth (85%). Their name derives from the term *chondrule*, which are striking spherical inclusions in those rocks. The origin of those is related to melting events in solids of the early solar system, the nature of which is the study of *cosmochemistry* (Cowley, 1995). *Carbonaceous chondrites* are 5% of all falling meteors, and are believed to be the earliest solids we know in the solar system. Inhomogeneities may have been created from (i) the initial decoupling from a triggering event, or from (ii) energetic-particle nuclear processing in the jet-wind phase of the newly-forming star. (See Chap. 6 and, e.g. Gaidos et al., 2009, for the case of our Solar system). Radioactive dating is an important tool in such studies.

(b) The fate of the ejecta of a stellar nucleosynthesis event is of concern at the next-larger scale. Stellar winds in late evolutionary stages of stars such as the *asymptotic giant* or *Wolf Rayet* phases, and also explosive events, novae and two kinds of supernovae (according to their different evolutionary tracks) involve different envelope masses, ejection energies, and dynamics. The astronomical display of such injection of fresh nuclei into interstellar space is impressive[6] throughout the early phases of the injection event; however, no real *mixing* with ambient interstellar gas occurs yet at this phase Ejected gas expands into the lower-pressure interstellar medium, but decelerates upon collisional interaction with interstellar atoms, and collisionless interactions with the magnetized plasma. This process is an important ingredient for the acceleration of cosmic rays. Once ejecta velocities have degraded to the velocity range of interstellar gas (~100-few km/s), the actual *mixing* process can become efficient. Cooling

[6] AGB stars form colorful planetary nebulae, massive-star winds form gas structures within the *HII-regions* created by the ionizing radiation of the same stars, and thus a similarly-rich variety of colorful filamentary structure from atomic recombination lines results. Supernova remnants are the more violent version of the same processes.

processes of gas in its different phases are key processes, and also incur characteristic astronomical signatures. ($H_\alpha$ radio emission, C[II] recombination in the IR, or FIR thermal emission of dust are important examples.) Radioactive isotopes are key sources of energy for the astronomical display (supernova light curves), and sensitive tracers of the nucleosynthesis conditions of these events.

(c) Co-evolving stellar groups and clusters provide an astrophysical object in its own right. The combined action of stars, successively reaching their individual wind phases and their terminating supernovae, shape the interstellar environment so that it may vary for each nucleosynthesis event. *Giant HII regions* and *Superbubbles* are the signposts of such 10–100 pc-sized activity, which can be seen even in distant galaxies (Oey et al., 2007). The evolution of disks in galaxies is determined by the processes on this scale: Formation of stars out of Giant Molecular Clouds, as regulated by *feedback* from the massive stars, as it stimulates further star formation, or terminates it, depending on gas dynamics and the stellar population. This is currently the frontier of the studies of cosmic evolution of galaxies (Calzetti and Kennicutt, 2009). Cumulative kinetic energy injection may be sufficient to increase size and pressure in a cavity generated in the interstellar medium, such that *blow-out* may occur perpendicularly to the galactic disk, where the pressure of ambient interstellar gas is reduced with respect to the galactic disk midplane. This would then eject gas enriched with fresh nucleosynthesis product into a galaxy's halo region through a *galactic fountain*. Only the fraction of gas below galactic escape velocity would eventually return on some longer time scale ($> 10^7$–$10^8$ years), possibly as *high-velocity clouds (HVCs)*. Long-lived ($\sim$My) radioactive isotopes contribute with age dating and radioactive tracing of ejecta flows.

(d) In a normal galaxy's disk, large-scale dynamics is set by differential rotation of the disk, and by large-scale regular or stochastic turbulence as it results from star formation and incurred wind and supernova activity (see (c)): This drives the evolution of a galaxy.[7] As a characteristic time scale for rotation we may adopt the solar orbit around the Galaxy's center of $10^8$ years. Other important large-scale kinematics is given by the spiral density waves sweeping through the disk of a galaxy at a characteristic pattern speed, and by the different kinematics towards the central galaxy region with its bulge, where a bar often directs gas and stellar orbits in a more radial trajectory, yet with a bar pattern speed that will differ from Keplerian circular orbits in general. Infalling clouds of gas from the galactic halo, but also gas streams from nearby galaxies and from intergalactic space will add drivers of turbulence in a galaxy's disk at this large-scale. The mixing characteristics of the interstellar medium therefore will, in general, depend on location and history within a galaxy's evolution. Radioactive

[7] Feedback from supermassive black holes is small by comparison, but may become significant in AGN phases of galaxies and on the next-larger scale (clusters of galaxies, see (e)). Galaxy interactions and *merging* events are also important agents over cosmic times, their overall significance for cosmic evolution is a subject of many current studies.

isotopes are part of the concerted abundance measurement efforts which help to build realistic models of a galaxy's chemical evolution (see Sect. 7.3.2).

(e) On the largest scale, the above gas streams into and away from a galaxy are the mixing agents on the intergalactic scale. Galactic *fountains* thus offer alternative views on above superbubble blow-out, and this may comprise a *galactic wind* ejected from galaxies (observed e.g. in starburst galaxies, see Heckman et al., 1990). Galaxies are part of the cosmic web and appear in coherent groups (and clusters). Hot gas between galaxies in such clusters can be seen in X-ray emission, elemental abundances can be inferred from characteristic recombination lines. Gas clouds between galaxies can also be seen in characteristic absorption lines from distant quasars, constraining elemental abundances in intergalactic space. The estimated budget of atoms heavier than H and He appears incomplete (the *missing metals* issue, Pettini (see 2004, for a review)), which illustrates that mixing on these intergalactic scales is not understood.

In Appendix A we comment on each of these scales and related astrophysical findings and issues.

## *7.3.2 Modeling the Chemical Evolution of a Galaxy*

Nucleosynthesis events produce new isotopes, which are mixed with ambient gas to then end up in new generations of stars, which again lead to nucleosynthesis events (see Fig. 7.4). This cycle began from *first stars* (*Population III stars*) created from almost *metal-free primordial* gas, and has since continued to form stars until today.[8] Star formation, evolution, and nucleosynthesis all vary with changing metal content. It is the challenge of *chemical evolution models* to account for the complex and various astrophysical processes in a suitably-summarizing description to represent the known astronomical constraints. Radioactivities contribute to those constraints, complementing the archeological memory of metal-poor stars in our Galaxy and various abundance measurements of specific objects throughout the universe.

### 7.3.2.1 The Concept

*Analytical treatments* of chemical evolution have been proposed and discussed 40 years ago, in first attempts to *relate the elemental-abundance distribution and their evolution in galaxies to the activity of star formation and its history* (Clayton, 1968; Cameron and Truran, 1971; Truran and Cameron, 1971; Audouze and Tinsley, 1976; Tinsley, 1980, and many others). Later, the physical processes involved received

[8] Young stars which have been created rather recently in the cosmic history are called *Population I*. They are the only stellar population which contains massive (hence short-lived) stars observable today.

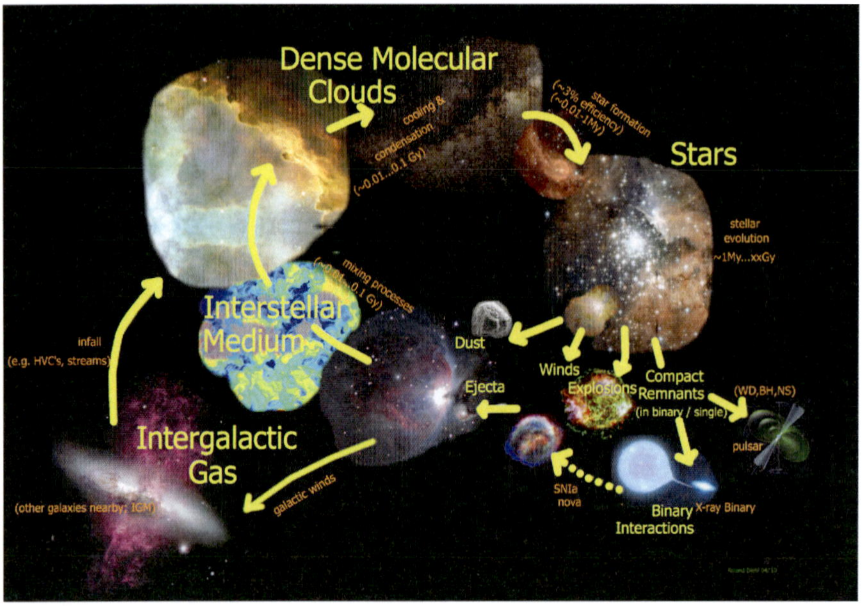


**Fig. 7.4** Illustration of the cycle of matter. Stars form from molecular clouds, and eventually return gas enriched with nucleosynthesis products into interstellar space

more sophisticated treatments, such as allowing for external gas flows, multiple and independent components of a galaxy, and more complex histories of how different stellar components inject their products (Clayton, 1988; Matteucci and Francois, 1989; Pagel, 1997; Chiappini et al., 1997; Prantzos and Silk, 1998; Boissier and Prantzos, 1999; François et al., 2004).[9] These models provide a framework, in which the rich variety of astronomical abundance data can be exploited to obtain a coherent and consistent description, and, e.g., describes the observed distribution of stars of different ages in the solar neighborhood.

In a galaxy, chemical evolution tracks the reservoirs of *gas* and *stars* over time. Key ingredients are:

(a) *Gas* is *consumed* by the process of star formation
(b) *Stars* evolve and *return gas* enriched with *metals*
(c) *Gas* (and *stars*) may be *lost* from the galaxy
(d) *Gas* (and *stars*) may be *accreted* from outside the galaxy

We trace these processes through relations among the different components. Mass conservation reads therefore:

[9] See also Appendix A for a specific and useful standard description by Clayton (1985, 1988).

$$m = m_{\text{gas}} + m_{\text{stars}} \tag{7.2}$$

which may be supplemented by *infall* and *outflow* terms.The populations of stars may usefully be subdivided into *luminous* ($l$) and *dead* ($c$ for 'compact') stars:

$$m_{\text{stars}} = m_{\text{l}} + m_{\text{c}} \tag{7.3}$$

The theories for astrophysical processes can be introduced to obtain a formalism linking different observational quantities:

1. The birth rate of stars can be introduced through theory of star formation, linking the birth rate, e.g. to the gas content of a galaxy.
2. The theory of stellar evolution allows to track the stellar population over time, as it grows from star birth, and as stars explode and/or compact stellar remnants are formed when stellar evolution reaches its end points.
3. The nucleosynthesis yields of stars in their different phases can be obtained from theories per source type, finally predicting how gas is enriched with metals over time.
4. Gas dynamics in galaxies can be modeled from theories of mixing to determine the composition of gas over time. Stars formed at later times thus inherit enriched compositions, with respect to stars formed earlier.
5. Feedback theories can be used to refine the simplistic star formation concept above, as the gas reservoir is modified. Star formation efficiency may have lower thresholds (requiring a critical mass or local density of gas), or a cut-off from a massive star enriching interstellar gas around such that it cannot form cool and dense cores as birth sites of new stars.

The observational quantities which are fed into this framework to obtain a constrained and more realistic description are:

(i) Number counts or densities of stars in intervals of metallicity. Stellar evolution theory can be used to convert this into a stellar population census versus age.
(ii) Abundances of elements or isotopes, for different locations or galaxy components (stars, gas), and for different ages (as inferred from the stellar observables).

Within the framework of the chemical evolution model, now the often uncertain parameters in the different kinds of theories listed above can be adjusted, to verify if a consistent description can be found, and appears plausible.

#### 7.3.2.2 The Role of Stars

Star formation is often prescribed linking star formation efficiency to mass density of interstellar gas, and to produce a universal spectrum of stellar masses. The *stellar birth rate* $B(m,t) = \Psi(t) \cdot \phi(m)$ links this *star formation rate* $\Psi(t)$ with the *Initial Mass Function* $\phi(m)$. An often used expression for the star formation rate is the

*Schmidt Law* (Schmidt, 1959), which states that the star formation rate is proportional to a power $N$ of the gas density:

$$\Sigma = A \cdot \sigma^N \tag{7.4}$$

It was shown that this holds for the vicinity of the Sun and nearby regions. Later, universality of this relationship was shown for a sample of different galaxies (Kennicutt, 1998b). Thus $\Sigma$ has been defined as the star formation rate in $M_\odot$ $y^{-1}$ $kpc^{-2}$, $\sigma$ being the gas mass expressed as a local surface density (integrated over the disk thickness) in $M_\odot/pc^2$. Star formation efficiency among different regions may be compared if the *specific* star formation rate is used, which normalizes the above rate by the gas mass of a region or galaxy. The steepness of the dependence of star formation on gas density (exponent $N$) has been empirically estimated as $1.4 \pm 0.15$) (Kennicutt, 1998b). Also, star formation has been inferred to occur with a lower threshold on gas density of a few $M_\odot/pc^2$. The normalization $A$ caters for different environments. Star formation characteristics for the Galaxy are discussed in Sect. 7.2.

The initial mass of a newborn star appears to follow a universal probability function with mass: $dN = \zeta(m)dm$ is the number of stars formed in a mass interval $dm$ from a probability function $\zeta(m)$; conveniently, this probability function is expressed in logarithmic mass intervals, and then called $\phi(m)$. For the high-mass end, i.e. $m > 0.5\,M_\odot$, a *Salpeter function* represents this distribution quite adequately (Kroupa, 2002). Note that this is the *initial* mass spectrum of stars in a cluster. This is different from the observed galactic-average mass spectrum. The relation between these two different mass spectra has recently been explained for the different scales of star clusters and galaxies, adopting the plausible process of most-massive cluster stars terminating star formation in a cluster, and then keeping track of stellar evolution (Weidner et al., 2010). This is an interesting way to ensure universality of the mass spectrum at birth, and also to naturally explain the observed minimum threshold for star formation. The Salpeter function reads:

$$\phi(m) \sim m^{-1.35} \tag{7.5}$$

while for lower mass stars, the relation is

$$\phi(m) \sim m^{-0.35} \tag{7.6}$$

This mass spectrum now can be combined with the star formation history and with stellar evolution as a function of initial mass $m$ to determine the *destruction* and *production* rates per isotope.

Stellar evolution depends on initial mass and metallicity, as analyzed in detail in stellar evolution theory (see Chap. 3). The time it takes a star of a given mass to evolve into late phases where significant and enriched material is returned to the interstellar gas varies between a few My for the most massive stars (approximately 100 $M_\odot$) and several Gy for solar-sized stars to enter their giant phase. Substantial

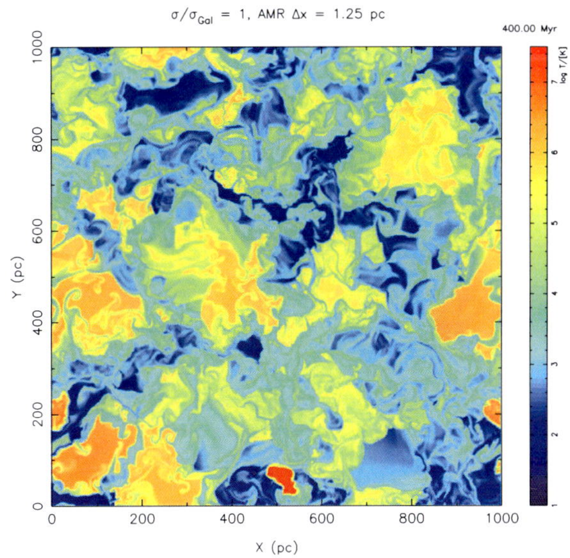


**Fig. 7.5** The interstellar medium is dynamic and evolves rapidly, driven by winds and supernova explosion. This graph shows the temperature distribution obtained from hydrodynamical simulations of 300 My of evolution in a cube of dimensions 1 kpc on each side (de Avillez and Breitschwerdt, 2004)

enrichment of interstellar gas is probably restricted to stars above masses of 3 $M_\odot$ (see Chap. 3).

Stellar evolution of single stars eventually leads to compact remnant stars (white dwarfs, neutron stars, or black holes, depending on the mass of the star), which locks up that part of stellar gas remaining at the end. Binary systems, however, open channels for re-cycling this locked-up stellar mass into the gas reservoir, through stellar explosions which disrupt the entire star in thermonuclear supernovae (SN Ia; see Chap. 5). Since the interstellar medium environment of different types of stars (low-mass stars, of massive stars, and of the delayed SNIa explosions) all will be different in density and dynamic state (turbulence), corresponding impacts are expected on mixing and formation of new stars nearby.

The *contribution of new material* into interstellar matter includes two astrophysical processes, the *mass ejection* by stars of different mass during their lifetime, and the *mixing* of ejecta with interstellar gas from which new stars may be formed. The detailed picture of how these processes occur has not been obtained; simulations such as shown in Fig. 7.5 (Breitschwerdt, 2001) illustrate typical studies. Their goals are an understanding of distribution and recycling of stellar material and of feedback from massive-star activity.

If complexities of stellar evolution and delayed ejections of newly enriched gas are to be avoided, one often assumes *instantaneous recycling*, i.e. the *stellar yield* is employed as a contribution to the gas at the time of star formation in such approximation. For contributions from massive stars this may be not too bad, but for AGB and other giant stars, or even more so for novae or thermonuclear supernovae, delays and extended ejection may be inadequately represented in such treatment. If only the complexities of enrichment of star-forming gas from previous stellar generations are to be avoided, *instantaneous mixing* is assumed as an approximation in all current formulations of chemical evolution. This may be overcome in descriptions of *dynamical* chemical evolution.

#### 7.3.2.3 The Roles of Gas and Dust

Gas and dust are spread out in interstellar space, subject to gravity from stars and gas of the galaxy components, and to energy injected by stars through radiation, winds, and explosions. The morphology of the interstellar medium is complex on scales below kpc (Fig. 7.5) – understanding of these structures and processes which drive them will be essential for our understanding of the evolution of galaxies. From the immediate surroundings of the Sun, we can map out (to some degree) clouds and hot cavities and their relation to groups of stars (Fig. 7.6), which generally confirms this picture from simulations. Star formation activity and its efficiency is regulated

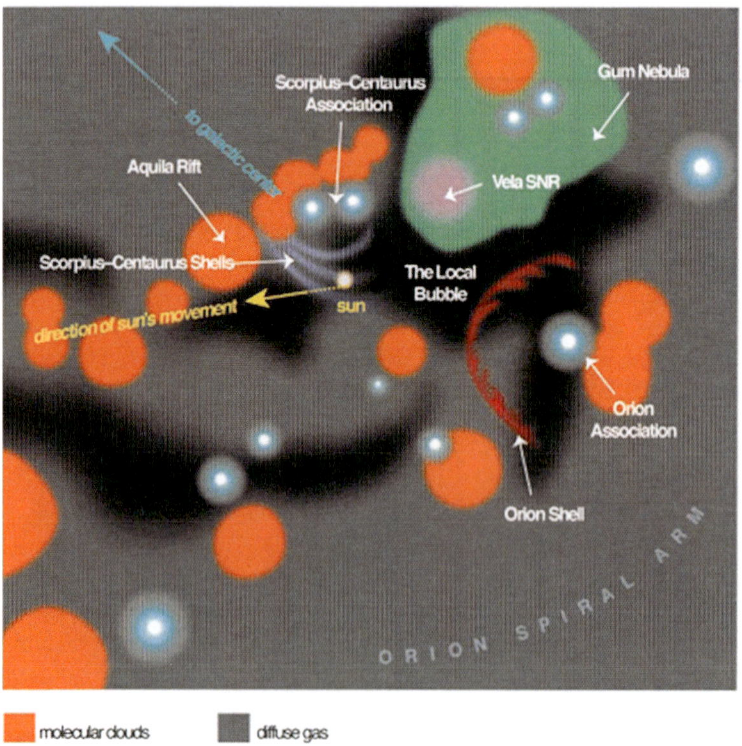


**Fig. 7.6** The interstellar medium in the solar vicinity has been mapped in more detail than can be done elsewhere. This figure has been assembled by Frisch et al. (2009), summarizing our current knowledge about prominent clouds and cavities within a few 100 kpc

by how energy is transported from the stellar population into the gas (e.g. Jappsen et al., 2005). Turbulent energy and its cascading has been understood to play a major role; the self-gravitation process as estimated through the *Jeans mass* provides crude guidance only. Star formation in Taurus is found to be faster and incompatible with self gravitation only, for example.

Dust formation is rather well modeled in AGB star envelopes (Sedlmayr and Patzer, 2004). For more massive stars, this has not been achieved; Wolf-Rayet winds are complex, clumpy, and very energetic. Exploding supernova envelopes are even more dynamic, and dust formation is largely unexplored.

Interstellar dust is modified in size and composition on its journey through interstellar space (see Jones, 2005, for a review). Once created, it rapidly grows by condensation of interstellar molecules, growing considerable ice mantles. Interstellar shocks, but also the intense radiation near massive stars, re-processes dust grains through partial or full evaporization of ice mantles, interstellar shocks enhance grain collisions and may incur sputtering of larger grains into smaller ones. Dust is a prime tracer for star forming environments, as radiation from stars heats dust to higher temperatures than the typical $\sim$10 K in normal interstellar space; thermal emission is observed and studied through infrared telescopes.

#### 7.3.2.4 Evolving Massive-Star Groups in the Galaxy

All above considerations concern a larger (representative, or averaged) region of a galaxy. But the formation of massive stars occurs in *groups*, from a parental giant molecular cloud coeval groups of hundreds to thousands of massive stars will be born. The stellar content, interstellar-gas enrichment, and dynamical state of restricted regions such as the ones resulting from the evolution of a single giant molecular complex may be different from the galactic average. It may also be more straightforward in its treatment, as approximations in theories may be more valid in this restricted context. What is expected for a group of massive stars, as they evolve from their birth through stellar evolution with terminating core-collapse supernovae?

For such regions with a more limited content of massive stars, a steady-state assumption may not apply, time histories are explicitly evaluated. As an example discussed here, the $^{26}$Al content of the interstellar medium around each group evolves with the heterogeneity of ejection given by the evolution of stars of different initial masses. Production of new stars will not continue steadily and will even terminate completely once feedback of energy from those stars will disrupt the parental cloud. The existing massive stars will eventually terminate their stellar evolution in core-collapse supernovae, and any further $^{26}$Al production in such region is left to probably less-efficient AGB stars and novae.

The $^{26}$Al production though hydrostatic and explosive burnings exhibits large variations of $^{26}$Al ejection for a massive star over its evolution (see Chap. 4 and Fig. 4.20). Initially, hydrostatic burning in the core may produce $^{26}$Al which is buried and partly decays. But as evolution towards giant phases develops strong stellar winds, the hydrogen envelope is gradually pealed off and interior burning products may be admixed into the wind in the Wolf-Rayet phase of stellar evolution. This typically sets in beyond 3 My for the most-massive stars, which then also are the first to explode as core-collapse supernovae beyond ~4 My and eject $^{26}$Al from late and explosive nuclear burning.

For a coeval group of stars, the characteristic $^{26}$Al content of surrounding interstellar medium as a function of time after birth is shown in Fig. 7.7 (from Voss et al. 2009; see also Plüschke et al., 2001b; Cerviño et al., 2000). Shading shows the 68 and 95% variances of $^{26}$Al amounts, variations derive from statistical variance due to the limited number of massive stars at each age interval. For this simulation, a group of 100 massive stars in the mass range 8–120 $M_\odot$ were traced, with their initial masses distributed according to the Salpeter mass function.

Figure 7.7 shows that the $^{26}$Al content varies by almost an order of magnitude between ~3 and 20 My after star birth. This may provide a diagnostic for groups of massive stars from their nucleosynthesis: If remaining present-day stars can be counted, the age of the entire population is constrained from $^{26}$Al gamma-ray brightness; if age is well-constrained, $^{26}$Al gamma-ray emission allows inferences on a possibly embedded population of stars which may have escaped detection. In any case, $^{26}$Al measurements provide a consistency check of our understanding of massive-star evolution and -nucleosynthesis.

With the typical time scales of massive-star evolution of about 1–10 My, the current model places origins of massive stars into dense parts of parental giant molecular clouds (see review by McKee and Ostriker, 2007; Zinnecker and Yorke, 2007). One of the most important questions in models of the star formation process is the release of energy and matter by massive stars and their effects on the surrounding clouds (Oey and Clarke, 2007). Such feedback on star formation in dense clouds can be either negative (termination of star formation through dispersal of the natal cloud) or positive (triggering of further star formation by compression of cloud material). Most massive-star clusters are believed to be dispersed due to the action of their stars within 10 My (e.g. Pfalzner, 2009). $^{26}$Al studies from specific regions

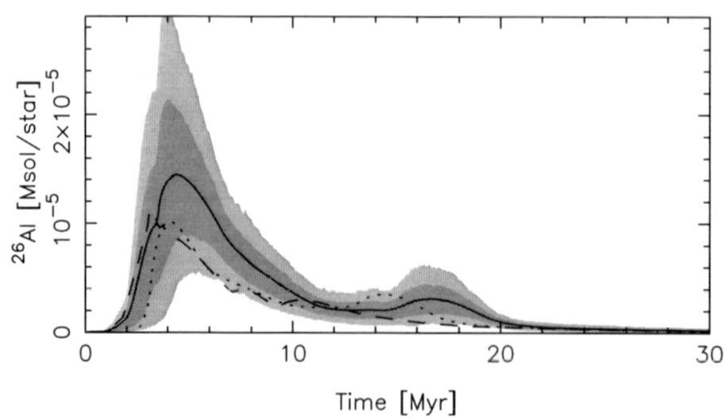


**Fig. 7.7** The $^{26}$Al content of the interstellar medium around a group of massive stars evolves with the heterogeneity of ejection given by the evolution of stars of different initial masses. This graph shows the simulation for a coeval group of massive stars born at $t = 0$ with an initial mass function of Salpeter slope (Voss et al., 2009)

hosting massive stars thus have a role in improving our understanding of the cycles of star formation and how energy and matter modify the vicinity of their birth sites, thus driving the evolution of galaxies.

## 7.4 $^{26}$Al in the Interstellar Medium

The long-lived isotopes $^{26}$Al and $^{60}$Fe are neither part of the *light-element* family (originating from primordial or cosmic-ray spallation nucleosythesis) nor of the *heavy-element* family (originating from probably explosive sites with extreme densities of matter, energy, and neutrons). $^{26}$Al and $^{60}$Fe most likely originate from stellar sources, hence from nuclear reactions inside stars with contributions from explosive nucleosynthesis. The explosions mainly help to eject stellar matter into interstellar space, but also constitute a commonly-occurring nucleosynthesis environment involved in $^{26}$Al and/or $^{60}$Fe production. But it is a plausible starting point to assume that rather normal stars are the sources of both $^{26}$Al and $^{60}$Fe (see Chap. 3, 4, 5).

$^{26}$Al is produced by proton captures on $^{25}$Mg (see e.g. Prantzos & Diehl, 1996, for a review on $^{26}$Al nuclear physics and astrophysics). Therefore, all hydrogen-burning environments are expected to process pre-existing $^{25}$Mg into $^{26}$Al. But (as shown by Prantzos & Diehl (1996) in their Fig. 3.1) other reactions will occur in those environments, and the questions controlling cosmic $^{26}$Al abundance are: Will $^{26}$Al survive in these environments? Will it find its way into interstellar space before it decays?

Candidate source regions inside stars in hydrostatic equilibrium (see Chap. 3) are the hydrogen layers (core or shell), where protons are abundant (and we may assume some $^{24,25}$Mg being present here in ~standard abundance), but also the Ne—O layers, where $^{25}$Mg nuclei are abundant from Ne-burning reactions, and protons may be produced by $(X, p)$ reactions. In the current Galaxy, the ambient-medium abundance of $^{25}$Mg may be assumed *solar*, i.e. $\log(^{25}\mathrm{Mg}) = 6.6$ (where $\log(\mathrm{H}) = 12$) (Asplund et al., 2009). For hotter environments, more abundant $^{24}$Mg may provide $^{26}$Al fuel, as it is processed by nuclear reactions into $^{25}$Mg through successive p capture reactions and intermediate $\beta$-decays. $^{24}$Mg has a solar abundance of $\log(\mathrm{Mg}) = 7.49$ (Asplund et al., 2009). Also in stellar explosions, novae (Chap. 5) and supernovae (Chap. 4), $^{26}$Al synthesis reactions may occur, as gas is heated by the explosion shock wave to temperatures where protons are released and may undergo nuclear reactions on ambient Mg nuclei. In interstellar space, spallation reactions of cosmic rays on heavier nuclei also may produce $^{26}$Al- or vice versa, heavy-nuclei cosmic rays are broken up by interactions with ambient gas to produce $^{26}$Al as a cosmic-ray component (see Chap. 10). It turned out that, from all these candidate sources, stellar sources such as Wolf-Rayet and possibly AGB stars, and core-collapse supernovae, very likely dominate above the other candidate sources in producing interstellar $^{26}$Al (see observational argument below, and theory considerations in Chaps. 3, 4 and 5).

$^{26}$Al may be destroyed within its production site, as it may undergo further proton or neutron captures, processing it into Manganese, more-massive Aluminium, and Silicon isotopes. Yet, the $^{27}$Al$(p, \alpha)^{24}$Na reaction will close a nuclear-reaction cycle, which avoids loss of the nucleus to $^{26}$Al abundance from such $^{26}$Al destruction, as long as the competing leakage reaction from that cycle $^{27}$Al$(p, \gamma)^{28}$Si proceeds at a low rate. Within such a cycle, an equilibrium abundance of $^{26}$Al would be established, balancing production and destruction reaction flows. In sufficiently-hot environments, $^{26}$Al may also be destroyed rather rapidly through thermal excitation and decay of $^{26}$Al$^m$: This energy level of $^{26}$Al at 228 keV excitation energy then becomes relevant (at $T \geq 0.1\ 10^9$ K), because $^{26}$Al decay from this excited and metastable state is rather rapid (within ~6 s) due to its 0+ spin nature.

Once transported away from the hot nuclear-reaction zone in a star, $^{26}$Al will $\beta$-decay with its 1.04 My lifetime (Fig. 1.3). The radioactive lifetime of isotopes against $\beta$-decay is a result of the nuclear structure of the isotope, specifically the angular momentum and parity of the ground state with respect to those of the potential decay product. In the case of $^{26}$Al, the ground state is a 5+ state, potential lower-lying states of the daughter isotope $^{26}$Mg are 2+ and 0+. By selection rules for transitions, the angular-momentum difference of 3 cannot be fulfilled with the positron, antineutrino, and a dipole electromagnetic transition. Therefore this decay is inhibited, leading to a lifetime of $1.042 \times 10^6$ years (or, equivalently, a half-life of 722,000 years) (from averaging over four independent measurements; the value recommended in the *Table of Isotopes* is $(1.034 \pm 0.046)$ My, Endt (1990)). In 99.76% of all decays, a gamma-ray photon of energy 1,808.63 keV ($\pm$0.07 keV uncertainty) is emitted. 82.1% of decays are $\beta^+$-decays which also emit a positron, the remainder of decays occur through capture of an electron from the atomic shell. Photon yields at 1,130 (2.4% of decays) and 2,938 keV (0.24 of decays) are weak by comparison with the 1,808.63 keV line. The photon yield from positron annihilation ($\simeq$1.6 photons of 511 keV, depending on annihilation conditions) is more indirect, as the positron's lifetime depends on gas and magnetic-field conditions and can reach the order of $10^5$ years (Jean et al., 2009).

$^{26}$Al produced in a supernova explosion will all be dispersed in the interstellar medium, as the synthesis occurs in outer regions (the O/Ne shell), far above the inner supernova core where some material may fall back onto the remnant star or black hole. $^{26}$Al produced hydrostatically in stars is ejected only in two possible ways: Either the supernova explosion occurs sufficiently soon after its production so that it still has not decayed, or else convection within the star may bring it up to the surface and a sufficiently-strong wind will carry it away from the star. Two types of star fulfill the latter conditions: In asymptotic-giant branch stars, ashes from hydrogen-shell burning are brought to the surface, the convective shell may be a *hot-bottom*-producer of $^{26}$Al (see Chap. 3). Alternatively, core hydrogen-burning ashes from the Helium shell are revealed in the strong winds of the massive-stars' Wolf-Rayet phase, in a later stage of this phase after the hydrogen envelope has been stripped (see Chap. 4).

The amount of $^{26}$Al ejected into the interstellar medium depends on many ingredients of the process from its production until release from the star. Typical values are $10^{-5}$–$10^{-4}$ M$_\odot$ (see Fig. 4.20). Several factors within stellar models play a role: Convective overturn may follow the *Schwarzschild criterion* where buoyancy of a homogeneous gas mixture is compared to restoring forces from pressure gradients (gravity and thermal), or the *Ledoux criterion* where the net effect of all restoring forces is evaluated. More complexity may be added by a chemical-composition gradient, which may add thermohaline convection or reduce convection overall; these effects are often implemented empirically through parameters for *semiconvection* (see Maeder & Meynet 2000). Stellar rotation incurs exchanges between local and global angular momenta inside a star, adding other origins of mixing processes. For ionized gas, magnetic fields also may be important in directing gas flows inside the star. The net result of these complex mixing processes finally determines the size of the convective stellar core and of the various burning shells. This, in turn, determines the amount of fuel available to nuclear fusion, as well as the transport time scales between hot regions (incurring fusion *and* destruction) and cooler regions where ashes may survive and be available for further transport and ejection.

Rotation has been shown in massive-star models to significantly enhance mixing, as a result enhancing the amounts of ejected $^{26}$Al (Palacios et al., 2005). Also, the abundant neutrinos from the collapsing stellar core may produce additional $^{26}$Al by spallating $^{26}$Mg nuclei in the Ne—O layers, the exact yield depending on the poorly known average neutrino energy (Woosley and Weaver, 1995) – this *neutrino process* that has not been included in many $^{26}$Al source models yet.

Important publications on $^{26}$Al yields from massive stars and their supernovae have been provided by Woosley and Weaver (1995), Timmes et al. (1995), Thielemann et al. (1996), Meynet et al. (1997), Palacios et al. (2005), Hirschi et al. (2005), and Limongi and Chieffi (2006). $^{26}$Al from AGB stars is addressed by Karakas and Lattanzio (2007). Nova contributions to $^{26}$Al are addressed by Jose and Hernanz (1998) and Gehrz et al. (1998).

The $^{60}$Fe isotope is another long-lived isotope, with radioactive lifetime of $3.8 \times 10^6$ years (Rugel et al. (2009); corresponding half-life is 2.62 My), also emitting gamma-rays upon decay which make it suitable for remote studies of cosmic nucleosynthesis. Gamma-ray energies are 1,173 and 1,332 keV from cascade transitions in the final $^{60}$Ni daughter nucleus of the $\beta$-decay in 99.85 and 99.98% of decays, respectively. A photon at 59 keV energy accompanies this $^{60}$Fe – $^{60}$Co – $^{60}$Ni – decay chain, from a transition in the primary daughter nucleus $^{60}$Co, but is emitted in 2% of $^{60}$Fe-decays only (most de-excitations occur through internal conversion), and thus relatively unimportant.

$^{60}$Fe is produced in the same massive-star zones as $^{26}$Al (Ne—O zone), both hydrostatically and explosively, by successive n-captures on $^{58}$Fe and $^{59}$Fe. It is also produced in the base of a massive-star's He-shell, by a less-vigorous *r process* during the explosion (Woosley and Weaver, 1995). Knowledge of the n-capture efficiency on unstable $^{59}$Fe, in addition to massive-star physics addressed in Chap. 4, make predicted interstellar amounts uncertain. But, stated in another view, measurements of $^{60}$Fe decay $\gamma$-rays from the same regions where $^{26}$Al $\gamma$-rays are seen will yield constraints on massive-star interiors from the comparisons of observed versus modeled $\gamma$-ray flux ratios.

Contrary to the case of $^{26}$Al, $^{60}$Fe is expected to be ejected *only* by the SN explosion and not by the stellar wind, since it is buried too deep (in the Ne—O shell) for ejection into interstellar space during any pre-supernova wind phase. It had been suggested, therefore, that detection of $^{60}$Fe in the Galaxy would help to decide whether WR stars or core-collapse SN are the major sources of observed $^{26}$Al. Stated more precisely, any massive-star region which is not in a steady state of production versus decay of these radioactivities from continued star formation, would show a variation of the $\gamma$-ray flux ratio from $^{26}$Al and $^{60}$Fe with age (see Voss et al., 2009, for model predictions) (see Sect. 7.5).

## *7.4.1 Observations of $^{26}$Al in the ISM*

Thirty years of gamma-ray measurements carried the celestial signal from $^{26}$Al decay through a history from *establishing proof of ongoing nucleosynthesis* in the Galaxy to *identifying $^{26}$Al origin* predominantly from massive stars into an *astronomical tool* for the study of massive-star/ISM interactions.

The first report of a gamma-ray line from a radioactive nucleus of cosmic origin arose from the HEAO-C instrument's 1979/1980 measurements, from a satellite mission carrying a Ge detector which located a line at 1,809 keV and plausibly attributed it to decay of live $^{26}$Al in the Galaxy's interstellar medium (Mahoney et al., 1984). $^{26}$Al radioactive decay within 1 My requires a source within the past million years, which is a rather brief recent time span in the Galaxy's history which spans 12 Gy or more. Therefore this is proof of currently-ongoing nucleosynthesis. The COMPTEL sky survey over 9 years then provided a sky image in this gamma-ray line, and had mapped structured $^{26}$Al emission, extended along the plane of the Galaxy (Plüschke et al., 2001a; Knödlseder et al., 1999; Oberlack et al., 1996; Diehl et al., 1995) (see Fig. 7.8), in broad agreement with earlier expectations of $^{26}$Al being produced throughout the Galaxy (Prantzos, 1993; Lingenfelter and Ramaty, 1978).

The COMPTEL detectors lacked the spectral resolution required for line identification and shape determination, with $\simeq$ 200 keV instrumental line width, compared to $\simeq$ 3 keV for Ge detectors. A balloon experiment provided therefore a surprise to the community, based on its Ge detectors: The $^{26}$Al line appeared significantly broadened to 6.4 keV, which implied a kinematic Doppler broadening of astrophysical origin of 540 km/s Naya et al. (1996). Considering the $1.04 \times 10^6$ year decay time of $^{26}$Al, such a large line width was hard to understand, as it requires either kpc-sized cavities around $^{26}$Al sources or major fractions of $^{26}$Al to be condensed on grains (Chen et al., 1997; Sturner and Naya, 1999).

The INTEGRAL satellite with its Ge-detector spectrometer SPI, launched in 2002, restored confidence in what was known about the morphology of the interstellar medium and its gas and dust compositions, through a precision measurement of the line width and the clear exclusion of significant broadening beyond a few tenths of keV, or, about 150 km/s, respectively (Diehl et al., 2003a, 2006a). With better statistics and the (modest, about 3°) imaging resolution, the SPI/INTEGRAL data even showed that the line centroid appeared to shift by a few tenths of keV

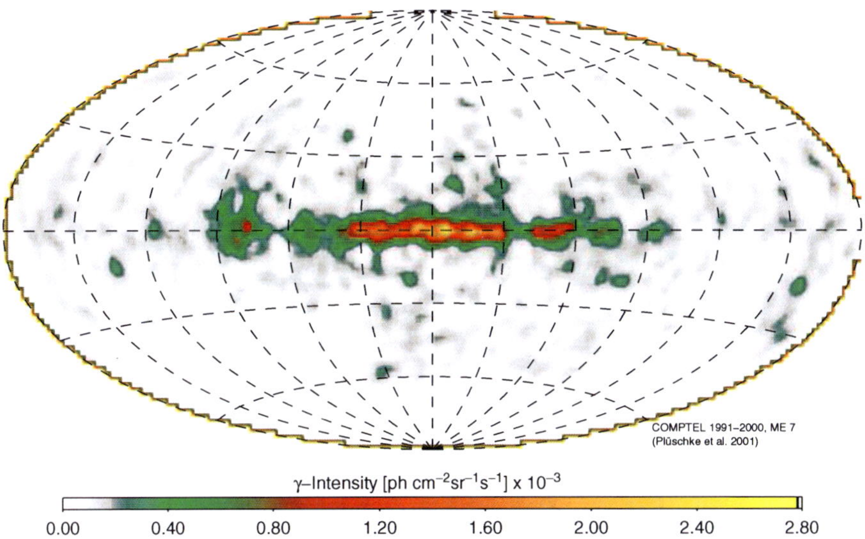


**Fig. 7.8** The $^{26}$Al sky as seen with the COMPTEL telescope. This image was obtained from measurements taken 1991–2000, and using a maximum-entropy regularization together with likelihood to iteratively fit a best image to the measured photons

with Galactic longitude as expected from the large-scale rotation of the Galaxy (see Fig. 7.11, Diehl et al. (2006a)). The INTEGRAL mission, too, was extended beyond the originally-planned 3+2 years of operation, and thus accumulated sufficient data along the plane of the Galaxy for a spatially-resolved $^{26}$Al line measurement separating different Galactic regions (Wang et al., 2009a; see Fig. 7.13). A dream of cosmic nucleosynthesis studies through radioactivity gamma-ray lines had come true: Astrophysical models of massive stars and their evolution could be confronted with observations of source aggregates in our Galaxy to learn about validity of such models. Clearly, now it would be desirable to extend this to more and different source groups and individual sources through improved instrumental sensitivity and spatial resolution, and to extend such studies to nearby galaxies, at least. But instrumental challenges are enormous (see Chap. 10), and such major advances can probably not be materialized within the next several decades (Boggs et al., 2006; Greiner et al., 2008) (see Chap. 10 and Fig. 10.4).

## 7.4.2 $^{26}$Al Throughout the Galaxy

### 7.4.2.1 Large-Scale Aspects

Observation of 1,808.63 keV gamma-rays from the decay of radioactive $^{26}$Al in the interstellar medium demonstrated that $^{26}$Al nucleosynthesis does occur in the present Galaxy. In HEAO-C data, it had been shown that an extended source

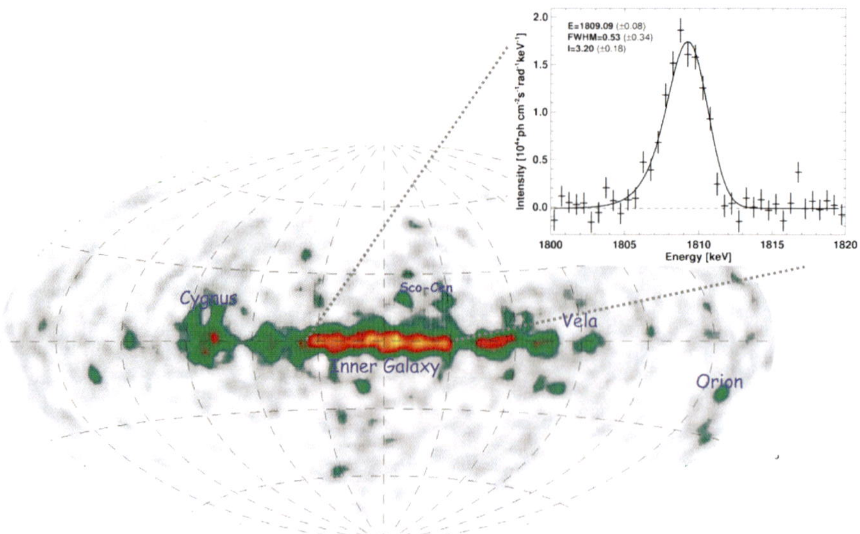


**Fig. 7.9** The $^{26}$Al line shape and intensity as seen with INTEGRAL's high-resolution spectrometer SPI (*top right*; Diehl et al. 2006), integrating over all sources from the inner Galaxy region. The COMPTEL image from Fig. 7.8 is shown for reference, with regions of special interest annotated

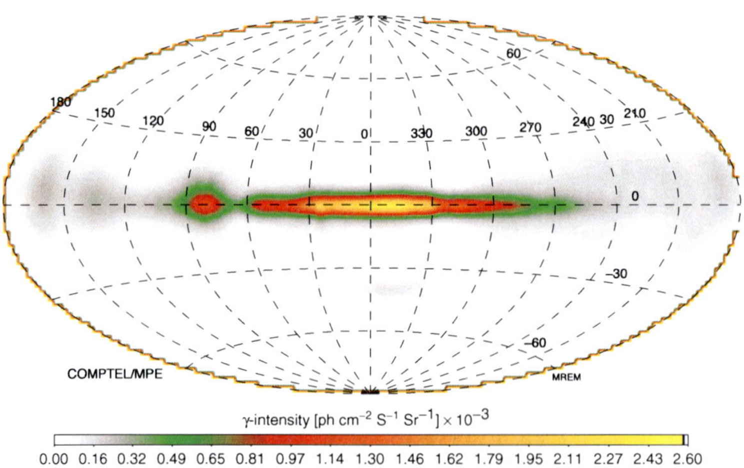


**Fig. 7.10** The $^{26}$Al sky as deconvolved from the same data as shown in Fig. 7.8, now using a multi-resolution expectation maximization based on likelihood and wavelet-domain hierarchical noise filtering

represents the observed signal better than assuming a point-like source concentration, e.g. in the Galaxy's central region. The imaging COMPTEL telescope then had provided an image of the $^{26}$Al gamma-ray sky at a spatial resolution of a few degrees (COMPTEL's point spread function (see Chap. 10) is 3.8°wide (FWHM) at an energy of 1.8 MeV). Actually, the images shown in Fig. 7.8 and 7.10 are two of many possible representations of the celestial distribution of emission, many of which would be consistent with the measured data. Facing a lack of tight constraints in image space from an instrument with a rather broad instrumental response function, the process of image extraction also must include statistical constraints, and, more importantly, prior-knowledge directions for the image. In the image shown, the two criterial are maximum likelihood (i.e., quality of the data fit to the image, which in the extreme fits all even noise features), and maximum image entropy, which prefers an image with least new information content. Alternative choices are to build up the image successively from largest to smaller scales, at each stage estimating statistical noise if the image were true already, and suppressing its effects in the iterative addition of next-finer spatial information. In this way one obtains a much smoother image (see Fig. 7.10).

Having established that there is $^{26}$Al emission throughout the disk of the Galaxy, two major astrophysical questions arise up front: (1) How much $^{26}$Al is there? (2) How does the $^{26}$Al fraction in interstellar gas vary over the Galaxy?

The somewhat irregular distribution of $^{26}$Al emission seen along the plane of the Galaxy provided a main argument for the idea that massive stars dominate the production of $^{26}$Al. Massive stars preferentially form in clusters; some of the nearby massive-star regions appear prominent in $^{26}$Al emission (e.g. in the Cygnus region; see below), while others do not. The massive-star census in the Galaxy is well known only out to distances of a few kpc from astronomical measurements of thermal emission from those stars. Large parts of the Galaxy are occulted and hence not accessible to such direct measurement, and one is left with considerable uncertainty about the Galaxy-wide distribution of massive stars. Probably, the *molecular ring* around the center of our Galaxy at a radial distance of 3–4 kpc from the center is a prominant birth site for massive stars, as are molecular clouds swept up along the Galaxy's spiral arms. A Galaxy-wide interpretation of the $^{26}$Al gamma-ray measurements would need such information, however, to resolve the distance uncertainty when assigning a measured flux along a line of sight to source intensities. The irregularity of the $^{26}$Al gamma-ray map hints at the possibility of localized efficient $^{26}$Al-producing regions. The total amount of $^{26}$Al in the Galaxy is thus rather uncertain.

Excess $^{26}$Mg found in meteorites (i.e. solid bodies from the solar system, formed some time between today and the initial cooling of the solar system's parental molecular cloud) shows that the hot disk-accretion phase of the presolar nebula apparently was characterized by an amount of radioactive $^{26}$Al (relative to the stable $^{27}$Al isotope) with a rather well-determined isotopic ratio $^{26}$Al/$^{27}$Al of $4.5 \times 10^{-5}$ (see details in Chap. 6). This is surprising, given that $^{26}$Al decays within about 1 My: The time it takes for a parental molecular cloud after decoupling from nucleosynthetically-enriched interstellar gas to form protostellar disks is much longer. Therefore, the solar system's $^{26}$Al/$^{27}$Al ratio has been interpreted as some kind of local and special in situ $^{26}$Al enrichment of the young solar nebula, such as by a nearby supernova or a AGB star event injecting fresh nucleosynthesis products at a *last moment*, or by enhanced cosmic-ray nucleosynthesis in the magnetically-active early Sun with its accretion disk (see details in Chap. 6). In that case, the solar value of $^{26}$Al content would be decoupled from the mean $^{26}$Al content of the interstellar medium in the Galaxy.

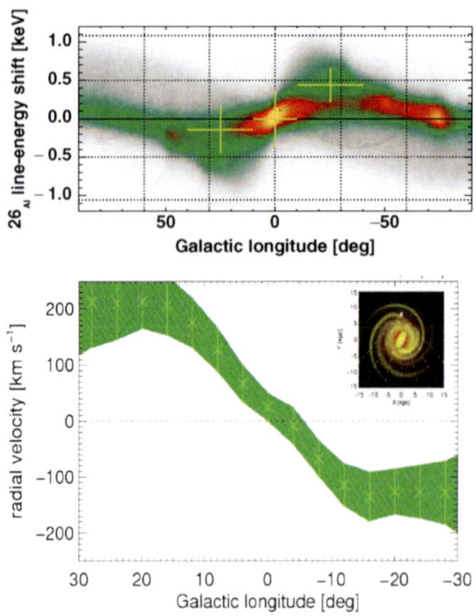


**Fig. 7.11** Kinematics of $^{26}$Al towards the inner Galaxy, from INTEGRAL/SPI measurements. *Top:* Indications of $^{26}$Al line position variation with Galactic longitude(Diehl et al., 2006a). The underlying colors show the expected *line brightness* and velocity shift versus longitude, data points are from first 1.5 years of INTEGRAL observations. *Bottom:* Longitude-velocity diagram, for hot ISM as traced through $^{26}$Al in the inner Galaxy. From 5 years of observation, the $^{26}$Al signal can be determined for narrower longitude intervals (here $\delta l = 16°, \delta b = 10°$), and shows the trend from the Galaxy's large-scale rotation. (Kretschmer et al., in prep. (2010))

There is one independent signature which can tell us about distances: The Doppler shift of spectral lines will encode the relative velocity of the source region with respect to the observer. Such information is widely used in disentangling brightness information along the observer's line of sight, for example in 21 cm mapping of atomic hydrogen thoughout the Galaxy (Hartmann and Burton, 1997). With Ge spectrometry of the $^{26}$Al gamma-ray line, an important step towards this direction was taken with the INTEGRAL mission (Fig. 7.11). These data show Doppler-shifts in the $^{26}$Al line consistent with expectation from the Galaxy's large-scale rotation. This is evidence that a substantial part or all of the observed $^{26}$Al gamma-ray brightness towards the inner Galaxy indeed originates there, rather than in specific regions more nearby. Therefore, we may regard as a rather secured first-order approximation a large-scale geometrical model of $^{26}$Al source distributions within the Galaxy, such as inferred from molecular cloud material, from free electrons, or from dust thermal emission, and normalize the projection of such a spatial distribution onto the sky with the observed $^{26}$Al gamma-ray line brightness to determine the total $^{26}$Al mass in our Galaxy. This has been done in several studies, based on COMPTEL and on INTEGRAL data. They all find an $^{26}$Al amount between 1.5 and 3 $M_\odot$, depending on data and models used. Diehl et al. (2006a) obtain (2.8 $\pm$ 0.8) $M_\odot$ of $^{26}$Al.

This $^{26}$Al mass determination based on observations sampling the *entire* Galaxy now allows a comparison with what would be expected from their candidate sources on theoretical grounds. For supernovae from core collapse of massive stars 2.1 ± 1.1 $M_\odot$ were predicted (Timmes et al., 1995); the earlier Wolf-Rayet phase was estimated to contribute a galactic total of 0.5 $M_\odot$ with a factor ~3 uncertainty (see Prantzos & Diehl, 1996, and references therein). Uncertainties are due to the incomplete census of WR stars in our Galaxy (van der Hucht, 2001) and yield uncertainties from metallicity dependencies. Recent models have assessed a Galactic contribution from WR stars of ~0.6–1.4 $M_\odot$ (Palacios et al., 2005), accounting for new insights on stellar rotation impacts and WR winds (see also Chap. 4). Models for these two candidate sources were best established, and total massive-star yields of about 2 $M_\odot$ are derived (Limongi and Chieffi, 2006). Models for novae and AGB stars were uncertain, by comparison. Estimates for novae ranged from 0.1 to 5 $M_\odot$ with large uncertainties, mainly from a lack of progenitor knowledge, and from lacking still an also quantitatively-realistic nova model. About 0.2 $M_\odot$ of Galactic $^{26}$Al may be due to classical novae, with again a factor ~3 uncertainty (see Jose and Hernanz, 1998, for a description of nova nucleosynthesis, but Chap. 5 presents an updated discussion in detail). Although AGB stars were once thought to be possibly important $^{26}$Al sources (e.g. Forestini and Charbonnel, 1997), their contribution is presently estimated to be below 1% of that of massive stars (see Chap. 3). Note that both AGB stars and novae are identified as $^{26}$Al producers from interstellar-grain inclusions in meteorites (see Clayton and Nittler, 2004, for a review, and Chaps. 3 and 5). Both these two types of sources are copious dust producers, more so than supernovae or WR stars. This illustrates that interstellar grain samples of nucleosynthesis are biased towards measuring dust-producing sources.

If we assume that massive stars are the dominating contributors, we may use their theoretical yield estimates per mass and integrate over the mass distribution function to compare with the observed $^{26}$Al amount. The normalization factor in such comparison is the Galactic supernova rate (see discussion in this Sects. 7.2 and 7.9). $^{26}$Al measurements, as interpreted within this framework of models, yields a core-collapse supernova rate of 1.9 ± 1.1 SN/century (Diehl et al., 2006a). This is consistent with other determinations, which are either extrapolated from counting trace objects in the solar vicinity and extrapolating, or inferences from other galaxies which are assumed to be similar to ours (see Appendix in Diehl et al., 2006a, for a detailed comparison and discussion of alternative methods). $^{26}$Al measurements, though limited in precision both from source model yields and the instrumental uncertainties, provide an alternative way to determine this characteristic quantity for our Galaxy, which is different in its systematics. Therefore, consistency of the core-collapse rate among methods is reassuring.

The global line width of the $^{26}$Al line will reflect Doppler broadening from the large-scale rotation within the Galaxy, but also from turbulent $^{26}$Al nuclei motions in the presumably-hot interstellar gas phase that $^{26}$Al may still reside in at the time of its decay, after its ejection from the source now typically traveling through interstellar space for $\simeq$ 1 My. Contrary to the balloon experiment based report of a

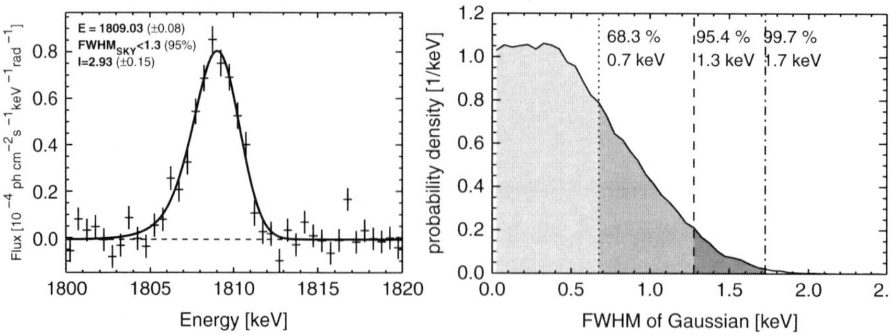


**Fig. 7.12** The $^{26}$Al line is constrained in width as rather narrow. INTEGRAL results from the Galaxy (*left*) show that any additional broadening beyond instrumental width is small (*right*: probability distribution for extra broadening) (Diehl et al., 2006b)

rather broad $^{26}$Al line with 5.4 keV broadening beyond instrumental resolution. But INTEGRAL's spectrometer confirms that the $^{26}$Al line is not significantly broadened beyond what is expected from the Galaxy's rotation and from typical ISM velocities in the 10-km/s range (Fig. 7.12).

Peculiarities of source regions in different evolutionary phases (Fig. 7.7) and localized bulk motion could also contribute to broadening of the line when integrating over the entire Galaxy as in Fig. 7.12. Therefore spatially-separated line shape measurements (see Fig. 7.13) are important to study $^{26}$Al line emission from specific massive-star groups. We now discuss a few such groups (see also Sect. 4.5 for the Vela region objects of interest).

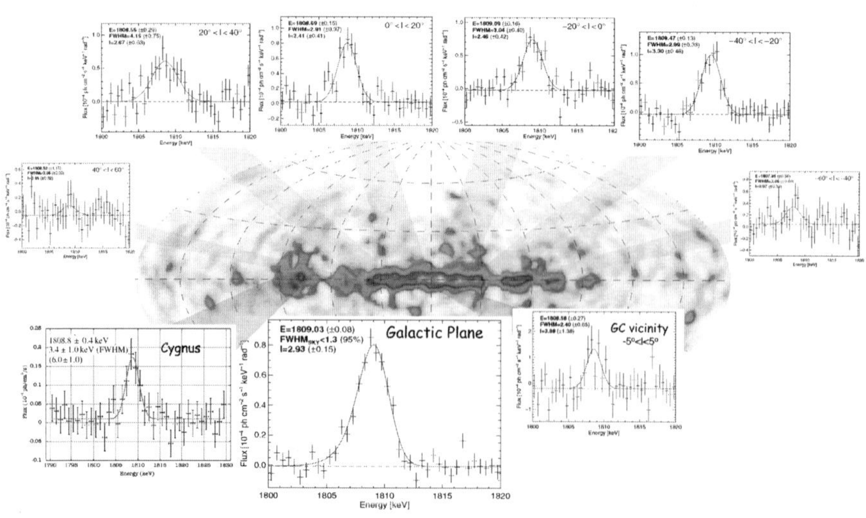


**Fig. 7.13** The $^{26}$Al line measurements from five years of INTEGRAL observations indicate that some variations of line position and width may occur for different regions along the plane of the Galaxy (Wang et al., 2009a)

#### 7.4.2.2 $^{26}$Al in the Cygnus Region

The $^{26}$Al gamma-ray map shows emission from the Cygnus region at Galactic longitudes around 30° as the clearest and most-prominent feature beyond the bright ridge in the inner Galaxy (see Figs. 7.8 and 7.10). Even considering imaging uncertainties, this emission is highly significant ($>10\sigma$ both in COMPTEL and INTEGRAL data). The INTEGRAL measurement of the $^{26}$Al line of the Cygnus region is shown in Fig. 7.13. Several studies have discussed astrophysical implications of these gamma-ray measurements (del Rio et al., 1996; Plüschke et al., 2002; Knödlseder et al., 2004; Martin et al., 2009).

Along this line of sight, there are 6 prominent OB associations at distances ranging from 0.7 to 2.5 kpc (Plüschke et al., 2002), plus about a dozen open clusters, some associated to those OB associations (Fig. 7.14). Their ages range from 2.5 to 7.5 My (Knödlseder et al., 2002). It appears that the Cygnus OB2 association dominates by far the stellar census of this *Cygnus complex*. Possibly associations Cyg OB1, 2, and 9 are related to OB2 and may originate from the same parental molecular cloud (Knödlseder et al., 2002) . Cyg OB2 may even be considered the most prominent case in our Galaxy of the extremely-rich superclusters, which appear prominent in other galaxies but are hard to recognize within our own Galaxy; about 120 stars in the high-mass range (20–120 $M_\odot$) have been identified to relate to Cyg OB2; the other associations typically are ten times smaller. The age and distance of Cyg OB2 is 2.5 My and 1.57 kpc, respectively.

Other prominent objects towards this line of sight include the Cygnus Loop supernova remnant, and a diffuse and extended structure of X-ray emitting (hot) gas called the Cygnus Superbubble. The Cygnus Loop is a young supernova remnant with an estimated age of 10,000 years, and relatively nearby at 540 pc distance (Blair et al., 2005), attributed to a $\sim$12 $M_\odot$ progenitor star. By itself, it is a

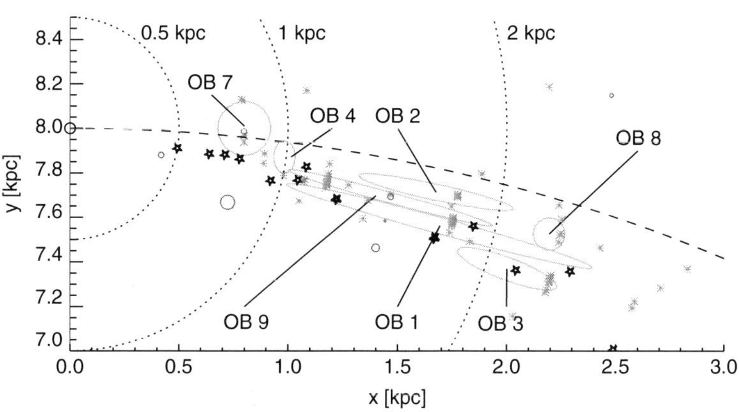


**Fig. 7.14** Top view of the Galactic plane, showing the position of the Sun and prominent objects towards Cygnus. The OB associations along the *line* of sight towards the Cygnus region cover a rather large range of distances (ellipses illustrate distance uncertainties; Plüschke et al. (2002)). The Cygnus OB2 group is by far the richest group of stars, and probably dominates $^{26}$Al production

candidate $^{26}$Al source, yet not more prominent than other massive stars if viewed at the characteristic $^{26}$Al time scale of 1 My. Its proximity and age makes it appear as a bright X-ray and radio source; its $^{26}$Al contribution appears unrelated to the Cygnus complex, however. The Cygnus Superbubble (Cash et al., 1980) is potentially more interesting for $^{26}$Al studies, as it is much more extended and thus may be related to the collective effects of many past supernovae and/or massive-star winds. It has been re-assessed from radio observations as being most likely a superposition of many smaller hot-gas filaments at a range of distances (Uyanıker et al., 2001). In the current picture of the Cygnus region, star formation has been ongoing for more than 10 My, and has led to a significant population of field and dispersed stars, in addition to the now-observed relatively-young OB associations. The large variations of visual magnitude extinction over small angular scales also indicates that the interstellar medium in the Cygnus complex is very heterogeneous and filamentary, with hot cavities bounded by dense remains of the parental molecular clouds (Lozinskaya et al., 2002; Comerón et al., 2008). Thus, one difficulty is to constrain the stellar population of a specific OB association (such as Cyg OB2) from all O stars seen towards this sightline, and to avoid inclusions of indirectly or even unrelated stars in such stellar budget (as discussed in detail by Comerón et al., 2008).

Because of this young age of the dominant source region of Cyg OB2, stellar evolution even for the most-massive stars should still not be completed, and contributions from core-collapse supernovae to $^{26}$Al production should be small or absent. Instead, Wolf-Rayet-wind ejected $^{26}$Al from hydrostatic nucleosynthesis may be assumed to dominate, currently originating from Cyg OB2 stars. In that case $^{26}$Al gamma-rays from the Cygnus region could disentangle the different $^{26}$Al production phases and regions within the same massive stars: In galactic-averaged analysis, one assumes a *steady state* situation of $^{26}$Al decay and production, such that the complete age range of stars is represented and contributes to $^{26}$Al production with its time-averaged numbers of stars per age interval and their characteristic $^{26}$Al ejection from either process (hydrostatic, or late-shell burning plus explosive; Limongi and Chieffi (see 2006, and Chap. 4 for a discussion of processes)). Here, the age of currently-ejecting massive star groups suggests that $^{26}$Al production is predominantly due to Wolf-Rayet wind ejection.

For a young and active region of massive-star action, one may plausibly assume that the interstellar medium would be peculiar and probably more dynamic than in a large-scale average. With the fine spectroscopic resolution of the INTEGRAL measurements, therefore initial hints for a broadened $^{26}$Al gamma-ray line were tantalizing. With better data, it turns out that the $^{26}$Al line seen from the Cygnus region is compatible with the laboratory energy (i.e. no bulk motion exceeding tens of km/s) and with instrumental line width (i.e. no excessive Doppler broadening beyond ~200 km/s (Martin et al., 2009). Note that $^{26}$Al ejection from Wolf Rayet winds would be with a velocity of ~1,500 km/s, decelerating as circumstellar gas would be swept up.

The total $^{26}$Al brightness towards the Cygnus direction in the longitude interval [70°, 96°] of $\sim 10^{-4}$ ph cm$^{-2}$ s$^{-1}$ shows $\sim 6 \times 10^{-5}$ ph cm$^{-2}$ s$^{-1}$ from the line of sight towards the Cygnus OB associations, and has been decomposed into a

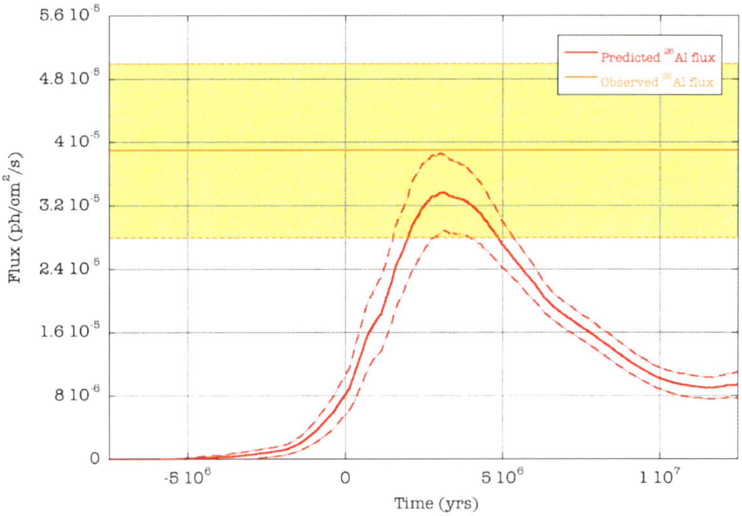


**Fig. 7.15** The time history of $^{26}$Al production in the Cygnus complex, as compared to the gamma-ray observations. Expectations from such populations synthesis are on the low side of observed $^{26}$Al gamma-rays, though consistent within uncertainties. The *shaded area* presents the range given by the $^{26}$Al gamma-ray data, the *dashed lines* bracket the uncertainty range of predictions from recent massive-star models through population synthesis. (Figure adapted from Martin et al., 2010)

contribution from the *Cygnus complex* of $\sim 3.9 \times 10^{-5}$ ph cm$^{-2}$ s$^{-1}$ (Martin et al., 2009). From the observed stars, accumulation of the expected $^{26}$Al production of massive-star groups as they evolve has always shown a problem of seeing 2–3 times more $^{26}$Al gamma-rays than predicted. While much of this discrepancy was due to the occultation of stars behind molecular clouds (Knödlseder et al., 2002), some underprediction ($\sim$25%) remains (Fig. 7.15), though now within uncertainties of data and models(Martin et al., 2010). The discrimination of the $^{26}$Al attributed to the Cygnus complex proper allows this more-specific comparison to expectations according to models (see Fig. 7.15) in this Fig. 7.15. Note that metallicity is one of the critical factors for Wolf-Rayet-phase contributions of $^{26}$Al, and hence models based on recently-reduced solar metallicity or improved values for the Cygnus region's metallicity may revise this picture in detail.

#### 7.4.2.3 $^{26}$Al in the Orion Region

The Orion region is the most-nearby region of massive stars, at a distance of $\sim$450 pc (Bally, 2008; Genzel and Stutzki, 1989). Its location towards the outer Galaxy and at Galactic latitudes around 20°is favorable, as potential confusion from other Galactic sources is negligable. The groups of massive stars, and in particular the Orion Nebula Cluster of stars, have been studied extensively, and are considered the prototype laboratory for astrophysical studies of *normal* massive-star activity. The dominating group of massive stars is the Orion OB1 association (Brown et al., 1994) with three major subgroups of different ages, one oldest subgroup *a* at 8–12 My, and two possibly coeval subgroups *b* (5–8 My) and *c* (2–6 My); subgroup *d* is the smallest and youngest at 1 My or below. Subgroup c hold most massive stars, about 45 in the mass range 4–120 $M_{\odot}$. These groups are located on the near side of the Orion A and B molecular clouds, which extend from 320 to 500 pc distance away from the Sun, and span a region of ∼120 pc perpendicular to our viewing direction.

Since at least two subgroups of the Orion OB1 association are in the age range where one expects ejection of $^{26}$Al (see Fig. 7.7), it is plausible to search for $^{26}$Al gamma-rays: At 400 pc, the output from 10 massive stars at $10^{-4}$ $M_{\odot}$ each would lead to a gamma-ray flux of $7.5 \times 10^{-5}$ph/cm$^2$/s, which should be detectable with current telescopes.

But with the COMPTEL imaging telescope, only faint hints for emission in the wider Orion region were noticed, at low surface brightness and apparently only at the level of typical noise (Fig. 7.16). Upon a closer inspection, however, a clear line at 1.8 MeV seen in spectra from those areas of the sky (typically not seen for other low-brightness regions in the map) was encouraging, and a model fit found the signal altogether being significant ($5\sigma$, Diehl (2002)).

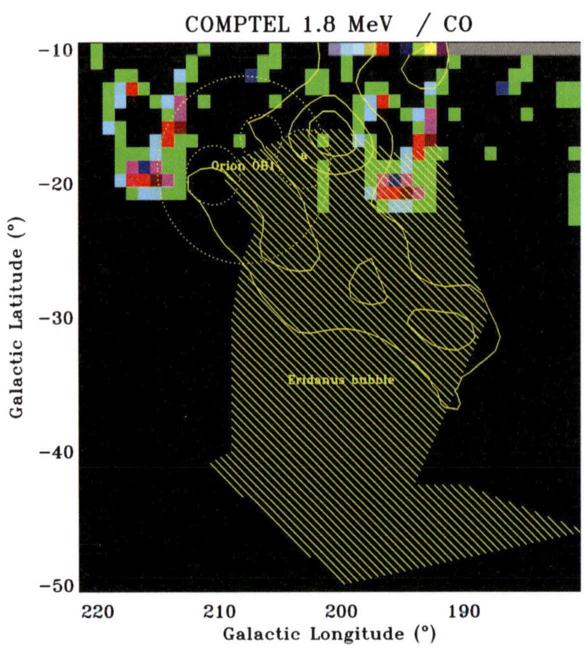


**Fig. 7.16** The $^{26}$Al distribution seen by COMPTEL towards Orion. The gamma-ray intensity map contours from COMPTEL measurements are inconsistent with a concentrated source, and, though uncertain in detail, suggest extended emission reaching out from the Orion clouds (*color pixels* in the map) and the OB1 subgroups (*circles*) towards high-latitude regions further away from the Galactic plane. The location of the interstellar cavity of Eridanus is indicated (*hatched*). From Diehl et al. (2003b)

The surprise (and reason why it was not easily seen) is that $^{26}$Al emission appears quite extended and not concentrated near the OB1 association (Fig. 7.16). Upon a closer look, this is not too surprising, however: Earlier X-ray studies and HI maps had revealed that a huge interstellar cavity was extending from the Orion molecular clouds towards the Sun, banana-shaped and extending over almost 300 pc (Burrows et al., 1993). Should the Orion OB association stars be connected and even be responsible for creating this cavity, then $^{26}$Al production from current-generation stars would find a pre-shaped cavity directing the flow of ejecta into it, rather than towards the dense remains of the parental molecular clouds on the far side of the OB association see sketch in Fig. 7.17. The oldest, and most-nearby, OB1 subgroup *a* thus may have created the Eridanus cavity on the near side of the Orion molecular clouds, triggering subsequent star formation (the later and more-distant OB1 subgroups) in the molecular cloud. The sketched scenario of Fig. 7.17 plausibly explains the offset of $^{26}$Al gamma-rays from their sources: fresh ejecta propagate further into the cavity, yet do not fill it as $^{26}$Al decays along with its flow.

Now with INTEGRAL's spectrometer interesting measurements would be the centroid (bulk motion towards the Sun?) and width ($^{26}$Al fractions moving within the hot cavity versus $^{26}$Al deposited at the cavity walls?). At the time of writing, no signal had been found; the exposure should have shown ($3\sigma$ significance) a line at the COMPTEL-suggested intensity $\sim 7 \times 10^{-5}$ ph cm$^{-2}$ s$^{-1}$ if the line were narrow. It is unclear if more Orion exposure will be obtained in the late INTEGRAL mission, as this region of the sky is largely devoid of other sources of interest to the broader INTEGRAL science community.

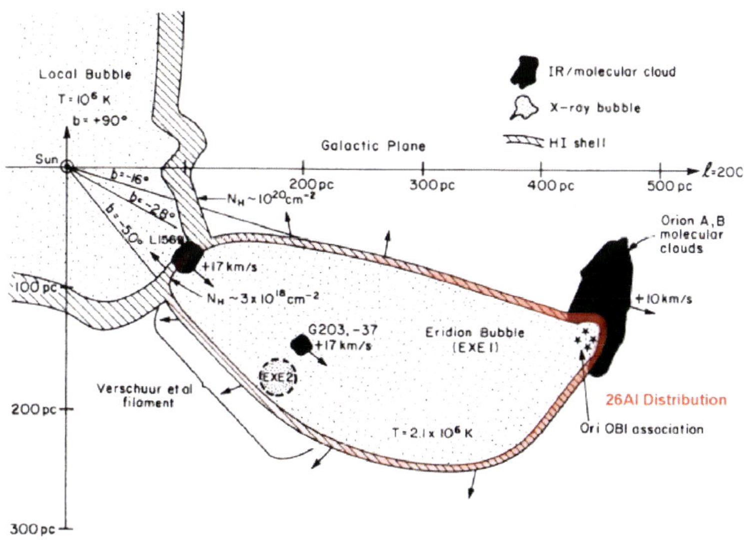


**Fig. 7.17** The $^{26}$Al distribution as it would emerge from ejecta of the Orion OB1 association, blown into the Eridanus cavity which extends from the Orion molecular clouds towards the Sun

#### 7.4.2.4 $^{26}$Al in the Sco-Cen Region

The environment of the Sun shows clear signatures of massive-star action. The interstellar medium near the Sun is characterized by a rather complex morphology (Frisch, 1995). The Local Bubble, a rather tenuous cavity, surrounds the Sun (Breitschwerdt et al., 1998), as seen in X-ray emission from hot gas in its interior, embedded cloudlets and interstellar gas appear streaming towards the local standard of rest from a preferred direction. In addition to the Eridanus cavity (Fig. 7.17), several loops/shells reminiscent of supernova remnants have been identified (de Geus, 1992). The Local Bubble is the most-nearby of such a cavity, and finding its origins is a prominent current science topic (Breitschwerdt and de Avillez, 2006; Welsh and Shelton, 2009).

The stellar association of Scorpius-Centaurus (Sco OB2) and its subgroups (Fig. 7.18, right) at a distance of about 100–150 pc (de Geus, 1992; de Zeeuw et al., 1999; Preibisch and Zinnecker, 1999) apparently plays a major role, making up a significant part of the massive stars in the solar vicinity (de Zeeuw et al., 1999). This association shows several subgroups of different ages (5, 16, and 17 My, with typical age uncertainties of 1–2 My; de Geus et al. (1989), Slesnick et al. (2008)). Its location in the sky up to 20°above the plane of the Galaxy is sufficiently-distinct from other Galactic source groups along the line of sight. The subgroup of Upper Sco has the right age for being dominant in terms of current massive-star mass ejection and supernova events (5 My, Voss et al. see also 2009). Signs of past supernova activity from this regions may, indirectly, be inferred from the morphology of the interstellar medium (Frisch, 1995), but also from estimates that the most massive star in Upper Sco presumably had $\sim 50\,M_{\odot}$ and thus may have exploded as a supernova about 1.5 Million years ago, and the pulsar PSR J1932+1059 may be its compact remnant (Hoogerwerf et al., 2000; Chatterjee et al., 2004). So, $^{26}$Al with its characteristic My time scale can help to shed light on the nature of such suggested massive-star activity from this nearby group.

Stellar subgroups of different ages would result from a star forming region within a giant molecular cloud if the environmental effects of massive-star action of a first generation of stars (specifically shocks from winds and supernovae) would interact

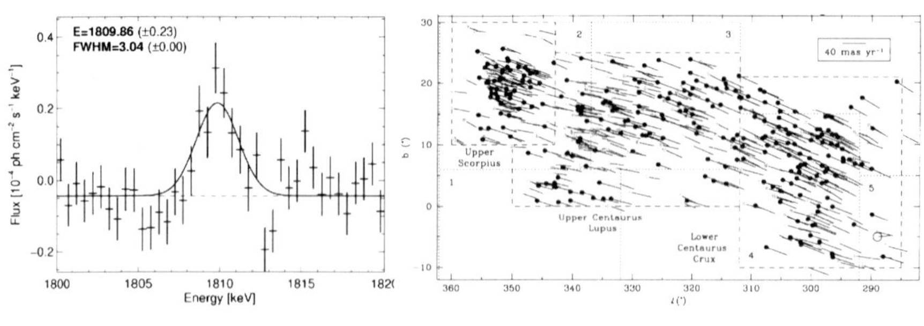


**Fig. 7.18** The $^{26}$Al signal disentangled from the Scorpius-Centaurus region with INTEGRAL (*left*; Diehl et al. 2010), and an illustration of the stars as tracked by Hipparcos in this nearby region (*right*, de Zeeuw et al., 1999)

with nearby dense interstellar medium, in a scenario of propagating or triggered star formation. Then later-generation ejecta would find the ISM pre-shaped by previous stellar generations. Such a scenario was proposed based on the different subgroups of the Scorpius-Centaurus Association (de Geus et al. 1989; Preibisch et al. 2002; Fig. 7.19) and the numerous stellar groups surrounding it (e.g. Preibisch and Mamajek, 2008; Fernández et al., 2008). Indications of recent star formation have been found in the L1688 cloud as part of the $\rho$ Oph molecular cloud, and may have been triggered by the winds and supernovae causing the $^{26}$Al we observe. The young $\rho$ Oph stars then could be interpreted as the latest signs of propagating star formation originally initiated from the oldest Sco-Cen subgroup in Upper Centaurus Lupus (Wilking et al., 2008). Many proposed scenarios of triggered star formation are only based on relatively weak evidence, such as the presence of Young Stellar Objects (YSOs) near shocks caused by massive stars. Positional evidence alone is not unequivocally considered to prove triggered star formation. Much more reliable conclusions can be drawn if the ages of the young stellar populations can be determined and compared to the moment in time at which an external shock from another star formation site arrived. Agreement of these timings would add convincing evidence for the triggered star formation scenario.

This may be the kind of activity that shaped the local interstellar medium (Frisch, 1995; Lallement et al., 2003), and specifically created the *Local Bubble*

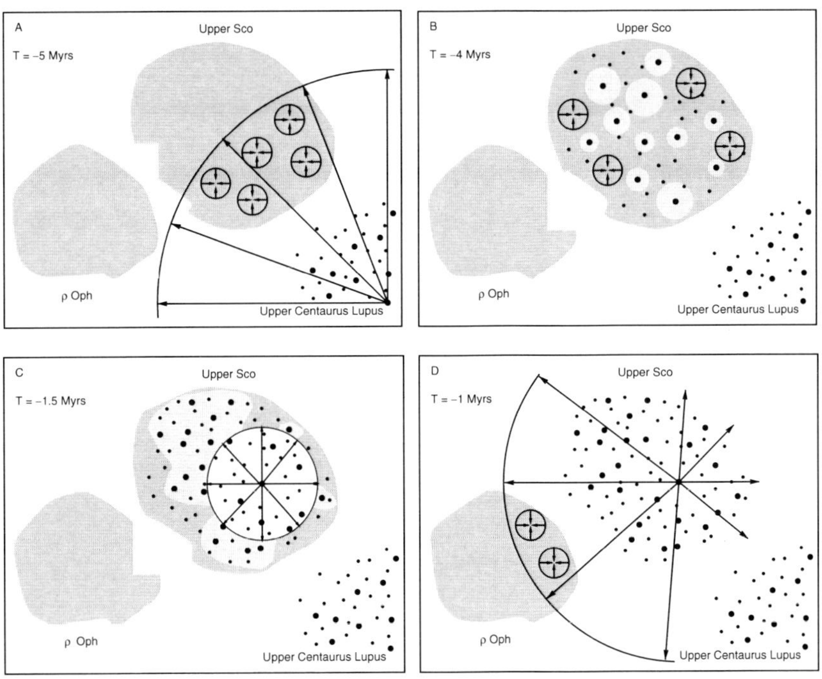


**Fig. 7.19** The objects in the Sco-Cen region, as they could have evolved in a scenario of *triggered star formation* (from Preibisch et al., 2002)

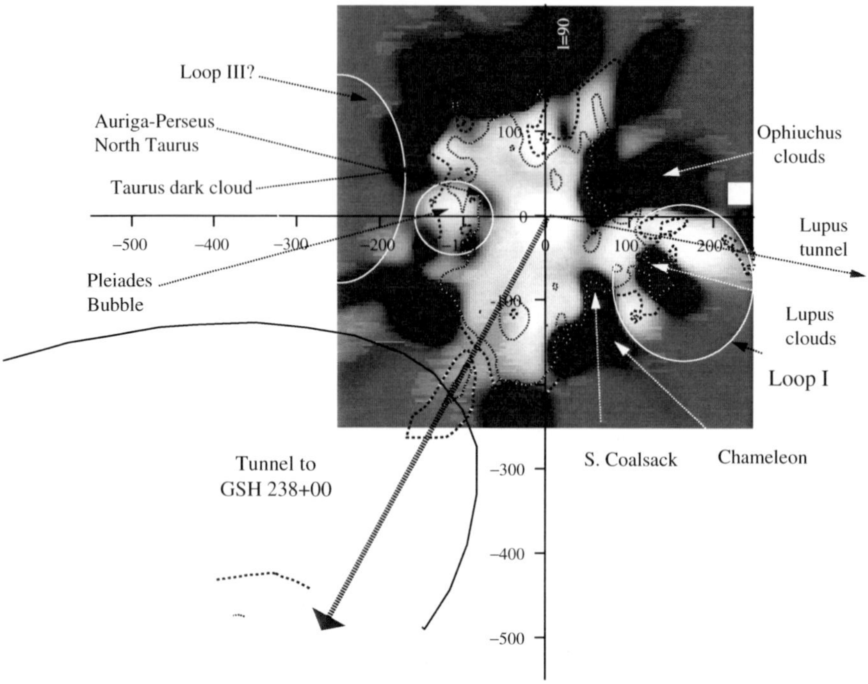


**Fig. 7.20** The local surroundings of the Sun in the plane of our Galaxy (NaI gas density, Lallement et al. (2003)). The vicinity of the Sun appears shaped by nearby energy injection, simultaneously squeezed from several supershells from more distant activity. The Sco-Cen association is located on the *righthand side* of the graph

(see Breitschwerdt et al., 1996, for a review). Also massive stars exploding very nearby will have an impact on terrestrial climate (Svensmark, 2007). Radioactive ejecta from such events may help to unravel the nearby cosmic climate (explosions, cosmic-ray intensity variations) on the time scale of My (see also $^{60}$Fe findings on Earth as discussed in Chap. 4).

Radioactive clocks such as $^{26}$Al can provide such information in this context in a most-direct way, being driven by physics which is rather independent of the environmental conditions of the dynamic ISM. The INTEGRAL signal which could be disentangled from this region of the sky (Fig. 7.18, left) provides a promising step to date the different Sco-Cen subgroups through their nucleosynthesis ejecta.

## 7.5 $^{60}$Fe and Massive-Star Structure

$^{60}$Fe radioactivity had been another candidate for nucleosynthesis investigations through cosmic gamma-ray observations. The decay chain of $^{60}$Fe (see Fig. 7.21) involves intermediate $^{60}$Co and produces two gamma-rays at 1,332 and 1,173 keV, respectively, which are well-known from laboratory work through the $^{60}$Co calibration sources. $^{60}$Fe may be produced through successive neutron capture

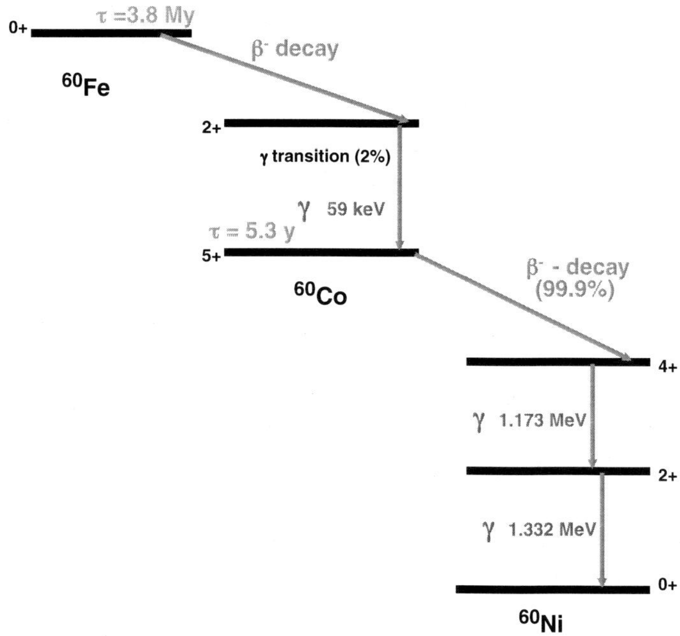


**Fig. 7.21** The $^{60}$Fe isotope decays with a radioactive lifetime of 3.8 My through $^{60}$Co to $^{60}$Ni. Note that per decay, two gamma-ray photons are obtained

reactions from the abundant and stable $^{56}$Fe isotope. Destruction may occur through radioactive decay, but also through further neutron capture reactions. $^{60}$Fe is a typical *s-process* isotope. Its cosmic sources therefore should be shell-burning massive stars, where both in the He shell and in the C shell candidate neutron source reactions are supposed to operate (Limongi and Chieffi, 2006; Timmes et al., 1995; Clayton, 1982) (see also Chaps. 3 and 4). But also in rare thermonuclear explosions, $^{60}$Fe could be abundantly produced (Woosley and Weaver, 1994).

Since massive stars are the most plausible sources of both $^{60}$Fe and $^{26}$Al, but their production occurs at quite different regions and burning episodes in those same stars, the determination of the ratio of their yields $r = Y_{60\text{Fe}}/Y_{26\text{Al}}$ should be a very sensitive global diagnostic of the validity of massive-star nucleosynthesis models (e.g. Woosley and Heger, 2007, and references therein). In steady-state approximation of current galactic nucleosynthesis (i.e. the galactic average synthesis rate of $^{26}$Al and $^{60}$Fe equals its decay rate), this production ratio and the total gamma-ray flux ratio accessible to gamma-ray telescopes are related through

$$\frac{I(^{60}\text{Fe})}{I(^{26}\text{Al})} = \frac{\dot{M}(^{60}\text{Fe})}{\dot{M}(^{26}\text{Al}} \cdot \frac{26}{60} \cdot 2 \tag{7.7}$$

where $\dot{M}(^{60}\text{Fe})$ is the total Galactic production in $M_{\odot}$/year (similarly for $^{26}$Al). The mass of each isotope maintained in steady state throughout the Galaxy then is

$$\langle M(^{60}\mathrm{Fe})\rangle = \dot{M}(^{60}\mathrm{Fe}) \cdot \tau(^{60}\mathrm{Fe}) \tag{7.8}$$

For determination of the integrated production rate, the yield in $^{60}$Fe per star of initial mass $M$ is weighted with the number of stars of mass $M$, the *initial mass function*, summing over the mass range for massive stars $[M_{\mathrm{low}}, M_{\mathrm{up}}]$. Normalization with the star formation rate cancels when the ratio for the $^{60}$Fe and $^{26}$Al isotopes is determined. This illustrates that much uncertainty related to unknowns of Galactic nucleosynthesis cancel for the isotope ratio $^{60}$Fe/$^{26}$Al, and a measurement of the production ratio for all massive stars is obtained, which only depends on nucleosynthetic yields per star of mass $M$ and the integration over the mass distribution of stars.[10]

Current models predict a gamma-ray flux ratio around 16% (Timmes et al., 1995; Woosley and Heger, 2007) or ~18 (±4)% (Limongi and Chieffi, 2006).[11]

$^{60}$Fe gamma-rays are hard to detect with current telescope sensitivities. RHESSI reported a marginal signal (2.6 $\sigma$ for the combined $^{60}$Fe lines at 1.173 and 1.332 MeV) (Smith, 2004) from the inner Galaxy, at the 10%-level of $^{26}$Al brightness; SPI aboard INTEGRAL obtains a similarly low value, around 10%, also at the 3$\sigma$-level from early analysis (Harris et al., 2005), but more convincingly the $^{60}$Fe lines were seen at 5$\sigma$ significance with 4 years of data (Wang et al., 2007)[12] (see Fig. 7.22), and reported a $^{60}$Fe to $^{26}$Al brightness ratio of 14 (±6)%. Obviously, $^{60}$Fe $\gamma$-ray intensity from the inner Galaxy remains substantially below its $^{26}$Al brightness. Though apparently observations and predictions agree, there are uncertainties in both (gamma-ray telescope data analysis; nuclear-physics issues, probably concerning the (uncertain) neutron capture reactions on unstable $^{59}$Fe; massive-star shell burnings). Therefore, it would be interesting to determine the $^{60}$Fe to $^{26}$Al ratio not only for integrated observations of the Galaxy, but more locally for specific regions of massive star nucleosynthesis.

Interesting confirmation of $^{60}$Fe being produced in Galactic nucleosynthesis in the solar neighborhood is obtained from two very different measurements: (a) Discovery of an excess $^{60}$Fe deposit trace in oceancrust material from the Pacific ocean, which is dated to about 2–3 My before today (see Knie et al., 2004, and discussion in Chaps. 4); and (b) inference of $^{60}$Fe in the early solar system, which may have even been enhanced over the interstellar value determined from current gamma-ray observations (see discussion in Chaps. 2 and 6 and Tachibana and Huss, 2003).

---

[10] The mass range commonly assumed is 8–120 M$_\odot$; but both mass limits are subject to some uncertainty (see discussion in Zinnecker and Yorke, 2007). Limongi and Chieffi (2006) discuss in detail the impact of varying the slope of the mass distribution function, and the upper mass limit for the integrated range of masses.

[11] A period of irritation occurred between 2002 and 2005, as nuclear cross sections and models were updated, and seemed to predict much higher ratios up to 1; see discussion in Woosley and Heger (2007) and Prantzos (2004).

[12] Instrumental background from $^{60}$Co produced locally in the satellite by cosmic rays may be a remaining concern. Wang et al. (2007) have accounted for such contribution within instrumental knowledge. Note that insufficient subtraction of instrumental contributions would make the Galactic $^{60}$Fe contribution even smaller.

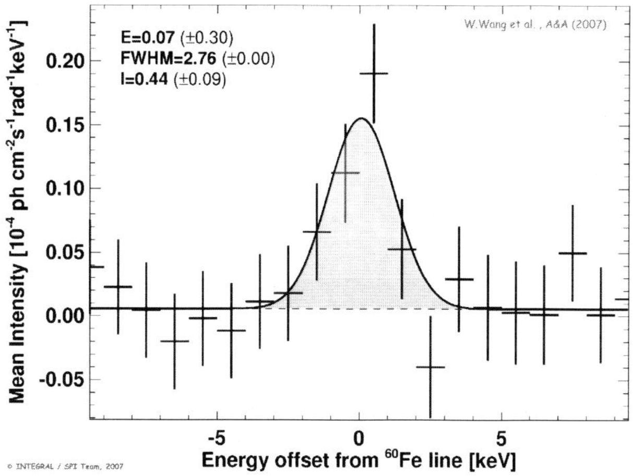


**Fig. 7.22** The $^{60}$Fe signal (both gamma-ray lines superimposed at their laboratory energies of 1,173 and 1,332 keV) as observed with INTEGRAL/SPI from the Galaxy (Wang et al., 2007)

## 7.6 $^{44}$Ti and Galactic Core-Collapse Supernovae

The rate of core-collapse supernovae in our Galaxy presumably is about one every 50 years, within a factor ~two (see Sect. 7.2, and Diehl et al., 2006a). Among core-collapse supernova nucleosynthesis products will plausibly be $^{44}$Ti (see Chap. 4), at a yield of $\sim 10^{-5}$ M$_\odot$ or thereabouts. $^{44}$Ti gamma-rays have clearly been measured from the 350-year-old Cas A supernova remnant (as reviewed by Vink, 2005; see Cas A discussion in Chap. 4). Indirectly, the late bolometric light curve of SN1987A also requires energy input of the kind which plausibly is obtained from $^{44}$Ti at an amount of even $0.5\text{–}2.0 \times 10^{-4}$ M$_\odot$ (Fransson and Kozma, 2002). In solar-system abundances, moreover, the $^{44}$Ca isotope sticks out in abundance among Ca isotopes between the most-abundant $\alpha$-nucleus $^{40}$Ca and typically low-abundant heavier Ca isotopes up to $^{48}$Ca; this is plausibly attributed to a unique additional production through the $\alpha$-multiple $^{44}$Ti nucleus and its subsequent decay.

Given apparent $^{44}$Ti production in core-collapse supernovae as observed (Cas A, SN87A), we should expect to see more, and especially younger Galactic supernova remnants in their $^{44}$Ti emission ($^{44}$Ti lifetime is 89 year). This is not so: Beyond Cas A, no other object in the Galaxy has been seen to emit $^{44}$Ti gamma-rays, despite reports about some candidates, see Dupraz et al. (1997). The Vela region candidate, which initially had been identified with SNR J0852-4642 (Iyudin et al., 1998), could not be confirmed; likewise other candidates in the Perseus region and south of the inner Galaxy. (Constraints from INTEGRAL's imaging survey are summarized by Renaud et al., 2006.) Is this small-number statistics? Or are we misled by the Cas A $^{44}$Ti data, and other types of sources are responsible for producing the $^{44}$Ca abundance seen in the solar system? (A rare subclass of SNIa had been discussed, with large $^{44}$Ti amounts per event for these *sub-Chandrasekhar* models; see Woosley and Weaver (1994); Shen et al. (2010)).

The et al. (2006) present a detailed discussion of the currently-believed source of cosmic $^{44}$Ca, $\alpha$-rich freeze-out in core-collapse supernovae. They show through simulations that the COMPTEL survey should have seen at least one or two sources of $^{44}$Ti gamma-rays in the inner Galaxy region, where massive stars are most abundant (Fig. 7.23). They simulated hypothetical galaxies with imprinted distributions of supernovae in space and in (uncertain) $^{44}$Ti yields, used then extinction models to evaluate supernova light which could have been witnessed from Earth in historical observations, and also evaluated if COMPTEL could have seen those hypothetical supernovae. Simulating a large sample of cases, they showed that statistical fluctuations are inadequate to explain the paucity of $^{44}$Ti sources along the plane of the Galaxy. $^{44}$Ti ejection must be a rather exceptional situation, so they conclude. Supported is such conclusion also by the rather high $^{44}$Ti amounts inferred for Cas A and SN1987A, in comparison to the 1D models discussed in Chap. 4. Parametric studies had supported the suspicion that $^{44}$Ti ejection would be more plausible in polar high-entropy regions, which are expected to develop in supernovae where spherical symmetry is not retained (Nagataki et al., 1998; Maeda and Nomoto, 2003). Rare asymmetric supernova explosions with rather high $^{44}$Ti ejections could then compensate for the low or absent $^{44}$Ti ejection in normal supernovae, where most of the

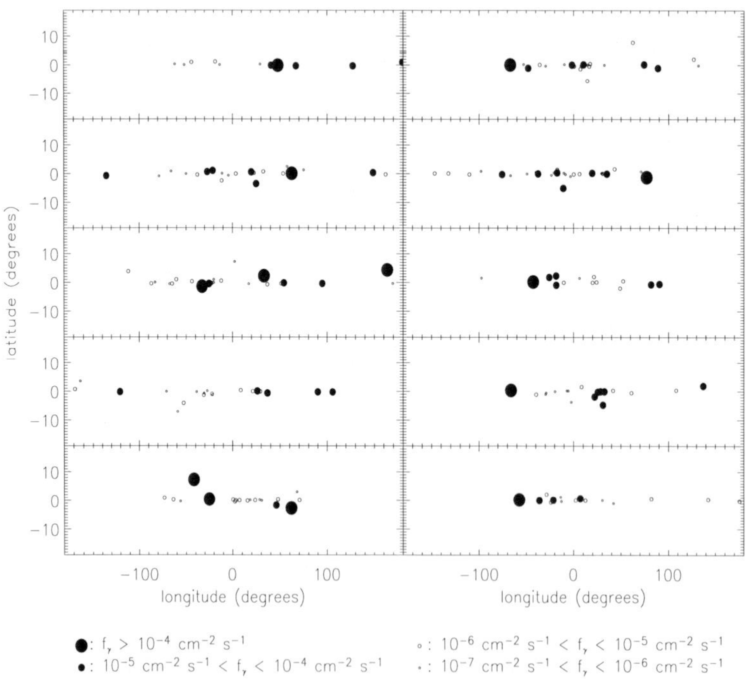


**Fig. 7.23** The $^{44}$Ti appearance of the sky, as expected from ten samples of Monte Carlo simulations (The et al., 2006). Here expectations from spatial distribution and $^{44}$Ti yields have been employed. The samples illustrate the expected statistical effects. A few candidate sources would normally be expected in *COMPTEL* data, therefore, if core-collapse supernovae all were drawn from the standard model employed here

$^{44}$Ti produced in the inner supernova region would remain burried in the compact remnant star rather than be ejected. Asymmetry of the explosion is indicated in Cas A (see discussion by Vink, 2005). This may be considered an exciting prospect, since extremely polar core-collapse explosions, the *collapsars*, are held responsible for bright gamma-ray burst explosions which can be seen to large (cosmological) distances.[13]

## 7.7 Positrons and their Annihilation

One of the most unambiguous manifestations of radioactivity is the emission of positrons from the $\beta^+$ decay of unstable nuclei located on the proton rich side of the valley of nuclear stability in the isotope chart. This is the case, e.g., for $^{26}$Al, $^{44}$Ti or $^{56}$Co, produced in massive stars and supernova explosions. If the lifetime of the unstable nucleus is larger than some timescale $\tau_T$, characterizing the *transparency* of the stellar production site to positrons, then a substantial fraction of the released positrons may escape that site and find its way into the interstellar medium. Most positrons from the decay of $^{44}$Ti ($\tau$ ~60 year) and $^{26}$Al ($\tau$ ~1 My) will obviously escape, while the fate of those released by short-lived $^{56}$Co ($\tau$ ~77 d) is not clear.

Positrons from radioactive decay are released with typical energies characteristic of the difference between nuclear energy levels, i.e. of the order of ~MeV. Once in the interstellar medium, those relativistic positrons slow down through various processes (e.g., Inverse-Compon scattering, synchrotron emission, Coulomb collsions, ionization losses) and finally annihilate. Annihilation with electrons takes place either directly, or after formation of a *positronium* atom (with the positron substituting the positively-charged atomic nucleus). In both cases, the characteristic $\gamma$-ray line at 511 keV is emitted, whereas in the latter case, a $\gamma$-ray continuum below 511 keV is also produced (see next section).

The 511 keV emission of interstellar e$^+$ annihilation was first detected from the general direction of the Galactic center in the early 1970s, by balloon borne experiments of low energy resolution (Johnson et al., 1972). It was unambiguously identified a few years later with high resolution Ge detectors (Leventhal et al., 1978). It is the first and most intense $\gamma$-ray line originating from outside the solar system that was ever detected. Its flux on Earth (~$10^{-3}$ cm$^{-2}$ s$^{-1}$), if combined with the distance to the Galactic center (~8 kpc) (assuming that annihilation occurs in the inner Galaxy), implies the annihilation of ~2 $10^{43}$ e$^+$ s$^{-1}$ (Sect. 7.7.2), releasing a power of ~$10^{37}$ erg/s or ~$10^4$ $L_\odot$ in $\gamma$-rays. Assuming a steady state, i.e. equality between production and annihilation rates of positrons, one should then seek for a source able to provide ~2 $10^{43}$ e$^+$/s. If the activity of that site were maintained at

[13] Only *long* GRBs are attributed to such collapsars. GRBs have been seen up to redshifts 8 and are thus the most-distant objects seen; understanding their nature will help to understand formation of stars at or before the epochs where galaxies formed.

such level during the $\sim 10^{10}$ year of the Galaxy's lifetime, a total amount of positrons equivalent to $\sim 3\ M_{\odot}$ would have been annihilated.

Imaging the Galaxy in annihilation $\gamma$-rays was considered to be the exclusive way to identify the cosmic $e^+$ sources (assuming that the spatial morphology of the $\gamma$-ray emission reflects the spatial distribution of the sources, i.e. that positrons annihilate close to their production sites). Because of the difficulties of imaging in the MeV region, progress was extremely slow in that field: Only in the 1990*s* the first hints on the spatial distribution of the 511 keV emission were obtained by the OSSE instrument aboard the Compton Gamma Ray Observatory (CGRO, Cheng et al., 1997; Purcell et al., 1997). The first reliable imaging of the 511 keV emission was obtained by the SPI coded-mask imaging instrument aboard ESA's INTEGRAL Gamma Ray Observatory: The emission is strongly concentrated in the inner Galaxy (the bulge, Knödlseder et al., 2005) and much weaker brightness is seen from the Galactic disk (Weidenspointner et al., 2008b). This emission morphology is strikingly-different than the ones seen in any other wavelength.

Several candidate sources of positrons were proposed over the years: radioactivity from $\beta^+$ decay of unstable nuclei produced in stellar explosions, high energy interactions occuring in cosmic rays or near compact objects (like pulsars and X-ray binaries) or the massive black hole in the Galactic center, etc. For a long time, radioactivity from $^{56}$Co produced in thermonuclear supernovae (SNIa) appeared as the most promising candidate, provided that just a few % of the positrons could escape the supernova remnant and annihilate in the interstellar medium. However, none of the candidate sources has a spatial pattern ressembling the one of the detected $\gamma$-ray emission. In particular, the release of the first year of SPI data, revealing the bulge but not yet the disk, prompted a series of *exotic* explanations involving dark matter particles, superconducting cosmic strings, etc. The confirmation of emission from the Galaxy's disk a few years later made such explanations much less likely or necessary, without completely eliminating them up to now. It is well possible, however, that rather the propagation of low energy (<MeV) positrons is the key to understanding the unusual morphology of the 511 keV emission.

In the following sections we present a brief overview of the radiative signatures of $e^+$ annihilation, the recent observations of the Galactic 511 keV emission, the pro's and con's of candidate $e^+$ sources, and the intricacies of $e^+$ propagation in the ISM.

## 7.7.1 Signatures of Positron Annihilation

The annihilation of a positron with an electron releases a total (rest-mass) energy of 1,022 keV in the form of two or more photons. Direct annihilation of a $e^-$–$e^+$ pair at rest produces two photons of 511 keV each. The situation is more complex in the case of positronium (*Ps*). Positronium has two basic states, depending on the relative orientations of the spins of the electron and the positron. The singlet state has antiparallel spins, total spin $S = 0$, is denoted as $^1S_0$ and is known as *para-positronium* (p-*Ps*). The triplet state has parallel spins, total spin $S = 1$, is denoted as $^3S_1$ and is known as *ortho-positronium* (o-*Ps*). From the $(2S + 1)$ spin degeneracy, it follows that *Ps* will be formed 1/4 of the time in the p-*Ps* state and 3/4 of the time in the o-*Ps* state. The energy difference between the two spin states (*hyperfine splitting*) is $8.4 \times 10^{-4}$ eV. Transitions between these states similar to the spin-flip transition in hydrogen, which produces the astrophysically-important 21 cm line of HI, are unimportant due to the short Ps lifetimes.

Spin and momentum conservation control the release of annihilation energy in the form of photons. Para-positronium annihilation releases two photons of 511 keV each in opposite directions (as in the case of direct $e^-$–$e^+$ annihilation). Ortho-positronium annihilation requires a final state with more than two photons from spin conservation; momentum conservation distributes the total energy of 1,022 keV among three photons producing a continuum of energies up to 511 keV (Fig. 7.24). The corresponding lifetimes before annihilation (in vacuum) are $1.2 \times 10^{-10}$ s for para-Ps and $1.4 \times 10^{-7}$ s for ortho-Ps.

If a fraction $f_{Ps}$ of the positrons annihilate via positronium formation, then the 3-photon $\gamma$-ray continuum of ortho-positronium will have an integrated intensity of

$$I_{3\gamma} \propto \frac{3}{4} 3\, f_{Ps} \tag{7.9}$$

The remaining fraction $1 - f_{Ps}$ will annihilate directly to 2 photons of 511 keV each. We should add to this the 2-photon contribution of the para-Positronium state, so that the 2-photon (511 keV line) intensity will be:

$$I_{2\gamma} \propto 2(1 - f_{Ps}) + \frac{1}{4} 2\, f_{Ps} = 2 - 1.5\, f_{Ps} \tag{7.10}$$

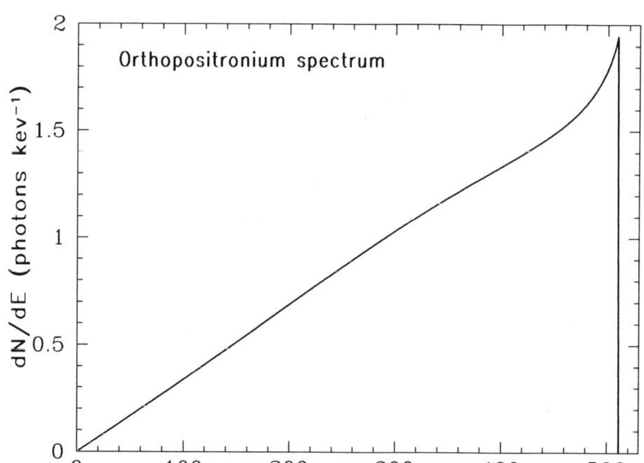


**Fig. 7.24** Spectrum of ortho-positronium annihilation with the three-photon continuum (from Ore and Powell, 1949).

By measuring the intensities of the 511 keV line and of the *Ps* continuum one can then derive the positronium fraction

$$f_{Ps} = \frac{8\, I_{3\gamma}/I_{2\gamma}}{9 + 6 I_{3\gamma}/I_{2\gamma}} \tag{7.11}$$

This quantity offers a valuable diagnostic of the physical conditions of the ISM where positrons annihilate, as these impact on the *Ps* formation efficiency.

## 7.7.2 Galactic Gamma-Ray Emission from Positron Annihilation

### 7.7.2.1 Imaging with INTEGRAL/SPI

Reliable imaging of the Galactic 511 keV emission was first obtained with ESA's INTEGRAL observatory (Winkler et al., 2003), launched in 2002. The SPI imaging spectrometer (Vedrenne et al., 2003) for the first time combines imaging with high-resolution spectroscopy. Around 0.5 MeV, SPI has a spatial resolution of 3°(FWHM) and a spectral resolution of ~2.1 keV (FWHM, at 0.5 MeV), allowing for a spatially resolved fine spectroscopy of the signal (including the underlying continuum emission).

The first 511 keV line and positronium continuum all-sky maps have been presented by Knödlseder et al. (2005) and Weidenspointner et al. (2006), respectively, based on approximately one year of SPI data. These two maps are identical within their uncertainties, which suggests that the positronium fraction does not vary over the sky. The images illustrate the remarkable predominance of the spheroidal component. OSSE data had suggested a relatively strong disk component, but the Galactic disk seemed to be completely absent in the first year SPI images. Model fitting indicated only a marginal signal from the Galactic disk, corresponding to a bulge-to-disk flux ratio $> 1$ (Knödlseder et al., 2005). This strong dominance of the Galactic bulge, unseen in any other wavelength, stimulated above-mentioned *unconventional* models involving dark matter.

After accumulating 5 years of INTEGRAL/SPI data the 511 keV line emission all-sky image revealed also fainter emission extending along the Galactic plane. With a much improved exposure with respect to the first year (in particular along the Galactic plane), 511 keV emission from the Galactic disk is now clearly detected (Weidenspointner et al. (2008b), Fig. 7.25). The detailed quantitative characterization of the different components of 511 keV emission requires parametrizing these in the form of (necessarily idealized) spatial emission models fitted to the data. No unique description emerges at present, since both the spheroid and the disk may have faint extensions contributing substantially to their total $\gamma$-ray emissivities. It turns out that the bulge emission is best described by combining a narrow and a broad Gaussian, with widths (FWHM, projected onto the sky) of 3° and 11°, respectively. Another, more extended component is needed to fit the data, a rather thick disk of vertical extent 7° (FWHM projected on the sky); but morphological modeling

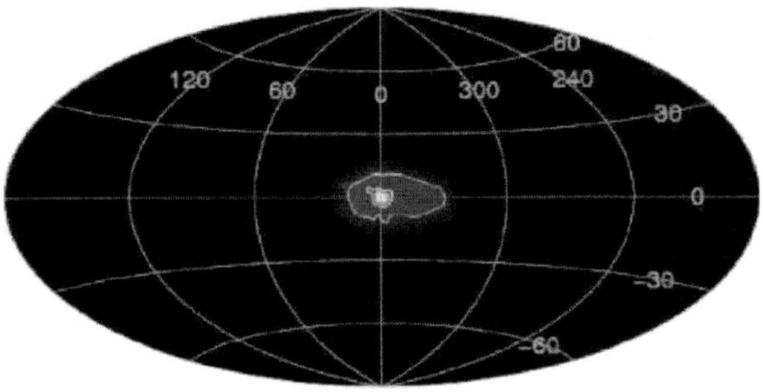


**Fig. 7.25** 511 keV line map derived from 5 years of INTEGRAL/SPI data. From Weidenspointner et al. (2008b)

**Table 7.5** Two model fits of the Galactic 511 keV emission (from Weidenspointner et al. (2008c)): fluxes, photon emissivities and $e^+$ annihilation rates (computed for a positronium fraction of $f_{ps}$=0.967, see Sect. 7.7.2.2). Notice that "thin" and "thick" disks have not the same meaning as in Sect. 7.2.1

| | $F_{511}$ ($10^{-4}$ ph cm$^{-2}$ s$^{-1}$) | $L_{511}$ ($10^{42}$/s) | $\dot{N}_{e^+}$ ($10^{42}$/s) |
|---|---|---|---|
| *Bulge + thick disk* | | | |
| Narrow bulge | $2.7^{+0.9}_{-0.4}$ | $2.3^{+0.8}_{-0.7}$ | $4.1^{+1.5}_{-1.2}$ |
| Broad bulge | $4.8^{+0.7}_{-0.4}$ | $4.1^{+0.6}_{-0.4}$ | $7.4^{+1.0}_{-0.8}$ |
| Thick disk | $9.4^{+1.8}_{-1.4}$ | $4.5^{+0.8}_{-0.7}$ | $8.1^{+1.5}_{-1.4}$ |
| Total | 17.1 | 10.9 | 19.6 |
| Bulge/Disk | 0.8 | 1.4 | 1.4 |
| *Halo + thin disk* | | | |
| Halo | $21.4^{+1.1}_{-1.2}$ | $17.4^{+0.9}_{-1.1}$ | $31.3^{+2.2}_{-2.6}$ |
| Disk | $7.3^{+2.6}_{-1.9}$ | $2.9^{+0.6}_{-0.6}$ | $5.2^{+1.1}_{-1.1}$ |
| Total | 28.7 | 20.3 | 36.5 |
| Halo/Disk | 2.9 | 6 | 6 |

gets quite vague here. This model is presently agreed as a baseline model for the Galactic 511 keV emission observed by SPI. It implies a total $e^+$ annihilation rate of $2 \times 10^{43}$ $e^+$/s and a spheroid/disk ratio of 1.4 (Table 7.5). It should be noted, however, that alternative models, involving extended components of low surface brightness (thus far undetected by SPI) are also possible. One such alternative (Weidenspointner et al., 2008c) involves a centrally condensed but very extended halo and a thinner disk (projected vertical extent of 4°), with a spheroid/disk ratio of 6 (Table 7.5).

With more SPI data being accumulated, more detailed constraints on the morphology of the disk emission can be made. So far, the flux in the disk component remains concentrated to longitudes $|l| < 50°$; no significant 511 keV line emission has yet been detected from beyond this interval. It appears, though, that currently-accumulated SPI data yield a flux from negative longitudes of the Galactic disk that is twice as large as the flux from an equivalent region at positive longitudes. The significance of this asymmetry is still rather low, about $\sim 4\sigma$. It should also be noted that a different analysis of the same SPI data finds no evidence for a disk asymmetry (Bouchet et al., 2008), although it cannot be excluded either. Several interpretational discussions of such asymmetry have been presented (e.g. Higdon et al., 2009); (Weidenspointner et al., 2008a), but are somewhat controversial. Clearly, clarifying the asymmetric or symetric nature of the disk profile should be a major aim of the 511 keV studies in the years to come.

#### 7.7.2.2 Spectroscopy with INTEGRAL/SPI

The excellent spectral resolution of SPI allowed for the first time to study the spectrum of the 511 keV emission in great detail and for different regions. The spectra of the Galactic spheroidal emission were analyzed by Churazov et al. (2005) and Jean et al. (2006), based on the first year of SPI data. The line displays no spectral shift, i.e. it has an energy $E = 511 \pm 0.08$ keV (Churazov et al., 2005) and it is composed of two spectral components (assumed, to first order, to be represented by Gaussians): a narrow line with a width of FWHM $= 1.3 \pm 0.4$ keV and a broad component with a width of FWHM $= 5.4 \pm 1.2$ keV (Fig. 7.26). The width of the broad line is in agreement with the broadening expected from positronium annihilation via charge exchange with hydrogen atoms. The narrow line component contains $\sim 2/3$ of the total annihilation line flux while the broad one makes up the remaining $\sim 1/3$ of the flux. Table 7.6 summarizes the results of the spectral analysis of the Galactic 511 keV emission after the first year of SPI data.

SPI also clearly detected the ortho-positronium continuum with an intensity that corresponds to a positronium fraction of $f_{Ps}$=97±2% (Jean et al. (2006); see Eq. (7.11)). This value is in good agreement with earlier measurements obtained by other instruments (97±3%, Kinzer et al. (1996)) and TGRS (94±4 %, Harris et al. (1998)).

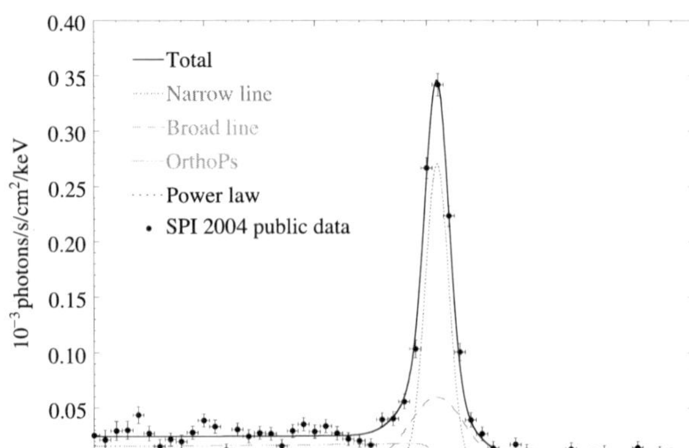


**Fig. 7.26** Fit of the spectrum of the annihilation emission measured by SPI with narrow and broad Gaussian lines and an ortho-positronium continuum. The power-law account for the Galactic diffuse continuum emission (from Jean et al. (2006))

**Table 7.6** Results of spectral analysis of Galactic 511 keV emission from the region within 8° from the Galactic center. $I_n$, $\Gamma_n$, $I_b$ and $\Gamma_b$ are the flux and width (FWHM) of the narrow and broad lines, respectively, $I_{3\gamma}$ is the flux of the ortho-positronium continuum and $A_c$ is the amplitude of the Galactic continuum at 511 keV. The first set of error bars refers to $1\sigma$ statistical errors and the second set to systematic errors (from Jean et al. (2006))

| Parameters | Measured values |
|---|---|
| $I_n$ ($10^{-3}$ ph cm$^{-2}$ s$^{-1}$) | $0.72 \pm 0.12 \pm 0.02$ |
| $\Gamma_n$ (keV) | $1.32 \pm 0.35 \pm 0.05$ |
| $I_b$ ($10^{-3}$ ph cm$^{-2}$ s$^{-1}$) | $0.35 \pm 0.11 \pm 0.02$ |
| $\Gamma_b$ (keV) | $5.36 \pm 1.22 \pm 0.06$ |
| $I_{3\gamma}$ ($10^{-3}$ ph cm$^{-2}$ s$^{-1}$) | $4.23 \pm 0.32 \pm 0.03$ |
| $A_c$ ($10^{-6}$ ph cm$^{-2}$ s$^{-1}$ keV$^{-1}$) | $7.17 \pm 0.80 \pm 0.06$ |

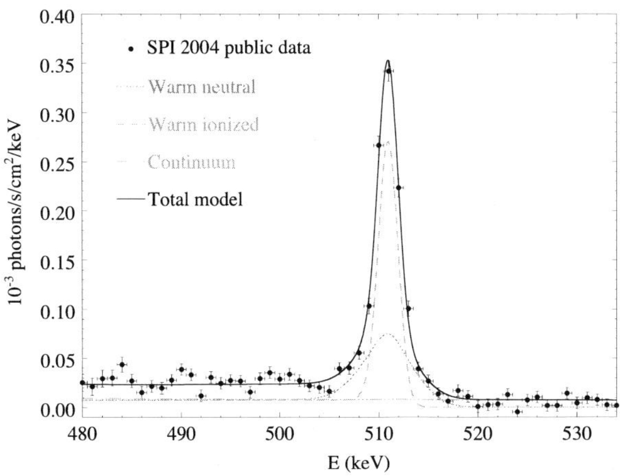


**Fig. 7.27** Best fit of the spectrum measured by SPI using the warm components of the ISM and the Galactic continuum. Contributions from the molecular, cold and hot components are not needed to explain the data (from Jean et al. (2006))

The shape of the annihilation line and the relative intensity of the ortho-positronium continuum are closely related to the physical conditions such as density, temperature and chemical abundances of the interstellar medium in which positrons annihilate. The results of the spectral analysis of the bulge emission suggest that positrons annihilate mostly in a warm medium ($T$ $\sim 10^4$ K) and a non-negligible ionized fraction: $>1\%$ according to Churazov et al. (2005), and up to 50% according to Jean et al. (2006). The latter adopted spectral models appropriate for $e^+$ annihilation in the different ISM phases of Table 7.3 and adjusted the phase fractions $f_i$ (with $i$={molecular, cold, warm neutral, warm ionized, hot}) so as to obtain the best fit to the spectral details measured by SPI (see Fig. 7.27). Since the bulge is dominated by hot gas (see Sect. 7.1), one may ask how positrons end up annihilating in sub-dominant (by volume) phases of the ISM. This immediately suggests that positron propagation may be crucial in understanding the 511 keV emission. Furthermore, the small width of the 511 keV line, along with the high positronium fraction, both suggest that positrons annihilate at very low energies ($<10$ keV).

#### 7.7.2.3 Relevant Observations at MeV Energies

The observed $\gamma$-ray emission from the decay of $^{26}$Al in the Galaxy (see Sect. 7.4) implies that $^{26}$Al provides an important contribution to the Galactic amount of positrons. The detected flux translates into a decay rate of $^{26}$Al which depends slightly on the adopted 3D distribution of $^{26}$Al in the Galaxy. The most recent analysis of SPI data results in a rate of $\dot{N}_{26}$= 4.3 $10^{42}$/s or 2.7 $M_{\odot}$/My (Wang et al., 2009b). Assuming a steady state, i.e. equality between production and decay rates, this is also the present production rate of $^{26}$Al in the Galaxy (Sect. 7.4).

Being predominantly a $\beta^+$-emitter (with a branching ratio of $f_{e^+,26}$=82%, see Fig. 1.3) $^{26}$Al is itself a source of positrons. The corresponding Galactic $e^+$ production rate is $\dot{N}_{e^+,26}= f_{e^+,26}\dot{N}_{26} \sim 3.5\ 10^{42}/s^1$. This constitutes a significant contribution to the total Galactic $e^+$ production rate (Table 7.5): 17% of the total $e^+$ annihilation rate and almost half of the (thick) disk in the baseline model, or 10% of the total and 70% of the thin disk in the Halo+thin disk model. We shall see that positrons from this and other $\beta^+$-decaying nuclei can readily explain the disk emissivity, while the bulge emissivity remains hard to explain.

Important complementary information on the energies of the annihilating positrons is obtained from the analysis of the observed continuum emission at somewhat higher energies (above 511 keV and into the MeV region) The reason is that positrons in several $e^+$ candidate sources are typically emitted at relativistic energies, in some cases even far above 1 MeV. They behave essentially like relativistic electrons of cosmic rays, producing bremsstrahlung and inverse-Compton emission while slowing down to thermal energies (eV) of the interstellar medium, where they eventually annihilate. But positrons may also annihilate *in flight* while still having relativistic energies, giving rise to a unique $\gamma$-ray continuum signature at energies above 511 keV (as the center-of-mass energy is transferred to annihilation photons, systematically increasing and broadening their typical energies in observer's frame). The shape and amplitude of this $\gamma$-ray emission depend on the injection spectrum of positrons and the corresponding total *in-flight* annihilation rate. For positrons injected at low energies (of the order of ~MeV, such as those released by radioactivity), the amplitude of the in-flight annihilation continuum above 1 MeV is quite small, while for sources injecting positrons at much higher energy (such as cosmic-ray positrons from pion decay), the annihilation $\gamma$-ray spectrum would extend up to GeV energies and include a considerable $\gamma$-ray flux. The high energy $\gamma$-ray continuum above 1 MeV therefore constrains the energy and the annihilation rate of relativistic positrons, when all other sources of such high energy emission are properly accounted for.

Diffuse Galactic continuum emission has been well-measured at least in the inner part of the Galactic disk (longitudes $-30° < l < 30°$) in the hard-X-ray through $\gamma$-ray regime by INTEGRAL, OSSE, COMPTEL, and EGRET Bouchet et al. (2008). It is is mostly due to various interactions of cosmic rays propagating with interstellar gas, which also produce secondary positrons (Strong et al., 2007). In view of the uncertainties still affecting the propagation parameters, soe (albeit very little) room is still left for a contribution of in-flight $e^+$ annihilation to the MeV continuum.

The constraints to the injection energy of positrons have been pointed out a long time ago by Agaronyan and Atoyan (1981). They showed that the positrons which are responsible for the Galactic 511 keV line cannot be produced in a steady state by the decay of the $\pi^+$ created in proton – proton collisions (between cosmic rays and the ISM) or else the in-flight annihilation emission should have been detected. A similar argument was used by Beacom and Yüksel (2006) and Sizun et al. (2006) to constrain the mass of the candidate dark matter particle which could be the source of positrons in the Galactic spheroid. If such particles produce positrons (in their decay or annihilation) at a rate which corresponds to the observed 511 keV emission, then their mass should be less than a few MeV, otherwise the kinetic energy of the created positrons would have been sufficiently high to produce a measurable $\gamma$-ray continuum emission in the 1–30 MeV range (Fig. 7.28). The same argument allows one to constrain the initial kinetic energy of positrons and thus to eliminate several classes of candidate sources, like e.g. pulsars, ms pulsars, magnetars, cosmic rays etc., as major positron producers, due to their high positron injection energies.

### *7.7.3 Candidate Positron Sources: A Critical Assessment*

The physical processes and candidate sources of $e^+$ production in the Milky Way have been reviewed recently in Prantzos et al. (2010), and we summarize them here. Positrons can be produced either by radioactivity in stars, novae and supernovae, or through various high energy processes in compact objects (production of positrons by cosmic rays, which are accelerated by supernovae, belongs to the latter class); processes involving dark matter (DM) annihilation, decay or de-excitation is a third possibility.

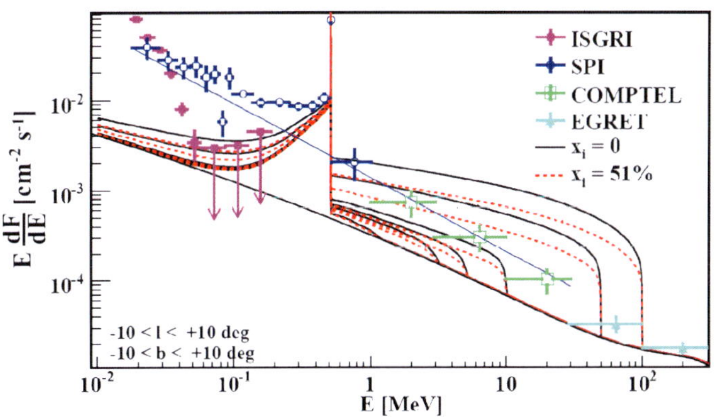


**Fig. 7.28** Spectrum of the inner Galaxy as measured by various instruments, compared to various theoretical estimates made under the assumption that positrons are injected at high energy: the four pairs of *curves* result from positrons injected at 100, 30, 10 and 3 MeV (from *top* to *bottom*) and correspond to positrons propagating in neutral (*solid*) or 50% ionized (*dotted*) media (from Sizun et al. (2006)). This constrains the injected positron energy (or, equivalently, the mass of decaying/annihilating dark matter particles) to a few MeV

Among the various astrophysical sources of positrons proposed so far, the only one known with certainty to release $e^+$ in the ISM is $\beta^+$ radioactivity of $^{26}$Al; the observed intensity of its characteristic 1.8 MeV emission in the Galaxy corresponds to ~3–4 $10^{42}$ $e^+$/s. A similar amount is expected from the decay of $^{44}$Ti, on the grounds of nucleosynthesis arguments (it is the parent nucleus of stable $^{44}$Ca). Both radionuclides are produced mostly in massive stars and their positrons should be released along the Galactic plane, as traced by the 1.8 MeV emission; they could thus account for the observed disk 511 keV emission.

Radioactivity of $^{56}$Co from SNIa was traditionally considered to be the major $e^+$ producer in the Galaxy. Both the typical $^{56}$Ni yield of a SNIa and the Galactic SNIa rate are rather well constrained, resulting in 5 $10^{44}$ $e^+$/s produced *inside* SNIa. If only $f_{esc}$ ~4% of them escape the supernova to annihilate in the ISM, the observed total $e^+$ annihilation rate can be readily explained. However, observations of two SNIa, interpreted in the framework of 1-D (stratified) models, suggest that the positron escape fraction is negligible *at late times*. On the other hand, both observations of early spectra and 3-D models of SNIa suggest that a sizeable fraction of $^{56}$Ni is found at high velocity (close to the surface), making – perhaps – easier the escape of $^{56}$Co positrons. SNIa remain a serious candidate, with a potential Galactic yield of $2 \times 10^{43}$ $e^+$/s. But the expected spatial distribution of SNIa in the Galaxy corresponds to a much smaller bulge-to-disk ratio than the one of the observed 511 keV profile.

There are several candidate sources producing positrons through high-energy processes: pulsars, millisecond pulsars, magnetars, microquasars, Low mass X-ray binaries (LMXRBs), and the Central massive black hole in the Milky Way. Among those sources, the most promising appear to be LMXRBs (or the microquasar variant of that class of sources) and the massive black hole at the Galactic center. There is no evidence that either of those sources produces positrons, however, and the $e^+$ yields evaluated by various authors are often (intentionally) optimistic rather than typical values. Furthermore, because of the current low activity of the central SMBH (much lower than the one of LMXRBs) it has to be assumed that the source was much more active in the past, thus dropping the assumption of *steady state* between $e^+$ production and annihilation, which is adopted in all other cases.

Each one of the candidate positron sources can be discussed in the light of the observational constraints, using three main criteria: (i) the total $e^+$ annihilation rate ($\sim 2 \times 10^{43}$/s), (ii) the typical energy of the injected positrons, or the equivalent mass of annihilation DM particles (<3–7 MeV) and (perhaps, most significantly) (iii) the morphology of the 511 keV emission (with a bulge/disk ratio B/D>1 in the case of a thin disk emission) (see Table 7.7). A fourth constraint, the longitudinally asymmetric disk emission, could be added to this list, once robustly established.

Assuming a steady state regime, the $e^+$ annihilation rate has to be equal to the *average* $e^+$ *production rate during the lifetime of* $e^+$ *in the ISM*. The only source definitely known to provide substantial amounts of $e^+$ at a well constrained rate is the radioactive decay of $^{26}$Al: 0.4 $10^{43}$ $e^+$/s. The decay of $^{44}$Ti probably provides another 0.3 $10^{43} e^+ s^{-1}$. GCRs probably provide 0.1 $10^{43}$ $e^+$ $s^{-1}$. Nova models

**Table 7.7** Properties of candidate positron sources in the Milky Way (From Prantzos et al. (2010)

| Source | Process | $E(e^+)$[a] (MeV) | $e^+$ rate[b] $\dot{N}_{e^+}$($10^{43}$/s) | Bulge/ Disk[c] $B/D$ | Comments |
|---|---|---|---|---|---|
| Massive stars: $^{26}$Al | $\beta^+$ | ~1 | 0.4 | <0.2 | $\dot{N}$, $B/D$: observationally inferred |
| Supernovae: $^{44}$Ti | $\beta^+$ | ~1 | 0.3 | <0.2 | $\dot{N}$: robust estimate |
| SNIa: $^{56}$Ni | $\beta^+$ | ~1 | *2* | <0.5 | Assuming $f_{e^+,esc}$=0.04 |
| Novae | $\beta^+$ | ~1 | *0.02* | <0.5 | Insuffiicent $e^+$ production |
| Hypern./GRB: $^{56}$Ni | $\beta^+$ | ~1 | ? | <0.2 | Improbable in inner MW |
| Cosmic rays | p-p | ~30 | 0.1 | <0.2 | Too high $e^+$ energy |
| LMXRBs | $\gamma-\gamma$ | ~1 | *2* | <0.5 | Assuming $L_{e^+}$ ~0.01 $L_{obs,X}$ |
| Microquasars | $\gamma-\gamma$ | ~ 1 | *1* | <0.5 | $e^+$ load of jets uncertain |
| Pulsars | $\gamma-\gamma$ | >30 | *0.5* | <0.2 | Too high $e^+$ energy |
| ms pulsars | $\gamma-\gamma$ | >30 | *0.15* | <0.5 | Too high $e^+$ energy |
| Magnetars | $\gamma-\gamma$ | >30 | *0.16* | <0.2 | Too high $e^+$ energy |
| Central black hole | p-p | High | ? | | |
| | $\gamma-\gamma$ | 1 | ? | | Requires $e^+$ diffusion to ~1 kpc |
| Dark matter | Annih. | 1 (?) | ? | | Light scalar particle, cuspy DM profile |
| | Deexcit. | 1 | ? | | Only cuspy DM profiles allowed |
| | Decay | 1 | ? | | Ruled out for all DM profiles |
| Observational constraints | | <7 | 2 | 1.4 | |

[a] Typical values are given.
[b] $e^+$ rates: in roman: observationally deduced or reasonable estimates; in italic: speculative (and rather closer to upper limits). ?: positron production rate unknown.
[c] Sources are simply classified as belonging to either young ($B/D$ <0.2) or old (<0.5) stellar populations.

(as constrained against several observables such as ejecta abundances, velocities etc.) may provide a $e^+$ yield from $^{22}$Na decay not be much below the reported value of $10^{41}$ $e^+$/s. The $e^+$ of all other candidate sources is entirely speculative at present. Values discussed above should be considered as optimistic rather than typical values. No useful observational constraints exist up to now on the $e^+$ yields of hypernovae/GRBs, pulsars, ms pulsars, magnetars, LMXRBs, microquasars, the MBH at the Galactic center or dark matter annihilation. SNIa remain an intriguing, but serious candidate, with a potential Galactic yield of $2 \times 10^{43}$ $e^+$/s.

Radioactive decay produces positrons of $E \leq 1$ MeV, naturally fulfilling the observational constraint on continuum $\gamma$ rays from in flight annihilation. The same applies to pair creation through $\gamma - \gamma$ collisions in the inner accretion disk or at the base of the jets of LMXRBs, microquasars and the SMBH at the Galactic center. Conversely, pair creation involving very high energy photons, as in e.g. pulsars or magnetars, will produce positrons of too high energy. The same holds for energetic p – p collisions in Galactic cosmic rays or in the baryonic jets of LMXRBs, microquasars and the Galactic SMBH. Those processes produce $e^+$ of >30 MeV, thus may be discarded as major $e^+$ sources in the Milky Way. Also, that same constraint limits the mass of putative decaying or annihilating DM particles to <10 MeV, while it does not constrain the mass of de-exciting DM particles.

None of the $e^+$ sources reproduces the large bulge-to-disk ratio $\geq 1$ ratio inferred from SPI data. The best-established $e^+$ sources, $\beta^+$-decay from $^{26}$Al and $^{44}$Ti produced in massive stars, yield a bulge-to-disk ratio $\leq 0.2$, as derived from the observed distribution of the 1.8 MeV $^{26}$Al line. The corresponding skymap and longitude profile (normalised to a disk emissivity of $0.7 \times 10^{43}$ $e^+$/s from radioactivity of $^{26}$Al and $^{44}$Ti) are displayed in Figs. 7.29 and 7.30, respectively, where they are compared to the SPI data of Weidenspointner et al. (2008b). Notice that similar profiles are expected for pulsars, magnetars, hypernovae and gamma-ray bursts (albeit with different normalisations).

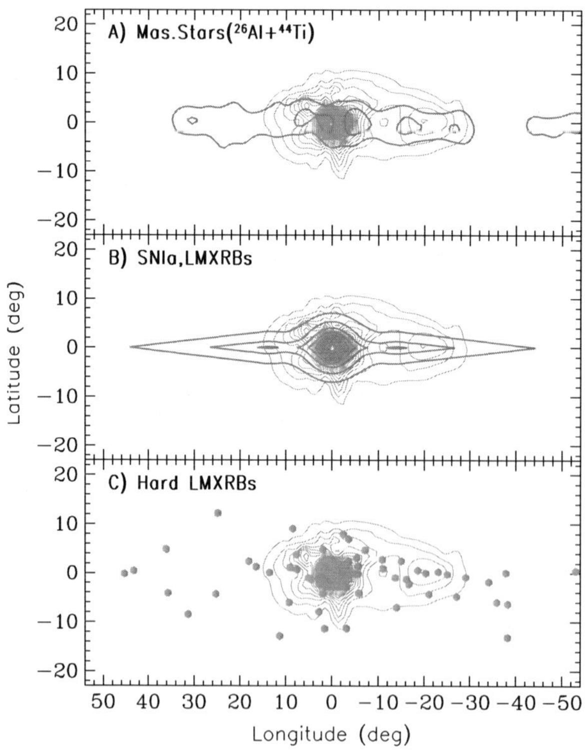


**Fig. 7.29** Maps of the Galactic 511 keV emission (flux in cm$^{-2}$ s$^{-1}$ sterad$^{-1}$), as observed from SPI (in all panels, *thin isocontours* from Weidenspointner et al. (2008b)) and from observationally based or theoretical estimates. (**a**) Observed $^{26}$Al (and, presumaby, $^{44}$Ti) map; (**b**) Accreting binary systems (SNIa and, presumably, LMXRBs); (**c**) Observed Hard LMXRBs (from Bird et al. (2007)). From Prantzos et al. (2010)

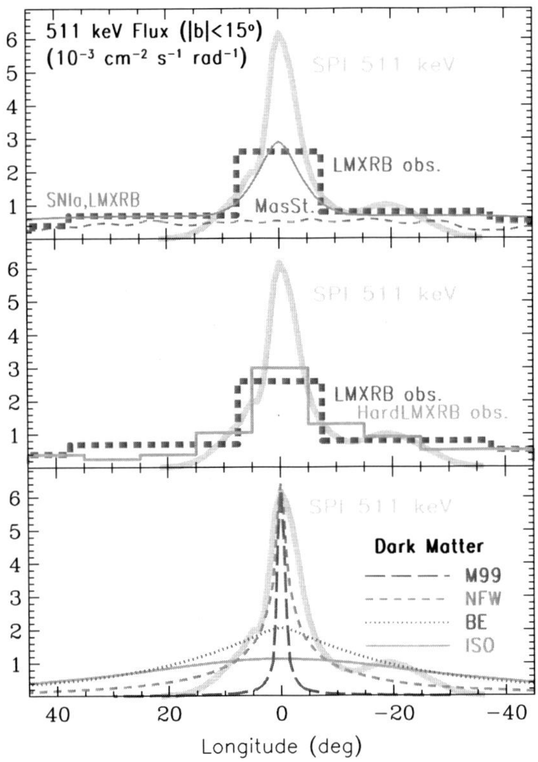


**Fig. 7.30** Intensity of 511 keV emission as a function of Galactic longitude. All fluxes are integrated for latitudes $|b| < 15°$. In all panels, the *thick solid curve* corresponds to SPI observations, i.e. the map of Fig. 7.29. (*Note*: SPI maps and fluxes are provided here for illustration purposes only; quantitative comparison of model predictions to data should only be made through proper convolution with the SPI response.) The *thick dotted histogram* (*top* and *middle*) are the observed longitude distribution of LMXRBs (from Grimm et al., 2002); the latter resembles closely the theoretically estimated longitude distribution of SNIa (*thin solid curve* in the *upper panel*), which has been normalised to a total emissivity of $1.6 \times 10^{43}$ $e^+$/s, with Bulge/Disk = 0.45 (maximum Bulge/Disk ratio for SNIa from Table 7.4). Also, in the upper panel, the *lower dashed curve* corresponds to the expected contribution of the $^{26}$Al and $^{44}$Ti $\beta^+$-decay from massive stars. The *thin solid histogram* in the *middle panel* is the observed longitude distribution of Hard LMXRBs (from Bird et al. (2007)) and it has the same normalization as the thick histogram. In the *bottom* panel, the SPI 511 keV profile is compared to profiles expected from dark matter annihilation. From Prantzos et al. (2010)

Binaries involving low mass stars, such as SNIa, novae and LMXRBs, are expected to have a steeper longitude profile, with a maximal bulge-to-disk ratio $\leq 0.5$ (assuming the bulge and disk masses of Sect. 7.2.3). Using data from Fig. 7.3 and Table 7.4 one may estimate an expected sky distribution and corresponding longitude profile of SNIa, also displayed in Figs. 7.29 and 7.30. Figure 7.30 indicates that the theoretically expected SNIa longitude profile ressembles the observed profile of LMXRBs (from Grimm et al. (2002)). This similarity reflects the fact that both classes of sources involve an old stellar population, which is more abundant in the inner Galaxy and the bulge. Novae are also expected, on those grounds, to have a similar distribution.

The upper panel of Fig. 7.30 clearly shows that

(a) The expected outer disk ($l$ >20°) contribution of massive star radioactivity ($^{26}$Al and $^{44}$Ti) is not yet detected, due to insufficient sensitivity; SPI/INTEGRAL or a future instrument should reveal that component, or else positrons diffuse far away from their sources.
(b) SNIa or LMXRBs or microquasars can explain only about half of the strong 511 keV emission from the inner Galaxy, assuming they produce as many positrons as in Table 7.7)
(c) Any remaining annihilation $\gamma$-ray emission requires a supplementary source (dark matter or the central SMBH, provided its positrons can diffuse to kpc distances and fill the bulge).

In the middle panel of Fig. 7.30 we compare the longitude profile of the observed 511 keV emission to the one of the hard LMXRBs (emitting in the 20–100 keV range), as seen with IBIS/INTEGRAL (from Bird et al. (2007)). Weidenspointner et al. (2008b) noticed that the latter distribution exhibits a pronounced asymmetry, with source number ratio $N(l < 20°)/N(l > 20°) = 1.7$, which matches well the asymmetry in the 511 keV flux reported in the same paper. They suggested then that hard LMXRBs may be at the origin of the disk emission.

We note that the study of the same SPI/INTEGRAL data by Bouchet et al. (2008) with different methods does not find significant disk asymmetry. Obviously, the important (and intriguing) observational result of Weidenspointner et al. (2008b) needs confirmation by further observations/analysis.

The morphology of the observed 511 keV emission provides also some interesting constraints in the case of dark matter particles as positron sources (under the assumption of negligible $e^+$ propagation) (as analysed in Ascasibar et al., 2006): (i) Particle candidates with velocity dependent cross section are excluded as the main source of 511 keV emission, (ii) Decaying dark matter cannot be the main source of low energy positrons, because the resulting flux profile is too flat, compared to SPI data. Notice that this latter feature is a generic property of all models involving decaying particles, where the positron production (and annihilation) rate is proportional to the DM density profile: even *cuspy* profiles, such as the *Navarro-Frenk-White* profile do not provide a $\gamma$-ray flux profile sufficiently peaked towards the inner Galaxy. Annihilating or de-exciting DM produces positrons at a rate proportional to the square of the DM density profile and leads to a much more peaked $\gamma$-ray profile. Light scalar annihilating particles remain a possible candidate, provided the dark matter halo is at least as cuspy as the Navarro-Frenk-White profile with $\gamma \sim 1$ (see bottom panel of Fig. 7.30); however, astrophysical evidence favors flatter DM halo profiles.

The main features of all these candidate $e^+$ sources are summarized in Table 7.7. The $e^+$ production rates of all those sources are extremely uncertain (except those of

$^{26}$Al, $^{44}$Ti and GCRs) and the values listed above should be considered as optimistic rather than typical ones. Only in the case of novae may the estimated production value be used to eliminate those sources as important $e^+$ producers. Source morphology and high energy of produced positrons appear to exclude pulsars, magnetars and Galactic cosmic rays as major contributors to the observed 511 keV emission from the bulge. Source morphology alone would exclude hypernovae and GRBs. The high energy of positrons disfavors ms pulsars, as well as p-p collisions from any source (micro-quasars, LMXRB jets, the central SMBH). This still leaves several potentially important $e^+$ source candidates, but none of them appears compatible with the observed morphology of 511 keV emission.

Thus, *if positrons annihilate near their sources*, one has to conclude that (i) either an unknown class of sources dominates $e^+$ production, or that (ii) positrons are produced by a combination of the sources of Table 7.7, e.g. (a) $^{26}$Al +$^{44}$Ti for the disk and dark matter for the bulge, or (b) $^{26}$Al +$^{44}$Ti+LMXRBs (or microquasars) for the disk *and* the bulge plus a contribution from the central SMBH for the inner bulge, or (c) some other combination.

Positron escape from the disk to the halo would obviously alleviate the morphology problem, by reducing the disk 511 keV emissivity and thus increasing the bulge-to-disk ratio of those classes of sources. However, although some of the basic physical processes underlying $e^+$ propagation are well understood, there is no clear global picture of how far positrons can propagate in the magnetized, turbulent interstellar medium of the Galaxy.

### 7.7.4 Implications of Positron Transport in the Interstellar Medium

Positrons injected from various sources interact with the surrounding medium either through collisions (collisional transport) or through interaction with plasma waves (collisionless transport). The collisional transport predicts distances explored by MeV particles that are large – of the order of 10 kpc/n ($n$ is the local density in $cm^{-3}$). The distances are shorter by a factor 0.75 if the tangled components of the magnetic field are added (Jean et al., 2009). Such distances are much larger than the typical sizes of the hot and warm phases of the ISM. Since the spectral analysis of the 511 keV emission suggests that positrons annihilate mostly in the warm phases, the only way to stop positrons is by collisionless processes.

Collisionless transport is governed by wave-particle interactions. In the case of a magnetized plasma, positrons are spiraling along the magnetic field lines. The gyroradius of a positron with Lorentz factor $\gamma$ is $r_g \sim 1.7\ 10^9\ B_{\mu G}^{-1}(\gamma^2 - 1)^{1/2}$ cm, where the local mean magnetic field $B_{\mu G}$ is expressed in $\mu G$. In a magnetized, turbulent, plasma, the most efficient of collisionless processes is scattering off magnetic fluctuations of size $r_B \simeq r_g$, which induce *resonant* pitch angle scattering of positrons (e.g. Kulsrud (2005), and references therein), or *non-resonant* interactions with fluctuations on scales just above $r_g$ (see e.g. Toptygin, 1985; Ragot, 2006). The resonant interactions involve either the particle gyro-motion around the mean magnetic field (*cyclotrotron resonance*) or the parallel motion of the particle along the field line (*Cherenkov resonance*). The Larmor radius of the resonant positron being small – of the order of $r_g \simeq 10^9$ cm and the positron gyromotion polarization being left-handed only MHD waves (Alfvén or fast magnetosonic) can fulfill the resonance conditions. Nonresonant compressible perturbations may also affect the positron mean free path, as it seems to be the case for sub-MeV electrons in the solar wind, where there are strong indications for MeV electron re-acceleration Ragot (1999). Moreover, *adiabatic deceleration* of positrons in jets or expanding shells (for example in SN remnants) results in positron cooling, even without Coulomb collisions; this occurs if the positron mean free path, which is dominated by $e^+$ scattering by waves, is shorter than the typical scale of bulk plasma motion.

To summarize, wave-particle interactions, both resonant and adiabatic non-resonant, could result in particle deceleration, but also *re-acceleration*, depending strongly on the local conditions. The transport of energetic (>GeV) Galactic cosmic rays is driven by such collisionless processes. But in the case of MeV positrons the situation is not clear, because there is no observational evidence on the level of intestellar turbulence at such small scales (although one may reasonably expect that it can be quite high in the vicinity of some positron sources, e.g. supernovae).

The propagation and transport of low energy positrons has important implications for the Galactic 511 keV emission (first raised by Prantzos, 2006): the morphology of the 511 keV emission does not necessarily reflect the morphology of the underlying $e^+$ source distribution. Positrons from SNIa are released in the hot and rarefied ionized medium, since the scaleheight of SNIa is considerably larger than the scaleheight of the cool, dense gas in the Galactic disk (see Fig. 7.3). The $e^+$ propagation distances are then quite large, allowing $e^+$ from the disk to annihilate far away from their sources (perhaps in the halo, where a low surface brightness emission should be expected); this fact may considerably reduce the bulge/disk ratio of 511 keV emission of any class of astrophysical $e^+$ sources, thus alleviating the morphology problem.

Furthermore, *if* the magnetic field of the Milky Way halo has a strong poloidal component, as suggested (see Fig. 7.31, *left*), then some positrons escaping the disk may be channelled into the bulge and annihilate there, enhancing even more the bulge/disk $e^+$ annihilation ratio (suggested by Prantzos, 2006). In that case, positrons from SNIa may suffice to cause *quantitatively both* the total observed $e^+$ annihilation rate ($\sim 2 \times 10^{43}$ $e^+$/s) *and* the corresponding bulge-to-disk ratio, *provided* that the escaping $e^+$ fraction from SNIa is ~3–4% . It is rather improbable (although it cannot yet be excluded) that the Galactic magnetic field has a large scale regular poloidal component; observations of external spirals suggest rather an X-shaped halo field (e.g. Fig. 7.31, *right*), in which case it would be difficult for disk positrons to find their way into the bulge. Still, the issue is of considerable interest to urge a better assessment of the poorly known global configuration of the Galactic magnetic field.

Higdon et al. (2009) suggested that positron propagation throughout the specific parts of our Galaxy may be all that is needed for understanding not only the

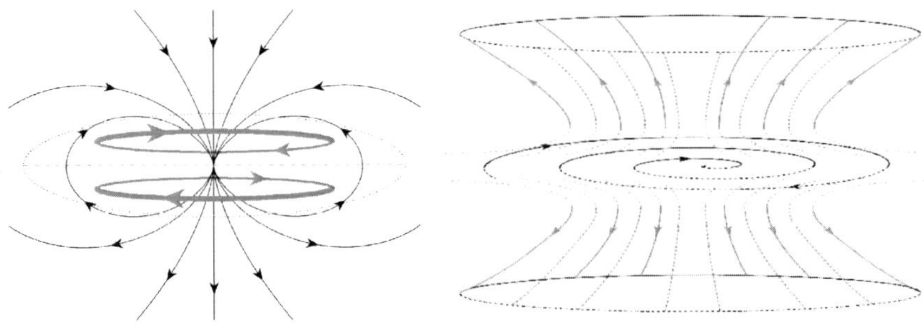

**Fig. 7.31** Possible configurations of the large scale magnetic field of the Milky Way. *Top*: As derived from Faraday polarization measurements of the MW, according to Han (2004). *Bottom*: Scetch of the observable components of the large scale magnetic field of the disk galaxy NGC253 (which shows, however, signs of starburst activity, unlike the Milky Way); the halo magnetic field is even and pointing outwards, whereas the *dotted parts* are not observed (from Heesen et al. (2009))

spatial morphology of the 511 keV emission, but also its spectral properties. They started from the rather bold assumption that radioactivity (from $^{26}$Al, $^{44}$Ti and, mostly, from $^{56}$Co) is the sole $e^+$ source in the Galaxy. They then considered (i) a fairly detailed description of the various phases of the ISM and (ii) a particular phenomenological model of collisionless scattering of MeV positrons by turbulent fluctuations of the ISM, that was used to describe energetic particle (electron or proton) propagation in the interplanetary medium. Depending on the nature of the medium, positrons are assumed to propagate either by diffusion along magnetic flux tubes (in ionized media, where turbulence cascades down to the gyroradius $r_g$) or by streaming with mean velocity $\beta c$ (in neutral media, where turbulence is quenched by ion-neutral collisions at scales $>> r_g$). Putting together these ingredients, Higdon et al. (2009) find excellent agreement with each and every observed property of the Galactic annihilation emission available so far: spatial morphology, i.e. bulge-to-disk ratio, spectral features (including a narrow and a broad 511 keV line) and even an asymmetry between fluxes from negative and positive longitudes. Their extremely precise *predictions* for the various properties of the 511 keV emission (which fit extremely well – to better than 10% – each and every observable) appear surprising, and may largely underestimate the various uncertainties of this complex astrophysical problem (see Prantzos et al., 2010 for a detailed criticism).

### *7.7.5 Summarizing the Positron Annihilation Puzzle*

More than 30 years after its discovery, the origin of positrons annihilating in the Galaxy remains unknown. Progress in the field will require advances in several directions:

(i) *Observations of 511 keV emission*: What is the true spatial distribution of the emission? How far the spheroid and disk extend? Are there yet undetected regions of low surface brightness? Is the disk emission asymmetric indeed? How do the 1.8 MeV and 511 keV disk emissions compare to each other? A much deeper exposure of the Galaxy and a better understanding of the backgrounds will be required to tackle those issues. Even if INTEGRAL's mission is extended to 2012, it seems improbable that it will be able to provide the answers; and no other mission in this energy range is scheduled at present.

(ii) *Physics of $e^+$ sources*: What is the $e^+$ escaping fraction in SNIa ? What is the SNIa rate in the inner (star forming) and in the outer (inactive) bulge? What are the $e^+$ yields, activity timescales, and spatial distribution in the inner bulge of LMXRBs or microquasars? How can the past level of activity of the central massive black hole be reliably monitored?

(iii) *Positron propagation:* What is the large scale configuration of the Galactic magnetic field? What are the properties of interstellar plasma turbulence and how they affect the positron transport? What are the dominant propagation modes of positrons and what is the role of re-acceleration?

The many faces of the Galactic 511 keV emission make this problem one of the most inriguing problems in high energy astrophysics today and for the years to come.

## 7.8 Radioactivities in Cosmic Rays

Along with stellar grains, Galactic Cosmic Rays (GCR) provide a sample of matter from outside the Solar system. Despite almost a century of active reasearch, the physics of GCR (concerning their sources, acceleration and propagation in the Galaxy) is not yet thoroughly understood. In particular, key questions regarding GCR physics are related to the timescales of various processes (acceleration in one event or in a series of events, confinement in the Galaxy, etc.). Radionuclides unstable to $\beta^{\pm}$ decay or *e*-capture, with laboratory lifetimes close to the timescales of interest for GCR studies, provide important probes of the aforementioned processes. After a short introduction to the observations (Sect. 7.8.1) and the physics (Sect. 7.8.2) of GCR, we present the main *radioactive probes* for the study of various GCR timescales in Sect. 7.8.3.

### *7.8.1 Observations of Cosmic Rays: Spectra and Composition*

GCR spectra measured in Earth's vicinity have the same overall shape for all chemical elements. Above a few GeV/nucleon, and up to $10^6$ GeV/nucleon, the GCR intensity $I(E)$ (in $\mathrm{cm}^{-2} s^{-1} st^{-1} GeV^{-1}$) is decribed by a practically featureless power-law as a function of kinetic energy $E$

$$I(E) = 1.8\,E^{\alpha} \tag{7.12}$$

with slope $\alpha$=-2.7. At $E \sim 10^6$ GeV/nucleon (the *knee*), the GCR spectrum steepens to $\alpha = -3$. At $E < 1$ GeV/nucleon, the GCR spectrum progressively flattens (with $\alpha$ becoming even positive) and its intensity varies, in a way which anticorrelates with the solar activity (*solar modulation*); the solar wind prevents the lowest energy GCR from entering the heliosphere. Correcting for that effect (*demodulating*) allows one to infer, through the time-averaged locally observed spectra, the *true* spectrum of GCR as they propagate in the ISM (Fig. 7.32). This allows one to evaluate the kinetic energy density of GCR $\varepsilon_{\mathrm{GCR}}$ as

$$\varepsilon_{\mathrm{GCR}} = 4\pi \int \frac{E_k}{\upsilon}\, I(E)\, dE \tag{7.13}$$

where $\upsilon$ is the velocity of a particle with kinetic energy $E$. The inferred local GCR energy density is $\varepsilon_{\mathrm{GCR}} \sim 1\,\mathrm{eV/cm^3}$. It is comparable to the local energy densities of starlight ($\varepsilon_{\mathrm{Light}}$), of ISM gas thermal motion ($\varepsilon_{\mathrm{Gas}} \sim n\kappa T$) and of the magnetic field ($\varepsilon_B \sim \frac{B^2}{8\pi}$), i.e. GCR constitute an important dynamical agent in the Galaxy.

As we shall see in Sect. 7.8.2, GCR are confined for a timescale $\tau_{\mathrm{Conf}} \sim 10^7$ year in the Galactic disk (before escaping to the halo and then to the intergalactic space), i.e. in a roughly cylindrical volume $V$ of radius $R \sim 15$ kpc and height $h \sim 4$ kpc. Assuming that $\varepsilon_{\mathrm{GCR}}$ is the same throughout that volume, one finds that the

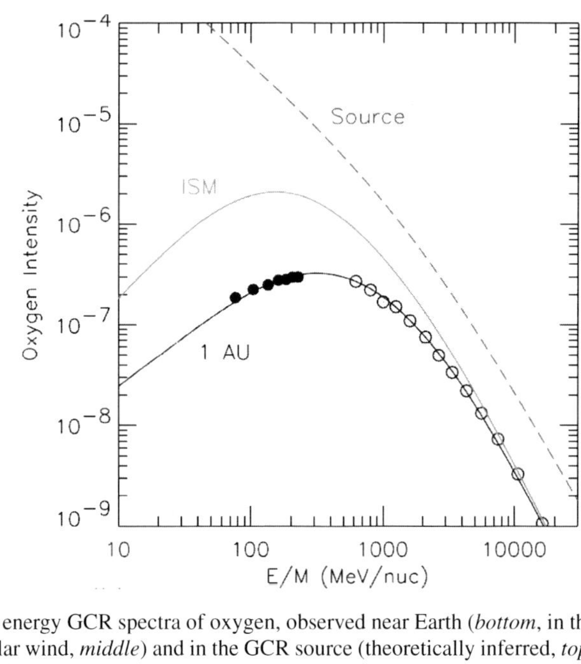


**Fig. 7.32** Low energy GCR spectra of oxygen, observed near Earth (*bottom*, in the ISM (demodulated for the solar wind, *middle*) and in the GCR source (theoretically inferred, *top*) (from Wiedenbeck et al., 2007)

total GCR power is $P_{GCR} = \varepsilon V/\tau_{Conf} \sim 2 \times 10^{41}$ erg/s. Taking into account the expected SN frequencies in the Milky Way ($f_{GalSN} \sim$ 3/century) and assuming a kinetic energy of $E_{SN} = 1.5 \times 10^{51}$ erg for each SN, one sees that $P_{GCR} \sim 0.1$ $P_{GalSN}$ (where $P_{GalSN} = f_{GalSN} E_{SN} \sim 1.5 \times 10^{42}$ erg/s). Those numbers suggest that SN accelerate GCR with an efficiency of the order of 10%.

The composition of arriving CR is best measured in the low energy range (0.1–1 GeV/nucleon) because in that region GCR intensity is highest and experimental techniques for particle identification have better resolution than at higher energies. In Fig. 7.33, the arriving GCR composition is compared to the solar system one. Both samples of matter are dominated by the elements H, He, C, O, Ne, Mg, Si and Fe. The overall similarity of the two abundance curves suggests that they have similar nucleosynthetic origin, i.e. from massive stars and supernovae (for nuclei in the C—Fe peak range). Volatile elements (He, O, Ne, S and Ar) are somewhat depleted relative to refractory Si, perhaps as a result of some fractionation process, favouring the acceleration of more refractory elements (see next section).

The most striking difference between GCR and solar abundances is the GCR overabundance of some nuclides which are relatively rare in the solar system composition. The atomic numbers of those rare nuclides are only a few units below those of the nuclides dominating the abundance curves. This is the case of e.g. Li, Be and B with respect to the slightly heavier C, N and O nuclei ($(\frac{\mathrm{LiBeB}}{\mathrm{CNO}})_{solar} \sim 10^{-6}$,

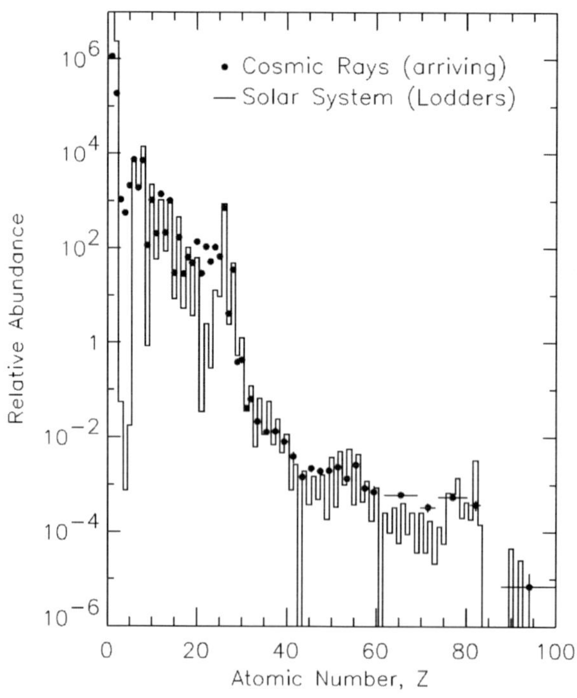


**Fig. 7.33** Elemental abundances in arriving GCR (*dots*), compared to the solar system ones (*histogram*) (from Wiedenbeck et al., 2007)

while $(\frac{\mathrm{LiBeB}}{\mathrm{CNO}})_{\mathrm{GCR}} \sim 10^{-1}$) or of Sc, Ti and V nuclei with respect to Fe peak nuclei (see Fig. 7.33). The relative overabundance of those rare nuclei in GCR has long been recognised to be the result of their *secondary* nature: they are produced by the nuclear fragmentation (*spallation*) of their heavy progenitors (*primaries*), during the GCR propagation in the ISM.

As we shall see in the next sections, the properties of GCR secondaries, both stable and radioactive, provide important information on the physics of Galactic cosmic rays.

## 7.8.2 The GCR Odyssey: Sources, Acceleration, Propagation

From the previous section one may infer that several steps are involved between the production of the GCR nuclides in stellar interiors and their detection near Earth: (1) stellar nucleosynthesis, (2) ejection by stellar winds and explosions, (3) elemental fractionation, (4) acceleration of primary GCR nuclides, by shocks due to SN and winds of massive stars, (5) propagation through the ISM of the Galaxy, (6) modulation at the heliospheric boundary and (7) detection of arriving GCR. In particular, GCR transport through the ISM has been studied with models of varying sophistication, which account for a large number of astrophysical observables (see the comprehensive review of Strong et al. (2007) and references therein).

To describe the composition data, less sophisticated models are sufficient, like e.g. the *leaky-box* model. In that model, GCR are assumed to fill homogeneously a cylindrical box (the Galactic disk) and their intensity in the ISM is assumed to be in a steady state (equilibrium), between several production and destruction processes. The former involve acceleration in GCR sources and production *in-flight* through fragmentation of heavier nuclides, while the latter include either physical losses from the *leaky box* (*escape* from the Galaxy) or losses in energy space: *fragmentation*, *ionization losses* and *radioactive decay*. Most of the physical parameters describing these processes are well known, although some spallation cross sections still suffer from considerable uncertainties. The many intricacies of GCR transport are encoded in a simple parameter, the *escape length* $\Lambda_{\mathrm{esc}}$ (in g/cm$^2$): it represents the average column density traversed by GCR nuclei before escaping the Galactic leaky box.

The abundance ratio of a secondary to a primary nuclide depends essentially on $\Lambda_{\mathrm{esc}}$. Observations of $\frac{\mathrm{LiBeB}}{\mathrm{CNO}}$ (and more recently of $\frac{ScTiV}{\mathrm{Fe}}$) in arriving GCR, interpreted in this framework, suggest a mean escape length $\Lambda_{\mathrm{esc}} \sim 7$ g/cm$^2$. In fact, the observed GCR secondary/primary ratios display some energy dependence, which translates into an energy dependent $\Lambda_{\mathrm{esc}}(E)$, going through a maximum at $E \sim 1$ GeV/nucleon and decreasing both at higher and lower energies. The observed energy dependence of that key phenomenological parameter can be interpreted in the framework of more sophisticated GCR transport models and provides valuable insight into the physics of GCR transport (role of turbulent diffusion, convection by a Galactic wind, re-acceleration, etc.); those same models can be used to infer the injection spectra of GCR at the source (see Jones et al., 2001; Strong et al., 2007).

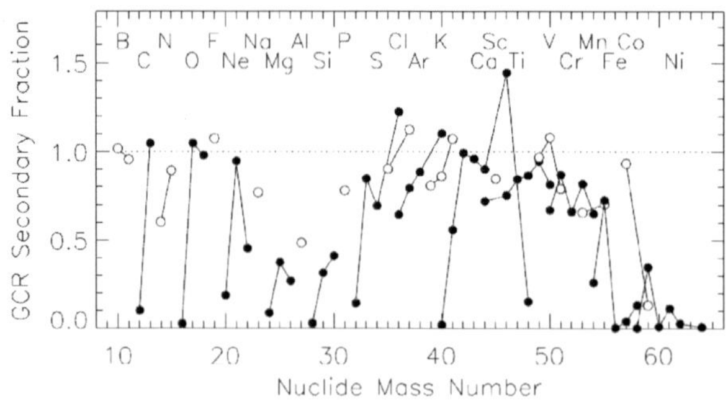


**Fig. 7.34** Secondary fractions in arriving GCR for isotopes from B to Ni, after the analysis of ACE/CRIS measurements. Nuclides with filled (*open*) symbols correspond to isotopes with even (*odd*) Z, while solid lines connect isotopes of the same element. From Wiedenbeck et al., 2007)

Once the key parameters of the leaky-box model are adjusted to reproduce the key secondary/primary ratios, the same formalism may be used in order to evaluate the secondary fractions (produced by fragmentation in-flight) of all GCR nuclides. Those fractions depend critically on the relevant spallation cross-sections (well known in most cases) and appear in Fig. 7.34. Fractions close to 1 imply an almost purely secondary nature while fractions close to 0 characterize primary nuclides (like, e.g. $^{12}$C, $^{16}$O, $^{24}$Mg, $^{56}$Fe etc.). The former are very sensitive to the adopted $\Lambda_{esc}$, contrary to the latter (Wiedenbeck et al., 2007).

The *source abundances of primary GCR* (GCRS) derived that way display both similarities and differences when compared to the solar ones. It was recognised quite early on, that the observed pattern of GCRS/Solar abundances shows some correlation with the First Ionisation Potential (FIP): elements of high FIP are depleted in GCRS relatively to low FIP ones. Since the latter are more easily ionized than the former, it is conceivable that they are also more easy to pre-accelerate (in some suitable environment) before SN shocks accerelerate them to relativistic GCR energies (Meyer, 1985). Although similar FIP-dependent abundance patterns are observed in the solar corona, this idea did not evolve in a fully self-consistent model for the origin of GCR.

Another, long standing, idea attracted considerable attention in the past 15 years. It invokes chemical volatility to explain fractionation in the GCR sources. Of course, volatility is somewhat related to FIP: high FIP elements (e.g. He, N, Ne, Ar) are, in general, volatile, where low FIP elements are, in general, refractories (e.e Mg, Si, Fe etc.), and condense readily into dust grains. Meyer et al. (1997) and Ellison et al. (1997) suggested that grains are efficiently accelerated in SN shocks, because of their high mass/charge ratio; subsequent sputtering of atoms from these fast grains creates a *pool* of *suprathermal* ions, that are further accelerated to relativistic GCR energies from SN shocks. Building on those ideas, Meyer et al. (1997) proposed a model explaining quantitatively the inferred pattern of GCRS/Solar abundances (Fig. 7.35). Despite its success, the model relies on several assumptions which lack empirical support (notice, however, that interplanetary dust particles display traces of grain acceleration, as reported by Westphal and Bradley (2004).

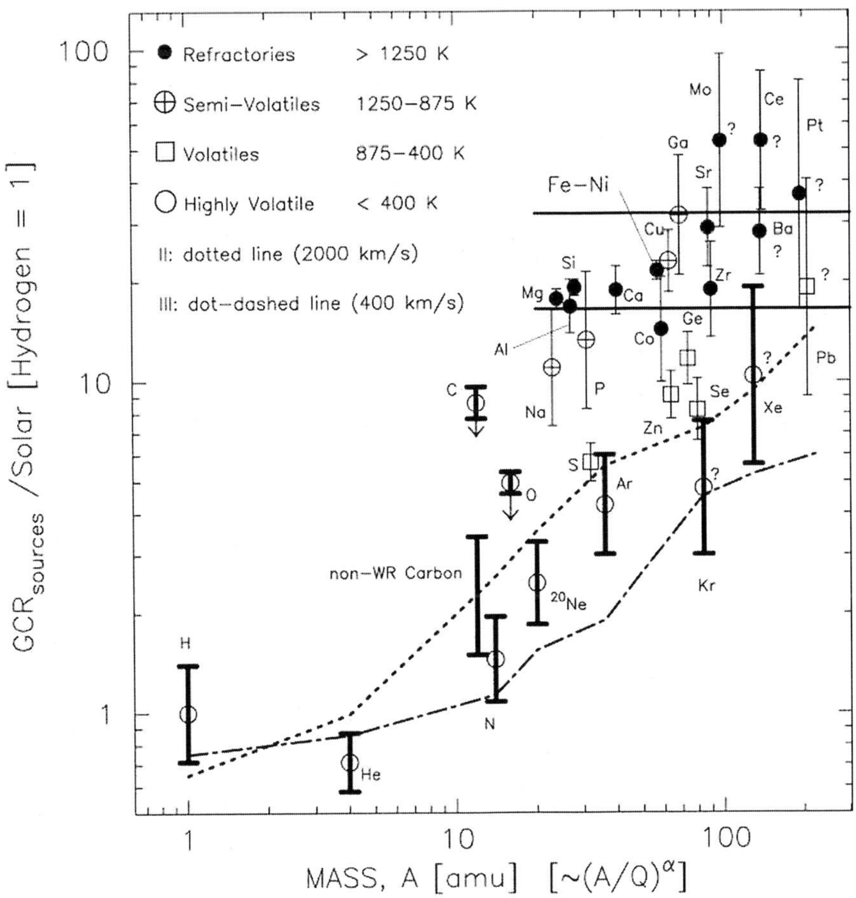


**Fig. 7.35** GCR source abundance relative to solar abundance vs. atomic mass number. All values are measured relative to GCR hydrogen at a given energy per nucleon. The elements are divided, on the basis of condensation temperature, into refractory, semivolatile, volatile, and highly volatile groups. The refractories are essentially completely locked in grains in the ISM, while the highly volatile ones are gaseous. The arrows on carbon and oxygen indicate that these elements have an additional source from $^{22}$Ne—C—O enriched Wolf-Rayet wind material. Model predictions for the abundances of volatile elements from a high Mach number shock model are shown with a *dotted line*, and those for a lower Mach number model with a *dot-dashed line*. The *horizontal solid lines* on the *right side* of the plot are limits on predicted abundance of iron and other refractory elements. The label on the abscissa [$\sim(A/Q)^{\alpha}$, where $\alpha$ is some unspecified constant] is a reminder that, for most ionization models, A/Q is a roughly monotonically increasing function of mass. The abundances of Kr, Xe, Mo, Ba, Ce, Pt, and Pb relative to Fe may contain systematic errors that are difficult to evaluate (indicated with a question mark) (from Ellison et al., 1997)

In view of the complexities introduced by fractionation, it appears safer to rely either on refractory primary elements (unaffected by fractionation) or to ratios of isotopes belonging to the same element (since fractionation effects cancel out), in order to probe the source composition of GCR. The ACE/CRIS observations allowed to determine the primary GCRS abundances of 19 refractory nuclides, belonging to Mg, Al, Si, Ca, Fe and Ni. As reviewed in Wiedenbeck et al. (2007) those nuclides have a solar composition (within uncertainties), with the possible exception of $^{58}$Fe. Since more than half of solar Fe come from a *long-lived* source (SNIa, Goswami and Prantzos, 2000), the GCRS composition has to originate from an extremely well mixed sample of the ISM (or of solar type stars).

Contrary to the refractory nuclides, the volatile $^{22}$Ne/$^{20}$Ne ratio in the GCR source has long been known to exceed its solar value by a factor of $\sim$5. This excess of $^{22}$Ne is attri uted to the contribution of Wolf-Rayet winds to GCRs, as originally suggested by Casse and Paul (1982) and quantitatively elaborated in subsequent studies (e.g. Prantzos et al., 1985).

It is not clear at present whether the totality of the GCRS composition originates from the same site (where the nucleosynthetic products of many stars and SN are well mixed) or whether it originates as a diversified sample of nuclei (produced from a variety of sources distributed all over the Galaxy), which are mixed well after their acceleration by SN shocks, during their $10^7$ year travel through the Milky Way. The latter case corresponds to the, now defunct, FIP-based paradigm of GCRS composition (Meyer, 1985). The former possibility is now embodied in the *super-bubble* paradigm, advocated in Higdon et al. (1998) and further supported by Binns et al. (2005) on the basis of the $^{22}$Ne/$^{20}$Ne analysis. GCR (both refractories produced by sputtered grains and volatiles) are accelerated – with higher efficiency than in the normal ISM – from the thermal pool of ions present in a super-bubble, which is created and chemically enriched by the action of dozens or hundreds of stars and SN. In that model, SN explosions have to occur $>10^5$ year apart frome each other, in order to allow for *e*-capture decay of $^{59}$Ni, which is absent in arriving GCR (see next section). The super-bubble paradigm has been criticised on several grounds in Prantzos (2007).

### 7.8.3 Radioactive Isotopes in Galactic Cosmic Rays

Radioactive nuclides present in GCR may provide key information about various timescales concerning the acceleration and propagation of cosmic rays in the Galaxy. Primary nuclides unstable against $\beta^{\pm}$ and $\alpha$ decays or spontaneous fission, are sensitive to the timescale between nucleosynthesis and arrival on Earth; however, U and Th have such large lifetimes that they are essentially stable during the propagation of GCR in the Galaxy and they cannot be used as probes of that timescale. Primary nuclides unstable against *e*-capture only, are sensitive to the timescale from their nucleosynthesis to acceleration, since after acceleration to GCR energies they can hardly capture an orbital electron and they propagate essentially without decay; $^{57}$Co and, especially, $^{59}$Ni belong to this category. Secondaries unstable against $\beta^{\pm}$

**Table 7.8** Decay modes and half-lives of key secondary radioactivities in GCR (from Yanasak et al., 2001b)

| Isotope | Decay mode | Mode half-life (year) |
|---|---|---|
| $^{10}$B | $\beta^-$ | $1.51 \times 10^6$ |
| $^{14}$C | $\beta^-$ | $5.70 \times 10^3$ |
| $^{26}$Al | $\beta^+$ | $8.73 \times 10^5$ |
| | $e$-capture | $8.45 \times 10^6$ |
| $^{36}$Cl | $\beta^-$ | $3.07 \times 10^5$ |
| | $e$-capture | $1.59 \times 10^7$ |
| $^{54}$Mn | $\beta^-$ | $(6.3 \pm 1.7) \times 10^5$ |
| | $e$-capture | 0.8548 |

and $\alpha$ decays or spontaneous fission can probe the residence timescale of GCR in the Galaxy, provided their half-lifes are comparable to that timescale; a handful of nuclides belong to this class (see Table 7.8). The GCR residence time, combined with the mean path length (obtained from the secondary/primary ratio) and the particle velocity, allows one to evaluate the mean density of the ISM traversed by GCR. Finally, secondary nuclides unstable against $e$-capture only, can be used as probes of re-acceleration effects, because $e$-capture is more likely at low energies. The next two subsections summarize the excellent review of Mewaldt et al. (2001) on radioactivities in GCR.

#### 7.8.3.1 Timescales of Cosmic-Ray Transport

Determination of the confinement (or residence or escape) timescale of GCR in the Galaxy is a key issue, because $\tau_{\rm Conf}$ determines the power required to sustain the energy density of GCR (see Sect. 7.8.1). All four nuclides of Table 7.8 have been measured in arriving GCR after the Ulusses and ACE results and the situation is summarized in Fig. 7.36. Confinement times are obtained in the framework of a Leaky box model with energy dependent escape length and the average value $\tau_{\rm Conf}$ is found to be $15.0 \pm 1.6$ Million year. In the Leaky box model, $\tau_{\rm Conf}=\Lambda_{\rm esc}/(\upsilon\rho)$, where $\upsilon = \beta c$ is the GCR velocity (at $<1$ GeV/nucleon, the effects of solar modulation have to be accounted for in the calculation of $\beta$). This allows one to evaluate the average ISM density traversed by GCR as $n = \rho/m_p$ =0.36 H atoms/cm$^3$ (Yanasak et al., 2001b). This is a factor of $\sim$3 lower than the canonical value of the local ISM density ($\sim$1 H atom/cm$^3$).

In the Leaky box model, where GCR intensities and ISM densities are uniform in the GCR propagation volume (and in time), such measurements can probe only the average density of the confinement region. The above result implies that GCR spend a large fraction of their confinement time in a volume of smaller average density than the one of the local gas, i.e. in the Galactic halo. In more realistic models, involving e.g. diffusion (see Strong et al., 2007), the aforementioned radioactivities probe a volume of radius $R$ which is limited by their mean life $\tau$, such that $R \sim (\gamma D \tau)^{1/2}$, where $D$ is the spatial diffusion coefficient and $\gamma = (1-\beta^2)^{-1/2}$. In that scheme, at

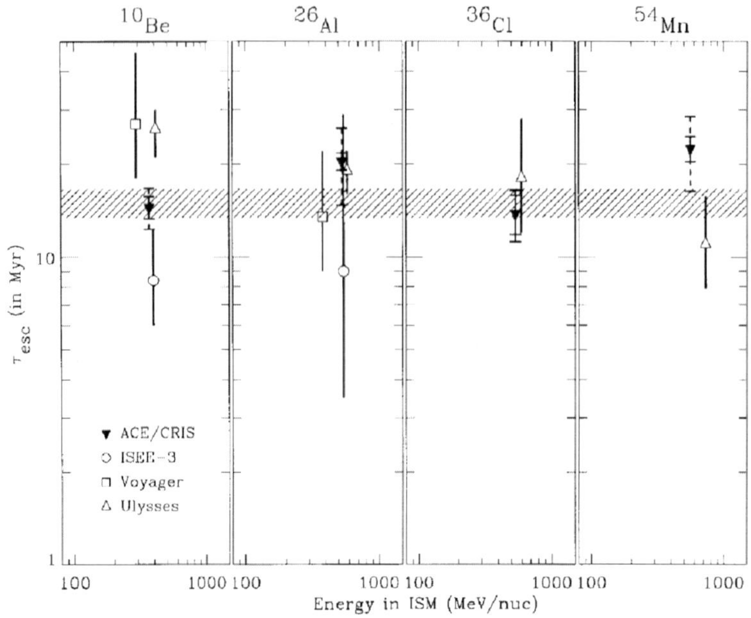


**Fig. 7.36** Confinement times obtained by ACE/CRIS and previous experiments. Uncertainties shown with *solid error bars* are 1 standard deviation statistical. The average value of the confinement time, $\tau_{esc} = 15.0 \pm 1.6$ My, indicated by the CRIS data for the four clock isotopes is shown as a hatched band (From Yanasak et al., 2001a)

1 GeV/nucleon, $^{10}$Be probes regions out to ∼400 pc (i.e. beyond the gaseous layer), while $^{14}$C probes the immediate vicinity of the solar system; however, its expected signal is lower than the background due to $^{14}$C produced inside the ACE/CRIS instrument. Notice that uncertainties in the derived ISM densities are dominated by uncertainties in fragmentation cross-sections, rather than by measurement uncertainties (Yanasak et al., 2001a).

Finally, in diffusion models of GCR propagation, the expected surviving fraction of secondary radioactivities depends on the assumed diffusion coefficient $D$. Comparing the ACE/CRIS measurements for $^{10}$Be, $^{26}$Al and $^{36}$Cl at 400 MeV/nucleon with theoretical predictions of Ptuskin and Soutoul (1998), values of $D \sim 2 \times 10^{28}$ cm$^2$/s are found (see Fig. 9 in Mewaldt et al. (2001)).

#### 7.8.3.2 Timescales of CR Acceleration: $^{59}$Ni

Radionuclides unstable to $e$-capture, if accelerated to >500 MeV/nucleon, become fully stripped of orbital electrons and cannot decay. This suggests that, if $f = X/D_X$ (the abundance fraction of such a nuclide $X$ relative to its decay product $D_X$) arriving on Earth >>1, then the timescale $\Delta t$ between the nucleosynthesis and acceleration of $X$ is smaller than its decay timescale $\tau_X$ by $e$-capture; inversely, an arriving fraction $f = X/D_X$ <<1 would imply that $\Delta t > \tau_X$. Three such radionuclides were identified by Soutoul et al. (1978) as potentially important in that respect: $^{56}$Ni, $^{57}$Co and $^{59}$Ni, with half-lives against $e$-capure from 6 days to $7.6 \times 10^4$ year. It is well known now that $^{56}$Ni is the dominant product of explosive nucleosynthesis in supernovae, however, GCR composition is dominated by its decay product $^{56}$Fe; thus, $\Delta t$ is much larger than 6 days.

The most interesting case is the one of $^{59}$Ni, which probes rather large values of the timescale $\Delta t$. Nucleosynthesis calculations by Woosley and Weaver (1995) suggest that ~2/3 of the $A = 59$ isobars is synthesized as $^{59}$Ni The analysis of ACE/CRIS data showed that very little (if any at all) $^{59}$Ni exists in arriving GCR, while its daughter nuclide $^{59}$Co is present at the level of $^{59}$Co/$^{60}$Ni~0.25 (Fig. 7.37). The absence of $^{59}$Ni from GCR implies that $\Delta t > 10^5$ year.

Such a long timescale between nucleosynthesis and GCR acceleration has important implications for our understanding of GCR composition. It implies that SN cannot accelerate their own ejecta: even in a rarefied medium, the SN ejecta are completely diluted in (and mixed up with) the interstellar gas in timescales as long as 105 year. If GCR are accelerated in super-bubbles, as suggested in (Higdon et al., 1998), then the timescale between two successive SN explosions in that environment has to be $> 10^5$ year (mildly constraining, except for the largest super-bubbles, with hundreds of SN).

There is a loop-hole to the above argument: while $e$-capture is essentially impossible at sufficiently high energies (>500 MeV/nucleon), it becomes possible at lower energies. The average energies of CRIS measurements (~300 MeV/nucleon) corresponds to ~450 MeV/nucleon outside the heliosphere. If $^{59}$Ni produced by

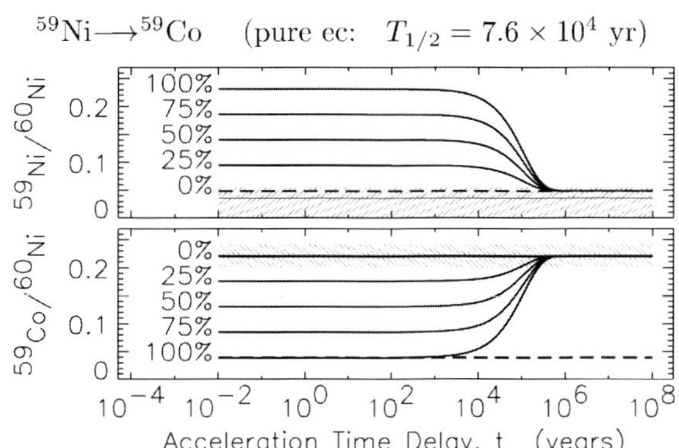


**Fig. 7.37** Calculated abundances at Earth of $^{59}$Ni/$^{60}$Ni (upper panel) and $^{59}$Co/$^{60}$Ni (*lower panel*) are shown as a function of the time delay between nucleosynthesis and GCR acceleration. Calculated abundances are a combination of a secondary component (*dashed lines*) produced by nuclear fragmentation during transport and a surviving primary component. The different *curves* correspond to different assumed fractional contributions of $^{59}$Ni in the primary $A = 59$ material, as indicated by the labels on the *curves*. The time dependences are the result of the exponential decay of the primary $^{59}$Ni into $^{59}$Co as the result of the electron-capture decay of $^{59}$Ni before acceleration. The hatched regions indicate the abundances measured with CRIS, including 1 $\sigma$ uncertainties (From Wiedenbeck et al., 1999)

SN is not directly accelerated to high energies, it may capture an electron and decay in flight, provided it spends $\sim 10^5$ year travelling at such low energies (before getting re-accelerated to higher ones). In that case, its absence in arriving GCR provides no information on the delay between its nucleosynthesis and acceleration.

This brings in the important issue of GCR re-acceleration: do GCR get their high energies in a single SN blast wave, or in a series of SN shock waves (*distributed acceleration*)? Secondary radionuclides, unstable against *e*-capture, with short enough timescales, can be used as probes of such processes: these include $^7$Be, $^{37}$Ar, $^{44}$Ti, $^{49}$V, $^{55}$Fe, with laboratory lifetimes extending up to 67 year. Comparing the measured abundance ratios of those nuclides to their stable daughter nuclei in various energy bands, one may, in principle, constrain the extent of re-acceleration. Unfortunately, the analysis of observations from various experiments (*Ulysses, Voyager, ACE*) has produced contradicting results up to now. However, as stressed in Mewaldt et al. (2001), meassurements of *e*-capture radionuclides in GCR are still in their infancy. The same holds for models to interpret the data, which suffer from uncertainties related to cross-sections (for fragmentation, *e*-capture and loss as a function of energy), the history of GCR acceleration and re-acceleration, inhomogeneities of the ISM, etc. Improvements in both measurements and interpretation will contibute to a much better understanding of the physics of GCR propagation.

## 7.9 Cosmic Star Formation and Supernova Rates

The star formation rate is the key driver of evolution of the ISM within galaxies, and it strongly influences their global evolution via energy, momentum, and chemical feedback from stellar winds and supernova explosions. A fully satisfactory first-principles theory of star formation does not yet exist (for recent reviews see Stahler and Palla, 2005; Zinnecker and Yorke, 2007; McKee and Ostriker, 2007), and consequently we also do not have a full understanding of galaxy evolution with its added complexities due to dark matter, cosmological evolution, and a quasi-random merger history. The determination of the cosmic star formation history, SFR(z), offers useful observational constraints on models of galaxy evolution, and is also an important goal for a variety of related phenomena. The reconstruction of the Star Formation History (SFH), i.e., the SFR as a function of space/time is possible via an analysis of the color-magnitude diagrams (CMDs) of selected stellar populations in the local Universe (e.g., Brown et al., 2006; Cole et al., 2007). Ironically, the Milky Way encounters greater challenges for this kind of study, due to the location of the observer within a heavily obscuring dust layer, than nearby galaxies do.

The cosmological framework is now better known than ever before, but our empirical knowledge of the evolution of the baryon budget from the dark ages to the present, in particular its chemical enrichment history, leaves much to be discovered (e.g., Pettini, 2004; Prochaska and Tumlinson, 2009; Hartmann et al., 2009). There are several observational fingerprints of the cosmic star formation history. Emission in the UV from a first generation of massive stars is responsible, at least in part, for the re-ionization of the Universe at z > 6, and the integrated light of evolving stellar populations generated a cosmic UV-Optical-IR background, which in turn is responsible for a significant optical depth to pair creation for photons in the TeV regime. Radioactivity produced in supernovae leads to the production of $\gamma$-rays. As the ejecta expand, escape of some fraction of these $\gamma$-rays, predominantly from SNIa, contributes substantially to the diffuse background in the MeV band, which also receives contributions from MeV-blazars (e.g., Ajello et al., 2009), $\gamma$-rays from cosmic ray interactions in the ISM of normal galaxies (Pavlidou and Fields, 2002), and possibly dark matter annihilations. Likewise, neutrinos produced in stellar collapses throughout the history of cosmic star formation add up to a diffuse neutrino background (Hartmann and Woosley, 1997), which is becoming detectable with new generation detectors (Horiuchi et al., 2009). Radioactivity is thus directly involved in the cosmic high-energy background, as some fraction of the energy stored in unstable isotopes does not power the supernova but is instead released via escaping photons. The question how star formation began, and how it progressed in various galaxy types is of importance in the context of many frontier problems of observational astronomy, and several reliable tracers are available to measure its cosmic rate. IR-UV continuum- and optical line emission can be used to determine the specific SFR in galaxies, and thereby an extinction corrected rate density. As it becomes increasingly difficult to find small, LMC-like star forming regions at the highest redshifts (the presently most distant, spectrally confirmed galaxy is IOK-1; J132359.8+272456 at $z = 6.4$; Iye et al., 2006) with surveys, direct probes of star formation via their associated Gamma-Ray Burst (GRB) activity (Woosley and Bloom, 2006) (see Vedrenne and Atteia (2010) for a general review of GRBs) are needed to accurately localize these sites of star formation, so that future follow-up with large aperture telescopes can trace and probe the cosmic star formation history (Hartmann, 2009).

If we use the Milky Way as a standard, with $SFR_{MW}$ $\sim$2.3 $M_\odot$/yr (see Sect. 7.1.3), and combine it with an approximate density of Milky-Way-like galaxies of order $n_{gal}$ $10^{-2}$/Mpc$^3$, the local rate density is about $2 \times 10^{-2}$ $M_\odot$/Mpc$^3$/year. How does this rate evolve with cosmic look-back time?

Figure 7.38 shows the so-called *Madau plot*: star formation rate density (per co-moving volume and time) vs. redshift, originally derived from observations of the Hubble Deep Field (Madau et al., 1996; Madau, 1997). Rates are obtained by converting rest frame continuum luminosity density to comoving SFR via various calibrations (see Moustakas et al., (2006) for discussions of uncertainties in such calibrations). The SFR-density to $z \sim 5$ indicated a maximum at $z$ $\sim$1–2, and a relatively fast decline beyond. Soon after this study (but see also Lilly et al., 1996), a flurry of activities led to improved estimates, with better low-$z$ calibrations and dust extinction taken into account. The SFR($z$) function was extended to higher redshifts, its overall normalization moved up substantially, and its high-$z$ tail flattened out. Recently, Hopkins and Beacom (Hopkins and Beacom, 2006) discussed acceptable SFR($z$) functions based on a careful review of the uncertainties in the normalization resulting from dust corrections, assumptions about the IMF, and uncertainties in the employed calibration relations. Despite significant variation evident from their

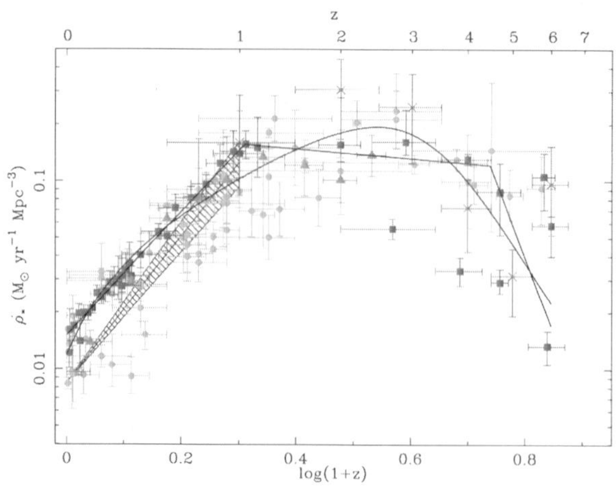


**Fig. 7.38** The cosmic star formation rate density as a function of redshift, as first shown in the influential papers by Madau and collaborators (Madau et al., 1996; Madau, 1997). The key features of SFR($z$) revealed at tha time are a surprisingly rapid increase with look-back time (redshift) and a peak in the $z \sim 1-2$ range. Adapted from Hopkins and Beacom (2006)

compilation, the SFR($z$) appears to be established to within a factor of two back to the epoch at which the reionization of the Universe was complete (at $z \sim 6$).

Extending the Madau plot to redshifts much larger than those reachable with supernovae is an important goal for observational cosmology. Directly tracing SFR($z$) back to the first stars is an even more daunting task, and first efforts to find supernovae from the first generation of stars have been undertaken (Frost et al., 2009). However, GRBs offer a unique and likely much easier way to accomplish this goal.

### *7.9.1 Observational Implications of the Evolving Star Formation Rate*

There are several multi-messenger signals associated with star formation and its associated stellar death we can probe with cumulative measures:

- *The Cosmic Gamma-Ray Background (CGB)*: Gamma-rays escaping from supernovae produce a cumulative background in the MeV regime (Fig. 7.39), which can be used to constrain the global star formation history (Watanabe et al., 2003). Although Type Ia supernovae have a lower rate than core collapse supernovae, the are the dominant supernova contributor to this MeV background, because they produce more radioactive nickel and have a larger $\gamma$-ray escape fraction due to the absence of a massive envelope. With reasonable assumptions about the SNIa delay time distribution (see Sect. 7.3.2 and A.2.1) and SFR($z$), the diffuse $\gamma$-ray background detected in the MeV band can be explained in principle.

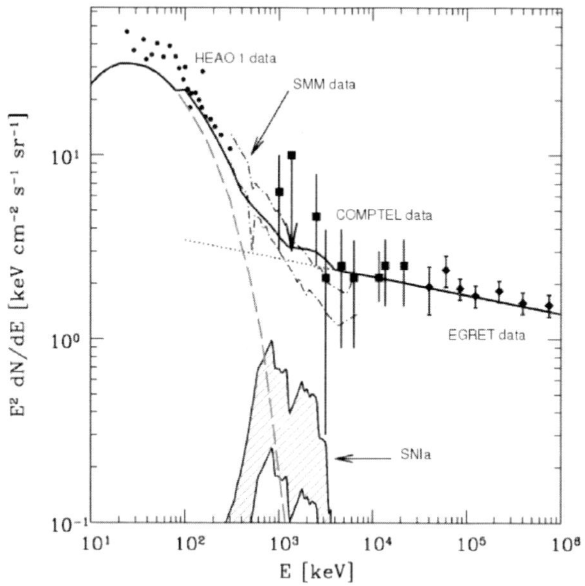


**Fig. 7.39** The diffuse, extragalactic $\gamma$-ray background in the MeV range. The gap between established contributions from Seyfert galaxies on the low-energy side, and from blazars on the high-energy side can be filled with $\gamma$-rays from supernovae, decaying dark matter, and a subset of MeV-blazars. SNIa contribute a significant fraction of the flux observed in the MeV band, but recent *Swift* observations suggest that the bulk of the emission in this regime may originate in a class of AGNs known as blazars. Adapted from Beacom (2006)

Seyfert galaxies do not contribute in this energy regime, but some blazars do, as established with the COMPTEL instrument aboard the Compton Gamma Ray Observatory (CGRO) (Bloemen et al., 1995). Recent observations with the BAT instrument on *Swift* suggest that a strongly evolving population of flat-spectrum radio quasars (FSRQs) may be the underlying source population that dominates the emission in this band (Ajello et al., 2009). The MeV background is thus explained by a mixture of photons from radioactive material from supernovae and photons originating from accretion flows in super-massive black holes at the centers of active galaxies. An accurate determination of SFR($z$) and continued observations of blazars will eventually allow accurate estimates of the ratio between these two contributing categories, and thus constrain the production of radioactive isotopes in supernovae.

- *The Cosmic Neutrino Background (CNB)*: The measurement of SFR($z$) is also responsible for a diffuse, extragalactic neutrino background in the 10 MeV-window, which may soon be detectable by Super-Kamiokande (Horiuchi et al., 2009). Supernova neutrinos probe the dense, hot interiors of proto- neutron stars at the heart of core-collapse SNe. In addition to direct SFR measures from rates of SNIa and SNII in the nearby universe (the limit of direct SN searches is approaching z 2, e.g., Riess et al. (2001)), the CGB and CNB offer two valuable, cumulative measurements of the SFR($z$) function. It must be mentioned that all these methods require a conversion from event rates to star formation rates, which inevitably involves the Initial Mass Function and various assumptions about its evolution and dependence on environment. Neutrinos from the Sun have revealed vital information on stellar models, and similarly the CNB may yield valuable constraints on the final stages of collapsing massive stars.

- *Extragalactic Background Light (EBL)*: The electromagnetic sky background from the radio- to the $\gamma$-ray regime is in part truly diffuse and in part composed of emission from unresolved populations of various sources. The truly diffuse Cosmic Microwave Background (CMB) is a relic of the early history of the universe, but all other bands measure integrated contributions from stars and accreting sources during the buildup of galaxies and clusters of galaxies. Knowledge of SFR($z$) is critical in the re-construction of this meta-galactic radiation field as a function of time (e.g., Kneiske et al., 2002; Primack et al., 2008). We can utilize the diffuse extragalactic background spectrum, and its evolution in time, as a tracer of cosmic evolution. Generations of stars produced a growing UV-Optical-IR component of the EBL, and reprocessing by dust in the ISM added further IR components. This background destroys photons of very high energies via pair creation, which allows us to probe EBL($z$) with appropriate sources. While TeV emission from nearby blazars at $z < 0.1$ is strongly suppressed, GeV emission from GRBs is suppressed as well, but only after propagation through much greater distances (e.g., Hartmann, 2007; Stecker, 2007). The opacity of the universe thus directly relates to the cosmic star formation history, and the accurate determination of SFR($z$) is an essential requirement for all models of photon – photon pair creation extinction corrections to be applied to GeV-TeV sources. NASA's Fermi mission is now creating a rapidly growing data base of GeV-spectra of nearby AGNs and more distant GRBs. The Italian AGILE satellite is contributing as well, and ground-based TeV observatories, H.E.S.S., MAGIC, and VERITAS, are complementing this with TeV data. The utilization of these data for studying black hole accretion in AGNs and relativistic jets in GRBs requires corrections for high-energy extinction via pair creation off the EBL, and thus implies the need for measuring SFR($z$). It is important to re-emphasize that the nearby universe quickly becomes opaque at TeV energies, but that the range of GeV photons is much larger, so that it is crucial to establish the photon SED in the Universe as a function of time, and not just the present-day spectrum, i.e., the SED of the EBL at $z = 0$. This task is thus connected to the task of measuring SFR($z$). The synergy between space-based GeV observations and ground-based TeV observations is leading to startling results on the EBL, raising questions on the transparency (or opaqueness) of the universe (Aharonian et al., 2008; Stecker and Scully, 2009). Recent reviews of high-energy astrophysics in the TeV regime can be found in Aharonian et al. (2008) and Hinton and Hofmann (2009). For a recent discussion of the contribution of stars, via SFR(z), to the EBL see Razzaque et al. (2009).

Cosmic star formation and galaxy evolution occupies a central role in the developing frontiers of cosmology. As is the case for galactic chemical evolution, star formation is the quintessential driver of galaxy evolution. In this section, we discussed various tracers of the cosmic star formation history, having by no means exhausted all the exciting aspects this field entails. We pointed to direct probes of cosmic chemical evolution with GRB afterglow spectroscopy, which can lead us through the era of reionization to the very first stars, and we discussed the various backgrounds resulting from cosmic star formation and stellar demise. Radioactivity is an integral part of this field, contributing in unique ways, but the field of cosmic chemical evolution is just emerging. We build on our experience with galactic chemical evolution, and complement it in many ways. The observational challenges of probing abundances at $z >> 1$ are tremendous, but judging from progress made in past decades we will continue to improve our basic understanding of how the baryons made in stars were transported and processed in the interstellar medium and in the large-scale filamentary structures of the cosmic web.

## References


Agaronyan, F.A., Atoyan, A.M.: On the origin of the galactic annihilation radiation. Sov. Astron. Lett. **7**, 395 (1981)

Aharonian, F., Buckley, J., Kifune, T., Sinnis, G.: High energy astrophysics with ground-based gamma ray detectors. Rep. Prog. Phys. **71**(9), 096901 (2008). doi:10.1088/0034-4885/71/9/096901

Ajello, M., Costamante, L., Sambruna, R.M., Gehrels, N., Chiang, J., Rau, A., Escala, A., Greiner, J., Tueller, J., Wall, J.V., Mushotzky, R.F.: The evolution of swift/BAT Blazars and the origin of the MeV background. ApJ **699**, 603–625 (2009). doi:10.1088/0004-637X/699/1/603, 0905.0472

Armstrong, J.W., Rickett, B.J., Spangler, S.R.: Electron density power spectrum in the local interstellar medium. ApJ **443**, 209–221 (1995). doi:10.1086/175515

Ascasibar, Y., Jean, P., Boehm, C., Knödlseder, J: Constraints on dark matter and the shape of the Milky Way dark halo from the 511-keV line. MNRAS **368**, 1695–1705 (2006). doi:10.1111/j.1365-2966.2006.10226.x, arXiv:astro-ph/0507142

Asplund, M., Grevesse, N., Sauval, A.J., Scott, P.: The chemical composition of the sun. ARA&A **47**, 481–522 (2009). doi:10.1146/annurev.astro.46.060407.145222, 0909.0948

Audouze, J., Tinsley, B.M.: Chemical evolution of galaxies. ARA&A **14**, 43–79 (1976). doi:10.1146/annurev.aa.14.090176.000355

Bally, J.: Overview of the Orion Complex. In: Reipurth, B. (ed.) Handbooks of Star Forming Regions, vol. I, pp. 459 (2008)

Beacom, J.F.: The cosmic stellar birth and death rates. New Astron Rev 50, 561–565 (2006). arXiv:astro-ph/0602101

Beacom, J.F., Yüksel, H.: Stringent constraint on galactic positron production. Phys. Rev. Lett. **97**(7), 071102 (2006). doi:10.1103/PhysRevLett.97.071102, arXiv:astro-ph/0512411

Bell, E.F., Zucker, D.B., Belokurov, V., Sharma, S., Johnston, K.V., Bullock, J.S., Hogg, D.W., Jahnke, K., de Jong, J.T.A., Beers, T.C., Evans, N.W., Grebel, E.K., Ivezic, Z., Koposov, S.E., Rix, H.W., Schneider, D.P., Steinmetz, M., Zolotov, A.: The accretion origin of the milky way's stellar halo. ArXiv e-prints 706, 0706.0004 (2007)

Bienaymé, O., Soubiran, C., Mishenina, T.V., Kovtyukh, V.V., Siebert, A.: Vertical distribution of galactic disk stars. A&A **446**, 933–942, arXiv:astro-ph/0510431 (2006)

Binns, W.R., Wiedenbeck, M.E., Arnould, M., Cummings, A.C., George, J.S., Goriely, S., Israel, M.H., Leske, R.A., Mewaldt, R.A., Meynet, G., Scott, L.M., Stone, E.C., von Rosenvinge, T.T.:

Cosmic-ray neon, wolf-rayet stars, and the superbubble origin of galactic cosmic rays. ApJ **634**, 351–364, arXiv:astro-ph/0508398 (2005)

Bird, A.J., Malizia, A., Bazzano, A., Barlow, E.J., Bassani, L., Hill, A.B., Bélanger, G., Capitanio, F., Clark, D.J., Dean, A.J., Fiocchi, M., Götz, D., Lebrun, F., Molina, M., Produit, N., Renaud, M., Sguera, V., Stephen, J.B., Terrier, R., Ubertini, P., Walter, R., Winkler, C., Zurita, J.: The third ibis/isgri soft gamma-ray survey catalog. ApJS **170**, 175–186 (2007)

Blair, W.P., Sankrit, R., Raymond, J.C.: Hubble space telescope imaging of the primary shock front in the Cygnus loop supernova remnant. AJ **129**, 2268–2280 (2005). doi:10.1086/429381

Bloemen, H., Bennett, K., Blom, J.J., Collmar, W., Hermsen, W., Lichti, G.G., Morris, D., Schoenfelder, V., Stacy, J.G., Strong, A.W.,Winkler, C.: New COMPTEL gamma-ray source (GRO J0516-609) near PKS 0506-612/0522-611: first evidence for "MeV Blazars". A&A **293**, L1–L4 (1995)

Boggs, S., Kurfess, J., Ryan, J., Aprile, E., Gehrels, N., Kippen, M., Leising, M., Oberlack, U., Wunderer, C., Zych, A., Bloser, P., Harris, M., Hoover, A., Klimenk, A., Kocevski, D., McConnell, M., Milne, P., Novikova, E.I., Phlips, B., Polsen, M., Sturner, S., Tournear, D., Weidenspointner, G., Wulf, E., Zoglauer, A., Baring, M., Beacom, J., Bildsten, L., Dermer, C., Hartmann, D., Hernanz, M., Smith, D., Starrfield, S.: The advanced compton telescope. In: Society of Photo-Optical Instrumentation Engineers (SPIE) Conference Series, Presented at the Society of Photo-Optical Instrumentation Engineers (SPIE) Conference, vol. 6266 (2006). doi:10.1117/12.670605

Boissier, S., Prantzos, N.: Chemo-spectral evolution of the Milky Way and of spiral disks. Ap&SS **265**, 409–410 (1999). doi:10.1023/A:1002181826858

Boissier, S., Prantzos, N., Boselli, A., Gavazzi, G.: The star formation rate in disc galaxies: thresholds and dependence on gas amount. MNRAS **346**, 1215–1230, arXiv:astro-ph/0309372 (2003)

Bouchet, L., Jourdain, E., Roques, J.P., Strong, A., Diehl, R., Lebrun, F., Terrier, R.: Integral spi all-sky view in soft gamma rays: a study of point-source and galactic diffuse emission. ApJ **679**, 1315–1326, 0801.2086 (2008)

Boulanger, F., Abergel, A., Bernard, J.P., Burton, W.B., Desert, F.X., Hartmann, D., Lagache, G., Puget, J.L.: The dust/gas correlation at high Galactic latitude. A&A **312**, 256–262 (1996)

Boulanger, F., Perault, M.: Diffuse infrared emission from the galaxy. I – Solar neighborhood. ApJ **330**, 964–985 (1988). doi:10.1086/166526

Breitschwerdt, D.: Modeling the local interstellar medium. Ap&SS **276**, 163–176 (2001)

Breitschwerdt, D., de Avillez, M.A.: The history and future of the Local and Loop I bubbles. A&A **452**, L1–L5 (2006). doi:10.1051/0004-6361:20064989, arXiv:astro-ph/0604162

Breitschwerdt, D., Egger, R., Freyberg, M.J., Frisch, P.C., Vallerga, J.V.: The local bubble origin and evolution. Space Sci. Rev. **78**, 183–198 (1996). doi:10.1007/BF00170805

Breitschwerdt, D., Freyberg, M.J., Truemper, J. (eds.): The Local Bubble and Beyond. Lecture Notes in Physics, vol. 506, Berlin Springer Verlag (1998). doi:10.1007/BFb0104684

Brown, A.G.A., de Geus, E.J., de Zeeuw, P.T.: The Orion OB1 association. 1: stellar content. A&A **289**, 101–120, arXiv:astro-ph/9403051 (1994)

Brown, T.M., Smith, E., Ferguson, H.C., Rich, R.M., Guhathakurta, P., Renzini, A., Sweigart, A.V., Kimble, R.A.: The detailed star formation history in the spheroid, outer disk, and tidal stream of the andromeda galaxy. ApJ **652**, 323–353 (2006). doi:10.1086/508015, arXiv:astro-ph/0607637

Burrows, D.N., Singh, K.P., Nousek, J.A., Garmire, G.P., Good, J.: A multiwavelength study of the Eridanus soft X-ray enhancement. ApJ **406**, 97–111 (1993). doi:10.1086/172423

Calzetti, D., Kennicutt, R.C.: The new frontier: galactic-scale star formation. PASP **121**, 937–941 (2009). doi:10.1086/605617, 0907.0203

Cameron, A.G.W., Truran, J.W.: The chemical evolution of the galaxy. J. Roy. Soc. Can. **65**, 1 (1971)

Case, G.L., Bhattacharya, D.: A new sigma -d relation and its application to the galactic supernova remnant distribution. ApJ **504**, 761, arXiv:astro-ph/9807162 (1998)

Cash, W., Charles, P., Bowyer, S., Walter, F., Garmire, G., Riegler, G.: The X-ray superbubble in Cygnus. ApJ **238**, L71–L76 (1980). doi:10.1086/183261

Casse, M., Paul, J.A.: On the stellar origin of the ne-22 excess in cosmic rays. ApJ **258**, 860–863 (1982)

Cerviño, M., Knödlseder, J., Schaerer, D., von Ballmoos, P., Meynet, G.: Gamma-ray line emission from OB associations and young open clusters. I. Evolutionary synthesis models. A&A **363**, 970–983, arXiv:astro-ph/0010283 (2000)

Chatterjee, S., Cordes, J.M., Vlemmings, W.H.T., Arzoumanian, Z., Goss, W.M., Lazio, T.J.W.: Pulsar parallaxes at 5 GHz with the very long baseline array. ApJ **604**, 339–345 (2004). doi:10.1086/381748, arXiv:astro-ph/0312044

Chen, W., Diehl, R., Gehrels, N., Hartmann, D., Leising, M., Naya, J.E., Prantzos, N., Tueller, J., von Ballmoos, P.: Implications of the broad 26Al 1809 keV line observed by GRIS. In: Winkler, C., Courvoisier, T.J.L., Durouchoux, P. (eds.) The Transparent Universe, vol. 382, pp. 105. ESA Special Publication (1997)

Cheng, L.X., Leventhal, M., Smith, D.M.: ApJ **481**, L43 (1997)

Chiappini, C., Matteucci, F., Gratton, R.: The chemical evolution of the galaxy: the Two-Infall model. ApJ **477**, 765 (1997). doi:10.1086/303726, arXiv:astro-ph/9609199

Churazov, E., Sunyaev, R., Sazonov, S., Revnivtsev, M., Varshalovich, D.: Positron annihilation spectrum from the Galactic Centre region observed by SPI/INTEGRAL MNRAS **357**, 1377 (2005)

Clayton, D.D.: Principles of Stellar Evolution and Nucleosynthesis. McGraw-Hill, New York (1968)

Clayton, D.D.: Cosmic radioactivity – a gamma-ray search for the origins of atomic nuclei. In: Essays in Nuclear Astrophysics, pp. 404. Cambridge University Press, Cambridge UK (1982)

Clayton, D.D.: Galactic chemical evolution and nucleocosmochronology: a standard model. In: Arnett, W.D., Truran, J.W. (eds.) Nucleosynthesis: challenges and new developments, pp. 65 (1985)

Clayton, D.D.: Nuclear cosmochronology within analytic models of the chemical evolution of the solar neighbourhood. MNRAS **234**, 1–36 (1988)

Clayton, D.D., Nittler, L.R.: Astrophysics with presolar stardust. ARA&A **42**, 39–78 (2004). doi:10.1146/annurev.astro.42.053102.134022

Cole, A.A., Skillman, E.D., Tolstoy, E., Gallagher, J.S., III, Aparicio, A., Dolphin, A.E., Gallart, C., Hidalgo, S.L., Saha, A., Stetson, P.B., Weisz, D.R., Leo, A.: A Late-blooming survivor of the epoch of reionization in the local group. ApJ **659**, L17–L20 (2007). doi:10.1086/516711, arXiv:astroph/0702646

Comerón, F., Pasquali, A., Figueras, F., Torra, J.: The outskirts of Cygnus OB2. A&A **486**, 453–466 (2008). doi:10.1051/0004-6361:200809917, 0806.1891

Cordes, J.M., Lazio, T.J.W.: NE2001.I. A New Model for the Galactic Distribution of Free Electrons and its Fluctuations. ArXiv Astrophysics e-prints astro-ph/0207156 (2002)

Cowley, C.: Introduction to Cosmochemistry. Cambridge University Press, Cambridge (1995)

Dame, T.M.: The distribution of neutral gas in the milky way. In: Holt, S.S., Verter, F. (eds.) Back to the Galaxy. American Institute of Physics Conference Series, vol. 278, pp. 267–278 (1993)

de Avillez, M.A., Breitschwerdt, D.: Volume filling factors of the ISM phases in star forming galaxies. I. The role of the disk-halo interaction. A&A **425**, 899–911 (2004). doi:10.1051/0004-6361:200400025, arXiv:astro-ph/0407034

de Geus, E.J.: Interactions of stars and interstellar matter in Scorpio Centaurus. A&A **262**, 258–270 (1992)

de Geus, E.J., de Zeeuw, P.T., Lub, J.: Physical parameters of stars in the Scorpio-Centaurus OB association. A&A **216**, 44–61 (1989)

del Rio, E., von Ballmoos, P., Bennett, K., Bloemen, H., Diehl, R., Hermsen, W., Knoedlseder, J., Oberlack, U., Ryan, J., Schoenfelder, V., Winkler, C.: 1.8 MeV line emission from the Cygnus region. A&A **315**, 237–242 (1996) de Zeeuw, P.T., Hoogerwerf, R., de Bruijne, J.H.J., Brown, A.G.A., Blaauw, A.: A HIPPARCOS census of the nearby OB associations. AJ **117**, 354–399 (1999). doi:10.1086/300682, arXiv:astroph/9809227

Diehl, R.: $^{26}$Al production in the Vela and Orion regions. New Astron. Rev. **46**, 547–552 (2002). doi:10.1016/S1387-6473(02)00199-9

Diehl, R., Dupraz, C., Bennett, K., Bloemen, H., HermsenW, Knoedlseder, J., Lichti, G., Morris, D., Ryan, J., Schoenfelder, V., Steinle, H., Strong, A., Swanenburg, B., Varendorff, M., Winkler, C.: COMPTEL observations of Galactic $^{26}$Al emission. A&A **298**, 445 (1995)

Diehl, R., Knödlseder, J., Lichti, G.G., Kretschmer, K., Schanne, S., Schönfelder, V., Strong, A.W., von Kienlin, A., Weidenspointner, G., Winkler, C., Wunderer, C.: SPI measurements of Galactic $^{26}$Al. A&A **411**, L451–L455 (2003a). doi:10.1051/0004-6361:20031347, arXiv:astro-ph/0309097

Diehl, R., Kretschmer, K., Plüschke, S., Cerviño, M., Hartmann, D.H.: Gamma-rays from massive stars in Cygnus and Orion. In: van der Hucht, K., Herrero, A., Esteban, C. (eds.) A Massive Star Odyssey: From Main Sequence to Supernova. IAU Symposium, vol. 212, pp. 706 (2003b)

Diehl, R., Halloin, H., Kretschmer, K., Lichti, G.G., Schönfelder, V., Strong, A.W., von Kienlin, A., Wang, W., Jean, P., Knödlseder, J., Roques, J.P., Weidenspointner, G., Schanne, S., Hartmann, D.H., Winkler, C., Wunderer, C.: Radioactive $^{26}$Al from massive stars in the galaxy. Nature **439**, 45–47, arXiv:astro-ph/0601015 (2006a)

Diehl, R., Halloin, H., Kretschmer, K., Strong, A.W., Wang, W., Jean, P., Lichti, G.G., Knödlseder, J., Roques, J.P., Schanne, S., Schönfelder, V., von Kienlin, A., Weidenspointner, G., Winkler, C., Wunderer, C.: $^{26}$Al in the inner Galaxy. Large-scale spectral characteristics derived with SPI/INTEGRAL. A&A **449**, 1025–1031 (2006b). doi:10.1051/0004-6361:20054301, arXiv:astroph/0512334

Diehl, R., Prantzos, N., von Ballmoos, P.: Astrophysical constraints from gamma-ray spectroscopy. Nucl. Phys. A **777**, 70–97 (2006c). doi:10.1016/j.nuclphysa.2005.02.155, arXiv:astroph/0502324

Dupraz, C., Bloemen, H., Bennett, K., Diehl, R., Hermsen, W., Iyudin, A.F., Ryan, J., Schoenfelder, V.: COMPTEL three-year search for galactic sources of ^44^Ti gamma-ray line emission at 1.157MeV. A&A **324**, 683–689 (1997)

Ellison, D.C., Drury, L.O., Meyer, J.: Galactic cosmic rays from supernova remnants. ii. shock acceleration of gas and dust. ApJ **487**, 197, arXiv:astro-ph/9704293 (1997)

Endt, P.M.: Energy levels of A = 21–44 nuclei (VII). Nucl. Phys. A **521**, 1–400 (1990). doi:10.1016/0375-9474(90)90598-G

Fernández, D., Figueras, F., Torra, J.: On the kinematic evolution of young local associations and the Scorpius-Centaurus complex. A&A **480**, 735–751 (2008). doi:10.1051/0004-6361:20077720, 0801.0605

Ferrière, K.M.: The interstellar environment of our galaxy. Rev. Mod. Phys. **73**, 1031–1066, arXiv:astro-ph/0106359 (2001)

Ferrière, K., Gillard, W., Jean, P.: Spatial distribution of interstellar gas in the innermost 3 kpc of our galaxy. A&A **467**, 611–627 (2007). doi:10.1051/0004-6361:20066992

Figer, D.F.: Massive star formation in the galactic center. ArXiv e-prints 803, 0803.1619 (2008)

Figer, D.F., Rich, R.M., Kim, S.S., Morris, M., Serabyn, E.: An extended star formation history for the galactic center from hubble space telescope nicmos observations. ApJ **601**, 319–339, arXiv:astro-ph/0309757 (2004)

Flynn, C., Holmberg, J., Portinari, L., Fuchs, B., Jahreiß, H.: On the mass-to-light ratio of the local galactic disc and the optical luminosity of the galaxy. MNRAS **372**, 1149–1160, arXiv:astroph/0608193 (2006)

Forestini, M., Charbonnel, C.: Nucleosynthesis of light elements inside thermally pulsing AGB stars: I. The case of intermediate-mass stars. A&AS **123**, 241–272 (1997). doi:10.1051/aas:1997348

François, P., Matteucci, F., Cayrel, R., Spite, M., Spite, F., Chiappini, C.: The evolution of the Milky Way from its earliest phases: constraints on stellar nucleosynthesis. A&A **421**, 613–621 (2004). doi:10.1051/0004-6361:20034140, arXiv:astro-ph/0401499

Fransson, C., Kozma, C.: Radioactivities and nucleosynthesis in SN 1987A. New Astron. Rev. **46**, 487–492 (2002). doi:10.1016/S1387-6473(02)00188-4, arXiv:astro-ph/0112405

Freudenreich, H.T.: A cobe model of the galactic bar and disk. ApJ **492**, 495, arXiv:astroph/9707340 (1998)

Frisch, P.C.: Characteristics of Nearby Interstellar Matter. Space Sci. Rev. **72**, 499–592 (1995). doi:10.1007/BF00749006

Frisch, P.C., Bzowski, M., Grün, E., Izmodenov, V., Krüger, H., Linsky, J.L., McComas, D.J., Möbius, E., Redfield, S., Schwadron, N., Shelton, R., Slavin, J.D., Wood, B.E.: The galactic environment of the sun: interstellar material inside and outside of the heliosphere. Space Sci. Rev. **146**, 235–273 (2009). doi:10.1007/s11214-009-9502-0

Frost, M.I., Surace, J., Moustakas, L.A., Krick, J.: A pilot search for population III supernova candidates in the Spitzer/IRAC dark field. ApJ **698**, L68–L71 (2009). doi:10.1088/0004-637X/698/1/L68, 0905.4494

Gaidos, E., Krot, A.N., Williams, J.P., Raymond, S.N.: $^{26}$Al and the formation of the solar system from a molecular cloud contaminated by Wolf-Rayet Winds. ApJ **696**, 1854–1863 (2009). doi:10.1088/0004-637X/696/2/1854, 0901.3364

Gehrz, R.D., Truran, J.W., Williams, R.E., Starrfield, S.: Nucleosynthesis in classical novae and its contribution to the interstellar medium. PASP **110**, 3–26 (1998). doi:10.1086/316107

Genzel, R., Stutzki, J.: The Orion molecular cloud and star-forming region. ARA&A **27**, 41–85 (1989). doi:10.1146/annurev.aa.27.090189.000353

Gillessen, S., Eisenhauer, F., Trippe, S., Alexander, T., Genzel, R., Martins, F., Ott, T.: Monitoring stellar orbits around the massive black hole in the galactic center. ArXiv e-prints 0810.4674 (2008)

Gilmore, G., Reid, N.: New light on faint stars. iii – galactic structure towards the south pole and the galactic thick disc. MNRAS **202**, 1025–1047 (1983)

Goldsmith, P.F.: Molecular clouds – an overview. In: Hollenbach, D.J., Thronson, H.A., Jr. (eds.) Interstellar Processes, Astrophysics and Space Science Library, vol. 134, pp. 51–70, D. Reidel Publ. Co., Dordrecht (1987)

Goswami, A., Prantzos, N.: Chemical evolution of intermediate mass nuclei in the solar neighbourhood and the halo of the galaxy. Bull. Astron. Soc. India **28**, 305–308 (2000)

Greggio, L.: The rates of type ia supernovae. i. Analytical formulations. A&A **441**, 1055–1078, arXiv:astro-ph/0504376 (2005)

Greiner, J., Iyudin, A., Kanbach, G., Zoglauer, A., Diehl, R., Ryde, F., Hartmann, D., Kienlin, A.V., McBreen, S., Ajello, M., Bagoly, Z., Balasz, L.G., Barbiellini, G., Bellazini, R., Bezrukov, L., Bisikalo, D.V., Bisnovaty-Kogan, G., Boggs, S., Bykov, A., Cherepashuk, A.M., Chernenko, A., CollmarW, Dicocco, G., Dröge, W., Gierlik, M., Hanlon, L., Horvath, I., Hudec, R., Kiener, J., Labanti, C., Langer, N., Larsson, S., Lichti, G., Lipunov, V.M., Lubsandorgiev, B.K., Majczyna, A., Mannheim, K., Marcinkowski, R., Marisaldi, M., McBreen, B., Meszaros, A., Orlando, E., Panasyuk, M.I., Pearce, M., Pian, E., Poleschuk, R.V., Pollo, A., Pozanenko, A., Savaglio, S., Shustov, B., Strong, A., Svertilov, S., Tatischeff, V., Uvarov, J., Varshalovich, D.A., Wunderer, C.B., Wrochna, G., Zabrodskij, A.G., Zeleny, L.M.: Gammaray burst investigation via polarimetry and spectroscopy (GRIPS). Exp. Astron. **23**, 91 (2008). doi:10.1007/s10686-008-9102-0

Grimm, H.J., Gilfanov, M., Sunyaev, R.: The milky way in x-rays for an outside observer. log(n)–log(s) and luminosity function of x-ray binaries from rxte/asm data. A&A **391**, 923–944, arXiv:astro-ph/0109239 (2002)

Groenewegen, M.A.T., Udalski, A., Bono, G.: The distance to the galactic centre based on population ii cepheids and rr lyrae stars. A&A **481**, 441–448, arXiv:0801.2652 (2008)

Guessoum, N., Jean, P., Gillard, W.: The lives and deaths of positrons in the interstellar medium. A&A **436**, 171–185 (2005). doi:10.1051/0004-6361:20042454, arXiv:astro-ph/0504186

Guibert, J., Lequeux, J., Viallefond, F.: Star formation in interstellar gas density in our galaxy. A&A **68**, 1–2 (1978)

Han, J.L.: The large-scale magnetic field structure of our galaxy: efficiently deduced from pulsar rotation measures. In: Uyaniker, B., Reich, W., Wielebinski, R. (eds.) The Magnetized Interstellar Medium, pp. 3–12 (2004)

Harris, M.J., Knödlseder, J., Jean, P., Cisana, E., Diehl, R., Lichti, G.G., Roques, J.P., Schanne, S., Weidenspointner, G.: Detection of $\gamma$-ray lines from interstellar 60Fe by the high resolution spectrometer SPI. A&A **433**, L49–L52 (2005). doi:10.1051/0004-6361:200500093, arXiv:astroph/0502219

Harris, M.J., Teegarden, B.J., Cline, T.L., Gehrels, N., Palmer, D.M., Ramaty, R., Seifert, H.: Transient gamma-ray spectrometer measurements of the positron annihilation spectrum from the galactic center. ApJ **501**, L55 (1998). doi:10.1086/311429, arXiv:astro-ph/9803247

Hartmann, D., Burton, W.B.: Atlas of Galactic Neutral Hydrogen. Cambridge Unversity, Cambridge (1997)

Hartmann, D.H.: Probing the extragalactic background with GLAST. In: Ritz, S., Michelson, P., Meegan, C.A. (eds.) The First GLAST Symposium, American Institute of Physics Conference Series, vol. 921, pp. 24–25 (2007). doi:10.1063/1.2757259

Hartmann, D., Kouveliotou, C., Piro, L., den Herder, J.W., Ohashi, T., Abel, T., Amati, L., Barthelmy, S., Beacom, J., Bloom, J., Bonamente, M., Branchini, E., Branduardi-Raymont, G., Bregman, J., Briggs, M., Bromm, V., Burkert, A., Burrows, D., Campana, S., Carilli, C., Cen, R., Coppi, P., Danforth, C., Diehl, R., Ettori, S., Fall, M., Fan, X., Fishman, G., Fields, B., Forman, C., Ghisellini, G., Galeazzi, M., Gehrels, N., Ghirlanda, G., Grindlay, J., Heger, A., Henry, P., Hermsen, W., Holland, A., Hughes, J., Kaastra, J., Kawai, N., Keel, B., Kelley, R., Kippen, M., Kusenko, A., Loeb, A., Madau, P., Matteucci, F., Mathews, G., Meegan, C., Meszaros, P., Mineo, T., Mitsuda, K., Molendi, S., Natalucci, L., Nomoto, K., O'Brien, P., O'Dell, S., Paerels, F., Pareschi, G., Petrosian, V., Prantzos, N., Primack, J., Prochaska, J., Ramirez-Ruiz, E., Ramsey, B., Rasmussen, A., Savaglio, S., Schaye, J., Snowden, S., Springel, V., Suto, Y., Tagliaferri, G., Takei, Y., Tawara, Y., Timmes, F., Townsely, L., Ubertini, P., van der Horst, A., Vink, J., Weisskopf, M., Wijers, R., Wilson-Hodge, C., Woosley, S., Yamasaki, N.: Reading the metal diaries of the universe: tracing cosmic chemical evolution. In: AGB Stars and Related Phenomenastro 2010: The Astronomy and Astrophysics Decadal Survey, Astronomy, vol. 2010, pp. 114 (2009)

Hartmann, D.H.: Tracing cosmic chemical evolution with GRBs. In: Meegan, C., Kouveliotou, C., Gehrels, N. (eds.) American Institute of Physics Conference Series, American Institute of Physics Conference Series, vol. 1133, pp. 464–466 (2009). doi:10.1063/1.3155952

Hartmann, D.H., Woosley, S.E.: The cosmic supernova neutrino background. Astropart. Phys. **7**, 137–146 (1997). doi:10.1016/S0927-6505(97)00018-2

Heckman, T.M., Armus, L., Miley, G.K.: On the nature and implications of starburst-driven galactic superwinds. ApJS **74**, 833–868 (1990). doi:10.1086/191522

Heesen, V., Beck, R., Krause, M., Dettmar, R.J.: Cosmic rays and the magnetic field in the nearby starburst galaxy ngc 253. i. The distribution and transport of cosmic rays. A&A **494**, 563–577, 0812.0346 (2009)

Higdon, J.C., Lingenfelter, R.E., Ramaty, R.: Cosmic-ray acceleration from supernova ejecta in superbubbles. ApJ **509**, L33–L36 (1998)

Higdon, J.C., Lingenfelter, R.E., Rothschild, R.E.: The galactic positron annihilation radiation and the propagation of positrons in the interstellar medium. ApJ **698**, 350–379 (2009)

Hinton, J.A., Hofmann, W.: Teraelectronvolt astronomy. ARA&A **47**, 523–565 (2009). doi:10.1146/annurev-astro-082708-101816

Hirschi, R., Meynet, G., Maeder, A.: Rotating massive stars: Pre-SN models and stellar yields at solar metallicity. Nucl. Phys. A **758**, 234–237 (2005). doi:10.1016/j.nuclphysa.2005.05.041, arXiv:astro-ph/0411670

Holmberg, J., Flynn, C.: The local surface density of disc matter mapped by hipparcos. MNRAS **352**, 440–446, arXiv:astro-ph/0405155 (2004)

Hoogerwerf, R., de Bruijne, J.H.J., de Zeeuw, P.T.: The origin of runaway stars. ApJ **544**, L133–L136 (2000). doi:10.1086/317315, arXiv:astro-ph/0007436

Hopkins, A.M., Beacom, J.F.: On the normalization of the cosmic star formation history. ApJ **651**, 142–154 (2006). doi:10.1086/506610, arXiv:astro-ph/0601463

Horiuchi, S., Beacom, J.F., Dwek, E.: Diffuse supernova neutrino background is detectable in Super-Kamiokande. Phys. Rev. D**79**(8), 083013 (2009). doi:10.1103/PhysRevD.79.083013, 0812.3157

Iye, M., Ota, K., Kashikawa, N., Furusawa, H., Hashimoto, T., Hattori, T., Matsuda, Y., Morokuma, T., Ouchi, M., Shimasaku, K.: A galaxy at a redshift $z = 6.96$. Nature **443**, 186–188 (2006). doi:10.1038/nature05104, arXiv:astro-ph/0609393

Iyudin, A.F., Schönfelder, V., Bennett, K., Bloemen, H., Diehl, R., Hermsen, W., Lichti, G.G., van der Meulen, R.D., Ryan, J., Winkler, C.: Emission from 44Ti associated with a previously unknown Galactic supernova. Nature **396**, 142–144 (1998). doi:10.1038/24106

Jappsen, A.K., Klessen, R.S., Larson, R.B., Li, Y., Mac Low, M.M.: The stellar mass spectrum from non-isothermal gravoturbulent fragmentation. A&A **435**, 611–623 (2005). doi:10.1051/0004-6361:20042178, arXiv:astro-ph/0410351

Jean, P., Gillard, W., Marcowith, A., Ferrière, K.: Positron transport in the interstellar medium. A&A **508**, 1099–1116 (2009). doi:10.1051/0004-6361/200809830, 0909.4022

Jean, P., Knödlseder, J., Gillard, W., Guessoum, N., Ferrière, K., Marcowith, A., Lonjou, V., Roques, J.P.: Spectral analysis of the Galactic e+ e− annihilation emission. A&A **445**, 579–589 (2006). doi:10.1051/0004-6361:20053765, arXiv:astro-ph/0509298

Johnson, W.N., III, Harnden, F.R., Jr., Haymes, R.C.: The spectrum of low-energy gamma radiation from the galactic-center region. ApJ **172**, L1 (1972). doi:10.1086/180878

Jones, A.P.: Dust formation, propagation and survival in the ISM. In: Wilson, A. (ed.) ESA Special Publication, ESA Special Publication, vol. 577, pp. 239–244 (2005)

Jones, F.C., Lukasiak, A., Ptuskin, V.S., Webber, W.R.: K-capture cosmic ray secondaries and reacceleration. Adv. Space Res. **27**, 737–741 (2001)

Jose, J., Hernanz, M.: Nucleosynthesis in classical novae: CO versus ONe white dwarfs. ApJ **494**, 680 (1998). doi:10.1086/305244, arXiv:astro-ph/9709153

Jurić, M., Ivezić, Ž, Brooks, A., Lupton, R.H., Schlegel, D., Finkbeiner, D., Padmanabhan, N., Bond, N., Sesar, B., Rockosi, C.M., Knapp, G.R., Gunn, J.E., Sumi, T., Schneider, D.P., Barentine, J.C., Brewington, H.J., Brinkmann, J., Fukugita, M., Harvanek, M., Kleinman, S.J., Krzesinski, J., Long, D., Neilsen, E.H., Jr., Nitta, A., Snedden, S.A., York, D.G.: The milky way tomography with sdss. i. Stellar number density distribution. ApJ **673**, 864–914 (2008)

Kalberla, P.M.W., Dedes, L.: Global properties of the HI distribution in the outer Milky Way. ArXiv e-prints 804, 0804.4831 (2008)

Karakas, A., Lattanzio, J.C.: Stellar models and yields of asymptotic giant branch stars. Publ. Astron. Soc. Aust. **24**, 103–117 (2007). doi:10.1071/AS07021, 0708.4385

Kennicutt, R.C., Jr.: Star formation in galaxies along the hubble sequence. ARA&A **36**, 189–232, arXiv:astro-ph/9807187 (1998a)

Kennicutt, R.C., Jr.: The global Schmidt law in star-forming galaxies. ApJ **498**, 541 (1998b). doi:10.1086/305588, arXiv:astro-ph/9712213

Kinzer, R.L., Purcell, W.R., Johnson, W.N., Kurfess, J.D., Jung, G., Skibo, J.: OSSE observations of positron annihilation in the Galactic plane. A&AS **120**, C317 (1996)

Kneiske, T.M., Mannheim, K., Hartmann, D.H.: Implications of cosmological gamma-ray absorption. I. Evolution of the metagalactic radiation field. A&A **386**, 1–11 (2002). doi:10.1051/0004-6361:20020211, arXiv:astro-ph/0202104

Knie, K., Korschinek, G., Faestermann, T., Dorfi, E.A., Rugel, G., Wallner, A.: $^{60}$Fe anomaly in a deep-sea manganese crust and implications for a nearby supernova source. Physical Rev. Lett. **93**(17), 171103 (2004). doi:10.1103/PhysRevLett.93.171103

Knödlseder, J., Cerviño, M., Le Duigou, J.M., Meynet, G., Schaerer, D., von Ballmoos, P.: Gammaray line emission from OB associations and young open clusters. II. The

Cygnus region. A&A **390**, 945–960 (2002). doi:10.1051/0004-6361:20020799, arXiv:astro-ph/0206045

Knödlseder, J., Dixon, D., Bennett, K., Bloemen, H., Diehl, R., Hermsen, W., Oberlack, U., Ryan, J., Schönfelder, V., von Ballmoos, P.: Image reconstruction of COMPTEL 1.8 MeV (26) AL line data. A&A **345**, 813–825, arXiv:astro-ph/9903172 (1999)

Knödlseder, J., Jean, P., Lonjou, V., Weidenspointner, G., Guessoum, N., Gillard, W., Skinner, G., von Ballmoos, P., Vedrenne, G., Roques, J.P., Schanne, S., Teegarden, B., Schönfelder, V., Winkler, C.: The all-sky distribution of 511 keV electron-positron annihilation emission. A&A **441**, 513–532 (2005). doi:10.1051/0004-6361:20042063, arXiv:astro-ph/0506026

Knödlseder, J., Valsesia, M., Allain, M., Boggs, S., Diehl, R., Jean, P., Kretschmer, K., Roques, J.P., Schönfelder, V., Vedrenne, G., von Ballmoos, P., Weidenspointner, G., Winkler, C.: SPI/INTEGRAL observation of 1,809 keV gamma-ray line emission from the CYGNUS X region. In: Schoenfelder, V., Lichti, G., Winkler, C. (eds.) 5th INTEGRAL Workshop on the INTEGRAL Universe, vol. 552, pp. 33, ESA Special Publication (2004)

Kretschmer et al., in preparation for A&A (2010)

Kroupa, P.: The initial mass function of stars: evidence for uniformity in variable systems. Science **295**, 82–91 (2002). doi:10.1126/science.1067524, arXiv:astro-ph/0201098

Kulkarni, S.R., Heiles, C.: The atomic component. In: Hollenbach, D.J., Thronson, H.A., Jr. (eds.) Interstellar Processes, Astrophysics and Space Science Library, vol. 134, pp. 87–122 (1987)

Kulsrud, R.M.: Plasma Physics for Astrophysics. Princeton University Press, Princeton (2005)

Lallement, R., Welsh, B.Y., Vergely, J.L., Crifo, F., Sfeir, D.: 3D mapping of the dense interstellar gas around the Local Bubble. A&A **411**, 447–464 (2003). doi:10.1051/0004-6361:20031214

Launhardt, R., Zylka, R., Mezger, P.G.: The nuclear bulge of the galaxy. iii. Large-scale physical characteristics of stars and interstellar matter. A&A **384**, 112–139, arXiv:astro-ph/0201294 (2002)

Leventhal, M., MacCallum, C.J., Stang, P.D.: Detection of 511 keV positron annihilation radiation from the galactic center direction. ApJ **225**, L11–L14 (1978). doi:10.1086/182782

Lilly, S.J., Le Fevre, O., Hammer, F., Crampton, D.: The Canada-France Redshift Survey: the luminosity density and star formation history of the Universe to Z approximately 1. ApJ **460**, L1 (1996). doi:10.1086/309975, arXiv:astro-ph/9601050

Limongi, M., Chieffi, A.: The nucleosynthesis of $^{26}$Al and $^{60}$Fe in solar metallicity stars extending in mass from 11 to 120 M<??>: The hydrostatic and explosive contributions. ApJ **647**, 483–500 (2006). doi:10.1086/505164, arXiv:astro-ph/0604297

Lingenfelter, R.E., Ramaty, R.: Gamma-ray lines – A new window to the Universe. Phys. Today **31**, 40–47 (1978)

Liszt, H.S., Burton, W.B.: The gas distribution in the central region of the Galaxy. III – A barlike model of the inner-Galaxy gas based on improved H I data. ApJ **236**, 779–797 (1980). doi:10.1086/157803

Lozinskaya, T.A., Pravdikova, V.V., Finoguenov, A.V.: Searches for the shell swept up by the Stellar Wind from Cyg OB2. Astron. Lett. **28**, 223–236 (2002). doi:10.1134/1.1467257, arXiv:astro-ph/0203015

Lyne, A.G., Manchester, R.N., Taylor, J.H.: The galactic population of pulsars. MNRAS **213**, 613–639 (1985)

Madau, P.: Cosmic star formation history. In: Holt, S.S., Mundy, L.G. (eds.) American Institute of Physics Conference Series, American Institute of Physics Conference Series, vol. 393, pp. 481–490 (1997). doi:10.1063/1.52821

Madau, P., Ferguson, H.C., Dickinson, M.E., Giavalisco, M., Steidel, C.C., Fruchter, A.: High-redshift galaxies in the Hubble Deep Field: colour selection and star formation history to z˜4. MNRAS **283**, 1388–1404, arXiv:astro-ph/9607172 (1996)

Maeda, K., Nomoto, K.: Bipolar supernova explosions: nucleosynthesis and implications for abundances in extremely metal-poor stars. ApJ **598**, 1163–1200 (2003). doi:10.1086/378948, arXiv:astro-ph/0304172

Maeder, A., Meynet, G.: The Evolution of Rotating Stars. A&A, 38, 143–190 (2000). doi 10.1146/annurev.astro.38.1.143

Mahoney, W.A., Ling, J.C., Wheaton, W.A., Jacobson, A.S.: HEAO 3 discovery of Al-26 in the interstellar medium. ApJ **286**, 578–585 (1984). doi:10.1086/162632

Maness, H., Martins, F., Trippe, S., Genzel, R., Graham, J.R., Sheehy, C., Salaris, M., Gillessen, S., Alexander, T., Paumard, T., Ott, T., Abuter, R., Eisenhauer, F.: Evidence for a long-standing top-heavy initial mass function in the central parsec of the galaxy. ApJ **669**, 1024–1041, arXiv:0707.2382 (2007)

Mannucci, F., Della Valle, M., Panagia, N., Cappellaro, E., Cresci, G., Maiolino, R., Petrosian, A., Turatto, M.: The supernova rate per unit mass. A&A **433**, 807–814, arXiv:astro-ph/0411450 (2005)

Martin, P., Knödlseder, J., Diehl, R., Meynet, G.: New estimates of the gamma-ray line emission of the Cygnus region from INTEGRAL/SPI observations. A&A **506**, 703–710 (2009). doi:10.1051/0004-6361/200912178

Martin, P., Knödlseder, J., Meynet, G., Diehl, R.: Predicted gamma-ray line emission from the Cygnus complex. A&A, **511**, 86 (2010). ArXiv e-prints 1001.1522 (2010)

Matteucci, F., Francois, P.: Galactic chemical evolution – Abundance gradients of individual elements. MNRAS **239**, 885–904 (1989)

McKee, C.F., Ostriker, E.C.: Theory of star formation. ARA&A **45**, 565–687 (2007). doi:10.1146/annurev.astro.45.051806.110602, 0707.3514

McKee, C.F., Williams, J.P.: The luminosity function of ob associations in the galaxy. ApJ **476**, 144 (1997)

Mewaldt, R.A., Yanasak, N.E., Wiedenbeck, M.E., Davis, A.J., Binns, W.R., Christian, E.R., Cummings, A.C., Hink, P.L., Leske, R.A., Niebur, S.M., Stone, E.C., Von Rosenvinge, T.T.: Radioactive clocks and cosmic-ray transport in the galaxy. Space Sci. Rev. **99**, 27–39 (2001)

Meyer, J.: The baseline composition of solar energetic particles. ApJS **57**, 151–171 (1985)

Meyer, J., Drury, L.O., Ellison, D.C.: Galactic cosmic rays from supernova remnants. i. A cosmicray composition controlled by volatility and mass-to-charge ratio. ApJ **487**, 182, arXiv:astroph/9704267 (1997)

Meynet, G., Arnould, M., Prantzos, N., Paulus, G.: Contribution of Wolf-Rayet stars to the synthesis of $^{26}$Al. I. The $\gamma$-ray connection. A&A **320**, 460–468 (1997)

Moustakas, J., Kennicutt, R.C. Jr., Tremonti, C.A.: Optical star formation rate indicators. ApJ **642**, 775–796 (2006). doi:10.1086/500964, arXiv:astro-ph/0511730

Nagataki, S., Hashimoto, M.A., Sato, K., Yamada, S., Mochizuki, Y.S.: The high ratio of 44Ti/56Ni in Cassiopeia A and the Axisymmetric Collapse-driven Supernova Explosion. ApJ **492**, L45 (1998). doi:10.1086/311089, arXiv:astro-ph/9807015

Nakanishi, H., Sofue, Y.: Three-dimensional distribution of the ISM in the Milky Way galaxy: I. The H I Disk. PASJ **55**, 191–202, arXiv:astro-ph/0304338 (2003)

Nakanishi, H., Sofue, Y.: Three-dimensional distribution of the ISM in the Milky Way galaxy: II. The molecular gas disk. PASJ **58**, 847–860, arXiv:astro-ph/0610769 (2006)

Naya, J.E., Barthelmy, S.D., Bartlett, L.M., Gehrels, N., Leventhal, M., Parsons, A., Teegarden, B.J., Tueller, J.: Detection of high-velocity $^{26}$Al towards the Galactic Centre. Nature **384**, 44–46 (1996). doi:10.1038/384044a0

Oberlack, U., Bennett, K., Bloemen, H., Diehl, R., Dupraz, C., Hermsen, W., Knoedlseder, J., Morris, D., Schoenfelder, V., Strong, A., Winkler, C.: The COMPTEL 1.809MeV all-sky image. A&AS **120**, C311 (1996)

Oey, M.S., Clarke, C.J.: Massive stars: feedback effects in the local universe. ArXiv Astrophysics e-prints arXiv:astro-ph/0703036 (2007)

Oey, M.S., Meurer, G.R., Yelda, S., Furst, E.J., Caballero-Nieves, S.M., Hanish, D.J., Levesque, E.M., Thilker, D.A., Walth, G.L., Bland-Hawthorn, J., Dopita, M.A., Ferguson, H.C., Heckman, T.M., Doyle, M.T., Drinkwater, M.J., Freeman, K.C., Kennicutt, R.C., Jr., Kilborn, V.A., Knezek, P.M., Koribalski, B., Meyer, M., Putman, M.E., Ryan-Weber, E.V., Smith, R.C., Staveley-Smith, L., Webster, R.L., Werk, J., Zwaan, M.A.: The survey for ionization in neutral gas galaxies. III. Diffuse, warm ionized medium and escape of ionizing radiation. ApJ **661**, 801–814 (2007). doi:10.1086/517867, arXiv:astro-ph/0703033

Olling, R.P., Merrifield, M.R.: Luminous and dark matter in the Milky Way. MNRAS **326**, 164–180 (2001). doi:10.1046/j.1365-8711.2001.04581.x, arXiv:astro-ph/0104465

Ore, A., Powell, J.L.: Three-photon annihilation of an electron-positron pair. Phys. Rev. **75**, 1696–1699 (1949). doi:10.1103/PhysRev.75.1696

Pagel, B.E.J.: Nucleosynthesis and Chemical Evolution of Galaxies. Nucleosynthesis and Chemical Evolution of Galaxies, p. 392. Cambridge University Press, Cambridge, UK (Oct 1997) ISBN 0521550610.

Palacios, A., Meynet, G., Vuissoz, C., Knödlseder, J., Schaerer, D., Cerviño, M., Mowlavi, N.: New estimates of the contribution of Wolf-Rayet stellar winds to the Galactic $^{26}$Al. A&A **429**, 613–624 (2005). doi:10.1051/0004-6361:20041757, arXiv:astro-ph/0409580

Pavlidou, V., Fields, B.D.: The guaranteed gamma-ray background. ApJ **575**, L5–L8 (2002). doi:10.1086/342670, arXiv:astro-ph/0207253

Pettini, M.: Element abundances through the cosmic ages. In: Esteban, C., García López, R., Herrero, A., Sánchez, F. (eds.) Cosmochemistry. The Melting Pot of the Elements, pp. 257–298 (2004)

Pfalzner, S.: Universality of young cluster sequences. A&A **498**, L37–L40, DOI 10.1051/0004-6361/200912056, 0904.0523 (2009)

Plüschke, S., Diehl, R., Schönfelder, V., Bloemen, H., Hermsen, W., Bennett, K., Winkler, C., McConnell, M., Ryan, J., Oberlack, U., Knödlseder, J.: The COMPTEL 1.809 MeV survey. In: Gimenez, A., Reglero, V., Winkler, C. (eds.) Exploring the Gamma-Ray Universe, vol. 459, pp. 55–58, ESA Special Publication (2001a)

Plüschke, S., Kretschmer, K., Diehl, R., Hartmann, D.H., Oberlack, U.: OB Associations as sources of galactic radioactivity. In: Deiters, S., Fuchs, B., Just, A., Spurzem, R., Wielen, R. (eds.) Dynamics of Star Clusters and the Milky Way, Astronomical Society of the Pacific Conference Series, vol. 228, pp. 544 (2001b)

Plüschke, S., Cerviño, M., Diehl, R., Kretschmer, K., Hartmann, D.H., Knödlseder, J.: Understanding $^{26}$Al emission from Cygnus. New Astron. Rev. **46**, 535–539 (2002). doi:10.1016/S1387-6473(02)00197-5

Pohl, M., Englmaier, P., Bissantz, N.: Three-dimensional distribution of molecular gas in the barred Milky Way. ApJ **677**, 283–291 (2008). doi:10.1086/529004, arXiv:0712.4264

Prantzos, N.: On the diffuse galactic emission at 511 keV and 1,809 keV. A&AS **97**, 119–126 (1993)

Prantzos, N.: Radioactive $^{26}$Al and $^{60}$Fe in the Milky Way: implications of the RHESSI detection of 60Fe. A&A **420**, 1033–1037 (2004). doi:10.1051/0004-6361:20035766, arXiv:astroph/0402198

Prantzos, N.: On the intensity and spatial morphology of the 511 kev emission in the milky way. A&A **449**, 869–878, arXiv:astro-ph/0511190 (2006)

Prantzos, N.: Origin and Evolution of the Light Nuclides, pp. 27. Springer, Dordrecht (2007)

Prantzos, N., Diehl, R.: Radioactive $^{26}$Al in the galaxy: observations versus theory. Phys. Rep. **267**, 1–69 (1996). doi:10.1016/0370-1573(95)00055-0

Prantzos, N., Arnould, M., Arcoragi, J.P., Casse, M.: Neutron-rich nuclei in cosmic rays and wolfrayet stars. In: Jones, F.C. (ed.) International Cosmic Ray Conference, International Cosmic Ray Conference, vol. 3, pp. 167–170 (1985)

Prantzos, N., Silk, J.: Star formation and chemical evolution in the MilkyWay: cosmological implications. ApJ **507**, 229–240 (1998). doi:10.1086/306327

Prantzos, N., Bohem, C., Bykov, A., Diehl, R., Ferrière, K., Guessoum, N., Jean, P., Knoedlseder, J., Marcowith, A., Moskalenko, I., Strong, A., Weidenspointner, G.: "The 511 keV emission from positron annihilation in the Galaxy" Rev Mod Phys, (2010) in press

Preibisch, T., Brown, A.G.A., Bridges, T., Guenther, E., Zinnecker, H.: Exploring the full stellar population of the upper scorpius OB Association. AJ **124**, 404–416 (2002). doi:10.1086/341174

Preibisch, T., Mamajek, E.: The nearest OB Association: Scorpius-Centaurus (Sco OB2), p. 235 (2008)

Preibisch, T., Zinnecker, H.: The history of low-mass star formation in the upper scorpius OB Association. AJ **117**, 2381–2397 (1999). doi:10.1086/300842

Primack, J.R., Gilmore, R.C., Somerville, R.S.: Diffuse extragalactic background radiation. In: Aharonian, F.A., Hofmann, W., Rieger, F. (eds.) American Institute of Physics Conference Series, American Institute of Physics Conference Series, vol. 1085, pp. 71–82 (2008). doi:10.1063/1.3076779

Prochaska, J.X., Tumlinson, J.: Baryons: What, When and Where? pp. 419–+ (2009). doi:10.1007/978-1-4020-9457-6-16

Ptuskin, V.S., Soutoul, A.: Cosmic-ray clocks. Space Sci. Rev. **86**, 225–238 (1998). doi:10.1023/A:1005071612125

Purcell, W.R., Cheng, L.X., Dixon, D.D.: OSSE Mapping of Galactic 511 kev Positron Annihilation Line Emission. ApJ **491**, 725 (1997)

Ragot, B.R.: Nongyroresonant pitch angle scattering. ApJ **518**, 974–984 (1999). doi:10.1086/307315

Ragot, B.R.: Nonresonant pitch-angle scattering of low-energy electrons in the SolarWind. ApJ **642**, 1163–1172 (2006). doi:10.1086/501159

Rattenbury, N.J., Mao, S., Sumi, T., Smith, M.C.: Modelling the galactic bar using ogle-ii red clump giant stars. MNRAS **378**, 1064–1078, arXiv:0704.1614 (2007)

Razzaque, S., Dermer, C.D., Finke, J.D.: The stellar contribution to the extragalactic background light and absorption of high-energy gamma rays. ApJ **697**, 483–492 (2009). doi:10.1088/0004-637X/697/1/483, 0807.4294

Renaud, M., Vink, J., Decourchelle, A., Lebrun, F., Terrier, R., Ballet, J.: An INTEGRAL/IBIS view of young Galactic SNRs through the 44Ti gamma-ray lines. New Astron. Rev. **50**, 540–543 (2006). doi:10.1016/j.newar.2006.06.061, arXiv:astro-ph/0602304

Riess, A.G., Nugent, P.E., Gilliland, R.L., Schmidt, B.P., Tonry, J., Dickinson, M., Thompson, R.I., Budavári, T., Casertano, S., Evans, A.S., Filippenko, A.V., Livio, M., Sanders, D.B., Shapley, A.E., Spinrad, H., Steidel, C.C., Stern, D., Surace, J., Veilleux, S.: The farthest known supernova: support for an accelerating universe and a glimpse of the epoch of deceleration. ApJ **560**, 49–71 (2001). doi:10.1086/322348, arXiv:astro-ph/0104455

Robin, A.C., Reylé, C., Derrière, S., Picaud, S.: A synthetic view on structure and evolution of the milky way. A&A **409**, 523–540, arXiv:astro-ph/0401052 (2003)

Robitaille, T.P., Whitney, B.A.: The present-day star formation rate of the milky way determined from spitzer-detected young stellar objects. ApJ **710**, L11–L15, 1001.3672 (2010)

Rugel, G., Faestermann, T., Knie, K., Korschinek, G., Poutivtsev, M., Schumann, D., Kivel, N., G¨unther-Leopold, I., Weinreich, R.,Wohlmuther, M.: New measurement of the Fe60 half-life. Phys. Rev. Lett. **103**(7), 072502 (2009). doi:10.1103/PhysRevLett.103.072502

Russeil, D., Adami, C., Georgelin, Y.M.: Revised distances of Northern HII regions. A&A **470**, 161–171 (2007)

Salpeter, E.E.: The luminosity function and stellar evolution. ApJ **121**, 161 (1955)

Sawada, T., Hasegawa, T., Handa, T., Cohen, R.J.: A molecular face-on view of the Galactic Centre region. MNRAS **349**, 1167–1178 (2004). doi:10.1111/j.1365-2966.2004.07603.x, arXiv:astroph/0401286

Scannapieco, E., Bildsten, L.: The type ia supernova rate. ApJ **629**, L85–L88, arXiv:astroph/0507456 (2005)

Schmidt, M.: The rate of star formation. ApJ **129**, 243 (1959). doi:10.1086/146614

Sedlmayr, E., Patzer, A.B.C.: Grain formation and dynamical atmosphere. In: Jorissen, A., Goriely, S., Rayet, M., Siess, L., Boffin, H. (eds.) EAS Publications Series, vol. 11, pp. 51–66 (2004). doi:10.1051/eas:2004003

Shen, K.J., Kasen, D., Weinberg, N.N., Bildsten, L., Scannapieco, E.: Thermonuclear .Ia Supernovae from Helium Shell Detonations: Explosion Models and Observables. ArXiv e-prints 1002.2258 (2010)

Sizun, P., Cassé, M., Schanne, S.: Continuum $\gamma$-ray emission from light dark matter positrons and electrons. Phys. Rev. D **74**(6), 063514 (2006). doi:10.1103/PhysRevD.74.063514, arXiv:astroph/0607374

Slesnick, C.L., Hillenbrand, L.A., Carpenter, J.M.: A large-area search for low-mass objects in upper scorpius. II. Age and mass distributions. ApJ **688**, 377–397 (2008). doi:10.1086/592265, 0809.1436

Smith, D.M.: Gamma-ray line observations with RHESSI. In: Schoenfelder, V., Lichti, G., Winkler, C. (eds.) 5th INTEGRAL Workshop on the INTEGRAL Universe, vol. 552, pp. 45, ESA Special Publication (2004)

Soutoul, A., Casse, M., Juliusson, E.: Time delay between the nucleosynthesis of cosmic rays and their acceleration to relativistic energies. ApJ **219**, 753–755 (1978)

Stahler, S.W., Palla, F.: The Formation of Stars. Wiley, New York (2005)

Stecker, F.W.: Digging for the truth: photon archeology with GLAST. In: Ritz, S., Michelson, P., Meegan, C.A. (eds.) The First GLAST Symposium, American Institute of Physics Conference Series, vol. 921, pp. 237–239 (2007). doi:10.1063/1.2757310

Stecker, F.W., Scully, S.T.: Is the universe more transparent to gamma rays than previously thought? ApJ **691**, L91–L94 (2009). doi:10.1088/0004-637X/691/2/L91, 0807.4880

Strong, A.W., Moskalenko, I.V., Ptuskin, V.S.: Cosmic-ray propagation and interactions in the galaxy. Ann. Rev. Nucl. Part. Sci. **57**, 285–327, arXiv:astro-ph/0701517 (2007)

Sturner, S.J., Naya, J.E.: On the nature of the high-velocity $^{26}$Al near the galactic center. ApJ **526**, 200–206 (1999). doi:10.1086/307979

Sullivan, M., Le Borgne, D., Pritchet, C.J., Hodsman, A., Neill, J.D., Howell, D.A., Carlberg, R.G., Astier, P., Aubourg, E., Balam, D., Basa, S., Conley, A., Fabbro, S., Fouchez, D., Guy, J., Hook, I., Pain, R., Palanque-Delabrouille, N., Perrett, K., Regnault, N., Rich, J., Taillet, R., Baumont, S., Bronder, J., Ellis, R.S., Filiol, M., Lusset, V., Perlmutter, S., Ripoche, P., Tao, C.: Rates and properties of type ia supernovae as a function of mass and star formation in their host galaxies. ApJ **648**, 868–883, arXiv:astro-ph/0605455 (2006)

Svensmark, H.: Cosmoclimatology: a new theory emerges. Astron. Geophys. **48**(1), 010000–1 (2007). doi:10.1111/j.1468-4004.2007.48118.x

Tachibana, S., Huss, G.R.: The initial abundance of 60fe in the solar system. ApJ **588**, L41–L44 (2003). doi:10.1086/375362

Thielemann, F.K., Nomoto, K., Hashimoto, M.A.: Core-collapse supernovae and their ejecta. ApJ **460**, 408 (1996). doi:10.1086/176980

Timmes, F.X., Woosley, S.E., Hartmann, D.H., Hoffman, R.D., Weaver, T.A., Matteucci, F.: $^{26}$Al and $^{60}$Fe from supernova explosions. ApJ **449**, 204 (1995). doi:10.1086/176046, arXiv:astroph/9503120

Tinsley, B.M.: Evolution of the stars and gas in galaxies. Fund. Cosmic Phys. **5**, 287–388 (1980)

Toptygin, I.N.: Cosmic Rays in Interplanetary Magnetic Fields. D.Reidel Publishing Co., Dordrecht (1985)

Truran, J.W., Cameron, A.G.W.: Evolutionary models of nucleosynthesis in the galaxy. Ap&SS **14**, 179–222 (1971). doi:10.1007/BF00649203

Uyanıker, B., Fürst, E., Reich, W., Aschenbach, B., Wielebinski, R.: The Cygnus superbubble revisited. A&A **371**, 675–697 (2001). doi:10.1051/0004-6361:20010387

van der Hucht, K.A.: The VIIth catalogue of galactic Wolf-Rayet stars. New Astron. Rev. **45**, 135–232 (2001). doi:10.1016/S1387-6473(00)00112-3

Vedrenne, G., Atteia, J.: Gamma-Ray Bursts (2010). doi:10.1007/978-3-540-39088-6

Vedrenne, G., Roques, J.P., Schönfelder, V., Mandrou, P., Lichti, G.G., von Kienlin, A., Cordier, B., Schanne, S., Knödlseder, J., Skinner, G., Jean, P., Sanchez, F., Caraveo, P., Teegarden, B., von Ballmoos, P., Bouchet, L., Paul, P., Matteson, J., Boggs, S., Wunderer, C., Leleux, P., Weidenspointner, G., Durouchoux, P., Diehl, R., Strong, A., Cassé, M., Clair, M.A., André, Y.: SPI: The spectrometer aboard INTEGRAL. A&A **411**, L63–L70 (2003). doi:10.1051/0004-6361:20031482

Vink, J.: Gamma-ray observations of explosive nucleosynthesis products. Adv Space Res **35**, 976–986 (2005). doi:10.1016/j.asr.2005.01.097, arXiv:astro-ph/0501645

Voss, R., Diehl, R., Hartmann, D.H., Cerviño, M., Vink, J.S., Meynet, G., Limongi, M., Chieffi, A.: Using population synthesis of massive stars to study the interstellar medium near OB associations. A&A **504**, 531–542 (2009). doi:10.1051/0004-6361/200912260, 0907.5209

Wang, W., Harris, M.J., Diehl, R., Halloin, H., Cordier, B., Strong, A.W., Kretschmer, K., Knödlseder, J., Jean, P., Lichti, G.G., Roques, J.P., Schanne, S., von Kienlin, A., Weidenspointner, G., Wunderer, C.: SPI observations of the diffuse $^{60}$Fe emission in the Galaxy. A&A **469**, 1005–1012 (2007). doi:10.1051/0004-6361:20066982, 0704.3895

Wang, W., Lang, M.G., Diehl, R., Halloin, H., Jean, P., Knödlseder, J., Kretschmer, K., Martin, P., Roques, J.P., Strong, A.W., Winkler, C., Zhang, X.L.: Spectral and intensity variations of Galactic $^{26}$Al emission. A&A **496**, 713–724 (2009a). doi:10.1051/0004-6361/200811175, 0902.0211

Wang, W., Lang, M.G., Diehl, R., Halloin, H., Jean, P., Knödlseder, J., Kretschmer, K., Martin, P., Roques, P., Strong, A.W., Winkler, C., Zhang, X.L.: Spectral and intensity variations of galactic 26^al emission. ArXiv e-prints 0902.0211 (2009b)

Watanabe, K., Hartmann, D.H., Leising, M.D., The, L.: Constraining the cosmic star formation rate with the MeV backgroung. Nucl. Phys. A **718**, 425–427 (2003). doi:10.1016/S0375-9474(03)00821-2, arXiv:astro-ph/0306161

Weidenspointner, G., Shrader, C.R., Knödlseder, J., Jean, P., Lonjou, V., Guessoum, N., Diehl, R., GillardW, Harris, M.J., Skinner, G.K., von Ballmoos, P., Vedrenne, G., Roques, J.P., Schanne, S., Sizun, P., Teegarden, B.J., Schönfelder, V., Winkler, C.: The sky distribution of positronium annihilation continuum emission measured with SPI/INTEGRAL. A&A **450**, 1013–1021 (2006). doi:10.1051/0004-6361:20054046, arXiv:astro-ph/0601673

Weidenspointner, G., Skinner, G., Jean, P., Knödlseder, J., von Ballmoos, P., Bignami, G., Diehl, R., Strong, A.W., Cordier, B., Schanne, S., Winkler, C.: An asymmetric distribution of positrons in the Galactic disk revealed by $\gamma$-rays. Nature **451**, 159–162 (2008a). doi:10.1038/nature06490

Weidenspointner, G., Skinner, G., Jean, P., Knödlseder, J., von Ballmoos, P., Bignami, G., Diehl, R., Strong, A.W., Cordier, B., Schanne, S., Winkler, C.: An asymmetric distribution of positrons in the galactic disk revealed by $\gamma$-rays. Nature **451**, 159–162 (2008b)

Weidenspointner, G., Skinner, G.K., Jean, P., Knödlseder, J., von Ballmoos, P., Diehl, R., Strong, A., Cordier, B., Schanne, S., Winkler, C.: Positron astronomy with spi/integral. New Astron. Rev. **52**, 454–456 (2008c)

Weidner, C., Kroupa, P., Bonnell, I.A.D.: The relation between the most-massive star and its parental star cluster mass. MNRAS **401**, 275–293 (2010). doi:10.1111/j.1365-2966.2009.15633.x, 0909.1555

Welsh, B.Y., Shelton, R.L.: The trouble with the Local Bubble. Ap&SS **323**, 1–16 (2009). doi:10.1007/s10509-009-0053-3, 0906.2827

Westphal, A.J., Bradley, J.P.: Formation of glass with embedded metal and sulfides from shockaccelerated crystalline dust in superbubbles. ApJ **617**, 1131–1141, arXiv:astro-ph/0410553 (2004)

Wiedenbeck, M.E., Binns, W.R., Christian, E.R., Cummings, A.C., Dougherty, B.L., Hink, P.L., Klarmann, J., Leske, R.A., Lijowski, M., Mewaldt, R.A., Stone, E.C., Thayer, M.R., von Rosenvinge, T.T., Yanasak, N.E.: Constraints on the time delay between nucleosynthesis and cosmic-ray acceleration from observations of <??>^59ni and <??>^59co. ApJ **523**, L61–L64 (1999)

Wiedenbeck, M.E., Binns, W.R., Cummings, A.C., Davis, A.J., de Nolfo, G.A., Israel, M.H., Leske, R.A., Mewaldt, R.A., Stone, E.C., von Rosenvinge, T.T.: An Overview of the Origin of Galactic Cosmic Rays as Inferred from Observations of Heavy Ion Composition and Spectra, pp. 415. Springer, Dordrecht (2007)

Wilking, B.A., Gagné, M., Allen, L.E.: Star Formation in the $\rho$ Ophiuchi Molecular Cloud, p. 351 (2008)

Winkler, C., Courvoisier, T.J.L., Di Cocco, G., Gehrels, N., Giménez, A., Grebenev, S., Hermsen, W., Mas-Hesse, J.M., Lebrun, F., Lund, N., Palumbo, G.G.C., Paul, J., Roques, J.P., Schnopper, H., Schönfelder, V., Sunyaev, R., Teegarden, B., Ubertini, P., Vedrenne, G., Dean, A.J.: The INTEGRAL mission. A&A **411**, L1–L6 (2003). doi:10.1051/0004-6361:20031288

Woosley, S.E., Bloom, J.S.: The supernova gamma-ray burst connection. ARA&A **44**, 507–556 (2006). doi:10.1146/annurev.astro.43.072103.150558, arXiv:astro-ph/0609142

Woosley, S.E., Heger, A.: Nucleosynthesis and remnants in massive stars of solar metallicity. Phys. Rep. **442**, 269–283 (2007). doi:10.1016/j.physrep.2007.02.009, arXiv:astro-ph/0702176

Woosley, S.E., Weaver, T.A.: Sub-Chandrasekhar mass models for Type IA supernovae. ApJ **423**, 371–379 (1994). doi:10.1086/173813

Woosley, S.E., Weaver, T.A.: The evolution and explosion of massive stars. II. Explosive hydrodynamics and nucleosynthesis. ApJS **101**, 181 (1995). doi:10.1086/ 192237

Yanasak, N.E., Wiedenbeck, M.E., Mewaldt, R.A., Davis, A.J., Cummings, A.C., George, J.S., Leske, R.A., Stone, E.C., Christian, E.R., von Rosenvinge, T.T., Binns, W.R., Hink, P.L., Israel, M.H.: Measurement of the secondary radionuclides $^{10}$Be, $^{26}$Al, $^{36}$Cl, $^{54}$Mn, and $^{14}$C and implications for the galactic cosmic-ray age. ApJ **563**, 768–792 (2001a). doi:10.1086/323842

Yanasak, N.E., Wiedenbeck, M.E., Mewaldt, R.A., Davis, A.J., Cummings, A.C., George, J.S., Leske, R.A., Stone, E.C., Christian, E.R., von Rosenvinge, T.T., Binns, W.R., Hink, P.L., Israel, M.H.: Measurement of the secondary radionuclides $^{10}$be, $^{26}$al, $^{36}$cl, $^{54}$mn, and $^{14}$c and implications for the galactic cosmic-ray age. ApJ **563**, 768–792 (2001b)

Zinnecker, H., Yorke, H.W.: Toward understanding massive star formation. ARA&A **45**, 481–563 (2007). doi:10.1146/annurev.astro.44.051905.092549, 0707.1279

Zoccali, M., Renzini, A., Ortolani, S., Greggio, L., Saviane, I., Cassisi, S., Rejkuba, M., Barbuy, B., Rich, R.M., Bica, E.: Age and metallicity distribution of the galactic bulge from extensive optical and near-ir stellar photometry. A&A **399**, 931–956, arXiv:astro-ph/0210660 (2003)

# Part IV
# Tools for the Study of Radioactivities in Astrophysics

At the foundations of science lie the instruments (detectors) which enable us to record data that encode information about objects under study. Also important are the methods used to organize our concepts and notions about these objects in the forms of astrophysical theories or phenomenological descriptions. These *tools* are mostly based on general physical or mathematical principles, but with increasing sophistication one often requires the development of specialized methods that may not be common in other areas of astronomy. In this last part of the book we address how theorists construct models of astrophysical objects and processes, exemplified by the core-collapse of massive stars as a key source of cosmic radioactivity. We further present the astrophysics of nuclear reactions and discuss them from the laboratory- and theory point of view. Finally, we present two specific classes of instruments used to measure cosmic radioactivities, one employed in terrestrial laboratories for the study of interstellar dust grains, another operating in space for the study of electromagnetic radiation from cosmic radioactive decays.

# Chapter 8
# Computer-Modeling of Stars

**M. Liebendörfer**

A human being experiences his immediate environment on the scale of meters, seconds and grams. These are also the natural scales of his actions. Thus, as soon as he starts to explore the laws of physics, he can easily move around masses at the scale of grams, objects on the scale of meters and perform experiments on the scale of seconds. On these scales, the experimentator has full control on the setup of an experiment and direct access to all degrees of freedom during the evolution of the experiment. This direct access is lost in experiments that explore the physics on scales that are many orders of magnitude smaller. The experimentator still has full control on the setup, for example, by putting a specific target into a properly designed accelerator beam. But the measurements are then limited to the far field, where only a superposition of the effects of the microscopic physics becomes detectable. The large number of degrees of freedom that may be present in the microscopic physics must be explored by clever variations of the experimental setup. Most astronomical observations are obviously also taken from the far field, because the distance to the observed source is so much larger than the length scale of the source. Hence, many degrees of freedom of the dynamics on the length scale of the source are only indirectly accessible for the observer. Moreover, it is not possible to efficiently manipulate and prepare matter outside the solar system in order to produce systematic variations in the setup as in terrestrial experiments.

In this situation, a computer model is a particularly useful tool. It implements known input physics and evolves sophisticated coupled systems in space-time in order to reproduce the observed far field response of a corresponding event in nature. In contrast to the real event, the data of the computer model is fully accessible and permits to study all degrees of freedom in detail. Of course there is no guarantee that the results represent an accurate image of the actual astrophysical source. Only the real experiment is capable of probing the laws of physics realised in nature, while the computer model is bound to elaborate the properties of the theory it is

M. Liebendörfer (✉)
Universität Basel, 4056 Basel, Switzerland, matthias.liebendoerfer@unibas.ch

Liebendörfer, M.: *Computer-Modeling of Stars*. Lect. Notes Phys. **812**, 439–459 (2011)
DOI 10.1007/978-3-642-12698-7_8 

built on. The strength of the computer simulation is that it demonstrates and extends the predictive power of the underlying theory. It is therefore essential to check the models for predictions that could be in reach of a targeted future observation, so that a specific model can be supported or falsified. The prediction of the production of radioactive nuclei that can be observed from far distances is one possibility.

Before an astrophysical model is constructed, it is of primary importance to perform order of magnitude estimates and to identify relevant and irrelevant processes. The presence of equilibrium conditions can further reduce the degrees of freedom. This information is then used to compose a sequence of relevant processes that form a scenario. In a second step, one substitutes the more complicated processes by simple approximate laws (for example power laws) that allow an analytical investigation of the most important aspects of the model. In stellar structure, for example, order of magnitude arguments show that the matter is in thermodynamical equilibrium. Hence, an equation of state can be defined that expresses the gas pressure, $p$, as a function of the local density, $\rho$, temperature and composition. In some regimes, the microscopic physics information in the equation of state is well approximated by an equation of state of the form $p = \kappa\rho^{1+1/n}$. Here, $\kappa$ is a constant and $n$ a parameter called the polytropic index. If this approximation holds, the equations of stellar structure reduce to the Lane-Emden equation (e.g. Chandrasekhar, 1967) which has analytical solutions for specific integer values of $n$. Scientific computing provides the tools to specify additional solutions for non-integer values of $n$. An almost arbitrary numerical precision can be obtained because all physical uncertainties remain hidden by the power law of the equation of state and the reduction of the star to spherical symmetry. In this example, the scenario and the simplification of the input physics have to be pre-defined manually before scientific computing can be used to determine specific solutions of a simple model.

It is numerically more challenging to let a computer model contribute at the stage of the order of magnitude estimates and the elaboration of the scenario. Before the irrelevant processes are excluded, the model is rich and complicated. Instead of judging processes in each fluid cell by hand, the finite differencing can be set up in a manner that filters out the irrelevant processes automatically and establishes equilibria accurately. A well-known example of this technique is provided by the numerical propagation of a shock front: Only a minority of computer codes resolves the microscopic width of the shock. Most codes just ensure the accurate conservation of mass, momentum and energy across the shock front to obtain the shock propagation speed and the thermodynamic conditions on both sides of the shock. Here, it is the numerical algorithm that dynamically performs the simplification of the model while the microscopic viscosity of the fluid does not enter the calculation whether it is implemented or not. Hence, one may enable a rich set of input physics and design the finite differencing such that fundamental laws of physics are fulfilled under all possible conditions. Then, the complexity of the model is limited by the scale on which unresolvable small-scale structures are dissipated in space and time. This scale depends on the numerical algorithm and the mesh topology rather than the investigated physics. Because the behaviour at the resolution threshold can greatly vary from numerical algorithm and grid to the next, we believe that it is important for the reliability of astrophysical models to be investigated with several different numerical tools that are simple and efficient enough to be broadly used.

Computer models of stellar evolution and explosion have a long tradition and have mostly been carried out in spherical symmetry (e.g. Henyey et al., 1959; Woosley et al., 2002; Colgate and White, 1966; Bethe, 1990). The restriction to spherical symmetry is a consequence of the fact that the hydrodynamic time scale is many orders of magnitude smaller than the lifetime of a star. Traditional stellar evolution codes assumed that the matter is in hydrostatic equilibrium and thus close to spherical. If one implements hydrodynamic equilibrium that adjusts slowly on the time scale of the thermodynamic evolution of the star, sufficiently large time steps can be taken to numerically cover the lifetime of the star. The crucial effect of mixing by convection and rotation on the evolution has been included by effective mixing parameters (e.g. Ludwig et al., 1999) and the inclusion of an effective evolution of angular momentum in centrifugally deformed layers (Meynet and Maeder, 1997; Langer et al., 1997) that can still be labelled and evolved by a unique radial parameter. Several computer codes have been built to follow the nucleosynthesis through the different stages of stellar evolution. The application of traditional simulation codes relied on instructed users that learned to handle and extend the codes within the different research groups. Publications therefore mostly focus on the extensions of the numerical methods and the properties of the new stellar models that resulted (Nomoto and Hashimoto, 1988; Maeder and Meynet, 1989; Woosley and Weaver, 1995). It is a more recent phenomenon that the different codes obtain a name and an identity as software package (Limongi et al., 2000; Rauscher et al., 2002; Young and Arnett, 2005; Eggenberger et al., 2008; Dotter and Paxton, 2009).

As soon as the iron core becomes gravitationally unstable at the end of the evolution of massive stars, it starts to collapse on the dynamical time scale. This marks the end of the applicability of stellar evolution codes that usually are not designed to describe dense matter with enhanced weak interactions. For a range of stellar masses, which for solar metallicity is believed to be within $\sim$ 8–40 $M_{\odot}$, this last phase of stellar evolution leads to the supernova explosion, which may eject matter into the interstellar medium after its processing by explosive nucleosynthesis. This phase is modelled by 'supernova codes' that traditionally have also assumed spherical symmetry – but for a different reason than the stellar evolution codes. The growing importance of general relativistic corrections around the dense neutron star at the center of the collapse and the large impact of neutrino physics and transport on the dynamics of the explosion makes it computationally very challenging to treat this phase in multidimensional simulations.

Hence, for both stellar evolution and stellar core collapse, realistic three-dimensional turnover and important fluid instabilities were for a long time only simulated in a phenomenological manner. A very simple estimate shows that this insatisfactory situation is about to change: In order to sample a dynamically interesting profile of a physical quantity, one needs a minimum resolution of $\sim$100 data points. If one requests that the evolution of these data points should not touch the scale on which unresolvable features are numerically dissipated, one needs another order of magnitude, i.e. a resolution $\sim$1,000 points in one dimension. For a three-dimensional simulation, this then leads to $\sim 1{,}000^3 = 10^9$ fluid cells. Assuming that each cell holds at least 10 real variables, this means that a three-dimensional astrophysical simulation has to evolve about 80 Gigabyte of data at each time step. As this level of resolution is now achievable on high-performance computer clusters, the link between input physics and astrophysical observables receives a new quality in the transition from generic computer models in spherical or axisymmetry to detailed three-dimensional models. This transition has already occurred in astrophysical fields that are dominated by hydrodynamics (e.g. cosmology) while it will occur over the next years in the fields that have to couple the hydrodynamics to radiative transfer. In stellar evolution, it becomes possible to perform dynamical global stellar models in three dimensions over short time intervals so that the parameterised long-term models can be gauged (Brun et al., 2004; Meakin and Arnett, 2007). Simulations of thermonuclear explosions that lead to the display of a type Ia supernova have also entered the three-dimensional aera (Fryxell et al., 2000; Zingale et al., 2005; Röpke et al., 2007). The ignition and detailed propagation of the burning front in the 3D geometry is the crucial ingredient of type Ia supernova simulations. Likewise, the simulations of stellar core collapse are ready to make the transition to three dimensions in spite of GR effects and the difficulties with multidimensional neutrino transport, because there is a global symmetry center that initially permits some simplifying approximations, because the neutrino spectra are continuous and smooth, and because the dynamical time scale and the neutrino propagation time scale differ by less than an order of magnitude (a more detailed discussion follows below).

In addition to these more application-specific code developments, there are several well-documented and publicly available general purpose multidimensional hydrodynamics codes. As starting points we might mention the following ones: GADGET is a cosmological N-body code based on the Smoothed Particle Hydrodynamics method and parallelised using the Message Passing Interface (MPI). It is a Lagrangean approach with a hierarchical tree for the non-local evaluation of Newtonian gravity (Springel et al., 2001). VH-1 is a multidimensional hydrodynamics code based on the Lagrangian remap version of the Piecewise Parabolic Method (PPM) (Blondin and Lufkin, 1993) that has been further developed and parallelised using MPI. ZEUS-2D (Stone and Norman, 1992a, b; Stone et al., 1992) and ZEUS-3D are widely used grid-based hydrodynamics codes for which a MPI-parallel ZEUS-MP version exists as well. It offers the choice of different advection schemes on a fixed or moving orthogonal Eulerian mesh in a covariant description. While these more traditional approaches distribute in the form of a software package that includes options to switch on or off, recent open source projects try to provide the codes in the form of a generic framework that can host a variety of different modules implementing different techniques. In this category we could mention the FLASH code (e.g. Calder et al., 2002) with its main focus on the coupling of adaptive mesh stellar hydrodynamics to nuclear burning and the WHISKY code (Giacomazzo and Rezzolla, 2007) as a recent general relativistic hydrodynamics code based on the CACTUS environment.

For the accurate prediction of yields of radiactive nuclei in astrophysical scenarios it would be necessary to perform accurate models of stellar evolution and the different types of stellar explosions. The involved physics has been discussed in Chapters 4 and 9. Here we try to give some additional information about the technical tools that enable these simulations. Rather than continuing the discussion for all of these domains, we select the problem of stellar core collapse and explosion, where we think that the technical uncertainties are largest. Nevertheless, many of the discussed considerations and difficulties of current multidimensional models are quite general and also apply to the other astrophysical applications.

## 8.1 Models of Core-Collapse Supernovae

### *8.1.1 Basic Physical Description*

The general scenario of a core-collapse supernova is described in Chap. 4. Supernova matter is described as a self-gravitating compressible fluid with negligible viscosity. The compactness of the gravitationally bound matter is sufficient to make general relativistic effects important. With respect to the electromagnetic interaction, the fluid is assumed to be locally charge neutral, but it is allowed to couple to dynamical magnetic fields as described by the equations of ideal magneto-hydrodynamics. With respect to the strong interaction, nuclear statistical equilibrium can be assumed in a large part of the computational domain. In this case, it is sufficient to specify the triple $(n_b, T, Y_e)$ in order to uniquely describe the local thermodynamic state and the nuclear composition. The three independent variables are the baryon density $n_b$, the temperature $T$, and the electron fraction $Y_e = (n_e - n_{e^+})/n_b$, where $n_e$ is the electron density and $n_{e^+}$ the positron density. The weak interaction rates are not guaranteed to be in equilibrium. In fact, in dense regions of the computational domain the reaction rates are orders of magnitude faster than the dynamical time scale, while in other parts the weak reaction rates are negligible. Among the key reactions are the capture of electrons, $e$, on protons, $p$, and the capture of positrons, $e^+$, on neutrons, $n$. These reactions produce electron flavour neutrinos, $\nu_e$, and antineutrinos, $\bar{\nu}_e$, and influence the structure and dynamics of the gravitationally bound fluid by changing its temperature and electron fraction. The inverse reactions must be considered as well:

$$\begin{aligned} e + p &\rightleftharpoons n + \nu_e \\ e^+ + n &\rightleftharpoons p + \bar{\nu}_e. \end{aligned} \tag{8.1}$$

At low density, the neutrinos have a long mean free path and escape from the star. At high density, the neutrinos are trapped due to a short mean free path. Being fermions, they can block the reactions (8.1) such that the effective rates of electron and positron capture are determined by the ability of neutrinos to diffuse out of the high-density regime. As the weak interaction rates increase with the square of the neutrino energy, one has to solve an energy-dependent radiative transfer problem. The radiative transfer problem is difficult to solve because the semi-transparent regime can be subject to turbulence and asymmetric convective turnover. Moreover, one has to consider the additional emission of $\mu$ flavour neutrinos and $\tau$ flavour neutrinos and their antineutrinos.

## 8.1.2 Basic Mathematical Description

With the ingredients summarised above, the mathematical description of the problem consists of an advection problem of the compressible fluid that is coupled to a spectral radiative transfer problem for the neutrinos. In principle, the description should be given in full general relativity. However, the general relativistic problem is only solved in rare cases (Liebendörfer et al., 2004; Ott, 2009). Most multi-dimensional supernova models with spectral neutrino transport use a non-relativistic framework with a modified gravitational potential that makes up for the most important general relativistic effects (Marek et al., 2006). The radiative transfer part is usually kept accurate to order $v/c$, where $v$ is the velocity of the fluid and $c$ the velocity of light.

Using Cartesian space coordinates $x_i$ and time $t$ in the laboratory frame, the hydrodynamic part can be written as

$$\frac{\partial \rho}{\partial t} + \frac{\partial}{\partial x_j}\left(\rho v_j\right) = 0 \tag{8.2}$$

$$\frac{\partial}{\partial t}\left(\rho v_i\right) + \frac{\partial}{\partial x_j}\left(v_i \rho v_j - b_i b_j\right) + \frac{\partial P}{\partial x_i} + \rho \frac{\partial \phi}{\partial x_i} = \rho \dot{v}_i \tag{8.3}$$

$$\frac{\partial E}{\partial t} + \frac{\partial}{\partial x_j}\left[\left(E + P\right) v_j - b_i v_i b_j\right] + \rho v_i \frac{\partial \phi}{\partial x_i} = \rho \dot{e} \tag{8.4}$$

$$\frac{\partial}{\partial t}\left(\rho Y_e\right) + \frac{\partial}{\partial x_j}\left(Y_e \rho v_j\right) = \rho \dot{Y}_e. \tag{8.5}$$

The mass density $\rho = m_b n_b$ relates to the baryon density $n_b$ by a constant conversion factor $m_b$ that represents an average mass per baryon. Einsteins convention is used to sum expressions with equal indices over the different spatial directions denoted by indices $i = 1 \ldots 3$. All equations are written in conservative form, i.e. in the general form

$$\frac{\partial U}{\partial t} + \frac{\partial F_j}{\partial x_j} = S, \tag{8.6}$$

where $U$ is the density of a conserved quantity, $\mathbf{F}$ the flux density of this quantity, and $S$ a local source density. The straight-forward finite differencing of a conservation equation guarantees for a fluid cell with volume $dV$ that the change of the conserved quantity in the cell, $UdV$, is exactly given by the in/out-flow of the quantity through the cell-boundary and the source or sink of the quantity integrated over the cell volume, $SdV$. Equation (8.2) expresses the conservation of baryon number $\rho dV = n_b m_b dV$. The conservation of momentum $\rho v_i dV$ is described in Eq. (8.3) and the conservation of energy $EdV = \left(\rho e + \frac{1}{2}\rho v^2 + \frac{1}{2}b^2\right) dV$ in Eq. (8.4). Equation (8.5) describes the conservation of the lepton number $\rho Y_e dV$. The pressure $P(\rho, T, Y_e)$ and the specific internal energy $e(\rho, T, Y_e)$ are provided by an equation of state (EoS). The right hand sides of Eqs. (8.3), (8.4) and (8.5) contain source terms. The terms $\rho\dot{v}_i$, $\rho\dot{e}$ and $\rho\dot{Y}_e$ stand for the exchange rates of momentum, energy, and lepton number, respectively, with the neutrino radiation field. The gravitational potential $\phi$, finally, is determined by the Poisson equation

$$\nabla^2\phi = 4\pi\rho. \tag{8.7}$$

If a magnetic field $\mathbf{B} = \sqrt{4\pi}\mathbf{b}$ is present, the momentum and energy equations (8.3) and (8.4) obtain contributions from the magnetic stress as indicated. The time-evolution of the magnetic field is described by

$$\frac{\partial \mathbf{b}}{\partial t} - \nabla \times (\mathbf{v} \times \mathbf{b}) = 0, \tag{8.8}$$

which guarantees that the magnetic field stays divergence free. Note that the 9 equations (8.2)–(8.8) determine the 9 unknowns $\rho$, $T$, $Y_e$, $v_i$, $b_i$. Only the three source terms in Eqs. (8.3), (8.4) and (8.5) remain open. They are determined by the neutrino transport equation that evaluates the weak interactions and the exchange of momentum, energy and lepton number between the fluid and the neutrino radiation field.

For each constituent of the fluid that cannot move with respect to the baryons, it is sufficient to assume thermal equilibium and to specify an abundance. Constituents like the neutrinos that are not always in thermal equilibrium and that are able to propagate with respect to the fluid are more difficult to describe. The distribution function $f\left(t, \mathbf{x}, \omega, \frac{\mathbf{p}}{p}\right)$ additionally accepts the neutrino energy $\omega$ and the direction of propagation $\frac{\mathbf{p}}{p}$. The neutrino mass is neglected in the following kinetic treatment of the neutrinos, ($|\mathbf{p}|\, c = pc = \omega$). The neutrino abundance is given by the integration of the distribution function over all neutrino energies $\omega$ and all propagation directions $\Omega$,

$$Y_\nu(t, \mathbf{x}) = \frac{4\pi}{n_b (hc)^3} \int f\left(t, \mathbf{x}, \omega, \frac{\mathbf{p}}{p}\right) \omega^2 d\omega d\Omega, \tag{8.9}$$

whereas the time evolution of the neutrino distribution function is described by the Boltzmann equation

$$\frac{\partial f}{\partial t} + \frac{\partial f}{\partial \mathbf{x}} \cdot \frac{\mathbf{p}}{m} + \frac{\partial f}{\partial \mathbf{p}} \cdot \mathbf{K} = \dot{f}_{\text{collision}}. \tag{8.10}$$

The first term is the time derivation of the neutrino distribution function, which we would like to calculate. The second term describes the advection of neutrinos in and out of the fluid cell. In contrast to the analogous second term in Eq. (8.6), it depends on the particle velocity $\mathbf{p}/m$, which for the massless neutrinos becomes the light speed $c$ instead of the fluid velocity. The third term describes the action of a force field $\mathbf{K}$ on the propagating neutrinos. The right hand side of Eq. (8.10) is given by the rate of local weak interactions between the neutrinos and the fluid. If this collision term is very large (i.e. if the mean free path of the neutrinos is very small), one can show that the appropriate integration over energy and angle of Eq. (8.10) will lead to an equation like Eq. (8.5) for the time evolution of the neutrino abundance in Eq. (8.9). The full system of equations that needs to be solved is now given by Eqs. (8.2) to (8.5) and Eqs. (8.7) and (8.8), coupled to Eq. (8.10) by the interaction rates on the right hand sides of the equations. The collision integral should at least implement the weak interactions listed in Bruenn (1985).

### *8.1.3 Scales and Challenges*

The large *density range* covered by a supernova model is typical for astrophysical applications. After the gravitational collapse on a time scale of ~100 ms (milliseconds), the central density in the compact remnant can reach up to $\rho_c \sim 10^{15}$ g/cm$^3$. At the outer boundary of the computational domain (at a radius of ~10,000 km in the progenitor star), densities of order $10^5$ g/cm$^3$ are typical. In the case of a successful explosion, the density between the compact remnant and the ejecta may drop to order 1 g/cm$^3$. The neutrinos decouple from matter at densities $\sim 10^{11}$–$10^{12}$ g/cm$^3$, depending on the neutrino energy and type. The location where the energy-dependent optical depth is of order unity is called the *neutrinosphere* (in analogy to the photosphere). The neutrino-opaque region inside the neutrinospheres forms the protoneutron star. During the simulation, the size of the *protoneutron star* shrinks from an initial radius ~70 to ~20 km due to the accumulation of mass and the emission of neutrinos. The protoneutron star has an initial mass of ~1.2 $M_\odot$ and accretes infalling outer layers until the supernova explosion is launched or the maxium stable mass is exceeded so that a black hole is formed. In order to resolve the steep density profile at the edge of the protoneutron star, a *resolution* of $\Delta x < 300$ m is desirable to investigate the protoneutron star itself. A resolution of $\Delta x \sim 1$ km might just be sufficient if one is only interested in the explosion on the surface of the protoneutron star. The sound speed in the compact remnant is $c_s \sim 0.3c$, which leads to *time steps* $\Delta t < 10^{-5}$ s if the algorithm has to obey the CFL condition $\Delta t < \Delta x/c_s$. On the one hand, this is a reasonable time step compared to the dynamical time scale $(G\rho_c)^{-\frac{1}{2}} \sim 0.2$ ms of the protoneutron star. On the other hand, the postbounce phase between the collapse and the explosion is thought to extend on a time scale of order seconds. The assumed neutrino diffusion time scale is consistent with the detection of neutrinos over several seconds from supernova SN1987A in the Magellanic cloud. However, even with time steps of order $\mu$s, the

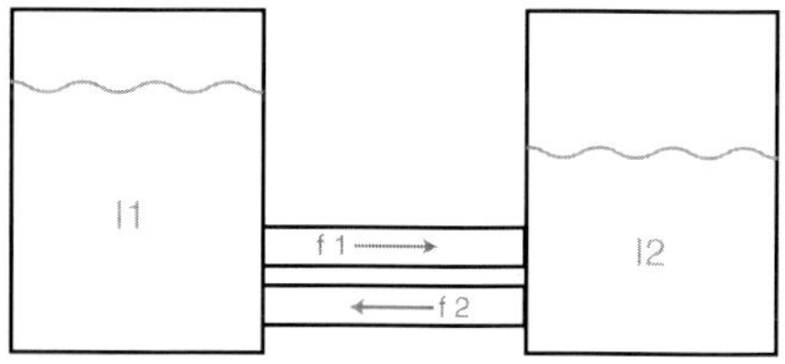


**Fig. 8.1** Two containers with content $I_1$ and $I_2$ exchanging a conserved quantity on a charactersitic time scale $\tau$

reaction rates in the neutrino-opaque regime are fast enough to require an *implicit finite differencing* to find the correct equilibrium value.

Since implicit finite differencing is an important and ubiquitous concept for the modelling of stars, we dedicate a paragraph to this powerful technique. Consider two containers with contents $I_1$ and $I_2$, respectively, as drawn in Fig. 8.1. The containers are able to exchange content by two channels having fluxes $|f_1| = I_1/\tau$ and $|f_2| = I_2/\tau$ in the direction indicated by the arrows. $\tau$ is an arbitrary time scale. With this, the system is described by the coupled system of equations

$$\frac{\partial I_1}{\partial t} = \frac{I_2 - I_1}{\tau}, \qquad \frac{\partial I_2}{\partial t} = -\frac{I_2 - I_1}{\tau}. \tag{8.11}$$

The system is readily solved analytically by investigating the time evolution of the total quantity $I_2 + I_1$ and the content difference $I_2 - I_1$. Now, imagine a computer code that solves the finite differenced system

$$\begin{aligned} I_1^{n+1} &= I_1^n + \frac{I_2^{n+k} - I_1^{n+k}}{\tau}\Delta t \\ I_2^{n+1} &= I_2^n - \frac{I_2^{n+k} - I_1^{n+k}}{\tau}\Delta t. \end{aligned} \tag{8.12}$$

The upper index indicates the time step number: $I^0$ is the initial container content, $I^1$ the value calculated after the first time step $\Delta t$, $I^2$ the value after the second time step, and so forth. If one sets $k = 0$ in Eq. (8.12) one obtains Euler forward finite differencing or explicit finite differencing. In this case, the values of $I^{n+1}$ are directly expressed by the values of $I^n$ of the previous time step as described in Eq. (8.12). Now let's start from the initial values $I_1^0 = 0.6$ and $I_2^0 = 0.4$. Thinking through few time steps shows clearly that a reasonable solution can only be expected when the time step $\Delta t$ is much smaller than the characteristic time scale $\tau$. If it is chosen larger, the solution diverges! This cannot happen with Euler backward or implicit finite differencing. For implicit finite differencing one sets $k = 1$ in Eq. (8.12). Because $I^{n+1}$ now appears on both sides of Eq. (8.12) one has to solve the system of equations first for $I^{n+1}$ and obtains

$$
\begin{aligned}
I_2^{n+1} + I_1^{n+1} &= I_2^n + I_1^n \\
I_2^{n+1} - I_1^{n+1} &= \left(I_2^n - I_1^n\right)\left(1 + \frac{2\Delta t}{\tau}\right)^{-1}
\end{aligned} \tag{8.13}
$$

One can now convince oneself that Eq. (8.13) is equivalent to the explicit version of Eq. (8.12) for very small time steps $\Delta t \ll \tau$, while the equilibrium configuration is obtained without difficulty for very large time steps $\Delta t \gg \tau$. Hence, an implicitly finite differenced equation puts automatically all processes into equilibrium that have a faster characteristic time scale than the time step. This technique is crucial for the successful evolution of the weak interaction rates, where many different characteristic time scales couple accross the divers thermodynamic conditions found in a supernova. However, this stability comes at the cost of solving the system for $I^{n+1}$, which may quickly become a limiting factor for the memory and CPU-time of the simulation.

Even if the *computational domain* has to extend to 10,000 km radius in order to capture most of the infalling layers during the postbounce evolution, the initially interesting dynamics is confined to a sphere of about 500 km radius (see e.g. Marek and Janka, 2009). This still gives a ratio of neutrino-opaque to neutrino-transparent volume of $50^3/500^3 = 10^{-3}$. At the first glance it is therefore surprising that most investigators choose a *comoving frame* description of the neutrino transport which is excessively inefficient in the neutrino-transparent regimes that constitute 99.9% of the computational domain. The reason is that the diffusive limit of the approach has to be very accurate in order to resolve the *small neutrino flux* $H$ at high neutrino density $J$ in the core, where ratios of $\frac{H}{cJ} \sim 10^{-4}$ can prevail. While the diffusion equation has no difficulty to resolve arbitrarily small fluxes (the flux in the comoving frame naturally scales with the mean free path), it seems close to impossible for a Boltzmann equation that is discretised in the laboratory frame to obtain equally accurate results.

With the neutrino-driven (Marek and Janka, 2009), the magneto-rotational (Bisnovatyi-Kogan et al., 2008), the acoustic (Burrows et al., 2007) and the phase-transition induced (Sagert et al., 2009) explosion mechanisms, several supernova explosion mechanisms are discussed as cause of the energetic stellar explosions. The difficulty to pinpoint a specific mechanism is not due to the lack of available energy to explode the star. The gravitational binding energy released during the collapse of the inner stellar core to the protoneutron star amounts to several $10^{53}$ erg, i.e. much more than the $\sim 10^{51}$ erg kinetic energy of a typical supernova explosion. The major part of the released energy is first stored as thermal or rotational energy in the protoneutron star and then emitted from there by the emission of neutrinos or build-up of magnetic fields. The difficulty is to reliably *quantify the energy transfer* that occurs from the small coupling of the large energy reservoir in the protoneutron star to its surface layers that have densities of $10^{10}$–$10^8$ g/cm$^3$ with much lower characteristic energies. In the neutrino-driven explosion mechanism the energy is transferred by the neutrinos. The energy deposition rate of the neutrinos depends on their spectra and on the propagation angles in the layers that are close to the energy-dependent neutrinospheres. In the magento-rotational explosion mechanism the rotational energy from the collapse is transferred via its coupling to magnetic fields, in the (magneto-)acoustic explosion mechanism the energy is transferred by (magneto-)acoustic waves, and in the phase-transition induced mechanism the energy is transferred by a shock that emerges from the protoneutron star.

A further difficulty is that there are *fluid instabilities* like the standing accretion shock instability (SASI Foglizzo et al., 2007) and large scale convective turnover from neutrino heating (Herant et al., 1994) that lead to flow patterns with large regions of outflow alongside of narrow downstreams that affect the efficiency of the energy transfer. Typical dynamical time scales in this less dense regime are of order of several 10 ms. Finally, the effects of *magnetic fields* in combination with rotation have been discussed over decades, but are only starting to be quantitatively explored by the advent of three-dimensional simulations that are based on the ideal magneto-hydrodynamic (MHD) equations. In combination with differential rotation, the magneto-rotational instability (MRI) might develop on the length scale of meters, which cannot be resolved in global simulations. The magnetic fields might grow over a longer postbounce time scale and lead to deformations of the fluid flow in the layers above the protoneutron star that are crucial for the energy transfer.

## 8.1.4 Solution Strategies

### 8.1.4.1 Spherically Symmetric Models

An important milestone has been reached by the traditional spherically symmetric approach. In this case, the degrees of freedom of the distribution function reduce to $f(t, r, \omega, \mu)$, where $r$ is the radial coordinate, $\mu = \cos\theta$ and $\theta$ is the angle between the propagation direction of the neutrino and the radial direction. For example, $\mu = 1$ means 'radially outward' propagation, $\mu = 0$ means "tangential" propagation and $\mu = -1$ means "radially inward" propagation. At first, one might be surprised that the neutrino distribution function has an angular dependence in spite of the assumption of spherical symmetry. This fact is understandable in the following way: Let's assume that the abundance of neutrinos is spherically symmetric. If all neutrinos propagate radially outward, the configuration is obviously spherically symmetric. But if all neutrinos propagate tangentially to a sphere around the origin and isotropically within the surface of the sphere, the configuration is also spherically symmetric. Hence, there are different spherically symmetric configurations that can be distinguished by the angle of the neutrino propagation with respect to the radial direction. The neutrino number density, for example, is then given by the integration of the neutrino distribution function over energy and the propagation angle

$$n_\nu(t, r) = \frac{4\pi}{(hc)^3} \frac{1}{2} \int_{-1}^{+1} \int_0^\infty f(t, r, \omega, \mu)\, \omega^2 d\omega d\mu. \tag{8.14}$$

The time evolution of the neutrino distribution function is described by the Boltzmann transport equation Lindquist (1966), Castor (1972), Mezzacappa and Bruenn (1993) and Liebendörfer et al. (2004)

$$\frac{df}{cdt}+\mu\frac{\partial f}{\partial r}+\left[\mu\left(\frac{d\ln\rho}{cdt}+\frac{3v}{cr}\right)+\frac{1}{r}\right]\left(1-\mu^2\right)\frac{\partial f}{\partial\mu}$$
$$+\left[\mu^2\left(\frac{d\ln\rho}{cdt}+\frac{3v}{cr}\right)-\frac{v}{cr}\right]\omega\frac{\partial f}{\partial\omega}=j\left(1-f\right)-\chi f$$
$$+\frac{1}{c\left(hc\right)^3}\left[\left(1-f\right)\int R^{\mathrm{in}}f'\omega'^2d\omega d\mu'-f\int R^{\mathrm{out}}\left(1-f'\right)\omega'^2d\omega d\mu'\right]. \tag{8.15}$$

Why does Eq. (8.15) look so complicated with respect to Eq. (8.10)? The transport terms on the left hand side of the Boltzmann equation would be equally simple as in Eq. (8.10) if the momentum phase space of the neutrino distribution function was described in the *laboratory frame*. However, most supernova modellers choose a description in the *comoving frame* because there are very sensitive regimes where the neutrinos have a small mean free path. There, the neutrinos diffuse with respect to the turbulent fluid and not with respect to the inertial frame. Moreover, the computationally expensive reaction rates on the right hand side of the Boltzmann equation are difficult to calculate and tabulate in the laboratory frame, where the emission angles depend on the fluid velocity. In the comoving frame, the propagation terms naturally vanish with increasing opacity and the weak interaction rates can establish an accurate thermal and weak equilibrium between the fluid and the trapped neutrinos. On the other hand, the neutrinos propagate between fluid cells with relative velocities so that their energy and propagation direction must be corrected for Doppler frequency shift and angular aberration. These velocity-dependent terms are responsible for the complicated appearance of Eq. (8.15). The collision integral on the right hand side of Eq. (8.15) describes weak interactions between the neutrinos and the fluid. Examples are the isotropic neutrino emissivity $j\left(\rho,T,Y_e,\omega\right)$ and absorptivity $\chi\left(\rho,T,Y_e,\omega\right)$ for the reactions Eq. (8.1) and the collision integral for the scattering of neutrinos with the nucleons or electrons in the fluid. The latter depends on scattering kernals that describe inscattering, $R^{\mathrm{in}}\left(\rho,T,Y_e,\omega,\mu,\omega',\mu'\right)$, and outscattering, $R^{\mathrm{out}}\left(\rho,T,Y_e,\omega,\mu,\omega',\mu'\right)$. Due to the fermionic nature of the neutrinos, the appropriate blocking factors $(1-f)$ must be applied. The solution of Eq. (8.15) determines the collision integral and therefore the exchange rates $\rho\dot{\mathbf{v}}$, $\rho\dot{e}$ and $\rho\dot{Y}_e$ that couple to Eqs. (8.3), (8.4) and (8.5). From the emission and absorption terms alone one gets for example (Mezzacappa and Bruenn, 1993)

$$\rho\dot{v}=\frac{2\pi}{(hc)^3}\int\left(j+\chi\right)f\omega^3d\omega\mu d\mu$$
$$\rho\dot{e}=\frac{2\pi c}{(hc)^3}\int\left[j-\left(j+\chi\right)f\right]\omega^3d\omega d\mu$$
$$\rho\dot{Y}_e=\frac{2\pi cm_b}{(hc)^3}\int\left[j-\left(j+\chi\right)f\right]\omega^2d\omega d\mu.$$

Successful comparisons of the results have been performed with alternative radiative transfer algorithms like the Variable Eddington Factor (VEF) method (Rampp and Janka, 2002; Liebendörfer et al., 2005) and the Multi-Group Flux-Limited Diffusion (MGFLD) approximation (Bruenn et al., 2001; Liebendörfer et al., 2004). However, models in spherical symmetry cannot treat the fluid instabilities and mixing that turn out to be crucial for the onset of the explosion and the further evolution of the expanding ejecta.

#### 8.1.4.2 Axisymmetric Models

In spherically symmetric models, only very peculiar progenitor stars (Kitaura et al., 2006) or exotic input physics (Sagert et al., 2009) can explain explosions. It is well-known that fluid instabilities between the protoneutron star and the standing accretion shock increase the absorption of neutrino energy behind the shock (Herant et al., 1994). Indeed, more progenitor stars are found to explode in axisymmetric models. However, the results of different groups have not yet converged: Models from the Florida/Oak Ridge group predict explosions ~300 ms postbounce (Messer et al., 2008), models from the Garching group predict weak explosions $\sim 600$ ms (Marek and Janka, 2009), and models from the Arizona group predict explosion $\sim 1,200$ ms (Burrows et al., 2007) based on a different explosion mechanism.

Models in axisymmetry have not been performed without approximations in the neutrino transport part. In attempts to solve a direct discretisation of the analogue of Eq. (8.15) (Livne et al., 2004), one had to drop the complicated but important velocity-dependent observer corrections, so that the first law of thermodynamics is violated for the trapped neutrinos. Implementations of the VEF method were more successful, but so far only in combination with the ray-by-ray approximation (Marek and Janka, 2009). The ray-by-ray approach treats the neutrino transport in separate angular wedges with periodic boundary conditions at their interfaces to the neighbor wedges. Lateral transport is only supported in the neutrino-opaque regime by a term that advects trapped neutrinos with the fluid. The MGFLD approximation has been implemented in true axisymmetry (Dessart et al., 2006) and in ray-by-ray approaches (Bruenn et al., 2009). The VEF and MGFLD approximations have both the important advantage that they are based on the solution of moments equations of the radiation field that can naturally resolve small neutrino fluxes in the diffusion limit. While both methods share the drawback that the implicit solution of moments equations is very inefficient in the neutrino-transparent regime, a traditional flux limiter in the MGFLD approach additionally looses the information about propagation directions outside the neutrinospheres (Ott et al., 2008). We believe that the VEF approach can successfully cope with most of the challenges listed in Sect. 8.1.3. However, due to the large computational cost it appears for the time being uneconomical to use this otherwise favourable approach in global three-dimensional models.

#### 8.1.4.3 Three-Dimensional Models

Even if axisymmetry supports crucial fluid instabilities in all these models, it still severely restricts the degrees of freedom of the fluid. Patterns of the accretion flow that include accretion tubes or funnels instead of cones and the three-dimensional spreading of non-axisymmetric modes cannot be investigated in axisymmetry. Also the Lorentz force, that acts on the fluid when magnetic fields are present, has an intrinsically three-dimensional nature. In search of a neutrino transport approximation that can be used as soon as possible to explore the rich physics of three-dimensional dynamical phenomena, we step back from the ideal that the numerical algorithm or method has to be uniform across the whole computational domain. There is no *best algorithm* without a clear specification what problem the algorithm has to solve. In global astrophysical models, many different problems have to be solved in different parts of the computational domain and hence, it is unlikely that there is *one* best algorithm to do this *everywhere*. In three-dimensional supernova models one should exploit that a lot of computational cost could be avoided by adaptive algorithms. A numerically well-resolved model that is based on good physical approximations is much easier to control and safer to interpret than a very comprehensive model that is numerically under-resolved or too inefficient to be sufficiently checked before the production phase. In this spirit, we start with a list of the most important technical requirements that characterise a useful supernova model:

1. The hydrodynamics algorithm must prevent that errors from the calculation of the gravitational potential propagate into the specific internal energy of the fluid and it must conserve baryon number, energy and momentum in order to be able to handle shocks and dissipate unresolved turbulence.
2. Reaction rates must be time-implicitly finite differenced and obey detailed balance in order to be able to find thermodynamic equilibrium where appropriate.
3. In the diffusion limit, mechanical work acting on the neutrino gas must change the neutrino temperature consistently with the first law of thermodynamics and the *discretised* equations must deliver diffusive fluxes that scale with the mean free path to arbitrary small numbers.
4. The neutrino spectrum must be retained when neutrinos are transported between fluid cells of different thermodynamic state and the flux factor in the transparent regime should rather depend on the distance to the neutrinosphere of appropriate neutrino energy than on irrelevant local conditions of the fluid.
5. In order to explore the large parameter space of possible initial configurations, the code must be capable to run over thousands of dynamical time scales within reasonable wallclock time.

In order to implement these requirements we developed the Isotropic Diffusion Source Approximation (IDSA) (Liebendörfer et al., 2009). It attempts to implement an *adaptive algorithm* in the sense that the separation into hydrodynamics and radiative transfer is not based on particle species as in Eqs. (8.2)–(8.5) and (8.15), but on the local opacity. One particle species is allowed to have a component that evolves in the hydrodynamic limit, while another component of the same particle species is treated by radiative transfer. The distribution function of the neutrinos is decomposed into a trapped neutrino component and a streaming neutrino component. The thermodynamics of trapped neutrinos is included in the standard hydrodynamics scheme that evolves the background matter. The neutrino losses due to diffusion are described by an additional source term, the *diffusion source*, that converts trapped neutrinos into streaming neutrinos. The obtained streaming neutrino emission rates are then integrated over space to obtain the neutrino flux. Finally, the neutrino flux is converted to a local streaming neutrino abundance Eq. (8.9) that interacts with matter and feeds back into the evaluation of the diffusion source for the next time step. The conversion from the neutrino flux to the neutrino abundance is performed by a geometric estimate of the flux factor based on the location of the neutrinospheres.

An *adaptive algorithm* can significantly boost the efficiency of a simulation because good approximations can be used locally, even if they do not hold in the full computational regime. For example, the diffusion limit can be used in the computational domain of large optical depth and ray-tracing can be used in the computational domain with negligible optical depth. The challenge of this approach is the implementation of a smooth and reasonably consistent transition between the different regimes of approximation. The 3D results shown in Fig. 4.10 have been produced by implementing the IDSA in the magnetohydrodynamics code FISH (Kaeppeli et al., 2009).

The theoretical understanding of the supernova explosion mechanism is crucial for the understanding of the stellar life cycle, the feedback of internal energy to the interstellar gas in star-forming regions, and the enrichment of the Galaxy with heavy elements. Supernovae are active in all observational windows and emit a broad spectrum of electro-magnetic waves, neutrinos, cosmic rays and gravitational waves. A quantitative understanding of the supernova explosion mechanism may grant observational access to matter under extreme conditions where new physics could be discovered. The knowledge of the input physics is continuously improving by terrestrial experiments and theoretical developments. Similarly, the data base of astronomical observations is growing due to the increased sensitivity of observational instruments and improved statistics. The computational model as link between the input physics and the astronomical observable is most productive if the uncertainties in the model are smaller than the uncertainties in the input physics or the observables. The main difficulty in quantifying the uncertainties in the understanding of the explosion mechanism is that the interaction between the neutrinos and matter constitute a 3D dynamical radiative transfer problem that requires the solution of a spectral transport equation for the neutrinos in a general relativistically curved environment where turbulence and mixing spans many orders of magnitude in density and time scales.

## 8.2 Models of Ejecta for Nucleosynthetic Yield Prediction

As discussed in Chaps. 4 and 9, the nucleosynthetic yields are determined by (i) the dynamics of the ejection process, (ii) the initial composition and thermodynamic conditions of the ejecta and (iii) the weak interactions between the ejecta and the neutrinos that are emitted from the neutrino spheres. In the remainder of this section we will discuss these ingredients and their uncertainties based on the supernova models outlined above.

### *8.2.1 The Dynamics of the Ejection Process*

The described uncertainties about the supernova explosion mechanism and the discussed technical difficulties to constrain the uncertainties by accurate three-dimensional computer models are responsible for the largest part of the uncertainties of the core-collapse supernova nucleosynthesis yields. Fortunately, not every detail of the multidimensional computer model impacts the ejection dynamics with similar significance. The first and perhaps most significant unknown is the delay time $\sim 500$ ms between the core bounce and the onset of the explosion. If this delay time is long, the innermost stellar layers that surround the collapsing iron core will accrete onto the protoneutron star and settle there before the explosion. The shock wave of the explosion then runs through the outer stellar layers performing explosive nucleosynthesis. The amount of mass that remains bound (including the protoneutron star) is called the *mass cut*. It determines which stellar layers form the innermost ejecta and has therefore a significant impact on the ejected supernova yields. Since it determines the total amount of ejected $^{56}$Ni, the mass cut can be inferred from the observed mass of $^{56}$Ni in the ejecta (Thielemann et al., 1996). A very long delay time between the core bounce and the supernova explosion increases the chance that the protoneutron star accretes more than its maximum stable mass so that the protoneutron star collapses to a black hole instead of driving an explosion. Thus, the unknown delay time determines in this way also the mass range of progenitor stars that will eject nucleosynthesis yields.

The second important unknown in current supernova models is the explosion energy. The otherwise most plausible neutrino-driven supernova mechanism (Bethe and Wilson, 1985) shows a tendency of producing too low explosion energies throughout the literature. This is mostly due to a negative feedback between the accretion rate and the neutrino heating: Whenever the explosion starts, the accretion rate is reduced. This leads to a significant decrease of the neutrinos that are emitted from the compressed accreted matter. This decrease in the neutrino luminosity from the neutrinospheres reduces the neutrino heating behind the shock and hinders the explosion to pick up as much energy as the observation of supernovae suggests. This problem is most severe in spherically symmetric approaches where accretion and explosion cannot coexist at the same time. In multidimensional models, it is plausible that narrow accreting downstreams maintain a high accretion luminosity while regions inbetween them are already expanding toward the explosion (Herant et al., 1994). Multidimensional simulations of the post-explosion phase show strong mixing by the Rayleigh-Taylor instabilities that occur when the cold outer stellar layers are accelerated from below by the hot ejecta (Kifonidis et al., 2003; Hammer et al., 2009). However, this late-time mixing determines rather the geometrical distribution of the yields than the nucleosynthesis itself.

### 8.2.2 The Thermodynamic Conditions of the Ejecta

The nucleosynthesis is mainly influenced by the initial composition and the time evolution of the thermodynamic conditions in the ejecta. Traditional yield predictions adopt the composition of the progenitor star as initial condition of the ejecta, infer the mass cut from the observed $^{56}$Ni mass and set artificially a kinetic piston or thermal energy (Aufderheide et al., 1991) into a hydrodynamical model in order to produce timelines of thermodynamic conditions in the ejecta that can be post-processed with large nuclear reaction networks (see Chap. 9). Individual fluid elements from different stellar layers are post-processed in this way until the reactions freeze out. The resulting yields are then obtained by collecting the abundances from the treated fluid elements. This approach is very clever, because the above-mentioned two big uncertainties of supernova models with respect to the mass cut and the explosion energy are circumvented by observational information. In a spherically symmetric model, the temperature of the matter at the time it is passed by the shock can even be estimated analytically from Chap. 4, Eq. 4.4. If one determines the thermodynamical conditions of the ejecta more accurately from a hydrodynamical simulation, one faces the problem that the thermodynamic evolution of the ejecta depends itself on the energy budget of the nuclear reactions that one would like to evaluate after the fact in a post-processing step. Fortunately, it is a good approximation to run the hydrodynamics together with an efficient small network of selected (mostly $\alpha$-)nuclei (Hix et al., 1998) that correctly deliver the overall energy budget of the nucleosynthesis to the thermodynamical evolution.

### 8.2.3 The Exposure of the Ejecta to Neutrinos

The large reaction networks that are used to determine the nucleosynthesis trace the evolution of the abundances of thousands of nuclei. Hence, they keep an accurate track of the composition – with one exception: the neutrinos. Because the neutrinos are not trapped in the ejecta, their abundance is not determined by local conditions, but by the distance to the neutrinospheres and the conditions at the neutrinospheres. The influence of the neutrinos on the nucleosynthesis is a third factor of uncertainty in the yield prediction that relates to the underlying supernova model. The exchange of neutrons into protons and vice versa by the reactions in Eq. (8.1) depends on the ratio of the neutrino to anti-neutrino abundances and their spectra (they are not in thermal equilibrium with the ejecta). These neutrino properties are determined at the neutrinospheres close to the surface of the protoneutron star.

While the mass cut and the explosion energy could be gauged by the observation, it is not possible to reliably derive the influence of neutrinos from observations. The only detection of neutrinos from supernova 1987A was statstically too scarce to pinpoint the neutrino energies. Fortunately the neutrino energies and spectra are less difficult to predict from the above-described supernova models than the multidimensional explosion dynamics: The neutrinos decouple from the matter at the surface of the protoneutron star where the geometry is still reasonably spherically symmetric. Hence, the spherical models described in Sect. 8.1.4 can be used to derive the neutrino abundances as function of radius and time in general relativity with sophisticated input physics. In this way, the neutrino luminosities can be chosen consistently with the progenitor model that serve for the ejection model. This is important because the neutrino luminosities and spectra can strongly vary between progenitor models of different main sequence mass. A compilation of luminosities and mean energies is given in (Liebendörfer et al., 2003). As discussed in Chap. 4, yield predictions have shown that the neutrinos have a significant impact on the nucleosynthesis of the innermost ejecta, where the neutrino flux and target density in the ejecta is largest (Woosley et al., 1990; Fröhlich et al., 2006b; Pruet et al., 2006).

How can the three uncertainties *dynamics*, *thermodynamic conditions* and *neutrino influence* further be reduced? In the mean time, the neutrino luminosities are also available from accurate axisymmetric supernova models (Marek and Janka, 2009). The main difference to the spherical case is that the multidimensional accretion flow permits time-variable fast downflows that contribute a time-variable accretion luminosity component to the total neutrino luminosities. Additionally, it is now interesting to not only use the neutrinos from the supernova model, but also the dynamics of the ejection process and the consistent thermodynamic condition of the ejecta. However, because there are only very few supernova models that explode naturally (Hoffman et al., 2008; Wanajo et al., 2009), one has to produce series of explosion models by artificially adding external energy to the neutrino heating in a parameterised form (Fröhlich et al., 2006a). As a next step, the axisymmetric supernova models discussed in Sect. 8.1.4 should be refined and brought to mutual agreement and convergence between the different groups so that the three-dimensional models with spectral neutrino transport that are constructed now along the lines of Sect. 8.1.4 find again a reliable reference. Then the new generation of three-dimensional models will lead to new insight: If they produce explosions that are consistent with the observations, they can serve as models to predict the nucleosynthesis yields in much more detail. If, on the other hand, they still do not produce sufficiently energetic explosions, one has for the first time proven missing physics in the models and can start a new quest for the supernova mechanism.

## References


Aufderheide, M.B., Baron, E., Thielemann, F.: Shock waves and nucleosynthesis in type II supernovae. ApJ **370**, 630–642 (1991). doi:10.1086/169849

Bethe, H.A.: Supernova mechanisms. Rev. Mod. Phys. **62**, 801–866 (1990). doi:10.1103/RevModPhys.62.801

Bethe, H.A., Wilson, J.R.: Revival of a stalled supernova shock by neutrino heating. ApJ **295**, 14–23 (1985). doi:10.1086/163343

Bisnovatyi-Kogan, G.S., Moiseenko, S.G., Ardelyan, N.V.: Different magneto-rotational supernovae. Astron. Rep. **52**, 997–1008 (2008). doi:10.1134/S1063772908120056

Blondin, J.M., Lufkin, E.A.: The piecewise-parabolic method in curvilinear coordinates. ApJS **88**, 589–594 (1993). doi:10.1086/191834

Bruenn, S.W.: Stellar core collapse – Numerical model and infall epoch. ApJS **58**, 771–841 (1985). doi:10.1086/191056

Bruenn, S.W., De Nisco, K.R., Mezzacappa, A.: General relativistic effects in the core collapse supernova mechanism. ApJ **560**, 326–338 (2001). doi:10.1086/322319, arXiv:astro-ph/0101400

Bruenn, S.W., Mezzacappa, A., Hix, W.R., Blondin, J.M., Marronetti, P., Messer, O.E.B., Dirk, C.J., Yoshida, S.: 2D and 3D core-collapse supernovae simulation results obtained with the CHIMERA code. J. Phys. Conf. Ser. **180**(1), 012018 (2009). doi:10.1088/1742-6596/180/1/012018

Brun, A.S., Miesch, M.S., Toomre, J.: Global-scale turbulent convection and magnetic dynamo action in the solar envelope. ApJ **614**, 1073–1098 (2004). doi:10.1086/423835, arXiv:astroph/0610073

Burrows, A., Livne, E., Dessart, L., Ott, C.D., Murphy, J.: Features of the acoustic mechanism of core-collapse supernova explosions. ApJ **655**, 416–433 (2007). doi:10.1086/509773, arXiv:astroph/0610175

Calder, A.C., Fryxell, B., Plewa, T., Rosner, R., Dursi, L.J., Weirs, V.G., Dupont, T., Robey, H.F., Kane, J.O., Remington, B.A., Drake, R.P., Dimonte, G., Zingale, M., Timmes, F.X., Olson, K., Ricker, P., MacNeice, P., Tufo, H.M.: On validating an astrophysical simulation code. ApJS **143**, 201–229 (2002). doi:10.1086/342267, arXiv:astro-ph/0206251

Castor, J.I.: Radiative transfer in spherically symmetric flows. ApJ **178**, 779–792 (1972). doi:10.1086/151834

Chandrasekhar, S.: An Introduction to the Study of Stellar Structure. Dover, New York (1967)

Colgate, S.A., White, R.H.: The hydrodynamic behavior of supernovae explosions. ApJ **143**, 626 (1966). doi:10.1086/148549

Dessart, L., Burrows, A., Ott, C.D., Livne, E., Yoon, S., Langer, N.: Multidimensional simulations of the accretion-induced collapse of white dwarfs to neutron stars. ApJ **644**, 1063–1084 (2006). doi:10.1086/503626, arXiv:astro-ph/0601603

Dotter, A., Paxton, B.: Evolutionary implications of the new triple-$\alpha$ nuclear reaction rate for low mass stars. A&A **507**, 1617–1619 (2009). doi:10.1051/0004-6361/200912998, 0905.2397

Eggenberger, P., Meynet, G., Maeder, A., Hirschi, R., Charbonnel, C., Talon, S., Ekstr¨om, S.: The Geneva stellar evolution code. Ap&SS **316**, 43–54 (2008). doi:10.1007/s10509-007-9511-y

Foglizzo, T., Galletti, P., Scheck, L., Janka, H.: Instability of a stalled accretion shock: evidence for the advective-acoustic cycle. ApJ **654**, 1006–1021 (2007). doi:10.1086/509612, arXiv:astroph/0606640

Fröhlich, C., Hauser, P., Liebendörfer, M., Martínez-Pinedo, G., Thielemann, F., Bravo, E., Zinner, N.T., Hix, W.R., Langanke, K., Mezzacappa, A., Nomoto, K.: Composition of the innermost core-collapse supernova Ejecta. ApJ **637**, 415–426 (2006a). doi:10.1086/498224, arXiv:astro-ph/0410208

Fröhlich, C., Martínez-Pinedo, G., Liebendörfer, M., Thielemann, F., Bravo, E., Hix, W.R., Langanke, K., Zinner, N.T.: Neutrino-induced nucleosynthesis of A¿64 nuclei: the $\nu\pi$ process. Phys. Rev. Lett. **96**(14), 142502 (2006b). doi:10.1103/PhysRevLett.96.142502, arXiv:astroph/0511376

Fryxell, B., Olson, K., Ricker, P., Timmes, F.X., Zingale, M., Lamb, D.Q., MacNeice, P., Rosner, R., Truran, J.W., Tufo, H.: FLASH: an adaptive mesh hydrodynamics code for modeling astrophysical thermonuclear flashes. ApJS **131**, 273–334 (2000). doi:10.1086/317361

Giacomazzo, B., Rezzolla, L.: WhiskyMHD: a new numerical code for general relativistic magnetohydrodynamics. Class. Quantum Grav. **24**, 235 (2007). doi:10.1088/0264-9381/24/12/S16, arXiv:gr-qc/0701109

Hammer, N.J., Janka, H., Mueller, E.: Three-dimensional simulations of mixing instabilities in supernova explosions. ArXiv e-prints 0908.3474 (2009)

Henyey, L.G., Wilets, L., Böhm, K.H., Lelevier, R., Levee, R.D.: A method for atomic computation of stellar evolution. ApJ **129**, 628 (1959). doi:10.1086/146661

Herant, M., Benz, W., Hix, W.R., Fryer, C.L., Colgate, S.A.: Inside the supernova: a powerful convective engine. ApJ **435**, 339–361 (1994). doi:10.1086/174817, arXiv:astro-ph/9404024

Hix, W.R., Khokhlov, A.M., Wheeler, J.C., Thielemann, F.: The quasi-equilibrium-reduced alpha-network. ApJ **503**, 332 (1998). doi:10.1086/305968, arXiv:astro-ph/9805095

Hoffman, R.D., Müller, B., Janka, H.: Nucleosynthesis in O-Ne-Mg supernovae. ApJ **676**, L127–L130 (2008). doi:10.1086/587621, 0712.4257

Kaeppeli, R., Whitehouse, S.C., Scheidegger, S., Pen, U., Liebendoerfer, M.: FISH: a 3D parallel MHD code for astrophysical applications. ArXiv e-prints 0910.2854 (2009)

Kifonidis, K., Plewa, T., Janka, H., Müller, E.: Non-spherical core collapse supernovae. I. Neutrino-driven convection, Rayleigh-Taylor instabilities, and the formation and propagation of metal clumps. A&A **408**, 621–649 (2003). doi:10.1051/0004-6361:20030863, arXiv:astroph/0302239

Kitaura, F.S., Janka, H., Hillebrandt, W.: Explosions of O-Ne-Mg cores, the Crab supernova, and subluminous type II-P supernovae. A&A **450**, 345–350 (2006). doi:10.1051/0004-6361:20054703, arXiv:astro-ph/0512065

Langer, N., Fliegner, J., Heger, A., Woosley, S.E.: Nucleosynthesis in rotating massive stars. Nucl. Phys. A **621**, 457–466 (1997). doi:10.1016/S0375-9474(97)00290-X

Liebendörfer, M., Mezzacappa, A., Messer, O.E.B., Martinez-Pinedo, G., Hix, W.R., Thielemann, F.: The neutrino signal in stellar core collapse and postbounce evolution. Nucl. Phys. A **719**, 144 (2003). doi:10.1016/S0375-9474(03)00984-9, arXiv:astro-ph/0211329

Liebendörfer, M., Messer, O.E.B., Mezzacappa, A., Bruenn, S.W., Cardall, C.Y., Thielemann, F.: A finite difference representation of neutrino radiation hydrodynamics in spherically symmetric general relativistic spacetime. ApJS **150**, 263–316 (2004). doi:10.1086/380191, arXiv:astroph/0207036

Liebendörfer, M., Rampp, M., Janka, H., Mezzacappa, A.: Supernova simulations with boltzmann neutrino transport: a comparison of methods. ApJ **620**, 840–860 (2005). doi:10.1086/427203, arXiv:astro-ph/0310662

Liebendörfer, M., Whitehouse, S.C., Fischer, T.: The isotropic diffusion source approximation for supernova neutrino transport. ApJ **698**, 1174–1190 (2009). doi:10.1088/0004-637X/698/2/1174, 0711.2929

Limongi, M., Straniero, O., Chieffi, A.: Massive stars in the range 13–25M.: evolution and nucleosynthesis. II. The solar metallicity models. ApJS **129**, 625–664 (2000). doi:10.1086/313424, arXiv:astro-ph/0003401

Lindquist, R.W.: Relativistic transport theory. Ann. Phys. **37**, 487–518 (1966). doi:10.1016/0003-4916(66)90207-7

Livne, E., Burrows, A., Walder, R., Lichtenstadt, I., Thompson, T.A.: Two-dimensional, timedependent, multigroup, multiangle radiation hydrodynamics test simulation in the core-collapse supernova context. ApJ **609**, 277–287 (2004). doi:10.1086/421012, arXiv:astro-ph/0312633

Ludwig, H., Freytag, B., Steffen, M.: A calibration of the mixing-length for solar-type stars based on hydrodynamical simulations. I. Methodical aspects and results for solar metallicity. A&A **346**, 111–124, arXiv:astro-ph/9811179 (1999)

Maeder, A., Meynet, G.: Grids of evolutionary models from 0.85 to 120 solar masses – Observational tests and the mass limits. A&A **210**, 155–173 (1989)

Marek, A., Dimmelmeier, H., Janka, H., Müller, E., Buras, R.: Exploring the relativistic regime with Newtonian hydrodynamics: an improved effective gravitational potential for supernova simulations. A&A **445**, 273–289 (2006). doi:10.1051/0004-6361:20052840, arXiv:astro-ph/0502161

Marek, A., Janka, H.: Delayed neutrino-driven supernova explosions aided by the standing accretion-shock instability. ApJ **694**, 664–696 (2009). doi:10.1088/0004-637X/694/1/664, 0708.3372

Meakin, C.A., Arnett, D.: Turbulent convection in stellar interiors. I. Hydrodynamic simulation. ApJ **667**, 448–475 (2007). doi:10.1086/520318, arXiv:astro-ph/0611315

Messer, O.E.B., Bruenn, S.W., Blondin, J.M., Hix, W.R., Mezzacappa, A.: Multidimensional, multiphysics simulations of core-collapse supernovae. J. Phys. Conf. Ser. **125**(1), 012010 (2008). doi:10.1088/1742-6596/125/1/012010

Meynet, G., Maeder, A.: Stellar evolution with rotation. I. The computational method and the inhibiting effect of the $\mu$-gradient. A&A **321**, 465–476 (1997)

Mezzacappa, A., Bruenn, S.W.: Stellar core collapse – a Boltzmann treatment of neutrinoelectron scattering. ApJ **410**, 740–760 (1993). doi:10.1086/172791

Nomoto, K., Hashimoto, M.: Presupernova evolution of massive stars. Phys. Rep. **163**, 13–36 (1988)

Ott, C.D.: TOPICAL REVIEW: the gravitational-wave signature of core-collapse supernovae. Class. Quantum Grav. **26**(6), 063001 (2009). doi:10.1088/0264-9381/26/6/063001, 0809.0695

Ott, C.D., Burrows, A., Dessart, L., Livne, E.: Two-dimensional multiangle, multigroup neutrino radiation-hydrodynamic simulations of postbounce supernova cores. ApJ **685**, 1069–1088 (2008). doi:10.1086/591440, 0804.0239

Pruet, J., Hoffman, R.D., Woosley, S.E., Janka, H., Buras, R.: Nucleosynthesis in early supernova winds. II. The role of neutrinos. ApJ **644**, 1028–1039 (2006). doi:10.1086/503891, arXiv:astroph/0511194

Rampp, M., Janka, H.: Radiation hydrodynamics with neutrinos. Variable Eddington factor method for core-collapse supernova simulations. A&A **396**, 361–392 (2002). doi:10.1051/0004-6361:20021398, arXiv:astro-ph/0203101

Rauscher, T., Heger, A., Hoffman, R.D., Woosley, S.E.: Nucleosynthesis in massive stars with improved nuclear and stellar physics. ApJ **576**, 323–348 (2002). doi:10.1086/341728, arXiv:astroph/0112478

Röpke, F.K., Hillebrandt, W., Schmidt, W., Niemeyer, J.C., Blinnikov, S.I., Mazzali, P.A.: A three-dimensional deflagration model for type Ia supernovae compared with observations. ApJ **668**, 1132–1139 (2007). doi:10.1086/521347, 0707.1024

Sagert, I., Fischer, T., Hempel, M., Pagliara, G., Schaffner-Bielich, J., Mezzacappa, A., Thielemann, F., Liebendörfer, M.: Signals of the QCD phase transition in core-collapse supernovae. Phys. Rev. Lett. **102**(8), 081101 (2009). doi:10.1103/PhysRevLett.102.081101, 0809.4225

Springel, V., Yoshida, N., White, S.D.M.: GADGET: a code for collisionless and gasdynamical cosmological simulations. New Astron. **6**, 79–117 (2001). doi:10.1016/S1384-1076(01)00042-2, arXiv:astro-ph/0003162

Stone, J.M., Mihalas, D., Norman, M.L.: ZEUS-2D: a radiation magnetohydrodynamics code for astrophysical flows in two space dimensions. III – the radiation hydrodynamic algorithms and tests. ApJS **80**, 819–845 (1992). doi:10.1086/191682

Stone, J.M., Norman, M.L.: ZEUS-2D: a radiation magnetohydrodynamics code for astrophysical flows in two space dimensions. I – the hydrodynamic algorithms and tests. ApJS **80**, 753–790 (1992a). doi:10.1086/191680

Stone, J.M., Norman, M.L.: ZEUS-2D: a radiation magnetohydrodynamics code for astrophysical flows in two space dimensions. II. The magnetohydrodynamic algorithms and tests. ApJS **80**, 791 (1992b). doi:10.1086/191681

Thielemann, F., Nomoto, K., Hashimoto, M.: Core-collapse supernovae and their ejecta. ApJ **460**, 408 (1996). doi:10.1086/176980

Wanajo, S., Nomoto, K., Janka, H., Kitaura, F.S., Müller, B: Nucleosynthesis in electron capture supernovae of asymptotic giant branch stars. ApJ **695**, 208–220 (2009). doi:10.1088/0004-637X/695/1/208, 0810.3999

Woosley, S.E., Hartmann, D.H., Hoffman, R.D., Haxton, W.C.: The nu-process. ApJ **356**, 272–301 (1990). doi:10.1086/168839

Woosley, S.E., Heger, A., Weaver, T.A.: The evolution and explosion of massive stars. Rev. Mod. Phys. **74**, 1015–1071 (2002). doi:10.1103/RevModPhys.74.1015

Woosley, S.E., Weaver, T.A.: The evolution and explosion of massive stars. II. Explosive hydrodynamics and nucleosynthesis. ApJS **101**, 181 (1995). doi:10.1086/192237

Young, P.A., Arnett, D.: Observational tests and predictive stellar evolution. II. Nonstandard models. ApJ **618**, 908–918 (2005). doi:10.1086/426131, arXiv:astro-ph/0409658

Zingale, M., Woosley, S.E., Rendleman, C.A., Day, M.S., Bell, J.B.: Three-dimensional numerical simulations of Rayleigh-Taylor unstable flames in type Ia supernovae. ApJ **632**, 1021–1034 (2005). doi:10.1086/433164, arXiv:astro-ph/0501655

# Chapter 9
# Nuclear Reactions

**M. Wiescher and T. Rauscher**

Nuclear reaction rates determine the abundances of isotopes in stellar burning processes. A multitude of reactions determine the reaction flow pattern which is described in terms of reaction network simulations. The reaction rates are determined by laboratory experiments supplemented by nuclear reaction and structure theory. We will discus the experimental approach as well as the theoretical tools for obtaining the stellar reaction rates. A detailed analysis of a reaction is only possible for a few selected cases which will be highlighted in this section. The bulk of nuclear reaction processes is however described in terms of a statistical model approach, which relies on global nuclear structure and reaction parameters such as level density and mass and barrier penetration, respectively. We will discuss a variety of experimental facilities and techniques used in the field, this includes low energy stable beam experiments, measurements at radioactive beam accelerators, and neutron beam facilities.

## 9.1 Nuclear Reactions in Astrophysical Environments

Nuclear reactions are the engine of stellar evolution and determine the overall production of the long-lived radioactive isotopes in a variety of nucleosynthesis patterns. A detailed understanding of the characteristic production and depletion rates within the framework of the nucleosynthesis process is crucial for reliable model predictions and the interpretation of the observed abundances.

There are several experimental and theoretical challenges in obtaining stellar reaction rates. The interaction energies in stellar environments extend from basically zero projectile energy up to only several MeV. This is especially challenging for the measurement of the relevant reaction cross sections which can be extremely small, especially for reactions with charged projectiles. This also makes theoretical

M. Wiescher (✉)
University of Notre Dame, Notre Dame, IN 46556, USA

T. Rauscher
Universität Basel, 4056 Basel, Switzerland

Wiescher, M. Rauscher, T.: *Nuclear Reactions*. Lect. Notes Phys. **812**, 461–489 (2011)
DOI 10.1007/978-3-642-12698-7_9 

predictions extremely difficult because several reaction mechanisms (see below) may compete and simpler approximations may only be of limited use.

Another challenge arises from the fact that nuclear burning at high temperature produces very short-lived isotopes which subsequently decay to long-lived and stable species. Current experimental technology can only access a fraction of those yet and is still limited in obtaining detailed information on their properties. The possibility to measure cross sections of reactions with such highly unstable nuclei is even more limited currently. Thus, investigations of nucleosynthesis in high-temperature environments largely rely on theoretical models, which not only have to treat the reaction mechanisms properly but also are required to predict nuclear properties far from stability.

Moreover, depending on the actual plasma conditions, reactions in an astrophysical plasma may proceed fundamentally differently from those in the laboratory. This is due to two effects. On one hand, laboratory nuclei are always part of atoms or molecules whereas astrophysical nuclear burning involves fully ionized nuclei immersed in a cloud of free electrons (and photons). The Coulomb charge of a nucleus is partially shielded by the surrounding electrons but this shielding (or screening) effect will be different for an atom or molecule and a plasma because of the different electron distribution and kinetics. While theoretical cross sections always imply bare nuclei, the values have to be appropriately converted (also based on a theoretical treatment of different screening mechanisms) for comparison to low-energy laboratory cross sections and for application in astrophysical plasmas. Additionally, the quantum mechanically and geometrically different electron distribution in a plasma directly affects electron capture reactions. For example, nuclei such as $^{7}$Be or $^{44}$Ti, decaying by capturing an electron from the K-shell of the atom under laboratory conditions will not be able to do this in a stellar plasma. Instead, electron capture inside a star involves capturing a free electron from the plasma, which is more unlikely and therefore the terrestrial half-life can be shorter than the one in a stellar environment (Iliadis, 2007; Johnson et al., 1992).

Finally, due to the high photon and matter densities in astrophysical environments, nuclei very quickly reach thermal equilibrium with the surroundings by excitation and de-excitation via photons and by collisions. In most cases, this happens on a shorter timescale than that of nuclear transformations (one of the exceptions being isomeric states). Consequently, the nuclei involved in the reactions occur not only in their ground states, as in the laboratory, but also their excited states are populated with a probability involving the Boltzmann factor. So far, reactions on excited states can only be treated theoretically. The population of excited nuclear states does not only depend on the plasma temperature but also on the structure of a nucleus. Nuclei with isolated levels reachable within a few keV projectile energy will exhibit pronounced thermal population already at low plasma temperature. This is especially important for modern s-process studies which require high accuracy knowledge of neutron capture rates. Thermal effects are also important in decays and neutrino reactions because the available phase space of the reaction products is altered, leading to a modification of the rate. For example, electron capture rates in the stellar core collapse are enhanced at temperatures $T > 1.5$ MeV because of the unblocking of low-lying neutron states by thermal excitation (Cooperstein and Wambach, 1984).

### 9.1.1 Reaction Networks and Thermonuclear Reaction Rates

The change of abundances $Y$ with time due to nuclear processes is traced by coupled differential equations. To be fully solvable, the number of equations $N$ must equal the number of involved nuclei acting as reaction partners and thus an equation matrix of size $N^2$ has to be solved. Such a set of coupled equations is called reaction network and can generally be written as

$$\dot{Y}_i = \frac{1}{\rho N_{\mathrm{A}}}\dot{n}_i = \frac{1}{\rho N_{\mathrm{A}}}\left\{\sum_j {}^1_iP_j\ {}_i\lambda_j + \sum_j {}^2_iP_j\ {}_ir_j + \sum_j {}^3_iP_j\ {}_i\hat{r}_j + \dots\right\}, \quad (9.1)$$

where $1 \le i \le N$ numbers the nucleus, ${}_i\lambda_j$ is the $j$th rate for destruction or creation of the $i$th nucleus without a nuclear projectile involved (this includes spontaneous decay, lepton capture, photodisintegration), and ${}_ir_j$ is the rate of the $j$th reaction involving a nuclear projectile and creating or destroying nucleus $i$. Similarly, we have three-body reactions where nucleus $i$ is produced or destroyed together with two other (or similar) nuclei. Reactions with more participants (denoted by ... above) are unlikely to occur at astrophysical conditions and are usually neglected. The quantities ${}^1_iP_j$, ${}^2_iP_j$, and ${}^3_iP_{jk}$ are positive or negative integer numbers specifying the amount of nuclei $i$ produced or destroyed, respectively, in the given process. As shown below, the rates $\lambda$, $r$, and $\hat{r}$ contain the abundances of the interacting nuclei. Rates of type $\lambda$ depend on one abundance (or number density), rates $r$ depend on the abundances of two species, and rates $\hat{r}$ on three.

Using abundances $Y$ instead of number densities $n = Y\rho N_{\mathrm{A}}$ (where $\rho$ is the plasma density) has the advantage that a change in the number of nuclei in a given volume due to density fluctuations is factored out and only changes by nuclear processes are considered. Using abundance changes, the total energy generation rate per mass due to nuclear reactions can easily be expressed as

$$\dot{\varepsilon} = -\sum_i \dot{Y}_i N_A M_i c^2, \quad (9.2)$$

with the rest masses $M_i c^2$ of the participating nuclei.

The rates ${}_i\lambda_j$ appearing in the first term of Eq. (9.1) are reactions per time and volume, and only contain the abundance $Y_j$. For example, ${}_i\lambda_j$ is simply $n_jL_j = Y_j\rho N_{\mathrm{A}}L_j$ for $\beta$-decays. The factor $L_j = (\ln 2)/{}^jT_{1/2}$ is the usual decay constant (with the unit 1/time) and is related to the half-life ${}^jT_{1/2}$ of the decaying nucleus $j$. It has to be noted that some decays depend on the plasma temperature and thus $L_j$ is not always constant, even for decays.

Two-body rates $r$ include the abundances of two interacting particles or nuclei. In general, target and projectile follow specific thermal momentum distributions $dn_1$ and $dn_2$ in an astrophysical plasma. With the resulting relative velocities $\mathbf{v}_1 - \mathbf{v}_2$, the number of reactions per volume and time, is given by

$$r_{12} = \int \hat{\sigma}(|\mathbf{v}_1 - \mathbf{v}_2|)|\mathbf{v}_1 - \mathbf{v}_2| dn_1 dn_2, \tag{9.3}$$

and involves the reaction cross section $\hat{\sigma}$ as a function of velocity/energy, the relative velocity $\vec{\mathbf{v}}_1 - \vec{\mathbf{v}}_2$ and the thermodynamic distributions of target and projectile $dn_1$ and $dn_2$. The evaluation of this integral depends on the type of particles (fermions, bosons) and distributions which are involved.

However, many two-body reactions can be simplified and effectively expressed similarly to one-body reactions, only depending on one abundance (or number density). If reaction partner 2 is a photon, the relative velocity is always $c$ and the quantities in the integral do not depend on $dn_1$. This simplifies the rate expression to

$$\lambda_1 = L_\gamma(T) n_1, \tag{9.4}$$

where $L_\gamma(T)$ stems from an integration over a Planck distribution for photons of temperature $T$. This is similar to the decay rates introduced earlier and therefore we replaced $r$ by $\lambda$ in our notation and can include this type of reaction in the first term of Eq. (9.1). A similar procedure is used for electron captures by protons and nuclei. Because the electron is about 2,000 times less massive than a nucleon, the velocity of the nucleus is negligible in the center-of-mass system in comparison to the electron velocity ($|\vec{\mathbf{v}}_{\text{nucleus}} - \vec{\mathbf{v}}_{\text{electron}}| \approx |\vec{\mathbf{v}}_{\text{electron}}|$). The electron capture cross section has to be integrated over a Fermi distribution of electrons. The electron capture rates are a function of $T$ and $n_e = Y_e \rho N_A$, the electron number density. In a neutral, completely ionized plasma, the electron abundance $Y_e$ is equal to the total proton abundance $Y_e = \sum_i Z_i Y_i$ and thus

$$\lambda_{\text{nucleus,ec}} = L_{\text{ec}}(T, \rho Y_e) n_{\text{nucleus}}. \tag{9.5}$$

Again, we have effectively a rate per target $L$ (with unit 1/time) similar to the treatment of decays earlier and a rate per volume including the number density of only one nucleus. We denote the latter by $\lambda$ and use it in the first term of Eq. (9.1). This treatment can be applied also to the capture of positrons, being in thermal equilibrium with photons, electrons, and nuclei. Furthermore, at high densities ($\rho > 10^{12}$g/cm$^3$) the size of the neutrino scattering cross section on nucleons, nuclei, and electrons ensures that enough scattering events occur to lead to a continuous neutrino energy distribution. Then also the inverse process to electron capture (neutrino capture) can occur as well as other processes like, e.g., inelastic scattering, leaving a nucleus in an excited state which can emit nucleons and $\alpha$ particles. Such reactions can be expressed similarly to photon and electron captures, integrating over the corresponding neutrino distribution.

In the following, we focus on the case of two interacting nuclei or nucleons as these reactions will be extensively discussed in Sects. 9.3 and 9.4. This will result in an actual two-body rate $r$ to be used in the second term of Eq. (9.1). Here, we mention in passing that Eq. (9.3) can be generalized to 3 and more interacting nuclear species by integrating over the appropriate number of distributions, leading to rates $\hat{r}$ and higher order terms in Eq. (9.1).

Turning our attention back to two-body reactions, we note that the velocity distributions can be replaced by energy distributions. Furthermore, it can be shown that the two distributions in Eq. (9.3) can be replaced by a single one in the center-of-mass system. This time the resulting expression describes a rate $r$ including two abundances (or number densities) and showing up in the second term of Eq. (9.1). The rate $r$ is defined as an interaction of two reaction partners with an energy distribution $\phi(E)$ according to the plasma temperature $T$ and a reaction cross section $\sigma^*(E)$ specifying the probability of the reaction in the plasma:

$$r = \frac{n_1 n_2}{1+\delta_{12}} \int_0^\infty \sigma^*(E)\phi(E)\,dE. \tag{9.6}$$

The factor $1/(1+\delta_{12})$ with the Kronecker symbol $\delta$ is introduced to avoid double counting. The nuclear cross section is defined as in standard scattering theory by

$$\sigma = \frac{\text{number of reactions target}^{-1}\ \text{s}^{-1}}{\text{flux of incoming projectiles}}. \tag{9.7}$$

However, in an astrophysical plasma, nuclei quickly (on the timescale of nuclear reactions and scattering) reach thermal equilibrium with all plasma components. This allows thermal excitation of nuclei which follows a Boltzmann law and gives rise to the *stellar* cross section

$$\sigma^* = \frac{\sum_x (2J_x+1)\sigma^x e^{-\frac{E_x}{kT}}}{\sum_x (2J_x+1) e^{-\frac{E_x}{kT}}} = \frac{\sum_x (2J_x+1)\sigma^x e^{-\frac{E_x}{kT}}}{G(T)}, \tag{9.8}$$

where the sum runs over all excited states $x$ of the target (for simplicity, here we assume the projectile, i.e. the second reaction partner, does not have excited states) with spin $J_x$ and excitation energy $E_x$. The quantity $G$ is the partition function of the nucleus. Cross sections $\sigma = \sigma^{x=0}$ measured in terrestrial laboratories do not include such thermal effects. At low temperature (e.g., for the $s$ process) the stellar enhancement factor SEF $= \sigma^*/\sigma$ will only differ from unity when there are excited states within a few keV above the reaction threshold. At the large temperatures reached in explosive burning, thermal enhancement can lead to a considerable deviation from the ground-state cross section.

Nuclei in an astrophysical plasma obey a Maxwell-Boltzmann (MB) distribution $\phi(E) = \phi_{\mathrm{MB}}$ and we obtain finally:

$$r = \frac{n_1 n_2}{1+\delta_{12}} < \sigma^* v > \tag{9.9}$$

$$< \sigma^* v > = (\frac{8}{\mu\pi})^{1/2}(kT)^{-3/2} \int_0^\infty E\sigma^*(E) e^{-\frac{E}{kT}}\, dE. \tag{9.10}$$

Here, $\mu$ denotes the reduced mass of the two-particle system and $< \sigma^* v >$ is the reaction rate per particle pair or *reactivity*.

As mentioned above the charge of the reaction partners can be screened. For most astrophysical conditions this can be included by introducing a screening factor $f_{\mathrm{screen}}$, modifying the above rate for bare nuclei (Iliadis, 2007; Salpeter and Van Horn, 1969)

$$r^{\mathrm{scr}} = f_{\mathrm{screen}} r. \tag{9.11}$$

The screening factor is derived from the plasma conditions of the specific stellar environment. At high densities and low temperatures screening factors can enhance reactions by many orders of magnitude and lead to *pycnonuclear ignition* (Yakovlev et al., 2006). However, note that the above factorization is not valid for vanishing temperatures when nuclei are trapped in a Coulomb lattice.

Forward and reverse rates are related. Applying the well-known reciprocity theorem for nuclear transitions (Blatt and Weisskopf, 1991) and further assuming that the reactands in the entrance channel $a$ as well as the reaction products in the exit channel $b$ are instantaneously thermalized (the *detailed balance* principle), the relation (Holmes et al., 1976; Iliadis, 2007)

$$< \sigma^* v >^{b\to a} = \frac{1+\delta_{b_1 b_2}}{1+\delta_{a_1 a_2}} \frac{G_{a_1} G_{a_2}}{G_{b_1} G_{b_2}} \left(\frac{\mu_a}{\mu_b}\right)^{3/2} e^{-\frac{Q}{kT}} < \sigma^* v >^{a\to b}, \tag{9.12}$$

relating the *stellar* reverse rate to the *stellar* forward rate. The latter has the reaction $Q$-value $Q$. For captures (forward channel $a$) and photodisintegrations (reverse channel $b$), Eq. (9.12) transforms to

$$L_\gamma = \frac{1}{1+\delta_{a_1 a_2}} \frac{G_{a_1} G_{a_2}}{G_b} \left(\frac{\mu_a kT}{2\pi\hbar^2}\right)^{3/2} e^{-\frac{Q}{kT}} < \sigma^* v >^{\mathrm{capture}}. \tag{9.13}$$

These expressions will not be valid anymore if any of the involved rates was derived from a laboratory cross section. They also imply that the detailed balance assumption is valid. Detailed balance can be violated in nuclei with long-lived isomeric states which are not populated or depopulated during regular reaction timescales. For these cases, reactions to separate final states have to be calculated and the (de)population of these states by photon transitions followed explicitly (Ward et al., 1980). Important examples for such nuclei are $^{26}$Al and $^{180}$Ta (Rauscher et al., 2002).

### 9.1.2 Reaction Equilibria

It is not always necessary to solve a full reaction network (Eq. 9.1) including all the rates. On one hand, simplifications can often be made by omitting slow reactions which will not contribute significantly during the timescale of the astrophysical event. These are, for example, charged-particle reactions on heavy targets in hydrostatic stellar burning. On the other hand, high temperature can establish reaction equilibria. When both forward and reverse reactions become sufficiently fast to reach equilibrium with abundances set at equilibrium values. The equilibrium abundances of nuclei can be derived by using the relations 9.12 and 9.13 in the network equation (9.1) and assuming $\dot{Y} = 0$. Somewhat depending on the density, for $T > 4 - 5$ GK all reactions are in a full *nuclear statistical equilibrium* (NSE) and the abundances are given by

$$Y_i = G_i \left(\rho N_{\mathrm{A}}\right)^{A_i - 1} \frac{A_i^{3/2}}{2_i^A} \left(\frac{2\pi\hbar^2}{m_{\mathrm{u}}kT}\right)^{\frac{3}{2}(A_i-1)} e^{\frac{B_i}{kT}} Y_{\mathrm{n}}^{N_i} Y_{\mathrm{p}}^{Z_i}, \tag{9.14}$$

$$\sum_i A_i Y_i = 1 \sum_i Z_i Y_i = Y_{\mathrm{e}} \tag{9.15}$$

with $Z_i$, $N_i$, $A_i$, and $B_i$ being the charge, neutron number, mass number, and the binding energy of the nucleus $i$, respectively, the atomic mass unit $m_{\mathrm{u}}$, and the abundances of free neutrons $Y_{\mathrm{n}}$, free protons $Y_{\mathrm{p}}$, and free electrons $Y_{\mathrm{e}}$. Here, it is assumed that reactions via the strong and electromagnetic interactions are in equilibrium while the weak interaction is not. Therefore, $Y_{\mathrm{e}}$ can still be time-dependent and thus also the resulting NSE abundances $Y_i$.

At $T < 4$ GK and/or low densities only some reactions may be in equilibrium while others are too slow. This gives rise to the so-called *quasi-statistical equilibrium* (QSE) where only groups of nuclei are equilibrated and those groups are connected by slower reactions which are not in equilibrium (Hix and Thielemann, 1999). Abundance ratios within a QSE group can be determined by application of Eq. (9.14) while the connecting, slow reaction determines the amount of matter in each group relative to the other groups at a given time. QSE occurs in low temperature, low density Si-burning and in O-burning of stars. Often, the slowest rate falling out of equilibrium first is that of the strongly density-dependent triple-$\alpha$ reaction.

A special case of a QSE is the *waiting-point approximation*, often used in r-process calculations (Cowan et al., 1991; Arnould et al., 2007). There, the network is reduced to neutron capture reactions and their reverse reactions, and $\beta^-$-decay (with possible release of neutrons). Assuming equilibrated capture and photodisintegration, QSE within an isotopic chain is obtained and the relative abundances are given by

$$\frac{Y_{i+1}}{Y_i} = n_{\mathrm{n}} \frac{G_{i+1}}{2G_i} \left(\frac{A_i + 1}{A_i}\right)^{3/2} \left(\frac{2\pi\hbar^2}{m_{\mathrm{u}}kT}\right)^{3/2} e^{\frac{Q_{\mathrm{ncap}}}{kT}}. \tag{9.16}$$

The neutron number density is denoted by $n_{\mathrm{n}}$ and the neutron capture $Q$-value is given by the neutron separation energy in nucleus $i+1$: $Q_{\mathrm{ncap}} = S_{\mathrm{n},i+1}$. The indices $i$ are ordered by increasing neutron number. The $\beta^-$-decays are much slower and not in equilibrium. Synthesis of the next element is delayed until the decay of the produced isotopes. Typically, only one or two nuclides have significant abundances in such an isotopic QSE chain, hence the name waiting-point approximation.

The advantage of using equilibria is that the rates – and thus the cross sections – do not have to be known explicitly. The resulting abundances are completely determined by basic nuclear properties and the conditions in the astrophysical environment.

## 9.2 Relevant Energy Range of Astrophysical Cross Sections

In the general calculation of the reaction rate according to Eq. (9.10) the nuclear cross section has to be known. Although the integration limits in Eq. (9.10) run from Zero to Infinity, significant contributions to the integral only come from a comparatively narrow energy range. This is due to the shape of the MB distribution, showing a peak around the energy $E_{\mathrm{MB}} = kT$ and quickly approaching very small values both towards $E = 0$ and $E \gg kT$. For a slowly varying cross section (as found, e.g., in non-resonant neutron-induced reactions), the relevant energy range is simply given by the peak of the distribution, $E_0 = E_{\mathrm{MB}}$ and its width $\Delta_0 = \Delta_{\mathrm{MB}}$. For partial waves higher than s-waves, the additional centrifugal barrier introduces a stronger energy dependence in the cross section and shifts the relevant range to slightly higher energy, i.e. $E_0 \approx 0.172 T_9 (\ell + 1/2)$ MeV and $\Delta_0 \approx 0.194 T_9 \sqrt{\ell + 1/2}$ for partial waves $\ell > 0$ (Rauscher et al., 1997; Wagoner, 1969). Charged-particle cross sections exhibit a strong energy dependence at energies close to and below the Coulomb barrier. They decrease by many orders of magnitude towards lower energy. Using the astrophysical S-factor

$$S(E) = \sigma E e^{2\pi\eta}, \tag{9.17}$$

with $\eta$ being the Sommerfeld parameter describing the barrier penetrability, most of the Coulomb suppression is taken out and $S(E)$ is easier to handle because it is varying less with energy than $\sigma$. Inserting definition 9.17 into Eq. (9.10) shows that the penetration factor causes a significant shift of the relevant energy range towards higher energy. The resulting energy window (the *Gamow window*, given by the *Gamow peak* appearing when folding the charged particle cross section with the MB distribution) can be approximated by Iliadis (2007); Rauscher et al. (1997)

$$E_0 = 0.12204 \left( Z_1^2 Z_2^2 \mu \right)^{1/3} T_9^{2/3} \quad \mathrm{MeV} \tag{9.18}$$

$$\Delta_0 = 4\sqrt{\frac{E_0 kT}{3}} = 0.23682 \left( Z_1^2 Z_2^2 \mu \right)^{1/6} T_9^{5/6} \quad \mathrm{MeV}. \tag{9.19}$$

Here, $T_9$ is the plasma temperature in GK. The idea of a single, relevant energy window is only viable for non-resonant cross sections or reactions with broad resonances. Strong, narrow resonances lead to fragmentation of the peak and split it up in several small energy ranges around the resonance energies, with decreasing weight towards higher energy.

It is important to note that Eq. (9.18) is not always valid. It is based on the assumption that the energy dependence of the cross section is mainly determined by the penetration of the projectile through the Coulomb barrier. However, the dependence is dominated by the one of the smallest width in the entrance or exit channel for resonant reactions or smallest *averaged* width in the case of Hauser-Feshbach compound reactions (see Sect. 9.3.2). This smallest width can also be the one of the exit channel, leading to a different maximum in the contribution to the reaction rate integral than estimated from Eq. (9.18). This is often the case in capture reactions when $\Gamma_{\mathrm{projectile}} \gg \Gamma_\gamma$ (Iliadis, 2007; Newton et al., 2007). Because of the weak energy dependence of the $\gamma$-width, there would not be a Gamow window. Effectively, however, the Gamow window is shifted to energies where $\Gamma_{\mathrm{projectile}}$ (which is strongly energy dependent) becomes smaller than $\Gamma_\gamma$. Since reaction rates at higher temperature are determined by cross sections at higher energy, the discrepancy between Eq. (9.18) and the true maximum of the integrand is more pronounced at high temperature than at low temperature. Therefore, the relevant energy range for reactions in explosive burning should be derived by a proper inspection of the product of the (predicted) cross sections and the MB distribution (Rauscher, 2010a). For other charged particle captures in astrophysics, often $\Gamma_{\mathrm{projectile}} \ll \Gamma_\gamma$ due to the low interaction energy implied by $E_{\mathrm{MB}} = kT = T_9/11.6045$ MeV, unless for light nuclei (with low Coulomb barrier). Regarding neutron captures, although $\Gamma_{\mathrm{n}} \gg \Gamma_\gamma$ will apply in most cases (unless very close to the reaction threshold), the shape of the integrand is mostly determined by the shape of the MB distribution and obviously not by any Coulomb penetration. Therefore, the relevant energy window for neutrons can still be estimated from the MB distribution as shown above.

## 9.3 Nuclear Reaction Models

Having determined the relevant energy range, the cross sections have to be predicted by reaction models or determined experimentally. As previously mentioned, often measurements for astrophysics prove difficult due to small cross sections or/and unstable nuclei involved. However, even if a measurement is feasible, the resulting cross section has to be corrected for effects of electron screening and thermal excitation of the target via theoretical models before being used to compute an astrophysical reaction rate.

Here, we provide a brief overview of approaches to predict low-energy cross sections of reactions involving the strong force. Decays and other reactions via the weak force are important but cannot be discussed due to limited space. The reader is referred to other sources, e.g. (Möller et al., 2003; Fuller et al., 1982; Vogel, 2006) and references therein. We also do not cover fission reactions which are important in extremely neutron-rich explosive environments where a $r$ process could occur and reach the region of fissionable nuclei. Current predictions of fission barriers, however, carry large uncertainties. For details, see e.g. (Cowan et al., 1991; Arnould et al., 2007; Panov et al., 2005, 2010; Goriely et al., 2009) and references therein. We also only discuss reactions between a nucleus and a nucleon or an $\alpha$-particle as the majority of reactions in astrophysics is of that type.

The interaction of a particle with a nucleus can excite few or many degrees of freedom, i.e. transfer energy to few (or none) or to many of the nucleons constituting the target nucleus. In nature, all interaction types are, in principle, possible but often only one will be dominating at a given interaction energy but with gradual transitions from one type to the other within certain energy intervals. For theory, it is simpler to consider extreme, idealized cases. Interdependence and interference effects between different reaction mechanisms, even if in principle understood, are very difficult to predict and especially so for the required large number of reactions with unstable nuclei required in astrophysics. In the following we introduce a selection of relevant reaction mechanisms considered in literature. The number of degrees of freedom which can be excited depends on the number of states or levels present in the system formed by projectile and target (Descouvemont and Rauscher, 2006). Therefore, it is helpful to distinguish between compound systems with low and high level densities.

### 9.3.1 Resonance and Potential Models

Low level-density systems exhibit no or only few, isolated resonances in the relevant energy range. These involve mostly light nuclei which have few, widely spaced excited states within several tens of MeV above the ground state and therefore also show only few resonances even when the separation energy of the projectile from the compound system is large. A similar situation also occurs for heavier nuclei with closed shells or heavier nuclei far off stability and close to the driplines where the projectile separation energy becomes very low (e.g. in neutron capture on extremely neutron-rich nuclei) and in consequence the compound system is formed at very low relative energy.

In principle, isolated resonances can be included by the *Breit-Wigner formula* (Blatt and Weisskopf, 1991)

$$\sigma^x = \frac{\omega^2}{4\pi} \frac{2J+1}{(2J_x+1)(2J_{\mathrm{proj}}+1)} \frac{\Gamma_a^x \Gamma_b}{(E-E_{\mathrm{res}})^2 + \frac{\Gamma_{\mathrm{tot}}^2}{4}}, \tag{9.20}$$

where $J$ and $E_{\mathrm{res}}$ refer to the spin and energy of the resonance, $\omega$ is the de Broglie wavelength, and $\Gamma_{\mathrm{tot}}$ is the total resonance width, including the entrance and exit widths $\Gamma_a^x$ and $\Gamma_b$ plus all other open channels. Note that the widths are energy dependent. For a narrow resonance, inserting the above in Eq. (9.10) yields

$$N_A \langle \sigma v \rangle = 1.54 \times 10^{11} \frac{1}{(\mu T_9)^{3/2}} \frac{2J+1}{(2J_x+1)(2J_{\mathrm{proj}}+1)} \frac{\Gamma_a^x \Gamma_b}{\Gamma_{\mathrm{tot}}} e^{-\frac{11.6045 E_{\mathrm{res}}}{T_9}} . \quad (9.21)$$

This gives the reactivity in units of $\mathrm{cm^3 s^{-1} mole^{-1}}$ when the widths and the resonance energy $E_{\mathrm{res}}$ are inserted in units of MeV (Iliadis, 2007). (Note that the above equations do not involve stellar cross sections. For a true stellar cross section and rate, a thermally weighted sum of target states has to be used, according to Eq. (9.8).) However, tails of resonances with the same $J$ may interfere and there may also be interference with a direct component (see below). Therefore, additional interference terms may have to be added (see, e.g., Rauscher and Raimann, 1996). Furthermore, location of the resonance and the widths have to be predicted from nuclear structure. Currently, this is not possible from first principles (except for the lightest nuclei) with the accuracy needed in applications. Therefore, this information usually has to come from experiments.

Instead of Breit-Wigner formulas and interference terms, often the *R-matrix method* (Lane and Thomas, 1958) is used. It is applied to parameterize experimentally known cross sections with as few parameters as possible, implicitly accounting for resonances and their interference. The R-matrix approach can be used to extrapolate the nuclear cross section from existing data to the Gamow range as long as nuclear structure information about resonance levels and non-resonant reaction contributions are included.

In addition to possible resonance contributions a direct capture process can occur. These are fast, one-step processes in which a captured particle directly enters the final state. Typical reaction timescales of direct processes are of the order of $10^{-22}$ s whereas compound reactions, distributing the energy among a large number of nucleons, take of the order of $10^{-16}$ s. Direct reactions include transfer processes where a particle exchange takes place between projectile and target nucleus, and capture processes in which the projectile is being fully captured by the target nucleus. These two reaction types can be treated in ab initio models, determining the cross sections from wave functions obtained by solving the Schrödinger equation using effective potentials.

For transfer reactions often the *Distorted Wave Born Approximation* (DWBA) (Satchler, 1983; Glendenning, 2004) is used, utilizing optical potentials to compute the cross sections from the overlap integral of distorted scattering wave functions and the bound state wave function. The DWBA implicitly assumes that elastic scattering is dominant while non-elastic contributions can be treated perturbatively.

On the other hand, capture reactions can be calculated with a simple potential model, which is a first-order approach involving an electromagnetic operator describing the emission of photons due to the dynamics in the movement of electric charges (Descouvemont and Rauscher, 2006). In the potential model the differential cross section is proportional to the matrix element defined by the overlap of the final state $\phi_\beta$ of the final nucleus and the initial state composed of the target wave function $\phi_\alpha$ and a (distorted) scattering wave of the projectile $\chi_\alpha$. This can be decomposed into an overlap function $S$ of the target and the final nucleus and a radial integral containing the scattered wave $\chi_\alpha^x(r)$, the bound state wave function of the projectile in the target $\phi_{a+A}$, and the radial form of the electromagnetic operator $\mathcal{O}_{EM}$ (Kim et al., 1987)

$$\frac{d\sigma^x}{d\Omega} \propto \left|\langle\phi_\beta|\mathcal{O}_{EM}|\chi_\alpha\phi_\alpha^x\rangle\right|^2 \propto S\left|\phi_{a+A}(r)\mathcal{O}_{EM}(r)\chi_\alpha^x(r)\,dr\right|^2. \tag{9.22}$$

The wave functions $\phi_{a+A}(r)$ and $\chi_\alpha^x(r)$ are obtained by solving the radial Schrödinger equation with appropriate effective potentials.

Both approaches, DWBA and potential model, require a renormalization of the resulting cross section through spectroscopic factors $S$, describing nuclear structure effects by the overlap of initial and final state of the system. These spectroscopic factors have to be obtained from nuclear structure models or by comparison with experiment (Satchler, 1983; Glendenning, 2004).

Microscopic reaction models are first principle methods, starting from effective nucleon-nucleon interactions and treating all nucleons in a Hamiltonian with exact antisymmetrization of the wave functions. Because of this, no artificial distinction between direct and resonant contributions has to be made. Unfortunately, such reaction models are limited to systems of few nucleons. Although the Quantum Monte Carlo method (Pieper and Wiringa, 2001) is promising, it is currently limited to $A \leq 10$ and not applicable to continuum states. Cluster models have been often used for light systems so far (see Descouvemont, 2003; Descouvemont and Rauscher, 2006, and references therein). They assume that the nucleons are grouped in clusters and use cluster wave functions defined in the shell model and computed with an adapted effective nucleon-nucleon force. The Resonating Group model (RGM) and the Generator Coordinate Method (CGM) are two equivalent implementations differing in the definition of the relative wave function of the clusters (Descouvemont, 2003; Descouvemont and Rauscher, 2006).

### 9.3.2 Statistical Model

In systems with high level density $\rho(J,\pi,E)$, individual resonances cannot be resolved anymore and an average over the overlapping resonances can be used instead (Fig. 9.1). Further assuming that the relative phases are randomly distributed, interferences will cancel and a simple sum of Breit-Wigner contributions can be replaced by a level-density weighted sum of averaged widths $\langle\Gamma\rangle$ over all spins $J$ and parities $\pi$ (Descouvemont and Rauscher, 2006; Gadioli and Hodgson, 1992)

$$\sigma^x(E) \propto \frac{1}{(2J_x+1)(2J_{\text{proj}}+1)} \tag{9.23}$$

$$\times \sum_{J,\pi} [(2J+1)\rho(J,\pi,E_c) \tag{9.24}$$

$$\times \left\langle \Gamma^{x}_{\text{pro}}(\{J_x, \pi_x\} \rightarrow \{J, \pi\}, E) \right\rangle \tag{9.25}$$

$$\frac{\left\langle \Gamma_{\text{b}} \left( \sum_{J_{\text{fin}}, \pi_{\text{fin}}, E_{\text{fin}}} (\{J, \pi\} \rightarrow \{J_{\text{fin}}, \pi_{\text{fin}}\}, E_{\text{fin}}) \right) \right\rangle}{\langle \Gamma_{\text{tot}} \rangle} \tag{9.26}$$

$$\times W(J, \pi, E_{\text{c}})] , \tag{9.27}$$

$$E_{\text{c}} = E + E_{\text{sep,pro}} - E_x , \tag{9.28}$$

$$E_{\text{fin}} = E_{\text{c}} - E_{\text{sep,fin}} - E_{x,\text{fin}} . \tag{9.29}$$

This is called the *Hauser-Feshbach* or *statistical model of compound reactions* (Hauser and Feshbach, 1952). Width fluctuation corrections $W$ account for non-statistical correlations but are only important close to channel openings (Ericson, 1960). The separation energy $E_{\text{sep,pro}}$ of the projectile in the compound system determines at which energy $E_{\text{c}}$ the compound system is formed. The averaged width of the exit channel $\langle \Gamma_{\text{b}} \rangle$ usually includes a sum over energetically possible final states at energy $E_{x,\text{fin}}$ or an integral over a level density of the final system when individual states are not known or numerous. For capture, compound and final system are identical. The averaged widths are related to transmission coefficients $\mathcal{T} = 2\pi\rho \langle \Gamma \rangle$. The latter are calculated from the solution of a (radial) Schrödinger

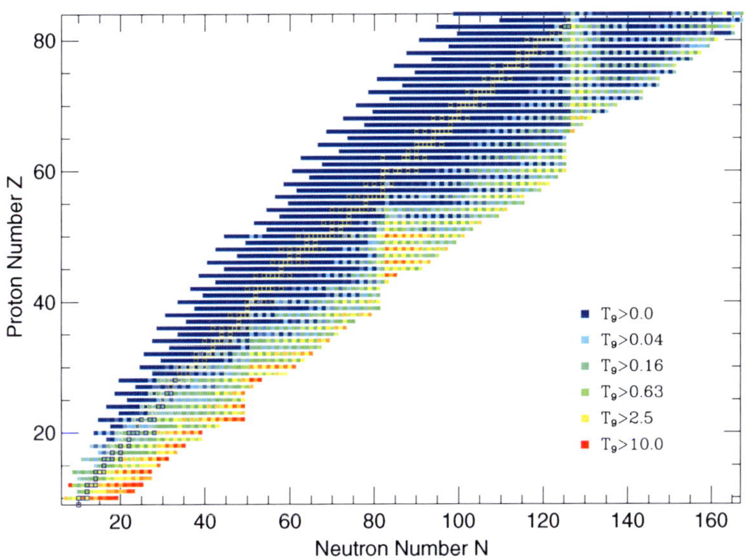


**Fig. 9.1** Applicability of the Hauser-Feshbach model to calculate astrophysical reaction rates for neutron-induced reactions: Shown are the stellar temperatures above which the rate can be calculated from Hauser-Feshbach cross sections (reprinted from Rauscher et al., 1997, with permission)

equation using an optical potential. (It is to be noted that these potentials differ from the ones employed for low-density systems described in Sect. 9.3.1.)

The challenge for nuclear astrophysics lies in the determination of globally applicable descriptions of low-energy optical potentials as well as level densities, masses (determining the separation energies), and spectroscopy (energies, spins, parities) of low-lying excited states, to be applied for a large number of nuclei at and far from stability. For details on the different properties and the remaining open problems in their treatment (see, e.g., Descouvemont and Rauscher, 2006; Rauscher et al., 1997; Arnould et al., 2007; Goriely et al., 2008; Rauscher, 2010b, and references therein). For a general discussion of the applicability of the statistical model, see Rauscher et al. (1997, 2000).

## 9.4 Experimental Facilities and Techniques

The experimental determination or verification of nuclear reaction rates requires a large variety of facilities and techniques. This is in particular true if one wants to establish experimentally reaction rates associated with the production of long-lived radioactive isotopes associated with galactic gamma sources. Nuclear astrophysics related experiments include low energy high intensity accelerator measurements with stable beams to study charged particle reactions of relevance for quiescent stellar burning which may possibly lead to the production of $^{7}$Be, $^{22}$Na and $^{26}$Al. High flux neutron beam studies to explore neutron induced reactions for the weak and main *s* process which can be associated with the production of long-lived radioactive isotopes such as $^{60}$Fe and $^{98}$Tc. Real and virtual photon beams are increasingly used for probing nuclear reactions associated with explosive nucleosynthesis events such as the *p* process but can also be used to probe indirectly neutron capture reactions associated with the *s* process. Intense radioactive beams are the primary tools for exploring nuclear reactions and decay mechanisms far of stability which are expected to occur in explosive stellar environments and may lead to the production of long-lived radioactive elements such as $^{18}$F, $^{26}$Al, $^{44}$Ti and $^{56}$Ni.

### *9.4.1 Low-Energy Facilities, Underground Techniques*

Low energy charged particle measurements belong to the most challenging experiments in nuclear astrophysics. The cross sections need to be measured at the extremely low energies associated with the Gamow range of quiescent stellar burning. This requires to determine the cross sections of proton capture reactions for hydrogen burning in main sequence stars at energies well below 100 keV. Measurements for helium burning in red giant stars need to be explored in the 200–500 keV range and heavy ion fusion reactions in subsequent stellar evolution phases need to be measured near 1–2 MeV center of mass energy. The cross sections are extremely low, typically in the femto-barn range, which requires a long time, in excess of days, to accumulate a statistical relevant amount of reaction yield data. Typical experimental techniques are summarized in the text book literature (Iliadis, 2007) and will not be discussed here.

The critical issue with low cross sections is that the yield of reaction events is low in the detectors measuring the characteristic gamma or particle radiation produced. This requires using high efficiency detector material with high resolution to separate the characteristic events from random background events. High beam intensity is desired to increase the event rate, however it may also increase beam induced background on target impurities and is limited by target stability.

The second critical issue is the background rate in the detector. There are typically three different kind of background, cosmic ray induced background in the detector environment, natural long-lived radioactivity in the detector material and the surrounding environment, and beam induced background on low Z target impurities and beam defining slits or apertures. This background must be reduced as far as possible to identify reaction events in the spectrum.

Cosmic ray induced background affects the spectra up to very high energies and makes it difficult to extract weak signals. That background is the most important to remove. Natural environmental background will be strong in an underground environment except for salt mine locations. But the characteristic $\gamma$ lines are mainly below 3 MeV and can be shielded locally. Neutron background is more difficult to absorb and needs special shielding arrangements. Beam induced background depends critically on the target as well as the choice and preparation of the target material. It is difficult to suppress and may require active shielding procedures.

This can be done by identifying the event electronically by its particular characteristics such as coincidence conditions in a particular decay sequence, pulse shape or timing conditions and reject the background events which do not fulfill these requirements. This can lead to active background suppression by up to three orders of magnitude (Runkle et al., 2005; Couture et al., 2008). While this clearly helps in many cases a more efficient background reduction is desired.

The high energy cosmic ray induced background can be most successfully suppressed by operating the experiments in a deep underground environment where the cosmic ray flux is heavily reduced. This was demonstrated with the installation of the LUNA accelerator facility at the Gran Sasso deep underground laboratory in Italy. The cosmic ray induced background was successfully removed and several critical reactions of the pp-chains and the CNO cycles were successfully measures in the or near the Gamow energy range (Costantini et al., 2009). As a consequence of this successful operation new underground accelerator facilities are being proposed or planned which would allow to cover reactions over a wider energy range than available at LUNA. This is of particular importance for an improved R-matrix analysis and extrapolation. Higher energies are also of great relevance for the underground measurements of $\alpha$ capture reactions and stellar neutron sources in helium burning. In particular it will also improve the chances for pursuing heavy ion fusion reaction studies towards lower energies. There are presently three major initiatives for the construction of new underground accelerator facilities. The proposal to establish an underground accelerator facility ELENA at the Boulby salt mine in the UK seeks to take advantage of the reduced level of neutron and natural activity in a salt environment. The disadvantage will be the reduced depth level compared to the Gran Sasso location. The second proposal is for the development of a two accelerator facility DIANA at the DUSEL underground laboratory at Homestake mine in South Dakota. The third proposal in debate is the construction of an accelerator facility in an abandoned train tunnel in the Pyrenees mountains at Canfranc, Spain. With these facilities the community hopes to address the new and critical questions about stellar reaction cross sections and provide the final answer on the nuclear engine of stellar evolution.

However, it has been demonstrated that alternative inverse kinematics methods are a very powerful tool in reducing the background. They are based on the technique of using a high intensity heavy ion beam on a hydrogen or helium gas target and separate the heavy ion recoil reaction products from the primary beam through a high resolution electromagnetic mass separator system from the primary beam. This method has been demonstrated to be successful at a number of different separator facilities such as DRAGON at TRIUMF, Vancouver (Vockenhuber et al., 2007) (Fig. 9.2) and ERNA at the Ruhr University Bochum (Di Leva et al., 2009).

The detection of the recoiling charged particle has a clear efficiency advantage compare to the gamma detection. The possible detection of the gamma rays in coincidence with the reaction products reduces dramatically the backgrounds. However,

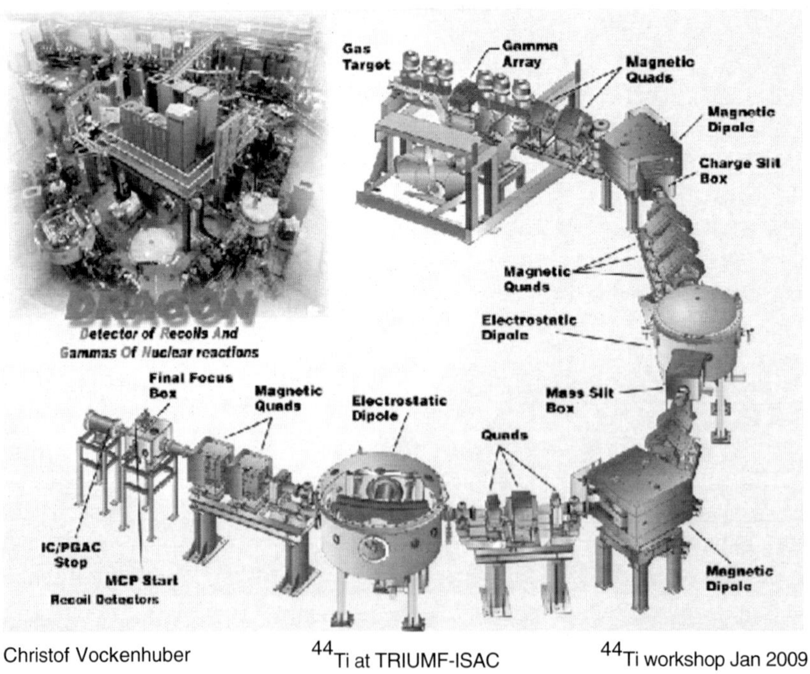


**Fig. 9.2** The *Dragon* facility in Vancouver, Canada, is an example of nuclear experiment facilities, now aimed at experiments for astrophysically-relevant reactions: Radioactive isotopes can be selected and accelerated to form projectiles for such reactions of interest

there are several experimental challenges associated with using recoil separators. At the low stellar energies, the energy spread and the angular aperture are much larger than the acceptance of any of the cited existing recoil separators. In order to measure an absolute cross section the transmission of the recoils should ideally be 100% or exactly known. It is also necessary to know precisely the charge state distribution of the recoil products. In addition, since the primary beam intensity is typically many orders of magnitude larger than the recoiling reaction products, a large spatial separation between the reaction products and the beam is required, which is difficult to realize for beams with a large energy spread. Therefore, solar fusion reactions are particularly challenging to measure with recoil separators and are typically used for higher energies and for the helium or heavy ion burning reactions.

Dedicated next generation separators for low energy nuclear astrophysics studies with stable ion beams coming on line are the ST.GEORGE facility in Notre Dame (Couder et al., 2008) and the modified and upgraded ERNA facility at CIRCE in Caserta, Italy. Both separators feature large angular and energy acceptance and will be equipped with high density gas jet targets, which will ensure a well defined interaction region.

## *9.4.2 Laboratory Neutron Sources*

Many of the observed or anticipated long lived radioactive isotopes in our galaxy are produced by neutron induced nucleosynthesis in the weak or main *s* process taking place in helium and carbon burning stellar environments. This includes $^{41}$Ca, $^{60}$Fe, $^{63}$Ni, but also more massive isotopes such as $^{98}$Tc and $^{99}$Tc and possibly numerous long lived isomers.

The study of neutron induced stellar reactions leading to the production of such isotopes requires high intensity neutron sources with a well defined energy distribution to determine the reaction cross sections at stellar energies of a few keV. Neutrons in that energy range can be produced in several ways. Nuclear reactions such as $^{7}$Li(p,n) or $^{3}$H(p,n) with high intensity proton beams provided by low-energy particle accelerators offer the possibility of tailoring the neutron spectrum to the energy range of interest; this has the advantage of low backgrounds. A particularly successful approach is to simulate a quasi-stellar neutron spectra in the laboratory. In bombarding thick metallic lithium targets with protons of 1,912 keV, the resulting neutrons exhibit a continuous energy distribution with a high-energy cutoff at $E_n$ = 106 keV and a maximum emission angle of 60 degrees. The angle-integrated spectrum corresponds closely to a Maxwell-Boltzmann distribution for kT = 25 keV (Ratynski and Käppeler, 1988). Hence, the reaction rate measured in such a spectrum yields immediately the proper stellar cross section.

Higher intensities can be achieved via photon production by bombarding heavy-metal targets with typically 50-MeV electron beams from linear accelerators. When these energetic neutrons are slowed down in a moderator, the resulting spectrum contains all energies from thermal energy to nearly the initial energy of the electron beam. Since the astrophysical relevant range corresponds to only a narrow window in this spectrum, background conditions are more complicated and measurements need to be carried out at larger neutron flight paths. In turn, the longer flight paths are advantageous for neutron-resonance spectroscopy with high resolution.

The most intense keV neutron flux is produced by spallation reactions. The LANSCE facility at Los Alamos is particularly suited for neutron TOF work due to the favorable repetition rate of only 12 Hz (Lisowski et al., 1990), and because the accumulation of a number of beam pulses in an external storage ring yields extremely intense neutron bursts. Accordingly, excellent signal-to-background ratios can be achieved. The n_ToF facility at CERN provides high intensity neutron beam pulses with a lower repetition rate of 0.4 Hz (Borcea et al., 2003). This has proved highly advantageous for a large number of experimental neutron capture studies along the *s*-process path.

The experimental methods for measuring (n,$\gamma$) cross sections fall into two groups: TOF techniques based on the detection of the prompt capture $\gamma$ rays and activation methods.

The TOF techniques can be applied in measurements of most stable nuclei but require a pulsed neutron source to determine the neutron energy via the flight time between target and detector. Capture events in the samples are identified by the prompt $\gamma$ ray cascade in the product nucleus.

The best signature for the identification of neutron capture events is the total energy of the capture gamma cascade, which corresponds to the binding energy of the captured neutron. Hence, accurate measurements of (n,$\gamma$) cross sections require a detector that operates as a calorimeter with good energy resolution and is insensitive to neutron exposure. In the gamma spectrum of such a detector, all capture events would fall in a line at the neutron binding energy (typically between 5 and 10 MeV), well separated from the gamma-ray backgrounds that are inevitable in neutron experiments. Such detectors have been successfully developed at the various laboratories using arrays of $4\pi$ $BaF_2$ scintillator detectors with a large number of independent detector modules (Heil et al., 2001) and have emerged as standard technology for these kinds of measurements.

A completely different approach to determining stellar (n,$\gamma$) rates is activation in a quasi-stellar neutron spectrum. Compared with the detection of prompt capture gamma rays, this method offers superior sensitivity, which means that much smaller samples can be investigated. Since it is also selective with respect to various reaction products, samples of natural composition can be studied instead of the expensive enriched samples required by the TOF techniques. However, the activation technique is restricted to cases where neutron capture produces an unstable nucleus, and it yields the stellar rate only for two thermal energies at kT = 25 and 52 keV. This method is however particulary powerful in obtaining cross sections for reactions producing long-lived radioactive materials which can be identified by their particular decay characteristics and signature. This activation technique has been used for a variety of measurements. The technique can be applied to short-lived products with half-lives in the millisecond range and allows for cross section measurements with uncertainties of a few percent.

### 9.4.3 Accelerator Mass Spectroscopy

Classical activation techniques require a characteristic decay signal associated with the decay pattern or the half-life of the produced radioactive isotope. This can be difficult in cases where no characteristic gamma or particle decay pattern exists or where the decay analysis of the $\beta$ decay signal is prohibited by high background activity. In these cases activation analysis through accelerator mass spectrometry (AMS) offers a powerful tool to measure cross sections through ultra-low isotope-ratio determination. The AMS method was successfully introduced for the study of the neutron-capture cross section of $^{62}$Ni(n,$\gamma$)$^{63}$Ni (Nassar et al., 2005), and extended to other neutron and charge-particle-induced reactions, such as $^{25}$Mg(p,$\gamma$)$^{26}$Al (Arazi et al., 2006) and $^{40}$Ca($\alpha$,$\gamma$)$^{44}$Ti (Nassar et al., 2006).

In these cases samples were either irradiated in a neutron spectrum resembling a stellar Maxwell Boltzmann distribution or by charged particles of well known energies. After the irradiation the samples must be chemically treated to extract the radioactive reaction products. This requires some time and limits AMS activation studies to more longer lived isotopes. Since isotopic and isobaric interferences may represent a major challenge in AMS measurements of irradiated samples, extensive background studies for these isotopes are always necessary prior to the irradiations in order to demonstrate that the required sensitivity can be reached. In AMS, negative ions are extracted from an ion source which have to pass a low energy mass spectrometer prior to entering a tandem accelerator. When passing the stripper, positive ions are produced while within this stripping process molecular isobars are destroyed. One positive charge state is selected with a second (high-energy) mass spectrometer system which is optimized for mass, charge and isobar separation trough possible combination of dipole magnet separators, Wien-filters, and more recently magnetic gas filled separators for improved isobar separation. With such a system the concentration ratio of the radioisotope is determined relative to a stable isotope by measuring the number of radionuclides relative to the current of the isotopic ions in front of the detector, after adjusting the injector magnet, terminal and Wien-filter voltage appropriately. By measuring relative to a standard sample of known isotopic ratio, factors like stripping yields and transmissions mostly cancel.

The difficulties with AMS experiments is in the chemical preparation of the sample and the sufficient separation of the extracted radioactive ions from background events. While AMS is a widely established method with many applications, the analysis of the very limited number of radioactive products from low cross section reactions remains challenging. Systematic studies are necessary to reduce possible uncertainties.

Dedicated AMS facilities with an established nuclear astrophysics program are the Vienna Environmental Research Accelerator (VERA) (Kutschera et al., 1997), the Center for Isotopic Research on Cultural and Environmental Heritage (CIRCE) in Caserta/Italy (Terrasi et al., 2007) or the Munich Tandem accelerator facility (Knie et al., 2000) which is optimized for the analysis of more massive radioactive isotopes. A new AMS program is presently being developed utilizing the Notre Dame tandem accelerator (Robertson et al., 2007).

### 9.4.4 Radioactive Beam Techniques

The development of radioactive accelerated beams for low energy nuclear astrophysics experiments has been one of the large challenges of the field. The experimental study of nuclear reactions and decay processes far of stability is necessary for the understanding of explosive nucleosynthesis processes such as the *rp* process in cataclysmic binary systems or the *r* process and *p* process in the supernova shock front. These processes can in particular contribute to the production of long-lived galactic radioactivity by primary reaction or also by secondary decay processes from the reaction path towards the line of stability.

For the purpose of studying the origin of long-lived radioactive isotopes in astrophysical environments radioactive beams are utilized in two ways, for producing long-lived targets by implantation for subsequent irradiation with neutron, charged particle or possibly intense photon beams or for direct reaction measurement in inverse kinematics on light ion target materials. The later approach requires well defined mono-energetic and intense radioactive beams and a detection system for light or heavy recoil reaction products.

The main challenge in this approach is to produce a sufficiently high intensity of radioactive beams which have to be produced on-line as a secondary reaction product. This requires high cross sections for the production process and high primary beam intensities. A variety of different approaches has been chosen in the past to optimize the production efficiency and maximize the intensity of the radioactive beams. A technique developed for small scale facilities is the selection of specific nuclear reactions tailored for the on-line production of radioactive beams at optimum conditions. The secondary particles can be used for subsequent nuclear reaction studies after blocking and separation from the primary beams (Kolata et al., 1989). The efficient separation of a suitably high intensity beam of radioactive species is the most challenging problem for this approach.

An alternative approach is the ISOL (Isotope Separation On-Line) technique where high energy protons are used to bombard heavy ion targets for producing a large number of radioactive species through spallation processes. These isotopes diffuse out of the target into an ion source for being charged and re-accelerated for secondary beam decay or reaction experiments. The method has been proven to be very powerful over the years but is limited to isotopes with lifetimes appreciably longer that the time necessary for the diffusion transport and ionization process. This can be different for different elements because of the associated chemical processes between the isotopes and the surrounding environment.

The third approach is based on the use of energetic heavy projectiles bombarding light or target nuclei fragmenting on impact. This fragmentation process generates a cocktail beam of many radioactive species which move forward with high velocity because the initial momentum of the primary particles is maintained. For experiments with a specific secondary particle, it must be selected by fragment separator systems which separate and focus the isotopes by magnetic fields and energy loss characteristics in heavy wedge materials. For nuclear astrophysics related experiments the fast beam particles need to be slowed down by energy loss in gas or solid material and re-accelerated to energies corresponding to the temperatures in the explosive stellar scenarios.

There is a number of laboratories which have focused on nuclear reaction studies with radioactive beams. The first fully operating radioactive beam laboratory based on the ISOL principle was the coupled cyclotron facility at Louvain la Neuve which did a number of successful radioactive beam studies of relevance for investigating the the production of $^{18}$F in novae (De Sérville et al., 2009). These measurements were complemented by measurements at the HRIBF facility at Oak Ridge using intense $^{18}$F beams (Chae et al., 2006). Both facilities produce the radioactive species by nuclear reactions on thin production targets, with the reaction products being transported into an ion source for producing and subsequently accelerating the secondary beam. The intensity is largely limited by target technology and beam transport and re-ionization efficiency.

The premier ISOL radioactive beam facility is ISAC at TRIUMF Canada. The primary 600 MeV proton beam is provided by the TRIUMF cyclotron. The reaction products are post-accelerated in an RFQ SC LINAC accelerator combination to energies of 0.3–3 MeV/u. ISAC has successfully performed a number of radioactive beam experiments of relevance for explosive hydrogen and helium burning. Most notable a direct study of $^{21}$Na(p,$\gamma$)$^{22}$Mg in inverse kinematics to probe the production mechanism of $^{22}$Na in Ne nova explosion environments (D'Auria et al., 2004). The facility also runs a successful program with stable beams which was utilized to investigate the production of $^{44}$Ti (Vockenhuber et al., 2007). Presently a number of studies associated with the production of the long-lived $\gamma$ emitter $^{26}$Al are being performed.

Other ISOL based radioactive beam facilities such as Spiral facility at GANIL in Caen, France or REX-ISOLDE at CERN have been used to perform interesting experiments for nuclear astrophysics but have been less concerned with the question of nuclear production mechanisms for long lived cosmic gamma emitters.

There have been a number of fast radioactive beam facilities with scientific programs in nuclear astrophysics primarily aimed at the study of nucleosynthesis processes far off stability. However the rapid new developments in fast beam physics promises a number of new experimental opportunities which can provide benefits for studying reactions associated with the production of long-lived gamma emitters in explosive nucleosynthesis events.

There are currently four major fragmentation facilities in the world: GANIL and GSI in Europe, NSCL/MSU in the US and RIKEN in Japan. They are all based on Heavy Ion accelerators which operate in complementary energy domains. Because of the high energy of the fragment products low energy reaction experiments for nuclear astrophysics are not possible but the development of indirect techniques to determine critical reaction or decay parameters has been the primary goal. In the context of long-lived isotopes of astrophysical interest a major contribution was the development of fast beams such as $^{8}$B at NSCL/MSU, RIKEN, and GSI for utilizing Coulomb dissociation techniques for probing critical reactions such as $^{7}$Be(p,$\gamma$)$^{8}$B. The NSCL and RIKEN also successfully developed a $^{44}$Ti beam for new measurements of its half-life (Görres et al., 1998). More half-life measurements of long-lived isotopes such as $^{60}$Fe are presently underway to re-evaluate these critical parameters.

## 9.5 Specific Experiments

The complexities of the experiments and the uncertainties in the experimental results affect the reliability of model predictions on the nucleosynthesis of long-lived radioactive species. In particular recent studies of critical nuclear reactions and decay processes exhibit considerable differences to earlier studies which so far have been the reference point for nucleosynthesis simulations and predictions for long-lived radioactive isotope abundances in stellar burning environments. It is therefore important to carefully evaluate the experimental results and clarify possible discrepancies and inconsistencies in the data. This section will discuss the present status of the experimental reaction rates and evaluate future opportunities to improve the existing data base.

### *9.5.1 Experiments with Stable Beams*

Many of the long-lived radioactive gamma emitters in our universe have been produced by radiative capture reactions on stable isotopes. The best known examples are $^{26}$Al, which is primarily formed by proton capture on stable $^{25}$Mg isotopes, $^{25}$Mg(p,$\gamma$)$^{26}$Al, and $^{44}$Ti which is most likely produced via alpha capture on stable $^{40}$Ca isotopes, $^{40}$Ca($\alpha$,$\gamma$)$^{44}$Ti. Extensive measurements using in-beam $\gamma$ spectroscopy techniques have been made for both reactions and have formed the basis for earlier reaction rate compilations.

The low energy reaction cross section of $^{25}$Mg(p,$\gamma$)$^{26}$Al is characterized by several resonances with energies between 30 and 400 keV. The reaction rate is directly correlated to the strengths $\omega\gamma$ of the resonances. The strengths for the resonances above 190 keV have been determined from the on-resonance thick target yield in radiative capture measurements (Elix et al., 1979; Iliadis et al., 1990). The strengths of lower energy resonances are estimated on the basis of single particle transfer reaction studies. Of particular importance are three resonances at 90 keV, 130 keV and at 190 keV which determine the reaction rate at temperatures typical for stellar hydrogen burning in AGB stars and nova explosions. Because the low energy radiative capture measurements have been handicapped by cosmic ray induced background, an alternative measurement was done using the AMS technique to analyze the number of $^{26}$Al reaction products after irradiation at resonance energies (Arazi et al., 2006). The experiment was successful and confirmed the resonance strengths of the known resonances at 304 keV, 347 keV, and 418 keV resonance energy. However the results indicated a substantially lower strength for the critical resonance at 190 keV. This would reduce the reaction rate by about a factor of five at the temperature range between 0.2 and 1.0 GK. This result introduced a large uncertainty in the reaction rate which motivated a new experimental study at LUNA in the Gran

Sasso laboratory using in-beam gamma spectroscopy techniques with a variety of high efficiency and high resolution gamma detector devices. The measurements confirmed earlier gamma spectroscopy studies of the strengths of higher energy resonances (Elix et al., 1979; Iliadis et al., 1990) tabulated in the NACRE compilation (Angulo et al., 1999). The new results are being prepared for publication. Parallel to the gamma spectroscopy measurement, the irradiated samples were analyzed for their $^{26}$Al content using AMS techniques. The AMS measurements were performed at the CIRCE facilities. Excellent agreement is demonstrated for the resonance at 304 keV, additional experiments are being pursued for lower energy resonances to address the inconsistencies in the strength determination for the 190 keV resonance.

The $^{40}$Ca$(\alpha, \gamma)^{44}$Ti reaction is considered to be one of the major production reactions for $^{44}$Ti in supernova shock front nucleosynthesis. The cross section for this radiative capture process has been explored in a number of in-beam gamma spectroscopy studies down to center of mass energies of 2.5 MeV (Simpson et al., 1971; Cooperman et al., 1977). The cross section is characterized by a large number of resonances and the initial reaction rate determinations were based on an analysis of resonance strengths. Despite the high level density in $^{44}$Ti, it was noted that the experimental reaction rate is substantially smaller than the reaction rate based on statistical model Hauser Feshbach predictions (Rauscher et al., 2000). The reaction was studied independently using a thick He-gas cell target and counting the long lived $^{44}$Ti reaction products by AMS techniques (Nassar et al., 2005) to determine the integral yield over an energy range of 1.7–4.2 MeV. The extracted reaction rate is substantially higher than the ones discussed in the literature (Rauscher et al., 2000). A more recent study of the reaction using inverse kinematics techniques was performed at the ISAC facility at TRIUMF, Vancouver, separating the $^{44}$Ti reaction products on-line with the DRAGON recoil separator. The measurements covered the energy range of 2.3–4.2 MeV (center of mass) in more than 100 small energy steps. The extracted yield was mostly interpreted as on-resonance resonance thick target yield and translated to a resonance strength. There are large uncertainties associated with this approach, in particular with the determination of the resonance energies, which have not been unequivocally determined in the experiment. In some cases several of the quoted resonances agree with previously identified states, in other cases it needs to be confirmed that the observed yields really correspond to additional resonances and do not originate from tail contributions of resonant yield curves associated with the different states. As far as the resonance levels which have been observed in both studies are concerned the published strengths are comparable to each other. Nevertheless the reaction rate suggested by Vockenhuber et al. (2007) is larger by more than a factor of two than the rates projected on the basis of the in-beam gamma spectroscopy measurements, but it is in agreement with the projections by Rauscher et al. (2000). The difference is mainly due to the difference in resonance numbers. While the resonance identification in previous work was based on a careful analysis of the particular gamma decay characteristics of the observed levels, the analysis of the recoil data is insufficient in providing information to differentiate between different resonances. It cannot be excluded that the number of identified states are overestimated; a more detailed gamma spectroscopy study with thin targets is therefore highly advisable to remove the existing uncertainties.

### 9.5.2 Experiments with Neutron Beams

A particularly interesting case is the origin of the long-lived gamma emitter $^{60}$Fe. Its characteristic $\gamma$-radioactivity has been observed with the INTEGRAL gamma ray telescope in supernova remnants near the solar system. These observations are complemented by recent AMS studies which suggest high $^{60}$Fe abundance in deep sea ferromanganese sediments (Knie et al., 2004). These $^{60}$Fe observations have been interpreted as indication for the existence of a recent ($\approx$3 million years) supernova event in the solar system vicinity. A more quantitative interpretation of the time and distance of the proposed supernova event requires a detailed knowledge of the nucleosynthesis history of $^{60}$Fe.

The radioactive $^{60}$Fe isotope is produced by a sequence of neutron capture reactions of stable iron isotopes such as $^{58}$Fe(n,$\gamma$)$^{59}$Fe(n,$\gamma$)$^{60}$Fe, the production rate and final abundance of the long-lived $^{60}$Fe depends on the reaction rate of these feeding processes as well as on the rate of the $^{60}$Fe(n,$\gamma$)$^{61}$ depletion reaction. No experimental information are available about the associated cross sections except for the neutron capture reaction $^{58}$Fe(n,$\gamma$)$^{59}$Fe. Present simulations of the $^{60}$Fe nucleosynthesis rely entirely of statistical model predictions of the neutron capture rates. Because of the relatively low level density in the associated $^{60}$Fe, $^{61}$Fe compound nuclei these model predictions are unreliable and need to be tested experimentally. This is underlined by the direct comparison between the experimental cross sections for neutron capture on the stable isotopes $^{56}$Fe, $^{57}$Fe, and $^{58}$Fe which were all measured through neutron activation techniques and theoretical Hauser Feshbach predictions which show considerable discrepancies in particular in the cases of $^{56}$Fe(n,$\gamma$)$^{57}$Fe and $^{57}$Fe(n,$\gamma$)$^{58}$Fe. For $^{58}$Fe(n,$\gamma$)$^{59}$Fe on the other hand , the agreement seems reasonable well but that cannot be extrapolated towards neutron captures on the more neutron rich Fe isotopes which are subject of the here proposed measurements.

Particularly important is the determination of the reaction rate of $^{59}$Fe(n,$\gamma$)$^{60}$Fe since it competes directly with the $^{59}$Fe $\beta$-decay which would by-pass the production of $^{60}$Fe. A direct measurement of this critical reaction in the traditional activation or time of flight spectroscopy technique is not feasible because the target is radioactive and only small amounts can be accumulated. these small amounts nevertheless produce a large background activity level, which would prohibit any of the described methods. The cross section for the ground state decay branch of $^{59}$Fe(n,$\gamma_0$)$^{60}$Fe can however be investigated using inverse $^{60}$Fe($\gamma$,n)$^{59}$Fe Coulomb dissociation techniques. The $^{60}$Fe beam can be produced by fragmentation of a heavy ion such as $^{64}$Ni on a light Be target at an energy of 500 MeV/u. The $^{59}$Fe recoil products, and the released reaction neutrons, as well as $\gamma$ rays can be detected with reasonable 200 keV resolution using a combination of a magnetic separator system and a neutron detector wall. This allows particle identification of all reaction products.

### 9.5.3 Experiments with Radioactive Beams or Targets

The depletion processes of long-lived radioactive isotopes includes the natural decay. Simulating this branch requires not only a good knowledge of the laboratory lifetime but also of the nature of the decay process since extreme environmental effects can change the decay rates drastically. In terms of $\beta$ decay, the decay can be accelerated through the decay of thermally excited states as in the case of $^{26}$Al. For decay through electron capture, the decay can be slowed down since the nuclei are completely ionized and the electrons have to be captured from the stellar plasma rather than from the inner K- or L-shell of the atom. This affects in particular the lifetime of $^{44}$Ti, which primarily decays by electron capture.

Often the depletion is primarily driven by nuclear reactions, such as $^{22}$Na(p,$\gamma$)$^{23}$Mg, $^{26}$Al(p,$\gamma$)$^{27}$Si, $^{44}$Ti($\alpha$,p)$^{47}$V, or $^{60}$Fe(n,$\gamma$)$^{61}$Fe, but also capture reactions on shorter-lived excited configurations of these nuclei are possible, such as $^{26}$Al*(p,$\gamma$)$^{27}$Si. There are two possibilities for experimental studies of the reaction cross sections. The first one is based on the production of highly enriched long-lived radioactive targets, which can be prepared through standard chemical target preparation techniques using externally bred radioactive material, or by implantation of radioactive ions at low energy ISOL facilities. The disadvantage of both techniques is that the actual $\gamma$ measurements have to be performed in a high radiation background environment produced by the sample itself.

Nevertheless, earlier measurements of reactions such as $^{22}$Na(p,$\gamma$)$^{23}$Mg (Seuthe et al., 1990) and $^{26}$Al(p,$\gamma$)$^{27}$Si (Buchmann et al., 1984) relied entirely on this approach. In both cases a large number of resonances were detected and the resonance strength determined for calculating the reaction rates. The results for $^{22}$Na(p,$\gamma$)$^{23}$Mg were confirmed by new direct measurements using improved target and detection techniques (Stegmüller et al., 1996), resulting in the observation of an additional low energy resonance at lower energies. Complementary spectroscopy techniques such as the study of the $\beta$-delayed proton decay of $^{23}$Al (Peräjärvi et al., 2000) and the heavy ion reaction induced $\gamma$ decay of proton unbound states in $^{23}$Mg (Jenkins et al., 2004) provided additional nuclear structure information which led to the reduction of uncertainties in the reaction rate.

The situation is similar with $^{26}$Al(p,$\gamma$)$^{27}$Si; after the initial study with radioactive targets (Buchmann et al., 1984). A number of transfer experiments (Schmalbrock et al., 1986; Vogelaar et al., 1996) providing complementary information about the threshold levels in $^{27}$Si not accessible to direct study by radiative capture measurements lead to an improved reaction rate for $^{26}$Al ground state capture. A first direct study of a lower energy resonance was successfully performed in inverse kinematics at the ISAC facility at TRIUMF using the DRAGON recoil separator (Ruiz et al., 2006). The resonance value is substantially smaller than the value quoted before (Vogelaar et al., 1996), which reduced the reaction rate slightly at temperatures anticipated for nova burning conditions.

Not included in the reaction rate calculations are possible contributions of proton capture on the thermally first excited state in $^{26}$Al (Runkle et al., 2001). Recently number of indirect measurements have been performed to explore the possible contribution to the total reaction rate of $^{26}$Al(p,$\gamma$)$^{27}$Si. Transfer reactions have been used to populate proton unbound states in $^{27}$Si measuring the subsequent proton decay to the ground state and the first excited state in $^{26}$Al (Deibel et al., 2009). This approach allows to determine the branching and the relative strength of the proton decays for each of the unbound states. This can be used to scale the reaction rate component for the proton capture on the first excited state.

Possible lower energy resonance contributions to the proton capture rates on the ground state (Lotay et al., 2009) and the excited state of $^{26}$Al (Lotay et al., 2009) have been explored by $\gamma$ spectroscopy techniques probing the proton unbound state in $^{27}$Si through heavy ion fusion evaporation reactions and measuring the $\gamma$ decay of proton unbound states. This is a particular efficient method to explore the levels near the threshold where proton decay is suppressed by the Coulomb barrier. The measurements provide critical information about spin and parity of the observed states but gives only limited information about the resonance strengths which is primarily determined by the proton decay strength.

The main reaction for the depletion of $^{60}$Fe in neutron rich environments is $^{60}$Fe(n,$\gamma$)$^{61}$Fe. The reaction rate used for nucleosynthesis simulations was for many years based on theoretical Hauser Feshbach model predictions. Recently an experiment has been performed at the FZK Karlsruhe in Germany to determine the stellar reaction cross section experimentally by neutron activation with the neutron beam resembling a quasi-stellar neutron spectrum (Uberseder et al., 2009). The activated $^{60}$Fe sample was prepared from PSI beamstop material. The cross section was determined from the characteristic $^{61}$Fe $\gamma$ activity relative to the amount of $^{60}$Fe nuclei in the target material. The latter was determined from the characteristic $^{60}$Fe $\gamma$ activity of the target sample. based on this the experimental results suggest a cross section which is twice as large as standard Hauser Feshbach predictions suggesting a much more rapid depletion of $^{60}$Fe in neutron rich environment than previously anticipated. The estimate of the number of $^{60}$Fe nuclei, however relied on adopting a half-life of $T_{1/2} = 1.49$ Gy (Kutschera et al., 1984). Recent work suggested that the half-live is considerably larger model $T_{1/2} = 2.62$ Gy (Rugel et al., 2009). This would translate into a considerably larger amount of $^{60}$Fe particles in the sample, suggesting a cross section which would be in fair agreement with the Hauser Feshbach predictions. New independent life time measurements for $^{60}$Fe are clearly necessary to address this issue and remove the uncertainty in the interpretation of the radiative capture data.

## References


Angulo, C., Arnould, M., Rayet, M., et al.: A compilation of charged-particle induced thermonuclear reaction rates. Nucl. Phys. A **656**, 3 (1999)

Arazi, A., Faestermann, T., Fernandez Niello, J.O., et al.: Measurement of $^{25}$Mg(p, $\gamma$)$^{26g}$Al resonance strengths via accelerator mass spectrometry. Phys. Rev. C **74**, 025802 (2006)

Arnould, M., Goriely, S., Takahashi, K.: The r-process of stellar nucleosynthesis: Astrophysics and nuclear physics achievements and mysteries. Phys. Rep. **450**, 97 (2007)

Blatt, J.M., Weisskopf, V.F.: Theoretical Nuclear Physics. Dover, New York (1991)

Borcea, C., Cennini, P., Dahlfors, M., et al.: Results from the commissioning of the n_TOF spallation neutron source at CERN. Nucl. Instr. Meth. A **513**, 524 (2003)

Buchmann, L., Hilgemeier, M., Krauss, A., et al.: The abundance of $^{26}$Al in the MgAl cycle. Nucl. Phys. A **415**, 93 (1984)

Chae, K.Y., Bardayan, D.W., Blackmon, J.C., et al.: First experimental constraints on the interference of $3/2^+$ resonances in the $^{18}$F(p, $\alpha$)$^{15}$O reaction. Phys. Rev. C **74**, 012801 (2006)

Cooperman, E.L., Shapiro, M.H., Winkler, H.: Helium burning of $^{40}$Ca. Nucl. Phys. A **284**, 163 (1977)

Cooperstein, J., Wambach, J.: Electron capture in stellar collapse. Nucl. Phys. A **420**, 591 (1984)

Costantini, H., Formicola, A., Imbriani, G., et al.: LUNA: a laboratory for underground nuclear physics. Rep. Prog. Phys **72**, 086301 (2009)

Couder, M., G.P.A. Berg, J. Görres, et al.: Design of the recoil mass separator St. George. Nucl. Instr. Meth. A **587**, 35 (2008)

Couture, A., Beard, M., Couder, M., et al.: Measurement of the $^{19}$F(p, $\gamma$)$^{20}$Ne reaction and interference terms from $E_{c.m.} = 200-760$ keV. Phys. Rev. C **77**, 015802 (2008)

Cowan, J.J., Thielemann, F.-K., Truran, J.W.: The r-process and nucleochronology. Phys. Rep. **208**, 267 (1991)

D'Auria, J. M., Azuma, R.E., Bishop, S., et al.: The $^{21}$Na(p, $\gamma$)$^{22}$Mg reaction from $E_{c.m.} = 200$ to 1103 keV in novae and X-ray bursts. Phys. Rev. C **69**, 065803 (2004)

De Sérville, N., Angulo, C., Coc, A., et al.: Low-energy $^{18}$F(p, $\alpha$)$^{15}$O cross section measurements relevant to nova $\gamma$-ray emission. Phys. Rev. C **79**, 015801 (2009)

Deibel, C.M., Clark, J.A., Lewis, R., et al.: Toward an experimentally determined $^{26m}$Al(p, $\gamma$)$^{27}$Si reaction rate in ONe novae. Phys. Rev. C **80**, 035806 (2009)

Descouvemont, P.: Theoretical Models for Nuclear Astrophysics. Nova Science Publications, New York (2003)

Descouvemont, P., Rauscher, T.: Cross section predictions for hydrostatic and explosive burning. Nucl. Phys. A **777**, 137 (2006)

Di Leva, A., Gialanella, L., Kunz, R., et al.: Stellar and Primordial Nucleosynthesis of $^{7}$Be: Measurement of $^{3}$He($\alpha$, $\gamma$)$^{7}$Be. Phys. Rev. Lett. **102**, 232502 (2009)

Elix, K., Becker, H.W., Buchmann, L., et al.: Search for low-energy resonances in $^{25}$Mg(p, $\gamma$)$^{26}$Al. Z. Phys. A **293**, 261 (1979)

Ericson, T.: Fluctuations of Cross Sections in the "Continuum" Region. Phys. Rev. Lett. **5**, 430 (1960)

Fuller, G.M., Fowler, W.A., Newman, M.J.: Stellar weak interaction rates for intermediate-mass nuclei. II – A = 21 to A = 60. ApJ **252**, 715 (1982)

Görres, J., Meißner, J., Schatz, H., et al.: Half-Life of $^{44}$Ti as a Probe for Supernova Models. Phys. Rev. Lett. **80**, 2554 (1998)

Gadioli, E., Hodgson, P.E.: Pre-Equilibrium Nuclear Reactions. Clarendon Press, Oxford (1992)

Glendenning, N.K.: Direct Nuclear Reactions. World Scientific Publishing, Singapore (2004)

Goriely, S., Hilaire, S., Koning, A.J.: Improved predictions of nuclear reaction rates with the TALYS reaction code for astrophysical applications. Astron. Astrophys. **487**, 767 (2008)

Goriely, S., Hilaire, S., Koning, A.J., Sin, M., Capote, R.: Towards a prediction of fission cross sections on the basis of microscopic nuclear inputs. Phys. Rev. C **79**, 024612 (2009)

Hauser, W., Feshbach, H.: The Inelastic Scattering of Neutrons. Phys. Rev. **87**, 366 (1952)

Heil, M., Reifarth, R., Fowler, M.M., et al.: A $4\pi$ $BaF_2$ detector for $(n, \gamma)$ cross-section measurements at a spallation neutron source. Nucl. Instr. Meth. A **459**, 229 (2001)

Hix, W.R., Thielemann, F.-K.: Silicon Burning. II. Quasi-Equilibrium and Explosive Burning. ApJ. **511**, 862 (1999)

Holmes, J.A., Woosley, S.E., Fowler, W.A., Zimmerman, B.A.: Tables of Thermonuclear-Reaction-Rate Data for Neutron-Induced Reactions on Heavy Nuclei. At. Data Nucl. Data Tables **18**, 305 (1976)

Iliadis, C.: Nuclear Physics of Stars. Wiley-VCH, New York (2007)

Iliadis, C., Schange, T., Rolfs, C., et al.: Low-energy resonances in $^{25}$Mg(p, $\gamma$)$^{26}$Al, $^{26}$Mg(p,$\gamma$)$^{27}$Al and $^{27}$Al(p,$\gamma$)$^{28}$Si. Nucl. Phys. A **512**, 509 (1990)

Jenkins, D.G., Lister, C.J., Janssens, R.V.F., et al.: Reevaluation of the $^{22}$Na(p, $\gamma$) Reaction Rate: Implications for the Detection of $^{22}$Na Gamma Rays from Novae. Phys. Rev. Lett. **92**, 031101 (2004)

Johnson, C.W., Kolbe, E., Koonin, S.E., Langanke, K.: The fate of $^7$Be in the Sun. ApJ. **392**, 320 (1992)

Kim, K.H., Park, M.H., Kim, B.T.: Radiative capture reaction $^7$Be(p, $\gamma$)$^8$B at low energies. Phys. Rev. C **23**, 363 (1987)

Knie, K., Faestermann, T., Korschinek, G., et al.: High-sensitivity AMS for heavy nuclides at the Munich Tandem accelerator. Nucl. Instr. Meth. B **172**, 717 (2000)

Knie, K., Korschinek, G., Faestermann, T., et al.: $^{60}$Fe Anomaly in a Deep-Sea Manganese Crust and Implications for a Nearby Supernova Source. Phys. Rev. Lett. **93**, 171103 (2004)

Kolata, J.J., Morsad, A., Kong, X.J., et al.: A radioactive beam facility using a large superconducting solenoid. Nucl. Instr. Meth. B **40**, 503 (1989)

Kutschera, W., Collon, P., Friedmann, H., et al.: VERA: A new AMS facility in Vienna. Nucl. Instr. Meth. B **123**, 47 (1997)

Kutschera, W., Billquist, P.J., Frekers, D., et al.: Half-life of $^{60}$Fe. Nucl. Instr. Meth. B **5**, 430 (1984)

Lane, A.M., Thomas, R.G.: R-Matrix Theory of Nuclear Reactions. Rev. Mod. Phys. **30**, 257 (1958)

Lisowski, P.W., Bowman, C.D., Russell, G.J., Wender, S.A.: The Los Alamos National Laboratory Spallation Neutron Sources. Nucl. Sci. Eng. **106**, 208(1990)

Lotay, G., Woods, P.J., Sewryniak, D., et al.: Identification of Key Astrophysical Resonances Relevant for the $^{26g}$Al(p, $\gamma$)$^{27}$Si Reaction in Wolf-Rayet Stars, AGB stars, and Classical Novae. Phys. Rev. Lett. **102**, 162502 (2009)

Lotay, G., Woods, P.J., Sewryniak, D., Carpenter, M.P., Janssens, R.V.E., Zhu, S.: $\gamma$-ray spectroscopy study of states in $^{27}$Si relevant for the $^{26m}$Al(p, $\gamma$)$^{27}$Si reaction in novae and supernovae. Phys. Rev. C **80**, 055802 (2009)

Möller, P., Pfeiffer, B., Kratz, K.-L.: New calculations of gross β-decay properties for astrophysical applications: Speeding-up the classical r process. Phys. Rev. C **67**, 055802 (2003)

Nassar, H., Paul, M., Ahmad, I., et al.: Stellar (n, $\gamma$) Cross Section of $^{62}$Ni. Phys. Rev. Lett. **94**, 092504 (2005)

Nassar, H., Paul, M., Ahmad, I., et al.: $^{40}$Ca($\alpha$, $\gamma$)$^{44}$Ti Reaction in the Energy Regime of Supernova Nucleosynthesis. Phys. Rev. Lett. **96**, 041102 (2006)

Newton, J.R., Iliadis, C., Champagne, A.E., Coc, A., Parpottas, Y., Ugalde, C.: Gamow peak in thermonuclear reactions at high temperatures. Phys. Rev. C **75**, 045801 (2007)

Panov, I.V., Kolbe, E., Pfeiffer, B., Rauscher, T., Kratz, K.-L., Thielemann, F.-K.: Calculations of fission rates for r-process nucleosynthesis. Nucl. Phys. A **747**, 633 (2005)

Panov, I.V., Yu Korneev, I., Rauscher, T., Martínez-Pinedo, G., Kelic-Heil, A., Zinner, N.T., Thielemann, F.-K.: Neutron-induced astrophysical reaction rates for translead nuclei. Astron. Astrophys. **513**, A61 (2010) arXiv:0911.2181

Peräjärvi, K., Siiskonen, T., Honkanen, A., Dendooven, P., et al.: Measurement of the IAS resonance strength in $^{23}$Mg. Phys. Lett. B **492**, 1 (2000)

Pieper, S.C., Wiringa, R.B.: Quantum Monte Carlo Calculations of Light Nuclei. Ann. Rev. Nucl. Part. Phys. **51**, 53 (2001)

Ratynski, W., Käppeler, F.: Neutron capture cross section of $^{197}$Au: A standard for stellar nucleosynthesis. Phys. Rev. C **37**, 595(1988)

Rauscher, T., Raimann, G.: Astrophysical reaction rates for $^{10}$B(p, $\alpha$)$^7$Be and $^{11}$B(p, $\alpha$)$^8$Be from a direct model. Phys. Rev. C **53**, 2496 (1996)

Rauscher, T., Thielemann, F.-K., Kratz, K.-L.: Nuclear level density and the determination of thermonuclear rates for astrophysics. Phys. Rev. C **56**, 1613 (1997)

Rauscher, T., Thielemann, F.-K.: Astrophysical Reaction Rates From Statistical Model Calculations. At. Data Nucl. Data Tables **75**, 1 (2000)

Rauscher, T., Heger, A., Hoffman, R.D., Woosley, S.E.: Nucleosynthesis in Massive Stars with Improved Nuclear and Stellar Physics. ApJ. **576**, 323 (2002)

Rauscher, T.: Relevant energy ranges for astrophysical reaction rates. Phys. Rev. C **81**, 045807 (2010a)

Rauscher, T.: Astrophysical Rates for Explosive Nucleosynthesis: Stellar and Laboratory Rates for Exotic Nuclei. Nucl. Phys. A **834**, 635 (2010b) arXiv:0911.0262

Rauscher, T., Thielemann, F.K., Görres, J., Wiescher, M.: Capture of alpha particles by isospin-symmetric nuclei. Nucl. Phys. A **675**, 695 (2000)

Robertson, D., Schmitt, C., Collon, Ph. et al.: A new AMS set up for nuclear astrophysics experiments. Nucl. Instr. Meth. B **259**, 669 (2007)

Rugel, G., Faestermann, T., Knie, K., et al.: New Measurement of the $^{60}$Fe Half-Life. Phys. Rev. Lett. **103**, 072502 (2009)

Ruiz, C., Parikh, A., José, J., Buchmann, L., et al.: Measurement of the $E_{c.m.} = 184$ keV Resonance Strength in the $^{26g}$Al(p, $\gamma$)$^{27}$Si Reaction. Phys. Rev. Lett. **96**, 252501 (2006)

Runkle, R.C., Champagne, A.E., Angulo, C., et al.: Direct Measurement of the $^{14}$N(p, $\gamma$)$^{15}$O S Factor. Phys. Rev. Lett. **94**, 082503 (2005)

Runkle, R.C., Champagne, A.E., Engel, J.: Thermal Equilibration of $^{26}$Al. ApJ **556**, 970 (2001)

Salpeter, E.E., Van Horn, H.M.: Nuclear Reaction Rates at High Densities. ApJ. **155**, 183 (1969)

Satchler, G.R.: Direct Nuclear Reactions. Clarendon, Oxford (1983)

Schmalbrock, P., Donoghue, T.R., Wiescher, M., Wijekumar, V., Browne, C.P., Rollefson, A.A., Rolfs, C., Vlieks, A.: Proton threshold states in $^{27}$Si and their implications on hydrogen burning of $^{26}$Al. Nucl. Phys. A **457**, 182 (1986)

Seuthe, S., Rolfs, C., Schröder, U., et al.: Resonances in the $^{22}$Na(p, $\gamma$)$^{23}$Mg reaction. Nucl. Phys. A **514**, 471 (1990)

Simpson, J.J., Dixon, W.R., Storey, R.S.: Study of $^{44}$Ti by the $^{40}$Ca($\alpha$, $\gamma$)$^{44}$Ti Reaction. Phys. Rev. C **4**, 443 (1971)

Stegmüller, F., Rolfs, C., Schmidt, S., et al.: $^{22}$Na(p, $\gamma$)$^{23}$Mg resonant reaction at low energies. Nucl. Phys. A **601**, 168 (1996)

Terrasi, F., Rogalla, D., De Cesare, N., et al.: A new AMS facility in Caserta/Italy. Nucl. Instr. Meth. B **259**, 14 (2007)

Uberseder, E., Reifarth, R., Schumann, D., et al.: Measurement of the $^{60}$Fe(n, $\gamma$)$^{61}$Fe Cross Section at Stellar Temperatures. Phys. Rev. Lett. **102**, 151101 (2009)

Vockenhuber, C., Ouellet, C.O., The, L.-S., et al.: Measurement of the $^{40}$Ca($\alpha$, $\gamma$)$^{44}$Ti reaction relevant for supernova nucleosynthesis. Phys. Rev. C **76**, 035801 (2007)

Vogel, P.: Neutrino nucleus cross section at low energies. Nucl. Phys. A **777**, 340 (2006)

Vogelaar, B., Mitchell, L.W., Kavanagh, R.W., et al.: Constraining $^{26}$Al + p resonances using $^{26}$Al($^{3}$He, d)$^{27}$Si. Phys. Rev. C **53**, 1945 (1996)

Wagoner, R.V.: Synthesis of the Elements Within Objects Exploding from Very High Temperatures. ApJ. Suppl. **18**, 247 (1969)

Ward, R.A., Fowler, W.A.: Thermalization of long-lived nuclear isomeric states under stellar conditions. ApJ. **238**, 266 (1980)

Yakovlev, D.G., Gasques, L.R., Afanasjev, A.V. et al.: Fusion reactions in multicomponent dense matter. Phys. Rev. C **74**, 035803 (2006)

# Chapter 10
# Instruments for Observations of Radioactivities

G. Kanbach and L. Nittler

## 10.1 Astronomical Telescopes

### *10.1.1 Measuring Radiation from Cosmic Radioactivity*

#### 10.1.1.1 General Considerations

Radioactivity is characterized by the emission of particles or photons that accompany the nuclear transformations of unstable isotopes. Direct observations of secondary particles (e.g. $\beta^{\pm}$, or $\alpha$) are only possible with in-situ measurements in the *local environment* of the solar system and often the detected particles are not very specific as to their parent nuclei. Transitions between energy levels of radioactive and excited nuclei however produce characteristic X- and $\gamma$-ray lines that can be detected from astronomical distances. Only one secondary *particle* resulting from radioactive decay, the positron, signals its presence in a characteristic $\gamma$-ray line: positrons annihilate with their anti-particle (electrons) and convert the pair's rest mass into a line at 511 keV (see Chap. 7). Nuclear energy levels range from 10s of keV upwards to energies of 10s of MeV with most important astrophysical lines in the range from about 100 keV to several MeV.

Detection of MeV photons must exploit the dominant interaction processes in this energy range: the photoelectric effect (*photoeffect*) and *Compton scattering*. Both these interaction processes and their cross-sections depend on photon energy and target material. The typical energy of transition from dominating photoeffect to Compton scattering is around the rest mass energy of the electron, i.e. 511 keV. For photon energies above the highest atomic binding energy the photoelectric cross section is given by $\sigma \propto Z^n E^{-3}$, where $Z$ is the atomic number (number of protons in the nucleus) and $n$ is an index with a value between 4 and 5. At lower photon energies characteristic line-structures appear in the cross section (K-, L-, M-edges, etc.)

G. Kanbach (✉)
MPE, 85748 Garching, Germany

L. Nittler
Carnegie Institution for Science, Washington DC 20005, USA

Kanbach, G., Nittler, L.: *Instruments for Observations of Radioactivities*. Lect. Notes Phys. **812**, 491–515 (2011)
DOI 10.1007/978-3-642-12698-7_10 

which indicate the atomic energy levels of the electrons released in an atomic transition. The experimental consequence is that a high-efficiency photoelectric detector should be made of high-Z material, and it will work best below a few 100 keV. Spectroscopy with such detectors however requires careful calibration, because the structures in the cross section will shape the instrumental response and appear in the measurements. At higher energies, incoherent scattering of photons with electrons, (*Thompson* and *Compton scattering*), lead to the release of an energetic electron and a secondary photon, and dominate photon interactions. The maximum of this interaction cross section is around 511 keV ($m_e c^2$). The high- and low-energy ends of the cross section behavior can be described by asymptotic limits in terms of the *Thompson* cross section

$$\sigma_T = \frac{8\pi}{3}\left(\frac{\alpha\hbar}{m_e c}\right)^2 \simeq 6.652 \times 10^{-25}\ \text{cm}^2 \tag{10.1}$$

and the photon energy in units of the electron rest mass, $\varepsilon = h\nu/m_e c^2$. In the *Thompson regime*, $\varepsilon \ll 1$, the total cross section is about $\sigma = \sigma_T(1 - 2\varepsilon \ldots)$. In the high energy limit (the *Klein Nishina* regime, $\varepsilon \gg 1$) the total cross section[1] is a decreasing function of energy:

$$\sigma = \frac{3}{8}\sigma_T \varepsilon^{-1}(1 + 2\ln\varepsilon) \tag{10.2}$$

#### 10.1.1.2 Instrument Types

Building an efficient detector for ~MeV gamma radiation however not only demands a good choice of detector material, but also requires that the detectors are sensitive to measure the secondary particles (electrons, positrons) released by the incoming energetic photons. This can be achieved in several ways:

- *Ionization chambers*: the detector is intrinsically capable of measuring the presence of ionization and to generate an electronic signal. Gas-filled ionization chambers, proportional counters, or Geiger counters were the original devices to detect high energy radiation. The small amounts of charge generated in the detector volume by a single photon is often amplified by a strong electric field generated from electrodes. This results in acceleration, collisional secondary ionization and the formation of a break-down cascade, that is more easily detectable. Variations of this principle led to gas-filled drift chambers or spark chambers that also allowed to locate the impact coordinates of primary photons. After semiconductors with sufficient volume had been developed (e.g. Si, Ge, Cadmium Zinc Telluride (CZT) or Cadmium Telluride (CdTe)) the principles and concepts of ionization chambers were transferred to the new solid state detectors. The higher densities and generally higher atomic numbers of solid state detectors led to much

[1] Detailed tables of photon cross sections for various elements and compounds are available at `http://physics.nist.gov/PhysRefData/Xcom/Text/XCOM.html`.

higher efficiencies for the conversion of $\gamma$-rays and, in the case of cryogenically cooled detectors, also to very fine energy resolution (Ge detectors).
- *Scintillation detectors*: the detector material emits light under the impact of ionization (*fluorescence, scintillation*). In addition to the scintillator material itself, these detectors require a photon detector to record the emitted light. Commonly used are *organic scintillators* (liquid or solid), scintillating *crystals* (e.g. NaI, CsI, BGO, $La_2Br_3$), or *noble gases* (e.g. liquid Xenon). Traditionally the detection of scintillation light was achieved with *photo-multiplier tubes (PMTs)*, which are low noise and very fast detectors. The typical quantum efficiency of a PMT photocathodes however is only around ~20%. More recently the readout of scintillation light is also done with *solid-state photodetectors* such as PIN diodes, Si drift detectors (SDD), and avalanche photo diodes (APD). The advantages of solid state readout are a higher detection efficiency, the small size and the possibility of complex arrangements (pixels, arrays) of Si detectors, in addition to the straightforward interface with modern readout electronics.

All high energy detectors in space are exposed to radiation of energetic particles (cosmic rays, radiation belts) and thus *photons of non-astronomical origin*. The structures of spacecraft and detector alike will become sources of local background under this irradiation. Direct nuclear interactions and the creation of radioactive isotopes may lead to intense levels of background, unless counter measures are taken. Direct and prompt interactions of incident charged particles can be rejected by enclosing the sensitive volumes in veto shield detectors. These *anti-coincidence* detectors are often made of plastic scintillator (e.g. *COMPTEL*, Schoenfelder et al. (1993), *EGRET*, Kanbach et al. (1989), *Fermi*, Atwood et al. (2009)) but massive shields of inorganic scintillators, like *CsI (SMM-GRS*, Forrest et al. (1980)) or *BGO (INTEGRAL-SPI*, Vedrenne et al. (2003)) are also employed to better cover the low energy range. There is a trade-off to be made between adding rejection capability for undesired events and adding mass which generates such undesired events.

To improve handling of the *intrinsic background* of a gamma-ray telescope, several options are available:

- The structural materials close to the sensitive detectors should be chosen specifically to ensure low activation by the dominant external particle environment (i.e., low cross sections for such reactions). For example, aluminum, which activates to radioactive $^{24}$Na, could be replaced by beryllium; hydrazine propellant, which thermalizes and captures neutron on protons, emitting 2.2 MeV photons, should be minimized.
- The sensitive parts of a detector could be mounted at a distance from the mass of the spacecraft, e.g. on a boom.
- the detection process should fully exploit the characteristics of gamma-ray interactions to discriminate against background through selections in the measured interaction details. Examples that were used include *pulse shape discrimination* to separate neutron and $\gamma$-ray interactions or specific coincidence trigger requirements.

- The *choice of orbit* of a low-energy gamma-ray telescope should avoid regions of elevated particle background in the Earth's magnetosphere (radiation belts, South Atlantic Anomaly). A low altitude, circular equatorial orbit provides the best environment for such telescopes, since it is also shielded against low energy solar particles.

Even with the above counter measures, the instrumental background of a nuclear line telescope will remain at a significant level and often dominates celestial signal even for strong sources (see Chap. 7). The spectrum of this background extends from several 100 keV up to about 5 MeV and shows a continuum and several strong complexes of nuclear lines. Detailed simulations are essential, and were developed based on the Monte-Carlo principle and detailed high-energy interaction physics software packages such as *GEANT*, e.g. by Weidenspointner et al. (2005). This allows to estimate the intensity and spectrum of instrumental background, for future and past missions. Simulations and trade-offs for options of telescope design and orbit can thus be investigated.

The usefulness of an astronomical telescope can be expressed as the level of *sensitivity* reachable during typical periods of observation. Sensitivity is given as the minimum source flux that can be detected significantly above the level of background. Generally the instrumental-background signature is spread out, and varies smoothly over the field of view, but it contains the spectral features mentioned above. In a telescope with temporal, angular and energy resolution, the relevant level of background is then the level contained in the temporal, angular and spectral resolution elements which would be populated by the signal from a real source (*point spread function*). The resolution elements of measurement and analysis, i.e. bins in a measured signal parameter, pixels in an image, or $\Delta E/E$ in a spectrum, should therefore correspond to the properties of the astronomical sources in order to optimize the sensitivity. Of course the combination of angular and spectral resolution will lead to best results, but as we discuss below, temporal resolution for transient or variable sources (flares, bursts) may well substitute a lack of angular resolution. We discuss now typical examples of successful instruments, that pioneered nuclear-line astrophysics.

### 10.1.2 Photon Collectors

*SMM-GRS*, the $\gamma$-ray spectrometer of the *Solar Maximum Mission (SMM)* was an actively shielded multicrystal scintillation $\gamma$-ray spectrometer with a wide field of view, sensitive to $\gamma$-rays in the energy range from about 300 keV to 100 MeV (Forrest et al., 1980). Its goal was to measure solar flares and during the mission from 1980 to 1989 a total of 185 flares were detected (Vestrand et al., 1999). The discrimination of signal and background in *SMM-GRS* was based on the time profile of the flares riding on top of a slowly varying background and on a model of the orbital background averaged over locations of similar magnetospheric conditions. *SMM-GRS* was therefore a typical photon collector with little angular reso-

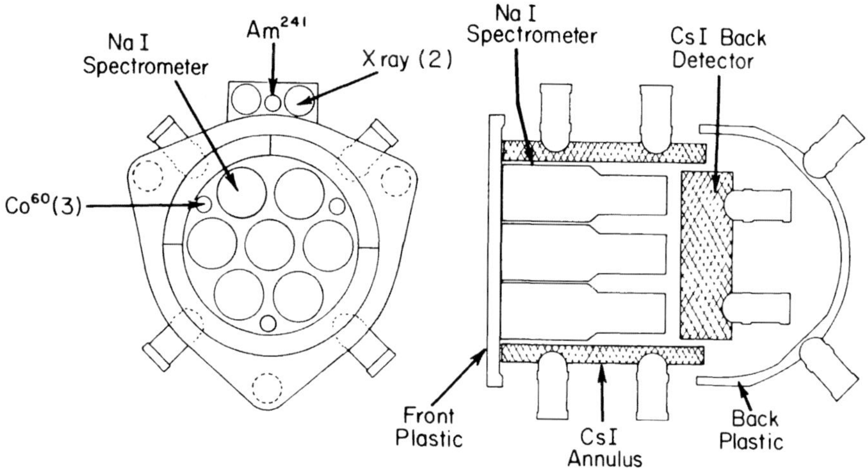


**Fig. 10.1** The *Gamma Ray Spectrometer* on the *Solar Maximum Mission* (Forrest et al., 1980). Seven 3" NaI detectors are enclosed in an anti-coincidence shield made of a plastic scintillator (*front* and *back*) and of CsI scintillators (*sides* and *back*)

lution. The $\gamma$-ray spectrometer, shown in Fig. 10.1, was composed of 7 cylindrical NaI(Tl) crystals with a diameter of 7.6 cm and a height of 7.6 cm. Each of these crystals was viewed by a PMT. The crystals were surrounded on the sides by a 2.5 cm thick CsI annulus. On the rear side a circular CsI crystal with a thickness of 7.6 cm and a diameter of 25 cm was placed. The 7 NaI (Tl) crystals and the back CsI crystal formed together the high-energy detector. In order to suppress charged particles the front and rear sides were covered with sheets of plastic scintillators thus shielding the main detector from all sides. The $\gamma$-ray spectrometer had a very wide field of view with a diameter of $\simeq$120°, an energy resolution of 7% at 662 keV and an effective area ranging from 20 to 200 cm$^2$ as a function of energy.

### *10.1.3 Imaging Instruments*

*COMPTEL* was the first truly imaging telescope for the MeV range and was in orbit for 9 years from 1991 to 2000 (Schoenfelder et al., 1993) aboard NASA's *Compton Gamma-Ray Observatory*. A photomontage of *COMPTEL* is shown in Fig. 10.2.

The upper detector, called D1, consists of 7 modules filled with the liquid scintillator NE 213A. Each module of 28 cm diameter and 8.5 cm thickness is viewed from the sides by 8 photomultiplier tubes. The total geometrical area of D1 is 4,300 cm$^2$. The lower detector, D2, separated from D1 by 1.5 m, consists of 14 modules of NaI(Tl)-scintillator. Each module of 28 cm diameter and 7.5 cm thickness is viewed from below by 7 photomultipliers. The total geometrical area of D2 is 8,600 cm$^2$. From the relative pulse heights of the photomultiplier tubes in each D1- and

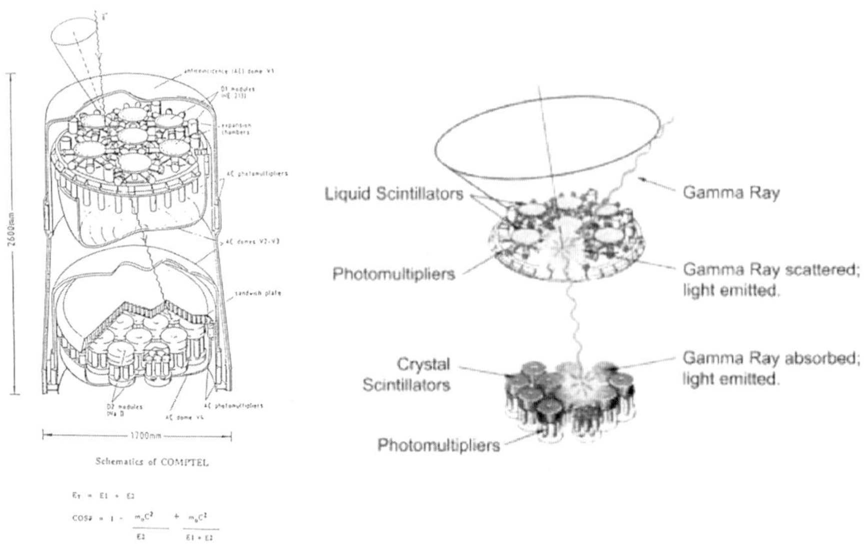


**Fig. 10.2** The *COMPTEL* instrument on the *Compton Gamma Ray Observatory CGRO* (Schoenfelder et al., 1993). A schematic drawing shows a typical scattering event and the cone with opening angle $\phi$ which contains the incident photon

D2-module the locations of the interactions are determined to within about 2 cm. The sum of the photomultiplier signals of each module provides the energy losses E1 and E2 . Both, D1 and D2 are completely surrounded by veto-domes of 1.5 cm thick plastic scintillator to reject charged particles. A valid event trigger is generated by a time-of-flight delayed coincidence between D1 and D2 and the absence of a veto signal. The positions of the interactions in D1 and D2 define the direction of the scattered photon and the Compton scattering formula

$$cos(\phi) = 1 - m_e c^2 \left( \frac{1}{E2} - \frac{1}{E1 + E2} \right)$$

is used to estimate the opening angle of a cone that contains the incident gamma ray (see schematic in Fig. 10.2)

*COMPTEL* covers the energy range 0.8–30 MeV. Within its large field-of view of about 1 steradian (64° FWHM), its angular resolution ranges from 1.7° to 4.4° FWHM (depending on energy). The energy resolution is in the range 5–8% FWHM, and the effective detection area is in the range 20–30 cm$^2$ (again depending on energy). The sensitivity of *COMPTEL* is significantly determined by the instrumental background. A substantial suppression is achieved by the combination of the charged particle veto domes, the time-of-flight measurement technique, pulse shape discrimination in D1, Earth horizon angle cuts, and proper event selection cuts. In spite of these significant countermeasures the actual total in-flight instrumental background was higher than expected prior to launch by about a factor of 4.

As a consequence, the sensitivity of *COMPTEL* was 2-times lower than expected. Fortunately, however, this loss in sensitivity could be compensated by the 4-times longer mission life-time (which originally was planned to be 2.25 years, only). The actually achieved point source sensitivity for deep observations ($T = 6 \times 10^6$ s) was $6.3 \times 10^{-5}$ ph cm$^{-2}$ s$^{-1}$ for continuum emission between 1 and 30 MeV (corresponding to about 3% of the Crab-flux) and $1.6 \times 10^{-5}$ ph cm$^{-2}$ s$^{-1}$ for line emission at 1.157 and 1.809 MeV ($^{44}$Ti and $^{26}$Al lines).

## *10.1.4 Spectrometry and Imaging*

As we have discussed above, angular resolution is an essential requirement to achieve astronomical sensitivity. When the spectrometer instrument for ESA's INTEGRAL mission was designed it was clear that the superb energy resolution of a Ge spectrometer had to be combined with a telescope that provides at least moderate angular resolution. In the Spectrometer for Integral, SPI, this was achieved with a coded-mask system (see Fig. 10.3). The SPI mask, which is situated 171 cm above the detection plane, features a HURA (Hexagonal Uniform Redundant Array) coded mask pattern with 120° symmetry enclosed within a diameter of 72 cm. Of the 127 individual cells (60 mm side to side) in the mask, 63 are opaque (3 cm thick blocks of Densimet (a tungsten alloy), opacity of about 90%) and 64 are transparent. The *shadowgram* projected onto the detector plane is measured with 19 Ge detectors. The reverse-electrode n-type Ge detectors also have a hexagonal shape with a side length of 3.2 cm and a height of 69.42 mm and are operated at a temperature of ~90 K cooled by a Sterling system. The detector is enclosed in a massive anti-coincidence shield of BGO scintillation detectors. The energy range extends over 20 keV–8 MeV with a typical energy resolution of 2.5 keV at 1.33 MeV. The coded mask system provides an angular resolution of about 2.5° with a positioning accuracy for a strong source of 10 arc min. The fully coded field-of-view has a diameter of 16°. Further details can be found in Vedrenne et al. (2003).

## *10.1.5 Perspectives of Cosmic Gamma-Ray Spectroscopy*

Large efforts are presently underway to develop the next generation of Compton telescopes. Some projects, such as the *Advanced Compton Telescope ACT* (Boggs et al., 2006), *GRIPS* (Greiner et al., 2009), and *GRI* (Knödlseder, 2007) are still in a study phase. Other experiments have been built as prototypes or balloon payloads. Among those are MEGA (Kanbach et al., 2003), the *Nuclear Compton Telescope NCT* (NCT Collaboration et al., 2007), and *TIGRE* (Zych et al., 2006). Scheduled for a space mission is currently only one system which employs a Compton scattering technique – the Soft Gamma Ray detector on the Japanese mission *Astro-H* (Takahashi et al., 2009). The NuSTAR mission employs X-ray techniques of mirror deflection; it is scheduled for launch in 2012, and will hopefully demonstrate that

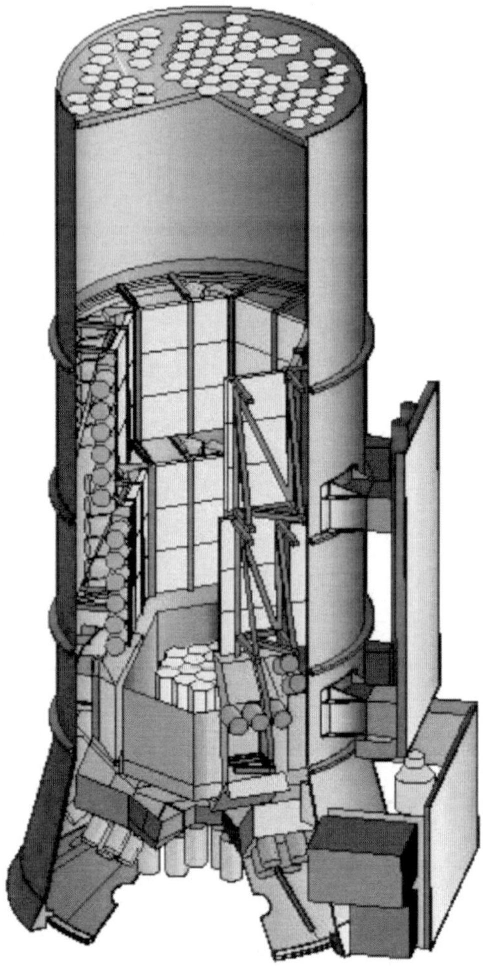

**Fig. 10.3** The *Spectrometer Instrument (SPI)* on ESA's *INTEGRAL* mission (Vedrenne et al., 2003). The $\gamma$-ray sky is projected through a coded mask onto an array of 19 Ge detectors. Deconvolution of the shadowgram allows an angular resolution of 2.7° and the Ge detectors provide an energy resolution of $E/\Delta E \sim 600$

at least the low-energy $^{44}$Ti decay lines at 68 and 78 keV can be measured with this technique.

From *COMPTEL* we have learned that the sky is rich in phenomena and objects that can be studied around 1 MeV. But it is also true, that with *COMPTEL* we could only see the tip of the iceberg. The achieved sensitivity was still modest. The next telescope would have to be more sensitive by one or two orders of magnitude. In the studies listed above different concepts of Compton telescopes are presently investigated and tested. Instead of scintillators also other detector materials are considered and tested, like silicon strip detectors, position sensitive germanium detectors,

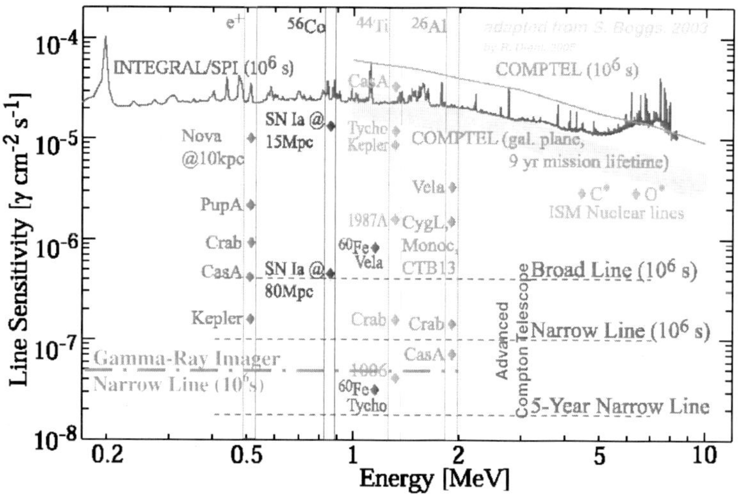


**Fig. 10.4** Next-generation gamma-ray telescopes will make several additional sources accessible to detailed comparisons between theoretical source models and data about isotopic abundances and their spatial distribution for these objects. The current *tip of the iceberg* source data samples will thus be substantiated. Still, object numbers remain modest, in comparison to other, more developed disciplines of astronomy (adapted from Boggs, 2006)

Cd-Te-detectors, liquid xenon gas detectors, and high pressure gas detectors. In order to achieve the required improvement in sensitivity, the detection efficiency has to be drastically increased and the background has to be reduced considerably. An increase of the efficiency by a factor of 5–10 is possible within reasonable experimental limitations. The levels of sensitivity expected for one example of a proposed experiment, the *Advanced Compton Telescope* (Boggs et al., 2006), are shown in Figure 10.4, together with brightness levels of several representative cosmic sources, such as novae, supernovae, and diffuse radioactivity lines. What is needed, is a decision that the still largely undeveloped MeV wavelength band is opened for astronomy with a sensitive new mission.

## 10.2 Analyzing Material Samples From and Within Meteorites

Radioactivities in meteorite samples are in general detected by separating individual atoms from a sample and weighing them with a device known as a mass spectrometer. For long-lived radionuclides, that is those with lifetimes significantly longer than the 4.6 billion year age of the Solar System, the nuclides themselves are detected, along with their decay products. In the case of short-lived isotopes created by cosmic-ray interactions with the meteorites during the time interval between ejection from a parent body and arrival on Earth, the parent isotopes can also be directly measured. For short-lived isotopes that existed in the early Solar System but have since fully decayed (*extinct* isotopes), only the decay products can now be determined. In most cases, stable isotopes of the parent and daughter elements must also be measured in order to derive reliable information. Depending on the sample and the isotopic system, a wide array of instrumentation and techniques can be used to derive information on radioactive elements in extraterrestrial samples. These are the subject of this section.

### *10.2.1 Measurement Principles and Techniques*

When a radionuclide decays within a solid sample, the isotopic compositions of both the parent and daughter elements are changed. By plotting isotopic and/or elemental ratios, one can infer either the age of the sample (time elapsed since the sample solidified to the point that the parent isotope was fixed in the structure) or, for extinct isotopes, the initial abundance of the parent isotope. This is illustrated in Fig. 10.5, which shows an example of *Rb-Sr* dating based on the decay of long-lived ($\tau \sim 7.2 \times 10^{11}$ year) $^{87}$Rb to $^{87}$Sr. Each plotted point represents a different mineral that obtained a different Rb/Sr elemental ratio upon crystallization of the rock. Initially all of the minerals had the same $^{87}$Sr/$^{86}$Sr ratio (open circles), but with time, some of the $^{87}$Rb decayed, moving the points to the up and left on the plot (filled symbols). The age of the rock can be calculated from the slope of the resulting line; this line is called an *isochron*.

For the case of extinct nuclides, the daughter isotope ratio is plotted versus a stable isotope of the parent element, and the slope of an isochron gives the initial ratio of the extinct nuclide to the stable isotope. For example, to detect extinct $^{26}$Al, one plots the measured $^{26}$Mg/$^{24}$Mg ratio versus the $^{27}$Al/$^{24}$Mg ratio and the slope gives the initial $^{26}$Al/$^{27}$Al ratio of the sample (see Chap. 6). Often, one seeks information

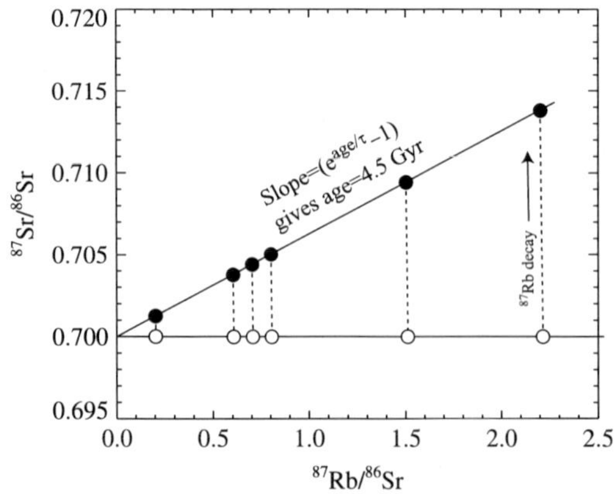


**Fig. 10.5** Schematic example of radiometric dating of rocks. *Open symbols* represent the initial isotopic compositions of a set of minerals that co-formed with varying Rb/Sr ratios. After 4.5 Gigayears (Gy), some of the long-lived $^{87}$Rb has decayed to $^{87}$Sr (*filled symbols*); the slope of the resulting *isochron* gives the age of the rock

on a single object with a single parent/daughter element ratio, for example presolar grains (Chap. 2). In such cases, isochrons cannot be constructed and assumptions regarding the initial stable isotopic composition must be made to determine initial abundances of radionuclides. Usually, such assumptions can be made with sufficient confidence for the problem at hand.

Note that there are many complications to radiometric age-dating and determination of extinct isotopes, both arising from the laboratory analyses (see below) and from the samples themselves. For example laboratory contamination can compromise analyses, especially for very-low-abundance elements. Atoms of either the parent and/or daughter elements may have diffused over time into or out of the sample, so that the measured ratios do not reflect the simple isochron behavior described above. Researchers have developed many techniques for identifying and overcoming such problems.

Mass spectrometers determine the elemental, isotopic or molecular composition of a sample by first converting the atoms or molecules of the sample into a beam of ions and then using electric and/or magnetic fields to disperse the beam in mass. Thus every mass spectrometer requires an ion source, a filter to separate masses and one or more ion detectors. Several examples of different types of mass spectrometers used for meteoritic radioactivity research are detailed in subsequent sections. However, we note here that a key distinction must be made between *bulk* analyses – those that are made on relatively large samples, usually chemically treated to concentrate the elements of interest, and in situ analyses – those that sample specific locations within a solid sample. For bulk analyses, the sample may be a single mineral grain a few 100 $\mu$m in size up to several grams of a whole meteorite. The sample is first dissolved in pure acids and other solvents and then chemical techniques (e.g. cation exchange columns) are used to purify the element(s) of interest. For in situ measurements, no chemical preparation is necessary; polished rock sections or dispersed mineral grains can be directly sampled by the instrumentation, down to sub-micron spatial scales.

Which technique is used for a specific problem depends on the problem itself, but is largely a question of sample size, abundances of the elements of interest and the required analytical precision. Because fundamentally this work involves counting atoms, the ultimate precision depends on the number of atoms that can be measured in a given sample. Since the magnitude of a radiogenic isotope effect depends on the parent/daughter ratio (e.g., Fig. 10.5), the lower this ratio is, the higher a precision is required and thus a larger number of atoms must be counted. In general, the highest precisions can be obtained with bulk techniques, but the attainable precision of modern in situ instrumentation has greatly increased in recent years, and many problems can now be addressed at smaller scales than was historically possible.

Almost all of the mass spectrometers described below use a magnetic-sector design for mass filtering the ion beam, often with an additional electrostatic analyzer (ESA) for double-focusing the beam in energy and mass. For example, Fig. 10.6 shows a schematic diagram of the NanoSIMS 50L ion microprobe (Cameca Instruments) used for many in situ measurements of presolar grains and other meteoritic components. In this double-focusing mass spectrometer, the ESA is used to disperse

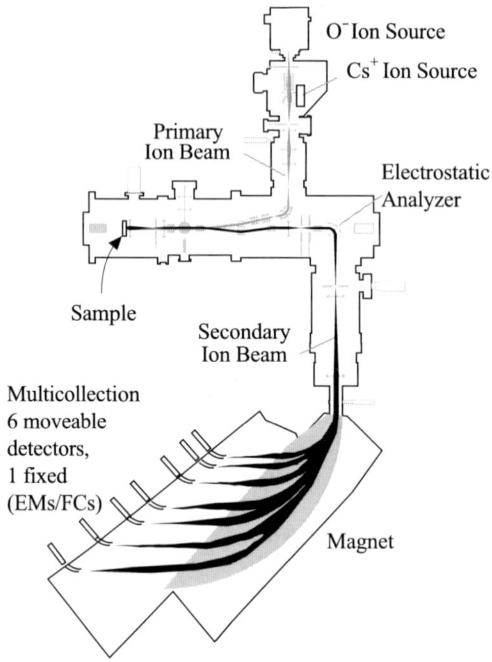


**Fig. 10.6** Schematic illustration of Cameca Instruments NanoSIMS 50L ion microprobe

the beam of ions from the sample according to kinetic energy and an electromagnet is used to disperse the beam in mass/charge ratio. The ESA and magnet are matched such that ions of a given mass/charge ratio are brought to the same focal point regardless of their kinetic energy.

The most common detectors used for meteoritic mass spectrometry are Faraday Cups (FCs) and electron multipliers (EMs). FCs measure the total charge deposited in a small conductive cup; they have essentially 100% detection efficiency but electronic noise requires minimum ion currents of $>10^4$–$10^5$/s for accurate measurements. In an EM, a single ion triggers an exponentially increasing cavalcade of electrons, which generates a measurable electronic pulse. EMs are thus used for low-counting-rate situations. However, in general the detection efficiency of an EM is less than unity, varies with element (and isotope) and changes with time as it is bombarded with more and more ions (*aging*). These problems limit the ultimate precision that may be attained by measurements that use EMs, compared to those that use FCs. Some mass spectrometers are built with a single ion detector. In these systems, the current in the magnet is repeatedly changed (*peak-jumping*) in order to focus different isotopes onto the detector. Repeated cycles of peak-jumping are used to build up statistics for the isotopes of interest and calculate ratios. In contrast, in a *multicollection* system (e.g., Fig. 10.6), the magnet spreads the ion beam out along its focal plane so that different isotopes are measured simultaneously in a set of detectors. Multicollection both decreases the time required for analyses and improves the accuracy of measurements compared to peak-jumping.

Other types of mass spectrometry, including accelerator mass spectrometry and gas-source mass spectrometry, while used for some meteorite research, are not used commonly for research related to the topics of this book. These are not discussed further here.

### 10.2.2 Bulk Techniques

The most commonly-used methods for bulk isotopic analysis of meteoritic components are Thermal Ionization Mass Spectrometry (TIMS) and Multi-Collector Inductively-Coupled Mass Spectrometry (MC-ICPMS). In TIMS, the element to be analyzed is chemically separated from the sample, and deposited from aqueous solution onto a metal (commonly W, Re, or Ta) filament. As the filament is heated by an electric current, atoms from the sample boil off and become ionized. The ions then pass through a magnetic mass spectrometer and are counted, usually by a multi-collection system of Faraday Cups. TIMS has the advantages of very high transmission of ions to the detectors as well as producing very stable beams, especially for elements with relatively low ionization potentials, like Mg, Cr, Fe, Ni, Sr, Pb and many others. For example, the original discovery of extinct $^{26}$Al in the Solar System was made by TIMS measurement of Mg isotopes (Gray, 1974; Lee et al., 1976). However, the thermal ionization processes introduces a mass fractionation, such that lighter atoms more easily get ionized, and the data must be corrected for this. Quantification of concentration ratios of different elements (needed to determine radiometric ages and abundances of extinct isotopes, see above) is accomplished by isotope dilution, where a known amount of artificially-enriched isotopic tracers are added to the sample and measured along with the isotopes of interest.

In bulk MC-ICPMS analysis, drops of a solution of purified element are sprayed into an Ar plasma at very high temperature. The plasma evaporates the droplets and ionizes the sample atoms with high efficiency and the ions are then passed through a multi-collector double-focusing mass spectrometer. MC-ICPMS has the advantages of high ionization efficiency for all elements, including some elements difficult to measure by TIMS (Lee and Halliday, 1995), as well as relatively rapid data collection. As in TIMS, mass fractionation effects are significant and must be corrected for, often by measuring standards with known isotopic composition.

### 10.2.3 In Situ Techniques

In addition to being used for bulk analyses, MC-ICPMS instruments can be equipped with a laser system for *situ* analysis. A UV laser (e.g. 193 nm) is focused into a 50–100 μm spot on a sample, ablating material that is then transferred into the Ar plasma. High precisions can be obtained, but this method has the disadvantage that all elements in the sample are transferred to the plasma and ionized, leading to the possibility of unresolved isobaric interferences in mass spectra. Nonetheless, the technique has been successfully applied to studies of extinct radioactivity in meteorites (Young et al., 2005).

The most widely used in situ mass spectrometric technique for meteoritic research is Secondary Ion Mass Spectrometry (SIMS, Fig. 10.6). In SIMS, a beam of primary ions is focused onto a solid sample in the presence of a strong electric field. The primary beam sputters atoms from the surface, some of which get ionized and transferred into a mass spectrometer. Either a $Cs^+$ beam is used to generate negative secondary ions of electronegative elements (e.g., C, O, S, etc.) or an $O^-$ beam is used to generate positive ions of electropositive species (e.g., Mg, Ti etc.). SIMS combines high sensitivity with high spatial resolution to allow isotopic ratios of even minor elements to be measured in very small samples and was crucial in the discovery and detailed isotopic characterization of presolar grains. Modern SIMS instruments, *e.g,* the Cameca NanoSIMS (Fig. 10.6) and ims-1280 ion probes, include very high transmission of secondary ions and multicollection detector systems, and in the case of the NanoSIMS, greatly improved spatial resolution over previous generations of instruments ($<100$ nm vs. $\sim 1\ \mu$m.). These characteristics allow higher-precision data to be acquired on smaller spatial scales than was ever possible before, allowing, for example, detailed isotopic characterization of submicron presolar grains (Nguyen and Zinner, 2004; Zinner et al., 2005) as well as high-precision $^{26}$Al systematics on small scales in meteoritic inclusions (Kita ct al., 2008).

An extraordinarily useful feature of SIMS instruments is the ability to produce mass-filtered images of a sample. Two types of SIMS imaging are commonly used in meteoritic research. In *direct imaging,* the instrument behaves as a microscope: the primary beam is defocused on the sample and the ion optics of the instrument transfer an image of the surface through the mass spectrometer onto an imaging detector (e.g. a micro-channel plate or a CCD). The spatial resolution of this technique is set by the ion optics and is typically $>500$ nm. In *scanning* or *raster* imaging, a focused beam is scanned over the sample with synchronized collection of secondary ions; spatial resolution is set by the size of the primary beam ($>50$ nm in the NanoSIMS). Both techniques are extremely useful for scanning large numbers of meteoritic grains to search for isotopically anomalous, presolar grains of stardust (Nguyen and Zinner, 2004; Nittler et al., 1994; Nagashima et al., 2004). Examples of NanoSIMS raster imaging are shown in Fig. 10.7. In the top panels, O isotopic ratio images of an area of a primitive meteorite clearly reveal two sub-micrometer grains with anomalous $^{17}$O and $^{18}$O abundances (circled). In the bottom panels, a SiC grain believed to have originated from a supernova is revealed to have monoisotopic $^{26}$Mg, from in situ decay of $^{26}$Al (the inferred $^{26}$Al/$^{27}$Al ratio for this grain is $\sim$0.5).

Additional in situ techniques used in meteoritic radioactivity research include laser-heating noble-gas analyses and Resonance Ionization Mass Spectrometry (RIMS). In the former, a sample is melted with a laser, releasing atoms of noble gases trapped within. These gases are collected, ionized by bombarding them with electrons and measured with a mass spectrometer. This technique has been successfully applied to He and Ne isotopes in individual presolar grains of graphite and

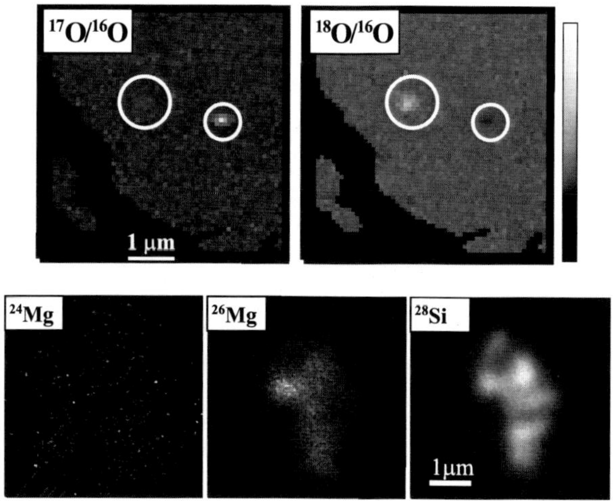


**Fig. 10.7** Example NanoSIMS isotopic images of presolar grains. *Top Panels*: Isotopic images of a $5 \times 5\ \mu m^2$ area of a primitive meteorite; *circles* indicate highly anomalous presolar grains surrounded by isotopically normal (solar-composition) material (Nguyen et al., 2010). *Bottom panels*: Isotopic images of a presolar SiC grain from a supernova. Mg in this grain is mono-isotopic $^{26}$Mg, due to decay of $^{26}$Al (Nittler et al., 2007)

SiC, in some cases providing evidence of extinct $^{22}$Na (Heck et al., 2007). RIMS uses lasers to ablate material from samples and then to selectively ionize atoms of a specific element; these are measured by means of a time-of-flight mass spectrometer. This technique has extraordinary sensitivity for the selected element and eliminates isobaric interferences that precludes measurement of some isotopes by SIMS. It has provided extremely useful isotopic data for trace elements within single presolar grains, including Zr, Mo, Ba, and Ru (Nicolussi et al., 1997; Savina et al., 2004).

A fundamental question for analysis of small samples like presolar grains is whether a given isotopic signature might be detectable in a given sample. Because in situ isotopic measurements are governed by Poisson statistics, the answer depends on the total number of atoms of the isotope of interest that can be detected, which clearly depends on the specific problem. The number of detectable atoms depends on the abundance of the element in the sample (determined by chemistry), the size of the sample (giving the total number of atoms) and the efficiency of the instrument (giving the fraction of atoms in a sample that can actually be detected and counted). For SIMS, the latter efficiency typically ranges from $10^{-7}$ to $10^{-2}$ and depends both on the low ionization probability of secondary ions during the sputtering process and the efficiency of transmitting ions through the mass spectrometer. As an example, let us consider the detection by SIMS of extinct $^{26}$Al in presolar SiC grains. A one-micron SiC grain contains some $5 \times 10^{10}$ atoms. Assuming an initial $^{26}$Al/$^{27}$Al ratio of $10^{-3}$, a typical Al concentration of 1% in presolar SiC, and a detection efficiency of $10^{-3}$, consuming an entire 1-μm grain would yield some 500 atoms of radiogenic $^{26}$Mg, corresponding to a statistical uncertainty of ~4%. Measuring only a fraction of the grain (to preserve it for additional isotopic analyses, for instance), or smaller grains, or grains with lower initial $^{26}$Al contents would correspondingly lower the number of detected atoms and decrease the precision. Moreover, whether the radiogenic signature is observable depends also on the amount of non-radiogenic Mg present in the grain. Clearly, similar considerations must be made for any given problem at hand.

### *10.2.4 Perspectives for Astronomy with Meteorite Samples*

As in many fields, the technology for chemical and isotopic analysis of geological and cosmochemical materials is rapidly evolving and improving and this will certainly continue into the future. Recent improvements in the stability of commercial instruments and advances in chemical treatments used to purify samples are now providing for extremely high precision measurements of bulk samples by TIMS and ICP-MS (Brennecka, 2010) and of intermediate-scale (tens of microns) in situ analyses by SIMS (Villeneuve et al., 2009). For analysis of tiny samples like presolar grains, significant effort is being expended to improve both sensitivity and spatial resolution. Current approaches are largely based on combining well-developed techniques. For example, SIMS has the advantage of achieving high spatial resolution, but is fundamentally limited by the low ionization probability for secondary ions during sputtering. In contrast, previous RIMS instruments have extraordinary sensitivity for certain elements, but limited spatial resolution due to the use of lasers to ablate atoms from samples. A new instrument under development at the University of Chicago (Davis et al., 2009) combines a very high resolution sputtering $Ga^+$ ion gun with multiple lasers for resonance ionization to achieve both sensitivity and spatial resolution beyond what is currently possible. As this machine comes on line in the next few years it is sure to increase the number of interesting elements measurable in presolar grains as well as to open other new avenues of extraterrestrial materials research. Another new instrument, the *MegaSIMS* constructed at the University of California, Los Angeles (McKeegan et al., 2009), takes another approach. In this instrument, the front end of a commercial SIMS instrument is used to produce secondary ions, which are then accelerated to MeV energies with a tandem accelerator, allowing for highly efficient suppression of molecular interferences, and thus allowing for higher sensitivity. Although developed specifically for solar-wind samples returned by NASA's Genesis mission, this instrument will surely be also used for exciting research on other samples where sensitivity is a priority, including presolar grains and extinct radioactivity in meteorites.

A fundamentally different technology, atom probe tomography, also holds great promise for isotopic and analysis of small extraterrestrial samples. In an atom probe, the sample is embedded in a very small, extremely sharp tip with a high potential applied to it. Applying a laser to the tip allows for single atoms to be extracted from the tip, one by one, and transmitted to a mass spectrometer. Reconstruction of atom trajectories allows the 3-dimensional reconstruction of the original positions and chemical identity of up to tens of millions of individual atoms. Technical limitations have thus far mostly limited the use of atom probe tomography to materials science applications, but recent attempts to apply the technique to presolar grains (Heck et al., 2010) suggest that it is likely to become a very useful tool for cosmochemistry in coming years.

## 10.3 Detection and Analysis of Cosmic Rays

The discovery of Galactic cosmic rays was inextricably tied to the development of sub-atomic particle physics in the early decades of the twentieth century. In particular, the discoveries in 1912–1914 by Hess and Kohlhörster, through high-altitude balloon flights, of a source of ionization that increases with altitude clearly indicated an extraterrestrial source of high-energy radiation (see historical review in Longair (1992)). Through the decades, increasingly more sophisticated instruments and experiments have been developed and deployed to determine the composition and energy spectra of cosmic rays. Because primary cosmic rays cannot penetrate the Earth's thick atmosphere to be detected at the ground, balloon-borne and space-based experiments remain crucial to progress in cosmic ray physics to the present day. However, due their very low flux, ground-based telescopes are still required to detect the highest-energy cosmic rays, through the *air showers* of secondary particles they produce as they traverse the atmosphere. This section describes some of the basic techniques used to detect and characterize GCRs both from the ground and at high altitudes.

### *10.3.1 Ground-Based Observations*

When Galactic cosmic rays enter the Earth's atmosphere, they interact with its atoms and molecules, generating a cascade of secondary particles. For most cosmic rays, these cascades overlap and lead to a relatively constant flux at ground level of secondary particles, dominated by muons. In contrast, air showers are caused by single primary cosmic rays with sufficient energy that their secondary particle cascades are detectable at ground level. Modern ground-based cosmic-ray telescopes are comprised of vast areas of particle and light detectors, designed to detect and characterize as many secondary particles as possible and hence reconstruct the characteristics of the original high-energy cosmic ray. For example, the largest such facility in the world, the Pierre Auger Observatory, is planned to ultimately include sites in Argentina and Colorado; the southern site is operational. It uses two common types of air shower detectors: an array of 1,600 water tank detectors spaced out over some 3,000 $km^2$ surrounded by four inward-looking *fluorescence detectors* (FDs) (The Pierre Auger Collaboration et al., 2007). The water tanks detect high-energy particles through the Cherenkov light they emit while passing through the tanks (essentially an electromagnetic sonic boom since the speed of light in the water is smaller than the particle velocities). The FDs consist of telescopes designed to detect UV fluorescent light from interaction of the shower particles with atmospheric nitrogen. Together, the two types of detectors allow reconstruction of the initial energy and direction of the primary cosmic ray. Similar techniques are used in other large air shower experiments, such as HiRes (Abbasi et al., 2004, 2008) and AGASA (Takeda et al., 2003).

### 10.3.2 High-Altitude and Space-Based Observations

As a charged particle passes through matter, it loses energy through ionization of atoms and molecules of the material. These ionization losses can cause significant damage to crystal structures and molecular chains. Many early studies of the composition of cosmic rays took advantage of this radiation damage in certain materials, such as many plastics as well as minerals in meteorites and lunar samples. The damaged areas have much higher chemical reactivity than the undamaged material. Thus, chemical etching of the materials reveals *tracks*, the nature (e.g., size) of which can be empirically related to the identities and energies of the incident particles (Fleisher et al., 1975). This technique was used starting in the late 1960s with balloon-borne plastic detectors to identify elements heavier than Fe, including heavy radioactive elements like U, in the primary GCRs (Blanford et al., 1969).

Because of limitations in the track technique (e.g., relatively poor resolution in charge identification), modern determinations of the composition of Galactic cosmic rays are based on electronic measurements of the total energy and energy loss rate as incident particles pass through various combinations of detectors. The rate of energy loss, $-dE/dx$, is proportional to the square of the particle's charge and depends also on its velocity as well as properties of the material (the Bethe-Bloch formula, Longair (1992)). If both $-dE/dx$ and the total kinetic energy can be independently determined, the charge and mass of the particle can be inferred. In practice, stacks of detectors are used to determine these parameters and additional detectors can be used to refine particle trajectories and exclude backgrounds. Examples of modern cosmic-ray experiments are given below.

Detectors used for cosmic-ray experiments commonly include: *solid-state detectors*, Si or Ge crystals in which the incident radiation releases electron-hole pairs that can be measured as an electrical pulse; *scintillation detectors*, crystals (e.g., NaI or certain plastics) in which light is produced by the incident radiation and detected by photomultiplier tubes; *Cherenkov detectors,* a type of scintillator for which particles above a threshold energy emit detectable light; and *transition radiation detectors,* stratified detectors with different indices of refraction; X-rays are emitted when particles cross the interfaces, with the amount of radiation dependent on the velocity of the particle.

A wealth of data on the isotopic and elemental composition of GCR nuclei has been obtained in the last decade by the Cosmic Ray Isotope Spectrometer (CRIS, Stone et al., 1998) on the *Advanced Composition Explorer* satellite, launched by NASA in 1997 and still operational as of this writing. A schematic diagram of the CRIS instrument is shown in Fig. 10.8 (from George et al., 2009). The instrument consists of a position sensitive *hodoscope* detector (HNX,Y) to determine particle trajectories and stacks of solid-state Si detectors (E1–E9) to measure energy and

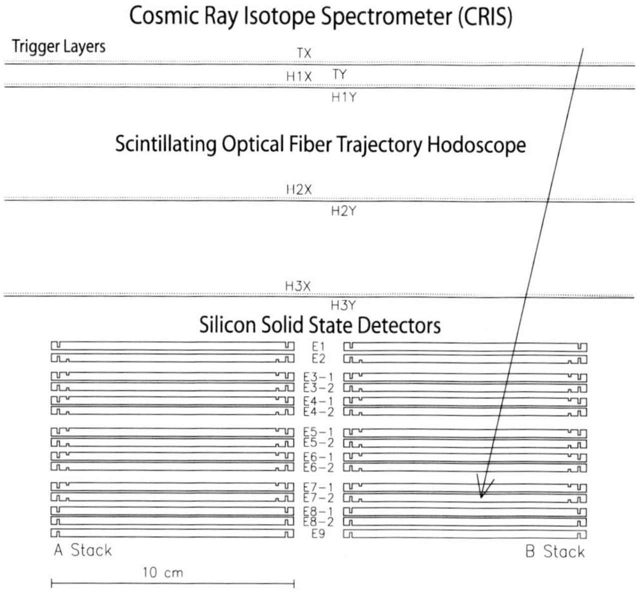


**Fig. 10.8** Schematic of CRIS instrument on the *Advanced Composition Explorer* satellite, from (George et al., 2009). *Arrow* indicates the trajectory of a single particle entering the instrument

energy loss rate. The hodoscope is constructed of layers of criss-crossed scintillating plastic fibers coupled to image-intensified CCD detectors. As a particle passes through the layers, the patterns of light observed on the CCDs can be inverted to determine the trajectory of the particle through the instrument with high precision. Additional layers of scintillating fibers at the top (TX,TY) serve as a trigger to signal the instrument that a particle has arrived. The arrow indicates the trajectory of a particle that enters the instrument and is finally stopped in the Si layer E7. For this event, the signals measured in detectors E1–E6 provide a measure of the energy loss rate (−dE/dx) whereas the total energy is found from that deposited in all seven detectors.

The process of particle discrimination in CRIS is illustrated in Fig. 10.9 (Stone et al., 1998), showing data obtained with a CRIS Si detector stack from a laboratory calibration experiment. Each data point represents the energy lost in detectors E1 through E3 plotted against the energy deposited in E4 for a single particle traversing the detector stack. Curves corresponding to the distinct isotopes of the indicated elements are clearly seen.

Other modern GCR composition instruments and experiments are designed around similar principles though with widely varying details. For example, the Trans-Iron Galactic Element Recorder (TIGER) experiment, flown on high-altitude balloons in Antarctica in 2001 and 2003, used a combination of scintillation and Cherenkov detectors to determine the composition of GCR with atomic number $26 < Z < 38$ (Rauch et al., 2009). To determine the composition of GCR at higher energies than studied by the experiments described above, the Transition Radiation

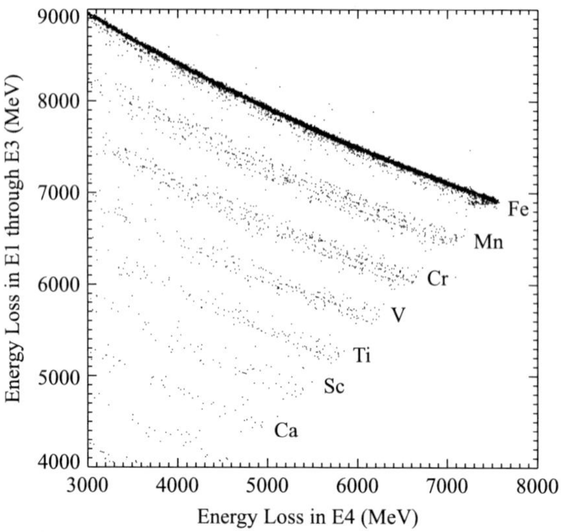


**Fig. 10.9** Calibration data for CRIS: plot of energy deposited in Si detectors E1–E3 versus that in detector E4. Particles of a given mass and charge plot along distinct curves, allowing isotope discrimination. Figure taken from Stone et al. (1998)

Array for Cosmic Energetic Radiation (TRACER) instrument uses a combination of Cherenkov, scintillator, gas-filled proportional counter and transition radiation detectors (Ave et al., 2008).

The instruments described above are designed to detect and characterize the composition of GCR nuclei. As discussed in Chap. 7, cosmic-ray positrons also provide important information regarding the origin and interactions of GCRs in the Galaxy. A recent report that the positron fraction of cosmic rays increases with increasing energy, inconsistent with purely secondary sources, was based on data from the PAMELA instrument on board the Russian Resurs-DK1 satellite, launched in 2006 (Adriani et al., 2009). PAMELA is broadly similar to other instruments designed to identify positrons (e.g., HEAT, Barwick et al. (1997)) and consists of scintillator detectors to measure time of flight and *dE/dx* for traversing particles and to distinguish upward-traveling electrons from downward-traveling positrons, a magnetic spectrometer to measure the rigidity (momentum per unit charge) of the particles, and an electromagnetic calorimeter consisting of interleaved layers of W absorbers and solid-state Si detectors. The calorimeter allows positrons to be distinguished from protons and anti-protons from electrons in the instrument.

## References

Abbasi, R.U., Abu-Zayyad, T., Amann, J.F., Archbold, G., Bellido, J.A., Belov, K., Belz, J.W., Bergman, D.R., Cao, Z., Clay, R.W., Cooper, M.D., Dai, H., Dawson, B.R., Everett, A.A., Fedorova, Y.A., Girard, J.H., Gray, R.C., Hanlon, W.F., Hoffman, C.M., Holzscheiter, M.H., Hüntemeyer, P., Jones, B.F., Jui, C.C., Kieda, D.B., Kim, K., Kirn, M.A., Loh, E.C.,

Manago, N., Marek, L.J., Martens, K., Martin, G., Matthews, J.A., Matthews, J.N., Meyer, J.R., Moore, S.A., Morrison, P., Moosman, A.N., Mumford, J.R., Munro, M.W., Painter, C.A., Perera, L., Reil, K., Riehle, R., Roberts, M., Sarracino, J.S., Sasaki, M., Schnetzer, S.R., Shen, P., Simpson, K.M., Sinnis, G., Smith, J.D., Sokolsky, P., Song, C., Springer, R.W., Stokes, B.T., Taylor, S.F., Thomas, S.B., Thompson, T.N., Thomson, G.B., Tupa, D., Westerhoff, S., Wiencke, L.R., Vanderveen, T.D., Zech, A., Zhang, X.: Measurement of the flux of ultrahigh energy cosmic rays from monocular observations by the high resolution fly's eye experiment. Phys. Rev. Lett. **92**(15), 151101-1 – 151101-4 (2004). doi:10.1103/PhysRevLett.92.151101, arXiv:astro-ph/0208243

Abbasi, R.U., Abu-Zayyad, T., Allen, M., Amman, J.F., Archbold, G., Belov, K., Belz, J.W., Ben Zvi, S.Y., Bergman, D.R., Blake, S.A., Brusova, O.A., Burt, G.W., Cannon, C., Cao, Z., Connolly, B.C., Deng, W., Fedorova, Y., Finley, C.B., Gray, R.C., Hanlon, W.F., Hoffman, C.M., Holzscheiter, M.H., Hughes, G., Hüntemeyer, P., Jones, B.F., Jui, C.C.H., Kim, K., Kirn, M.A., Loh, E.C., Maestas, M.M., Manago, N., Marek, L.J., Martens, K., Matthews, J.A.J., Matthews, J.N., Moore, S.A., O'Neill, A., Painter, C.A., Perera, L., Reil, K., Riehle, R., Roberts, M., Rodriguez, D., Sasaki, N., Schnetzer, S.R., Scott, L.M., Sinnis, G., Smith, J.D., Sokolsky, P., Song, C., Springer, R.W., Stokes, B.T., Thomas, S.B., Thomas, J.R., Thomson, G.B., Tupa, D., Westerhoff, S., Wiencke, L.R., Zhang, X., Zech, A.: First observation of the Greisen-Zatsepin-Kuzmin suppression. Phys. Rev. Lett. **100**(10), 101101-1 – 101101-4 (2008). doi:10.1103/PhysRevLett.100.101101, arXiv:astro-ph/0703099

Adriani, O., Barbarino, G.C., Bazilevskaya, G.A., Bellotti, R., Boezio, M., Bogomolov, E.A., Bonechi, L., Bongi, M., Bonvicini, V., Bottai, S., Bruno, A., Cafagna, F., Campana, D., Carlson, P., Casolino, M., Castellini, G., de Pascale, M.P., de Rosa, G., de Simone, N., di Felice, V., Galper, A.M., Grishantseva, L., Hofverberg, P., Koldashov, S.V., Krutkov, S.Y., Kvashnin, A.N., Leonov, A., Malvezzi, V., Marcelli, L., Menn, W., Mikhailov, V.V., Mocchiutti, E., Orsi, S., Osteria, G., Papini, P., Pearce, M., Picozza, P., Ricci, M., Ricciarini, S.B., Simon, M., Sparvoli, R., Spillantini, P., Stozhkov, Y.I., Vacchi, A., Vannuccini, E., Vasilyev, G., Voronov, S.A., Yurkin, Y.T., Zampa, G., Zampa, N., Zverev, V.G.: An anomalous positron abundance in cosmic rays with energies 1.5–100 GeV. Nature **458**, 607–609 (2009). doi:10.1038/nature07942, 0810.4995

Atwood, W.B., Abdo, A.A., Ackermann, M., Althouse, W., Anderson, B., Axelsson, M., Baldini, L., Ballet, J., Band, D.L., Barbiellini, G., Bartelt, J., Bastieri, D., Baughman, B.M., Bechtol, K., Bédérède, D., Bellardi, F., Bellazzini, R., Berenji, B., Bignami, G.F., Bisello, D., Bissaldi, E., Blandford, R.D., Bloom, E.D., Bogart, J.R., Bonamente, E., Bonnell, J., Borgland, A.W., Bouvier, A., Bregeon, J., Brez, A., Brigida, M., Bruel, P., Burnett, T.H., Busetto, G., Caliandro, G.A., Cameron, R.A., Caraveo, P.A., Carius, S., Carlson, P., Casandjian, J.M., Cavazzuti, E., Ceccanti, M., Cecchi, C., Charles, E., Chekhtman, A., Cheung, C.C., Chiang, J., Chipaux, R., Cillis, A.N., Ciprini, S., Claus, R., Cohen-Tanugi, J., Condamoor, S., Conrad, J., Corbet, R., Corucci, L., Costamante, L., Cutini, S., Davis, D.S., Decotigny, D., DeKlotz, M., Dermer, C.D., de Angelis, A., Digel, S.W., do Couto e Silva, E., Drell, P.S., Dubois, R., Dumora, D., Edmonds, Y., Fabiani, D., Farnier, C., Favuzzi, C., Flath, D.L., Fleury, P., Focke, W.B., Funk, S., Fusco, P., Gargano, F., Gasparrini, D., Gehrels, N., Gentit, F., Germani, S., Giebels, B., Giglietto, N., Giommi, P., Giordano, F., Glanzman, T., Godfrey, G., Grenier, I.A., Grondin, M., Grove, J.E., Guillemot, L., Guiriec, S., Haller, G., Harding, A.K., Hart, P.A., Hays, E., Healey, S.E., Hirayama, M., Hjalmarsdotter, L., Horn, R., Hughes, R.E., Jóhannesson, G., Johansson, G., Johnson, A.S., Johnson, R.P., Johnson, T.J., Johnson, W.N., Kamae, T., Katagiri, H., Kataoka, J., Kavelaars, A., Kawai, N., Kelly, H., Kerr, M., Klamra, W., Knödlseder, J., Kocian, M.L., Komin, N., Kuehn, F., Kuss, M., Landriu, D., Latronico, L., Lee, B., Lee, S., Lemoine-Goumard, M., Lionetto, A.M., Longo, F., Loparco, F., Lott, B., Lovellette, M.N., Lubrano, P., Madejski, G.M., Makeev, A., Marangelli, B., Massai, M.M., Mazziotta, M.N., McEnery, J.E., Menon, N., Meurer, C., Michelson, P.F., Minuti, M., Mirizzi, N., Mitthumsiri, W., Mizuno, T., Moiseev, A.A., Monte, C., Monzani, M.E., Moretti, E., Morselli, A., Moskalenko, I.V., Murgia, S., Nakamori, T., Nishino, S., Nolan, P.L., Norris, J.P., Nuss, E., Ohno, M., Ohsugi, T., Omodei,

N., Orlando, E., Ormes, J.F., Paccagnella, A., Paneque, D., Panetta, J.H., Parent, D., Pearce, M., Pepe, M., Perazzo, A., Pesce-Rollins, M., Picozza, P., Pieri, L., Pinchera, M., Piron, F., Porter, T.A., Poupard, L., Rainò, S., Rando, R., Rapposelli, E., Razzano, M., Reimer, A., Reimer, O., Reposeur, T., Reyes, L.C., Ritz, S., Rochester, L.S., Rodriguez, A.Y., Romani, R.W., Roth, M., Russell, J.J., Ryde, F., Sabatini, S., Sadrozinski, H., Sanchez, D., Sander, A., Sapozhnikov, L., Parkinson, P.M.S., Scargle, J.D., Schalk, T.L., Scolieri, G., Sgrò, C., Share, G.H., Shaw, M., Shimokawabe, T., Shrader, C., Sierpowska-Bartosik, A., Siskind, E.J., Smith, D.A., Smith, P.D., Spandre, G., Spinelli, P., Starck, J., Stephens, T.E., Strickman, M.S., Strong, A.W., Suson, D.J., Tajima, H., Takahashi, H., Takahashi, T., Tanaka, T., Tenze, A., Tether, S., Thayer, J.B., Thayer, J.G., Thompson, D.J., Tibaldo, L., Tibolla, O., Torres, D.F., Tosti, G., Tramacere, A., Turri, M., Usher, T.L., Vilchez, N., Vitale, V., Wang, P., Watters, K., Winer, B.L., Wood, K.S., Ylinen, T., Ziegler, M.: The Large Area Telescope on the Fermi Gamma-Ray Space Telescope Mission. ApJ **697**, 1071–1102 (2009). doi:10.1088/0004-637X/697/2/1071, 0902.1089

Ave, M., Boyle, P.J., Gahbauer, F., Höppner, C., Hörandel, J.R., Ichimura, M., Müller, D., Romero-Wolf, A.: Composition of primary cosmic-ray nuclei at high energies. ApJ **678**, 262–273 (2008). doi:10.1086/529424, 0801.0582

Barwick, S.W., Beatty, J.J., Bhattacharyya, A., Bower, C.R., Chaput, C.J., Coutu, S., de Nolfo, G.A., Knapp, J., Lowder, D.M., McKee, S., Mueller, D., Musser, J.A., Nutter, S.L., Schneider, E., Swordy, S.P., Tarle, G., Tomasch, A.D., Torbet, E., The HEAT Collaboration: Measurements of the cosmic-ray positron fraction from 1 to 50 GeV. ApJ **482**, L191–L194 (1997). doi:10.1086/310706, arXiv:astroph/9703192

Blanford, G.E., Friedlander, M.W., Klarmann, J., Walker, R.M., Wefel, J.P., Wells, W.C., Fleischer, R.L., Nichols, G.E., Price, P.B.: Observation of trans-iron nuclei in the primary cosmic radiation. Phys. Rev. Lett. **23**, 338–342 (1969). doi:10.1103/PhysRevLett.23.338

Boggs, S.E.: The advanced Compton telescope mission. New Astron. Rev. **50**, 604–607 (2006). doi:10.1016/j.newar.2006.06.076, arXiv:astro-ph/0608532

Boggs, S., Kurfess, J., Ryan, J., Aprile, E., Gehrels, N., Kippen, M., Leising, M., Oberlack, U., Wunderer, C., Zych, A., Bloser, P., Harris, M., Hoover, A., Klimenk, A., Kocevski, D., McConnell, M., Milne, P., Novikova, E.I., Phlips, B., Polsen, M., Sturner, S., Tournear, D.,Weidenspointner, G., Wulf, E., Zoglauer, A., Baring, M., Beacom, J., Bildsten, L., Dermer, C., Hartmann, D., Hernanz, M., Smith, D., Starrfield, S.: The advanced Compton telescope. In: Society of Photo-Optical Instrumentation Engineers (SPIE) Conference Series, Presented at the Society of Photo-Optical Instrumentation Engineers (SPIE) Conference, vol. 6266 (2006). doi:10.1117/12.670605

Brennecka, G.A.: A complication in determining the precise age of the solar system. Tech. rep. (2010)

Davis, A.M., Stephan, T., Veryovkin, I.V., Pellin, M.J., Savina, M.R.: The ion nanoprobe: a new instrument for studying the isotopic and elemental composition of the solar system and beyond at the few-nanometer scale. In: Lunar and Planetary Institute Science Conference Abstracts, vol. 40, Abstract 2472. Houston (2009)

Fleisher, R.L., Price, P.B., Walker, R.M.: Nuclear Tracks in Solids: Principles and Applications. University of California Press, Berkeley (1975)

Forrest, D.J., Chupp, E.L., Ryan, J.M., Cherry, M.L., Gleske, I.U., Reppin, C., Pinkau, K., Rieger, E., Kanbach, G., Kinzer, R.L.: The gamma ray spectrometer for the Solar Maximum Mission. Sol. Phys. **65**, 15–23 (1980). doi:10.1007/BF00151381

George, J.S., Lave, K.A., Wiedenbeck, M.E., Binns, W.R., Cummings, A.C., Davis, A.J., de Nolfo, G.A., Hink, P.L., Israel, M.H., Leske, R.A., Mewaldt, R.A., Scott, L.M., Stone, E.C., von Rosenvinge, T.T., Yanasak, N.E.: Elemental composition and energy spectra of galactic cosmic rays during solar cycle 23. ApJ **698**, 1666–1681 (2009). doi:10.1088/0004-637X/698/2/1666

Gray, C.M.: Excess 26Mg in the Allende meteorite. Nature **251**, 495–497 (1974). doi:10.1038/251495a0

Greiner, J., Iyudin, A., Kanbach, G., Zoglauer, A., Diehl, R., Ryde, F., Hartmann, D., Kienlin, A.V., McBreen, S., Ajello, M., Bagoly, Z., Balasz, L.G., Barbiellini, G., Bellazini, R., Bezrukov, L.,

Bisikalo, D.V., Bisnovaty-Kogan, G., Boggs, S., Bykov, A., Cherepashuk, A.M., Chernenko, A., Collmar, W., Di Cocco, G., Dröge, W., Gierlik, M., Hanlon, L., Horvath, I., Hudec, R., Kiener, J., Labanti, C., Langer, N., Larsson, S., Lichti, G., Lipunov, V.M., Lubsandorgiev, B.K., Majczyna, A., Mannheim, K., Marcinkowski, R., Marisaldi, M., McBreen, B., Meszaros, A., Orlando, E., Panasyuk, M.I., Pearce, M., Pian, E., Poleschuk, R.V., Pollo, A., Pozanenko, A., Savaglio, S., Shustov, B., Strong, A., Svertilov, S., Tatischeff, V., Uvarov, J., Varshalovich, D.A., Wunderer, C.B., Wrochna, G., Zabrodskij, A.G., Zeleny, L.M.: Gamma Ray Burst investigation via Polarimetry and Spectroscopy (GRIPS). Exp. Astron. **23**, 91–120 (2009). doi:10.1007/s10686-008-9102-0

Heck, P.R., Marhas, K.K., Hoppe, P., Gallino, R., Baur, H., Wieler, R.: Presolar He and Ne isotopes in single circumstellar SiC grains. ApJ **656**, 1208–1222 (2007). doi:10.1086/510478

Heck, P.R., Pellin, M.J., Davis, A.M., MArtin, I., Renaud, L., Benbalagh, R., Isheim, D., Seidman, D.N., Hiller, J., Stephan, T., Lewis, R.S., Savina, M.R., Mane, A., Elam, J., Stadermann, F.J., Zhao, X., Daulton, T.L., Amari, S.: Atom-probe tomographic analyses of presolar silicon carbide grains and meteoritic nanodiamonds – first results on silicon carbide. In: Lunar and Planetary Institute Science Conference Abstracts, vol 41, p. Abstract #2112. Houston (2010)

Kanbach, G., Andritschke, R., Bloser, P.F., Schopper, F., Schoenfelder, V., Zoglauer, A.: Concept study for the next generation medium-energy gamma-ray astronomy mission: MEGA. In: Truemper, J.E., Tananbaum, H.D. (eds.) Society of Photo-Optical Instrumentation Engineers (SPIE) Conference Series, Presented at the Society of Photo-Optical Instrumentation Engineers (SPIE) Conference, 24–28 August 2002, Waikoloa, Hawaii, USA, vol. 4851, pp. 1209–1220 (2003). doi:10.1117/12.461348

Kanbach, G., Bertsch, D.L., Favale, A., Fichtel, C.E., Hartman, R.C., Hofstadter, R., Hughes, E.B., Hunter, S.D., Hughlock, B.W., Kniffen, D.A., Lin, Y.C., Mayer-Hasselwander, H.A., Nolan, P.L., Pinkau, K., Rothermel, H., Schneid, E., Sommer, M., Thompson, D.J.: The project EGRET (Energetic Gamma Ray Experiment Telescope) on NASA's Gamma-Ray Observatory GRO. Space Sci. Rev. **49**, 69–84 (1989). doi:10.1007/BF00173744

Kita, N.T., Ushikubo, T., Fournelle, J., Knight, K.B., Mendybaev, R.A., Davis, A.M., Richter, F.M.: Internal isochron of CAI using high precision SIMS Mg isotope analyses. Geochim. Cosmochim. Acta Suppl. **72**, A 477 (2008)

Knödlseder, J.: GRI: The gamma-ray imager mission. Adv. Space Res. **40**, 1263–1267 (2007). doi:10.1016/j.asr.2007.07.036, 0705.1754

Lee, D., Halliday, A.N.: Hafnium-tungsten chronometry and the timing of terrestrial core formation. Nature **378**, 771–774 (1995). doi:10.1038/378771a0

Lee, T., Papanastassiou, D.A., Wasserburg, G.J.: Demonstration of Mg-26 excess in Allende and evidence for Al-26. Geophys. Res. Lett. **3**, 41–44 (1976). doi:10.1029/GL003i001p00041

Longair, M.S.: High Energy Astrophysics. Vol. 1: Particles, Photons and Their Detection, pp. 436. Cambridge University Press, Cambridge, UK. ISBN 0521387736 (Mar 1992)

McKeegan, K.D., Kallio, A.P., Heber, V., Jarzebinski, G., Mao, P.H., Coath, C.D., Kunihiro, T., Wiens, R., Allton, J., Burnett, D.S.: Oxygen isotopes in a genesis concentrator sample. In: Lunar and Planetary Institute Science Conference Abstracts, vol. 40, Abstract 2494. Houston (2009)

Nagashima, K., Krot, A.N., Yurimoto, H.: Stardust silicates from primitive meteorites. Nature **428**, 921–924 (2004). doi:10.1038/nature02510

NCT Collaboration, Boggs, S., Chang, Y.: The Nuclear Compton Telescope (NCT): scientific goals and expected sensitivity. Adv. Space Res. **40**, 1281–1287 (2007). doi:10.1016/j.asr.2007.07.012

Nguyen, A.N., Zinner, E.: Discovery of ancient silicate stardust in a meteorite. Science **303**, 1496–1499 (2004). doi:10.1126/science.1094389

Nguyen, A.N., Nittler, L.R., Stadermann, F.J., Stroud, R.M., Alexander, C.M.O'D.: Coordinated analyses of presolar grains in the Allan Hills 77307 and Queen Elizabeth Range 99177 meteorites. ApJ, in press (2010)

Nicolussi, G.K., Davis, A.M., Pellin, M.J., Lewis, R.S., Clayton, R.N., Amari, S.: s-process zirconium in presolar silicon carbide grains. Science **277**, 1281–1283 (1997). doi:10.1126/science.277.5330.1281

Nittler, L.R., Hoppe, P., Stroud, R.M.: Elemental heterogeneity in an isotopically homogeneous SiC aggregate from a supernova. In: Lunar and Planetary Institute Science Conference Abstracts, vol. 38, Abstract 2321. Houston (2007)

Nittler, L.R., Alexander, C.M. O'D., Gao, X., Walker, R.M., Zinner, E.K.: Interstellar oxide grains from the Tieschitz ordinary chondrite. Nature **370**, 443–446 (1994). doi:10.1038/370443a0

Rauch, B.F., Link, J.T., Lodders, K., Israel, M.H., Barbier, L.M., Binns, W.R., Christian, E.R., Cummings, J.R., de Nolfo, G.A., Geier, S., Mewaldt, R.A., Mitchell, J.W., Schindler, S.M., Scott, L.M., Stone, E.C., Streitmatter, R.E., Waddington, C.J., Wiedenbeck, M.E.: Cosmic ray origin in OB associations and preferential acceleration of refractory elements: evidence from abundances of elements 26Fe through 34Se. ApJ **697**, 2083–2088 (2009). doi:10.1088/0004-637X/697/2/2083, 0906.2021

Savina, M.R., Davis, A.M., Tripa, C.E., Pellin, M.J., Gallino, R., Lewis, R.S., Amari, S.: Extinct technetium in silicon carbide stardust grains: implications for stellar nucleosynthesis. Science **303**, 649–652 (2004). doi:10.1126/science.3030649

Schoenfelder, V., Aarts, H., Bennett, K., de Boer, H., Clear, J., Collmar, W., Connors, A., Deerenberg, A., Diehl, R., von Dordrecht, A., den Herder, J.W., Hermsen, W., Kippen, M., Kuiper, L., Lichti, G., Lockwood, J., Macri, J., McConnell, M., Morris, D., Much, R., Ryan, J., Simpson, G., Snelling, M., Stacy, G., Steinle, H., Strong, A., Swanenburg, B.N., Taylor, B., de Vries, C., Winkler, C.: Instrument description and performance of the imaging gamma-ray telescope COMPTEL aboard the compton gamma-ray observatory. ApJS **86**, 657–692 (1993). doi:10.1086/191794

Stone, E.C., Cohen, C.M.S., Cook, W.R., Cummings, A.C., Gauld, B., Kecman, B., Leske, R.A., Mewaldt, R.A., Thayer, M.R., Dougherty, B.L., Grumm, R.L., Milliken, B.D., Radocinski, R.G., Wiedenbeck, M.E., Christian, E.R., Shuman, S., Trexel, H., von Rosenvinge, T.T., Binns, W.R., Crary, D.J., Dowkontt, P., Epstein, J., Hink, P.L., Klarmann, J., Lijowski, M., Olevitch, M.A.: The cosmic-ray isotope spectrometer for the advanced composition explorer. Space Sci. Rev. **86**, 285–356 (1998). doi:10.1023/A:1005075813033

Takahashi, T., Kelley, R., Mitsuda, K., Kunieda, H., Petre, R., White, N., Dotani, T., Fujimoto, R., Fukazawa, Y., Hayashida, K., Ishida, M., Ishisaki, Y., Kokubun, M., Makishima, K., Koyama, K., Madejski, G.M., Mori, K., Mushotzky, R., Nakazawa, K., Ogasaka, Y., Ohashi, T., Ozaki, M., Tajima, H., Tashiro, M., Terada, Y., Tsunemi, H., Tsuru, T.G., Ueda, Y., Yamasaki, N., Watanabe, S., The ASTRO-H Team: The ASTRO-H mission. In: Kawai, N., Mihara, T., Kohama, M., Suzuki, M. (eds.) Astrophysics with All-Sky X-Ray Observations, Proceedings of the RIKEN Symposium, 10–12 June, 2008. RIKEN, and JAXA Suzuki Umetaro Hall, RIKEN Wako, Saitama, Japan. pp. 356–361 (2009)

Takeda, M., Sakaki, N., Honda, K., Chikawa, M., Fukushima, M., Hayashida, N., Inoue, N., Kadota, K., Kakimoto, F., Kamata, K., Kawaguchi, S., Kawakami, S., Kawasaki, Y., Kawasumi, N., Mahrous, A.M., Mase, K., Mizobuchi, S., Morizane, Y., Nagano, M., Ohoka, H., Osone, S., Sasaki, M., Sasano, M., Shimizu, H.M., Shinozaki, K., Teshima, M., Torii, R., Tsushima, I., Uchihori, Y., Yamamoto, T., Yoshida, S., Yoshii, H.: Energy determination in the Akeno Giant Air Shower Array experiment. Astropart. Phys. **19**, 447–462 (2003). doi:10.1016/S0927-6505(02)00243-8, arXiv:astroph/0209422

The Pierre Auger Collaboration, Abraham, J., Abreu, P., Aglietta, M., Aguirre, C., Allard, D., Allekotte, I., Allen, J., Allison, P., Alvarez, C., Alvarez-Muñiz, J., Ambrosio, M., Anchordoqui, L., Andringa, S., Anzalone, A., Aramo, C., Argirò, S., Arisaka, K., Armengaud, E., Arneodo, F., et al.: Correlation of the highest-energy cosmic rays with nearby extragalactic objects. Science **318**, 938–943 (2007). doi:10.1126/science.1151124, 0711.2256

Vedrenne, G., Roques, J., Schönfelder, V., Mandrou, P., Lichti, G.G., von Kienlin, A., Cordier, B., Schanne, S., Knödlseder, J., Skinner, G., Jean, P., Sanchez, F., Caraveo, P., Teegarden, B., von Ballmoos, P., Bouchet, L., Paul, P., Matteson, J., Boggs, S., Wunderer, C., Leleux, P., Weidenspointner, G., Durouchoux, P., Diehl, R., Strong, A., Cassé, M., Clair, M.A., André, Y.: SPI: the spectrometer aboard INTEGRAL. A& A **411**, L63–L70 (2003). doi:10.1051/0004-6361:20031482

Vestrand, W.T., Share, G.H., Murphy, R.J., Forrest, D.J., Rieger, E., Chupp, E.L., Kanbach, G.: The solar maximum mission atlas of gamma-ray flares. ApJS **120**, 409–467 (1999). doi:10.1086/313180

Villeneuve, J., Chaussidon, M., Libourel, G.: Homogeneous distribution of 26Al in the solar system from the Mg isotopic composition of chondrules. Science **325**, 985–988 (2009). doi:10.1126/science.1173907

Weidenspointner, G., Harris, M.J., Sturner, S., Teegarden, B.J., Ferguson, C.: MGGPOD: a Monte Carlo suite for modeling instrumental line and continuum backgrounds in gamma-ray astronomy. ApJS **156**, 69–91 (2005). doi:10.1086/425577, arXiv:astro-ph/0408399

Young, E.D., Simon, J.I., Galy, A., Russell, S.S., Tonui, E., Lovera, O.: Supra-canonical 26Al/27Al and the residence time of CAIs in the solar protoplanetary disk. Science **308**, 223–227 (2005). doi:10.1126/science.1108140

Zinner, E., Nittler, L.R., Hoppe, P., Gallino, R., Straniero, O., Alexander, C.M.O'D.: Oxygen, magnesium and chromium isotopic ratios of presolar spinel grains. GeoChim. Cosmochim. Acta **69**, 4149–4165 (2005). doi:10.1016/j.gca.2005.03.050

Zych, A.D., O'Neill, T.J., Bhattacharya, D., Trojanowski, C., Wijeratne, S., Teichegaeber, C., Mathews, M.: TIGRE prototype gamma-ray balloon instrument. In: Society of Photo-Optical Instrumentation Engineers (SPIE) Conference Series, Presented at the Society of Photo-Optical Instrumentation Engineers (SPIE) Conference, "Hard X-Ray and Gamma-Ray Detector Physics VIII" on the "Optics & Photonics 2006" SPIE Symposium, August 15–17, 2006, San Diego, CA, vol. 6319 (2006). doi:10.1117/12.683838

# Part V
# Epilogue

The previous Chapters presented in detail how astrophysical studies of cosmic radioactivities proceed via measurements and theory development. Now we take a moment to offer an outlook and perspective of this field. As the book approaches its end, we append a few items, such as a glossary of a few key terms and a history of selected milestones that marked progress (from the Editors point of view), which the student may find useful when reviewing the material gathered in this book.

# Chapter 11
# Perspectives

**R. Diehl, D.H. Hartmann, and N. Prantzos**

It is more than a century now since the phenomenon of radioactivity has been discovered. As noticed by D. D. Clayton in his introduction to Chap. 2, E. Rutherford was the first to point out that radioactive elements were produced in the Cosmos: In 1929 he had suggested that the long lived actinides Uranium and Thorium were synthesised in the Sun about 100 Million years ago and were somehow transported to Earth. It took another 20 years before another scientist suggested that radioactivity occurs *now* in the Universe and has observable effects: in 1950, L. Borst noticed that the exponential decrease of the luminosity of Type I supernovae (with a characteristic timescale of 55 days) is a strong indication for a radioactive source. But he misidentified that source with $^{7}$Be-decay. Two years later, P. Merrill identified absorption lines of short-lived Technetium in the spectra of S-stars, a clear indication that radioactivity (and nucleosynthesis) occurs in the Universe today.

Since then, the phenomenon of radioactivity evolved to a major astrophysical tool, allowing astronomers to

- determine the ages of stars and of our Galaxy
- probe the physical processes occuring deep inside supernova envelopes
- infer the current rate of nucleosynthesis in the Milky Way
- understand (or, at least, constrain) the timescales of the acceleration of Galactic cosmic rays and of their propagation through the interstellar medium
- infer the physical conditions inside the stars producing some of the extinct radioactivities found in meteoritic dust grains
- constrain the environment of the proto-solar nebula and the activity of the young Sun
- (and many more, as described in this book)

R. Diehl (✉)
Max Planck Institut für extraterrestrische Physik, 85748 Garching, Germany

D.H. Hartmann
Clemson University, Clemson SC 29634-0978, USA

N. Prantzos
Institut d'Astrophysique, 75104 Paris, France

Diehl, R. et al.: *Perspectives*. Lect. Notes Phys. **812**, 519–523 (2011)
DOI 10.1007/978-3-642-12698-7_11 

Radioactivity is the decay of unstable nuclei, therefore it is intimately related to nuclear astrophysics and stellar nucleosynthesis, i.e. the production of such nuclei in various cosmic environments. Obviously, the steady progress in our understanding of stellar structure, evolution and explosions, and of nuclear reactions in stellar conditions, was necessary in order to turn radioactivity into a powerful astrophysical tool.

The current state of the art in modelling nucleosynthesis in single stars of low and high mass, as well as in stellar explosions occuring in binary systems, is presented in Chaps. 3, 4 and 5 of this book, respectively. Present day models are much more sophisticated (including, e.g., mass loss and rotation) than their earlier counterparts and their results are, presumably, closer to reality. Despite their sophistication, however, such models can hardly be considered as *realistic*, since key ingredients, such as convection, mass loss and rotation, can only be treated in a parametrised way at present; this is also the case for models of both core collapse and thermonuclear supernovae. Moreover, the exploration of geometrical (3D) effects has only recently started and will undoubtely bring some surprises. Preliminary results are also currently obtained on the effects of the stellar magnetic field. In one case, at least, i.e. positron escape from SNIa, such effects are known to play a critical role.

The book's Chap. 6 briefly describes the exciting aspects of *our* Sun and the solar system, which have been connected to measurements and insights on radioactive isotopes all along. The Sun has recently taught us to retain modesty and critical views: Neutrino measurements were in striking conflict with solar nuclear reaction models, and could only recently be reconciled again through the now-established phenomena of neutrino flavors and neutrinos with non-zero mass. Refined (3D) models of the solar photosphere have significantly reduced *metallicity* in our cosmic standard for elemental abundances. Both these recent insights tell us that it is prudent scientific endeavor to question and consolidate seemingly-obvious details, before taking giant leaps and declare physics and astrophysics *understood*. The early history of the solar system is now again debated with great enthusiasm, making use of more precise observations and theoretical models of radioactivity signatures in our solar system bodies, the only place where we can study cosmic materials directly. We open the curtain to cosmic chemical evolution with measurements of $^{60}$Fe on Earth, and with $^{26}$Al from nearby stars helping us to understand the local interstellar environment. Detail matters.

This book's Chap. 7 then is a mixture of complex astrophysical problems and observational constraints which come into play once single objects are put into a greater context of a galaxy. The interstellar medium connects stellar sources across cosmic time intervals, in its transport properties for kinetic energy and matter. Our models for these processes are first-order and simple, still. They help us to explain the coarse aspects of galaxy evolution. But again, more detailed astrophysical understanding will be required for a *realistic* model of how stars and supernovae feed back their newly-produced isotopes and their violence from stellar winds and explosions into next generations of stars to form, and to carry on the cosmic cycle of matter. All great themes of current astrophysics necessarily are involved here: Nucleosynthesis yields for entire populations of source types, their occurrence rates over a galaxy's evolution, the feedback, mixing, and transport processes across the many degrees of freedom for energy. Observing electromagnetic radiation from cosmic objects is a great tool to study distant physical processes. Its limitations become obvious in view of the complexities of a galaxy's evolution.

The book's Chaps. 8, 9 and 10 provide a glimpse of the variety of *tools* which are involved in today's astrophysical work; we concentrate here on tools which most-directly relate to cosmic radioactivities. Nuclear physics made great strides to establish the concepts of nature as atomic nuclei are held together and interact; in recent years, much attention turned towards nuclear-structure details which matter in cosmic environments, and involves experiments with radioactive beams or targets. This field evolves back to focus more towards fundamental science, away from the power-generation and weapons applications of earlier years, as demonstrated in Chap. 9. Supernovae still challenge even most-advanced modern computing facilities, from their huge dynamic ranges in space and time domains. Chapter 8 gives a snapshot of how physics and mathematics ideas are employed to obtain computer simulations which are approaching reality, within such technical limitations; this is a prominent example of a study of *complex systems*. The instruments to obtain measurements from cosmic radioactivities then are highlighted in Chap. 10. We face very different categories of equipment, from space telescopes for penetrating gamma-rays through cosmic-ray probes and to sophisticated laboratory mass spectroscopy for minute samples of cosmic material extracted from meteorites. Each of these experimental fields is in the hands of small groups of experts, only a handful of laboratories world-wide working in each of these fields. Advances are very sensitive to specific conditions under which such small laboratories operate in their countries. Neither large science communities nor large industrial applications fuel technological evolutions here, rather the passion of experimental physicists is the driver of progress.

While the theoretical insights of D.D. Clayton laid the foundations of *astronomy with radioactivities* in the 1960s, the field of $\gamma$-ray line astronomy was mainly driven by observations. The discovery of the 1.8 MeV $\gamma$-ray line of $^{26}$Al came somewhat as a surprise (as discussed in Chap. 2), and the one of the 511 keV line (the first $\gamma$-ray line originating from outside the solar system that was ever detected) was also unexpected. Equally unexpected was the detection of the $^{56}$Co lines from SN1987A about 6 months earlier than predicted from spherically symmetric models of the supernova explosion. It was the improved angular resolution of the COMPTEL imaging gamma-ray telescope within a large field of view and the multi-year CGRO mission that made possible the identification of massive stars as major $^{26}$Al sources in the 1990s. Similarly, the improved spectral capabilities of SPI/INTREGRAL (and its energy range, including the few-hundred keV range) allowed astronomers to perform the first reliable all-sky map of the Galactic 511 keV emission, which revealed and consolidated its surprisingly large intensity ratio of the Galaxy's bulge to its disk – the current puzzle both for cosmic-ray propagation and for Galactic positron sources.

These discoveries, but also the detections of $^{44}$Ti in Cas A and of $^{60}$Fe in the Galaxy's interstellar medium, boosted an intense activity both in theoretical astrophysics and in nuclear laboratory experiments in the past three decades, generating hundreds of papers on various aspects of cosmic radioactivities. Also the cosmic environment of the Sun and the important diagnostics of radioactivities received a boost in the past decade, from more precise measurements, and not least from the discovery of $^{60}$Fe radioactivity on Earth.

Compared to other fields in astrophysics, the ratio of theoretical implications to observational data has been extremely high in the case of $\gamma$-ray line astronomy. It should be noticed that some of these implications were totally *unforseen* by the pioneers of the field. For instance, long-lived radioactivities, such as $^{26}$Al, can be used not only to probe the interior physics of their stellar sources, but also the physics of the Galaxy at large: star formation rate, locations of high-mass star forming sites, distribution of supernova ejecta out of the Galactic plane, etc. Similarly, the properties of positron annihilation emission can help to probe the physics of cosmic ray propagation and the interstellar medium in a new way, or the configuration of the galactic magnetic field (see Chap. 7).

One might think then that such a prolific field can only have a bright future. Laboratory equipment for the analysis of meteoritic inclusions and their isotopic abundances proceeds to ever smaller grains and precision. However, the scarcity of the astronomical data from gamma-ray studies tells a different story. Ideas and concepts promise to dig deeper into the nuclear universe, yet by modest increments, compared to other disciplines of astronomy. It will take a considerable effort to increase the number of sources where our theories can be challenged and expanded; even an increase by a modest factor of a few will require a multi-national space mission. No mission dedicated to $\gamma$-ray line astronomy is on the horizon at present. Current excitement focuses on questions of cosmology and dark components of the universe, and on breaking grounds for new astronomy through gravitational waves and neutrinos. So, proposed gamma-ray telescope projects may stimulate more excitement and support in new space programs only beyond the next decade. At least, interesting new data are imminent from the NuSTAR X-ray telescope (planned for a 2012 launch) for the $^{44}$Ti lines at 68 and 78 keV, for Cas A and other candidate young SNR in the Galaxy and for SN1987A. Other isotopic information may become available from superb resolutions now reaching into isotopic line shifts for molecular lines in the radio regime, and even for atomic lines in the optical. X-ray lines may help to constrain elemental abundances in hot gas of the intergalactic medium, in addition to hot ISM in SNR. It is clear that a considerable effort will be required in all those fields before robust theoretical predictions become available on the yields of various radioactivities (verified most-directly by the intensities of the corresponding $\gamma$-ray lines from radioactive decay). Valuable new insights on cosmic radioactivities may arise from spin-offs from other fields, such as high-resolution spectroscopy resolving isotopic information, or astro-particle advances on cosmic ray details, or solid-state research connected to materials science.

Similarly, significant effort is required in theoretical and associated fields of physics: Refined descriptions of chemical evolution accounting for various galaxy components, a better understanding of nuclear structure derived from nuclear reaction experiments and theories, and more realistic models of stellar explosions.

All these elements are needed in order to improve our *view and theoretical understanding* of the cosmic sources of new isotopes which have been detected up to now. This book aims to help students of astrophysics to better understand the role of cosmic radioactivities in relation to their specific interests. It will hopefully also help build up resources and excitement to further advance our understanding of the origins of the cosmic elements – one of the *big questions* agreed upon in the science community.

# Appendix A
# Annotations on Chemical Evolution

## A.1 Chemical Evolution: Analytical Descriptions

D.D. Clayton
Clemson University, Clemson SC 29634-0978, USA

### A.1.1 The Clayton Standard Model

One of the most useful families of analytic solutions of galactic chemical evolution with radioactivity is defined by choosing the integrable function $d\theta/dt$ used to integrate Eq. (2.18) to be

$$\frac{d\theta}{dt} = f(t)/M_G(t) = \frac{k}{(t + \Delta)} \tag{A.1}$$

which can be integrated easily to obtain $\theta(t)$. Constants $k$ and $\Delta$ are simply two parameters of this family of solutions that allow form-fitting of the ratio $f(t)/M_G(t)$. The constant $k$ is simply an integer. The parameter $\Delta$ has dimensions of time and prevents the ratio $k/(t+\Delta)$ from extreme temporal behavior near $t = 0$. Physical meaning can be assigned by examining the inverse of Eq. (A.1). The ratio $M_G(t)/f(t) = \tau_f$ equals the time that would be required at the current rate of mass infall $f(t)$ onto the disk for such infall to replenish the current mass of the disk. That replenishment time is equal to $(t+\Delta)/k$. Larger values of the integer $k$ correspond to higher rates of galactic infall and shorter replenishment times for the disk mass. The units of time can be chosen for convenience. Instead of seconds, one may prefer Gyr, in which case the infall rate $f(t)$ is mass per Gyr, and $t$ and $\Delta$ are expressed in Gyr. The units of mass can conveniently be expressed as multiples of an initial galactic gas mass $M_G(0)$ when the model is first applied. All functions of galactic chemical evolution can be analytically expressed with this standard model (Clayton, 1985). Each quantity has its own time dependence, one which can be chosen to mimic what our galaxy is believed to have done. To express this succinctly, define a new timelike variable $x = (t + \Delta)/\Delta$ which measures time in units of the parameter $\Delta$. Then

Diehl, R. et al., (eds.): *Appendices*. Lect. Notes Phys. **812**, 525–558 (2011)
DOI 10.1007/978-3-642-12698-7 

$$\text{Infall}: f(t) = \frac{(kM_G(0)}{\Delta}\mathrm{x}^{k-1}\, e^{-\omega t} \tag{A.2}$$

$$\text{Gas mass}: \ M_G(t) = M_G(0)\mathrm{x}^k\, e^{-\omega t} \tag{A.3}$$

$$\text{ISM Mass fraction } Z: \ Z - Z_0 = \frac{y\,\omega\Delta}{(k+1)}(x - x^{-\mathrm{k}}) \tag{A.4}$$

$$\text{Radioactive}: Z_\lambda = y\,\omega\, e^{-\lambda t}\, \mathrm{x}^{-k} I_k(t,\lambda) \tag{A.5}$$

The polynomials $I_k(t, \lambda)$ are given by Clayton (1985, 1988). The first four are, with $x = (t + \Delta)/\Delta$ and $d = (\Delta\,\lambda)^{-1}$ :

$$k = 0 : \lambda I_0 = e^{\lambda t} - 1 \tag{A.6}$$

$$k = 1 : \lambda I_1 = e^{\lambda t}(x - d) - (1 - d) \tag{A.7}$$

$$k = 2 : \lambda I_2 = e^{\lambda t}(x^2 - 2dx + 2d^2) - (1 - 2d + 2d^2) \tag{A.8}$$

$$k = 3 : \lambda I_3 = e^{\lambda t}(x^3 - 3dx^2 + 6d^2 x - 6d^3) - (1 - 3d + 6d^2 - 6d^3) \tag{A.9}$$

Graphs of all functions are displayed by Clayton (1985), who first found this family of solutions now called the Clayton Standard Model. It is a standard for demonstrating the behavior of radioactivity within galactic chemical evolution rather than for the physical issues of galactic evolution. Researchers are advised to confirm that these functions do indeed satisfy the differential equations of galactic chemical evolution and also to generate graphs of the solutions. Its usefulness lies in providing exact solutions of the equations of galactic chemical evolution with star formation rate proportional to mass of interstellar gas. The shapes of the functions $f(t)$ and $M_G(t)$ can be fitted to astrophysically motivated shapes for them by choosing the parameters $k$, $\Delta$ and the initial values $Z_0$ and $M_G(0)$. These analytic solutions present more reliable evaluations of the age of the elements by nuclear cosmochronology than can be inferred from parameterized integrations of Eq. (2.13).

Chronological contraints on the age of the galaxy can best be discussed in terms of *the remainder* for each radioactive isotope. The remainder $r_{\lambda(t)}$ is defined as the ratio of the abundance at time $t$ of a radioactive isotope to the value it would have had were it stable instead of radioactive. The remainder $r_{\lambda(t)}$ is the analogue of the function $\exp -\lambda t$ for an ensemble of radioactive nuclei. Clayton (1985) showed that if initial metallicity is neglected, the remainder in the Clayton Standard Model is $r(t) = \exp(-\lambda t)[(k+1)/\Delta][x^{-}kI_k(\text{lambda}, t)/(x - x^{-}k)]$. Clayton (1985, 1988) display graphs of the remainders for $^{235}$U, $^{238}$U, $^{232}$Th and $^{187}$Re and how those remainders limit the age of the galaxy.

Understanding this family of solutions, which are easier to master than are the general equations, will make more detailed models of galactic chemical evolution easier to grasp as a result. The analytic models yield simple expressions for the birthdate spectrum for solar nuclei within each model. The birthdate spectrum of solar nuclei is the birthdate spectrum of nuclei existing in the ISM when the sun formed. For this standard model the rate of growth of stable solar atoms (those actually existing in the ISM gas when the sun formed) can be shown to have been

$$\frac{dB_\odot}{dt} = y\,\omega \left[\frac{(t+\Delta)}{(t_\odot+\Delta)}\right]^k \tag{A.10}$$

where the birthdates of solar nuclei range from the beginnings of nucleosynthesis at $t=0$ to its final contributions at the time $t=t_\odot$ when the sun formed from the ISM. Eq. (A.10) shows that the historical rate of increase of the solar concentration of element $Z$ among the actual solar (think "red") atoms was only $y\omega[\Delta/(t_\odot+\Delta)]^k$ at $t=0$, but rose to a full $y\,\omega$ at $t=t_\odot$. This increase by a factor $[(t_\odot+\Delta)/\Delta]^k$ reflects a bias against old birthdates among the solar atoms precisely because atoms born early tend to become locked up within stars and are not available for incorporation into the solar system. Low-$Z$ infall $f(t)$ exacerbates the effect. That bias can be rather large. Just as an example, if in units Gigayears (Gy), $t_\odot = 10$, $\Delta = 2$, and $k = 2$, the increasing bias toward recent nucleosynthesis in the spectrum of solar nuclei is $[(t_\odot+\Delta)/\Delta]^k = 6^2 = 36$. The earliest galactic nucleosynthesis is depleted with respect to that just prior to solar birth by this factor 36! All studies of nuclear cosmochronology must digest this big effect. For the "closed box" model $k=0$ so that $dB_\odot/dt = y\,\omega$ for all ages; that is, the birthrate spectrum is flat despite the exponential decline of the galactic nucleosynthesis rate.

The birthrate spectrum of Eq. (A.10) uses different normalization than was used in the discussion surrounding Eqs. (2.11) and (2.13). Integration of Eq. (A.10) over all times gives the ISM *concentration* (grams of that element per gram of solar matter) rather than the numbers of atoms. Contrast with the discussion preceeding Eq. (2.17). These alternative normalizations of the birthrate spectrum need be no obstacle to the reader, because the two concepts are linearly related.

A big advantage of the standard model is that within that family of solutions it is possible to evaluate the cosmochronometers. Doing so is more realistic than using simple parameterized assumptions, a procedure common in the historic literature. By plotting the ratio of Eqs. (A.5) to (A.4), one can construct the ratio of a radioac tive abundance to that of a stable one in the average ISM. Use the concept *remainder* in doing this. The remainder for any radioactive species is defined as its abundance at the time of solar birth divided by the abundance it would have had were it stable. The remainder ratio is independent of the yield $y$ for that nucleus. It depends only on the galactic model. Examination of such time-dependent abundance growth curves reveals that nucleosynthesis began between 12 and 18 Gy ago. The major uncertainty, which is still rather large, in this result appears to be the degree to which the beta decay of $^{187}$Re is speeded within stars (Yokoi et al., 1983). Such thermal increase of the $^{187}$Re decay rate helps explain the large amount of cosmoradiogenic $^{187}$Os in solar abundances without requiring an extremely old galaxy. The reader can best consult the papers cited above as well as a review (Cowan et al., 1991). Huss et al. (2009) provide a modern list of known extinct radioactivity within the framework of the standard model described above. What these considerations establish is that nuclear cosmochronology depends heavily upon the history of galactic chemical evolution. Attempts to find a radioactive age for our elements without such interplay are doomed. Many historical attempts to do so have been invalidated by these analytic models.

When one applies the standard model to our own galaxy, one must use values for the galactic parameters that approximate the Milky Way. In particular, astronomers have shown that today's interstellar medium near the sun has a mass $M_G(t)$ that is approximately 0.1 times the total local mass density of stars. If the galactic age today is taken to be $A_G$=12 Gy, Table 1, taken from Huss et al. (2009), shows for the specific choice $\Delta$=0.1 Gy the values of $\omega$ and of $k$ that will yield that observed gas fraction locally today. Similar constraints exist for each value for the parameter $\Delta$. Other astronomical observations offer related insights into the approximate values of the parameters that simulate our galaxy.

Another result of great importance for extinct radioactivity can be seen best within the standard family of models. An especially useful limit of Eq. (A.5) can be obtained for short-lived radioactivity, whenever $\lambda$ is small, by expanding Eq. (A.5) for small $\lambda$. The concentration of a radioactive nucleus is

$$Z_\lambda = y\,\omega\left(\lambda + \frac{k}{(t+\Delta)}\right)^{-1} \tag{A.11}$$

whereas a stable nucleus obeys Eq. (A.4). If the parameters are such that $x \gg x^{-k}$, as is almost always the case at the time $t = t_\odot$, the ratio of an extinct radioactive species to a stable species in the ISM is larger by the factor $(k+1)$ than one might estimate without taking galactic chemical evolution into account. As an example, consider the interstellar ratio of radioactive $^{26}$Al to stable $^{27}$Al. The formula

$$\frac{Z(^{26}\mathrm{Al})}{Z(^{27}\mathrm{Al})} = \frac{y(26)}{y(27)}\,(k+1)\,\frac{\tau_{26}}{t_\odot} \tag{A.12}$$

can be derived from such a ratio (Clayton et al., 1993). This result is larger by the factor $(k+1)$ than an estimate using only Eq. (2.10), which had traditionally been used in oversimplified discussions of the steady-state amount of short-lived interstellar radioactivity. The extra factor $(k+1)$ became important when a large mass of interstellar $^{26}$Al was detected by gamma-ray astronomy.

These analytic models, like all homogeneous models of galactic chemical evolution, can at best yield only an average expectation for the ISM. The true ISM is inhomogeneous in space and the nucleosynthesis rate is sporadic in time rather than maintaining its steady average. Nonetheless, the models are important in laying out properties of radioactive abundances that would be true for a rapidly mixed ISM and a smooth rate of galactic nucleosynthesis. One interprets the observations against this backdrop of expectation.

### A.1.2 The Closed-Box Model

Perhaps the earliest serious attempt at analytic results in galactic evolution was the idea of a disk of constant mass. Such a model would have no mass-increasing infall ($f = 0$) and so was called a *closed box* (Schmidt, 1963). Many nice conclusions can be drawn for such a model; but unfortunately, the evidence is that the galactic disk has grown in mass, not only by infall of residual matter but also by mergers of dwarf satellite companions of the Milky Way with it. Infall $f(t)$ includes such galactic cannibalism; but if such dwarf galaxies are not low in metallicity, they pose a serious complication. The infall rate has strong effect on nuclear cosmochronology, as Tinsley (1977) noted early. Clayton's standard linear model (above) sheds illumination on closed-box models, even when the star formation rate is not proportional to the mass of disk gas. From Eqs. (A.1) to (A.2) one sees that $f = 0$ arises from the choice $k = 0$. From Eq. (A.3) the mass of gas then declines exponentially according to $M_G(t) = M_G(0)e^{-\omega t}$; and so therefore also does the star formation rate in linear models (see Eq. (2.16) and following).

For the closed-box model the metallicity grows as $Z = y\,\omega\,t$ (Eq. (A.4) with $k = 0$). The metallicity growth is linear in time, a situation that has erroneously been called "constant nucleosynthesis". The galactic nucleosynthesis rate in such models actually declines exponentially in proportion to the gas mass; but that exponential rate of galactic nucleosynthesis is diluted by being admixed into an exponentially declining mass of ISM gas. The result is that $Z$ increases by equal amounts in equal time intervals. This model therefore clarifies as simply as possible the distinction between "nucleosynthesis rate" and "birthdate spectrum". Indeed, Eq. (A.10) confirms that the birthdate spectrum is $dB/dt = y\,\omega$ in that model with $k = 0$.

It is recommended that the student master the closed-box linear model as the clearest example of the distinction between *nucleosynthesis rate* and *birthdate spectrum*. Once that distinction is grasped, more complicated theoretical treatments will be understood more easily.

## A.2 Complexities in Descriptions of Chemical Evolution

R. Diehl
MPE Garching, Germany

D.H. Hartmann
Clemson University, USA, and

N. Prantzos
Institut d'Astrophysique de Paris, France

### *A.2.1 Approximations Related to Sources*

Here we add further comments to illustrate the complex path from above analytical model to a realistic description of chemical evolution within a galaxy.

Chemical evolution frameworks are used to describe specific elemental or isotopic abundances in our Galaxy in terms of the action of stars and gas flows, and compare these with measurements (e.g. from spectroscopy of stellar photospheres from stars of different ages). This has been done over many years (e.g. Cayrel et al., 2004; Barklem et al., 2005), successfully confirming our general understanding of chemical evolution. As such, this framework is not a *theory*, and the variety of observational constraints of the description are mostly *indirect*. We have star counts, and need to derive distances, ages, masses. We have abundances, and need to derive the time of their sampling, their biases, plus distances, and how representative they can be. The evolution of elemental abundances over Galactic history is taken as the observational dataset to be represented by the model. Then, stars attributed to different Galactic components (such as bulge and disk and thick versus thin disk; see Sect. 7.2) are probes to see if the overall model treatment is consistent (e.g. Schönrich and Binney, 2009). Interstellar gas abundances have also been used to add constraints; in this case only the *current* abundances are accessible. A few additional comments:

- Since *stellar age* is not accessible to direct measurement, it has been common to use the abundance of a major *metal* as a proxy, based on a relationship between metallicity and age. Plausibly this relation would be linear for an element which is produced with constant yield over the star formation period. Such an *age-metallicity relation* had been proposed for solar neighborhood stars (Edvardsson et al., 1993). It turned out later that this was an artifact of the selection method of equal numbers of stars per logarithmic metallicity range (Andersen et al., 2005) (see also Feltzing et al., 2001). Avoiding this age uncertainty, it has become common to use the Fe abundance relative to Hydrogen as a reference scale for chemical evolution (even if not quite scaling with time). One can test the validity of chemical evolution models by comparing their predicted evolutions of elemental ratios in space and time. For extragalactic regions, Oxygen is often preferred, as a tracer related to age. Since both Fe and O deplete onto dust grains, noble gases have also been used to represent a scale of time.
- Nucleosynthesis from other sources such as Type Ia supernovae and novae are believed to be caused by binary accretion accumulating a threshold mass onto a white dwarf star. Since a white dwarf by itself is an *end product* of stellar evolution, and the threshold mass accumulation will require accretion over time scales on the order of Gy, the *delays* caused by those sources are probably more significant than the time scale of interstellar mixing. Therefore, the non-instantaneous recycling has generally been considered a secondary (or non-) issue for chemical evolution models, considering major unknowns in a galaxy's mass budget over time and star formation history.

- Forming a star is taken as an instantaneous and spontaneous event, occurring at some probability related to the local interstellar gas density. The underlying concept is that a *Jeans mass* is the critical unit which defines when gravitationally-bound interstellar cloud matter becomes unstable to its own gravity and will collapse into a newborn star. We have learned now that not only is the formation of molecular clouds a dynamic process driven by interstellar turbulence, but also that gravitational collapse towards a new star will typically also involve gravitationally-unbound parts of a molecular cloud, as turbulent motions may exceed local gravitational energy and thus drive compression (Clark et al., 2008; Krumholz and McKee, 2005). This implies that there is a causal connection to the turbulent energy in interstellar gas and its star forming efficiency, rather than being determined by gas density alone. Furthermore, we have learned that gravitational collapse under self-gravity of an interstellar cloud which is more massive stan the *Jeans mass* cannot form massive stars exceeding about 20 $M_\odot$, as the radiation emerging from the forming star produces radiation pressure which counteracts and prevents accretion. Here, instabilities in the accretion are initiated by turbulences, and result in a complex mixture of outflow and infall with a net growth in mass of the central star, often accompanied by a second star forming within this unstable accretion disk (Krumholz et al., 2009). The entire process of forming such massive stars extends over times in excess of $10^4$ y, and growth of the mass of the star continues beyond the time of central hydrogen ignition. This refined picture of the onset of stellar evolution for more massive stars is not accounted for in conventional chemical evolution treatments. But its test or validation would first and also demand a proper handling of interstellar mixing within the model, for clear conclusions.

## A.2.2 Feedback and Mixing of Nucleosynthesis Ejecta

In Sect. 7.3 we described the astrophysical issues of mixing nucleosynthesis ejecta on cosmic timescales and different spatial scales. How much do we know about mixing on these various scales?

(a) The planetary-system scale is constrained by observations and theory to $\simeq 10^5$ to $10^6$ years. *Short-lived radioactivities* are excellent clocks to probe any mixing processes within this scale, with radioactive-decay times ranging from 0.15 My ($^{41}$Ca) through My ($^{26}$Al, $^{60}$Fe and 10 My ($^{107}$Pd, $^{182}$Hf) to beyond 100 My ($^{244}$Pu with 115 and $^{146}$Sm with 148 My). The method here is age dating of the formation of solids of different types as found in meteorites, and the variability of radioactive-daughter abundances for a specific class of solid, which then would constrain inhomogeneities within the solar system at different early epochs.

(b) Ejecta of a stellar nucleosynthesis event as being blown from a hot wind or explosion into surrounding interstellar gas has been addressed since the 1970s.

Weaver et al. (1977) presented a calculation of the evolution of a hot spherical cavity into homogeneous interstellar medium including radiative cooling, which includes the key concepts and processes underlying all treatments of ejecta mixing till today. Shull and Saken (1995) mainly address the stages observable as supernova remnants, attempting to provide a model for the astronomical display in electromagnetic radiation. The formation of a shock as high-velocity ejecta meet interstellar gas is the dominant characteristic of models. Instabilities developing at the interface region will be the key to effective mixing with ambient gas in the late phase, the *dilution* phase of such a diffusing stellar remnant. On sub-parsec scales, there are indications from the solar vicinity that dust and gas compositions are decoupled; chemical mixing is unclear at small scales (Slavin and Frisch, 2007; Quitte et al., 2010), and our common assumption of total interstellar medium being the sum of gas and dust abundances may not be realistic.

(c) Beyond diffusion of an individual nucleosynthesis event into its (assumed homogeneous) surroundings, the collective action of such events has received detailed treatment in theories of interstellar turbulence and its drivers (Klessen, 2004). Here a comprehensive hydrodynamical treatment of the above through numerical models is the tool to determine the thermodynamic parameters of interstellar gas in 3 dimensions over My time scales. We have learned that the ISM is rather dynamic and not at all characterized by *equilibria*, although when averaging over sufficiently-large volumes and/or times, one can distinguish more abundant from less abundant state variables and thus can understand this resemblance of *phases* as defined by McKee and Ostriker (1977) (Breitschwerdt, 2004). Inclusion of magnetic fields appears relevant on these spatial scales, especially since much of the volume is filled with hot, ionized plasma. From this we learned that the actual parameter values obtained for densities and temperatures depend on the strength and orientations (i.e. homogeneous versus irregular) of magnetic fields, as obviously does the 3D morphology (e.g. Balsara and Kim, 2005). – Combined with the previous case, these scales of stellar remnants or remnants of stellar groups will determine how rapidly or slow fresh material ejected by stellar sources will be recycled into a next generation of stars. Many theoretical treatments of such *chemical evolution* (see below) adopt a simplifying *instantaneous recycling* approximation, for the lack of better knowledge, although it always has been clear that this is an *unrealistic* assumption (Tinsley, 1975), and may be actually as little *instantaneous* as $10^8$–$10^9$ years (Malinie et al., 1993), if spatial chemical inhomogeneities are attributed to recycling delays. For the case of blow-out into the halo, Spitoni et al. (2009) have found that ejecta blown into halo regions partly fall back to the disk of the galaxy within 0.5–1 kpc of their natal massive-star groups, yet with a delay of between 36 and 133 My, depending on the size of the massive star group and its location in inner or outer regions of a galaxy's disk.

(d) Morphologies within galaxies have been studied by many groups, in an attempt to constrain inhomogeneities and thus mixing (e.g. Elmegreen et al., 2003; Oey and Clarke, 1997). As a general finding, star formation appears to be concentrated in *clumps* with a typical maximum size on the order of a kpc, as a result of the effects of galactic rotation and star forming efficiency, which both do not allow for larger coherent star forming complexes. In the evolution of galaxies, the current consensus it that streams of matter from outside a galaxy play a key role. Depending on how exactly such *mass-addition* to a galaxy will proceed, the conservation of kinetic energy and angular momentum clearly will control how a galaxy's disk will be fed with turbulent energy, and on which scales. Cold gas streams attracted by the dark matter halo may direct gas towards galaxies (Dekel et al., 2009), and gas stream densities are sufficiently high for cooling to be effective and maintaining gas temperatures in the $10^4$–$10^5$ K regime. In this view, infalling gas perturbs the disk of galaxies and incurs star formation, unless the bulge is already massive enough to direct inflow directly towards it. Lacking better knowledge, the star formation rate is assumed to equal the gas inflow rate – it is here where knowledge about interstellar mixing and feedback to the star formation rate and efficiency will be crucial to make such models more realistic. There are also alternate views on the drivers of galaxy evolution. Undebated is a major role of a supermassive black hole. But material inflow in the forms of gas and stars is discussed in merger models, and in the form of hot inflows. Within the Galaxy, it became clear that abundance variations across the Galaxy are more complex than assumed in large-scale views: Measurements of stellar abundances in the Scutum arm region where the long Galactic bar ends show abundances which have a remarkable deviation in $\alpha$/Fe ratios, while otherwise being solar-like (Davies et al., 2009). This suggests that (a) the assumption of a simple overall Galactocentric radial trend appears an oversimplification, as the Galactic bar's influence may be significant and extending from the center to radial distances of about 5 kpc, and (b) azimuthal abundance variations may be significant, due to the influence of the bar and spiral arms. – The inner region of our Galaxy also provides an interesting laboratory for large-scale mixing and the material ending up in star formation. Here, gas motions are peculiar (from CO observations, see Dame et al., 1987), and mixing occurs in a way different from the disk of the Galaxy, where spiral-wave stimulations and interstellar turbulence derived from this dominates on a larger scale (larger than individual associations, say 100 pc). Najarro et al. (2009) find that $\alpha$-elements appear significantly enriched with respect to Fe in the central star clusters of our Galaxy. This could mean that massive-star activity in this central region is largely increased. It could also mean that the seed material for later stellar generations is not the same and is taken from a different enrichment history than typical for the thin disk of the Galaxy, e.g. from bulge gas only. Note that the large-scale trend in the disk runs towards $\alpha$/Fe decrease towards the center (Rolleston et al., 2000; Smartt et al., 2001), while abundance ratios of $\alpha$-elements over Fe are higher in the bulge of the Galaxy (Feltzing and Gilmore, 2000).

(e) Intergalactic mixing is, at present, rather uncertain, except for galaxy clusters (see above). The observations of intergalactic gas regions in absorption line spectroscopy suggest that there are *missing metals* (Pettini, 2004, and references therein), which is to be understood as only 20% of the metals inferred from the estimated star formation history of the universe and their corresponding massive-star production is seen in such absorption-line measurements. From another viewpoint, this *missing metal* issue reinforces above predictions that most of the metal production in the universe eventually is blown out of galaxies in the form of hot gas, residing at intergalactic temperatures of $10^7$ K or thereabouts, thus escaping direct detection. Therefore, *mixing* of ejecta from galaxies into intergalactic space is one of the frontiers in studying *cosmic* chemical evolution – The exchange of gas from galaxies with the intergalactic medium is key to our understanding of cosmic chemical evolution and the evolution of galaxies. Note that (i) galaxies form from gas streaming towards the gravitational potential set by the large-scale distribution of dark matter, that (ii) cold gas streams into/towards galaxies regulates the star formation activity in galaxies once the initial burst of star formation has decayed and much of the galaxy's gas has been dispersed by massive-star activity, and that (iii) gas from galaxies is mixed into the IGM through Galactic winds and fountains, and through stripping from mergers or cometary-like interaction of galaxies with the IGM. The Galaxy's hot corona is revealed in recombination emission from highly-ionized species, such as OVI and OVII (Sembach et al., 2003). High-velocity clouds have been recognized in HI 21 cm surveys due to their peculiar (*forbidden*) velocities, and have been assessed to provide a continuous and significant feeding of the Galaxy's gas content with material from outside the Galaxy (Sembach et al., 2003). It is debated how much of this material is fallback from Galactic gas ejected though massive-star activity at previous epochs, and how large the delay of fallback is; on the other hand, extragalactic gas also may constitute a significant fraction of the infalling gas. Baumgartner and Breitschwerdt (2009) have studied the efficiency of supernovae in driving galactic outflows. The activity of galactic winds and fountains appears confined to a range of order 100 kpc around galaxies. At larger distances, the IGM appears generally mixed and thermalized, as the application of the virial theorem adequately describes the state of coronal gas observed in galaxy clusters through X-rays. In fact, as soon as the central region of a galaxy cluster is avoided, galaxy clusters appear to display a rather smooth profile of temperatures and abundances of IGM, which reflects the early nucleosynthesis activity within a galaxy cluster in an averaged sense. Rasmussen et al. (2006) discuss details of coronal gas around galaxies, and their dynamics and impact on galaxy evolution. Somerville (2008) discuss that our understanding of star formation in galaxies is influenced significantly by the accretion and energy outflows of supermassive black holes. The black-hole activity is largely self-regulated, according to these models, but influences the cosmic variance of star forming activity in galaxies. Somerville (2008) show that the $\Lambda$CDM model cannot reproduce the cosmic star formation history as observed, based on their understanding of the role of SMBHs, and may have to be modified to reduce the power in smaller spatial scales after the big bang. Yet, QSO absorption systems' abundances are not consistent with either model and may either be biased, or taken as evidence of incompleteness of the model used.

Comparing spectra for different quasars at redshifts 2.5–6 suggest that there is no significant chemical evolution in this redshift range (Maiolino et al., 2003; Maiolino, 2009). This may be taken as an indication that the quasar activity itself is related to a galaxy history of very rapid and intense star forming activity processing most of the galactic gas into enriched material, before then the quasar activity sets in, and slows down chemical evolution.

### *A.2.3 Describing a Complex Source Environment*

- The interstellar medium is now understood as highly dynamic, and not consisting of localized, spatially-confined phases in pressure equilibrium (as originally proposed by McKee and Ostriker, 1977). The large-scale (kpc) average properties of interstellar medium may still be well described by global thermodynamic properties resembling such equilibrium (de Avillez and Breitschwerdt, 2005). But on scales below few hundred pc, substantial deviations from equilibrium values are the rule rather than the exception (de Avillez and Breitschwerdt, 2004).
- Refined chemical evolution descriptions distinguish the evolution of *different galaxy components separately* (Chiappini et al., 2003). For example, a somewhat complex *star formation history* in the Galaxy appears, if one assumes that early-on intense star formation was responsible for generating a *thick disk* stellar population, but terminated when most of the gas was consumed (typical time scale Gy); upon return (and infall) of gas, a second, less intense but more sustained, star forming activity sets in within the *thin disk* of the Galaxy, continuing as local feedback allows. With such framework, a rather successful representation of many elemental abundance trends could be obtained. On the other hand, alternative proposals relate the thick-disk stars to accretion processes following collisions or close encounters with other galaxies, which are likely during the $\sim$12 Gy history of our Galaxy. Aiming for another possible explanation, recent studies (Schönrich and Binney, 2009) have shown that inclusion of *radial mixing of gas and stars* in the disk of the Galaxy by spiral-wave dynamics can account for the observed abundance patterns in thin and thick disk also with a simple, exponentially-decaying star formation history alone. This example may illustrate how mixing and other chemical evolution model ingredients may interplay, and re-emphasize a concerted effort to constrain each of the model inputs in many possible ways.
- Looking in more detail at abundance history data and their models for some elements (e.g. Ti), we must also conclude that none of the models to date is capable of reproducing the full variety of spectroscopically obtained elemental abundance data for Galactic stars. Interstellar mixing time scales are just one of the many unknowns in these models. But it is necessary to constrain each of the processes involved by observations, as interdependencies among the different uncertainties, e.g. of star formation history and recycling times, could result in erroneous conclusions on each of them.

## References


Andersen, J., Nordström, B., Mayor, M.: The solar neighborhood in four dimensions. In: Barnes, T.G., III, Bash, F.N. (eds.) Cosmic Abundances as Records of Stellar Evolution and Nucleosynthesis, Astronomical Society of the Pacific Conference Series, vol. 336, pp. 305 (2005)

Balsara, D.S., Kim, J.: Amplification of interstellar magnetic fields and turbulent mixing by supernova-driven turbulence. II. The role of dynamical chaos. ApJ **634**, 390–406 (2005). doi:10.1086/452626, arXiv:astro-ph/0504065

Barklem, P.S., Christlieb, N., Beers, T.C., Hill, V., Bessell, M.S., Holmberg, J., Marsteller, B., Rossi, S., Zickgraf, F., Reimers, D.: The Hamburg/ESO R-process enhanced star survey (HERES). II. Spectroscopic analysis of the survey sample. A&A **439**, 129–151 (2005). doi:10.1051/0004-6361:20052967, arXiv:astro-ph/0505050

Baumgartner, V., Breitschwerdt, D.: Metal enrichment of the intracluster medium: SNdriven galactic winds. Astron. Nachr. **330**, 898 (2009). doi:10.1002/asna.200911258, 0908.1309

Breitschwerdt, D.: Self-consistent modelling of the interstellar medium. Ap&SS **289**, 489–498 (2004). doi:10.1023/B:ASTR.0000014982.31688.bf, arXiv:astro-ph/0303237

Cayrel, R., Depagne, E., Spite, M., Hill, V., Spite, F., François, P., Plez, B., Beers, T., Primas, F., Andersen, J., Barbuy, B., Bonifacio, P., Molaro, P., Nordström, B.: First stars V – abundance patterns from C to Zn and supernova yields in the early galaxy. A&A **416**, 1117–1138 (2004). doi:10.1051/0004-6361:20034074, arXiv:astro-ph/0311082

Chiappini, C., Romano, D., Matteucci, F.: CNO evolution: Milky Way, dwarf galaxies and DLAs. Ap&SS **284**, 771–774 (2003). doi:10.1023/A:1024008903413

Clark, P.C., Bonnell, I.A., Klessen, R.S.: The star formation efficiency and its relation to variations in the initial mass function. MNRAS **386**, 3–10 (2008). doi:10.1111/j.1365-2966.2008.13005.x, 0803.4053

Clayton, D.D.: Galactic chemical evolution and nucleocosmochronology: a standard model. In: Arnett, W.D., Truran, J.W. (eds.) Nucleosynthesis: Challenges and New Developments, pp 65. University of Chicago Press, Chicago (1985)

Clayton, D.D.: Nuclear cosmochronology within analytic models of the chemical evolution of the solar neighbourhood. MNRAS **234**, 1–36 (1988)

Clayton, D.D., Hartmann, D.H., Leising, M.D.: On Al-26 and other short-lived interstellar radioactivity. ApJ **415**, L25–L29 (1993). doi:10.1086/187024

Cowan, J.J., Thielemann, F., Truran, J.W.: Radioactive dating of the elements. ARA&A **29**, 447–497 (1991). doi:10.1146/annurev.aa.29.090191.002311

Dame, T.M., Ungerechts, H., Cohen, R.S., de Geus, E.J., Grenier, I.A., May, J., Murphy, D.C., Nyman, L., Thaddeus, P.: A composite CO survey of the entire Milky Way. ApJ **322**, 706–720 (1987). doi:10.1086/165766

Davies, B., Origlia, L., Kudritzki, R., Figer, D.F., Rich, R.M., Najarro, F., Negueruela, I., Clark, J.S.: Chemical abundance patterns in the inner galaxy: the scutum red supergiant clusters. ApJ **696**, 2014–2025 (2009). doi:10.1088/0004-637X/696/2/2014, 0902.2378

de Avillez, M., Breitschwerdt, D.: From large to small scales: global models of the ISM. Ap& SS **289**, 479–488 (2004). doi:10.1023/B:ASTR.0000014981.50094.8f, arXiv:astro-ph/0310633

de Avillez, M.A., Breitschwerdt, D.: Global dynamical evolution of the ISM in star forming galaxies. I. High resolution 3D simulations: effect of the magnetic field. A&A **436**, 585–600 (2005). doi:10.1051/0004-6361:20042146, arXiv:astro-ph/0502327

Dekel, A., Birnboim, Y., Engel, G., Freundlich, J., Goerdt, T., Mumcuoglu, M., Neistein, E., Pichon, C., Teyssier, R., Zinger, E.: Cold streams in early massive hot haloes as the main mode of galaxy formation. Nature **457**, 451–454 (2009). doi:10.1038/nature07648, 0808.0553

Edvardsson, B., Andersen, J., Gustafsson, B., Lambert, D.L., Nissen, P.E., Tomkin, J.: The chemical evolution of the galactic disk – Part one – analysis and results. A&A **275**, 101 (1993)

Elmegreen, B.G., Elmegreen, D.M., Leitner, S.N.: A turbulent origin for flocculent spiral structure in galaxies. ApJ **590**, 271–283 (2003). doi:10.1086/374860, arXiv:astro-ph/0305049

Feltzing, S., Gilmore, G.: Age and metallicity gradients in the Galactic Bulge. A differential study using HST/WFPC2. A&A **355**, 949–965 (2000)

Feltzing, S., Holmberg, J., Hurley, J.R.: The solar neighbourhood age-metallicity relation – Does it exist? A&A **377**, 911–924 (2001). doi:10.1051/0004-6361:20011119, arXiv:astro-ph/0108191

Huss, G.R., Meyer, B.S., Srinivasan, G., Goswami, J.N., Sahijpal, S.: Stellar sources of the shortlived radionuclides in the early solar system. GeoChim. Cosmochim. Acta **73**, 4922–4945 (2009)

Klessen, R.S.: The relation between interstellar turbulence and star formation. Master's thesis, Astrophysikalisches Institut Potsdam, Potsdam (2004)

Krumholz, M.R., McKee, C.F.: A general theory of turbulence-regulated star formation, from spirals to ultraluminous infrared galaxies. ApJ **630**, 250–268 (2005). doi:10.1086/431734, arXiv:astro-ph/0505177

Maiolino, R.: Star formation and interstellar medium in the most distant AGNs. In: Wang, W., Yang, Z., Luo, Z., Chen, Z. (eds.) Astronomical Society of the Pacific Conference Series, vol. 408, pp. 235 (2009)

Maiolino, R., Juarez, Y., Mujica, R., Nagar, N.M., Oliva, E.: Early star formation traced by the highest redshift quasars. ApJ **596**, L155–L158 (2003). doi:10.1086/379600, arXiv:astro-ph/0307264

Malinie, G., Hartmann, D.H., Clayton, D.D., Mathews, G.J.: Inhomogeneous chemical evolution of the Galactic disk. ApJ **413**, 633–640 (1993). doi:10.1086/173032

McKee, C.F., Ostriker, J.P.: A theory of the interstellar medium – three components regulated by supernova explosions in an inhomogeneous substrate. ApJ **218**, 148–169 (1977). doi:10.1086/155667

Najarro, F., Figer, D.F., Hillier, D.J., Geballe, T.R., Kudritzki, R.P.: Metallicity in the galactic center: the quintuplet cluster. ApJ **691**, 1816–1827 (2009). doi:10.1088/0004-637X/691/2/1816, 0809.3185

Oey, M.S., Clarke, C.J.: The superbubble size distribution in the interstellar medium of galaxies. MNRAS **289**, 570–588, arXiv:astro-ph/9703104 (1997)

Pettini, M.: Element abundances through the cosmic ages. In: Esteban, C., García López, R., Herrero, A., Sánchez, F. (eds.) Cosmochemistry. The Melting Pot of the Elements, pp. 257–298. Cambridge University Press, Cambridge, UK (2004)

Rasmussen, J., Ponman, T.J., Mulchaey, J.S., Miles, T.A., Raychaudhury, S.: First results of the XI Groups Project: studying an unbiased sample of galaxy groups. MNRAS **373**, 653–665 (2006). doi: 10.1111/j.1365-2966.2006.11023.x

Rolleston, W.R.J., Smartt, S.J., Dufton, P.L., Ryans, R.S.I.: The Galactic metallicity gradient. A&A **363**, 537–554 (2000)

Schmidt, M.: The rate of star formation. II. The rate of formation of stars of different mass. ApJ **137**, 758 (1963). doi:10.1086/147553

Schönrich, R., Binney, J.: Origin and structure of the Galactic disc(s). MNRAS **399**, 1145–1156 (2009). doi:10.1111/j.1365-2966.2009.15365.x, 0907.1899

Sembach, K.R., Wakker, B.P., Savage, B.D., Richter, P., Meade, M., Shull, J.M., Jenkins, E.B., Sonneborn, G., Moos, H.W.: Highly ionized high-velocity gas in the vicinity of the galaxy. ApJS **146**, 165–208 (2003). doi:10.1086/346231, arXiv:astro-ph/0207562

Shull, J.M., Saken, J.M.: Noncoeval star formation, starbursts, and the growth of supershells in OB associations. ApJ **444**, 663–671 (1995). doi:10.1086/175638

Slavin, J.D., Frisch, P.C.: The chemical composition of interstellar matter at the solar location. Space Sci. Rev. **130**, 409–414 (2007). doi:10.1007/s11214-007-9186-2

Smartt, S.J., Venn, K.A., Dufton, P.L., Lennon, D.J., Rolleston, W.R.J., Keenan, F.P.: Chemical abundances in the inner 5 kpc of the Galactic disk. A&A **367**, 86–105 (2001). doi:10.1051/0004-6361:20000479, arXiv:astro-ph/0009157

Somerville, R.S.: The co-evolution of galaxies, black holes, and AGN in a hierarchical universe. In: Kodama, T., Yamada, T., Aoki, K. (eds.) Astronomical Society of the Pacific Conference Series, vol. 399, pp. 391 (2008)

Spitoni, E., Matteucci, F., Recchi, S., Cescutti, G., Pipino, A.: Effects of galactic fountains and delayed mixing in the chemical evolution of the Milky Way. A&A **504**, 87–96 (2009). doi:10.1051/0004-6361/200911768, 0906.3400

Tinsley, B.M.: Nucleochronology and Chemical Evolution. APJ **198,** 145–150 (1975). doi:10.1086/153586

Tinsley, B.M.: Chemical evolution in the solar neighborhood. III – time scales and nucleochronology. ApJ **216**, 548–559 (1977). doi:10.1086/155496

Weaver, R., McCray, R., Castor, J., Shapiro, P., Moore, R.: Interstellar bubbles. II – structure and evolution. ApJ **218**, 377–395 (1977). doi:10.1086/155692

Yokoi, K., Takahashi, K., Arnould, M. (1983) The Re-187 – Os-187 chronology and chemical evolution of the Galaxy. A&A **117**, 65–82

# Appendix B
# Radionuclides and Their Stellar Origins

**M. Lugaro and A. Chieffi**

We supplement this book, and in particular the discussion of stellar nucleosynthesis presented in Chap. 3, by a list of the unstable isotopes at which branching points that become relevant in the *s*-process reaction chain in AGB stars are activated. For sake of clarity and a better understanding it is advisable to go through the list with a chart of the nuclides at hand. For each branching point a brief description of its operation and its relevance in the study of the *s* process in AGB stars is presented. All listed isotopes suffer $\beta^-$ decay, unless specified otherwise. It should also be noted that usually in *s*-process conditions nuclear energy metastable levels higher than the ground state are not populated, thus the effect of these states does not need to be included in the study of branching points, except for the special cases reported in the list (see also Ward, 1977).

Branching factors (see Sect. 3.5.2) for each branching point can be calculated for a given temperature, density, and neutron density conditions referring to Takahashi and Yokoi (1987) for the $\beta$ decay rates, and to Rauscher and Thielemann (2000) for the neutron capture cross section, unless advised otherwise in the description below.[1]

**$^{35}$S** This branching point may lead to production of the rare neutron-rich $^{36}$S, whose abundance can be observed in stars via molecular lines, and may be measured in sulphur-rich meteoritic materials (for discussion and models see Mauersberger et al., 2004).

**$^{36}$Cl and $^{41}$Ca** These are both long-living nuclei produced and destroyed – mostly via $(n, p)$ and $(n, \alpha)$ channels – via neutron captures in AGB stars, and discussed in detail in Sect. 3.6.4. While $^{36}$Cl behaves as as stable nucleus during the *s* process, the half life of $^{41}$Ca against electron captures has a

M. Lugaro (✉)
Monash University, Victoria 3800, Australia

A. Chieffi
I.N.A.F., 00133 Roma, Italy

[1] Maria Lugaro thanks Roberto Gallino and Franz Kaeppeler for communicating the passion for branching points and for help with this section.

strong temperature and density dependence, which could make it act as a branching point and most importantly prevent its survival instellar environments as in the case, e.g., of the other long-living nucleus $^{205}$Pb.

$^{45}$**Ca** This branching point may lead to production of the rare neutron-rich $^{46}$Ca, which could be measured in Ca-rich meteoritic material.

$^{59}$**Fe** This important branching point leads to the production of the long-living radioactive nucleus $^{60}$Fe. See Sect. 3.6.3 for a detailed description and AGB model results.

$^{63}$**Ni** The half life of this nucleus decreases from 100 year to $\simeq$ 12 year at 300 MK. The associated branching point affects the production of the rare neutron-rich $^{64}$Ni as well as the $^{65}$Cu/$^{63}$Cu ratio.

$^{64}$**Cu** The half life of this nucleus is short, of the order of a few hours, however, this isotope is a branching point on the $s$-process paths as it has comparable $\beta^+$ and $\beta^-$ decay rates. The branching point may affect the production of $^{64}$Ni and $^{65}$Cu.

$^{65}$**Zn** This nucleus suffers $\beta^+$ decay and the branching point may affect the production of $^{65}$Cu.

$^{71}$**Ge** This nucleus suffers $\beta^+$ decay and the branching point may affect the production of $^{71}$Ga.

$^{79}$**Se** This branching point may lead to production of the long-living radioactive isotope $^{81}$Kr. This production occurs when the temperature increases in the AGB thermal pulse, and the half life of $^{79}$Se decreases from the terrestrial half life of 65,000 year to roughly 4 year at 300 MK due to population of the shorter-living isomeric state (Klay and Käppeler, 1988, and see Sect. 3.6.5). Operation of this branching point also affects the $^{81}$Br/$^{79}$Br ratio.

$^{80}$**Br** The half life of this nucleus is short, of the order of minutes, however, it is a branching point on the $s$-process paths as it can decay both $\beta^+$ and $\beta^-$, with the $\beta^-$ roughly ten times faster than the $\beta^+$ channel. It can affect the production of $^{81}$Kr.

$^{81}$**Kr** This nucleus is too long living ($T_{1/2}$ = 0.23 Million years, down to 2,300 year at temperature 300 MK) to act as a branching point during the $s$ process and rather behaves as a stable nucleus. Its production during the $s$ process is discussed in detail in Sect. 3.6.5. Its radiogenic decay leads to $^{81}$Br.

$^{85}$**Kr** The relatively long half life of $^{85}$Kr (11 year) allows this branching point to activate already at low neutron densities, $> 5 \times 10^8$ n/cm$^3$. The actual operation of this branching point is complicated by the fact that roughly 40% of $^{84}$Kr$(n, \gamma)$ reactions during the $s$-process result in the production of the isomeric state of $^{85}$Kr. Approximately 80% of these nuclei quickly decay into $^{85}$Rb, with a half live of 4.5 h, while the remaining 20% relax into the ground state. The production of $^{87}$Rb, a very long-living isotope with half live of 48 Gigayears (Gy) and a magic number of neutrons $N = 50$, has traditionally been attributed to the activation of the branching point at $^{85}$Kr (Lambert et al., 1995; Abia et al., 2001). van Raai et al. (2008) showed that the activation of the branching point at $^{85}$Kr mostly results in the production of $^{86}$Kr, a nucleus with a magic number of neutrons $N = 50$ and a very small neutron capture cross section of only $\simeq$3.4 mbarn. $^{86}$Kr is thus more likely to accumulate than to capture the further neutron that would allow the production of $^{87}$Rb. The importance of the production of $^{86}$Kr in meteoritic SiC grains and the *s* process is discussed in Sect. 3.5.5.

$^{86}$**Rb** The branching point at $^{86}$Rb is activated at relatively high neutron densities, above $10^{10}$ n/cm$^3$, being the half life of this nucleus 18.7 days, and it leads directly to the production of $^{87}$Rb. The importance of $^{87}$Rb in *s* process observations and models is discussed in Sect. 3.5.4.

$^{89,90}$**Sr** and $^{91}$**Y** The branching point at $^{89}$Sr may produce the unstable $^{90}$Sr, also a branching point producing $^{91}$Sr, which quickly decays into unstable $^{91}$Y. This is also a branching point, producing $^{92}$Y, which quickly decays into stable $^{92}$Zr. The final result of the operation of this chain of branching points is to decrease the production of $^{90}$Zr and $^{91}$Zr, with respect to that of $^{92}$Zr. This point is discussed by Lugaro et al. (2003), in relevance to the Zr isotopis ratios measured in meteoritic silicon carbide (SiC) grains from AGB stars.

$^{93}$**Zr** This nucleus is too long-living ($T_{1/2}$ = 1.5 Million years) to act as a branching point during the *s* process and rather behaves as a stable nucleus (see Sect. 3.5.2 and Fig. 3.10), with an experimentally determined neutron-capture cross section (Macklin, 1985b). Its production during the *s* process is discussed in detail in Sect. 3.6.5. Its radiogenic decay produces most of the solar abundance of $^{93}$Nb. (A small fraction of $^{93}$Nb is also contributed by the radiogenic decay of $^{93}$Mo ($T_{1/2}$ = 3, 500 years) which is not on the main *s*-process path but can be produced by neutron-capture on the relatively abundant *p*-only $^{92}$Mo, 15% of solar Mo).

$^{95}$**Zr** This important branching point can lead to production by the *s* process of $^{96}$Zr if $N_n > 5 \times 10^8$ n/cm$^3$. Zr isotopic ratios have been estimated in MS and S stars via molecular lines and measured in meteoritic SiC grains, providing constraints on the neutron density in the thermal pulses. This point is further discussed in Sects. 3.5.4 and 3.5.5.

$^{94,95}$**Nb** The half life of $^{94}$Nb decreases from terrestrial 20,000 year to $\simeq$ 0.5 year at 100 MK and $\simeq$ 9 days at 300 MK. This branching point can produce the unstable $^{95}$Nb, which is also a branching point producing the unstable $^{96}$Nb, which quickly decays into stable $^{96}$Mo. Via the operation of the $^{94}$Nb branching point the $^{94}$Mo nucleus is skipped during the *s*-process chain, this nucleus is in fact classified as a *p*-only nucleus.

$^{99}$**Tc** The half life of $^{99}$Tc is 0.21 Million years, and decreases to 0.11 Million years at 100 MK and to 4.5 year at 300 MK. Thus, the neutron-capture path of the branching point is mostly open, producing $^{100}$Tc, which quickly decays into $^{100}$Ru, thus skipping $^{99}$Ru (Fig. 3.10). Then, radiogenic decay of $^{99}$Tc produces $^{99}$Ru. The production of $^{99}$Tc is discussed in detail in Sect. 3.6.5, and mentioned in Sect. 3.6.5.1 in relation to $^{99}$Ru in meteoritic SiC grains.

$^{107}$**Pd** This nucleus is too long-living ($T_{1/2}$ = 6.5 Million years, down to $\simeq$ 700 year at 300 MK) to act as a branching point during the *s* process and rather behaves as a stable nucleus, with an experimentally determined neutron-capture cross section (Macklin, 1985a). Its production during the *s* process is discussed in detail in Section 3.6.5. Its radiogenic decay is responsible for production of $^{107}$Ag.

$^{128}$**I** The decay half life of this nucleus is too short to allow for neutron captures, however, there is a marginal branching point here due to the fact that $^{128}$I has both $\beta^+$ and $\beta^-$ decay channels. The $\beta^+$ channel has significant temperature and density dependence and represents only a few percent of the decay rate. Nevertheless, this branching point has been investigated in detail because it affects the precise determination of relative abundances of the two *s*-only isotopes $^{128}$Xe and $^{130}$Xe, and because the timescale for its activation of the order of 25 minutes. is comparable to that of the convective turn-over timescale of the material inside AGB thermal pulses of hours (Reifarth et al., 2004).

$^{133}$**Xe** May lead to production of the $^{134}$Xe. Of interest in relation to the Xe–S component from SiC grains in primitive meteorites, as discussed in Sect. 3.5.5.

$^{134,135,136,137}$**Cs** The chain of branching points at the Cs isotopes is of particular interest because it affects the isotopic composition of the *s*-process element Ba and in particular the relative abudances of the two *s*-only nuclei $^{134}$Ba and $^{136}$Ba, as it is discussed in Sect. 3.5.5 in relation to Ba data from meteoritic SiC grains. The branching point at $^{134}$Cs allows production of the long-living isotope $^{135}$Cs (see Sect. 3.6.5 for model results). The half lifes of both $^{134}$Cs and $^{135}$Cs have a strong theoretical temperature dependence, decreasing by orders of magnitude in stellar conditions. Specifically for the long-living $^{135}$Cs, $T_{1/2}$ varies from terrestrial of 2 Million years down to $\simeq$ 200 year at 300 MK, while its neutron-capture cross section has been experimentally determined (Jaag et al., 1997). The branching point at $^{135}$Cs can produce the unstable $^{136}$Cs, which is also a branching point producing the unstable $^{137}$Cs. With a constant half life of $\simeq$ 30 year, this is also a branching point producing the unstable $^{138}$Cs, which quickly decays into stable $^{138}$Ba.

$^{141}$**Ce** This branching point may lead to production of the neutron-rich $^{142}$Ce, thus skipping the *s*-only $^{142}$Nd and affecting the Nd isotopic ratios, which are measured in SiC stardust grains (Gallino et al., 1997).

$^{142,143}$**Pr** The branching point at $^{142}$Pr is affected by the temperature dependence of the $\beta^-$ half life of $^{142}$Pr, which increases to $\simeq$4 days at 300 MK from the terrestrial 19 h. The neutron-capture branch may produce the unstable $^{143}$Pr, which is also a branching point producing the unstable $^{144}$Pr, which quickly decays into $^{144}$Nd. The operation of this chain of branching points may affect the isotopic composition of Nd because $^{142}$Nd and $^{143}$Nd are skipped by the neutron-capture flux and their abundances are decreased.

$^{147}$**Nd** This branching point may lead to the production of the neutron-rich "r-only" $^{148}$Nd, which is of interest in relation to stellar SiC grain Nd data (Gallino et al., 1997).

$^{147,148}$**Pm** The branching point at $^{147}$Pm is affected by the strong temperature dependence of the $\beta^-$ decay of this nucleus, where the half life decreases from the terrestrial value of 2.6 year down to $\simeq$ 1 year at 300 MK. The neutron-capture cross section of this nucleus is experimentally determined (Reifarth et al., 2003). When the branching is open, it produces the unstable $^{148}$Pm, a branching point that may lead to production of $^{149}$Pm, which quickly decays into $^{149}$Sm. The operation of this chain of branching points affects the isotopic composition of Sm, by skipping $^{147}$Sm and the *s*-only $^{148}$Sm. This is of interest in relation to stellar SiC grain Sm data (Gallino et al., 1997).

$^{151}$**Sm** The operation of this branching point is affected by the temperature dependence of the $\beta^-$ decay rate of $^{151}$Sm, where the half life of this nucleus decreases from 93 year to $\simeq$ 3 year at 300 MK. Its operation changes the $^{153}$Eu/$^{151}$Eu ratio, which can be measured in stars (Sect. 3.5.4) and in SiC stardust grains (Sect. 3.5.5). Note that $^{151}$Sm is one of few radioactive nuclei acting as branching points on the *s*-process path for which an experimental determination of the neutron capture cross section is available (Abbondanno et al., 2004; Wisshak et al., 2006).

$^{153}$**Sm** This branching point can produce the neutron-rich $^{154}$Sm and affect the $^{153}$Eu/$^{151}$Eu ratio.

$^{152}$**Eu** This nucleus suffers both $\beta^-$ and $\beta^+$ decays, with rates showing a strong temperature dependence covering several orders of magnitude variation in stellar conditions. The $\beta^+$ decay rate also has a strong dependence on density. The operation of this branching point, in combination with that at $^{151}$Sm, makes possible the production of the rare *p*-only isotope $^{152}$Gd by the *s* process.

$^{154,155}$**Eu** The decay rate of $^{154}$Eu has a strong temperature dependence, with its half life decreasing from 8.8 year down to $\simeq$11 days at 300 MK. If activated, it leads to production of the unstable $^{155}$Eu, a branching point also with a temperature dependence, and an experimentally determined neutron-capture cross section (Jaag and Käppeler, 1995), which may produce $^{156}$Eu, which quickly decays into $^{156}$Gd. The operation of this chain of branching points affects the isotopic composition of Gd, which is a refractory element present in stellar SiC grains (Yin et al., 2006).

$^{153}$**Gd** This nucleus suffers $\beta^+$ decay with a temperature dependence, where the terrestrial half life of 239 days increases with increasing the temperature by up to an order of magnitude in AGB stars conditions. The operation of this branching point may affect the $^{153}$Eu/$^{151}$Eu ratio.

$^{163}$**Dy** and $^{163,164}$**Ho** The nucleus $^{163}$Dy is stable in terrestrial conditions, but it can become unstable inside stars: at 300 MK the half life of this isotope becomes $\simeq$ 18 days. Thus, a branching can open on the *s*-process path, leading to the production of the unstable $^{163}$Ho via $\beta^-$ decay of $^{163}$Dy. In this conditions, the $\beta^+$ half life of $^{163}$Ho (which also has a strong temperature and density dependence) is $\simeq$ 12 year, so another branching can open on the *s*-process neutron capture path. Neutron captures on $^{163}$Ho lead to production of the unstable $^{164}$Ho, which has fast $\beta^-$ and $\beta^+$ channels, both temperature dependent. The $\beta^-$ channel can eventually lead to the production of $^{164}$Er, a *p*-only nucleus, which may thus have a *s*-process component in its cosmic abundance.

$^{169}$**Er** This branching point may lead to the production of the neutron-rich $^{170}$Er.

$^{170,171}$**Tm** The branching point at $^{170}$Tm may produce the unstable $^{171}$Tm, which is also a branching point (with a temperature dependence) producing the unstable $^{172}$Tm, which quickly decays into $^{172}$Yb. By skipping $^{171,172}$Yb during the *s*-process flux, these branching points affect the isotopic composition of Yb, which is a refractory element present in meteoritic stellar SiC grains (Yin et al., 2006).

$^{176}$**Lu** A branching point at $^{176}$Lu is activated because of the production of the short-living (half life of $\simeq$ 4 hours.) isomeric state of $^{176}$Lu via neutron captures on $^{175}$Lu. The situation is further complicated because, at around 300 MK, the isomeric and the ground state of $^{176}$Lu are connected via the thermal population of nuclear states that can act as mediators between the two. Hence, the half life of the $^{176}$Lu system can decrease at such temperatures by orders of magnitude. This branching point is of importance for the production of the very long-living ground state of $^{176}$Lu (half life of 380 Gy) and of the stable $^{176}$Hf, which are both *s*-only isotopes, shielded by $^{176}$Yb against *r*-process production. Hence, the relative solar abundances of these two isotopes need to be matched by *s* process in AGB stars. For details and models see Heil et al. (2008) and Mohr et al. (2009).

$^{177}$**Lu** This branching point may lead to production of the unstable $^{178}$Lu, which quickly decays into $^{178}$Hf, thus decreasing the abundance of $^{177}$Hf.

$^{179}$**Hf**, $^{179,180}$**Ta** A branching point at $^{179}$Hf may be activated on the *s*-process path because this stable nucleus becomes unstable in stellar conditions (as in the case of $^{163}$Dy) with a $\beta^-$ half life of $\simeq$ 40 year at 300 MK. This may allow the production of the unstable $^{179}$Ta, which is also a branching point with a temperature-dependent $\beta^+$ decay rate, which may lead to the production of $^{180}$Ta, the least abundant nucleus in the solar system (Käppeler et al., 2004), as a few percent of neutron captures on $^{179}$Ta lead to production of the very long-living isomeric state of $^{180}$Ta, instead of the ground state, which suffers fast $\beta^+$ and $\beta^-$ decays. As in the case of $^{176}$Lu, the ground and the isomeric states of $^{180}$Ta can be connected via the thermal population of nuclear states that act as mediators between the two. It is still unclear if the cosmic abundance of $^{180}$Ta is to be ascribed to the *s* process or to nucleosynthetic processes in supernovae connected to neutrino fluxes.

$^{181}$**Hf** This branching point may lead to production of the long-living radioactive nucleus $^{182}$Hf (one of the few radioactive isotopes with an experimentally determined neutron-capture cross section available, Vockenhuber et al., 2007) whose decay into $^{182}$W is of extreme importance for early solar system datation. The half life of $^{181}$Hf is relatively long in terrestrial conditions (42 days), but decreases down to 2 days at 300 MK, when the $^{22}$Ne($\alpha$,n)$^{25}$Mg reaction is activated, so the actual production of $^{182}$Hf in AGB stars is predicted to be relatively marginal (see also Sect. 3.6.5).

$^{182,183}$**Ta** The branching point at $^{182}$Ta is temperature dependent and may produce the unstable $^{183}$Ta, also a branching point, producing $^{184}$Ta, which quickly decays into the stable $^{184}$W. These branching points may affect the isotopic composition of W, which is a refractory element that is present in stellar SiC grains (Avila et al., 2008).

$^{185}$**W** This branching point may produce $^{186}$W, and affect the isotopic composition of W as well as the $^{186}$Os/$^{188}$Os ratio. Its signature may be seen in data from stellar SiC grains for W and Os (Humayun and Brandon, 2007). Note that $^{185}$W is one of few radioactive nuclei acting as branching points on the *s*-process path for which an experimental determination of the neutron capture cross section is available, even thought only via indirect $(\gamma, n)$ studies, which have rather large uncertainties of about 30% (Sonnabend et al., 2003; Mohr et al., 2004).

$^{186}$**Re** This isotope decays in $\simeq$ 89 h, and has both $\beta^-$ and $\beta^+$ decay channel. The $\beta^-$ decay channel is faster by one to two orders of magnitude depending on the temperature, which also affect the $\beta^+$ decay rate. This branching point can affect the production of $^{186}$Os, $^{186}$W, and the very long-living $^{187}$Re, whose slow decay into $^{187}$Os is used as a cosmological clock (see discussion in Chap. 2).

$^{191}$**Os** This branching point has a mild temperature dependence whereby the half life of $^{191}$Os decreases with the temperature from the terrestrial 15 days to $\simeq$8 days at 300 MK. If activated, the neutron-capture branch can decrease the *s*-process abundances of $^{191}$Ir and $^{192}$Pt and lead to production of $^{192}$Os, thus affecting the isotopic composition of Os, which is measured in meteoritic materials (Brandon et al., 2005), and $^{193}$Ir.

$^{192}$**Ir** This branching point can produce $^{193}$Ir, and affect the *s*-process production of the rare proton-rich $^{192}$Pt. A few percent of the decay rate of $^{192}$Ir is made by $\beta^+$ decays.

$^{193}$**Pt** This isotope decays $\beta^+$ with a half life of $\simeq$ 50 year, which may affect the production of $^{193}$Ir.

$^{204}$**Tl** This branching point has a strong temperature dependence with its half life decreasing from the terrestrial value of $\simeq$ 3.8 year to $\simeq$ 7 days at 300 MK, leading to production of the *s*-only $^{204}$Pb.

$^{205}$**Pb** This nucleus is long-living in terrestrial conditions ($T_{1/2}$ = 15 Million years), but its half life against electron captures has a strong temperature and density dependence, which affects its survival in stellar environments, as in the case of $^{41}$Ca. Its production during the *s*-process is discussed in detail in Sect. 3.6.5. Its radiogenic decay is responsible for production of $^{105}$Tl.

$^{210}$**Bi** This temperature-dependent branching point may lead to production of the unstable $^{211}$Bi, which $\alpha$ decays into $^{207}$Tl, which quickly decays $\beta^+$ into $^{207}$Pb.

$^{210}$**Po** May produce $^{211}$Po, which quickly $\alpha$ decays into $^{207}$Pb. The $\alpha$ decay of $^{210}$Po, and $^{211}$Bi above, represent the chain of reactions that terminates the *s* process (Clayton and Rassbach, 1967; Ratzel et al., 2004).

To complete the picture we list nuclei that could be classified as potential *s*-process branching points, given that their terrestrial half life is greater than a few days, however, they do not open during the *s* process because their half life decreases with the temperature to below a few days. These are: $^{103}$Ru, $^{123}$Sn, $^{124}$Sb, $^{156}$Eu, $^{160,161}$Tb, $^{175}$Yb, $^{198}$Au, and $^{205}$Hg. Finally, we point out the special case of $^{157}$Gd, a stable nucleus which becomes unstable at stellar temperatures, but not enough to open a branching point on the $s$-process path in AGB stars.

## References


Abbondanno, U., Aerts, G., Alvarez-Velarde, F., Álvarez-Pol, H., Andriamonje, S., Andrzejewski, J., Badurek, G., Baumann, P., Bečcvář, F., Benlliure, J., Berthoumieux, E., Calviño, F., Cano-Ott, D., Capote, R., Cennini, P., Chepel, V., Chiaveri, E., Colonna, N., Cortes, G., Cortina, D., Couture, A., Cox, J., Dababneh, S., Dahlfors, M., David, S., Dolfini, R., Domingo-Pardo, C., Duran, I., Embid-Segura, M., Ferrant, L., Ferrari, A., Ferreira-Marques, R., Frais-Koelbl, H., Furman, W., Goncalves, I., Gallino, R., Gonzalez-Romero, E., Goverdovski, A., Gramegna, F., Griesmayer, E., Gunsing, F., Haas, B., Haight, R., Heil, M., Herrera-Martinez, A., Isaev, S., Jericha, E., Käppeler, F., Kadi, Y., Karadimos, D., Kerveno, M., Ketlerov, V., Koehler, P., Konovalov, V., Krtièka, M., Lamboudis, C., Leeb, H., Lindote, A., Lopes, I., Lozano, M., Lukic, S., Marganiec, J., Marrone, S., Martinez-Val, J., Mastinu, P., Mengoni, A., Milazzo, P.M., Molina-Coballes, A., Moreau, C., Mosconi, M., Neves, F., Oberhummer, H., O'Brien, S., Pancin, J., Papaevangelou, T., Paradela, C., Pavlik, A., Pavlopoulos, P., Perlado, J.M., Perrot, L., Pignatari, M., Plag, R., Plompen, A., Plukis, A., Poch, A., Policarpo, A., Pretel, C., Quesada, J., Raman, S., Rapp, W., Rauscher, T., Reifarth, R., Rosetti, M., Rubbia, C., Rudolf, G., Rullhusen, P., Salgado, J., Soares, J.C., Stephan, C., Tagliente, G., Tain, J., Tassan-Got, L., Tavora, L., Terlizzi, R., Vannini, G., Vaz, P., Ventura, A., Villamarin, D., Vincente, M.C., Vlachoudis, V., Voss, F., Wendler, H., Wiescher, M., Wisshak, K.: Neutron capture cross section measurement of 151Sm at the CERN neutron time of flight facility (n TOF). Phys. Rev. Lett. **93**(16), id. 161103 (2004). doi:10.1103/PhysRevLett.93.161103

Abia, C., Busso, M., Gallino, R., Domínguez, I., Straniero, O., Isern, J.: The 85Kr s-process branching and the mass of carbon stars. ApJ **559**, 1117–1134 (2001). doi:10.1086/322383, arXiv:astroph/0105486

Avila, J.N., Ireland, T.R., Holden, P., Gyngard, F., Bennett, V., Amari, S., Zinner, E.: Tungsten Isotopic Compositions in Presolar Silicon Carbide Grains. Meteoritics and Planetary Science Supplement **43**, 5120 (2008)

Brandon, A.D., Humayun, M., Puchtel, I.S., Leya, I., Zolensky, M.: Osmium isotope evidence for an s-process carrier in primitive chondrites. Science **309**, 1233–1236 (2005). doi:10.1126/science.1115053

Clayton, D.D., Rassbach, M.E.: Termination of the s-process. ApJ **148**, 69–85 (1967). doi:10.1086/149128

Gallino, R., Busso, M., Lugaro, M.: Neutron capture nucleosynthesis in AGB stars. In: Bernatowicz, T.J., Zinner, E. (eds.) American Institute of Physics Conference Series, vol. 402, pp. 115–153 (1997). doi:10.1063/1.53327

Heil, M., Winckler, N., Dababneh, S., Käppeler, F., Wisshak, K., Bisterzo, S., Gallino, R., Davis, A.M., Rauscher, T.: 176Lu/176Hf: a sensitive test of s-process temperature and neutron density in AGB stars. ApJ **673**, 434–444 (2008). doi:10.1086/523892

Humayun, M., Brandon, A.D.: s-process implications from osmium isotope anomalies in chondrites. ApJ **664**, L59–L62 (2007). doi:10.1086/520636

Jaag, S., Käppeler, F.: Stellar (n,$\gamma$) cross section of the unstable isotope 155Eu. Phys. Rev. C **51**, 3465–3471 (1995). doi:10.1103/PhysRevC.51.3465

Jaag, S., Käppeler, F., Koehler, P.: The stellar (n,$\gamma$) cross section of the unstable 135Cs. Nucl. Phys. A **621**, 247–250 (1997). doi:10.1016/S0375-9474(97)00247-9

Käppeler, F., Arlandini, C., Heil, M., Voss, F., Wisshak, K., Reifarth, R., Straniero, O., Gallino, R., Masera, S., Travaglio, C.: Stellar neutron capture on 180Tam . II. Defining the s-process contribution to nature's rarest isotope. Phys. Rev. C **69**(5), id. 055802 (2004). doi:10.1103/PhysRevC.69.055802

Klay, N., Käppeler, F.: β-decay rate of 79mSe and its consequences for the s-process temperature. Phys. Rev. C **38**, 295–306 (1988). doi:10.1103/PhysRevC.38.295

Lambert, D.L., Smith, V.V., Busso, M., Gallino, R., Straniero, O.: The chemical composition of red giants. IV. The neutron density at the s-process site. ApJ **450**, 302–317 (1995). doi:10.1086/176141

Lugaro, M., Davis, A.M., Gallino, R., Pellin, M.J., Straniero, O., Käppeler, F.: Isotopic compositions of strontium, zirconium, molybdenum, and barium in single presolar SiC grains and asymptotic giant branch stars. ApJ **593**, 486–508 (2003). doi:10.1086/376442

Macklin, R.L.: Neutron capture measurements. Nucl. Sci. Eng. **89**, 79 (1985a)

Macklin, R.L.: Neutron capture measurements on radioactive $^{93}$Zr. Ap& SS **115**, 71–83 (1985b). doi:10.1007/BF00653828

Mauersberger, R., Ott, U., Henkel, C., Cernicharo, J., Gallino, R.: The abundance of $^{36}$S in IRC+10216 and its production in the Galaxy. A&A **426**, 219–227 (2004). doi:10.1051/0004-6361:20040451, arXiv:astro-ph/0407003

Mohr, P., Bisterzo, S., Gallino, R., Käppeler, F, Kneissl, U., Winckler, N.: Properties of the 5-state at 839 keV in Lu176 and the s-process branching at A = 176. Phys. Rev. C **79**(4), id. 045804 (2009). doi:10.1103/PhysRevC.79.045804, 0903.3897

Mohr, P., Shizuma, T., Ueda, H., Goko, S., Makinaga, A., Hara, K.Y., Hayakawa, T., Lui, Y., Ohgaki, H., Utsunomiya, H.: s-process branching at $^{185}$W revised. Phys. Rev. C **69**(3), id. 032801 (2004). doi:10.1103/PhysRevC.69.032801, arXiv:astro-ph/0409642

Ratzel, U., Arlandini, C., Käppeler, F., Couture, A., Wiescher, M., Reifarth, R., Gallino, R., Mengoni, A., Travaglio, C.: Nucleosynthesis at the termination point of the s process. Phys. Rev. C **70**(6), id. 065803 (2004). doi:10.1103/PhysRevC.70.065803

Rauscher, T., Thielemann, F.K.: Astrophysical reaction rates from statistical model calculations. At. Data Nucl. Data Tables **75**, 1–351 (2000). doi:10.1006/adnd.2000.0834, arXiv:astro-ph/0004059

Reifarth, R., Arlandini, C., Heil, M., Käppeler, F., Sedyshev, P.V., Mengoni, A., Herman, M., Rauscher, T., Gallino, R., Travaglio, C.: Stellar neutron capture on promethium: implications for the s-process neutron density. ApJ **582**, 1251–1262 (2003). doi:10.1086/344718

Reifarth, R., Käppeler, F., Voss, F., Wisshak, K., Gallino, R., Pignatari, M., Straniero, O.: 128Xe and 130Xe: testing He-shell burning in asymptotic giant branch stars. ApJ **614**, 363–370 (2004). doi:10.1086/422206, arXiv:astro-ph/0405065

Sonnabend, K., Mengoni, A., Mohr, P., Rauscher, T., Vogt, K., Zilges, A.: Determination of the (n, $\gamma$) reaction rate of unstable 185W in the astrophysical s-process via its inverse reaction. Nucl. Phys. A **718**, 533–535 (2003). doi:10.1016/S0375-9474(03)00835-2

Takahashi, K., Yokoi, K.: Beta-decay rates of highly ionized heavy atoms in stellar interiors. At. Data Nucl. Data Tables **36**, 375–409 (1987). doi:10.1016/0092-640X(87)90010-6

van Raai, M.A., Lugaro, M., Karakas, A.I., García-Hernández, D.A.: Rubidium and zirconium production in massive AGB stars. In: Guandalini, R., Palmerini, S., Busso, M. (eds.) Evolution and Nucleosynthesis in AGB Stars, American Institute of Physics Conference Series, vol. 1001, pp. 146–153 (2008). doi:10.1063/1.2916957

Vockenhuber, C., Dillmann, I., Heil, M., Käppeler, F., Winckler, N., Kutschera, W., Wallner, A., Bichler, M., Dababneh, S., Bisterzo, S., Gallino, R.: Stellar (n,$\gamma$) cross sections of Hf174 and radioactive Hf182. Phys. Rev. C **75**(1), id. 015804 (2007). doi:10.1103/PhysRevC.75.015804

Ward, R.A.: The importance of long-lived isomeric states in s-process branching. ApJ **216**, 540–547 (1977)

Wisshak, K., Voss, F., Käppeler, F., Krtièka, M., Raman, S., Mengoni, A., Gallino, R.: Stellar neutron capture cross section of the unstable s-process branching point 151Sm. Phys. Rev. C **73**(1), id. 015802 (2006). doi:10.1103/PhysRevC.73.015802

Yin, Q.Z., Lee, C.T.A., Ott, U.: Signatures of the s-process in presolar silicon carbide grains: barium through hafnium. ApJ **647**, 676–684 (2006). doi:10.1086/505188

# Appendix C
# Milestones in the Science of Cosmic Radioactivities

- **1869** Mendeleyev organizes the chemical elements according to their properties
- **1895** Röntgen discovers X-rays
- **1896** Becquerel discovers radioactivity of Uranium
- **1929** Rutherford estimates the age of element origins from the $^{235}$U/$^{238}$U ratio
- **1930** Dirac predicts the anti-electron (positron)
- **1932** Anderson discovers the positron
- **1932** Chadwick discovers the neutron
- **1939** Bethe describes the CN cycle in stars
- **1946** Hoyle suggests iron synthesis in stars (e process)
- **1947** Suess correlates nuclear magic numbers with isotopic abundances
- **1948** Haxel, Jensen, & Suess and Mayer advance shell model of magic numbers
- **1950** Fowler demonstrates that p-capture by $^{12}$C yields radioactive $^{13}$N
- **1950** Borst advances radioactivity as the power of supernova luminosity
- **1952** Merrill discovers radioactive Tc in stellar atmospheres
- **1952** Salpeter introduces triple-alpha reaction through quasiequilibrium $^{8}$Be
- **1953** Hoyle predicts the 7.7 MeV resonance in $^{12}$C
- **1954** Hoyle advances nucleosynthesis (C to Ni) in massive stars
- **1956** Suess and Urey suggest solar abundances of chemical elements
- **1957** B2FH review and systematize stellar nucleosynthesis
- **1957** Cameron lectures on Stellar Evolution and Nucleogenesis
- **1958** Feynman and Gell-Mann theory of Fermi interaction for beta decay
- **1960** Reynolds discovers excess $^{129}$Xe from extinct $^{129}$I in meteorites
- **1960** Fowler and Hoyle introduce $^{238}$U/$^{232}$Th cosmochronology
- **1961** Clayton develops time-dependent s-process description
- **1962** Cameron suggest supernova injection of $^{26}$Al at solar birth
- **1963** Wallerstein & Greenstein find metal-poor red giant stars
- **1964** Clayton proposes cosmoradiogenic chronologies of Re*Os and U*Pb
- **1964** Reynolds & Turner discover Xe-X excess in four heavy Xe isotopes
- **1964** Bahcall studies electron capture in stellar interiors
- **1965** Seeger, Fowler, & Clayton develop time-dependent *r* process description
- **1965** Clayton & Craddock consider $\gamma$-ray lines from *r* process ejecta
- **1966** Colgate & White advance hydrodynamic models of SNII
- **1967** Fowler assembles experimentally guided thermonuclear reaction rates

- **1968** Clayton discusses Si-burning as a quasiequilibrium process
- **1969** Colgate & McKee suggest $^{56}$Ni as the power source of SNIa
- **1969** Clayton, Colgate & Fishman predict $^{56}$Co and $^{44}$Ti $\gamma$-lines from supernovae
- **1969** Clayton & Silk predict cosmic background gamma-rays from $^{56}$Co decay
- **1969** Arnett predicts significant $^{26}$Al yield in explosive carbon burning
- **1971** Reynolds shows that extinct $^{244}$Pu fission Xe exists in meteorites
- **1971** Clayton predicts $^{60}$Fe gamma-ray lines from integrated supernova history
- **1972** Black discovers $^{22}$Ne-rich gas in meteorites
- **1972** Johnson, Harnden & Haymes discover galactic positron annihilation
- **1972** Starrfield et al. describe radiogenic luminosity of novae
- **1973** Clayton proposes $^{56}$Co and $^{44}$Ti as a galactic positronium source
- **1973** R. Clayton, Grossman, & Mayeda discover $^{16}$O-rich Al-rich inclusions
- **1974** Clayton & Hoyle consider $\gamma$-ray line emission from novae
- **1974** Gray & Compston find excess $^{26}$Mg in Al-rich minerals in meteorites
- **1975** Clayton predicts $^{44}$Ca-rich Ca from extinct $^{44}$Ti in Ti-rich presolar minerals
- **1975** Clayton proposes $^{22}$Na $\gamma$-ray lines from young supernovae
- **1975** Clayton & Ward predict *s*-process produced Xe detectable in stardust
- **1976** Clayton & Hoyle predict anomalous nova dust with extinct $^{22}$Na and $^{26}$Al
- **1976** Ward, Newman, & Clayton analyze radioactive *s*-process branching
- **1976** Clayton, Dwek, & Woosley consider proton irradiation in the solar disk
- **1977** Wasserburg shows initial $^{26}$Al/$^{27}$Al = 5 $10^{-5}$ in Al-rich inclusions
- **1977** Cameron & Truran suggest supernova trigger to inject $^{26}$Al etc in early Sun
- **1977** Clayton proposes model of fossil $^{26}$Mg excess from $^{26}$Al
- **1978** Kelly & Wasserburg discover extinct $^{107}$Pd in an iron meteorite
- **1978** Srinivasan & Anders find *s*-process Xe in meteorites
- **1978** Truran et al. estimate $\gamma$-ray lines from $^{22}$Na in novae
- **1981** Clayton predicts $^{49}$Ti excess from $^{49}$V in supernova dust
- **1982** Mahoney et al. discover interstellar $^{26}$Al gamma-rays with HEAO-C (launched 1978)
- **1983** Yokoi, Takahashi, & Arnould calculate speedup of $^{187}$Re decay
- **1985** Clayton proposes models for mean ISM radioactivity and cosmochronology
- **1985** Launch of the Solar Maximum Mission (SMM) satellite
- **1985** Share & Leising measure integrated $^{26}$Al mass in the ISM with SMM
- **1986** von Ballmoos et al. confirm $^{26}$Al $\gamma$-rays from galactic center region
- **1987** Butcher measures galaxy age by Th/Nd in old dwarfs
- **1987** Kamiokande and IMB find 19 prompt neutrinos from SN1987A
- **1987** Anders isolates stardust by dissolving meteorites in acid
- **1987** Zinner studies stardust isotopically using a mass spectrometer
- **1987** Bernatowicz et al. isotopically study SiC grains from red giants
- **1988** Balloon-borne Germanium spectrometers record $^{56}$Co lines in SN1987A
- **1988** Matz, Share, & Chupp detect $^{56}$Co gamma-rays from SN1987A with SMM
- **1989** Hudson, et al. measure initial solar plutonium from $^{244}$Pu/$^{238}$U= 3 $10^{-3}$
- **1990** Leising & Share find $^{56}$Co gamma-rays from SN1987A with SMM
- **1990** Tueller et al. measure $\gamma$-ray line profiles in SN1987A with GRIS
- **1990** Woosley et al. study $\nu$ process in massive stars; $^{26}$Al enhancement

- **1991** Launch of the Compton Gamma Ray Observatory (CGRO)
- **1991** GALLEX and SAGE neutrino observatories detect pp neutrinos
- **1991** Cowan et al. survey radioactive dating of the elements
- **1991** Meyer & Howard consider $\gamma$-ray signatures of $r$-process events
- **1992** Diehl et al. observe galactic plane $^{26}$Al with COMPTEL
- **1992** Kurfess et al. detect $^{57}$Co in SN1987A with OSSE
- **1992** Clayton et al. propose delayed radioactive power for SNII
- **1993** Weaver & Woosley nucleosynthesis dependence on $^{12}C(\alpha, \gamma)^{16}O$
- **1993** Starrfield et al. estimate yields of $^{22}$Na and $^{26}$Al in novae
- **1993** Timmes et al. estimate galactic chemical evolution of $^{26}$Al
- **1993** Matthews & Schramm consider galaxy-merger effects on cosmochronology
- **1994** Iyudin et al. detect $^{44}$Ti gamma-rays in Cas A with COMPTEL
- **1995** Liu & Dalgarno propose SN1987A CO dust formation by radioactive dissociation
- **1995** Diehl et al. present COMPTEL map of galactic $^{26}$Al gamma-rays
- **1996** Srinivasan et al. find $^{41}Ca/^{40}Ca = 1.5 \times 10^{-8}$ in meteorites
- **1996** Naya et al. show highly Doppler-broadened $^{26}$Al line with GRIS
- **1996** delRio et al. detect $^{26}$Al gamma-rays from the Cygnus region with COMPTEL
- **1996** Hoppe et al. and Nittler et al. find extinct $^{26}$Al and $^{44}$Ti in SiC X-grains
- **1997** Timmes et al. use $^{26}$Al to constrain the galactic star formation rate
- **1997** Dupraz et al. publish COMPTEL search for $^{44}$Ti gamma-ray sources
- **1998** Oberlack et al. publish COMPTEL all-sky $^{26}$Al $\gamma$-ray line image
- **1998** Hernanz et al. estimate radioactivities produced in novae
- **1999** Clayton, Liu & Dalgarno discuss how radioactivity affects graphite dust
- **1999** Korschinek, Knie et al. find live $^{60}$Fe in terrestrial ocean sediments
- **1999** Truran studies nucleocosmochronology with $r$-process radioactivities
- **2000** Plüschke/Cerviño et al. apply population synthesis to $^{26}$Al in Cygnus
- **2000** Meyer & Clayton consider radioactivities in the early solar system
- **2002** Hoppe & Besmehn discover $^{49}$Ti-rich titanium owing to extinct $^{49}$V decay
- **2002** Diehl et al. detect $^{26}$Al emission from Orion with COMPTEL
- **2003** Diehl et al. use INTEGRAL/SPI to show narrow width of $^{26}$Al line
- **2003** Knödlseder et al. publish INTEGRAL/SPI all-sky map of positron annihilation
- **2005** Jean; Churazov et al. show annihilation to occur in warm-ionized ISM
- **2005** Smith/Harris et al. find evidence for $^{60}$Fe with RHESSI / INTEGRAL
- **2006** The et al. discuss implications of the paucity of $^{44}$Ti sources
- **2006** Diehl et al. resolve Doppler shifts of the $^{26}$Al line from Galactic rotation
- **2007** Wang et al. establish galactic $^{60}$Fe gamma-ray lines with INTEGRAL/SPI
- **2008** Arpesella et al. detect solar $^{7}$Be neutrinos with BOREXINO
- **2008** Weidenspointner et al. find galactic positron annihilation to be asymmetric
- **2009** Diehl et al. detect $^{26}$Al from the Sco-Cen association with INTEGRAL/SPI
- **2009** Martin et al. constrain massive-star models with $^{26}$Al in Cygnus
- **2010** Kretschmer et al. trace inner-Galaxy hot ISM with the l-v map of $^{26}$Al

Additional historic pointers in the area of gamma-ray astronomy can be found at http://heasarc.gsfc.nasa.gov/docs/history/.

# Appendix D
# Glossary: Key Terms in Astronomy with Radioactivities

*Astronomy with Radioactivity* uses terminologies from very different fields: Theoretical astrophysics of stars, stellar interiors, and stellar explosions is intertwined with theories of the interstellar medium and galaxy evolution. Observations from stellar-photosphere spectroscopy through radio, IR, and X- and gamma-ray spectroscopy span a wide range, additionally cosmic-ray composition measurements and laboratory mass spectroscopy of small inclusions in meteorites are discussed. Nuclear-physics experiments and theories for nuclear levels and reaction cross sections are deeply involved as well. Here we provide a glossary of a few terms that appear often throughout the book.

- *Abundances*. The abundance of a chemical element may be measured by mass-fraction, or number-fraction. For example, the mass-fraction abundance of oxygen in water is about 90%, while the number-fraction is only 33% because 1 atom in 3 in water is an oxygen atom. On the cosmic scale, the mass-fraction of hydrogen and helium are about 74 and 23–25% respectively, while their relative abundance ratio by number is about 10:1. Abundances are often quoted relative to solar values (although not all values of the solar abundances come directly from measurements of the solar spectrum). The solar mass fraction of all elements other than hydrogen and helium (the metallicity) is slightly larger than 1%. Another notation often employed is the bracket $[X/H] = \log(X/H)/\log(X/H)_\odot$, i.e., the logarithmic (base 10) value of the number ratio of species X to hydrogen, relative to that ratio in the Sun. For example, stars in the Milky Way with the lowest metallicity exhibit $[Fe/H] = -5$, i.e. $10^{-5}$ solar iron abundance.
- *BATSE*. One of the four gamma ray telescopes aboard CGRO. BATSE was designed for burst sources and transient sources of gamma rays.
- *Chemical Evolution*. The change with time of relative abundances of elemental (and isotopic) species in gas and stars of galaxies. The study of this abundance evolution, by elemental or isotopic mass fraction, as a function of cosmic time (often expressed in redshift for the distant/early universe) in terms of its astrophysical agents (stars and gas flows) is called the science of chemical evolution (though no chemistry is involved). When treating all species collectively, one often follows the overall metallicity, instead of individual abundances. One distinguishes galactic chemical evolution (GCE) and cosmic chemical evolution

(CCE) when referring to changes *within* a galaxy, or for the universe as a whole, respectively.

- *Chemodynamics.* When chemical evolution is combined with dynamic aspects, such as stellar motions or hydrodynamic feedback of massive stars on the ISM, one speaks of chemo-dynamics, thus generalizing CE to CD.
- *COMPTEL.* One of the four gamma ray telescopes aboard CGRO. The gamma-ray detectors in COMPTEL are counters of events of double Compton scattering. Their geometrical arrangement fixes the kinematics of the incident gamma ray, and thereby allows (coarse) imaging as well as spectroscopy.
- *Compton Gamma Ray Observatory (CGRO).* One of NASAs great observatories, launched by shuttle Atlantis in 1991, carrying four experiments (BATSE, COMPTEL, EGRET, and OSSE) to study gamma rays over a wide range of energies.
- *Core collapse (Supernova).* Once a star has consumed its nuclear fuel and cannot release nuclear binding energy in its core from nuclear reactions, it will not counteract to gravitational pressure which forces contraction. An important alternative internal energy reservoir for stellar interiors is kinetic energy of electrons confined into a small-volume stellar core, as degeneracy and the Pauli exclusion principle enforce electron energies up to the *Fermi energy*. If electrons are forced to Fermi energies reaching nuclear energies, atomic nuclei may capture such electrons in a nuclear weak transition. Such reduction of electron degeneracy pressure is the cause of core collapse in stars of masses around 10 $M_{\odot}$. Such core collapses, when they occur from a rotating massive star, are believed to be the origins of the *long-duration* subclass of *Gamma Ray Bursts*. Gamma-Ray Bursts are extremely bright flashes, and may arise from stellar core collapses as early as those have been formed, i.e. out to redshifts 10–20, well beyond where galaxies can be observed. Such remote light beacons allow us to study elemental abundances in the remote and young universe, from absorption line analysis (similar to the Fraunhofer lines in the Solar spectrum).
- *Cosmic Rays.* Particles which penetrate the Galaxy and have high energies above $\sim$MeV, so that their interaction with matter causes the generation of secondary-particle avalanches. Cosmic rays mostly consist of high-energy protons, about 1% are electrons, and a small fraction are heavier atomic nuclei up to Fe. The energy spectrum reaches up to $10^{21}$ eV, where one (sub-atomic!) cosmic-ray particle alone has an energy comparable to a speedy tennis ball. The origins of cosmic rays is among one of nature's great mysteries and subject to a field of astrophysics and astro-particle-physics. Supernovae, pulsars, and active galaxies play a role. We distinguish Galactic cosmic rays (GCR) from extragalactic cosmic rays, from Solar Energetic Particles (SEPs), due to their (sometimes uncertain) origins, as inferred from indirect arguments.
- *Extinct Radioactivity.* Solid objects retain a record of having once contained an abundant radioactive nucleus, which is no longer present owing to the age of the solid greatly exceeding the halflife of the radioactive nucleus. The radioactivity is therefore now extinct. The initial activity is recorded by an excess abundance of the daughter isotope of the radioactive nucleus. The effect is largest in solids having a large abundance of the parent element and a small abundance

of the daughter element. Many such examples exist, and they are called extinct radioactivity.

- *Gamma Ray Lines*. A gamma-ray line is a flux of gamma-ray photons having the specific energy of a nuclear transition between two nuclear energy levels in a specific nucleus. That flux reveals the presence of those excited nuclear states. For example, the 847 keV gamma ray transition in the $^{56}$Fe nucleus occurs when abundant radioactive $^{56}$Co decays populate the upper $^{56}$Fe level of the 847 keV transition, which then rapidly and spontaneously emits the 847 keV photon.
- *Gamow Peak*. In thermonuclear reactions inside stars energy dependent cross sections, $\sigma(E)$, must be convolved with the thermal probability density of particle velocities (or, equivalently, their relative energy, E), which is given by the characteristic Maxwell-Boltzmann, $f_{\mathrm{MB}}(E) \propto E\exp(-E/kT)$, where k is the Boltzmann constant, and the energy, $E$, is evaluated in the center-of-mass reference frame. Charged particle reactions rely on tunneling through the Coulomb barrier between the interacting particles, which causes the cross section to depend on energy exponentially, $\sigma(E) \propto \exp(-2\pi\eta)$, where the Sommerfeld parameter, $\eta$, depends on nuclear masses, charges, and scales with energy as $\eta \propto 1/\sqrt{E}$. Reactions rates are small at low $E$, as the cross section is then small, and also small at high $E$, as the Maxwell-Boltzmann distribution then indicates a small number of particles at such energies. The evaluation of thermonuclear reaction rates (see Chap. 9) in thermal environments (such as stellar interiors) is thus given by $< \sigma v > \propto \int_0^\infty dE\ E\ f_{\mathrm{MB}}(E)\sigma(E)$. The competition between nuclear physics ($\sigma(E)$) and thermal physics ($f_{\mathrm{MB}}(E)$) results in a product of functions in the integrand that peaks at a characteristic energy $E_0 \sim 1$ keV–1 MeV, depending on the burning stages involved (i.e., depending on the charges of the key reactants).
- *Initial Mass Function*. A probability density function (pdf), $\Phi(m)$, describing the likelihood of forming stars in the mass range $(m,\ m + dm)$ in the range from ml (the lower mass limit of stars, determined by the condition of stable core hydrogen burning) to mu (the upper mass limit, set by the limiting factor of radiation driven pulsational instabilities above which stars can not settle on a stable configuration). The commonly adopted range for these parameters is ml = 0.1 $M_\odot$, and mu = 100 $M_\odot$. A frequently adopted pdf is the Salpeter IMF, $\Phi(m) \propto m^{-\alpha}$, with $\alpha = 2.35$. As a pdf, $\Phi$ is normalized to unity when integrating over the indicated range of stellar masses. The average stellar mass is given by $< m >= \int_{m_l}^{m_u} dm\Phi(m)\, m$, which evaluates to $< m > \sim 0.5\ M_\odot$. If all stars above $m_{SN}$ = 10 $M_\odot$ end their lives as core collapse supernovae (ccSNe), the fraction of all stars that undergo core collapse becomes $f_{SN} = \int_{m_{SN}}^{m_u} dm\Phi(m) \sim 0.001$.
- *INTEGRAL*. An ESA mission called INTErnational Gamma-Ray Astrophysics Laboratory. Launched in 2002 for a planned 3-year mission, extended to at least end 2012, due to scientific success and uniqueness of its main instruments.
- *Massive Star*. Stars may have masses from 0.1 to 100–1000 times that of the Sun. Low-mass stars are most abundant (see *Initial Mass Function*). A star massive enough so that it can ignite nuclear burnings beyond Helium burning is called *massive*. Often one draws a sharper line and uses Carbon burning as a criterion, as such a star then undergoes rapid evolution due to substantial neutrino

energy losses and develops into a gravitational collapse, called a *core-collapse supernova*.

- *OSSE*. One of the four gamma ray telescopes aboard CGRO. The gamma-ray detector in OSSE was a NaI scintillator. The quantity of photo energy produced by the scintillator measures the energy of the incident gamma ray.
- *Quasiequilibrium*. During Silicon Burning, very hot $^{28}$Si nuclei do not combine by nuclear fusion; rather they melt by photoejection of protons, neutrons and alpha particles. Those free particle densities assume a steady state in which they are globally recaptured by coexisting nuclei at the same rate at which they are photoejected from that distribution of nuclei. That steady state is called quasiequilibrium, and accounts for the intermediate-mass abundance distribution of nuclei between $^{44}$Ca and $^{60}$Ni. It provides a good description of the nucleosynthesis in that mass range. Its discovery in 1967–1968 filled the last major gap in Hoyle's theory of nucleosynthesis in supernovae. This process is of vital significance for AwR, because it provides the nucleosynthesis source for radioactive $^{44}$Ti, $^{49}$V, $^{52,53}$Fe, and $^{56,57}$Ni, whose decays are prominent for gamma-ray line astronomy and for isotopes in supernova stardust (SUNOCONs). The alpha-rich freezeout occurs when the temperature is sufficiently high to break down all $^{28}$Si into primarily alpha particles, so that the reassembly during cooling also builds more nuclei in this mass range. This variant of quasiequilibrium is especially prolific in production of $^{44}$Ti, $^{56}$Ni and $^{57}$Ni. The transition from the dominance of stable nuclei (primarily $Z = N$) below $Z = 21$ to radioactive positron emitters (also primarily $N = Z$) above $Z = 21$ occurs because abundances are restricted by the gas having low excess number of neutrons, so that $Z = N$ nuclei must dominate the abundances; however, owing to the positive Coulomb energy between nuclear protons, those nuclei having $Z > 21$ have their most stable isotopes at $N = Z + 4$ rather than at $N = Z$. Therefore, the synthesis at $Z = N$ undergoes positron emission after the explosion to reach $Z = N - 4$. For example, stable $^{40}$Ca ($Z = N = 20$) is the most abundant isobar at $A = 40$, whereas radioactive $^{44}$Ti ($Z = 22$, $N = 22$) decays later to $^{44}$Ca, the most abundant isobar at $A = 44$.
- *Population Synthesis*. Clusters of stars are prime examples of systems of stars that are believed to be coeval, or nearly so. The evolution of such a system of stars, all born at the same time, is also known as a simple stellar population (SSP). Given an initial mass function (IMF) that describes how stars are statistically distributed in their initial mass, m, the theory of stellar evolution allows an evaluation of various quantities relevant to this SSP as a function of time. For example, the rate of supernovae of a given type can be modeled and the amount of energy returned to the ISM from these supernovae and the mass loss preceding them (e.g., STARBURST99). Such studies are relevant for the study of the properties of the ISM in galaxies as a function of time. Galaxies have a continuous star formation history, so that the delta-function burst represented by a SSP has to be convolved with an assumed or calculated SFR(t). In Chapter 7 we discuss this in the context of radioactivities ($^{26}$Al, $^{60}$Fe) in nearby star forming regions. Further applications are color evolution of star clusters and galaxies as a whole,

and simulations of compact source populations. The latter must also include a treatment of binary stars. Population synthesis relies heavily on input from the theories of stellar evolution in single and multiple star system, but it does not address the issues of star formation itself.

- *Presolar Grain*. Inclusions in meteorites which show an clearly-unusual isotopic abundance signature (i.e. large factors beyond the spread seen among meteoritic samples) are attributed to an origin outside the solar system. This implies that they must have formed from material that is a different mix than what made the solar-system bodies. Since meteorites formed when the solar system was in its infancy, these anomalous grains must have been formed before the solar system was established. It may be more appropriate to call these grains *stardust*, as such grain formation continues to take place throughout the Galaxy.
- *Radioactivity*. Atomic nuclei which are not in their most-stable configurations of nucleons will eventually change their internal arrangement towards a more stable configuration. These nucleonic configurations are dictated by the laws of quantum physics. Transitions are made possible by the nuclear forces. Upon state transitions, secondary particles may be ejected, which causes the *harmful* effects to biological life. Radioactivity changes the type of isotope. Radioactive decay of an isotope occurs at a random moment in time, which can be estimated (though not predicted) from the characteristic decay time of each unstable isotope. Emission of secondaries from radioactive decay is thus independent of temperature or ionization state of a particular atom, therefore carries unique information not distorted by such environmental parameters. The physical unit measuring radioactivity is the *Becquerel*, counting decays per second, and named after Henri Antoine Becquerel, who discovered radioactivity in 1896.
- *Radioactive Beam Facilities*. Particle accelerators for performance of nuclear-reaction experiments, where the accelerated particles themselves are unstable. This is necessary for the study of nuclear reactions relevant for astrophysics, as in cosmic sites many reaction partners will have a composition deviant from stable nuclei as we know them.
- *Radiogenic Luminosity*. Radioactivity keeps a gas hot when it would otherwise cool, so it radiates when it would otherwise be dark. The energy input to the gas from radioactivity is first degraded to heat. Radiogenic luminosity occurs in explosive objects (supernovae and novae) because they contain sufficient radioactivity and would otherwise rapidly cool by expansion. A related phenomenon, radiogenic excitation, occurs when the non-thermal radiations from the radioactivity cause some species to have higher excitation than it would otherwise have at the ambient temperature. Examples of the latter are (1) He+ emission lines from a gas too cool to normally contain He+ ions; (2) small abundance of the CO molecule in a gas having abundant C and O at a temperature low enough that thermal equilibrium would favor the CO molecule.
- *Secondary-ion mass spectrometry*. Called SIMS for short, the charged secondary ions liberated by sputtering are subjected to electric and magnetic fields that determine their masses accurately enough to count distinct isotopes of the common elements lighter than iron.

- *Solar Abundances*. See Abundances
- *Stardust*. Stardust is a scientific word for a solid mineral grain that condensed thermally from the hot but slowly cooling gases leaving a star. Their mineral structures and high-temperature stability both attest to condensation from hot gases. Their extreme non-solar isotopic compositions are uniform throughout each grain. SiC is $^{13}$C-rich and mildly $^{29,30}$Si-rich compared to the sun, whereas oxides are predominantly $^{17}$O-rich. Typical sizes 0.1–1 mm are not visible to the naked eye. They are solid pieces of stars. Most known examples are extracted from the meteorites, whose accumulation within the accretion disk collected also the stardust. Those extracted from meteorites are necessarily from a star that lived prior to the birth of the sun. Stardust is recognized by its extreme isotopic composition, measured by secondary-ion mass spectrometry, and contains many extinct radioactive nuclei.
- *SUNOCON*. Constructed from the words SUperNOva CONdensate, SUNOCON is a scientific word for one specific type of stardust of exceptional importance to nucleosynthesis, supernova structure and AwR. SUNOCONs condense thermally within the hot, deep interiors of young supernovae while they are expanding explosively and cooling adiabatically. Beginning times for condensation are after a few months and end after a few years. Their mineral structures include the most refractory of high-temperature minerals, graphite (C), TiC, SiC, TiN. Their extreme isotopic compositions indicate interior supernova shells, as do their large contents of extinct radioactivities, $^{44}$Ti, $^{26}$Al, $^{49}$V, $^{95,97}$Zr and others. SUNOCONs are specifically not dust that condenses around supernovae, or as the ejecta interacts with external matter, or as H-containing supernova dust, or as dust aggregates.
- *Thermonuclear Supernova*. A compact and degenerate white dwarf star cannot expand fast enough once nuclear Carbon burning is efficiently ignited in its interior. The flame spreads so rapidly that the release of nuclear energy exceeds the gravitational binding energy of such a white dwarf, and thus the entire star is disrupted. It is unclear in detail how such Carbon ignition might occur. Thermonuclear supernovae are believed to be the cause of the observed class of *Type Ia* supernovae. These typically produce large amounts of radioactive $^{56}$Ni, which makes them shine brightly so they can be seen from cosmological distances. They can be empirically calibrated in their absolute brightness, which is the basis for their role in studying the expansion history of the universe.
- *Yield*. The amount of material (usually in units of solar masses) returned to the ISM by novae, supernovae, stellar winds, etc., as a function of specific source parameters, such as the initial mass of the progenitor star. For example, the yield of $^{26}$Al is a sensitive function of mass, and, after weighting by the IMF, is of order $10^{-4}$ $M_\odot$ per star massive enough to end its life as a supernova. Yields are required for the calculation of galactic or cosmic chemical evolution models.

# Index